Recent Progress in

HORMONE RESEARCH

The Proceedings of the Laurentian Hormone Conference

VOLUME 42

RECENT PROGRESS IN HORMONE RESEARCH

*Proceedings of the
1985 Laurentian Hormone Conference*

Edited by
ROY O. GREEP

VOLUME 42

PROGRAM COMMITTEE

G. D. Aurbach
J. D. Baxter
J. C. Beck
J. H. Clark
H. Friesen
R. O. Greep
P. A. Kelly

I. A. Kourides
A. R. Means
J. E. Rall
N. B. Schwartz
J. L. Vaitukaitis
W. W. Vale

ACADEMIC PRESS, INC.
Harcourt Brace Jovanovich, Publishers
Orlando San Diego New York Austin
Boston London Sydney Tokyo Toronto

Copyright © 1986 by Academic Press, Inc.
ALL RIGHTS RESERVED.
NO PART OF THIS PUBLICATION MAY BE REPRODUCED OR
TRANSMITTED IN ANY FORM OR BY ANY MEANS, ELECTRONIC
OR MECHANICAL, INCLUDING PHOTOCOPY, RECORDING, OR
ANY INFORMATION STORAGE AND RETRIEVAL SYSTEM, WITHOUT
PERMISSION IN WRITING FROM THE PUBLISHER.

ACADEMIC PRESS, INC.
Orlando, Florida 32887

United Kingdom Edition published by
ACADEMIC PRESS INC. (LONDON) LTD.
24–28 Oval Road, London NW1 7DX

LIBRARY OF CONGRESS CATALOG CARD NUMBER: Med. 47-38

ISBN 0–12–571142–5

PRINTED IN THE UNITED STATES OF AMERICA

86 87 88 89   9 8 7 6 5 4 3 2 1

CONTENTS

List of Contributors and Discussants ix

Preface xiii

1. Neurons with Multiple Messengers with Special Reference to Neuroendocrine Systems
   Tomas Hökfelt, Barry Everitt, Björn Meister, Tor Melander, Martin Schalling, Olle Johansson, Jan M. Lundberg, Anna-Lena Hulting, Sigbritt Werner, Claudio Cuello, Hugh Hemmings, Charles Ouimet, Ivar Walaas, Paul Greengard, and Menek Goldstein 1
   Discussion by Beck, Chen, Cohen, Drucker, Goltzman, Greep, Greer, Hoffman, Hökfelt, Kronenberg, Lakoski, Morgan, Ojeda, Pomerantz, Rice, and Saffran 66

2. Regulation of Gene Expression by Androgens in Murine Kidney
   J. F. Catterall, K. K. Kontula, C. S. Watson, P. J. Seppänen, D. Funkenstein, E. Melanitou, N. J. Hickok, C. W. Bardin, and O. A. Jänne 71
   Discussion by Atlas, Bremner, Callard, Catterall, Cutler, Geller, Granner, Hershman, Hökfelt, Horton, Knobil, Kronenberg, Melner, Nicoll, Odell, Rall, Saffran, Singh, Walters, and Ying 103

3. Insulin Regulates Expression of the Phosphoenolpyruvate Carboxykinase Gene
   Daryl K. Granner, Kazuyuki Sasaki, Teresa Andreone, and Elmus Beale .. 111
   Discussion by Callard, Chen, Clark, Gorski, Granner, Kelly, Kronenberg, Lederis, Lefevre, Means, Melner, Murphy, Nicoll, Rall, and Ying 137

4. Molecular Characterization of Fibroblast Growth Factor: Distribution and Biological Activities in Various Tissues
   Andrew Baird, Frederick Esch, Pierre Mormède, Naoto Ueno, Nicholas Ling, Peter Böhlen, Shao-Yao Ying, William B. Wehrenberg, and Roger Guillemin 143
   Discussion by Baird, Chen, Friesen, Goltzman, Keefe, Laherty, Melner, Meserve, Nekola, Nicoll, Papkoff, Pomerantz, Rall, Ryan, Sherman, and Veldhuis 200

5. Atrial Natriuretic Factor: A New Hormone of Cardiac Origin
   Steven A. Atlas 207
   Discussion by Anogianakis, Atlas, Callard, Campbell, Carey, Friesen, Geller, Katz, Kronenberg, Lederis, New, Nicoll, Rice, Rittmaster, Scoggins, Sherman, and Vecsei 243

6. Recent Progress in the Control of Aldosterone Secretion
   Robert M. Carey and Subha Sen 251
   Discussion by Albertson, Atlas, Carey, Geller, Gill, Hershman, Morris, New, Nicoll, Nolin, Osathanondh, Raj, Rittmaster, Scoggins, Segre, Tait, and Vecsei 289

v

7. Evolution of a Model of Estrogen Action
   Jack Gorski, Wade V. Welshons, Dennis Sakai, Jeffrey Hansen, Jane Walent, Judy Kassis, James Shull, Gary Stack, and Carolyn Campen 297
   Discussion by *Anogianakis, Bradlow, Callard, Clark, Cohen, Cutler, Gorski, Grumbach, Kronenberg, Lakoski, Lefevre, Nolin, Rice, and Walters* 322

8. Determinants of Puberty in a Seasonal Breeder
   Douglas L. Foster, Fred J. Karsch, Deborah H. Olster, Kathleen D. Ryan, and Steven M. Yellon 331
   Discussion by *Anogianakis, Cutler, Foster, Hoffman, Keefe, Kulin, Peter, Robinson, Schwartz, Urbanski, Walters, and Ying* 378

9. The Onset of Female Puberty: Studies in the Rat
   Sergio R. Ojeda, Henryk F. Urbanski, and Carol E. Ahmed 385
   Discussion by *Eckstein, Kulin, Ojeda, Pescovitz, Terasawa, and Yoshinaga* .. 440

10. Pubertal Growth: Physiology and Pathophysiology
    Gordon B. Cutler, Jr., Fernando G. Cassorla, Judith Levine Ross, Ora H. Pescovitz, Kevin M. Barnes, Florence Comite, Penelope P. Feuillan, Louisa Laue, Carol M. Foster, Daniel Kenigsberg, Manuela Caruso-Nicoletti, Hernan B. Garcia, Mercedes Uriarte, Karen D. Hench, Marilyn C. Skerda, Lauren M. Long, and D. Lynn Loriaux 443
    Discussion by *Albertson, Cutler, Drucker, Grumbach, Ho, Kaplan, Keefe, Kelch, Nicoll, Osathanondh, Raj, Rall, and Thorner* 465

11. Seasonal Breeding in a Marsupial: Opportunities of a New Species for an Old Problem
    C. H. Tyndale-Biscoe, L. A. Hinds, and S. J. McConnell 471
    Discussion by *Callard, Cohen, Horton, Karch, MacDonald, McConnell, Raj, Rice, Thorner, Tyndale-Biscoe, and Yoshinaga* 507

12. Interactions of Catecholamines and GnRH in Regulation of Gonadotropin Secretion in Teleost Fish
    Richard E. Peter, John P. Chang, Carol S. Nahorniak, Robert J. Omeljaniuk, M. Sokolowska, Stephen H. Shih, and Roland Billard 513
    Discussion by *Anogianakis, Aurbach, Campbell, Carey, Cutler, Dobyns, Hershman, Horton, Kulin, McConnell, Nekola, Peter, Schwartz, Spitz, Thorner, Veldhuis, and Ying* 545

13. Evolutionary Aspects of the Endocrine and Nervous Systems
    Derek LeRoith, George Delahunty, Gaye Lynn Wilson, Charles T. Roberts, Jr., Joshua Shemer, Celeste Hart, Maxine A. Lesniak, Joseph Shiloach, and Jesse Roth... 549
    Discussion by *Aurbach, Blizzard, Cohen, Greer, Horton, Kronenberg, LeRoith, New, Nicoll, Odell, Peter, Pomerantz, and Sherman* 582

14. Physiological and Clinical Studies of GRF and GH
    Michael O. Thorner, Mary Lee Vance, William S. Evans, Robert M. Blizzard, Alan D. Rogol, Ken Ho, Denis A. Leong, Joao L. C. Borges, Michael J. Cronin, Robert M. MacLeod, Kalman Kovacs, Sylvia Asa, Eva Horvath, Lawrence Frohman, Richard Furlanetto, Georgeanna Jones Klingensmith, Charles Brook, Patricia Smith, Seymour Reichlin, Jean Rivier, and Wylie Vale.. 589

Discussion by *Blizzard, Cohen, Geller, Greer, Hershman, Hoffman, Kaplan, Kronenberg, New, Nolin, Rittmaster, Saffran, Samaan, Schriock, Segre, Spitz, Thorner, and VanderLaan*................................... 632

15. Structure and Expression of the Human Parathyroid Hormone Gene
    Henry M. Kronenberg, Tetsuya Igarashi, Mason W. Freeman, Tomoki Okazaki, Stephen J. Brand, Kristine M. Wiren, and John T. Potts, Jr. 641
    Discussion by *Aurbach, Brandi, Fitzpatrick, Goltzman, Kronenberg, and Melner*.. 661

16. Studies of the Multiple Molecular Forms of Bioactive Parathyroid Hormone and Parathyroid Hormone-Like Substances
    David Goltzman, Hugh P. J. Bennett, Michael Koutsilieris, Jane Mitchell, Shafaat A. Rabbani, and Marie F. Rouleau............................. 665
    Discussion by *Aurbach, Campbell, Cohen, Draper, Fitzpatrick, Goltzman, Kelly, Kulin, Odell, Rice, Samaan, Segre, and Stewart*................... 697

17. Nephrogenous Cyclic AMP, Adenylate Cyclase-Stimulating Activity, and the Humoral Hypercalcemia of Malignancy
    J. W. Godsall, W. J. Burtis, K. L. Insogna, A. E. Broadus, and A. F. Stewart 705
    Discussion by *Aurbach, Carey, Goltzman, Katz, Kelley, Kronenberg, Rice, Samaan, Segre, and Stewart* 743

Index 751

LIST OF CONTRIBUTORS AND DISCUSSANTS

C. E. Ahmed
B. D. Albertson
T. Andreone
G. Anogianakis
S. Asa
S. A. Atlas
G. D. Aurbach
A. Baird
C. W. Bardin
K. M. Barnes
E. Beale
J. C. Beck
H. P. J. Bennett
R. Billard
R. M. Blizzard
P. Böhlen
J. L. C. Borges
H. L. Bradlow
S. J. Brand
M. L. Brandi
W. J. Bremner
A. E. Broadus
C. Brook
W. J. Burtis
G. Callard
I. Callard
G. T. Campbell
C. Campen
R. M. Carey
M. Caruso-Nicoletti
F. G. Cassorla
J. F. Catterall
J. P. Chang
T. T. Chen
J. H. Clark
S. Cohen
F. Comite
M. J. Cronin
C. Cuello
G. B. Cutler, Jr.
G. Delahunty
B. M. Dobyns
M. W. Draper
W. D. Drucker
B. Eckstein

F. Esch
W. S. Evans
B. Everitt
P. Feuillan
L. A. Fitzpatrick
C. M. Foster
D. L. Foster
M. W. Freeman
H. G. Friesen
L. Frohman
D. Funkenstein
R. Furlanetto
H. B. Garcia
J. Geller
J. R. Gill
J. W. Godsall
M. Goldstein
D. Goltzman
J. Gorski
D. K. Granner
P. Greengard
R. Greep
M. A. Greer
M. M. Grumbach
R. Guillemin
J. Hansen
C. Hart
H. Hemmings
K. D. Hench
J. M. Hershman
N. J. Hickok
L. A. Hinds
K. Ho
R. A. Hoffman
T. Hökfelt
T. Horton
E. Horvath
A.-L. Hulting
T. Igarashi
K. L. Insogna
O. A. Jänne
O. Johansson
S. L. Kaplan
F. J. Karch
J. Kassis

M. S. Katz
D. L. Keefe
R. P. Kelch
P. A. Kelly
T. M. Kelly
D. Kenigsburg
G. J. Klingensmith
E. Knobil
K. K. Kontula
M. Koutsilieris
K. Kovacs
H. M. Kronenberg
H. Kulin
R. F. Laherty
J. M. Lakoski
L. Laue
K. Lederis
Y. A. Lefevre
D. A. Leong
D. LeRoith
M. A. Lesniak
N. Ling
L. M. Long
D. L. Loriaux
J. M. Lundberg
T. Maack
G. J. MacDonald
R. M. MacLeod
S. J. McConnell
A. Means
B. Meister
T. Melander
E. Melanitou
M. H. Melner
L. A. Meserve
J. Mitchell
R. O. Morgan
P. Mormède
D. J. Morris
L. Murphy
C. S. Nahorniak
M. V. Nekola
M. I. New
C. S. Nicoll
J. M. Nolin
W. D. Odell
S. R. Ojeda
T. Okazaki
D. H. Olster
R. J. Omeljaniuk

R. Osathanondh
C. Ouimet
H. Papkoff
O. H. Pescovitz
R. E. Peter
D. K. Pomerantz
J. T. Potts, Jr.
S. A. Rabbani
M. Raj
J. E. Rall
S. Reichlin
B. F. Rice
R. Rittmaster
J. Rivier
C. T. Roberts, Jr.
J. Robinson
A. D. Rogol
J. L. Ross
J. Roth
M. F. Rouleau
K. D. Ryan
R. Ryan
M. Saffran
D. Sakai
N. Samaan
K. Sasaki
M. Schalling
E. A. Schriock
N. B. Schwartz
B. A. Scoggins
G. Segre
S. Sen
P. J. Seppänen
J. Shemer
M. R. Sherman
S. H. Shih
J. Shiloach
J. Shull
P. Singh
M. C. Skerda
P. Smith
M. Sokolowska
I. M. Spitz
G. Stack
A. F. Stewart
D. R. Stewart
J. F. Tait
E. Terasawa
M. O. Thorner
C. H. Tyndale-Biscoe

CONTRIBUTORS

N. Ueno
H. F. Urbanski
M. Uriarte
W. Vale
M. L. Vance
W. VanderLaan
P. Vecsei
J. D. Veldhuis
I. Walaas
J. Walent

M. Walters
C. S. Watson
W. B. Wehrenberg
W. V. Welshons
S. Werner
G. L. Wilson
K. M. Wiren
S. M. Yellon
S.-Y. Ying
K. Yoshinaga

PREFACE

This volume of *Recent Progress in Hormone Research* (the thirteenth and last for which I shall serve as editor) is another in this serial publication based on the proceedings of the Laurentian Hormone Conference held in September 1985 in Banff, Alberta, Canada. The topics discussed were of timely and varied interest. They covered the latest developments in research on such important topics as chemical neuroanatomy of the brain based on the histochemistry of brain neuronal transmitters and peptides; hormonal regulation of gene transcription and expression; steroid hormone receptors and their action; the atrial naturietic factor; the fibroblast growth factor; neuroendocrinology of puberty in rodents, seasonal breeders, and humans; comparative endocrinology including studies on marsupials, teleosts, and microorganisms; and parathyroid hormone, bioactive forms, gene expression, and malignancy-associated hypercalcemia. These subjects have been covered by leading experts in their special fields, and readers will find up-to-the-minute information in every instance.

I wish to convey my gratitude and that of the Program Committee to the following persons who served so ably as chairpersons of the several sessions of this conference: Drs. Anthony R. Means, Robert J. Ryan, J. Edward Rall, Adolfo de Bold, James H. Clark, Jr., Selna L. Kaplan, Charles S. Nicoll, Robert M. Blizzard, and Gerald D. Aurbach. Our appreciation and admiration are due to Lucy Felicissimo and Linda Carsagnini for their prompt and skillful transcription of the taped discussions. It is also a pleasure to acknowledge our grateful appreciation of the fine relations we have had with the staff of Academic Press over the past years.

I am pleased and proud to wish my successor, Dr. James H. Clark, the same sense of satisfaction that I have derived from fostering this renowned publication.

Roy O. Greep

Neurons with Multiple Messengers with Special Reference to Neuroendocrine Systems[1]

TOMAS HÖKFELT,* BARRY EVERITT,† BJÖRN MEISTER,*
TOR MELANDER,* MARTIN SCHALLING,* OLLE JOHANSSON,*
JAN M. LUNDBERG,‡ ANNA-LENA HULTING,§ SIGBRITT WERNER,§
CLAUDIO CUELLO,** HUGH HEMMINGS,‡ CHARLES OUIMET,‡
IVAR WALAAS,‡ PAUL GREENGARD,‡ AND MENEK GOLDSTEIN‡‡

*Departments of Histology and ‡ Pharmacology, Karolinska Institutet, § Department of Endocrinology, Karolinska Hospital, Stockholm, Sweden, † Department of Anatomy, University of Cambridge, Cambridge, England, ** Department of Pharmacology, McGill University, Montreal, Canada, ‡ Laboratory of Molecular and Cellular Neuroscience, Rockefeller University and ‡‡ Department of Psychiatry, New York University Medical Center, New York, New York

I. Introduction

Two systems are responsible for communication within higher organisms, the endocrine system and the nervous system. Although distinct differences between them are obvious, many similarities can also be found. For example, during the last decades it has become increasingly clear that they often share the same type of chemicals as messenger molecules. Moreover, the endocrine and the nervous system are deeply interconnected, and it is today difficult to approach neurobiological or endocrinological problems without taking both systems into account.

Transmission of messages within the nervous system was long assumed to occur via electrical events. In the beginning of this century, however, the first experimental evidence was presented that chemical messengers may be involved (Elliot, 1905). Elliott's studies were focused on the peripheral nervous system and implicated adrenaline as a possible transmitter molecule. The demonstration of chemical transmission in the central nervous system was achieved later. Thus, although this type of transmission was advocated by outstanding scientists (see Dale, 1935), it was not until 1954 that the first unequivocal evidence for central chemical transmission was presented in the form of acetylcholine mediating this function in recurrent inhibition of Renshaw cells (see Eccles *et al.*, 1954).

[1] The Gregory Pincus Memorial Lecture.

Subsequently evidence from the biochemical studies of Vogt (1954) suggested the possibility that noradrenaline may be a transmitter in the central nervous system, a view that was strongly supported by histochemical studies using the formaldehyde-induced fluorescence method of Falck and Hillarp (Carlsson *et al.*, 1962; Dahlström and Fuxe, 1964, 1965; Fuxe, 1965a,b). Also the noradrenaline precursor dopamine rapidly took a position as transmitter candidate (Carlsson *et al.*, 1958). In the next decade attention was focused on amino acids including γ-aminobutyric acid (GABA), glycine, and glutamate (see Fonnum, 1978).

Chemical messengers released from neurons was one of the basic ideas developed by Ernst and Berta Scharrer and Bargmann in their description of neurosecretory cells, which released their chemicals into the blood stream (Bargmann and Scharrer, 1951; Scharrer and Scharrer, 1954). Such neurosecretory neurons were present not only in the magnocellular paraventricular and supraoptic nuclei projecting to the posterior pituitary, but were also the types of neurons which subsequently were shown to produce the messenger molecules involved in the control of anterior pituitary hormone secretion, i.e., the hypothalamic-releasing and inhibitory hormones transported in the portal vessels to the pituitary gland according to the concept of Harris (1955).

The first identification of a neurosecretory product was achieved when Du Vigneaud *et al.* (1954) succeeded in purifying and sequencing oxytocin and vasopressin and showed them to be peptides consisting of nine amino acids.

Peptides as messenger molecules received renewed interest, when Guillemin, Schally, Vale, and their collaborators in a series of remarkable achievements were able to demonstrate that the postulated hypothalamic inhibitory and releasing hormones were peptides: thyrotropin-releasing hormone (TRH) (Burgus *et al.*, 1970; Nair *et al.*, 1970), luteinizing hormone-releasing hormone (LHRH) (Amoss *et al.*, 1971; Schally *et al.*, 1971), somatostatin (Brazeau *et al.*, 1973), corticotropin-releasing hormone (CRF) (Spiess *et al.*, 1981; Vale *et al.*, 1981), and growth hormone-releasing factor (GRF) (Brazeau *et al.*, 1982; Guillemin *et al.*, 1982; Rivier *et al.*, 1982; Spiess *et al.*, 1982, 1983; Böhlen *et al.*, 1984). This development has been covered in the program of the Laurentian Hormone Conferences and is thoroughly documented in the series of *Recent Progress in Hormone Research* (Van Dyke *et al.*, 1955; Greer, 1956; Guillemin, 1964, 1977; McCann and Ramirez, 1964; Barraclough, 1966; Martini *et al.*, 1968; Schally *et al.*, 1968; Gual *et al.*, 1972; Kastin *et al.*, 1972; Reichlin *et al.*, 1972; Porter *et al.*, 1973; Knobil, 1974, 1980; Yen *et al.*, 1975; Vale *et al.*, 1975, 1983; Kaplan *et al.*, 1976; Wilber *et al.*, 1976; Labrie *et al.*, 1978; Krieger *et al.*, 1980; Straus *et al.*, 1981; Leeman *et al.*, 1982; Guillemin *et al.*, 1984; Bloom *et al.*, 1985; Gershengorn, 1985).

The radioimmunological and immunohistochemical analysis with antisera raised against these hypothalamic peptides revealed a surprisingly wide distribution not only within hypothalamic nuclei outside the basal hypothalamus, but also in extrahypothalamic areas including the spinal cord and the peripheral nervous system (see Hökfelt *et al.*, 1980a; Snyder, 1980; Krieger, 1983; books edited by Krieger *et al.*, 1983; and Björklund and Hökfelt, 1985). These findings raised the possibility that such peptides, as well as several others, for example substance P (Otsuka and Takahashi, 1977; Pernow, 1983), could have functions beyond those of hypothalamic hormones, for example, as transmitters or modulators (see Otsuka and Takahashi, 1977; Snyder, 1980; Pernow, 1983). The number of such peptide candidates has increased dramatically during the last 10 years. Moreover, it has been recognized that families of peptides exist which arise sometimes from the same and sometimes from different genes within the same cell. An example of this is the discovery of several substance P-like compounds belonging to the tachykinin family (Kimura *et al.*, 1983; Kangawa *et al.*, 1983; Minamino *et al.*, 1984; Nawa *et al.*, 1983; Tatemoto *et al.*, 1986).

The abundance of peptides has raised the question of whether or not they occur in separate systems, distinguished from the ones shown to contain, for example, acetylcholine, noradrenaline, and 5-hydroxytryptamine. This question is particularly relevant in the periphery, where a large proportion of the neurons, e.g., in the sympathetic ganglia, have long been positively identified as noradrenergic. In fact, the first evidence for coexistence was observed in sympathetic ganglia, where noradrenaline-containing cells could be demonstrated to contain somatostatin-like immunoreactivity (Hökfelt *et al.*, 1977a). Subsequently, it has been recognized that coexistence situations embracing a classical transmitter and peptide represent a common phenomenon and can be encountered both in the peripheral and central nervous system. Moreover, in this search it has been found that other types of coexistence situations occur, for example, several peptides in a neuron without an identified classical transmitter, or more than one classical transmitter.

The occurrence of multiple messengers in a cell has been known for a long time. Thus, many endocrine cells in the gastointestinal tract have been shown to contain both a peptide hormone and a biogenic amine, and these cells have been regarded as belonging to a common system, the APUD system (Pearse, 1969; Owman *et al.*, 1973). Moreover, in invertebrates the coexistence of several transmitters was early observed (see Osborne, 1983). More recently the adrenal medullary gland cells have been shown to contain peptides, for example enkephalin-like peptides (Fig. 1A–C) (Schultzberg *et al.*, 1978a,b; Viveros *et al.*, 1979; Lewis *et al.*, 1980, 1981; Stern *et al.*, 1980), opening up possibilities of complex

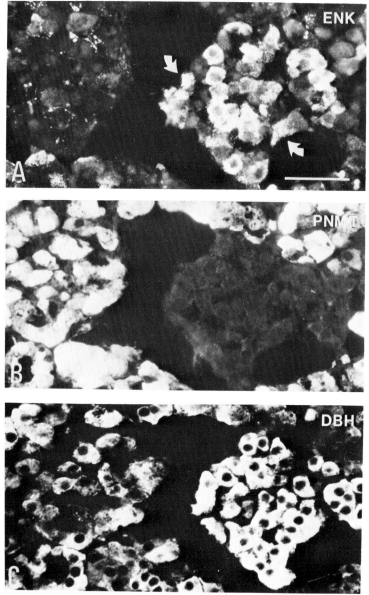

FIG. 1. Immunofluorescence micrographs of the adrenal medulla after incubation with antiserum to methionine-enkephalin (ENK) (A), phenylethanolamine-*n*-methyl transferase (PNMT) (B), and dopamine β-hydroxylase (DBH) (C). Strong ENK-like immunoreactivity (ENK-LI) can be seen in an island of cells (arrows in A), and these cells are PNMT negative but strongly DBH positive. Thus, the ENK-LI is present in noradrenaline cells. Note ENK-positive nerve endings. Bar indicates 50 μm. From Schultzberg *et al.* (1978b).

interactions (Fig. 2). Neuronal coexistence has been discussed in several articles (Burnstock, 1976; Smith, 1976; Hökfelt *et al.*, 1980a,b, 1982a, 1984a; Potter *et al.*, 1981; Cuello, 1982; Osborne, 1983; Chan-Palay and Palay, 1984).

The functional significance of coexistence of multiple messengers is still unclear, both with regard to endocrine cells as well as neurons. However, within a fairly short time data have accumulated indicating that neurons

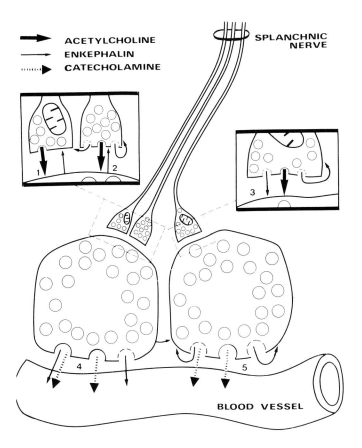

FIG. 2. Schematic illustration of enkephalin (ENK)-immunoreactive structures in the adrenal gland. Many gland cells contain ENK-LI which can be released into the blood or alternatively extracellular space, perhaps to act on receptors on adrenal gland cells or on presynaptic nerve endings. More recent evidence suggests that peptide and catecholamine are costored in the same vesicles (Viveros *et al.*, 1979). In addition, ENK-LI is present in nerve endings which disappear after section of the splanchnic nerve. ENK-LI may at least in part be present together with acetylcholine in the same nerves (Kondo *et al.*, 1985). From Schultzberg *et al.* (1978b).

may, in fact, release several messengers and that they may interact in a meaningful way. This information has been obtained in studies on, for example, the cat salivary gland (see Lundberg *et al.*, 1982d; Lundberg and Hökfelt, 1983) and frog sympathetic ganglia (Jan and Jan, 1983). The analysis of the central nervous system represents a much more difficult task and only fragmentary information on the functional meaning of coexistence is available. Moreover, and somewhat surprisingly, coexistence in endocrine cells, although established almost 20 years ago, has provided little information with regard to function. This may be due to the fact that presumed receptors for such hormones are widely spread and that possible interactions are difficult to analyze (see Fig. 2). Here, the nervous system provides a better model, since the target cells are defined by the neuronal projections and their terminal ramifications. It may be anticipated that the neuroendocrine system with its hypothalamic neurosecretory cells and with its well-defined targets, i.e., the hormone-producing cells in the anterior pituitary, also is accessible to functional analysis of effects possibly caused by multiple messengers released into the portal vessels.

In the present article we first want to discuss briefly the methodology involved in establishing coexistence, then summarize some of the present knowledge of various aspects on coexistence in general as revealed in studies on the peripheral and central nervous system and finally focus on hypothalamic systems presumably involved in the regulation of anterior pituitary hormone secretion.

II. Aspects on Methodology

A. DEMONSTRATION OF MULTIPLE ANTIGENS IN CELLS

Immunohistochemistry (see Coons, 1958; Sternberger, 1979) offers many advantages when attempting to characterize single neurons with regard to possible content of multiple antigens. At least three different approaches can be used as indicated in Fig. 3.

With the *"adjacent section" method* two or more serial sections are cut and incubated with antisera raised to different antigens (for examples, see Figs. 1, 25, and 27). This method can only be used for analysis of cell bodies, and the results are critically dependent on section thickness. Ideally sections down to 1 μm thickness should be used, but this can mostly be achieved only on plastic embedded material (see e.g., Berod *et al.*, 1984). With paraffin embedding 2–5 μm can be cut; in the cryostat it is difficult to obtain sections below 5 μm thickness. The advantage of this approach is that no interference between different antisera can occur and

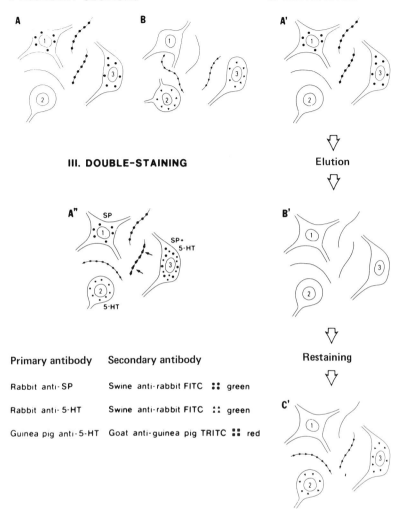

FIG. 3. Schematic illustration of three different approaches (I–III) to visualize two antigens in the same cell. (I) Thin adjacent sections (A and B) are incubated with antiserum to substance P (SP) and 5-hydroxytryptamine (5-HT), respectively. Cells 1–3 can be recognized in both sections, and one (No. 3) has both antigens, whereas cells 1 and 2 contain SP- and 5-HT-LI, respectively. Note that with this approach coexistence in nerve terminals cannot be demonstrated in view of their small size. (II) With the elution/restaining technique the same section (A') is incubated with SP antiserum, the staining pattern is photographed, and the section is eluted. After control incubation with FITC-conjugated antibodies, which should reveal complete lack of fluorescence (B'), the section is restained with 5-HT antiserum and the staining pattern photographed (C') and compared with the previous one. Again, it can be established that cell 3 contains both antigens. (III) With the double-staining technique a section (A'') is incubated with a mixture of SP antiserum raised in rabbit and 5-HT antiserum raised in guinea pig. The secondary antibodies are labeled with green fluorescent, FITC-conjugated swine anti-rabbit antibodies, and red fluorescent, TRITC-conjugated goat anti-guinea pig antibodies. Coexistence in cell 3 can be established by switching between proper filter combinations.

thus cross-reactivity problems will not be encountered. The disadvantages are mainly difficulties in unequivocally establishing the identity between cell profiles in the adjacent sections, and that coexistence in small structures such as nerve endings cannot be analyzed (see Figs. 25 and 27).

A second approach is the *"elution-restaining"* method as first described by Nakane (1968). Here the staining pattern obtained with one antibody is photographed, and the sections are then eluted in acid solutions, for example with acid potassium permanganate ($KMnO_4$) as recommended by Tramu *et al.* (1978). The sections are then processed for a second antiserum, and the new staining pattern photographed and compared with the first one (for example, see Figs. 18, 19, 21, and 26). It is important to carry out proper controls, i.e., to incubate the section with fluorescein isothiocyanate (FITC)-conjugated secondary antibody after elution, to validate complete removal of primary antibodies. The method offers the advantage of a safe establishment of identity, but the elution procedure may damage antigens in the sections so that restaining can not be carried out. In our experience, this approach can only be used with a limited number of antisera. A further disadvantage is the fact that all staining patterns of the first antiserum have to be photographed before elution, regardless of how successful the outcome is with the second incubation procedure.

When antisera raised in different species are available, a third method can be employed, the *"direct double staining."* The section is incubated with a mixture of two antisera raised, for example, in guinea pig and rabbit, respectively. As secondary antibody one can then use, for example, red fluorescent tetramethylrhodamine isothiocyanate (rhodamine, TRITC)-conjugated goat anti-guinea pig antibodies and green fluorescent FITC-conjugated swine anti-rabbit antibodies. If a proper balance can be obtained between the two antisera, and cross-reactivity can be excluded, this approach offers many advantages, since coexistence can be analyzed directly in the microscope by switching between proper filter combinations, and the findings can be documented immediately (see Figs. 19 and 23). Under favorable conditions, the direct double labeling technique offers possibilities to demonstrate coexistence in fibers.

If more than two antigens are to be analyzed, one may use, as stated above, the adjacent section method with thin sections. A further possibility is to combine, for example, the direct double labeling technique with the elution-restaining method (see Fig. 19).

B. SPECIFICITY—SENSITIVITY

Immunohistochemistry represents a uniquely powerful histochemical method but has inherent problems, especially with regard to specificity and sensitivity. *Specificity* problems arise with all immunological approaches, but are especially prominent for immunohistochemical methods, where often high concentrations of antibody are used. With novel improvements, e.g., the peroxidase–antiperoxidase technique of Sternberger and collaborators (1970) and its modifications, a very high sensitivity can be achieved with antibody dilutions of 10,000 or more, which reduces the risk for unspecific staining and cross-reactivity. However, it cannot be excluded, even at these dilutions, that the antibodies react with peptides or proteins containing a similar amino acid sequence. In fact, as discussed above, recent studies indicate that families of structurally similar peptides often exist. In this case sequence-specific antisera can perhaps be valuable in order to differentiate between such compounds.

Sensitivity also represents a serious problem in immunohistochemistry. Factors such as quality of antiserum, fixation, and species to be studied are important. To visualize peptides stores in cell bodies, it is often necessary to pretreat animals with colchicine, a mitosis inhibitor which arrests axonal transport and in this way causes accumulation of products synthesized in the cell body (see Dahlström, 1968, 1971; Hökfelt and Dahlström, 1971). Negative results should therefore be interpreted with caution and future experiments will probably demonstrate occurrence of certain peptides and transmitters in more systems than so far assumed.

C. TRACING OF TRANSMITTER IDENTIFIED PATHWAYS

An important task in the analysis of neuronal populations is to define their projections. Powerful techniques are now available for this purpose, including retrograde tracing methods (see Cowan and Cuenod, 1975). Moreover, retrograde tracing can be combined with immunohistochemistry, making possible mapping of transmitter-specific pathways (see Hökfelt *et al.*, 1983b; Skirboll *et al.*, 1984). With this approach retrograde tracers such as horseradish peroxidase (HRP) or fluorescent dyes (see Kuypers and Huisman, 1984) are injected into a certain brain area, and after retrograde transport to the cell bodies, they are visualized in the microscope. When combined with transmitter histochemistry, the sections are processed for immunohistochemistry and are analyzed for colocalization of retrograde tracer and immunohistochemical marker. This method is difficult in the median eminence because of problems of administering the retrograde tracer into this fragile structure with risks for diffusion into surrounding tissues and/or the third ventricle. However, some groups have succeeded in this task (Weigand and Price, 1980; Lechan *et al.*, 1982). We have attempted another approach, based on the fact that the median eminence is located outside the blood–brain barrier (Hökfelt *et al.*, 1985a). Thus, the retrograde tracer Fast Blue is injected intravenously, and nerve endings in the median eminence take up the dye, which is retrogradely transported to their parent cell bodies in hypothalamic (and perhaps other) regions. Using proper lesion controls, afferent projections to the median eminence can be defined in this way. Subsequently, immunohistochemistry using various antisera can be used to define the transmitter of the retrogradely labeled cells. In Fig. 4 this approach is schematically illustrated using antiserum to the enzyme tyrosine hydroxy-

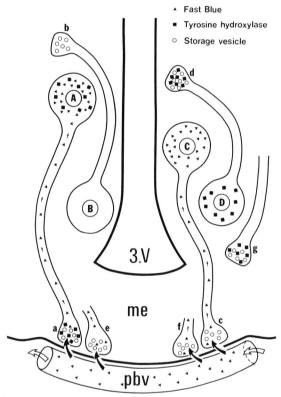

FIG. 4. Schematic illustration of our approach to try to elucidate the origin of nerve endings in the external layer of the median eminence (me). The retrograde tracer Fast Blue is injected intravenously and is taken up from the portal blood vessels (pbv) into nerve endings in the median eminence lying outside the blood–brain barrier and transported retrogradely to the cell bodies (A, C). After photography, the section is processed for indirect immunofluorescence histochemistry using, for example, antiserum to tyrosine hydroxylase (TH), a marker for dopamine neurons. Some neurons contain both Fast Blue and TH-LI (A), whereas another neuron (D) contains only TH-LI. The conclusion can therefore be drawn that one dopamine neuron (A) projects to external layer of the median eminence. 3.V, Third ventricle.

lase as marker for analysis of dopaminergic projections to the external layer of the median eminence (see also Fig. 24).

III. General Characteristics of Coexistence

A. DIFFERENT TYPES OF COEXISTENCE

Many examples of coexistence of a classical transmitter with one or more peptides have been encountered in the central (Table I) and peripheral (not shown) nervous system. Thus, coexistence situations can be found involving the catecholamines dopamine, noradrenaline, or adrenaline, as well as 5-hydroxytryptamine, acetylcholine, or GABA. In contrast no or only a few examples have been observed involving glutamate and glycine. This may reflect the lack of a good histochemical marker for these two amino acids. A second type of coexistence are neurons containing two or more peptides, but apparently lacking a classical transmitter. This group includes hypothalamic neurosecretory cells such as the neurons in the paraventricular and supraoptic nuclei, as well as many of the parvocellular systems projecting to the median eminence. Again, this may reflect lack of marker(s) for classical transmitter(s). A third example are neurons containing more than one classical transmitter, for example GABA plus 5-hydroxytryptamine (Belin *et al.*, 1983) or GABA plus dopamine (Everitt *et al.*, 1984b). Finally, ATP has been considered a transmitter candidate and may coexist, e.g., with noradrenaline and neuropeptide Y (NPY) (see Burnstock, 1985).

B. DIFFERENTIAL COEXISTENCE

A general feature encountered in the histochemical analysis of coexistence systems has been that a certain type of combination is limited to subpopulations of systems. For example, coexistence of 5-hydroxytryptamine and substance P has not been observed in the mesencephalic but only in the medullary raphe nuclei (Hökfelt *et al.*, 1978a). Furthermore, in medulla oblongata the proportion of neurons containing both 5-hydroxytryptamine and substance P, and sometimes TRH, varies considerably among different subnuclei (Johansson *et al.*, 1981).

Similar findings have been made in the analysis of NPY-LI in various central catecholamine cell groups (Everitt *et al.*, 1984a; for definition of nomenclature, see Dahlström and Fuxe, 1964; Hökfelt *et al.*, 1984d). Thus, numerous noradrenaline neurons of the A1 group as well as most adrenaline neurons of the C1, C2, and C3 (but not C2d) groups contain this peptide, and it is present in about 25% of the neurons in locus coeru-

TABLE I
Immunohistochemical Evidence for Coexistence of Classical Transmitters and Peptides in the Central Nervous System (Selected Cases)

Classical transmitter	Peptide[a]	Brain region (species)	References
Dopamine	CCK	Ventral mesencephalon (rat, cat, mouse, monkey, man?)	Hökfelt et al. (1980c,d, 1985b)
	Neurotensin	Ventral mesencephalon (rat)	Hökfelt et al. (1984a)
		Hypothalamic arcuate nucleus (rat)	Ibata et al. (1983); Hökfelt et al. (1984a)
Norepinephrine	Enkephalin	Locus coeruleus (cat)	Charnay et al. (1982); Léger et al. (1983)
	NPY	Medulla oblongata (man, rat)	Hökfelt et al. (1983); Everitt et al. (1984a)
	Vasopressin	Locus coeruleus (rat)	Everitt et al. (1984a)
	Neurotensin	Locus coeruleus (rat)	Caffé and van Leeuwen (1983)
Epinephrine	NPY	Medulla oblongata (rat)	Hökfelt et al. (1984a)
	Substance P	Medulla oblongata (rat)	Everitt et al. (1984a)
	CCK	Medulla oblongata (rat)	Lorenz et al. (1985)
	Substance P	Solitary tract nucleus (rat)	Hökfelt et al. (1985b)
5-HT		Medulla oblongata (rat, cat)	Chan-Palay et al. (1978); Hökfelt et al. (1978); Chan-Palay (1979); Johansson et al. (1981); Lovick and Hunt (1983)
	TRH	Medulla oblongata (rat)	Johansson et al. (1981)
	Substance P+TRH	Medulla oblongata (rat)	Johansson et al. (1981)
	CCK	Medulla oblongata (rat)	Mantyh and Hunt (1984)
	Enkephalin	Medulla oblongata, pons (cat)	Glazer et al. (1981); Hunt and Lovick (1982)
		Area postrema (rat)	Armstrong et al. (1984)

ACh	Enkephalin	Superior olive (guinea pig)	Altschuler et al. (1983)
		Spinal cord (rat)	Kondo et al. (1985)
	Substance P	Pons (rat)	Vincent et al. (1983)
	VIP	Cortex (rat)	Eckenstein and Baughman (1984)
	Galanin	Basal forebrain (rat)	Melander et al. (1985)
	CGRP	Medullary motor nuclei (rat)	Takami et al. (1985)
GABA	Motilin(?)	Cerebellum (rat)	Chan-Palay et al. (1981)
	Somatostatin	Thalamus (cat)	Oertel et al. (1983)
		Cortex, hippocampus (rat, cat, monkey)	Hendry et al. (1984); Jirikowski et al. (1984); Schmechel et al. (1984); Somogyi et al. (1984)
	CCK	Cortex (cat, monkey)	Hendry et al. (1984); Somogyi et al. (1984)
	NPY	Cortex (cat, monkey)	Hendry et al. (1984)
	Galanin	Hypothalamus (rat)	Melander et al. (1985)
	Enkephalin	Retina (chicken)	Watt et al. (1984)
		Ventral pallidum (rat)	Zahm et al. (1985)
	Opioid peptide	Basal ganglia (rat)	Oertel and Mugnaini (1984)
Glycine	Neurotensin	Retina (turtle)	Weiler and Ball (1984)

[a] This column contains the peptide against which the antiserum used for immunohistochemistry was raised. The exact structure of the peptide coexisting with the classical transmitter has, in most cases, not been defined.

leus (Holets *et al.*, 1985). So far no NPY-like immunoreactivity (NPY-LI) has been observed in dopamine neurons, but some of them contain instead cholecystokinin (CCK)-LI (Hökfelt *et al.*, 1980a,b) or neurotensin-LI in the ventral mesencephalon and in the hypothalamus (Hökfelt *et al.*, 1984c). Neurotensin-LI and CCK-LI may in addition be present in adrenaline neurons of the C2d group in the dorsal strip of the solitary tract nucleus. It appears therefore as if central neurons containing a classical transmitter can be subdivided into subclasses on the basis of the presence or absence of a certain peptide.

Another striking example in this respect has been observed in prevertebral ganglia in the guinea pig, where three distinct subpopulations of neurons have been found, containing somatostatin-, NPY-, and vasoactive intestinal polypeptide (VIP)-LI, respectively, which are located in different territories of the ganglion (Lundberg *et al.*, 1982c; Lindh *et al.*, 1986). They seem to have different targets for their projections and different functions (Furness *et al.*, 1983; Costa and Furness, 1984). Also in the central nervous system evidence is available for differential projections of subpopulations of transmitter-identified neurons containing a certain peptide. Thus, noradrenergic locus coeruleus neurons containing NPY seem to project preferentially to hypothalamic areas, whereas noradrenergic neurons projecting to the cortex and spinal cord more often lack this peptide (Holets *et al.*, 1985). Thus, the type of combination of messenger molecules may be related to function and/or projection site. Whether or not the peptide is of any direct importance for establishing connectivity is at present inconclusive.

C. SPECIES VARIABILITY

The occurrence and distribution of coexistence situations in different species have been studied only to a limited extent and mainly among mammalian species. In many cases the same coexistence situations can be found in several species. For example the 5-hydroxytryptamine/substance P/TRH systems in the medulla oblongata projecting to the spinal cord have been seen in the rat, cat, monkey, and man. Also dopamine/CCK coexistence has been observed in rat, mouse, cat, and monkey, although distinct differences in their distribution within the ventral mesencephalon and their projections have been encountered (see Hökfelt *et al.*, 1985b). In contrast, no dopamine/CCK coexistence was observed in guinea pig. In the peripheral nervous system somatostatin is frequently encountered in sympathetic noradrenergic neurons in the rat and guinea pig, but not in the cat (Lundberg *et al.*, 1982c). These findings suggest that coexistence combinations often are maintained in various mammalian species, but that distinct differences also are encountered.

IV. Mechanisms for Release of Coexisting Messengers

Analysis of the involvement of sympathetic and parasympathetic nerves containing several messengers in the salivary gland has given evidence for differential release of peptide and classical transmitter (see Lundberg, 1981; Lundberg *et al.*, 1982a; Lundberg and Hökfelt, 1983). Various mechanisms may underly such a differential release, but we have focused on a possible relation to subcellular storage sites. It is well known that most nerves contain at least two types of vesicles, small synaptic vesicles with a diameter of 500 Å and larger vesicles, often containing an electron-dense core and having a diameter of about 1000 Å. Early electron microscopic studies using potassium permanganate as fixative (Richardson, 1966) indicated that, e.g., both in peripheral and central noradrenergic neurons small and large vesicles may store noradrenaline (Fig. 5A) (Hökfelt, 1968; Ouimet *et al.*, 1981). Most immunocytochemical studies on the fine structural localization of peptides have revealed that they are stored in the large type vesicles (Fig. 5B) (see Pickel, 1985). However, with the latter technique classical transmitters, such as 5-hydroxytryptamine, only appear in the large vesicles (see e.g., Pelletier *et al.*, 1981). Immunohistochemistry may therefore not be a reliable technique for determining the true subcellular localization of messenger molecules.

Against this background we have employed subcellular fractionation and centrifugation techniques to compare the distribution of coexisting compounds. Two tissues were analyzed, the submandibular salivary gland of the cat, where acetylcholine and VIP coexist in parasympathetic nerves, and rat vas deferens, where noradrenaline and NPY are present together in the sympathetic nerves. The latter tissue has been thoroughly investigated in previous studies, and, using vas deferens of castrated rats, it is possible to obtain fairly pure fractions containing small and large vesicles, respectively (Fried *et al.*, 1978). The results indicate that in both tissues the peptide (VIP and NPY, respectively) can be observed in the high-density fraction, whereas the classical transmitters appear both in heavy and light fractions (Fig. 6). This suggests that peptides are exclusively stored in the large vesicles, whereas acetylcholine and noradrenaline are present in both small and large vesicles (Figs. 6 and 7) (Lundberg *et al.*, 1981; Fried *et al.*, 1985). This provides a morphological basis for differential release, which according to our hypothesis is frequency coded (see Lundberg and Hökfelt, 1983). Thus, at low frequencies small vesicles

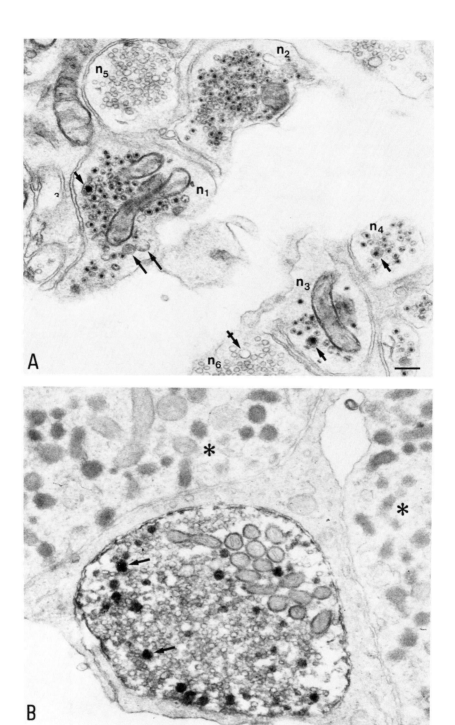

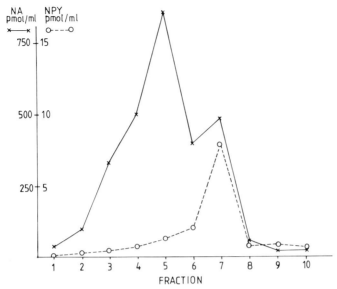

FIG. 6. Subcellular distribution of noradrenaline (full line) and neuropeptide Y (broken line) in a density gradient of rat vas deferens. On the Y axis, the total recovered sedimentable substance given as pmol/ml after centrifugation at 145,000 g_{max} for 45 minutes is shown. On the X axis the density gradient fractions 1–10 corresponding to increasing concentrations of sucrose are indicated (1 = 0.26 M; 10 = 1.2 M). For further details see Fried *et al.* (1985).

FIG. 5. Electron micrographs of nerve endings in the autonomic nervous system. (A) shows material fixed with potassium permanganate for optimal demonstration of amine content in small and large dense-core vesicles. (B) represents an enkephalin (ENK)-immunoreactive nerve ending in the adrenal medulla demonstrated with peroxidase–antiperoxidase immunohistochemistry. (A) In the permanganate-fixed material both small and large (arrows) vesicles contain a dense core suggesting presence of noradrenaline in several nerve endings (n_{1-4}). Note variable intensity of the dense core suggesting different amine levels in the large vesicles (short arrows show highly dense core, long arrows show vesicles with a low density). Almost all small vesicles contain a dense core. Two nerve endings ($n_{5,6}$) contain only empty vesicles and probably belong to cholinergic fibers. Note also empty large vesicle (crossed arrow). (B) The ultrastructural appearance of the ENK-positive nerve ending is characteristic of most immunohistochemical stainings for neuronal peptides revealing an electron-dense core only in the large vesicles (arrows). Asterisks denote adrenal medullary cells. Bar indicates 0.2 μm.

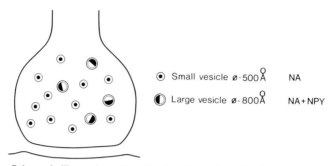

FIG. 7. Schematic illustration of localization of noradrenaline (NA) and neuropeptide Y (NPY) in nerve endings in the vas deferens. The small vesicles contain only NA, whereas the large vesicles contain both NA and NPY-LI.

are selectively activated resulting in release of only the classical transmitter; at high nerve pulse activity the large vesicles will release their content of classical transmitter and peptide (Fig. 8).

V. Coexistence and Plasticity

Coexistence situations seem to represent fairly stable phenomena in the sense that they can be demonstrated in a reproducible manner with immunohistochemical techniques. An important question is, however, whether or not the phenotype of coexistence can be influenced. It is well known from elegant studies from several laboratories that during development, peripheral neurons may change their expression of transmitters both *in vitro* and *in vivo* (Patterson, 1978; Potter *et al.*, 1983; Black *et al.*, 1984). In recent studies we have provided evidence that this can occur also in adult animals *in vivo* under certain experimental conditions, for example, in transplantation experiments in the central nervous (Foster *et al.*, 1985) as well as in iris of the rat (Björklund *et al.*, 1985). We would briefly like to describe the latter experiments.

The rat iris contains a supply of sympathetic noradrenergic nerves, a proportion of which contain NPY-LI, as well as cholinergic nerves (without a known peptide) and sensory neurons containing, for example, substance P. Sympathectomy by removal of the superior cervical ganglia causes a rapid and virtually complete disappearance of noradrenergic nerves as demonstrated with formaldehyde-induced fluorescence and this disappearance is irreversible (Malmfors, 1965). When long-term sympathectomized irides were analyzed with immunohistochemistry using the enzyme tyrosine hydroxylase as a marker for noradrenergic neurons, positive networks could be seen and NPY-immunoreactive fibers were

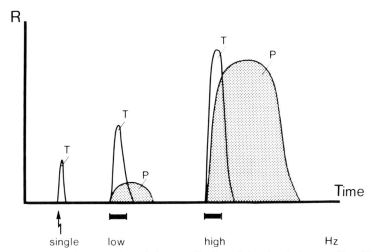

FIG. 8. Schematic illustration of the contribution of the classical transmitter (T) and coexisting peptide (P) to the functional response (R) upon various degrees of activation by electrical stimulation, ranging from single impulse to low and high frequencies (Hz). Single nerve impulses preferentially release the classical transmitter, resulting in a rapid response of short duration. At high frequencies peptides are released and cause a response with slower onset and a longer duration. From Lundberg and Hökfelt (1983).

also encountered (Björklund et al., 1985). The appearance of these fibers was time dependent, being first visible a few days after sympathectomy. These fibers were largely absent after parasympathectomy, i.e., removal of the ciliary ganglion. These findings indicate that cholinergic nerves under these experimental conditions can acquire sympathetic markers, i.e., the enzyme tyrosine hydroxylase and the peptide NPY (Björklund et al., 1985). Furthermore, these findings in general terms suggest possibilities of plasticity in the expression of phenotype also in adult animals and that they involve not only markers for the classical transmitter but also that a coexisting peptide can be expressed in parallel.

VI. Functional Aspects on Coexistence

The functional significance of the possible release of several messengers from the same nerve endings is still only incompletely understood. Our best evidence in this respect stems from studies on the cat salivary gland and rat vas deferens (see Lundberg, 1981; Lundberg et al., 1982c; Lundberg and Hökfelt, 1983) as well as from studies on autonomic ganglia in the bull frog (Jan and Jan, 1983). Moreover, results from studies in the central nervous system have given us some clues that mechanisms similar to those seen in the periphery also may operate here. In the following we would like to briefly go through some of these results.

A. POSTSYNAPTIC INTERACTIONS

The cat salivary gland represents a classical physiological model for analysis of nervous control of secretion and blood flow. It receives a parasympathetic innervation containing acetylcholine and VIP (Lundberg *et al.*, 1979) and two populations of sympathetic fibers containing noradrenaline and noradrenaline plus NPY, respectively (Lundberg *et al.*, 1982c). The latter nerves preferentially innervate blood vessels, whereas the noradrenergic nerves apparently lacking NPY are found around secretory elements (Fig. 9).

Extensive experiments by Lundberg, Änggård, and collaborators (see Lundberg *et al.*, 1982a) have shown that peptide and classical transmitter interact in a cooperative way and enhance each others effects on blood flow and secretion. They have been able to demonstrate that both peptides and classical transmitters are released upon electrical stimulation of the parasympathetic (Fig. 10) and sympathetic nerves, respectively. Functional analysis has revealed that acetylcholine induces both secretion and increase in blood flow (Fig. 10), whereas VIP has no apparent effect on secretion but can by itself increase blood flow. VIP can, however, potentiate acetylcholine-induced secretion, and acetylcholine and VIP infused together produce additive effects on blood flow. It has long been recognized that, whereas secretion induced by electrical stimulation can be completely blocked by atropine, suggesting a pure cholinergic

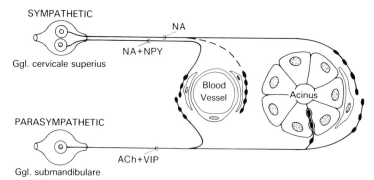

FIG. 9. Schematic illustration of the autonomic innervation of the cat submandibular salivary gland. Both the sympathetic and the parasympathetic neurons represent coexistence systems. The noradrenergic neurons innervating blood vessels contain a neuropeptide Y-like peptide and the cholinergic neurons contain vasoactive intestinal polypeptide-LI. From Lundberg and Hökfelt (1983).

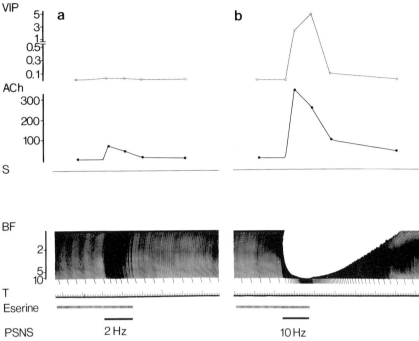

FIG. 10. Effects of low- (2 Hz) (a) and high- (10 Hz) (b) frequency stimulation of the submandibular branches of the caudal lingual nerve (parasympathetic nerve stimulation, PSNS) on secretion (S) and blood flow (BF) after atropine pretreatment. The output of the VIP (fmol min^{-1} g^{-1}) and acetylcholine (fmol min^{-1} g^{-1}) in the venous effluent from the gland is also shown. Physostigmine (eserine) has been given by local intraarterial infusion as indicated by the bar. Note that atropine completely blocks salivary secretion both at high- and low-frequency stimulation, whereas blood flow is affected only at low-frequency stimulation. No certain increase in release of VIP and in blood flow can be recorded at 2 Hz stimulation, whereas a marked increase in both parameters is seen at 10 Hz. Acetylcholine release can be recorded at both frequencies.

mediation, this drug cannot fully block vasodilation (Fig. 10). It seems likely that this atropine-resistant vasodilation is induced by VIP. A potentiating effect of VIP on acetylcholine-induced secretion may be explained by binding experiments, demonstrating that VIP can markedly change the binding characteristics for cholinergic agonists to membrane fragments from the salivary gland (Lundberg *et al.*, 1982b).

In the sympathetic system both messengers, noradrenaline and NPY, cause vasoconstriction (Fig. 11) (Lundberg and Tatemoto, 1982). Noradrenaline causes a rapid, short lasting effect, whereas NPY induces a slow, long lasting decrease in blood flow. Together they give a compound response, similar to the one seen after high-frequency stimulation (Fig.

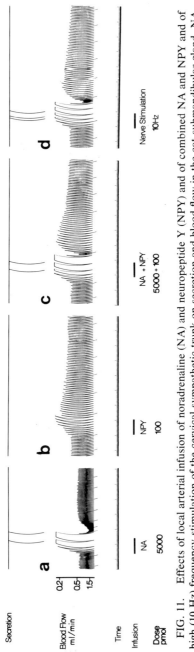

FIG. 11. Effects of local arterial infusion of noradrenaline (NA) and neuropeptide Y (NPY) and of combined NA and NPY and of high (10 Hz) frequency stimulation of the cervical sympathetic trunk on secretion and blood flow in the cat submandibular gland. NA causes a rapid response of short duration (a), NPY a response with a slow onset and of longer duration (b), and the combined administration of the two compounds (c) results in a response very similar to the one seen after stimulation at high frequency (d).

11). Thus, with regard to circulatory effects the coexisting peptides in both the parasympathetic and sympathetic nerves cooperate with their respective classical transmitter (Fig. 8). The classical transmitters are released already at low impulse activity, causing rapid effects with a short duration on blood flow and secretion. In contrast, the effects induced by the peptides have a slow onset and long duration and seem to be induced at high stimulation frequencies (see Lundberg, 1981) or alternatively when nerves are stimulated with trains of impulses (Edwards *et al.*, 1982).

Indirect evidence suggests that multiple messengers may also interact postsynaptically in a cooperative way in the central nervous system. Focusing on motor mechanisms in the spinal cord, Barbeau and Bedard (1981) have demonstrated that intravenously administered TRH can markedly activate the stretch reflex in chronically spinalized rats. The same effects could be observed with the 5-hydroxytryptamine precursor, 5-hydroxytryptophan, as shown earlier by Andén *et al.* (1964). In this model the effects of TRH could be blocked by a 5-hydroxytryptamine antagonist. TRH and 5-hydroxytryptamine coexist in medullary neurons (Table I) known to project to the spinal cord, giving rise to nerve endings in the ventral horns. Thus the two compounds are present in, and possibly released from, the same nerve endings and may act on the same or on two closely related (coupled?) receptor sites, possibly on motoneurons (Fig. 12). A strong cooperativity between TRH and 5-hydroxytryptamine at the spinal cord level has also been observed in studies of sexual behavior (Hansen *et al.*, 1983). In this study 5-hydroxytryptamine and TRH were administered intrathecally (Yaksh and Rudy, 1976), alone or together. When given separately, 5-hydroxytryptamine and TRH produced no or very small effects, but when administered together they caused a marked increase in two parameters recorded in the sexual behavior model, the mount and intromission latencies. These findings are in agreement with a cooperative interaction of these two compounds, possibly released from the same nerve endings in the ventral horn of the spinal cord (Fig. 12).

B. PRESYNAPTIC INTERACTIONS

Evidence for presynaptic interaction is available from studies both in the periphery and in the central nervous system. Thus, in the rat vas deferens, which contains noradrenaline and NPY in sympathetic nerves, NPY in a dose-dependent manner inhibits the electrically induced contraction of vas deferens. This effect seems to be due to inhibition of release of noradrenaline (Allen *et al.*, 1982; Lundberg *et al.*, 1982d; Ohhashi and Jacobowitz, 1983; Stjärne and Lundberg, 1984).

An apparent inhibitory action of a peptide on transmitter release has

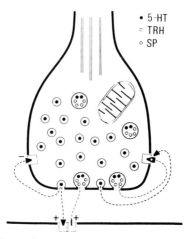

FIG. 12. Schematic illustration of a nerve ending in the ventral horn of the spinal cord containing 5-hydroxytryptamine (5-HT), thyrosin-releasing hormone (TRH), and substance P (SP). The small vesicles with a diameter of about 500 Å contain exclusively 5-HT, whereas the large vesicles with a diameter of about 1000 Å contain all three compounds. 5-HT acts both on a postsynaptic excitatory receptor and a presynaptic inhibitory receptor. TRH may act on a postsynaptic receptor closely related to the 5-HT receptor. SP may act on the presynaptic inhibitory 5-HT receptor to block this effect of 5-HT.

also been encountered in the central nervous system, i.e., in the caudate putamen of the cat. Here dopamine and CCK-LI coexist in nerve endings arising from neurons in the substantia nigra of the ventral mesencephalon (Hökfelt et al., 1985b). The effects of CCK peptides were analyzed on basal and electrically evoked tritium overflow from slices after preloading with tritiated dopamine. Sulfated, but not unsulfated CCK-7 and CCK-8 inhibited tritium overflow in very low concentrations. These effects were observed both on basal outflow as well as after electrical stimulation (Fig. 13). We have speculated that the inhibitory effect exerted by the two peptides in the two different systems discussed above may serve to prevent excessive release of the catecholamines, i.e., the main transmitters.

Mitchell and Fleetwood-Walker (1981) have described a different type of presynaptic interaction in the spinal cord. They analyzed the effect of TRH and substance P on potassium induced 5-hydroxytryptamine release from slices of the spinal cord. Neither peptide influenced potassium-induced tritium outflow. However, when cold 5-hydroxytryptamine was added to the bath in a concentration which inhibits 5-hydroxytryptamine release, substance P, but not TRH counteracted the inhibition of tritium outflow caused by 5-hydroxytryptamine. This indicates that substance P

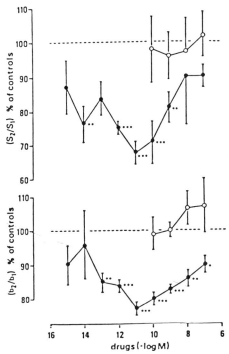

FIG. 13. Effects of sulfated CCK-8 (dots) and the nonsulfated form of CCK-8 (circles) on basal and electrically evoked tritium outflow slices of cat caudate nucleus previously labeled with [^{3}H]dopamine. After preincubation with [^{3}H]dopamine, slices were superfused with Krebs medium containing BSA and bacitracin and stimulated twice each at 2 Hz for 2 minutes (S_1, S_2). Drugs were added 30 minutes before S_2. Drug effects on basal tritium efflux are expressed as the ratio of the fractional tritium outflow of the fraction before S_2 to that before drugs were added (b_2/b_1), as a percentage of controls. The effect on electrically evoked tritium overflow is expressed as the ratio of tritium overflow evoked by S_2 to that evoked by S_1 as a percentage of controls. Stars indicate level of significance. Means ± SEM of four to six experiments. From Markstein and Hökfelt (1984).

act on the presynaptic inhibitory 5-hydroxytryptamine receptor and may cause an enhanced 5-hydroxytryptamine release (Fig. 12). In this way substance P acts to strengthen 5-hydroxytryptamine transmission in the spinal cord. As discussed above TRH which is costored together with 5-hydroxytryptamine and substance P also enhances 5-hydroxytryptamine transmission, but via a postsynaptic effect. It is therefore possible that two peptides released together with 5-hydroxytryptamine both cooperate with the amine to enhance transmission, but in two different ways, substance P by a presynaptic action, TRH by a postsynaptic action (Fig. 12).

VII. Peptide–Peptide Interactions

As discussed above there is evidence for neurons containing several peptides, but apparently no classical transmitter. We would like to discuss here one example, which may suggest some new principles on how multiple messengers may interact. The system in question is a population of primary sensory neurons, which early were demonstrated to contain substance P-like immunoreactivity (Takahashi *et al.*, 1974; Hökfelt *et al.*, 1975). More recently evidence has been presented that they contain another member of the tachykinin family, neurokinin A (neurokinin-α, substance K, neuromedin L) (Kimura *et al.*, 1983; Kangawa *et al.*, 1983; Nawa *et al.*, 1983; Minamino *et al.*, 1984), as well as the recently discovered calcitonin gene-related peptide (CGRP) (Amara *et al.*, 1982; Rosenfeld *et al.*, 1983a,b). Thus, several groups have described coexistence of CGRP- and substance P-LI in such primary sensory neurons (Gibson *et al.*, 1984; Wiesenfeld-Hallin *et al.*, 1984; Gibbins *et al.*, 1985; Lee *et al.*, 1985; Lundberg *et al.*, 1985). We have analyzed some possible types of interactions of these compounds, both at their central branches in the spinal cord as well as in the periphery, where their sensory processes terminate.

Central interaction was studied using the intrathecal administration technique of Yaksh and Rudy (1976) by injecting substance P, CGRP, or both drugs combined at lumbar levels. It is known that substance P injected in this way causes a characteristic, caudally directed biting and scratching behavior (Hylden and Wilcox, 1981; Piercey *et al.*, 1981; Seybold *et al.*, 1982). Our studies confirmed this to be a dose-dependent behavior with a relative short duration of a few minutes. CGRP alone in doses up to 20 μg did not cause any observable effects. However, when substance P (1 or 10 μg) and CGRP (20 μg) were given together, the biting and scratching behavior was markedly prolonged and lasted up to 40 minutes (Fig. 14). The mechanism underlying this dramatic effect has been analyzed by Terenius and collaborators (see Le Greves *et al.*, 1985). They found that CGRP is a potent inhibitor of a substance P endopeptidase. It may therefore be that two peptides released from the same nerve endings interact in a cooperative way, whereby one peptide (CGRP) inhibits breakdown of its comessenger (substance P).

The possible interactions of tachykinins and CGRP in peripheral tissues seem to be of a different type. When analyzing for the effects of these peptides on various parameters such as blood pressure, blood flow, extravasation, and insufflation pressure, distinct and differential effects were observed for the various compounds. As shown in Fig. 15 all compounds decrease blood pressure. Substance P, but not CGRP, causes extravasation and increased insufflation pressure in the lungs, whereas

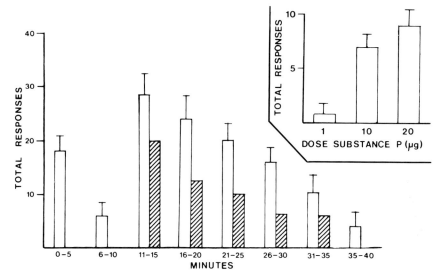

FIG. 14. Time course of biting/scratching responses for five rats injected with 10 μg SP + 20 μg CGRP (open bars) and one rat injected with 1 μg SP + 20 μg CGRP (hatched bars). The response lasts for up to 40 minutes. Inset shows average number of biting/scratching responses during the first 2 minutes after injection of various doses of substance P. From Wiesenfeld-Hallin et al. (1984).

	Extravasation (Evan's Blue)	Blood pressure	Insufflation pressure	Heart rate	Ventricular rate	Ventricular tension
Saline	0	0	0	0	0	0
Substance P	↑↑↑	↓↓	↑↑	0	0	0
Neurokinin A (⌁, Substance K, Neuromedin L)	↑↑	↓↓	↑↑↑	↓↓	↓↓	↓↓
Calcitonin Gene Related Peptide	0	↓↓↓	0	↑↑↑	↑↑↑	↑↑

FIG. 15. Schematic illustration of the effects of the tachykinins substance P and neurokinin A (substance K) and of calcitonin gene-related peptide on various parameters such as extravasation, blood pressure, insufflation pressure, heart rate, ventricular rate, and ventricular tension. These compounds are present in the same primary sensory neurons. Note the differential effects of the various compounds on recorded parameters. Data taken from Lundberg et al. (1985).

CGRP, but not substance P, markedly affects heart rate, ventricular rate, and ventricular tension. It may be speculated that the specificity of action in these cases is obtained by the distribution of receptors. For example, the heart muscles may have receptors preferentially for CGRP, but may lack substance P receptors. Such a receptor-mediated specificity would not require a neuron to be able to release CGRP and substance P differentially, but specificity would be obtained by the presence or absence of receptors.

VIII. Coexistence in Neuroendocrine Systems

Research during recent years has demonstrated that coexistence of multiple messengers is particularly frequent in neuroendocrine systems. Thus, the magnocellular neurons projecting from the supraoptic and paraventricular nuclei, as well as the parvocellular neurons located in the parvocellular part of the paraventricular nucleus and in the arcuate nucleus and some other hypothalamic regions, often contain more than one bioactive peptide. Perhaps the highest numbers of messengers have been detected in the magnocellular systems, where more than half a dozen peptides can be visualized with immunohistochemical techniques (see Brownstein and Mezey, 1986). We will here focus on two other areas: (1) the arcuate nucleus as the final integrator for output of neuroendocrine information from the brain to the anterior pituitary gland, and (2) a population of neurons in the parvocellular paraventricular nucleus projecting to the median eminence involved in control of adrenocorticotropin(ACTH) and prolactin secretion. We will also mention the possibility that the luteinizing hormone-releasing hormone (LHRH) neurons may contain a further compound involved in the control of luteinizing hormone secretion.

A. COEXISTENCE IN THE ARCUATE NUCLEUS

Early work by the Hungarian pioneers in neuroendocrine research (see Szentagothai *et al.*, 1962) demonstrated projections from the arcuate nucleus to the external layer of the median eminence, suggesting that this nucleus may be of importance in the control of anterior pituitary hormone secretion. In fact, such a tuberoinfundibular system represents a further type of neurosecretory neuron, first demonstrated by the Scharrers and Bargmann (Bargmann and Scharrer, 1951; Scharrer and Scharrer, 1954), and they could be the morphological substrate for the neurons producing and releasing the hypothalamic factors, postulated by Green and Harris (see Harris, 1955) to carry information between brain and anterior pituitary via the portal blood vessels. Using the formaldehyde-induced fluorescence technique (see Falck et al., 1962), Fuxe (1964) was able to chemically define the first tuberoinfundibular system as containing dopamine (Fuxe, 1964; Fuxe and Hökfelt 1966), strongly suggesting a role for this catecholamine in the central control of anterior pituitary hormone secretion. The existence of the tuberoinfundibular dopamine neurons was subsequently confirmed using immunohistochemistry and antiserum to the catecholamine-synthesizing enzyme tyrosine hydroxylase (Fig. 16) (Hökfelt et al., 1976). It is now generally accepted that dopamine represents an important prolactin release inhibitory factor (see MacLeod and Lehmeyer, 1974), but that these dopamine neurons also seem to be involved in parallel in control of other neuroendocrine events (see Fuxe and Hökfelt, 1969; Hökfelt and Fuxe, 1972).

With the purification and identification of several hypothalamic releasing and inhibitory hormones by Guillemin, Schally, Vale, and their collaborators, new avenues opened up to characterize the basal hypothalamus

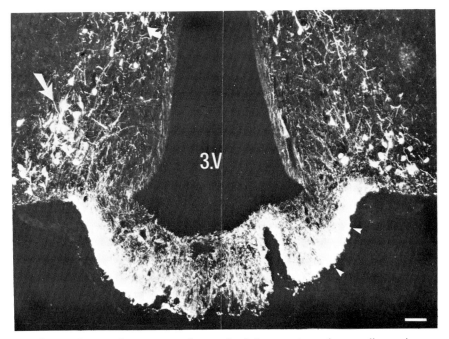

FIG. 16. Immunofluorescence micrograph of the arcuate nucleus–median eminence complex after incubation with antiserum to tyrosine hydroxylase. Positive cell bodies are seen in the arcuate nucleus, some located in the dorsal medial aspect (small arrow) and another group in the ventrolateral part of the nucleus (large arrow). Note dense network of fluorescent fibers mainly in the lateral parts of external layer (arrowheads). 3.V, Third ventricle. Bar indicates 50 µm.

with immunohistochemical techniques. Thus, the analysis of the distribution of LHRH, TRH, and somatostatin revealed numerous positive nerve endings in the external layer of the median eminence, but with distinctly differential distribution patterns (for refs., see Hökfelt et al., 1978b). None of these three systems, however, seemed to originate in the arcuate nucleus, but had their cell bodies outside the basal hypothalamus (Fig. 17). Also the recently discovered CRF (Spiess et al., 1981; Vale et al., 1981) has its cell bodies outside the basal hypothalamus, namely in the parvocellular paraventricular nucleus giving rise to a dense fiber network

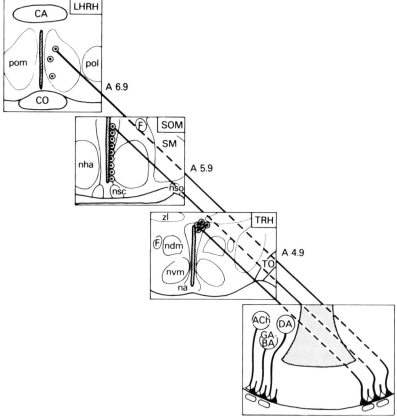

FIG. 17. Schematic illustration of the origin of the three hypothalamic hormones LHRH, somatostatin (SOM), and TRH, all of which project to the external layer of the median eminence. Thus, these major releasing and inhibitor factors have their origin outside the basal hypothalamus. Instead, the arcuate nucleus contains neurons with classical transmitters such as dopamine (DA), acetylcholine (ACh), and GABA. A more detailed account on the type of cells in the arcuate nucleus is given in Fig. 22.

in the external layer of the median eminence (see below). Finally, the most recent characterized hypothalamic hormone, growth hormone-releasing factor (GRF) (Böhlen *et al.*, 1984; Brazeau *et al.*, 1982; Guillemin *et al.*, 1982; Spiess *et al.*, 1983) turned out to be the first hypothalamic factor identified in cell bodies in the arcuate nucleus and in adjacent areas, giving rise to a dense fiber network in the external layer of the median eminence (for references, see below). It has been shown that virtually all these GRF-positive neurons also contain tyrosine hydroxylase-LI (Meister *et al.*, 1985a,b; Okamura *et al.*, 1985) and therefore in part are identical to the early discovered dopamine cells. For this reason we have undertaken an extensive analysis of the occurrence and distribution of peptides and classical transmitters in the arcuate complex with focus on possible coexistence situations, especially with the dopamine-synthesizing enzyme tyrosine hydroxylase (Everitt *et al.*, 1985). In a subsequent study we intend to determine which of these arcuate neuron populations project to the external layer of the median eminence by using combined retrograde tracing and immunohistochemistry (see Methods) (Hökfelt *et al.*, 1983b, 1985a; Skirboll *et al.*, 1984). This will help to understand at what level—pituitary, median eminence, and/or brain—these systems may exert a possible influence.

B. CLASSICAL TRANSMITTERS IN THE ARCUATE NUCLEUS

Immunohistochemical studies using antibodies to tyrosine hydroxylase have confirmed and extended earlier formaldehyde fluorescence studies and shown that dopamine neurons can be subdivided into two groups, a dorsal group consisting of small, strongly tyrosine hydroxylase-positive cell bodies and a ventrolateral group of larger cells with a less intense immunostaining (Fig. 16) (Hökfelt *et al.*, 1976, 1984c,d; Chan-Palay *et al.*, 1984; van den Pol *et al.*, 1984; Everitt *et al.*, 1985). The exact identity of the catecholamine(?) (amino acid?) in the latter cell group is at present under investigation (see Meister *et al.*, 1985b).

Biochemical analyses of glutamate decarboxylase (GAD), the GABA-synthesizing enzyme, combined with monosodium glutamate-induced lesions have suggested the presence of GABA neurons in the arcuate nucleus with projections to the median eminence (Walaas and Fonnum, 1968). The introduction of immunohistochemical analysis of GABA neurons with antiserum raised to GAD opened up the possibility of defining these neurons more exactly. In fact, GAD-positive neurons have been demonstrated in the arcuate nucleus (Vincent *et al.*, 1982a; Tappaz *et al.*, 1983; Mugnaini and Oertel, 1985), which could give rise to GABA fibers in the external layer of the median eminence, demonstrated with both auto-

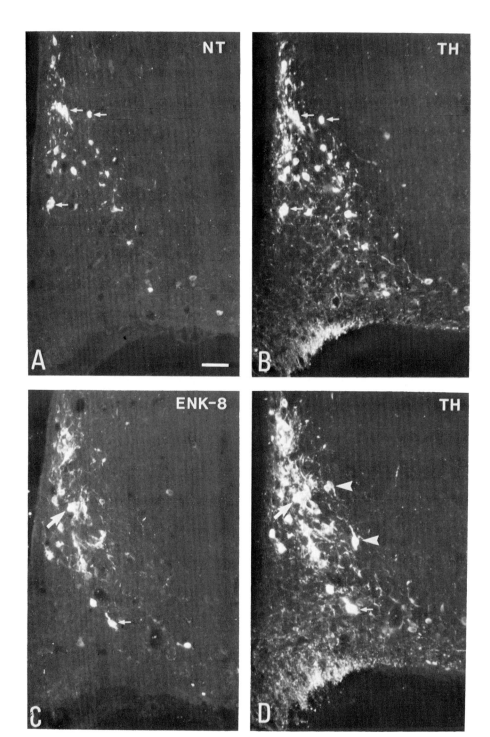

radiographic (Tappaz *et al.*, 1980) and immunohistochemical techniques (Perez de la Mora, 1981; Vincent *et al.*, 1982; Tappaz *et al.*, 1983). The number of GAD-immunoreactive cell bodies is, however, not large, and they exhibit a low GAD immunoreactivity. This, combined with the presence of dense GAD fiber networks obscuring GAD-positive cells, leaves the possibility open that the studies carried out so far have revealed only a fraction of all GABA neurons in this area. Elution-restaining experiments have revealed a considerable degree of coexistence between GAD- and tyrosine hydroxylase-positive neurons (Everitt *et al.*, 1984b, 1985), suggesting the presence of arcuate neurons producing two classical transmitters, GABA and dopamine.

Neurons containing a third classical transmitter, acetylcholine, may also be present in the arcuate nucleus although the evidence is less solid. Carson *et al.* (1977) have demonstrated acetylcholinesterase-positive cell bodies, which disappear after monosodium glutamate treatment. This lesion also caused a marked reduction in levels of the acetylcholine-synthesizing enzyme (ChAT), both in the arcuate nucleus and median eminence. In spite of the fact that several antisera against ChAT have been available for some years, it has so far not been possible to demonstrate ChAT-positive neurons in the arcuate nucleus to our knowledge. The exact definition of a cholinergic component in the arcuate–median eminence complex therefore awaits further experimental evidence.

C. PEPTIDES IN THE ARCUATE NUCLEUS

The presence of *neurotensin-* (Carraway and Leeman, 1973, 1975) positive neurons in the arcuate nucleus has been described in several studies (Uhl *et al.*, 1979; Kahn *et al.*, 1981; Jennes *et al.*, 1982; Ibata *et al.*, 1980, 1984a,b; Hökfelt *et al.*, 1984b). A porportion of these neurotensin-positive cells are identical to tyrosine hydroxylase-immunoreactive neurons located both in the dorsal and ventral aspects of the arcuate nucleus (Fig. 18A and B) (Ibata *et al.*, 1983; Hökfelt *et al.*, 1984b). Neurotensin-positive fibers are also present in the external layer of the median eminence. To what extent all these fibers originate from arcuate neurons is at the moment uncertain, since neurotensin-LI has also been described in parvocellular neurons (Sawchenko *et al.*, 1984). Neurotensin has been shown

FIG. 18. Immunofluorescence micrographs of two sections (A,B and C,D) of the arcuate nucleus incubated with antiserum to neurotensin (NT) (A) and enkephalin octapeptide (ENK-8) (C), followed by elution and restaining with tyrosine hydroxylase (TH) antiserum (B,D). Most NT- and ENK-8-immunoreactive cells also contain TH-LI (arrows). Arrowheads in D point to TH-positive/ENK-8-negative cells. Bar indicates 50 μm.

to exert variable effects on prolactin secretion. Thus, intraventricular injections inhibit prolactin secretion (Vijayan and McCann, 1979), in agreement with the inhibitory role of the coexisting dopamine on this pituitary hormone. Intravenous injections, on the other hand, elevate prolactin serum levels (Rivier *et al.*, 1977).

A further peptide present in the arcuate nucleus is *galanin* (Tatemoto *et al.*, 1983), which has been shown to be present with the highest numbers in its ventral part but also extending into the dorsal portion of the nucleus (Fig. 19B and E) (Melander *et al.*, 1986). Many of these galanin-positive cells also contain tyrosine hydroxylase-LI (Fig. 19A and D). A dense galanin-positive network is present in the external layer of the median eminence, possibly originating in the arcuate nucleus (Fig. 19B). So far no functional role for galanin in neuroendocrine events has been reported. It has been demonstrated that *GRF* (for references, see above) is located in the arcuate nucleus extending into adjacent areas (Fig. 19C and F) (Bloch *et al.*, 1983a,b; 1984; Bugnon *et al.*, 1983; Fellmann *et al.*, 1983; Jacobowitz *et al.*, 1983; Lechan *et al.*, 1984; Merchenthaler *et al.*, 1984; Smith *et al.*, 1984; Sawchenko *et al.*, 1985). It has also been shown that the vast majority of these GRF-immunoreactive cell bodies contain tyrosine hydroxylase-LI (Fig. 19A and D). Furthermore, a dense GRF-positive fiber network can be seen in the external layer of the median eminence, partly overlapping with the tyrosine hydroxylase-positive fibers (Fig. 19A and C).

These findings raise the possibility that dopamine participates in the regulation of growth hormone secretion either at the median eminence level or in the anterior pituitary. Such a possible function has been analyzed in many studies, but the interactions seem to be complex and controversial. Thus, dopamine, L-dopa, and dopamine agonists seem to cause growth hormone release and increase plasma concentrations of the hormone (Müller, 1973; Vijayan *et al.*, 1979a). Furthermore, dopamine receptor antagonists have been shown to reduce plasma growth hormone levels (Müller, 1973). Intraventricular infusions of dopamine and dopamine agonists elevate plasma growth hormone levels (Vijayan *et al.*, 1978). The site of action of dopamine is at the present not clear, but it may inhibit release of somatostatin (Fig. 20A) (Arimura and Fishback, 1981). Inhibition of secretion has also been suggested by Andersson *et al.* (1977, 1983). They demonstrated that intravenous injection of rat growth hormone caused a reduction of catecholamine turnover in the median and lateral palisade zone of the median eminence and suggested that this represents a short feed back phenomenon by which growth hormone inhibits its own secretion by reduction of dopamine synthesis and release. This effect could very well be exerted via an action on somatostatin

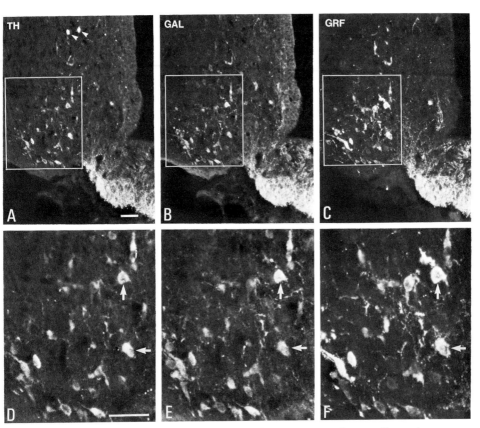

FIG. 19. Immunofluorescence micrographs of the arcuate nucleus–median eminence complex of one single section demonstrating distribution of tyrosine hydroxylase (TH) (A,D)-, galanin (GAL) (B,E)-, and growth hormone-releasing factor (GRF) (C,F)-like immunoreactivities. The section has first been double-stained with monoclonal antibodies to TH (A, D) using red fluorescent rhodamine-labeled secondary antibodies and with rabbit antiserum to GAL using green fluorescent FITC-conjugated secondary antibodies (B, E). The patterns in these micrographs can thus be viewed by switching between the two proper filter combinations. In a second step, the first two antisera have been eluted with acid potassium permanganate and the section restained with antiserum to GRF. Note that in the ventrolateral part of the arcuate nucleus (D–F; boxes in A–C), most cell bodies (arrows) contained all three compounds. Arrowheads point to two cells in the dorsal aspects apparently containing only TH-LI. Bars indicate 50 μm.

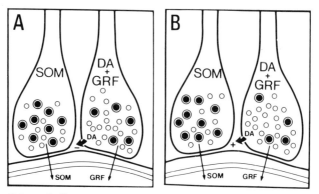

FIG. 20. Schematic illustration of possible types of interaction between a nerve ending containing growth hormone-releasing factor (GRF) and perhaps dopamine (DA) and another nerve ending containing somatostatin (SOM). GRF and somatostatin are released into the portal vessels, but dopamine may be released into extracellular space influencing adjacent somatostatin nerve endings, either inhibiting (A) or stimulating (B) release of this peptide. Dopamine may also be released into the portal vessels (not shown). For further details, see text.

release. However, *in vitro* studies have indicated that somatostatin may enhance dopamine release (Chihara *et al.*, 1979). To what extent these *in vitro* effects are significant for the *in vivo* situation remains to be seen, however. In conclusion, further studies are clearly needed to explain and understand the possible coexistence of dopamine and GRF in hypothalamic neurons projecting into the external layer of the median eminence. This includes characterization of the tyrosine hydroxylase-LI in the ventral group and to what extent these neurons in fact produce dopamine.

The occurrence and distribution of neurons containing *opioid peptides* within the arcuate nucleus are complex, with evidence for the presence of products from all three opioid peptide precursors, proenkephalin A, proenkephalin B, and proopiomelanocortin (POMC) (see also McGinty and Bloom, 1983). A similar abundance of opioid peptides has so far been observed only in one other region, the solitary tract nucleus (see Khachaturian *et al.*, 1985). The presence of met-enkephalin (met-ENK)- (Hughes *et al.*, 1975) positive cell bodies has been described in several studies (Hökfelt *et al.*, 1977; Sar *et al.*, 1978; Wamsley *et al.*, 1980; Kachaturian *et al.*, 1983, 1985; Krukoff and Calaresu, 1984; Merchenthaler *et al.*, 1985; Williams and Dockray, 1983). These cells are located in both the dorsal and ventral parts, but slightly medial to the main group of tyrosine hydroxylase-positive neurons. No evidence for coexistence of this immunoreactivity with tyrosine hydroxylase-LI has been obtained. Cell bodies with a similar distribution could also be observed with antiserum raised against the heptapeptide met-ENK-Arg-Phe, a sequence present in the proenkephalin A precursor (Stern *et al.*, 1979, 1981) confirming the studies of Williams and Dockray (1983). These neurons may in fact be identical to the met-ENK neurons described above, although this has not been directly established. Met-ENK-Arg-Gly-Leu (Kilpatrick *et al.*, 1981), a further product of proenkephalin A, was seen in numerous cells in the ventromedial and dorsal parts of the nucleus and the majority of these cells were also tyrosine hydroxylase positive (Fig. 18C and D). One more proenkephalin A product, metorphamide (Matsuo *et al.*, 1983; Weber *et al.*, 1983), is also present in the arcuate nucleus, but with a completely different distribution. Thus, numerous small cells can be seen in the medial, ventral part of the nucleus, close to the third ventricle. These cells are distinctly different from the tyrosine hydroxylase-positive ones, but are identical to NPY-immunoreactive neurons (Everitt *et al.*, 1985). Neurons containing dynorphin (1-13)-LI, a sequence within the proenkephalin B precursor (Goldstein *et al.*, 1979, 1980), can also be observed throughout the arcuate nucleus (Vincent *et al.*, 1982b; Weber *et al.*, 1982; Watson *et al.*, 1982, 1983) and a few of these contain tyrosine hydroxylase-LI (Everitt *et al.*, 1985).

Finally, numerous earlier studies have demonstrated the presence of neurons containing products of the POMC precursor in the arcuate nucleus (Watson *et al.*, 1977; Bloch *et al.*, 1978, 1979; Bugnon *et al.*, 1979; Sofroniew, 1979; Finley *et al.*, 1981; see Khachaturian *et al.*, 1985). All these neurons intermingle with the ventrolateral tyrosine hydroxylase-positive cell group, but no evidence for coexistence has been presented, either in studies using formaldehyde-induced fluorescence (Bloch *et al.*, 1979; Ibata *et al.*, 1980) or in studies using tyrosine hydroxylase antiserum (Everitt *et al.*, 1985). In conclusion, immunohistochemical analysis has revealed products from all three opioid peptide precursors in the arcuate nucleus with distinct and differential distribution patterns (see also McGinty and Bloom, 1983). Two of these peptides, met-ENK-Arg-Gly-Leu in a high proportion and dynorphin (1-13) in very low numbers, coexist with tyrosine hydroxylase-LI. It is here impossible to review the extensive literature concerned with the effects of opioid peptides and morphine on neuroendocrine parameters (see, e.g., Meites *et al.*, 1972; Rossier, 1982). A wide variety of effects have been described on several of the pituitary hormones and the present results may provide a morphological substrate for many types of interactions.

The medial, periventricular part of the arcuate nucleus contains two major neuron populations as judged from immunohistochemical analysis. Thus, a large proportion of these cells contain *NPY-LI* (Fig. 21A). These cells have earlier been described using an antiserum raised against avian and/or bovine pancreatic polypeptide (APP; Lorén *et al.*, 1979; Olschowka *et al.*, 1981; Card *et al.*, 1983), but recent evidence (see Lundberg

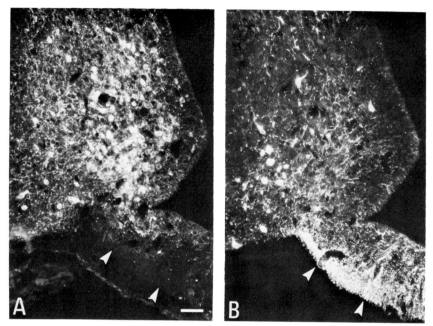

FIG. 21. Immunofluorescence micrographs of a section of the arcuate nucleus incubated first with antiserum to neuropeptide Y (NPY) (A) and after elution with tyrosine hydroxylase (TH) antiserum (B). Note that NPY- and TH-positive cells represent two distinct cell populations. No NPY fibers reach the external layer of the median eminence (arrowheads). Bar indicates 50 μm.

et al., 1984) suggests that the peptide demonstrated is NPY, a peptide discovered by Tatemoto and Mutt (Tatemoto, 1982; Tatemoto *et al.*, 1982).

The same neuron population also contains a metorphamide-like peptide as described above. It has been postulated that a neuron population with a similar distribution contains FMRFamide, a peptide isolated from ganglia of the clam (Price and Greenberg, 1977). In fact, several antibodies described in the literature show FMRF-positive neurons with a distribution similar to the APP/NPY-positive ones (Hökfelt *et al.*, 1983a; Chronwall *et al.*, 1984b). Our antiserum has been analyzed and compared with other antisera raised against pancreatic polypeptide-like peptides, and evidence has been presented that FMRFamide-LI, in fact, could represent crossreactivity with an NPY-like peptide (Lundberg *et al.*, 1984). It is our opinion that this may be the case in the arcuate nucleus. This is supported by studies using another antiserum considered to be specific for FMRFamide (Dockray, 1985), which exhibits a completely different

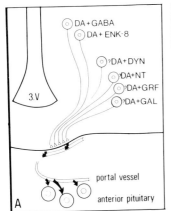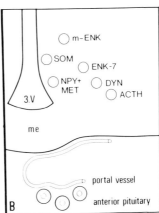

FIG. 22. Schematic illustration of arcuate neurons containing tyrosine hydroxylase-LI together with peptide-LI (A) and of other types of arcuate peptide neurons (B). The latter do not project to the external layer of the median eminence. Whether or not the ventral tyrosine hydroxylase-positive cell group in fact produces dopamine remains to be shown.

staining pattern. In fact, with this antiserum no or only single cells can be observed in the arcuate nucleus, although strongly staining cell bodies were observed inter alia in other hypothalamic nuclei (Everitt et al., 1985).

The second population of medially located arcuate cells contain *somatostatin*-LI, and these cells are located in a periventricular position mainly dorsal to the NPY-positive cell group (Kawano et al., 1982; Johansson et al., 1984; Vincent et al., 1985). In no case has coexistence with tyrosine hydroxylase-positive neurons been demonstrated. Chronwall et al. (1984a) have, however, described a very small proportion of somatostatin- plus NPY-positive cell bodies. A schematic illustration of various transmitters, peptides, and coexistence situations in the arcuate nucleus is shown in Fig. 22.

D. DARPP-32 IN TANYCYTES IN THE ARCUATE NUCLEUS–MEDIAN EMINENCE COMPLEX

The ultrastructural analysis of the basal hypothalamus revealed early that special types of glial cells, so-called tanycytes, constitute an important population of cells and cell processes in the median eminence (see e.g., Kobayashi and Matsui, 1969). Thus, many ependymal cells in the ventral aspects of the third ventricle give rise to processes which traverse the arcuate nucleus and run to the basal, ventral surface of the brain and into the median eminence. Moreover, many tanycytes are located within the median eminence and have processes which reach the ventricular surface as well as the external layer of the median eminence abutting on the portal blood vessels. Recently, a phosphoprotein has been discovered with an apparent molecular weight of 32,000, which is regulated by dopamine and adenosine 3,5-monophosphate, DARPP-32 (Hemmings *et al.*, 1984; Ouimet *et al.*, 1984; Walaas and Greengard, 1984). Using monoclonal antibodies to DARPP-32, Ouimet *et al.* (1984) demonstrated a wide distribution of this protein in the brain, mainly in areas enriched in dopamine nerve terminals, strongly supporting the view that this protein is present in dopamine-sensitive neurons. Of special interest in the context of the present article was the demonstration that tanycytes in the basal hypothalamus contained DARPP-32-LI (Ouimet *et al.*, 1984). This raises a clear possibility that dopamine released from processes of tuberoinfundibular dopamine neurons may exert at least part of its actions on tanycytes, in addition to having actions at the pituitary level as prolactin inhibitory factory (see MacLeod and Lehmeyer, 1974; Neill, 1980). A close relation between tyrosine hydroxylase-containing dopamine nerve endings in the external layer of the median eminence and DARPP-32-containing tanycytes can be demonstrated using the double labeling technique with antisera to these two compounds (Fig. 23A and B) as well as with LHRH-containing fibers (Fig. 23C and D).

These findings suggest a completely different type of interaction between dopamine and other elements in the external layer of the median eminence. Thus, we have earlier proposed that tanycytes may be involved in regulation of release of hypothalamic factors by controlling the space in the superficial area abutting onto the portal vessels (Hökfelt, 1973). Thus, by changing their shape, it would be possible for tanycytes to occupy a more or less extensive surface area, and in this way to influence the space that can be occupied by nerve endings containing hypothalamic factors. In an extensive series of articles we have earlier proposed that the inhibitory role of dopamine on LH secretion is exerted via an axo-axonic interaction, i.e., dopamine inhibits the release of LHRH at the median eminence level and in this way reduces LH levels in serum (Fuxe and Hökfelt, 1969; Hökfelt and Fuxe, 1972). The present model involving tanycytes may offer an alternative type of interaction. Thus, we have speculated that dopamine may influence the tanycytes to expand along the ventral surface and in this way limit the access of LHRH nerve endings to a secretion position and consequently decrease the amounts of LHRH released in the portal blood resulting in decreased LH levels (Hökfelt, 1973). The presence of a dopamine-sensitive phosphoprotein in the tanycytes, as first shown by Ouimet *et al.* (1984) and as confirmed in the present study, provides further indirect support for the hypothesis and makes it more interesting to pursue this line of research.

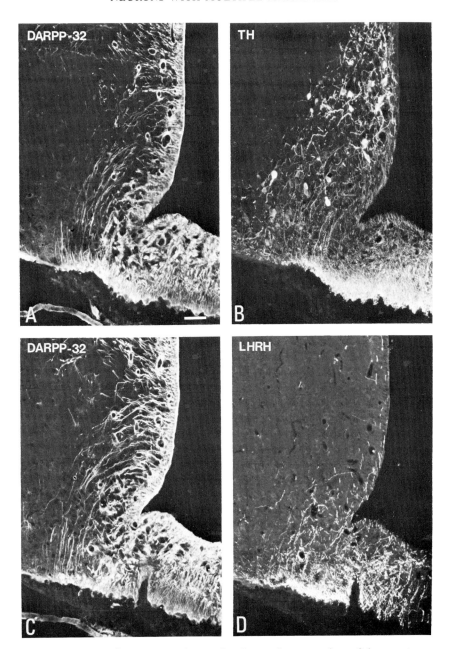

FIG. 23. Immunofluorescence micrographs of two adjacent sections of the arcuate nucleus–median eminence complex double stained with antisera to DARPP-32 and tyrosine hydroxylase (TH) (A,B) and to DARPP-32 and luteinizing hormone-releasing hormone (LHRH) (C,D). For further details, see text. Bar indicates 50 μm.

E. COMMENTS ON THE POSSIBLE FUNCTIONAL SIGNIFICANCE OF MULTIPLE MESSENGER SYSTEMS IN THE ARCUATE NUCLEUS

The picture sketched above, based on the immunohistochemical analysis of occurrence and distribution of various compounds in the arcuate nucleus, provides a complex view of the nucleus. It is not possible here to analyze these findings in detail. One can indeed consider whether or not the immunohistochemical demonstration of various immunoreactivities necessarily indicates a functional involvement, and it cannot be excluded that we are visualizing compounds which never participate in neuronal and neuroendocrine events under physiological conditions. In fact, an interesting question is to what extent the genomic control of peptide expression is sufficiently precise to exclude production of peptides which are not used by the neuron. On the other hand, it does not seem legitimate to rule out the present findings just because of the high degree of complexity. The analysis summarized here, and presented more in detail in the paper by Everitt *et al.* (1985), demonstrates an extensive heterogeneity and diversity of putative messenger molecules in the nervous system, which is not unique to the arcuate nucleus, but has been observed in many other brain regions, for example in the dorsal vagal complex, the dorsal horn of the spinal cord, the amygdaloid complex, and other hypothalamic nuclei. The histochemical findings may provide a basis for designing experiments to elucidate a possible role of various peptides and biogenic amines. It will also be important to establish exactly which of these systems project to the external layer of the median eminence, and thus can release their messengers into the portal vessels. By combining retrograde tracing with indirect immunofluorescence, it is possible to define such projections and their transmitter and/or peptide content (Fig. 24A and B). Furthermore, techniques are now available to search for the presence of these compounds in the blood of portal vessels as well as to analyze in detail their action on anterior pituitary hormone-producing cells.

IX. The CRF/PHI/Opiod Peptide Neurons: A Parvocellular Paraventricular Infundibular Coexistence System[2]

The parvocellular part of the paraventricular nucleus has emerged as a complex and important projection nucleus to the external layer of the median eminence (see Swanson and Sawchenko, 1983). We will not discuss this nucleus extensively and refer to the review article just men-

---

[2] It has come to our attention that PHI antisera may cross-react with CRF (Berkenbosch, Linton, and Tilders, *Neuroendocrinology*, submitted). We are at present investigating this issue.

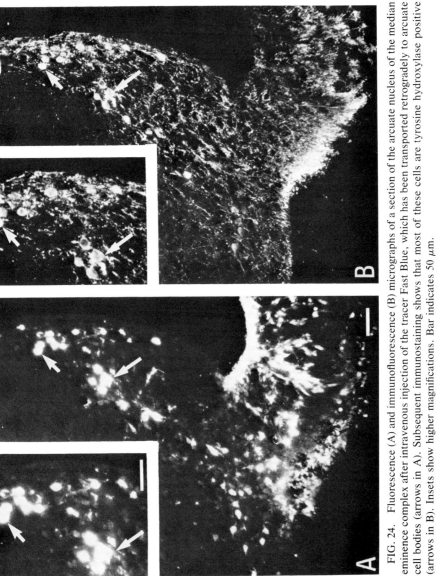

FIG. 24. Fluorescence (A) and immunofluorescence (B) micrographs of a section of the arcuate nucleus of the median eminence complex after intravenous injection of the tracer Fast Blue, which has been transported retrogradely to arcuate cell bodies (arrows in A). Subsequent immunostaining shows that most of these cells are tyrosine hydroxylase positive (arrows in B). Insets show higher magnifications. Bar indicates 50 μm.

tioned, but would like to concentrate on one subpopulation of neurons exhibiting an interesting coexistence situation. The discovery of the CRF (for references, see above) resulted in rapid production of powerful antisera and immunohistochemical description of CRF neurons, located inter alia in the parvocellular part of the paraventricular nucleus (Fellmann *et al.*, 1982; Bloom *et al.*, 1982; Olschowska *et al.*, 1982; Antoni *et al.*, 1983; Cummings *et al.*, 1983; Kawata *et al.*, 1983; Paull and Gibbs, 1983; Swanson *et al.*, 1983). Using an antiserum raised against the recently discovered peptide HI-27 [the peptide (P) having NH_2-terminal histidine (H) and COOH-terminal isoleucine (I) amide and 27 amino acid residues] (Tatemoto and Mutt, 1981), immunoreactive cell bodies have been observed in the parvocellular part of the paraventricular nucleus with a distribution very similar to the one seen for CRF. There are dense networks of PHI-immunoreactive fibers in the external layer of the median eminence with an overlapping distribution to CRF-positive ones (Fig. 25A and B). Finally, it is well known that this nucleus also contains enkephalin-immunoreactive neurons with projections to the external layer of the median eminence (Fig. 25C) (Hökfelt *et al.*, 1977b; Sar *et al.*, 1978; Wamsley *et al.*, 1980). Using the elution-restaining technique of Tramu *et al.* (1978), it could be demonstrated that CRF- and PHI-like material is present in the same cell bodies (Fig. 26A–D), and that a proportion of these cells also contain enkephalin-like immunoreactivity (Hökfelt *et al.*, 1982b, 1983c). It therefore seems likely that these three compounds are also present in the same nerve endings of the median eminence, although this has so far not been directly established by experimental evidence. Overlapping PHI-, CRF-, and ENK-immunoreactive fiber networks are also present in the human median eminence (Fig. 27A–C), suggesting coexistence of these peptides.

These findings indicate that a population of paraventricular-infundibular neurons both in rat and man may release at least the three compounds, CRF, PHI, and an opioid peptide. It is well known that the secretion of adrenocorticotropin hormone (ACTH) is regulated mainly by CRF. Prolactin release from the anterior pituitary gland has been assumed to be controlled primarily by an inhibitory factory, probably dopamine (see Neill, 1980). It has, however, been suggested that other messengers may be involved, for example, vasoactive intestinal polypeptide (VIP) (Kato *et al.*, 1978; Ruberg *et al.*, 1978; Shaar *et al.*, 1979; Vijayan *et al.*, 1979). In view of the structural similarity between VIP and PHI, and in view of the recent demonstration by Itoh *et al.* (1983) using recombinant DNA technology that VIP and PHI (PHM) are formed from the same gene, it may be anticipated that PHI exerts prolactin-releasing activity. In fact, it has been demonstrated that PHI releases prolactin from anterior pituitary cells (Fig. 28) (Samson *et al.*, 1983; Werner *et al.*, 1983).

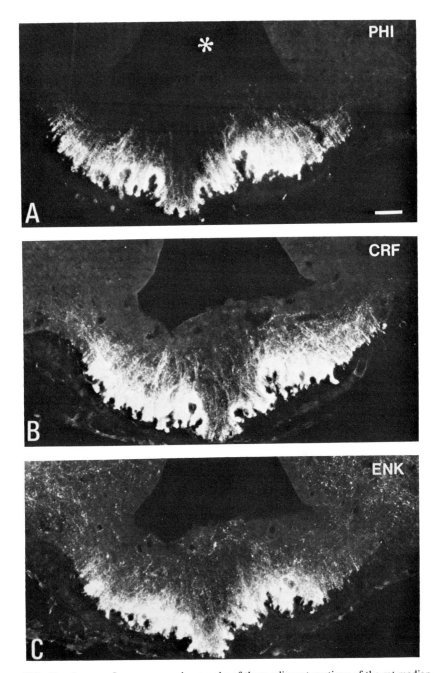

FIG. 25. Immunofluorescence micrographs of three adjacent sections of the rat median eminence incubated with antiserum to peptide HI (PHI) (A), corticotropin-releasing factor (CRF) (B), and met-enkephalin (ENK) (C). All three antisera reveal intense staining patterns in the external layer of the median eminence with a very similar distribution, suggesting a possible colocalization. Note that enkephalin immunoreactive fibers in addition are present in the arcuate nucleus. Asterisk indicates third ventricle. Bar indicates 50 μm.

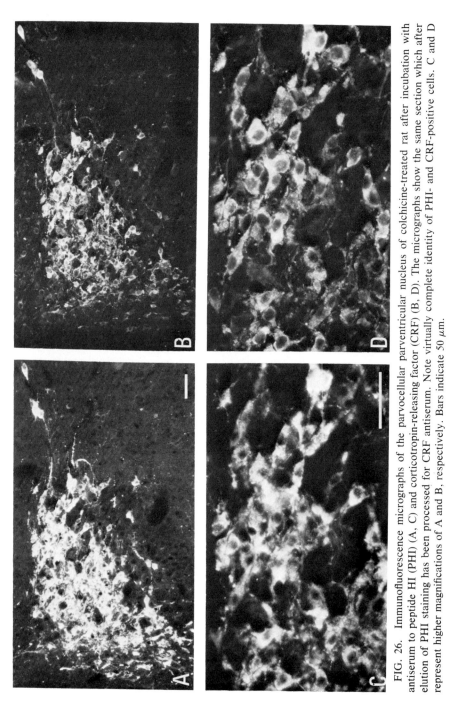

FIG. 26. Immunofluorescence micrographs of the parvocellular parventricular nucleus of colchicine-treated rat after incubation with antiserum to peptide HI (PHI) (A, C) and corticotropin-releasing factor (CRF) (B, D). The micrographs show the same section which after elution of PHI staining has been processed for CRF antiserum. Note virtually complete identity of PHI- and CRF-positive cells. C and D represent higher magnifications of A and B, respectively. Bars indicate 50 μm.

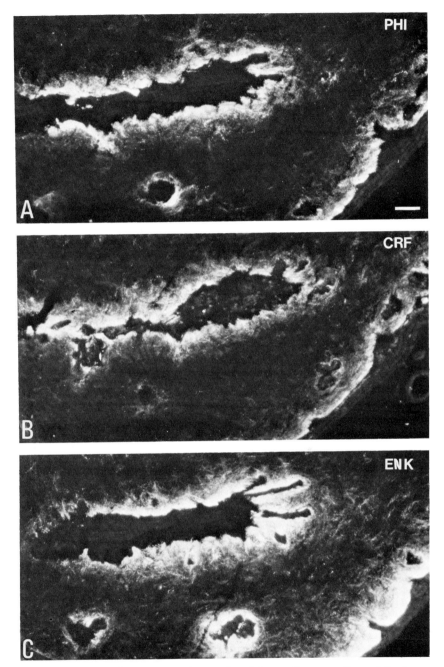

FIG. 27. Immunofluorescence micrographs of three adjacent sections of the human median eminence after incubation with antiserum to peptide HI (PHI) (A), corticotropin-releasing factor (CRF) (B), and met-enkephalin (ENK) (C). Note similar distribution of all three peptides, suggesting colocalization in the same fibers, as was seen also for the rat (Fig. 25). Bar indicates 50 μm.

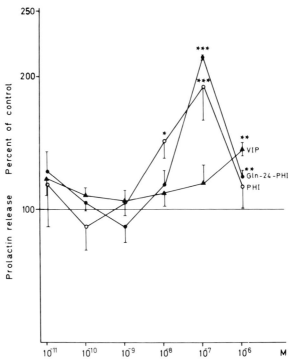

FIG. 28. Effects of peptide HI (PHI) (open circles), Gln-24-PHI (filled circles), and vasoactive intestinal polypeptide (VIP) (triangles) on prolactin secretion from dispersed rat anterior pituitary cells in culture. Prolactin values are expressed as percentage of respective basal release. From Werner et al. (1983).

It is known that certain types of stress cause release of both ACTH and prolactin (Ganong, 1963; Meites et al., 1963; Brown and Martin, 1974; Harms et al., 1975). It may be speculated that such a concomitant release of these two hypothalamic hormones is controlled by simultaneous release of PHI (and VIP) and CRF from the same nerve endings in the median eminence (Fig. 29). Moreover, the action of PHI on prolactin release may be strengthened by the concomitant release of the third peptide present in the same nerve endings, the enkephalin-like peptide (Fig. 29). It has been demonstrated that opioid peptides facilitate prolactin secretion, possibly by inhibiting release of dopamine (Bruni et al., 1977; Cocchi et al., 1977; Cusan et al., 1977; Ferland et al., 1977; Shaar et al., 1977; Meites et al., 1979). Concomitant release of an enkephalin peptide will therefore inhibit release of dopamine, a prolactin-inhibitory factor (MacLeod and Lehmayer, 1974), and reduce the inhibitory tone on prolactin secretion. Thus, PHI and the opiate peptide may enhance prolactin secretion by two different mechanisms—PHI by directly stimulating the

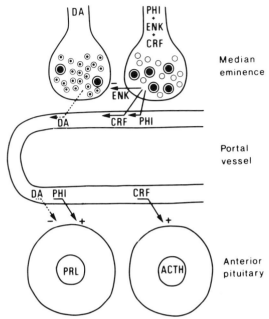

FIG. 29. Schematic illustration of two nerve endings in the external layer of the median eminence containing, respectively, dopamine (DA) and peptide HI (PHI) plus an opiate peptide (ENK) plus corticotropin-releasing factor (CRF). CRF and PHI are released into the portal vessels and stimulate secretion of, respectively, ACTH and prolactin. The opiate peptide may inhibit dopamine release, and in this way attenuate the inhibitor tone of dopamine on prolactin secretion.

lactotropes and the opiate peptide by inhibiting release of dopamine from nerve endings in the external layer of the median eminence (Fig. 29).

LEUKOTRIENES AND LHRH NEURONS

Recently a new group of bioactive compounds belonging to the arachidonic acid family has been discovered (see Samuelsson, 1983). They are present in certain peripheral tissues, for example leukocytes and are assumed to be involved in inflammatory and allergic reactions (see Samuelsson, 1983). Leukotrienes are formed from arachidonic acid via 5- lipoxygenation to an unstable epoxide LTA_4, which can be enzymatically hydrolyzed to LTB_4 and subsequently, by addition of the tripeptide glutathione to LTC_4 (Murphy et al., 1979; Morris et al., 1980). Furthermore, by removing, respectively, glutamic acid and glycine from the glutathione molecule, LTD_4 and LTE_4 can be formed (Örning and Hammarström,

1980; Bernström and Hammarström, 1981). Using a number of biochemical techniques including high-performance liquid chromatography, radioimmunoassay, and bioassay, evidence for formation of LTC_4, LTD_4, and LTE_4 in brain tissue has been presented (Lindgren et al., 1984; see also Dembinska-Kieć et al., 1984; Moskowitz et al., 1984). The biosynthesis of LTC_4 was studied in vitro in different regions of the brain exhibiting a regional variation with the highest capacity in the hypothalamus and median eminence. More recently, receptor binding studies have been performed and high-affinity binding sites have been demonstrated for LTC_4 in homogenates of rat central nervous system with a K_D of about 30 nmol (Schalling et al., 1986a). Furthermore, preliminary autoradiographic studies have shown that $[^3H]LTC_4$ binding exhibits a differential distribution pattern in sections of rat brain and that these patterns are different from the ones seen for LTB_4 (Schalling et al., 1986b). Taken together with iontophoretic studies in the cerebellum, showing that LTC_4 causes a long lasting activation of Purkinje cells (Palmer et al., 1980, 1981), these findings suggest that LTC_4 may have a messenger role in the nervous system.

Attempts have also been carried out to study possible leukotriene-containing systems with immunohistochemistry. Thus, using an antibody raised against LTC_4 conjugated to bovine serum albumin (Aeringhaus et al., 1982), sections from rat brain and other species have been analyzed using the immunofluorescence technique. Numerous fluorescent cell bodies were observed in many parts of the brain, but in control experiments it was not possible to abolish the staining patterns, except for the median eminence and adjacent areas (Lindgren et al., 1984). In the latter region the LTC_4-positive fibers overlapped LHRH-containing ones, and elution-restaining experiments showed identity of the fibers (Hulting et al., 1985). Against this background the effect of LTC_4 was studied on LH secretion. It could be demonstrated that LTC_4 in the picomolar range can release LH, but not growth hormone from dispersed rat anterior pituitary cells in vitro (Hulting et al., 1984, 1985). In contrast, LTD_4 had no effect on release of either luteinizing hormone or growth hormone. These findings indicate that release of luteinizing hormone from gonadotropes may be under control of more than one compound, perhaps an LTC_4-like compound, in addition to the conventional LHRH releaser.

X. Concluding Remarks

The basic message of this article is to convey the idea that neurons of different types in the periphery and in the central nervous system, including neurosecretory cells, produce, store, and perhaps release more than one messenger molecule. As shown in Fig. 30 this adds to past concepts

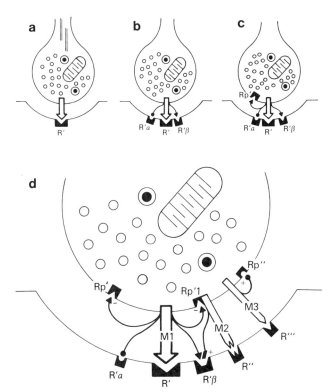

FIG. 30. Schematic illustration of the development of concepts of the transmitter–receptor relation. Initially a transmitter was assumed to react with a single type of postsynaptic receptor (a). It was then recognized that the same transmitter could act at different types of postsynaptic receptors (α-, β-receptors, etc.) (b). More recently it has been realized that there also exist presynaptic receptors (autoreceptors) (c). In (d) the idea of multiple messengers is illustrated. Different ways in which these compounds could interact are indicated. From Lundberg and Hökfelt (1983).

of chemical transmission a further step, whereby multiple messenger molecules may interact at the pre- and postsynaptic sites in the transmission process. It may be speculated (Fig. 31) that the present situation with regard to coexistence in nerve endings can be seen in a phylogenetic perspective, as outlined by Le Roith, Roth, and collaborators (Le Roith et al., 1986; Roth et al., 1986). Thus, coexisting messengers in primitive neurons are stored in the same vesicles, as they are in mammalian endocrine cells today. With the demand for faster communication, new types of vesicles, small synaptic vesicles, developed storing and releasing exclusively classical transmitters, being present in addition to the larger vesicles storing both classical transmitter and peptide(s). Interestingly,

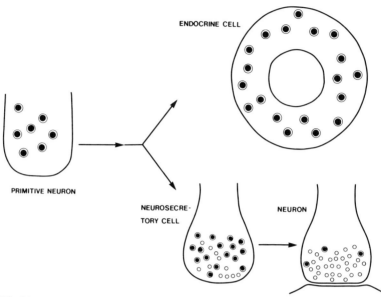

FIG. 31. Hypothetical development of messenger storage sites during phylogeny. In primitive neurons one type of large vesicle (Westfall and Kinnemon, 1984) may store both small messenger molecules such as biogenic amines and peptides. A similar situation is encountered in endocrine cells of today's mammalian species. Neurons have in addition developed a new type of storage vesicle, small synaptic vesicles, which store only classical transmitter. The larger vesicles contain both classical transmitter and peptides. See text.

the neurosecretory cells, representing an intermediate between endocrine cells and neurons, contain a higher proportion of large dense core vesicles than neurons releasing their messengers at more or less well defined synapses (Fig. 31).

Different types of compounds may be involved in such coexistence situations including classical transmitters, peptides, ATP, and perhaps members of the arachidonic acid family. The coexistence phenomenon has been documented extensively with immunohistochemical techniques, but knowledge about the functional significance and the mechanism of interaction between the putative messengers is still scarce. Studies in the peripheral nervous system suggest that classical transmitters and peptides are, indeed, coreleased and may interact in a cooperative way on effector cells. However, other types of interactions have been observed; for example, a coexisting peptide may inhibit the release of the classical transmitter. Therefore, interaction between different messengers released from the same nerve endings may be of several types and may, in a general sense, provide mechanisms for differential responses and for increasing the amount of information transferred at synapses.

It is important to note that the compounds discussed above do not necessarily have to be involved in the process of transmitting nerve impulses between neurons or to effector cells at synapses, but may also have other roles, for example, to exert trophic effects or induce other types of long-term events in neurons and effector cells. It has recently been shown that substance P exerts growth stimulatory effects on smooth muscle cells (Nilsson et al., 1985). Indeed, it may be argued that the fact that peptides are present in neurons, which already contain a transmitter substance, may provide a basis for "allowing" peptides to exert other types of functions.

Multimessenger transmission may represent a principle for increasing capacity for information transfer in the nervous system, a capacity which already appears enormous when considering just the number of neurons and their nerve endings in the mammalian nervous system. It may therefore be questioned why the nervous system should require multiple messengers. However, our neuronal machinery has a hard-to-imagine capacity and, in spite of the apparent redundancy of neurons and nerve endings, transmission via multiple messengers may be necessary to achieve the outstanding performance of our brain.

The number of peptides dealt with in this article is large, but many additional neuropeptides are present in other regions of the nervous system. Moreover, new peptides are discovered. An exciting development is the description by Seeburg and colleagues (Nikolics et al., 1985; Phillips et al., 1985) of a novel polypeptide with prolactin release-inhibiting activity present within the precursor for human gonadotropin-releasing hormone. It may, however, be wise to express a word of caution at this moment. There is, indeed, a discrepancy between the extensive immunohistochemical information and the in many cases meager physiological evidence for a functional role(s) of coexisting messengers. The importance of peptides is today in many cases difficult to evaluate, and it cannot be excluded that their role is considerably less significant than that of classical transmitters. Peptides may have played an important role in lower species, but may have been replaced as important functional messengers by "small molecule" transmitters. In fact, it cannot be excluded that peptides are sometimes carried along more or less as "silent passengers."

ACKNOWLEDGMENTS

This research was supported by grants from the Swedish Medical Research Council (04X-2887), Knut and Alice Wallenbergs Stiftelse, Petrus and Augusta Hedlunds Stiftelse, and Magnus Bergvalls Stiftelse. We thank Ms. W. Hiort, Ms. S. Nilsson, and Ms. A Peters for excellent technical assistance and Ms. E. Björklund for skillful help in preparing the manuscript.

Some results in this review article are taken from ongoing collaborative research with colleagues as indicated by the papers referred to (see reference list). We thank in particular Professor Jan Fahrenkrug, Department of Clinical Chemistry, Bispbjerg Hospital, Copenhagen, Denmark, for VIP and PHI antisera, Professor Lars Terenius, Department of Pharmacology, Uppsala University, Uppsala, Sweden, for enkephalin and NPY antisera, Dr. Åke Rökaeus and Dr. Elvar Theodorsson-Norheim, Department of Pharmacology, Karolinska Institutet, Stockholm, Sweden, for neurotensin and galanin antisera, respectively, and Dr. Wylie Vale, The Salk Institute, San Diego, California, for CRF and GRF antisera.

REFERENCES

Aehringhaus, U., Wölbling, R. H., König, W., Patrono, C., Peskar, B. M., and Peskar, B. A. (1982). *FEBS Lett.* **146**, 111–114.

Allen, J. M., Tatemoto, K., Polak, J. M., Hughes, J., and Bloom, S. R. (1982). *Neuropeptides* **3**, 71–77.

Altschuler, R. A., Parakkal, M. H., and Fex, J. (1983). *Neuroscience* **9**, 621–630.

Amara, S. G., Jonas, V., Rosenfeld, M. G., Ong, E. S., and Evans, R. M. (1982). *Nature (London)* **298**, 240–244.

Amoss, M., Burgus, R., Blackwell, R., Vale, W., Fellows, R., and Guillemin, R. (1971). *Biochem. Biophys. Res. Commun.* **44**, 205–210.

Andén, N.-E., Jukes, M., and Lundberg, A. (1964). *Nature (London)* **202**, 1222–1223.

Andersson, K., Fuxe, K., Eneroth, P., Gustafsson, J.-Å., and Skett, P. (1977). *Neurosci. Lett.* **5**, 83–89.

Andersson, K., Fuxe, K., Eneroth, P., Isaksson, O., Nyberg, F., and Roos, P. (1983). *Eur. J. Pharmacol.* **95**, 271–275.

Antoni, F. A., Palkovits, M., Makara, G. B., Linton, E. A., Lowry, P. J., and Kiss, J. Z. (1983). *Neuroendocrinology* **36**, 415–423.

Arimura, A., and Fishback, J. B. (1981). *Neuroendocrinology* **33**, 246–256.

Armstrong, D. M., Miller, R. J., Beaudet, A., and Pickel, V. M. (1984). *Brain Res.* **310**, 269–278.

Barbeau, H., and Bédard, P. (1981). *Neuropharmacology* **20**, 477–481.

Bargmann, W., and Scharrer, B. (1951). *Am. Sci.* **39**, 255–259.

Barraclough, C. A. (1966). *Recent Prog. Horm. Res.* **22**, 503–529.

Belin, M. F., Nanopoulos, D., Didier, M., Aguera, M., Steinbusch, H., Verhofstad, A., Maitre, M., and Pujol, J. F. (1983). *Brain Res.* **275**, 329–339.

Bernström, K., and Hammarström, S. (1981). *J. Biol. Chem.* **256**, 9579–9582.

Berod, A., Chat, M., Paut, L., and Tappaz, M. (1984). *J. Histochem. Cytochem.* **32**, 1331–1338.

Björklund, A., and Hökfelt, T. (eds.) (1985). "Handbook of Chemical Neuroanatomy. Vol. 4: GABA and Neuropeptides in the CNS, Part I." Elsevier, Amsterdam.

Björklund, H., Hökfelt, T., Goldstein, M., Terenius, L., and Olson, L. (1985). *J. Neurosci.* **5**, 1633–1643.

Black, I. B., Adler, J. E., Dreyfus, C. F., Jonakait, G. M., Katz, D. M., LaGamma, E. F., and Markey, K. M. (1984). *Science* **225**, 1266–1270.

Bloch, B., Bugnon, C., Fellmann, D., and Lenys, D. (1978). *Neurosci. Lett.* **10**, 147–152.

Bloch, B., Bugnon, C., Fellmann, D., Lenys, D., and Gouget, A. (1979). *Cell. Tissue Res.* **204**, 1–15.

Bloch, B., Brazeau, P., Bloom, F., and Ling, N. (1983a). *Neurosci. Lett.* **37**, 23–28.
Bloch, B., Brazeau, P., Ling, N., Böhlen, P., Esch, F., Wehrenberg, W. B., Benoit, R., Bloom, F., and Guillemin, R. (1983b). *Nature (London)* **301**, 607–609.
Bloch, B., Ling, N., Benoit, R., Wehrenberg, W. B., and Guillemin, R. (1984). *Nature (London)* **307**, 272–274.
Bloom, F. E., Battenberg, E. L. F., Rivier, J., and Vale, W. (1982). *Peptides* **4**, 43–48.
Bloom, F. E., Battenberg, E., Ferron, A., Mancillas, J. R., Milner, R. J., Siggins, G., and Sutcliffe, G. (1985). *Recent Prog. Horm. Res.* **41**, 339–361.
Böhlen, P., Brazeau, P., Bloch, B., Ling, N., Gaillard, R., and Guillemin, R. (1984). *Biochem. Biophys. Res. Commun.* **114**, 930–936.
Brazeau, P., Vale, W., Burgus, R., Ling, N., Butcher, M., Rivier, J., and Guillemin, R. (1973). *Science* **179**, 77–79.
Brazeau, P., Ling, N., Böhlen, P., Esch, R., Ying, S. Y., and Guillemin, R. (1982). *Proc. Natl. Acad. Sci. U.S.A.* **79**, 7009–7913.
Brown, G. M., and Martin, J. B. (1974). *Psychosom. Med.* **36**, 241–247.
Brownstein, M., and Mezey, E. (1986). *In* "Coexistence of Neuronal Messengers. A New Principle in Chemical Transmission" (*Prog. Brain Res.*) (T. Hökfelt, K. Fuxe, and B. Pernow, eds.). Elsevier, Amsterdam (in press).
Bruni, J. A., Van Vugt, D., Marshall, S., and Meites, J. (1977). *Life Sci.* **21**, 461–466.
Bugnon, C., Bloch, B., Lenys, D., Gouget, A., and Fellmann, D. (1979). *Neurosci. Lett.* **14**, 43–48.
Bugnon, C., Gouget, A., Fellmann, D., and Clavequin, M. C. (1983). *Neurosci. Lett.* **38**, 131–137.
Burgus, R., Dunn, T. F., Desiderio, D., Ward, D. N., Vale, W., and Guillemin, R. (1970). *Nature (London)* **226**, 321–325.
Burnstock, G. (1976). *Neuroscience* **1**, 239–248.
Burnstock, G. (1985). *TINS* **8**, 5–6.
Caffé, A. R., and van Leeuwen, F. W. (1983). *Cell Tissue Res.* **233**, 23–33.
Card, J. P., Brecha, N., and Moore, R. (1983). *J. Comp. Neurol.* **217**, 123–136.
Carlsson, A., Lindqvist, M., Magnusson, T., and Waldeck, B. (1958). *Science* **127**, 471.
Carlsson, A., Falck, B., and Hillarp, N.-Å. (1962). *Acta Physiol. Scand.* **56** (Suppl. 56), 1–28.
Carraway, R., and Leeman, S. E. (1973). *J. Biol. Chem.* **248**, 6854–6861.
Carraway, R., and Leeman, S. E. (1975). *J. Biol. Chem.* **250**, 1907–1911.
Carson, K. A., Nemeroff, C. B., Rone, M. S., Youngblood, W. W., Prange, A. J., Hanker, J. S., and Kizer, J. S. (1977). *Brain Res.* **129**, 169–173.
Chan-Palay, V. (1979). *Anat. Embryol.* **156**, 241–254.
Chan-Palay, V., and Palay, S. L. (eds.) (1984). "Coexistence of Neuroactive Substances in Neurons." Wiley, New York.
Chan-Palay, V., Jonsson, G., and Palay, S. L. (1978). *Proc. Natl. Acad. Sci. U.S.A.* **75**, 1582–1586.
Chan-Palay, V., Nilaver, G., Palay, S. L., Beinfeld, M. C., Zimmerman, E. A., Wu, J.-Y., and O'Donohue, T. L. (1981). *Proc. Natl. Acad. Sci. U.S.A.* **78**, 7787–7791.
Chan-Palay, V., Zaborszky, L., Köhler, C., Goldstein, M., and Palay, S. L. (1984). *J. Comp. Neurol.* **227**, 467–496.
Charnay, Y., Léger, L., Dray, F., Bérod, A., Jouvet, M., Pujol, J. F., and Dubois, P. M. (1982). *Neurosci. Lett.* **30**, 147–151.
Chihara, K., Arimura, A., and Schally, A. V. (1979). *Endocrinology* **104**, 1656–1662.
Chronwall, B. M., Chase, T. N., and O'Donohue, T. L. (1984a). *Neurosci. Lett.* **52**, 213–217.

Chronwall, B. M., Olschowka, J. A., and O'Donohue, T. L. (1984b). *Peptides* **5**, 569–584.
Cocchi, D., Santagostino, A., Gil-Ad, I., Ferri, S., and Müller, E. (1977). *Life Sci.* **20**, 2041–2046.
Coons, A. H. (1958). *In* "General Cytochemical Methods" (J. F. Danielli, ed.), pp. 399–422. Academic Press, New York.
Costa, M., and Furness, J. B. (1984). *Neuroscience* **13**, 911–919.
Cowan, W. M., and Cuenod, M. (eds.) (1975) "The Use of Axonal Transport for Studies of Neuronal Connectivity." Elsevier, Amsterdam.
Cross, B. A., Dyball, R. E. J., Dyer, R. G., Jones, C. W., Lincoln, D. W., Morris, J. F., and Pickering, B. T. (1975). *Recent Prog. Horm. Res.* **31**, 243–286.
Cuello, A. C. (ed.) (1982). "Co-transmission." MacMillan, London.
Cummings, S., Elde, R., Ells, J., and Lindall, A. (1983). *J. Neurosci.* **3**, 1355–1368.
Cusan, L., Dupont, A., Kledzik, G. S., Labrie, F., Coy, D. H., and Schally, A. V. (1977). *Nature (London)* **268**, 544–547.
Dahlström, A. (1968). *Eur. J. Pharmacol.* **5**, 111–113.
Dahlström, A. (1971). *Acta Neuropathol. Suppl.* **5**, 226–237.
Dahlström, A., and Fuxe, K. (1964). *Acta Physiol. Scand.* **62** (Suppl. 232), 1–55.
Dahlström, A., and Fuxe, K. (1965). *Acta Physiol. Scand.* **64** (Suppl. 247), 5–36.
Daikoku, S., Okamura, Y., Kawano, H., Tsuruo, Y., Maegawa, M., and Shibasaki, T. (1985). *Cell Tissue Res.* **240**, 575–584.
Dale, H. (1935). *Proc. R. Soc. Med.* **28**, 319–332.
Dembinska-Kieć, A., Simmet, T., and Peskar, B. A. (1984). *Eur. J. Pharmacol.* **99**, 57–62.
Dockray, G. J. (1985). *J. Neurochem.* **45**, 152–158.
Du Vigneaud, V., Lawler, H. C., and Popenoe, E. A. (1953). *J. Am. Chem. Soc.* **75**, 4880–4881.
Du Vigneaud, V., Ressler, C., Swan, J. M., Roberts, C. W., Katsoyannis, P. G., and Gordon, S. (1954). *J. Am. Chem. Soc.* **75**, 4879–4880.
Eccles, J. C., Fatt, P., and Koketsu, K. (1954). *J. Physiol. (London)* **126**, 524–526.
Eckenstein, F., and Baughman, R. W. (1984). *Nature (London)* **309**, 153–155.
Edwards, A. V., Järhult, J., Andersson, P. O., and Bloom, S. R. (1982). *In* "Systemic Role of Regulatory Peptides" (S. R. Bloom, J. M. Polak, and E. Lindenlaub, eds.), pp. 145–148. Schattauer, Stuttgart.
Elliott, T. R. (1905). *J. Physiol. (London)* **32**, 401–407.
Everitt, B. J., Hökfelt, T., Terenius, L., Tatemoto, K., Mutt, V., and Goldstein, M. (1984a). *Neuroscience* **11**, 443–462.
Everitt, J. E., Hökfelt, T., Wu, J.-Y., and Goldstein, M. (1984b). *Neuroendocrinology* **39**, 189–191.
Everitt, B. J., Meister, B., Hökfelt, T., Melander, T., Terenius, L., Rökaeus, Å., Theodorsson-Norheim, E., Dockray, G., Edwardson, J., Cuello, A. C., Elde, R. P., Goldstein, M., Hemmings, H., Ouimet, Ch., Walaas, I., Greengard, P., Vale, W., Weber, E., and Wu, J.-Y. (1985). *Brain Res. Rev.*, in press.
Falck, B., Hillarp, N. Å., Thieme, G., and Torp, A. (1962). *J. Histochem. Cytochem.* **10**, 348–354.
Fellmann, D., Bugnon, C., Gouget, A., and Cardot, J. (1982). *C.R. Séances Soc. Biol.* **176**, 511–516.
Fellmann, D., Gouget, A., and Bugnon, C. (1983). *C.R. Acad. Sci. Paris, Série III* **296**, 487–492.
Ferland, L., Fuxe, K., Eneroth, P., Gustafsson, J.-Å., and Skett, P. (1977). *Eur. J. Pharmacol.* **43**, 89–90.
Finley, J. C. W., Lindström, P., and Petrusz, P. (1981). *Neuroendocrinology* **33**, 28–42.

Fonnum, F. (ed.) (1978). "Amino Acids as Chemical Transmitters." NATO Advanced Study Institutes Series. Series A: Life Sciences. Plenum, New York.
Foster, G. A., Schultzberg, M., Björklund, A., Gage, F. H., and Hökfelt, T. (1985). In "Neural Grafting in the Mammalian CNS" (A. Björklund and U. Stenevi, eds.), pp. 179–189. Elsevier, Amsterdam.
Fried, G., Lagercrantz, H., and Hökfelt, T. (1978). Neuroscience 3, 1271–1291.
Fried, G., Terenius, L., Hökfelt, T., and Goldstein, M. (1985). J. Neurosci. 5, 450–458.
Furness, J. B., Costa, M., Emson, P. C., Håkanson, R., Moghimzadeh, E., Sundler, F., Taylor, J. L., and Chance, R. E. (1983). Cell Tissue Res. 234, 71–92.
Fuxe, K. (1964). Z. Zellforsch. 61, 710–724.
Fuxe, K. (1965a). Z. Zellforsch. 65, 573–596.
Fuxe, K. (1965b). Acta Physiol. Scand. 64 (Suppl. 247), 39–85.
Fuxe, K. and Hökfelt, T. (1966). Acta Physiol. Scand. 66, 243–244.
Fuxe, K., and Hökfelt, T. (1969). In "Frontiers in Neuroendocrinology" (W. F. Ganong and L. Martini, eds.), pp. 47–96. Oxford Univ. Press, London.
Ganong, W. F. (1963). In "Advances in Neuroendocrinology" (A. V. Nalbandov, ed.), pp. 92–149. Univ. of Illinois Press, Urbana, Illinois.
Gershengorn, M. C. (1985). Recent Prog. Horm. Res. 41, 607–646.
Gibbins, I. L., Furness, J. B., Costa, M., MacIntyre, I., Hillyard, C. J., and Girgis, S. (1985). Neurosci. Lett. 57, 125–130.
Gibson, S. J., Polak, J. M., Bloom, S. R., Sabate, I. M., Mulderry, P. M., Ghatei, M. A., McGregor, G. P., Morrison, J. F. B., Kelly, J. S., Evans, R. M., and Rosenfeld, M. G. (1984). J. Neurosci. 4, 3101–3111.
Glazer, E. J., Steinbusch, H., Verhofstad, A., and Basbaum, A. I. (1981). J. Physiol. (Paris) 77, 241–245.
Goldstein, A., Tachibana, S., Lowney, L. I., Hunkapiller, M., and Hood, L. (1979). Proc. Natl. Acad. Sci. U.S.A. 76, 6666–6670.
Goldstein, A., and Ghazarossian, V. E. (1980). Proc. Natl. Acad. Sci. U.S.A. 77, 6207–6210.
Greer, M. A. (1956). Recent Prog. Horm. Res. 12, 67–98.
Gual, C., Kastin, A. J., and Schally, A. V. (1972). Recent Prog. Horm. Res. 28, 173–200.
Guillemin, R. (1964). Recent Prog. Horm. Res. 20, 89–121.
Guillemin, R. (1977). Recent Prog. Horm. Res. 33, 1–20.
Guillemin, R., Brazeau, P., Böhlen, P., Esch, F., Ling, N., and Wehrenberg, W. B. (1982). Science 218, 585–587.
Guillemin, R., Brazeau, P., Böhlen, P., Esch, F., Ling, N., Wehrenberg, W. B., Bloch, B., Mougin, C., Zeytin, F., and Baird, A. (1984). Recent Prog. Horm. Res. 40, 233–286.
Hansen, S., Svensson, L., Hökfelt, T., and Everitt, B. J. (1983). Neurosci. Lett. 42, 299–304.
Harms, P. G., Langlier, P., and McCann, S. M. (1975). Endocrinology 96, 475–478.
Harris, G. (1955). "Neural Control of the Pituitary Gland." Arnold, London.
Hemmings, H. C., Nairn, A. C., Aswad, D. W., and Greengard, P. (1984). J. Neurosci. 4, 99–110.
Hendry, S. H. C., Jones, E. G., DeFelipe, J., Schmechel, D., Brandon, C., and Emson, P. C. (1984). Proc. Natl. Acad. Sci. U.S.A. 81, 6526–6530.
Hisano, S., Kawano, H., Nishiyama, T., and Daikoku, S. (1982). Cell Tissue Res. 224, 303–314.
Hökfelt, T. (1968). Z. Zellforsch. 91, 1–74.
Hökfelt, T. (1973). Acta Physiol. Scand. 89, 606–608.
Hökfelt, T., and Dahlström, A. (1971). Z. Zellforsch. 119, 460–482.

Hökfelt, T., and Fuxe, K. (1972). In "Brain-Endocrine Interaction. Median Eminence: Structure and Function" (K. M. Knigge, D. E. Scott, and A. Weindl, eds.), pp. 181–223. S. Karger, Basel.
Hökfelt, T., Kellerth, J.-O., Nilsson, G., and Pernow, B. (1975). Brain Res. 100, 235–252.
Hökfelt, T., Johansson, O., Fuxe, K., Goldstein, M., and Park, D. (1976). Med. Biol. 54, 427–453.
Hökfelt, T., Elfvin, L.-G., Elde, R., Schultzberg, M., Goldstein, M., and Luft, R. (1977a). Proc. Natl. Acad. Sci. U.S.A. 74, 3587–3591.
Hökfelt, T., Elde, R., Johansson, O., Terenius, L., and Stein, L. (1977b). Neurosci. Lett. 5, 25–31.
Hökfelt, T., Ljungdahl, Å., Steinbusch, H., Verhofstad, A., Nilsson, G., Brodin, E., Pernow, B., and Goldstein, M. (1978a). Neuroscience 3, 517–538.
Hökfelt, T., Elde, R., Fuxe, K., Johansson, O., Ljungdahl, Å., Goldstein, M., Luft, R., Efendic, S., Nilsson, G., Terenius, L., Ganten, D., Jeffcoate, S. L., Rehfeld, J., Said, S., Perez de la Mora, M., Possani, L., Tapia, R., Teran, L., and Palacios, R. (1978b). In "The Hypothalamus" (S. Reichlin, R. J. Baldessarini, and J. B. Martin, eds.), pp. 69–122. Raven, New York.
Hökfelt, T., Johansson, O., Ljungdahl, Å., Lundberg, J. M., and Schultzberg, M. (1980a). Nature (London) 284, 515–521.
Hökfelt, T., Lundberg, J. M., Schultzberg, M., Johansson, O., Ljungdahl, Å., and Rehfeld, J. (1980b). In "Neural Peptides and Neuronal Communication" (E. Costa and M. Trabucchi, eds.), pp. 1–23. Raven, New York.
Hökfelt, T., Rehfeld, J. F., Skirboll, L., Ivemark, B., Goldstein, M., and Markey, K. (1980c). Nature (London) 285, 476–478.
Hökfelt, T., Skirboll, L., Rehfeld, J. F., Goldstein, M., Markey, K., and Dann, O. (1980d). Neuroscience 5, 2093–2124.
Hökfelt, T., Vincent, S., Hellsten, L., Rosell, S., Folkers, K., Markey, K., Goldstein, M., and Cuello, C. (1981). Acta Physiol. Scand. 113, 571–573.
Hökfelt, T., Lundberg, J. M., Skirboll, L., Johansson, O., Schultzberg, M., and Vincent, S. R. (1982a). In "Cotransmission" (A. C. Cuello, ed.), pp. 77–126. Macmillan, London.
Hökfelt, T., Fahrenkrug, J., Tatemoto, K., Mutt, V., and Werner, S. (1982b). Acta Physiol. Scand. 116, 469–471.
Hökfelt, T., Lundberg, J. M., Tatemoto, K., Mutt, V., Terenius, L., Polak, M., Bloom, S. R., Sasek, C., Elde, R., and Goldstein, M. (1983a). Acta Physiol. Scand. 117, 315–318.
Hökfelt, T., Skagerberg, G., Skirboll, L., and Björklund, A. (1983b). In "Handbook of Chemical Neuroanatomy, Vol. 1: Methods in Chemical Neuroanatomy" (A. Björklund and T. Hökfelt, eds.), pp. 228–285. Elsevier, Amsterdam.
Hökfelt, T., Fahrenkrug, J., Tatemoto, K., Mutt, V., Werner, S., Hulting, A.-L., Terenius, L., and Chang, K. J. (1983c). Proc. Natl. Acad. Sci. U.S.A. 80, 895–898.
Hökfelt, T., Lundberg, J. M., Tatemoto, K., Mutt, V., Terenius, L., Polak, J., Bloom, S., Sasek, C., Elde, R., and Goldstein, M. (1983d). Acta Physiol. Scand. 117, 315–318.
Hökfelt, T., Johansson, O., and Goldstein, M. (1984a). Science 225, 1326–1334.
Hökfelt, T., Everitt, B. J., Theodorsson-Norheim, E., and Goldstein, M. (1984b). J. Comp. Neurol. 222, 543–559.
Hökfelt, T., Johansson, O., and Goldstein, M. (1984c). In "Handbook of Chemical Neuroanatomy, Vol. 2: Classical Transmitters in the CNS, Part I" (A. Björklund and T. Hökfelt, eds.), pp. 157–276. Elsevier, Amsterdam.

Hökfelt, T., Mårtensson, R., Björklund, A., Kleinau, S., and Goldstein, M. (1984d). In "Handbook of Chemical Neuroanatomy, Vol. 2: Classical Transmitters in the CNS, Part I" (A. Björklund and T. Hökfelt, eds.), pp. 277–379. Elsevier, Amsterdam.
Hökfelt, T., Dalsgaard, C.-J., Meister, B., and Everitt, B. J. (1985a). Neurosci. Lett. Suppl. p. 565.
Hökfelt, T., Skirboll, L., Everitt, B. J., Meister, B., Brownstein, M., Jacobs, T., Faden, A., Kuga, S., Goldstein, M., Markstein, R., Dockray, G., and Rehfeld, J. (1985b). Ann. N.Y. Acad. Sci. 448, 255–274.
Holets, V., Hökfelt, T., Terenius, L., Rökaeus, Å., and Goldstein, M. (1986). Neuroscience, submitted.
Hughes, J., Smith, T. W., Kosterlitz, H. W., Fothergill, L. H., Morgan, B. A., and Morris, H. R. (1975). Nature (London) 258, 577–579.
Hulting, A.-L., Lindgren, J.-Å., Hökfelt, T., Heidvall, K., Eneroth, P., Werner, S., Patrono, C., and Samuelsson, B. (1984). Eur. J. Pharmacol. 106, 459–460.
Hulting, A.-L., Lindgren, J.-Å., Hökfelt, T., Eneroth, P., Werner, S., Patrono, C., and Samuelsson, B. (1985). Proc. Natl. Acad. Sci. U.S.A. 52, 3834–3838.
Hunt, S. P., and Lovick, T. A. (1982). Neurosci. Lett. 30, 139–145.
Hylden, J. L. K., and Wilcox, G. L. (1981). Brain Res. 217, 212–215.
Ibata, Y., Watanabe, K., Kinoshita, H., Kubo, S., Sano, N., Yanaihara, C., and Yanaihara, N. (1980). Neurosci. Lett. 17, 185–189.
Ibata, Y., Fukui, K., Okamura, H., Kawakami, T., Tanaka, M., Obata, H. L., Isuto, T., Terubayashi, H., Yanaihara, C., and Yanaihara, N. (1983). Brain Res. 269, 177–179.
Ibata, Y., Kawakami, F., Fukui, K., Okamura, H., Obata-Tsuto, H. L., Tsuto, T., and Terubayashi, H. (1984a). Peptides Suppl. 1, 5, 109–120.
Ibata, Y., Kawakami, F., Fukui, K., Obata-Tsuto, H. L., Tanaka, M., Kubo, T., Okamura, H., Morimoto, N., Yanaihara, C., and Yanaihara, N. (1984b). Brain Res. 302, 221–230.
Itoh, N., Obata, K., Yanaihara, N., and Okamoto, H. (1983). Nature (London) 304, 547–549.
Jacobowitz, D. M., and Olschowka, J. A. (1982). Peptides 3, 569–590.
Jacobowitz, D. M., Schulte, H., Chrousos, G. P., and Loriaux, D. L. (1983). Peptides 4, 521–524.
Jan, Y. N., and Jan, L. Y. (1983). TINS 6, 320–325.
Jennes, L., Stumpf, W. E., and Kalivas, P. W. (1982). J. Comp. Neurol. 210, 211–224.
Jennes, L., Stumpf, W. E., Bissette, G., and Nemeroff, C. B. (1984). Brain Res. 308, 245–253.
Jirikowski, G., Reisert, I., Pilgrim, Ch., and Oertel, W. H. (1984). Neurosci. Lett. 46, 35–39.
Johansson, O., Hökfelt, T., Pernow, B., Jeffcoate, S. L., White, N., Steinbusch, H. W. M., Verhofstad, A. A. J., Emson, P. C., and Spindel, E. (1981). Neuroscience 6, 1857–1881.
Johansson, O., Hökfelt, T., and Elde, R. (1984). Neuroscience 13, 265–339.
Kahn, D., Abrams, G., Zimmerman, E. A., Carraway, R., and Leeman, S. E. (1981). Endocrinology 107, 47–53.
Kangawa, K., Minamino, N., Fukuda, A., and Matsuo, H. (1983). Biochem. Biophys. Res. Commun. 114, 533–540.
Kaplan, S. L., Grumbach, M. M., and Aubert, M. L. (1976). Recent Prog. Horm. Res. 32, 161–234.
Kastin, A. J., Gual, C., and Schally, A. V. (1972). Recent Prog. Horm. Res. 28, 201–216.
Kato, Y., Iwasaki, Y., Iwasaki, J., Abe, H., Yanaihara, N., and Imura, H. (1978). Endocrinology 103, 554–558.

Kawano, H., Daikoku, S., and Saito, S. (1982). *Brain Res.* **242**, 227–232.
Kawata, M., Hashimoto, K., Takahara, J. and Sand, Y. (1983). *Cell Tissue Res.* **230**, 239–246.
Khachaturian, H., Lewis, M. E., and Watson, S. J. (1983). *Neurology* **220**, 310–320.
Khachaturian, H., Lewis, M. E., Schäfer, M. K.-H., and Watson, S. J. (1985). *TINS* **8**, 111–119.
Kilpatrick, D. L., Jones, B. N., Kojima, K., and Udenfriend, S. (1981). *Biochem. Biophys. Res. Commun.* **13**, 698–705.
Kimura, S., Okada, M., Sugita, Y., Kanazawa, I., and Munekata, E. (1983). *Proc. Jpn. Acad.* **59B**, 101–104.
Knobil, E. (1974). *Recent Prog. Horm. Res.* **30**, 1–36.
Knobil, E. (1980). *Recent Prog. Horm. Res.* **36**, 53–88.
Koboyashi, H., and Matsui, T. (1969). *In* "Frontiers in Neuroendocrinology" (W. F. Ganong and L. Martini, eds.), pp. 3–46. Oxford Univ. Press, London and New York.
Kondo, H., Kuramoto, H., Wainer, B. H., and Yanaihara, N. (1985). *Brain Res.* **335**, 309–314.
Krieger, D. T. (1983). *Science* **222**, 975–985.
Krieger, D. T., Liotta, A. S., Brownstein, M. J., and Zimmerman, E. A. (1980). *Recent Prog. Horm. Res.* **36**, 277–336.
Krieger, D. T., Brownstein, M. J., and Martin, J. B. (eds.) (1983). "Brain Peptides." Wiley, New York.
Krukoff, T. L., and Calaresu, F. R. (1984). *Peptides* **5**, 931–936.
Kuypers, H. G. J. M., and Huisman, A. M. (1984). *Adv. Cell Neurobiol.* **5**, 307–340.
Labrie, F., Drouin, J., Ferland, L., Lagacé, L., Beaulieu, M., De Léan, A., Kelly, P. A., Caron, M. G., and Raymond, V. (1978). *Recent Prog. Horm. Res.* **34**, 25–81.
Lechan, R. M., Nestler, J. L., and Jacobson, S. (1982). *Brain Res.* **245**, 1–15.
Lechan, R. M., Lin, H. D., Ling, N., Jackson, I. M., Jacobson, S., and Reichlin, S. (1984). *Brain Res.* **309**, 55–61.
Lee, Y., Kawai, Y., Shiosaka, S., Takami, K., Kiyama, H., Hillyard, C. J., Girgis, S., MacIntyre, I., Emson, P. C., and Tohyama, M. (1985). *Brain Res.* **330**, 194–196.
Leeman, S. E., Aronin, N., and Ferris, C. (1982). *Recent Prog. Horm. Res.* **38**, 93–129.
Léger, L., Charnay, Y., Chayvialle, J. A., Bérod, A., Dray, F., Pujol, J. F., Jouvet, M., and Dubois, P. M. (1983). *Neuroscience* **8**, 525–546.
Le Grevés, P., Nyberg, F., Terenius, L., and Hökfelt, T. (1985). *Eur. J. Pharmacol.* **115**, 309–311.
LeRoith, D., Roth, J., Lesniak, M. A., dePablo, F., Bassas, L., and Collier, E. (1986). *Recent Prog. Horm. Res.*, in press.
Lewis, R., Stern, A. S., Kimura, S., Rossier, J., Stein, S., and Udenfriend, S. (1980). *Science*, **208**, 1459–1461.
Lewis, R., Stein, A. S., Kilpatrick, D., Gerber, L., Rossier, J. J., Stein, S., and Udenfriend, S. (1981). *Neuroscience* **1**, 80–82.
Lindgren, J. Å., Hökfelt, T., Dahlén, S. E., Patrono, C., and Samuelsson, B. (1984). *Proc. Natl. Acad. Sci. U.S.A.* **81**, 6212–6216.
Lindh, B., Hökfelt, T., Elfvin, L.-G., Terenius, L., Elde, R., and Goldstein, M. (1986). *J. Neurosci.*, in press.
Lorén, I. J., Alumets, R., Håkanson, R., and Sundler, F. (1979). *Cell Tissue Res.* **200**, 179–186.
Lorenz, R. G., Saper, C. B., Wong, D. L., Ciaranello, R. D., and Loewy, A. D. (1985). *Neurosci. Lett.* **55**, 255–260.

Lovick, T. A., and Hunt, S. P. (1983). *Neurosci. Lett.* **36,** 223–228.
Lundberg, J. M. (1981). *Acta Physiol. Scand.* **112** (Suppl. 496), 1–57.
Lundberg, J. M., and Hökfelt, T. (1983). *TINS* **6,** 325–333.
Lundberg, J. M., and Tatemoto, K. (1982). *Acta Physiol. Scand.* **116,** 393–402.
Lundberg, J. M., Hökfelt, T., Schultzberg, M., Uvnäs-Wallensten, K., Köhler, C., and Said, S. (1979). *Neuroscience* **4,** 1539–1559.
Lundberg, J. M., Fried, G., Fahrenkrug, J., Holmstedt, B., Hökfelt, T., Lagercrantz, H., Lundgren, G., and Änggård, A. (1981). *Neuroscience* **6,** 1001–1010.
Lundberg, J. M., Hedlund, B., Änggård, A., Fahrenkrug, J., Hökfelt, T., Tatemoto, K., and Bartfai, T. (1982a). *In* "Systemic Role of Regulatory Peptides" (S. R. Bloom, J. M. Polak, and E. Lindenlaub, eds.), pp. 93–119. Schattauer, Stuttgart.
Lundberg, J. M., Hedlund, B., and Bartfai, T. (1982b). *Nature (London)* **295,** 147–149.
Lundberg, J. M., Hökfelt, T., Änggård, A., Terenius, L., Elde, R., Markey, K., and Goldstein, M. (1982c). *Proc. Natl. Acad. Sci. U.S.A.* **79,** 1303–1307.
Lundberg, J. M., Terenius, L., Hökfelt, T., Martling, C. R., Tatemoto, K., Mutt, V., Polak, J., Bloom, S., and Goldstein, M. (1982d). *Acta Physiol. Scand.* **116,** 477–480.
Lundberg, J. M., Terenius, L., Hökfelt, T., and Tatemoto, K. (1984). *J. Neurosci.* **4,** 2376–2386.
Lundberg, J. M., Franco-Cereceda, A., Hua, X., Hökfelt, T., and Fischer, J. A. (1985). *Eur. J. Pharmacol.* **108,** 315–319.
McCann, S. M. (1982). *In* "Neuroendocrine Perspectives" (E. E. Müller and R. M. MacLeod, eds.), Vol. 1, pp. 1–22. Elsevier, Amsterdam.
McCann, S. M., and Ramirez, V. D. (1964). *Recent Prog. Horm. Res.* **20,** 131–170.
McCann, S. M., Vijayan, E., Mangat, H., and Nego-Vilar, A. (1982). *In* "Brain Peptides and Hormones" (R. Collu *et al.*, eds.), pp. 125–135. Raven, New York.
McCann, S. M., Lumpkin, M. D., Mizunuma, H., Khorram, O., and Samson, W. K. (1984). *Peptides* **5** (Suppl. 1), 3–7.
McGinty, J. F., and Bloom, F. (1983). *Brain Res.* **278,** 145–153.
MacLeod, R. M., and Lehmeyer, J. E. (1974). *Endocrinology* **94,** 1077–1085.
Malmfors, T. (1965). *Acta Physiol. Scand.* **64** (Suppl. 248), 1–93.
Mantyh, P. W., and Hunt, S. P. (1984). *Brain Res.* **291,** 49–54.
Markstein, R., and Hökfelt, T. (1984). *J. Neurosci.* **4,** 570–575.
Martini, L., Fraschini, F., and Motta, M. (1968). *Recent Prog. Horm. Res.* **24,** 439–485.
Matsuo, H., Miyata, A., and Mizuno, K. (1983). *Nature (London)* **305,** 721–723.
Meister, B., Hökfelt, T., Vale, W. W., and Goldstein, M. (1985a). *Acta Physiol. Scand.* **124,** 133–136.
Meister, B., Hökfelt, T., Vale, W. W., Sawchenko, P. E., Swanson, L., and Goldstein, M. (1985b). *Neuroendocrinology* **42,** 237–247.
Meites, J., Nicoll, C. S., and Talwalker, P. K. (1963). *In* "Advances in Neuroendocrinology (A. V. Nalbandov, ed.), pp. 238–277. Univ. of Illinois Press, Urbana, Illinois.
Meites, J., Lu, K. H., Wuttke, W., Welsch, C. W., Nagasawa, H., and Quadri, S. K. (1972). *Recent Prog. Horm. Res.* **28,** 471–516.
Meites, J., Bruni, J. F., Van Vugt, D. A., and Smith, A. F. (1979). *Life Sci.* **24,** 1325–1336.
Melander, T., Staines, W. A., Hökfelt, T., Rökaeus, Å., Eckenstein, F., Salvaterra, P. M., and Wainer, B. H. (1985). *Brain Res.* **360,** 130–138.
Melander, T., Hökfelt, T., and Rökaeus, Å. (1986). *J. Comp. Neurol.*, in press.
Merchenthaler, I., Vigh, S., Schally, A. V., and Petrusz, P. (1984). *Endocrinology* **114,** 1082–1085.
Merchenthaler, I., Maderdrut, J. L., Dockray, G. J., Altschuler, R. A., and Petrusz, P. (1986). *Brain Res.*, in press.

Minamino, N., Kangawa, K., Fukuda, A., and Matsuo, H. (1984). *Neuropeptides* **4**, 157–166.
Mitchell, R., and Fleetwood-Walker, S. (1981). *Eur. J. Pharmacol.* **76**, 119–120.
Morris, H. R., Taylor, G. W., Piper, P. J., and Tippins, J. R. (1980). *Nature (London)* **285**, 104–106.
Moskowitz, M. A., Kiwak, K. J., Hekimian, K., and Levine, L. (1984). *Science* **224**, 886–889.
Mugnaini, E., and Oertel, W. H. (1985). *In* "Handbook of Chemical Neuroanatomy" (A. Björklund and T. Hökfelt, eds.), Vol. 4, pp. 436–608. Elsevier, Amsterdam.
Müller, E. E. (1973). *Neuroendocrinology* **11**, 338–369.
Murphy, R. C., Hammarström, S., and Samuelsson, B. (1979). *Proc. Natl. Acad. Sci. U.S.A.* **76**, 4275–4279.
Nair, R. M. G., Barrett, J. F., Bowers, C. Y., and Schally, A. V. (1970). *Biochemistry* **9**, 1103–1106.
Nakane, P. K. (1968). *J. Histochem. Cytochem.* **16**, 557–560.
Nawa, H., Hirose, T., Takashima, H., Inayama, S., and Nakanishi, S. (1983). *Nature (London)* **306**, 32–36.
Neill, J. D. (1980). *In* "Frontiers in Neuroendocrinology" (L. Martini and W. F. Ganong, eds.), pp. 124–155. Raven, New York.
Nicolics, K., Mason, A. J., Szönyi, E., Rachmandran, J., and Seeburg, P. (1985). *Nature (London)* **316**, 511–517.
Nilsson, J., von Euler, A. M., and Dalsgaard, C.-J. (1985). *Nature (London)* **315**, 61–63.
Oertel, W. H., and Mugnaini, E. (1984). *Neurosci. Lett.* **47**, 233–238.
Oertel, W. H., Graybiel, A. M., Mugnaini, E., Elde, R. P., Schmechel, D. E., and Kopin, I. J. (1983). *J. Neurosci.* **3**, 1322–1332.
Ohhashi, T., and Jacobowitz, D. M. (1983). *Peptides* **4**, 381–386.
Okamura, H., Murakami, S., Chihara, K., Nagatsu, I., and Ibata, Y. (1985). *Neuroendocrinology* **41**, 177–179.
Olschowka, J. A., O'Donohue, T. L., and Jacobowitz, D. M. (1981). *Peptides* **2**, 309–331.
Olschowka, J. A., O'Donohue, R. L., Mueller, G. P., and Jacobowitz, D. M. (1982). *Peptides* **3**, 995–1015.
Örning, L., and Hammarström, S. (1980). *J. Biol. Chem.* **255**, 8023–8026.
Osborne, N. N. (1983). *In* "Dale's Principle and Communication between Neurones" (N. N. Osborne, ed.), pp. 83–94. Pergamon, Oxford.
Otsuka, M., and Takahashi, T. (1977). *Annu. Rev. Pharmacol. Toxicol.* **17**, 425–439.
Ouimet, C. C., Patrick, R. L., and Ebner, F. F. (1981). *J. Comp. Neurol.* **195**, 284–304.
Ouimet, C. C., Miller, P. E., Hemmings, H., Walaas, S. I., and Greengard, P. (1984). *J. Neurosci.* **4**, 111–124.
Owman, Ch., Håkanson, R., and Sundler, F. (1973). *Fed. Proc., Fed. Am. Soc. Exp. Biol.* **32**, 1785–1791.
Palmer, M. R., Mathews, R., Murphy, R. C., and Hoffer, B. J. (1980). *Neurosci. Lett.* **18**, 173–180.
Palmer, M. R., Mathews, W. R., Hoffer, B. J., and Murphy, R. C. (1981). *J. Pharmacol. Exp. Ther.* **219**, 91–96.
Patterson, P. H. (1978). *Annu. Rev. Neurosci.* **1**, 1–17.
Paull, W. K., and Gibbs, F. P. (1983). *Histochemistry* **78**, 303–316.
Pearse, A. G. E. (1969). *J. Histochem. Cytochem.* **17**, 303–313.
Pease, P. C. (1962). *Anat. Res.* **142**, 342.
Pelletier, G., Steinbusch, H. W., and Verhofstad, A. (1981). *Nature (London)* **293**, 71–72.
Perez de la Mora, M., Possani, L. D., Tapia, R., Teran, L., Palacios, R., Fuxe, K., Hökfelt, T., and Ljungdahl, A. (1981). *Neuroscience* **6**, 875–895.

Pernow, B. (1983). *Pharmacol. Rev.* 35, 85–141.
Phillips, H. S., Nicolics, K., Branton, D., and Seeburg, P. (1985). *Nature (London)* 316, 542–545.
Pickel, V. (1985). *In* "Handbook of Chemical Neuroanatomy" (A. Björklund and T. Hökfelt, eds.), Vol. 4, pp. 72–92. Elsevier, Amsterdam.
Piercey, M. F., Dobry, P. J. K., Schroeder, L. A., and Einspahr, F. J. (1981). *Brain Res.* 210, 407–412.
Porter, J. C., Mical, R. S., Ben-Jonathan, N., and Ondo, J. G. (1973). *Recent Prog. Horm. Res.* 29, 161–194.
Potter, D. D., Furshpan, E. J., and Landis, S. C. (1981). *Neurosci. Commun.* 1, 1–9.
Potter, D. D., Furshpan, E. J., and Landis, S. C. (1983). *Fed. Proc., Fed. Am. Soc. Exp. Biol.* 42, 1626–1632.
Price, D. A., and Greenberg, M. I. (1977). *Science* 197, 670–671.
Reichlin, S., Martin, J. B., Mitnick, M. A., Boshans, R. L., Grimm, Y., Bollinger, J., Gordon, J., and Malacara, J. (1972). *Recent Prog. Horm. Res.* 28, 229–277.
Richardson, K. C. (1966). *Nature (London)* 210, 756.
Rivier, C., Brown, M., and Vale, W. (1977). *Endocrinology* 100, 751–754.
Rivier, J., Spiess, J., Thorner, M. O., and Vale, W. (1982). *Nature (London)* 300, 276–278.
Rökaeus, Å., Melander, T., Hökfelt, T., Lundberg, J. M., Tatemoto, K., Carlquist, M., and Mutt, V. (1984). *Neurosci. Lett.* 47, 161–166.
Rosenfeld, M. G., Mermod, J.-J., Amara, S. G., Swanson, L. W., Sawchenko, P. E., Rivier, J., Vale, W. W., and Evans, R. M. (1983a). *Nature (London)* 304, 129–135.
Rosenfeld, M. G., Amara, S. G., Birnberg, N. C., Mermod, J. J., Murdoch, G. H., and Evans, R. M. (1983b). *Recent Prog. Horm. Res.* 39, 305–347.
Rossier, J. (1982). *In* "Frontiers in Neuroendocrinology" (W. F. Ganong and L. Martini, eds.), Vol. 7, pp. 191–209. Raven, New York.
Roth, J., LeRoith, D., Lesniak, M. A., dePablo, F., Bassas, L., and Collier, E. (1986). *In* "Coexistence of Neuronal Messengers: A New Principle in Chemical Transmission" (*Prog. Brain Res.*) (T. Hökfelt, K. Fuxe, and B. Pernow, eds.). Elsevier, Amsterdam (in press).
Ruberg, M., Rotsztejn, W., Arancibia, S., Besson, J., and Enalbert, A. (1978). *Eur. J. Pharmacol.* 51, 319–320.
Samson, W. K., Lumpkin, M. D., McDonald, J. K., and McCann, S. M. (1983). *Peptides* 4, 817–819.
Samuelsson, B. (1983). *Science* 220, 568–575.
Sar, M., Stumpf, W. E., Miller, R. J., Chang, K., and Cuatrecasas, P. (1978). *J. Comp. Neurol.* 182, 17–38.
Sawchenko, P. E., Swanson, L. W., and Vale, W. W. (1984). *J. Neurosci.* 4, 1118–1129.
Sawchenko, P. E., Swanson, L. W., Rivier, J., and Vale, W. W. (1985). *J. Comp. Neurol.* 237, 100–115.
Schalling, M., Neil, A., Terenius, L., Hökfelt, T., Lindgren, J.-Å., and Samuelsson, B. (1986a). *Eur. J. Pharmacol.*, in press.
Schalling, M., Hökfelt, T., Lindgren, J.-Å., and Samuelsson, B. (1986b). In preparation.
Schally, A. V., Arimura, A., Bowers, C. Y., Kastin, A. J., Sawano, S., Redding, R. W. (1968). *Recent Prog. Horm. Res.* 24, 497–581.
Schally, A. V., Arimura, A., Baba, Y., Nair, R. M. G., Matsuo, J., Redding, T. W., Debeljuk, L., and White, W. F. (1971). *Biochem. Biophys. Res. Commun.* 43, 393–399.
Scharrer, E., and Scharrer, B. (1954). *Recent Prog. Horm. Res.* 10, 183–240.
Schmechel, D. E., Vickrey, B. G., Fitzpatrick, D., and Elde, R. P. (1984). *Neurosci. Lett.* 47, 227–232.

Schultzberg, M., Hökfelt, T., Lundberg, J. M., Terenius, L., Elfvin, L.-G., and Elde, R. (1978a). *Acta Physiol. Scand.* **103**, 475–477.
Schultzberg, M., Lundberg, J. M., Hökfelt, T., Terenius, L., Brandt, J., Elde, R., and Goldstein (1978b). *Neuroscience* **3**, 1169–1186.
Seybold, V. S., Hylden, J. L. K., and Wilcox, G. L. (1982). *Peptides* **3**, 49–54.
Shaar, C. J., Fredrickson, R. C. A., Dininger, N. B., and Jackson, L. (1977). *Life Sci.* **21**, 853–860.
Shaar, C. J., Clemens, J. A., and Dininger, N. B. (1979). *Life Sci.* **25**, 2071–2074.
Skirboll, L., Hökfelt, T., Norell, G., Philipson, O., Kuypers, J. G. J. M., Bentivoglio, M., Catsman-Berrevoets, C. E., Visser, T. J., Steinbusch, H., Verhofstad, A., Cuello, A. C., Goldstein, M., and Brownstein, M. (1984). *Brain Res. Rev.* **8**, 99–127.
Smith, A. D. (1976). *In* "Neuron Concept Today" (J. Szentágothai, J. Hámori, and E. S. Vizi, eds.), pp. 49–61. Akadémiai Kiadó, Budapest.
Smith, R. M., Howe, P. R. C., Oliver, J. R., and Willoughby, J. O. (1984). *Neuropeptides* **4**, 109–115.
Snyder, S. (1980). *Science* **209**, 976–983.
Sofroniew, M. V. (1979). *Am. J. Anat.* **154**, 283–289.
Somogyi, P., Hodgson, A. J., Smith, A. D., Nunzi, M. G., Gorio, A., and Wu, J.-Y. (1984). *Neuroscience* **14**, 2590–2603.
Spiess, J., Rivier, R., Rivier, C., and Vale, W. (1981). *Proc. Natl. Acad. Sci. U.S.A.* **78**, 6517–6521.
Spiess, J., Rivier, J., Thorner, M., and Vale, W. (1982). *Biochemistry* **21**, 6037–6040.
Spiess, J., Rivier, J., and Vale, W. W. (1983). *Nature (London)* **303**, 532–535.
Stern, A. S., Lewis, R. V., Kimura, L. S., Rossier, J., Gerber, L. D., Brink, L., Stein, S., and Udenfriend, S. (1979). *Proc. Natl. Acad. Sci. U.S.A.* **76**, 6680–6683.
Stern, A. S., Lewis, R. V., Kimura, J., Rossier, J., Stein, S., and Udenfriend, S. (1980). *Arch. Biochem. Biophys.* **205**, 606–613.
Stern, A. S., Jones, B. N., Shively, J. E., Stein, S., and Udenfriend, S. (1981). *Proc. Natl. Acad. Sci. U.S.A.* **78**, 1962–1966.
Sternberger, L. A. (1979). "Immunocytochemistry." Wiley, New York.
Sternberger, L. A., Hardy, Jr., P. H., Cuculis, J. J., and Meyer, H. G. (1970). *J. Histochem. Cytochem.* **18**, 315–324.
Stjärne, L., and Lundberg, J. M. (1984). *Acta Physiol. Scand.* **120**, 477–479.
Straus, E., Ryder, S. W., Eng, J., and Yalow, R. S. (1981). *Recent Prog. Horm. Res.* **37**, 447–471.
Swanson, L. W., and Sawchenko, P. E. (1983). *Annu. Rev. Neurosci.* **6**, 275–325.
Swanson, L. W., Sawchenko, P. E., Rivier, J., and Vale, W. W. (1983). *Neuroendocrinology* **36**, 165–186.
Szentágothai, J., Flerko, B., Mess, B., and Halász, B. (1962). "Hypothalamic Control of the Anterior Pituitary." Akadémiai Kiadó, Budapest.
Takahashi, T., Konishi, S., Powell, D., Leeman, S. E., and Otsuka, M. (1974). *Brain Res.* **73**, 59–69.
Takami, K., Kawai, Y., Shiosaka, S., Lee, Y., Girgis, S., Hillyard, C. J., MacIntyre, I., Emson, P. C., and Tohyama, M. (1985). *Brain Res.* **328**, 386–389.
Tappaz, M. L., Aguera, M., Belin, M. F., and Pujol, J. F. (1980). *Brain Res.* **186**, 379–391.
Tappaz, M. L., Wassef, M., Oertel, W. H., Paut, L., and Pujol, J. F. (1983). *Neuroscience* **9**, 271–287.
Tatemoto, K. (1982). *Proc. Natl. Acad. Sci. U.S.A.* **79**, 5485–5489.
Tatemoto, K., and Mutt, V. (1981). *Proc. Natl. Acad. Sci. U.S.A.* **78**, 6603–6607.
Tatemoto, K., Carlquist, M., and Mutt, V. (1982). *Nature (London)* **296**, 659–660.

Tatemoto, K., Rökaeus, A., Jörnvall, H., McDonald, R. J., and Mutt, V. (1983). *FEBS Lett.* **164**, 124-128.
Tatemoto, K., Lundberg, J. M., Jörnvall, M., and Mutt, V. (1986). *Nature (London)* (submitted).
Tramu, G., Pillez, A., and Leonardelli, J. (1978). *J. Histochem. Cytochem.* **26**, 322-324.
Uhl, G. R., Goodman, R. R., and Snyder, S. H. (1979). *Brain Res.* **167**, 77-91.
Vale, W., Brazeau, P., Rivier, C., Brown, M., Boss, B., Rivier, J., Burgus, R., Ling, N., and Guillemin, R. (1975). *Recent Prog. Horm. Res.* **31**, 365-393.
Vale, W., Spiess, J., Rivier, C., and Rivier, J. (1981). *Science* **213**, 1394-1397.
Vale, W., Rivier, C., Brown, M. R., Spiess, J., Koob, G., Swanson, L., Bilezikjian, L., Bloom, F., and Rivier, J. (1983). *Recent Prog. Horm. Res.* **39**, 245-270.
van den Pol, A. N., Herbst, R. S., and Powell, J. F. (1984). *Neuroscience* **13**, 1117-1156.
van Dyke, H. B., Adamsons, K., Jr., and Engel, S. L. (1955). *Recent Prog. Horm. Res.* **19**, 1-35.
Vijayan, E., and McCann, S. M. (1979). *Endocrinology* **105**, 64-68.
Vijayan, E., Krulich, L., and McCann, S. M. (1978). *Neuroendocrinology* **26**, 174-185.
Vijayan, E., Samson, W., Said, S. I., and McCann, S. M. (1979). *Endocrinology* **104**, 53-57.
Vincent, S. R., Hökfelt, T., and Wu, J. Y. (1982a). *Neuroendocrinology* **34**, 117-125.
Vincent, S. R., Hökfelt, T., Christensson, I., and Terenius, L. (1982b). *Neurosci. Lett.* **33**, 185-190.
Vincent, S. R., Satoh, K., Armstrong, D. M., and Fibiger, H. C. (1983). *Nature (London)* **306**, 688-691.
Vincent, S. R., McIntosh, C. H. S., Buchan, A. M. J., and Brown, J. C. (1985). *J. Comp. Neurol.* **238**, 169-186.
Viveros, O. H., Diliberto, E. J., Hazum, E., and Chang, K.-J. (1979). *Mol. Pharmacol.* **16**, 1101-1108.
Vogt, M. (1954). *J. Physiol.* (London) **123**, 451-481.
Walaas, S. I., and Fonnum, F. (1978). *Brain Res.* **153**, 549-562.
Walaas, S. I., and Greengard, P. (1984). *J. Neurosci.* **4**, 111-124.
Wamsley, J. K., Young, W. S., III, and Kuhar, M. J. (1980). *Brain Res.* **190**, 153-174.
Watson, S. J., Barchas, J. D., and Li, C. H. (1977). *Proc. Natl. Acad. Sci. U.S.A.* **74**, 5155-5158.
Watson, S. J., Akil, H., Fischli, W., Goldstein, A., Zimmerman, E., Nilaver, G., and van Wimersma Greidanus, Tj. B. (1982). *Science* **216**, 85-87.
Watson, S. J., Khachaturian, H., Taylor, L., Fischli, W., Goldstein, A., and Akil, H. (1983). *Proc. Natl. Acad. Sci. U.S.A.* **80**, 891-894.
Watt, C. B., Su, Y. T., and Lam, D. M.-K. (1984). *Nature (London)* **311**, 761-763.
Weber, E., Roth, K. A., and Barchas, J. D. (1982). *Proc. Natl. Acad. Sci. U.S.A.* **79**, 3062-3066.
Weber, E., Esch, F. S., Böhlen, P., Paterson, S., Corbett, A., McKnight, A. T., Kosterlitz, H., Barchas, J. D., and Evans, C. J. (1983). *Proc. Natl. Acad. Sci. U.S.A.* **80**, 7362-7366.
Weigand, S. J., and Price, J. L. (1980). *J. Comp. Neurol.* **192**, 1-19.
Weiler, R., and Ball, A. K. (1984). *Nature (London)* **311**, 759-761.
Werner, S., Hulting, A.-L., Hökfelt, T., Eneroth, P., Tatemoto, K., Mutt, V., Maroder, L., and Wünsch, E. (1983). *Neuroendocrinology* **37**, 476-478.
Westfall, J., and Kinnemon, J. C. (1984). *Tissue Cell* **16**, 355-365.
Wiesenfeld-Hallin, Z., Hökfelt, T., Lundberg, J. M., Forssmann, W. G., Reinecke, M., Tschopp, F. A., and Fischer, J. A. (1984). *Neurosci. Lett.* **52**, 199-204.

Wilber, J. F., Montoya, E., Gendrich, R., Plotnikoff, N. P., White, W. F., Renaud, L., and Martin, J. B. (1976). *Recent Prog. Horm. Res.* **32**, 117–153.
Williams, R. G., and Dockray, G. J. (1983). *Neuroscience* **9**, 563–586.
Yaksh, T. L., and Rudy, T. A. (1976). *Physiol. Behav.* **17**, 1031–1036.
Yen, S. S. C., Lasley, B. L., Want, C. F., Leblanc, H., and Siler, T. M. (1975). *Recent Prog. Horm. Res.* **31**, 321–357.
Young, W. S. III, and Kuhar, M. J. (1979). *Brain Res.* **179**, 255–270.
Young, W. S. III, and Kuhar, M. J. (1980). *Proc. Natl. Acad. Sci. U.S.A.* **77**, 1696–1700.
Zahm, D. S., Zaborszky, L., Alones, V. E., and Heimer, L. (1985). *Brain Res.* **325**, 317–321.
Zamboni, I., and de Martino, C. (1967). *J. Cell Biol.* **148a**, 35.

DISCUSSION

W. D. Drucker. Dr. Hökfelt, have you been able to use your technique to study experimental models of disease? I am thinking for example of spontaneous Cushing's syndrome in dogs; are there any abnormalities or differences in the intensity of neurotransmitter staining? Alternatively, could your techniques be used to see how pharmacologic agents may be influencing neurotransmitters? We could use the example of Dr. Krieger's discovery that amelioration of Cushing's syndrome in some patients follows the use of cyproheptadine. We are really never sure with this compound if it is inhibiting serotonin specifically or where it might be acting anatomically.

T. Hökfelt. Thank you for bringing up two important questions. With regard to the first question concerning immunohistochemical analysis of pathological material, we have so far done very few studies. For example, we have looked postmortem at the brain of some children with the sudden infant death syndrome (SIDS) and of some schizophrenic patients but we have not been able to see any deficiencies. In order to be able to record effects, it would have to be major changes in levels, or fairly extensive degenerations. In reality changes may be much more subtle in those diseases, for example, changes in turnover or receptor sensitivity, and those would be very hard to discover with our technique. To study human material is a major problem for many reasons (postmortem delay, size of tissue, proper controls, etc.) and the sections I showed you are of course, the best we have. We cannot expect strong staining in the majority of cases, but we will continue our efforts in this area.

With regard to your second question concerning histochemical quantification, we have carried out such work for many years with Dr. Fuxe, originally subjectively estimating levels of catecholamines, for example, in the hypothalamus under various experimental conditions. In this work the formaldehyde-induced fluorescence technique was used. Today this is done using microspectrophotometry whereby standards with known amine concentrations can be inserted in the sections. In this way you can get a fairly accurate idea even of absolute levels.

To do the same thing with immunohistochemistry is in principle possible but several methodological problems are encountered. For example, each individual animal has to be perfused with fixative, which introduces variability. In contrast, with the formaldehyde method discussed above, up to 20 rat brains can be processed under identical conditions. In most experiments on changes in amine levels, synthesis inhibitors have been used and the decline in aminine stores has been studied in control and experimental groups. We have no good possibilities of manipulating peptide synthesis in a similar way. In fact immunohistochemistry is not a good method for measuring peptide levels. In our experiments, we have had difficulties in recording changes, even when peptide stores were reduced to 50% of normal. We have, however, been able to demonstrate changes in peptide levels after administration of drugs such as reserpine and 6-hydroxydopamine, which deplete, e.g., neuropeptide Y, which is costored with noradrenaline, and cysteamine can be shown to reduce somatostatin-like immunoactivity.

D. K. Pomerantz. I was intrigued by the data you showed in which the leukotriene caused LH release from the anterior pituitary cells. Do you have, or does anyone have data to show whether or not the LTC_4 can release follicle-stimulating hormone or, for that matter, any other anterior pituitary hormone?

T. Hökfelt. We have not been able to obtain consistent effects for other pituitary hormones.

A. Hoffman. Could you give us some of your data concerning the colocalization of neurotransmitters within the granules, and do you believe that when two neurotransmitters are colocalized within a neuron that this always implies cosecretion?

T. Hökfelt. If you have messengers of different types in a neuron, you would like to have some means of causing differential release, especially in cases of colocalization of a classical transmitter and a peptide. Costorage of peptides in hypothalamus, for example, CRF and PHI, as discussed in my presentation may not necessarily have to be released differentially. Findings in the peripheral nervous system mainly by Dr. Lundberg suggest a differential release in the sense that at low frequencies only the classical transmitter is released, whereas at high frequencies the peptide is released in addition. How is this achieved? We have considered the possibility that the two compounds have different storage sites and have, therefore, with Dr. Fried studied the subcellular distribution of noradrenaline and neuropeptide Y, which are costored in sympathetic nerves in the vas deferens of rats. In the electron microscope, nerve endings generally have two types of storage vesicles, small synaptic vesicles with a diameter of about 500 Å and, in lower numbers, a larger type of vesicle. According to our subcellular fractionation studies, the small vesicles contain only the amine, whereas the large vesicles have both amine and peptide. Our hypothesis therefore is that at low frequency only the small synaptic vesicles with the amine are activated whereas at high nerve activity the large vesicles release their content, i.e., amine plus peptide.

R. O. Morgan. The association between GnRH and leukotriene in stimulating LH secretion seems to imply that both are not so much neurotransmitters as they are local hormones. Have you also considered the possibility that GnRH may stimulate leukotriene production and that leukotriene C may actually be an intracellular messenger as well in the gonadotrope?

T. Hökfelt. That is a possibility. We have tried to demonstrate formation of LTC_4 in the pituitary, but so far without success. Therefore our hypothesis is that LTC_4 is produced in the median eminence and transported via portal vessels to the anterior pituitary. Of course, LTC_4 formation can occur, but in too low amounts to be detectable with our technique.

M. A. Greer. There is quite a bit of evidence that the hormone content of intracellular vesicles can be discharged by the classical exocytotic mechanism. There also is evidence that at least within the cell there can be differential release of individual components of vesicles, going back and forth in intracellular vesicles in the adrenal medulla for instance. I wonder if there is any direct evidence that there necessarily has to be exocytotic discharge of the entire secretory vesicle. As I understand from your data, you have made diagrams showing localization of different materials in discrete secretory vesicles. Is it possible that several substances might be within the same vesicle and that differential discharge could be by some mechanism that is not clear yet rather than the entire contents being popped out? Is that possible or is there compelling evidence against such a hypothesis?

T. Hökfelt. The proportion of messenger (transmitter) released from a vesicle is a controversial question. We do not have any data that would allow us to draw conclusions on that issue with respect to classical transmitter versus peptide.

S. R. Ojeda. If lipoxygenase activity behaves similarly to cyclooxygenase activity then it is very surprising that the hypothalamus would accumulate LTC_4 in such large amounts as to permit the detection of LTC_4 so clearly by immunohistochemistry. I noticed that you stressed the fact that you conjugated LTC_4 to BSA for the immunization; have you preabsorbed with BSA to control the specificity of your histochemical reaction?

T. Hökfelt. I agree with you that accumulation of LTC_4 in nerve endings seems an unlikely phenomenon, and as I tried to show, we are very uncertain about the identity of the compound which we see immunohistochemically. What got us interested in LTC_4 was that it, in contrast to other arachidonic metabolites such as prostaglandins, contains a peptide. We cannot exclude the possibility that this peptide moiety, which is glutathione, or another, unknown peptide is responsible for the immunoreaction.

S. R. Ojeda. Of course, if you perform a biochemical measurement of LTC_4 produced by brain tissue, apparently the production measure is exceedingly low, right?

T. Hökfelt. Yes.

D. Goltzman. I have a question about your elegant studies in which you injected a dye into the circulation and then localized it in cell bodies within nervous tissues through retrograde transport. I am wondering how you determined that retrograde transport was not occurring as a result of interaction with neurons in circumventricular organs other than the median eminence such as the area postrema of the subfornica organ?

T. Hökfelt. We have made a lesion of the median eminence, after which no retrogradely labeled neurons were seen, e.g., in the paraventricular nucleus or in the anterior periventricular area. The lesion also involves chiefly part of the arcuate nucleus, however, and we plan to make hypothalamic deafferentations to exclude transport from other regions outside the blood–brain barrier.

B. Rice. Is there a connection between the intensity of the staining and some functions, e.g., in adrenalectomized animals do you find higher staining activities than in an animal treated with high doses of glucocorticoid. Does the staining really represent just storage or does it represent activity.

T. Hökfelt. The physiological meaning of what we are seeing with immunohistochemistry is a completely open question. It is only when we try to connect these results to some physiological experiments that meaningful information can be obtained. Although I have not discussed this, such experiments have been carried out, e.g., on the cat salivary gland, mainly by my colleagues, Drs. Lundberg and Änggård, that show that these coexisting peptides have marked effects on blood flow through the gland. It may, however, be that peptides in other tissues have very little important function or no function at all. One could, for example, see this from an evolutionary point of view. In primitive organisms peptides could have been important factors in endocrine cells and primitive neurons. During evolution they might have been substituted by fast acting transmitters and the peptide may now be carried along as silent "passengers" without function.

With regard to your specific question there are studies, e.g., at NIH and the Salk Institute, in which the experiments you mentioned, for example, adrenalectomy, have been carried out. In fact increases in CRF levels could be shown in the cell bodies of the paraventricular neurons. Thus, adrenalectomy caused activation and increased CRF production, which could be recorded immunohistochemically.

R. Greep. In that connection it might be interesting to study the effect of down regulation on LHRH neurons that are unresponsive for a long time after excessive stimulation.

T. Hökfelt. We are now about to do such experiments, including analysis of the various stages of the cycle.

S. Cohen. I was intrigued by one of your earlier slides in which you started out with biochemistry, progressed through histology and histochemistry, and ended up with physiology. This reminded me of the time that a physiologist accused me of spending too much time during my biochemical lectures to medical students on the physiology of the endocrine glands and not enough on their biochemistry. I asked him what he considered the distinction. He said, "If it takes place within the cell it's physiological, if it takes place outside the cell it's biochemistry."

Do you think that in your studies, when you look at an object, you might be looking at different hormonal activities such as the β-lipoprotein of the brain that can be broken up into three hormones, ACTH, the enkephalin, and melatonin? I wonder if in your studies you may not have similar groupings of material so that it would show up as a single unit using one technique but would show up as of distinct hormones with a different technique.

T. Hökfelt. With my slide I just wanted to show that histochemistry and histology may contribute to the analysis of biological events by forming a link between biochemistry and physiology. Thus, histochemistry will provide information on the exact neuronal localization of hormones and transmitters which could be helpful in the design of physiological experiments.

S. Cohen. What I was thinking of was whether a compound viewed histochemically represents a distinct compound or hormone or whether it could represent a group of hormones.

T. Hökfelt. As I noted specificity is a serious problem in immunohistochemistry. As an example you may take the tachykinins. We now know that several substance P-like peptides exist in the nervous system. With most antisera it is difficult to distinguish between them histochemically. However, using sequence-specific antibodies it may be possible to improve the situation.

H. Kronenberg. Earlier you showed us how, among other things, peptides could be released from one neuron and act on another neuron. That being the case, how do you deal with the assumption that localization of the peptide defines the neuronal source of the peptide, as opposed to the possibility that in some cases the peptide got to one neuron from another neuron.

T. Hökfelt. This is one of my main points in trying to explain why it is so difficult to elucidate the functional role of peptides in the central nervous system. If you inject substance P into, for example, the dorsal horn of the spinal cord, which contains three different tachykinin systems, it is difficult to know to which of the three types of receptors this effect is related—to the receptor for the descending systems, for the primary sensory neurons, or the local neurons. Thus, the overlap of several systems containing the same peptide, sometimes coexisting with a classical transmitter, complicates the functional analysis.

T. Chen. I am fascinated by your model showing that neurotransmitters can act back on the presynaptic neuron. I guess this is called autoregulation. From a cell biologist's point of view, I presume that the neuropeptides are synthesized in the same classical pathway as the membrane receptors, so my question is how does the cell prevent the neurotransmitters from interacting with the newly synthesized receptors during their journey to the cell surface? Presumably they all go through rough endoplasm reticulum before the peptides are packaged while the receptors are incorporated into the plasma membrane.

T. Hökfelt. How the cell can keep different compounds apart is one of the great mysteries.

M. Saffran. Analogies can often be dangerous, but I wonder if something can be learned by looking at the similarities between the great neuroendocrine integrator in the hypothalamus and the visceral integrator of metabolism in the liver? The hepatocyte will make, store, and release small molecules such as glucose and make, store, and release larger molecules such as the blood clotting factors, albumin, and other protein characters. These are all made in the same cell, or stored in the same cell, they are released in response to different signals, so that nature has provided another model that can perhaps be applied to the study of hypothalamic cells.

J. M. Lakoski. Several laboratories have recently shown that classical neurotransmitters, in terms of their localization, appear to have some differences in sexual dimorphic nuclei, for example, serotonin in the hypothalamus. Do you see any evidence for sexual dimorphism in terms of peptide distribution and have the studies you presented today been done solely on male rats?

T. Hökfelt. Our studies were essentially done on male rats. Other immunohistochemical studies, e.g., by a Dutch group, have shown sex differences in the distribution of vasopressin in some brain areas, such as the septum.

J. C. Beck. You focused entirely on the single component of the physiological explanation for your observations. Is there any evidence for the coexistence or cohabitation of the traditional neurotransmitters with peptides and now the leukotrines may be playing a modulating role on the biosynthesis of the traditional neurotransmitter? One of the major epidemics sweeping the world is dementias of the Alzheimer's type and the locus ceruleus has been identified as a possible basic pathological site. I wonder if you have had an opportunity to look at this region of the brains of Alzheimer's disease patients?

T. Hökfelt. No, we have so far not studied such brains. With regard to possible intracellular actions of peptides, studies by Dr. K. Hole and Dr. M. Zigmond indicate that peptides such as cholecystokinin and secretin can effect the enzyme tyrosine hydroxylase in the brain and sympathetic ganglia. This suggests the possibility that the peptide can be involved in intracellular regulatory events.

Regulation of Gene Expression by Androgens in Murine Kidney

J. F. CATTERALL, K. K. KONTULA, C. S. WATSON, P. J. SEPPÄNEN,
B. FUNKENSTEIN, E. MELANITOU, N. J. HICKOK, C. W. BARDIN, AND
O. A. JÄNNE

The Population Council and The Rockefeller University, New York, New York

I. Introduction

The phenotypic expression of steroid hormone action in different target organs varies widely, yet the initial steps in the mechanism of action of all steroids are similar. These hormones probably enter the cell by passive diffusion through the plasma membrane after which they interact with soluble receptor molecules that are specific for each steroid class (O'Malley and Means, 1974; Katzenellenbogen, 1980; Bardin and Catterall, 1981). Whether this interaction occurs in the cytoplasm followed by translocation of the receptor–steroid complex to the nucleus (see Schrader, 1984) or primarily in the nucleus (King and Greene, 1984; Welshons *et al.*, 1984) does not change the basic concept that the receptor–steroid complex has a greater affinity for relevant sites on chromatin than the steroid-free receptor. These chromatin binding regions or "acceptor" sites may include DNA sequences which have been shown to specifically bind steroid receptors *in vitro* (Payvar *et al.*, 1981, 1983; Mulvihill *et al.*, 1982; Compton *et al.*, 1983; Renkawitz *et al.*, 1984). Although DNA binding alone is sufficient to account for interaction of the receptor–steroid complex with a specific DNA sequence *in vitro*, one or more nuclear proteins must be involved in determination of tissue- and cell-specific responses of genes to hormones *in vivo* (Spelsberg *et al.*, 1984). Recent studies have identified nuclear proteins with properties consistent with tissue-specific regulation of the prolactin gene (White *et al.*, 1985). It remains to be proven that these are indeed tissue-specific regulators and the mode of action of such regulatory proteins in steroid hormone action also awaits further elucidation. Whatever the precise nature of the interaction between steroid–receptor complexes and chromatin acceptor sites, the major result of this interaction is regulation of specific gene transcription (Ringold *et al.*, 1977; McKnight and Palmiter, 1979; Swaneck *et al.*, 1979;

Brock and Shapiro, 1983a). It is clear, however, that in other cases, steroid hormone treatment affects mRNA stability (Brock and Shapiro, 1983b). Both mechanisms have been documented in the regulation of gene expression by androgenic steroids (Derman, 1981; Page and Parker, 1982; Berger *et al.*, 1986).

The mechanism of androgen action is less well understood than that for other classes of steroids such as estrogens, progestins, and glucocorticoids. There are at least two major reasons for this: the main approaches to study steroid hormone action have recently been the analysis of the structure and function of steroid-regulated genes and the purification and characterization of steroid hormone receptors. Androgen-regulated gene products in general are not of the high abundance class. Therefore, recombinant DNA technology was first applied to those steroid-regulated systems that produced more readily obtainable gene products (O'Malley *et al.*, 1979; Yamamoto *et al.*, 1983; Groner *et al.*, 1984). Similarly, androgen receptors have proven to be unusually difficult to purify owing to their lability and low concentration in target tissues. Recent advances in the stabilization and purification of androgen receptors (Isomaa *et al.*, 1982; Chang *et al.*, 1983) should facilitate their more complete characterization.

One line of investigation that has been particularly useful for the study of androgen action is the identification of genetic mutations that cause androgen insensitivity. Testicular feminization (TF) was the first inherited disorder for which the phenotype, male pseudohermaphroditism, could be attributed to deficient receptors. However, studies of the clinical features of the syndrome alone could not distinguish among several possible mechanisms including defects in steroid metabolism. It was subsequently shown that the defect was in an early step in the pathway of androgen action (French *et al.*, 1966), but studies on the rat model of the syndrome (Stanley and Gumbreck, 1964; Stanley *et al.*, 1973) were required to prove the correlation between the phenotype and abnormality of the androgen receptor (Bullock and Bardin, 1970, 1972). Similar correlations were subsequently made in the mouse (Bullock *et al.*, 1971) and in man (Keenan *et al.*, 1974).

The animal models of the TF syndrome differed in that the rat exhibited a reduced androgen response while the response in the Tfm/Y mouse was completely abolished. These alternatives were assumed to represent syndromes analogous to incomplete and complete testicular feminization as described in man (Bardin *et al.*, 1973). This led to the conclusion that development of the male phenotype was directly proportional to the level of functional androgen receptors (i.e., the TF rat contained a reduced level of the receptor and the Tfm/Y mouse had no functional receptors (Bardin and Catterall, 1981). This view proved to be too simplistic be-

TABLE I
Androgen Receptor Status in Various Forms of Androgen Resistance

Phenotype	Receptor status			
	Absent	Abnormal	Normal	Reduced concentration
Complete TF[a]	+	+	+	−
Incomplete TF	−	+	+	+
Partial resistance	−	+	+	+
Infertility	−	+	+	+

[a] TF, Testicular feminization; (+) indicates that a given receptor status is associated with the phenotype, whereas (−) means that association of the phenotype with the indicated receptor status has not been observed. These data were compiled from Aiman and Griffin (1982), Griffin *et al.* (1982), Griffin and Wilson (1980), Verhoeven and Wilson (1979), and Warne *et al.* (1983).

cause, as clinical studies continued, it became clear that androgen insensitivity in man was associated with a wide variety of receptor abnormalities (Table I). For instance, complete TF can be associated with a lack of functional receptor, a receptor that exhibits abnormal binding, or an apparently normal receptor complement. On the other end of the spectrum, infertile, but otherwise normal men can exhibit abnormal, apparently normal, or a reduced level of apparently normal receptors. Thus, the direct correlation of functional receptor level and androgen resistance phenotype was untenable.

It seems likely that, insofar as the male phenotype is represented by the sum of the expression of androgen-induced (or -attenuated) genes, whatever mechanism causes differential regulation of gene expression by the hormone would also be responsible for phenotypic variation. In an attempt to study this variation at the molecular level, we have used the expression of several genes regulated by androgens in mouse kidney as representative of the male phenotype. In this article, we present data on the characterization of the genes for ornithine decarboxylase, β-glucuronidase, and kidney androgen-regulated protein. In addition, we will show that each of these genes exhibits a unique pattern of response to androgen treatment. The data support the hypothesis that gene regulation based on differential sensitivity to androgen–receptor complexes can explain tissue-specific phenotypic variation in response to the hormone.

II. Mouse Kidney Model System

Androgen action in the mouse kidney is mediated by the androgen receptor protein as is androgen-regulated differentiation of the secondary sex organs of the male genital tract. However, mouse kidney offers several features which simplify the study of androgenic effects. For instance, analysis of murine renal tissue before and after androgen treatment has shown that the hormone increases cell size but not cell number. Even under conditions in which androgens increase renal enzyme activities as much as 1000-fold, DNA replication is stimulated minimally or not at all (Kochakian and Harrison, 1962; Henningsson et al., 1978). This is in contrast to the male reproductive tract in which both hypertrophy and hyperplasia occur in response to male sex hormones. Testosterone is the major effector of the androgen-induced phenotype in murine kidney. Unlike many other androgen-responsive cells, testosterone is not reduced to 5α-dihydrotestosterone (DHT) in renal cells, which have low 5α-reductase activity, the enzyme that catalyzes the conversion of testosterone to DHT (Wilson et al., 1977; Bardin and Catterall, 1981). These two features (absence of hyperplasia and negligible 5α-reductase activity) serve to simplify the interpretation of data on primary effects of androgens in target cells.

In addition to these advantages, many enzyme markers for androgen action in the mouse kidney have been described (Bardin et al., 1978; Swank et al., 1978). In contrast to some target cells where steroid treatment induces large amounts of a few specific proteins (O'Malley and Means, 1974; Parker et al., 1978; Higgins and Burchell, 1978), androgens act on the mouse kidney to induce many enzymes but none predominates as a percentage of total protein concentration. Two-dimensional gel electrophoretic analysis of total soluble protein fraction of mouse kidney after testosterone treatment revealed no major changes in the complement of proteins detected in untreated animals (Swank et al., 1978). Comparison of renal proteins in the lysosomal and mitochondrial compartments by gel electrophoresis before and after androgen administration showed that three proteins were significantly induced (Bardin et al., 1978). These relatively abundant proteins were designated T_1, T_2, and T_3 and have molecular weights of 43,000, 60,000, and 54,000, respectively. However, their identities and functions remain unknown.

The best studied markers for androgen action in murine kidney are β-glucuronidase (Swank et al., 1978; Bardin and Catterall, 1981) and ornithine decarboxylase (Pajunen et al., 1982). Although the responses of the genes for these two enzymes do not represent the two extremes in this tissue, they illustrate the broad range of androgen effects which must be accounted for in mechanistic models of androgen action. Ornithine decarboxylase activity increases rapidly after androgen treatment becoming maximal within 24–48 hours of a single dose of testosterone (Pajunen et

TABLE II
Characteristics of Androgen-Regulated mRNAs in Mouse Kidney[a]

mRNA	Size (kb)	Abundance (%)	Fold induction
β-Glucuronidase	2.6	0.02	30[b]
KAP	0.9	4.0	4
MAK	1.6, 2.2	0.2[c]	8[d]
MK 908	1.5, 2.5	0.5[c]	8[d]
Ornithine decarboxylase	2.2, 2.7	0.1[c]	10[d]

[a] The data are adapted from the following references: β-glucuronidase, Catterall and Leary (1983), Palmer et al. (1983); KAP, Toole et al. (1979), Watson et al. (1984); MAK, Snider et al. (1985); MK 908, Berger et al. (1981); ornithine decarboxylase, Berger et al. (1984), Jänne et al. (1984), Kontula et al. (1984).
[b] The fold induction values, at least for β-glucuronidase and ornithine decarboxylase mRNAs, are dependent on the strain of animals used (see text).
[c] Based upon combined expression of both mRNA species.
[d] The two mRNA species in each case are coordinately regulated.

al., 1982). β-Glucuronidase activity, on the other hand, rises more slowly and requires 15–21 days of chronic androgen stimulation to reach its peak activity (Swank et al., 1973). The effect of androgens on steady-state levels of the messenger RNAs (mRNA) for these two enzymes has also been studied (Catterall and Leary, 1983; Palmer et al., 1983; Kontula et al., 1984; Jänne et al., 1984). Other androgen-responsive mRNAs in murine kidney have been identified by cell-free translation of mRNA from treated and untreated animals (Toole et al., 1979) and by differential hybridization of complementary DNAs (cDNAs) prepared from these mRNA populations to renal cDNA libraries (Berger et al., 1981; Snider et al., 1985). The androgen-responsive mRNAs for which cDNA probes are available are shown in Table II. The similarities in properties between MK908 mRNA (Berger et al., 1981) and the MAK mRNA (Snider et al., 1985) indicate that they may actually be identical gene products. The availability of these mRNA markers and their broad range of responses to androgenic stimulation facilitate the usefulness of the mouse kidney as a model system for androgenic control of male sex differentiation.

III. Three Androgen-Responsive Genes with Different Properties

In our studies of the effects of testosterone on the mouse kidney, three genes were chosen for intense study because of their unique properties. Each of these is first discussed in detail before their different responses to androgens are presented.

A. KIDNEY ANDROGEN-REGULATED PROTEIN GENE

Analysis of kidney mRNA, before and after androgen administration, by translation *in vitro*, revealed a relatively abundant testosterone-induced mRNA species that gave rise to a 20,000-Da peptide (Toole *et al.*, 1979). The peptide was designated kidney androgen-regulated protein, or KAP. The function of this protein and its site of synthesis in the kidney are unknown. Hybridization of KAP cDNA with mRNA from various tissues showed that only kidney had significant amounts of KAP mRNA (Toole *et al.*, 1979). Annealing of KAP cDNA with genomic DNA suggested that this protein was encoded by a unique gene which supported the idea that KAP was a single protein species rather than several peptides of similar molecular weight (Toole *et al.*, 1979). Northern blot hybridization using cloned KAP cDNA indicated that KAP mRNA was approximately 850 nucleotides in length (Watson *et al.*, 1984). KAP mRNA is of the relatively abundant class, representing 4–5% of total poly(A) RNA in the kidney of androgen-stimulated animals. It is induced in 2–6 hours by a single dose of testosterone and reaches its peak concentration of 3- to 4-fold over control levels in 24–48 hours. The constitutive amount of KAP mRNA in normal female animals is relatively high and its induction by testosterone small in relation to other androgen-regulated genes in this tissue (Berger *et al.*, 1981; Catterall and Leary, 1983; Kontula *et al.*, 1984).

Because KAP was discovered by cell-free translation of renal mRNA, nothing is known about the properties of the protein *in vivo*. In order to study characteristics and regulation of the KAP protein, a specific antibody should be available. However, to purify this protein from mouse renal tissue would be difficult indeed, as the only physical property on which to base an assay for its purification is its size, which may not be the same for the native protein and the cell-free translation product.

To circumvent these problems, we have deduced a partial amino acid sequence for KAP from the DNA sequence of cloned complementary DNA (unpublished data). A 22-residue peptide was selected from the available 120-amino acid sequence, synthesized and coupled to bovine serum albumin for immunization of rabbits (Fig. 1). The antibodies were tested for their ability to immunoprecipitate the 20,000-Da KAP peptide following cell-free translation of renal mRNA from androgen-treated mice. Figure 2A shows that the antibody raised against the 22-residue peptide efficiently and specifically precipitated a labeled protein with the characteristics of KAP. Kidney mRNA isolated from testosterone-treated animals produced approximately 4-fold more immunoprecipitable peptide than that from untreated female mice (Fig. 2A). The androgen-elicited

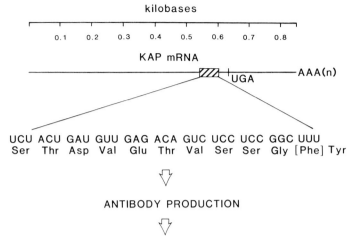

FIG. 1. Scheme for the preparation of KAP-specific antibody from synthetic KAP peptide. KAP mRNA was isolated and its sequence determined from complementary DNA clones. A 22-residue peptide based on the derived amino acid sequence was synthesized, purified, and coupled to bovine serum albumin (BSA) for immunization of rabbits. UGA, Translation stop codon; $AAA_{(n)}$, 3'-terminal poly(A) tail; hatched box, position of the derived amino acid sequence on which the synthetic peptide was based. The sequence shown is a C-terminal segment of the 22-amino acid peptide. The complete peptide sequence is NH_2-Thr-Pro-Ala-Ala-Asn-Lys-Gln-Asn-Ser-Glu-Phe-Ser-Thr-Asp-Val-Glu-Thr-Val-Ser-Ser-Gly-Tyr-COOH in which the C-terminal Tyr replaced the encoded Phe and was used for coupling to the carrier protein.

increase in biologically active KAP mRNA concentration was similar to that determined for KAP mRNA sequences by Northern blot hybridization analysis (Fig. 2B). These data indicated that the antibody raised against the synthetic peptide was specific for KAP. The antiserum was next used to develop a radioimmunoassay for KAP using kidney cytosol from androgen-treated animals as a standard and ^{125}I-labeled synthetic peptide as a tracer. Radioimmunoassay data showed that the concentration of KAP increased 3- to 4-fold upon androgen treatment similar to the mRNA increase, as assayed by both Northern blot analysis and translation *in vitro* followed by immunoprecipitation (Fig. 2). Assays of other tissues (or biological fluids) showed that, in addition to the soluble protein fraction of the kidney, KAP was found in the submaxillary gland and in the urine. A small amount was also detectable in liver. This pattern of expression is interesting in that both submaxillary gland and liver are androgen target organs (Wilson *et al.*, 1977; Derman, 1981), while the presence of KAP in the urine suggests that it may be secreted by the

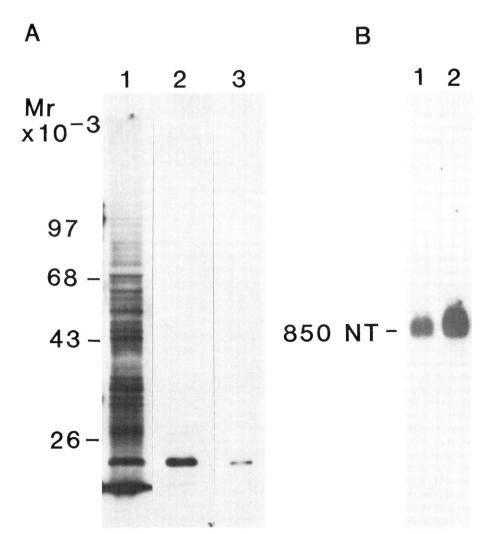

FIG. 2. (A) Products of cell-free translation of total kidney poly(A)-containing mRNA analyzed by SDS–polyacrylamide gel electrophoresis (10% total acrylamide concentration). Lane 1: total [^{35}S]methionine-labeled peptide products from translation of kidney poly(A)-RNA from testosterone-treated female mice. Lane 2: same as lane 1 after immunoprecipitation with antibody raised against the synthetic KAP peptide (see legend to Fig. 1). Lane 3: immunoprecipitation of translation products from renal poly(A)-RNA from normal female mice. (B) Northern blot analysis of murine kidney poly(A)-containing RNA isolated from intact females (lane 1) and testosterone-treated females (lane 2). RNA (8 μg) was denatured and analyzed according to the method of Thomas (1980) as previously described (Watson et al., 1984). NT, Nucleotides.

kidney like other androgen-induced proteins. Toole et al. (1979) previously reported that the submaxillary gland of male mice contained less than one KAP mRNA molecule per cell compared with 15,000 copies per cell in the kidney. The mRNA and protein measurements were done in different strains of mice and may, therefore, simply reflect a strain difference in the expression of KAP. Alternatively, androgen-dependent synthesis of KAP in submaxillary gland may be independent of an increase in the steady-state mRNA level. It seems unlikely that KAP is synthesized elsewhere and concentrated in the submaxillary gland as no immunoreactive protein was found in the serum.

B. ORNITHINE DECARBOXYLASE GENE

Ornithine decarboxylase is one of the rate-controlling enzymes in the biosynthesis of polyamines and catalyzes conversion of L-ornithine to putrescine. It is a constitutive, pyridoxal 5′-phosphate-dependent enzyme that is probably present in all cells and tissues (Pegg and McCann, 1982); Tabor and Tabor, 1984). Ornithine decarboxylase concentration is very low in quiescent cells, but its amount increases many fold in a variety of tissues within a few hours of exposure to trophic stimuli, such as hormones, drugs, tissue regeneration, and growth factors (Jänne et al., 1978; Pegg and McCann, 1982; Tabor and Tabor, 1984). Even though these stimuli may elevate the enzyme activity 1000-fold, ornithine decarboxylase still represents only a minute fraction of total cellular protein ranging from 0.0001 to 0.04% of the soluble protein in a maximally induced tissue. Interestingly, renal cells of androgen-treated mice seem to be the richest source of ornithine decarboxylase, although the physiological significance of the high enzyme activity in murine kidney is not yet understood.

Androgen induction of ornithine decarboxylase is steroid specific and occurs with relatively rapid kinetics. As illustrated in Fig. 3A, 5α-dihydrotestosterone administration for 5 days elicited a 400- to 600-fold increase in the enzyme activity, while other classes of steroids produced little or no effect (Pajunen et al., 1982). Additional studies showed that testosterone is as potent as DHT in increasing the enzyme activity and that stimulation of ornithine decarboxylase activity in murine kidney is always accompanied by a parallel change in the enzyme protein concentration (Seely and Pegg, 1983; Isomaa et al., 1983). These specificity studies indicate that androgens are the dominant steroidal regulators of ornithine decarboxylase gene expression in murine kidney, rather than glucocorticoids and estrogens as is the case in rat kidney (Brandt et al., 1972; Nicholson et al., 1976). A single dose of testosterone increases the activity and the enzyme protein concentration in a parallel fashion with

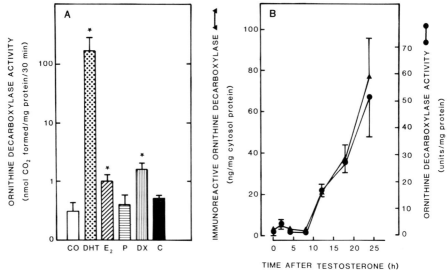

FIG. 3. Steroid specificity (A) and time course (B) of the induction of ornithine decarboxylase in murine kidney by androgens. (A) Castrated male mice were treated for 5 days with daily doses of 5α-dihydrotestosterone (DHT, 1 mg/day), estradiol-17β (E_2, 20 μg/day), progesterone (P, 1 mg/day), dexamethasone (DX, 100 μg/day), and cortisol (C, 1 mg/day). The animals were killed 24 hours after the last steroid dose (*$p < 0.05$, compared to vehicle-treated, castrated control mice, CO). (B) Female animals were given a single intraperitoneal dose of testosterone (10 mg) and killed at indicated time intervals. Data adapted from Pajunen et al. (1982) and Isomaa et al. (1983).

peak values at 18–24 hours of steroid administration (Fig. 3B). The extent and duration of this induction are dose dependent and related to the residence time of nuclear androgen receptors (Pajunen et al., 1982). The induction represents de novo synthesis of the enzyme protein, as it can be blocked by concomitant administration of cycloheximide.

The most striking characteristic of ornithine decarboxylase is its very rapid turnover in mammalian tissues and the rapid changes that occur in the amount of this enzyme after exposure to a variety of effectors in vivo (Jänne et al., 1978; Pegg and McCann, 1982; Tabor and Tabor, 1984). The turnover rate of mammalian ornithine decarboxylase is faster than that of any other mammalian enzyme; the half-life in mouse kidney is approximately 15 minutes (Seely and Pegg, 1983; Isomaa et al., 1983). It is of interest to note that the two other rate-controlling enzymes in polyamine biosynthesis (S-adenosylmethionine decarboxylase and spermidine/spermine N^1-acetyltransferase) also have relatively rapid turnover rates (Pegg and McCann, 1982). Androgen administration in pharmacological doses and for several days changes the turnover rate of ornithine decarboxylase

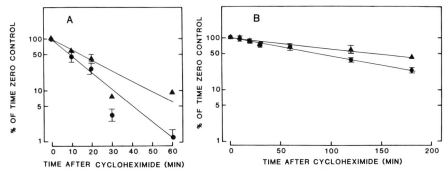

FIG. 4. Stabilization of ornithine decarboxylase in murine kidney by androgen treatment. Intact (A) and androgen-treated male mice (B) were administered cycloheximide (50 mg/kg body weight) at time 0 and killed at indicated time intervals thereafter. (●) Catalytic enzyme activity; (▲) immunoreactive enzyme protein concentration. The calculated half-lives for the catalytically active and immunoreactive enzymes were 9 and 16 and 90 and 140 minutes for intact and androgen-treated male mice, respectively. Androgen treatment with testosterone-releasing implants (200 μg/day) lasted for 7 days. From Isomaa et al. (1983).

(Fig. 4); the half-life increases from 10–15 to about 100–150 minutes. This change in the half-life applies to both the catalytically active and the immunoreactive enzyme protein (Seely and Pegg, 1983; Isomaa et al., 1983). The mechanisms responsible for slowing down the turnover rate of ornithine decarboxylase by androgens in murine kidney are currently unknown as is the physiological significance of this phenomenon.

Ornithine decarboxylase has been purified to an apparent homogeneity from a variety of sources including the murine kidney. The enzyme protein has a subunit molecular weight of about 50,000 (Fig. 5) and is apparently a dimeric molecule under physiological conditions. However, the 50,000-Da subunits are catalytically active, since they bind the active site-directed inhibitor, α-difluoromethylornithine. Charge heterogeneity similar to that shown in Fig. 5 for murine ornithine decarboxylase subunits (Isomaa et al., 1983) has been detected on two-dimensional gel electrophoresis for this enzyme from a variety of sources (McConlogue and Coffino, 1983; Mitchell and Wilson, 1983; Choi and Scheffler, 1983). The two (or more) forms of the enzyme may represent polypeptides with different degrees of posttranslational modifications such as phosphorylation; however, they are all catalytically active enzymes since each form is capable of binding α-[^{3}H]difluoromethylornithine (Fig. 5). Alternatively, the various forms of the enzyme could be encoded by different mRNAs (Hickok et al., 1986) and their charge heterogeneity be a consequence of dissimilarity in primary amino acid sequences.

Mouse kidney ornithine decarboxylase mRNA has been shown by

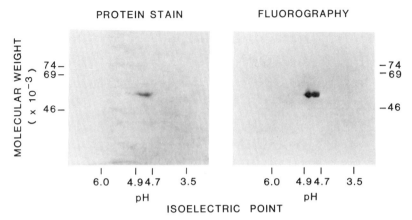

FIG. 5. Two-dimensional gel electrophoretic analysis of ornithine decarboxylase purified from kidneys of androgen-treated mice. (Left) Visualization of the enzyme protein (7 μg) by staining with Coomassie blue. (Right) Fluorographic detection of the enzyme after covalent labeling with α-[³H]difluoromethylornithine. From Isomaa et al. (1983).

Northern blot analysis to be composed of two species with approximate molecular sizes of 2.2 and 2.7 kilobases (Kontula et al., 1984; Berger et al., 1984; Jänne et al., 1984). The two mRNAs are constitutively expressed in kidneys of female mice and their accumulation is regulated by androgens in a parallel fashion (Kontula et al., 1984; Hickok et al., 1986). It is of note, however, that the expression of the 2.2- and 2.7-kb mRNA species shows genetic variation among inbred strains of mice; their ratio varied from 1:20 to 1:3 among the eight mouse strains studied (Melanitou et al., 1986). There are several mechanisms by which a single gene could code for multiple mRNAs, including the use of different promoters, the presence of multiple polyadenylylation/termination signals, and alternative splicing that leads to inclusion or deletion of exons and introns. By cloning and sequencing the cDNAs corresponding to the 3'-termini of the two ornithine decarboxylase mRNAs, we have shown that the 3'-nontranslated region of the 2.7-kb mRNA contains two potential polyadenylylation signals (AAUAAA), and utilization of the second AAUAAA signal for poly(A) addition seems to be the reason for the presence of the 2.7-kb ornithine decarboxylase mRNA (Hickok et al., 1986). This conclusion was confirmed by Northern blot analysis using hybridization probes isolated from defined regions of the ornithine decarboxylase cDNAs (Fig. 6).

Although the alternative use of two different polyadenylylation/termination signals within a single ornithine decarboxylase gene could explain the presence of the two mRNA species, additional results suggested that

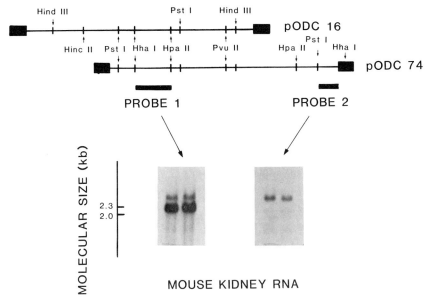

FIG. 6. Northern blot analysis of ornithine decarboxylase mRNAs with specific hybridization probes. Poly(A)-containing RNA was isolated from kidneys of androgen-treated mice, duplicate samples (6 μg RNA) were subjected to agarose gel electrophoresis in the presence of 2.2 M formaldehyde and transferred to nitrocellulose. Indicated fragments were cleaved from plasmids pODC16 and pODC74 carrying sequences complementary to ornithine decarboxylase mRNA, isolated, nick-translated, and used as hybridization probes. From Hickok et al. (1986).

more than one gene may be expressed. This is possible, since there are multiple ornithine decarboxylase genes in the murine genome (McConlogue et al., 1984; Berger et al., 1984; Kahana and Nathans, 1985), as illustrated by the Southern blot of mouse DNA in Fig. 7. Comparison of the partial nucleotide sequences of the two cDNAs corresponding to the 2.2- and 2.7-kb ornithine decarboxylase mRNAs revealed 5 mismatches within their 759-base pair overlap suggesting that they originate from two very similar, yet different ornithine decarboxylase genes (Hickok et al., 1986). It remains to be elucidated whether the two mRNAs present in murine kidney have any bearing on the charge heterogeneity observed for the enzyme protein isolated from the same tissue (Fig. 5). In addition to this microheterogeneity, two forms of the enzyme have been reported, which differ in their heat sensitivity, half-life *in vivo*, and androgen inducibility in mouse kidney (Loeb et al., 1984). The presence of the two mRNA species appears to be unrelated to these forms of the enzyme, since the relative amounts of the ornithine decarboxylase mRNAs did not correlate

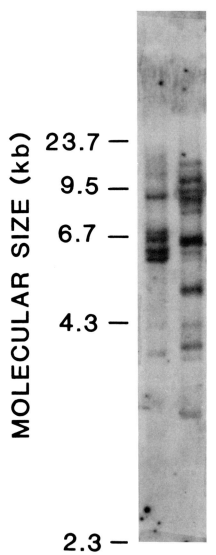

FIG. 7. Southern blot analysis of mouse liver DNA. Aliquots of high-molecular-weight DNA (10 μg) were cleaved with restriction endonucleases EcoRI (E) or BamHI (B), subjected to agarose gel electrophoresis and transferred to nitrocellulose. Ornithine decarboxylase gene sequences were visualized by hybridization with nick-translated pODC16 DNA carrying sequences complementary to ornithine decarboxylase mRNA.

with those of the enzyme forms. Furthermore, both mRNA species were induced to the same extent by androgens in contrast to only the heat-labile enzyme form being androgen regulated (Loeb et al., 1984).

Ornithine decarboxylase gene expression in murine kidney seems to be under genetic control. As mentioned above, the relative amounts of the 2.2- and 2.7-kb mRNA species vary among the 8 inbred strains of mice studied (Melanitou et al., 1986). This variation is apparently unrelated to the genetic regulation of the magnitude of androgen responsiveness of the ornithine decarboxylase gene(s). The data shown in Fig. 8 for two inbred mouse strains (A/J and C57BR/cdJ) exemplify the characteristics of this latter genetic variation in ornithine decarboxylase gene expression. Kidneys of untreated females do not have significantly different enzyme activities or mRNA concentrations; however, the renal enzyme concentrations in untreated male mice of C57BR/cdJ strain is over 10-fold higher than that of A/J males, which, in turn, is barely higher than in the female mice (Fig. 8a).

A blunted response in ornithine decarboxylase activity to exogenous testosterone is also seen in the A/J animals (Fig. 8A), although the difference between the two strains is not as marked as at physiological androgen concentrations. This genetically regulated androgen responsiveness was clearly evident at the level of ornithine decarboxylase mRNA accumulation (Fig. 8B). A similar strain-dependent variation in ornithine decarboxylase gene expression was found in several additional mouse strains, and it did not correlate either with nuclear androgen receptor concentrations or with serum testosterone levels in male mice (Melanitou et al., 1986). We infer from these data that it is the ornithine decarboxylase gene(s), whose androgen regulation is under a genetic control, rather than a genetic regulation of androgen production and nuclear androgen receptor concentrations. As will be discussed below, the genetic regulation of ornithine decarboxylase gene(s) is, however, distinctly different from that of the β-glucuronidase gene.

C. β-GLUCURONIDASE GENE

β-Glucuronidase is an acid hydrolase that is present in lysosomes and endoplasmic reticulum of virtually all tissues of the mouse. Murine kidney β-glucuronidase is unique in its regulation by androgens. The enzyme is synthesized in proximal tubule cells and secreted into urine, which, along with rat preputial gland (Tulsiani and Keller, 1975), is one of the richest sources of β-glucuronidase. The enzyme functions in the metabolism of mucopolysaccharides and possibly in the modification of certain glucuronide conjugates such as those of steroid hormones. β-Glucuronidase

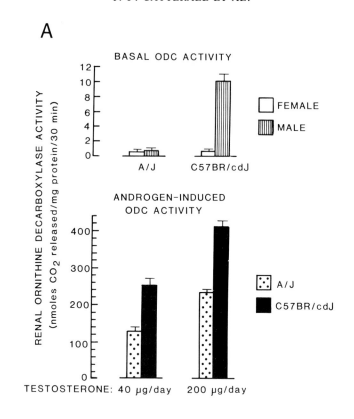
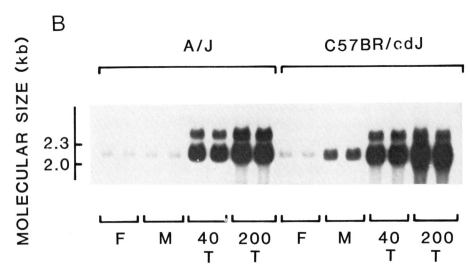

deficiency in humans is associated with the lysosomal storage disease mucopolysaccharidosis type VII (Sly et al., 1973; Hall et al., 1973). The enzyme is a tetramer of four identical subunits ($M_r=72,000-75,000$) and occurs in two forms, each with a unique subcellular localization. The most common form, designated L, is the lysosomal enzyme the subunits of which differ slightly from their counterparts in the microsomal membrane form, called X, in apparent size and charge (Tomino et al., 1975; Lusis and Paigen, 1977). However, these two forms of the enzyme are the products of the same gene (Lalley and Shows, 1974). In the microsomal compartment, each of the subunits of the X form can bind a protein, called egasyn, resulting in forms M_1-M_4 in which one to four molecules of egasyn are noncovalently associated with the X tetramer, anchoring it to the membrane (Lusis and Paigen, 1977). All six forms of the enzyme can be separated electrophoretically (Ganschow and Bunker, 1970).

β-Glucuronidase exhibits substrate specificity for β-D-glucuronide linkages, however, it will cleave such a bond associated with a wide variety of conjugate groups. Because of this versatility, very sensitive assays for β-glucuronidase activity have been established, using several "reporter" molecules that can be easily detected and quantitated after cleavage of the glucuronide linkage. The most useful of these assays involves the fluorogenic substrate, 4-methylumbelliferyl-β-D-glucuronide. Assays of enzyme activity in urine and renal homogenates provided a simple and sensitive measure of the effects of androgens on β-glucuronidase. Paigen and his colleagues used the results of such assays to establish the β-glucuronidase gene as one of the best characterized genetic loci in mammals (Paigen, 1979).

The genetic studies have led to the identification of four genes in the β-glucuronidase genetic complex, [Gus] (Table III). The β-glucuronidase structural gene, Gus-s, has three well-characterized alleles, $-s^a$, $-s^b$, and $-s^h$ which code for electrophoretic variants ($-s^a$ and $-s^b$) or an enzyme stability variant ($-s^h$). Recently, two new alleles have been identified and designated $-s^{cs}$ and $-s^{cl}$ (Pfister et al., 1985). Each of these alleles codes for

---

FIG. 8. Androgen regulation of ornithine decarboxylase (ODC) activity (A) and ornithine decarboxylase mRNA accumulation (B) in two inbred strains of mice. (A) The enzyme activity was measured in kidneys of mice of A/J and C57BR/cdJ strains in the indicated groups. Testosterone treatment was accomplished by subcutaneous implants releasing 40 and 200 μg steroid/day. The values are mean ± SE for 7–13 animals in each group. (B) Northern blot analysis of poly(A)-containing RNA from kidneys of A/J and C57BR/cdJ mice. Duplicate RNA samples (6 μg) prepared from pooled kidneys of 5–10 animals were subjected to agarose gel electrophoresis (see legend to Fig. 6) and hybridized with nick-translated pODC16 DNA. Testosterone (T) treatment was performed at indicated daily doses (in μg) for 7 days. F, Female; M, male. Adapted from Melanitou et al. (1986).

thermostability variants and the enzyme coded for by -s^{cl} may also lack an antigenic determinant present in enzymes encoded by other alleles. In addition to *Gus-s*, three regulatory alleles have been described (Table III). These have all been shown to be closely linked to *Gus-s*. The *Gus-t* locus regulates the temporal development of enzyme activity in various tissues. This locus is unique among the regulators of the [*Gus*] complex and proximal regulatory sites in mammals in general because, unlike the others, it acts in trans (Herrup and Mullen, 1977; Meredith and Ganschow, 1978; Lusis *et al.*, 1983). A second regulatory locus of the [*Gus*] complex, *Gus-u*, controls the synthesis of β-glucuronidase in all tissues at all stages of development (Paigen, 1979). This systemic regulator exhibits cis action. Finally, a third regulatory locus, *Gus-r*, acts in cis and controls the response of the structural gene to androgen stimulation. Mouse strains generally fall into one of two groups based on their responses to androgens. In one group (containing the *Gus-r*^{b} allele), β-glucuronidase activity develops slowly over a 3-week period reaching a plateau at approximately 21 days of androgen administration. The second group (which carry the *Gus-r*^{a} allele) expresses the enzyme more rapidly and reaches a higher maximum level during the induction. Other alleles of *Gus-r* have been identified which are associated with the alleles of *Gus-s* (above) and carry the same haplotype designations (-r^h, -r^{cs}, and -r^{cl}). No recombination has been observed between the *Gus-s* and *Gus-r* loci.

Expression of β-glucuronidase gene along with other androgen-controlled genes is also regulated by genetic loci remote from the [*Gus*] complex on mouse chromosome 5. The response of *Gus-s* to androgenic stimulation through *Gus-r* is abolished in Tfm/Y mice. The *Tfm* locus comprises the structural gene (or a regulator thereof) for the androgen receptor and is located on the X chromosome. Apparently, this mutation reduces the level of functional androgen receptors below a threshold at which it can be recognized by the regulatory site associated with the [*Gus*] complex, perhaps *Gus-r* itself (see below).

TABLE III
β-*Glucuronidase Genetic Complex [Gus]*[a]

Designation	Function	Chromosome
Gus-s	Structural gene	5
Gus-r	cis-regulatory gene	5
Gus-t	Temporal trans regulator	5
Gus-u	Systemic cis regulator	5

[a] The data are compiled from Paigen (1979) and Lusis *et al.* (1980, 1983).

Of the regulatory sites and other loci that affect *Gus-s* gene function, the *Gus-r* locus is of most interest due to its androgen-dependent function. *Gus-r* is also likely to be the most experimentally accessible with currently available technology. This notion is based on our assumption that, like other steroid-dependent regulatory sites, the *Gus-r* locus will be found near the site of initiation of transcription of the *Gus-s* gene. If this assumption is correct, then cloning and characterization of the *Gus-s* transcription unit and sequences flanking it will provide the means to identify and characterize *Gus-r*. The direct cloning of *Gus-r* would be very difficult since there is no *Gus-r* encoded gene product. It is also possible that *Gus-r* is located within *Gus-s*, perhaps as part of an intervening sequence (Slater *et al.*, 1985). Alternatively, *Gus-r* may represent a tissue- and gene-specific enhancer-like sequence, possibly acting at a distance from *Gus-s*. In the latter two cases, characterization of *Gus-r* is also approachable by cloning and sequencing of *Gus-s* and its flanking DNA on mouse chromosome 5. For instance, cloned gene fragments can be fused with a gene such as bacterial chloramphenicol acetyltransferase and assayed for expression of *Gus-r*-like function in cultured cells which contain androgen receptors.

The physical map of β-glucuronidase cDNA shown in Fig. 9 has been constructed from three separate sources of data. First, cDNA clones covering approximately 1.7 kb of the 2.6-kb mRNA sequence have been identified and sequenced. Second, the extreme 3'-end sequence, which was lacking in the cDNA clones isolated, was determined from a genomic DNA clone (see below) from the same mouse strain. The genomic sequence beyond the 3'-end of the cloned cDNA sequence contained an AATAAA polyadenylylation site (Fitzgerald and Shenk, 1981) which was followed 16 bp downstream by A_{14} (unpublished data). The 3'-end of the cDNA in Fig. 9 corresponds to the beginning of this poly(A) sequence.

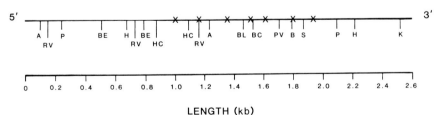

FIG. 9. Restriction endonuclease map of β-glucuronidase (*Gus-s*ᵃ) cDNA. X marks the locations of seven intervening sequences. Data for the presence or placement of intervening sequences 5' to the *Hin*dIII site (H) at 0.6–0.7 kb are presently unavailable. Restriction enzyme sites are A, *Ava*I; B, *Bam*HI; BC, *Bcl*I; BE, *Bst*EII; BL, *Bal*I; H, *Hin*dIII; HC, *Hin*cII; K, *Kpn*I; P, *Pst*I; PV, *Pvu*II; RV, *Eco*RV; S, *Sst*I.

The presence of the poly(A) sequence in the gene was unexpected and its origin is unknown. Poly(A) sequences in genes coding for mRNA are characteristic of processed pseudogenes (Vanin, 1984). However, two other considerations make this an unlikely explanation. First, the cloned *Gus-s*[a] gene contains intervening sequences (see below) and is, therefore, not completely processed. Second, there is no evidence for more than one β-glucuronidase gene in the mouse genome. Genetic evidence (Paigen, 1979) and DNA hybridization experiments (unpublished results) in fact suggest the presence of a single gene. The appropriate spacing between the AATAAA and the A_{14} sequences strongly suggests that this segment represents the 3'-end of the mature β-glucuronidase mRNA sequence. However, it is also possible that these sequences have no role in polyadenylylation of the β-glucuronidase mRNA and that they are simply contained within a larger 3'-noncoding segment. An alternative explanation is that the *Gus-s* gene contains an encoded poly(A) sequence, but the short length of the segment in this case resembles a product of reverse transcription rather than a functional poly(A) tail.

The third source of the *Gus-r*[a] restriction map was Southern analysis of full-length cDNA synthesized from partially purified mRNA (Fig. 10). In this case, the cDNA was prepared by the method of Okayama and Berg (1982) as modified by Gubler and Hoffman (1983). The cDNA was cleaved with various restriction enzymes and examined by the Southern procedure (Southern, 1975) using a fragment containing the first 75 bp of the cloned cDNA as hybridization probe. Enzymes were used that did not cleave within the region represented by the probe so that only the fragment that spanned the junction between the cloned and unmapped segments was detected in each digest.

Genomic DNA clones were isolated from a library (provided by Dr. R. M. Perlmutter, California Institute of Technology) prepared by partial *Eco*RI cleavage of mouse (BALB/c) liver DNA and insertion into the Charon 4A bacteriophage vector. *Gus-s*-containing phage isolates were plaque purified and the *Gus-s* inserts were subcloned into pBR322. Comparison of the restriction mapping data from the subclones, genomic DNA, and the cDNA clones led to the identification of seven intervening sequences (Fig. 9). Seven exon regions have been precisely mapped, but at least one additional exon remains to be mapped at the 5'-end. Completion of the *Gus-s*[a] allele map will require either the preparation of a cDNA clone complementary to the entire mRNA or analysis of 5'-end containing genomic fragments by hybridization directly with β-glucuronidase mRNA followed by S1-nuclease mapping or electron microscopy. The extent of the mRNA sequence still to be accounted for within the gene is 0.5–0.7 kb and may be encoded in more than one additional exon.

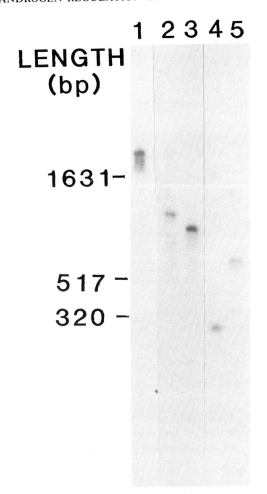

FIG. 10. Southern blot restriction analysis of the 5'-end of β-glucuronidase cDNA not represented in cloned fragments. Double-stranded cDNA was prepared from partially purified β-glucuronidase mRNA (Catterall and Leary, 1983) by the method of Gubler and Hoffman (1983). Aliquots of cDNA (approximately 20 ng) were digested with restriction enzymes and subjected to electrophoresis on a 1.2% agarose mini-gel (95 × 75 mm) for 2 hours at 100 V. Restriction sites 5' to the cloned segments were localized by hybridization to a labeled fragment comprising the 5'-most 75 nucleotides of the cloned cDNA. Lane 1: undigested double-stranded cDNA. Lanes 2–5 contained cDNA digested with AvaI, AvaI + PstI, BstEII, and EcoRV, respectively.

Characterization of the *Gus-r* locus is also possible by a functional analysis, in which its control of the expression of *Gus-s* to androgenic regulation is monitored. *Gus-r* was shown to regulate the increase in β-glucuronidase mRNA "activity" in frog oocytes injected with mouse renal mRNA (Paigen *et al.*, 1979). More direct measurements of β-glucuronidase mRNA by hybridization to cDNA probes showed that *Gus-r* acts by controlling steady-state mRNA levels (Palmer *et al.*, 1983; Watson and Catterall, 1986). Although the levels of β-glucuronidase mRNA in untreated females was about the same in animals with the [*Gus*]^A and [*Gus*]^B haplotypes, the amounts induced by physiological and pharmacological testosterone doses were significantly greater in [*Gus*]^A animals (Fig. 11). In order to further characterize the differences in accumulation of β-glucuronidase mRNA in these haplotypes, a complete time course of testosterone stimulation was carried out using a more directly quantifiable assay. Table IV shows the results of an assay, in which 10 μg poly(A)-containing mRNA prepared from kidneys of [*Gus*]^A or [*Gus*]^B animals was bound covalently to aminophenylthioether (APT) paper and hybridized with a radiolabeled cDNA probe. The results revealed that *Gus-r* was responsible for a 9- to 12-fold variation in the steady-state level of β-glucuronidase mRNA in response to a 10- to 15-day continuous testosterone administration. After 15 days of treatment, the *Gus-r*^{a} animals had reached a peak value 32.5-fold higher than that prior to treatment, while

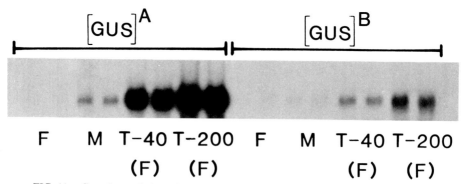

FIG. 11. Genetic regulation of androgen-induced β-glucuronidase mRNA accumulation. Poly(A)-RNA samples (6 μg) from inbred strains homozygous for the *Gus-r*^{a} or *Gus-r*^{b} regulatory allele in various hormonal states were analyzed by Northern blot hybridization with a β-glucuronidase cDNA plasmid. [*Gus*]^A, A haplotype which carries the *Gus-r*^{a} allele conferring high inducibility of the β-glucuronidase gene; [*Gus*]^B, B haplotype carrying the *Gus-r*^{b} allele which confers low inducibility of the β-glucuronidase gene. F, Female; M, male; T-40, 40 μg testosterone per day; T-200, 200 μg testosterone per day. From Catterall and Jänne (1986).

TABLE IV
Androgen-Induced Accumulation of
β-Glucuronidase mRNA in [Gus]^A and [Gus]^B
Haplotypes

Days of Treatment	Relative β-glucuronidase mRNA concentration[a]	
	[Gus]^A	[Gus]^B
0	1.0	1.0
0.5	3.9	0.7
1	4.3	1.5
2	6.9	1.7
3	11.0	2.3
4	10.3	2.4
5	12.0	1.7
7	15.0	2.2
10	31.8	2.5
15	32.5	3.6
20	24.2	3.8

[a] Data presented indicate the fold increase in the relative β-glucuronidase mRNA concentration when the untreated female level (day 0) is arbitrarily set at a value of 1.0. Animals were treated with Silastic implants that released 120 μg testosterone/day. The values represent the mean of three experiments each including poly(A)-containing RNA from pooled kidneys of 8–20 animals. Detection of β-glucuronidase mRNA sequences was carried out by the APT filter binding assay. Adapted from Watson and Catterall (1986).

animals carrying the Gus-r^b allele had attained a plateau level of 3.6-fold over untreated controls. While these results and others (Palmer et al., 1983) make it clear that Gus-r acts at the level of mRNA accumulation, the extent to which it affects transcription of the mRNA as opposed to mRNA stability remains unclear. The cis action of Gus-r is most simply interpreted as the positive regulation of transcription of Gus-s through the binding of a regulatory protein, perhaps the androgen receptor itself. However, preliminary experiments have detected no increase in the rate of Gus-s transcription (unpublished results) as determined by the nuclear run-off assay (McKnight and Palmiter, 1979). It is interesting to note that three other androgen-regulated genes in the mouse kidney have been shown not to be regulated by a transcriptional mechanism using this same assay (Berger et al., 1986).

IV. Differential Regulation of Gene Expression by Androgens

Androgens act on the genital tract as well as many extragenital tissues in mediating the development of the male phenotype (Bardin and Catterall, 1981). However, as discussed above, the complexity of this process is illustrated by the fact that a single clinical syndrome may display a variety of phenotypes as is the case with the various receptor abnormalities associated with the same or different forms of androgen resistance (Table I). It is thus evident that variation in the functional androgen receptor concentration alone is insufficient to explain masculinization of target cells and that the mechanisms by which androgens control development of the male phenotype involve a number of individually regulated events.

The male phenotype in a given target tissue can be interpreted as the consequence of the sum of the expression of the androgen-regulated genes in that tissue. The wide variation in responses to testosterone of the three genes described above is illustrated in Fig. 12. In this study, a single dose (10 mg) of testosterone was administered and changes in mRNA accumulation were followed over a 48-hour period. Both KAP and ornithine decarboxylase mRNA concentrations begin to increase within 6 hours of hormone treatment. In addition, each of these genes reached its maximum induction by 24 hours, however, the levels achieved in the two cases were different. KAP mRNA accumulation increased approximately 2-fold while that for ornithine decarboxylase mRNA was 8-fold. Unlike these two, the relative concentration of β-glucuronidase mRNA did not increase significantly until 24 hours after the hormone and the level reached during the 48-hour test period was less than 2-fold over the control value. This variation in the response to testosterone of these genes was not due to a lack of androgen receptor in the renal nuclei as shown in Fig. 12. Nuclear androgen receptor concentration increases rapidly after testosterone administration and was at its maximal level by the first time point measured (30 minutes). While the nuclear androgen receptor concentration decreased by approximately 50% during the ensuing 48-hour period, at no time did it return to the pretreatment level.

Table V shows the nuclear androgen receptor concentration in kidneys of normal females, males, and androgen-treated males. The number of receptors per cell nucleus is relatively low in murine kidney compared to other androgen target tissues such as the ventral prostate of the rat. The development of the normal male phenotype in the mouse is accompanied by a surprisingly small sex difference in renal nuclear androgen receptor concentrations (Table V). The receptor level in the nuclei can be signifi-

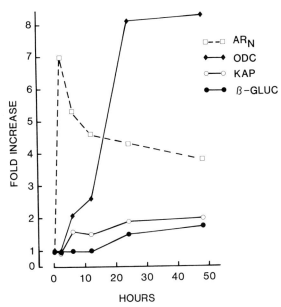

FIG. 12. Changes in mRNA accumulation and nuclear androgen receptor concentration after a single dose of testosterone. Testosterone (10 mg) was administered by intraperitoneal injection to female NCS mice. At the indicated times poly(A)-mRNA was prepared and analyzed by Northern blotting. Autoradiographs of each transfer filter were scanned with a densitometer. The untreated female level was assigned the value of 1. Androgen receptor measurements were made on nuclear extracts prepared from identically treated animals by an exchange assay (Isomaa et al., 1982). Nuclear androgen receptor (AR_N) values represent the mean of three experiments. Kidneys from four animals were pooled at each time point in all experiments. KAP, Kidney androgen-regulated protein; ODC, ornithine decarboxylase; β-GLUC, β-glucuronidase. From Catterall et al. (1985).

cantly increased by testosterone administration, reaching 5 times the normal male concentration. This latter value represents binding of virtually all receptors to chromatin of the renal cells, as the total number of receptors in kidneys of intact females and males or androgen-treated mice is approximately the same (Isomaa et al., 1982). The expressions of the three renal genes at physiological testosterone concentrations are shown in Fig. 13. In each case, a significantly higher steady-state mRNA accumulation was exhibited in normal males than normal females. Therefore, only a 2-fold increment in nuclear androgen receptor concentration seems to be responsible for the normal male phenotypic response of these genes. By driving virtually all available androgen receptor into the nucleus with exogenous testosterone administration, the accumulation of β-glucuroni-

TABLE V
Nuclear Androgen Receptor Concentrations in Kidneys of
Intact Females and Males, and in Testosterone-Implanted
Male Mice

Animal group	Androgen receptor concentration[a]	
Intact females	150 ± 60	(13)
Intact males	280 ± 100	(6)
Testosterone-implanted males[b]	1540 ± 390	(5)

[a] The receptor concentrations (mean ± SD) are expressed as molecules/cell assuming that the receptors are evenly distributed within renal cells with 6 pg DNA/cell. The number of animals per group is shown in parentheses.

[b] The animals were treated with testosterone-releasing implants (200 µg steroid/day) for 7 days.

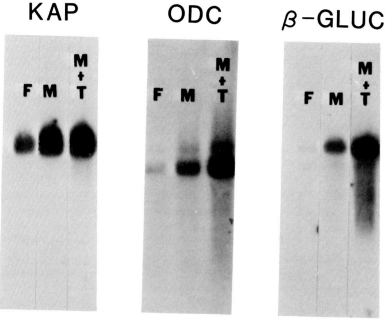

FIG. 13. Concentrations of the three mRNAs in intact females (F) and males (M), and in androgen-treated male mice (M+T). Renal poly(A)-RNA was prepared, subjected to agarose gel electrophoresis, and hybridized with appropriate cloned cDNAs. The autoradiograms were exposed to Kodak XAR film at −70°C for various periods of time to achieve comparable signals for each mRNA species. The sizes of the mRNAs are kidney androgen-regulated protein (KAP), 0.9 kb; ornithine decarboxylase (ODC), 2.2 and 2.7 kb; β-glucuronidase (β-GLUC), 2.6 kb. From Catterall et al. (1985).

dase and ornithine decarboxylase mRNAs was further increased over the male levels. However, KAP mRNA accumulation did not change from the male values during this treatment (Fig. 13).

These data suggest that the relatively small number of nuclear androgen receptors present in intact males is sufficient to maximally stimulate the gene for KAP but not the genes for ornithine decarboxylase and β-glucuronidase. One possible explanation for these results is that interaction between the androgen receptor and each gene is unique, being characterized by different affinities for interaction or by requirement for binding of different amounts of the receptor to produce each response. Such a hypothesis is attractive in that it can explain variations in phenotype that involve only a subset of or even a single response(s).

In order to further test this hypothesis, experiments were conducted under conditions in which receptor concentrations are markedly reduced thus limiting the development of the androgen-induced phenotype. Kidney poly(A)-containing RNA from testicular-feminized (Tfm/Y) mice before and after treatment with testosterone was analyzed for variation in steady-state mRNA accumulation for each gene (Fig. 14). Interestingly, while β-glucuronidase and ornithine decarboxylase mRNA accumulation was not induced at all by testosterone in these receptor-deficient mutants, KAP mRNA exhibited a partial response to the hormone. Although this was a surprising result, it fits the above hypothesis, if one concludes that the KAP gene either has higher affinity for the androgen receptor or

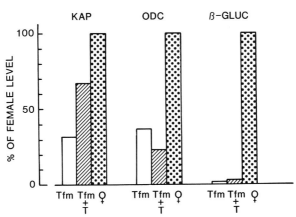

FIG. 14. Accumulation of the three mRNA species in response to testosterone in testicular feminized (Tfm) mice. The animals were treated for 4 days with Silastic rods releasing 200 μg testosterone (T) per day. Analysis of mRNA accumulation was performed as described in the legend to Fig. 12, and the values are given as percentages of those in nontreated, normal female mice. KAP, Kidney androgen-regulated protein; ODC, ornithine decarboxylase; β-GLUC, β-glucuronidase. From Catterall et al. (1985).

requires the binding of fewer androgen receptor molecules for activation. An identical argument would explain the lack of further stimulation of KAP gene expression during treatment of male animals with exogenous testosterone (Fig. 13), as the occupied androgen receptor concentration in intact males would have been sufficient for a maximal stimulation of this renal gene. Alternative explanations are also possible. For instance, the KAP response in Tfm/Y animals may be mediated by another receptor or by other hormones. While all alternatives for such an effect cannot easily be eliminated, additional studies showed that there was no increase in KAP mRNA level after treatment of normal control or Tfm/Y animals with 5β-dihydrotestosterone that functions through mechanisms different from those mediated by the normal androgen receptor. Furthermore, estradiol-17β and dexamethasone were ineffective in stimulating KAP gene expression, although an increase in KAP mRNA accumulation occurred upon progestin treatment (unpublished data).

If the KAP gene is more sensitive to nuclear hormone–receptor complex, then one would postulate that KAP mRNA accumulation would be less affected than the other two genes by treatment with antiandrogens. This is based on the assumption that fewer receptors are required for androgen induction of the KAP gene than the ornithine decarboxylase or β-glucuronidase genes. The nonsteroidal antiandrogen, flutamide, that

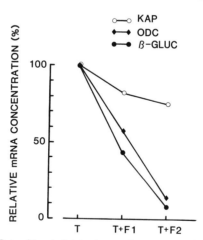

FIG. 15. Effects of flutamide administration on the response of the three genes to testosterone. In each case, the values in female animals treated with testosterone (T) (40 μg steroid/day) represent 100%. Flutamide was administered concomitantly with testosterone at two different doses (F1, 150 μg/day; F2, 650 μg/day). All treatments were for 8 days. Analysis of mRNA accumulation was carried out as described in the legend to Fig. 12. KAP, Kidney androgen-regulated protein; ODC, ornithine decarboxylase; β-GLUC, β-glucuronidase.

binds to the androgen receptor and forms biologically inactive receptor–ligand complexes (Kontula et al., 1985) was used to test this hypothesis. Animals were treated with a submaximal dose of testosterone (40 µg/day) alone and in combination with two different doses of flutamide (150 or 650 µg/day). As predicted, while β-glucuronidase and ornithine decarboxylase mRNA levels were reduced to one-tenth of the testosterone-induced values, KAP mRNA concentration remained at 76% of the induced level (Fig. 15). Administration of flutamide alone was unable to stimulate expression of any of the three genes, thus further confirming the concept that androgen receptor–flutamide complexes are biologically inactive (Kontula et al., 1985).

V. Summary and Conclusions

The murine kidney provides an unique experimental system to study the regulation of gene expression by androgens. Several genes in this tissue respond in a noncoordinate fashion to stimulation by testosterone. The three genes used in our studies exhibit diverse characteristics in both their structure and the regulation of their expression by androgens. The ornithine decarboxylase genes form a family of several members, while those for KAP and β-glucuronidase appear to be unique in the mouse genome. The product of the KAP gene is representative of the abundant class of cellular mRNAs, whereas β-glucuronidase and ornithine decarboxylase mRNAs are of the nonabundant type. KAP and ornithine decarboxylase mRNAs are induced with similar kinetics after hormone administration, but the extent of induction of KAP mRNA is much lower. The kinetics of β-glucuronidase mRNA induction differs greatly from the other two, requiring 15 days of steroid exposure to reach its peak value compared to 24–48 hours needed for full stimulation of KAP and ornithine decarboxylase mRNAs. Also, the KAP gene appears to be more sensitive to the concentration of functional androgen receptors than the β-glucuronidase or ornithine decarboxylase genes. The only unifying trait among these genes is the fact that they respond to androgens in a steroid receptor-dependent manner.

The male phenotype of a given cell consists of a diverse set of responses to a single stimulus, the male sex hormone. Analysis of androgenic control of specific gene expression should, therefore, lead to insights into the underlying mechanisms controlling development of this phenotype. The data presented support the concept that the expression of the male phenotype can be explained as the sum of individual, gene-specific interactions between the androgen–receptor complex and regulatory sites associated with the genes. Our studies did not address the question of a temporal and/or causal linkage of the individual responses; however, it is likely that this type of relationship exists in the development of the male phenotype. It has been suggested by others (Palmiter *et al.*, 1981) that gene-specific responses to steroids can be explained by multiple receptor binding sites creating dissimilar thresholds of response for different genes. Our data are consistent with such a mechanism. However, we prefer a model based upon differential binding affinity between chromatin and the androgen receptor, because this mechanism can also explain variations in only a single trait of a phenotype, such as loss of fertility in otherwise normal men (see Table I). Modulation of the affinity of the hormone–receptor complex for a specific binding site on chromatin could occur via a number of mechanisms, such as structural alterations in the receptor protein or the DNA sequence, changes in packaging of chromatin, and presence of tissue-specific factors regulating this interaction. The fact that changes in the concentration of the androgen receptor result in variations of the phenotypic expression is clearly illustrated in this work.

Details of the molecular mechanisms by which androgens modulate gene expression are unknown. While androgens have been shown to accelerate the rate of gene transcription (Compere *et al.*, 1981; Derman, 1981), increases in this rate were not sufficient to account for the androgen induction of the steady-state levels of mRNAs for proteins C1, C2, and C3 in the rat ventral prostate (Page and Parker, 1982) or of KAP, MK 908, and ornithine decarboxylase mRNAs in mouse kidney (Berger *et al.*, 1986). Transcription analyses were performed under conditions which allow RNA polymerases engaged in transcription to elongate their nascent transcripts but not to reinitiate. Nucleotide incorporation into specific gene products is, therefore, proportional to the rate of RNA chain initiation that prevailed *in vivo* prior to the start of the assay *in vitro*. Although the above results are very suggestive, their validation by more direct measurements awaits additional studies. Methods that allow direct measurement of absolute transcription rates (Brock and Shapiro, 1983a) or of mRNA turnover rates must be applied to the models for androgen regulation of gene expression, in order to determine with certainty the ultimate mechanisms by which male sex hormones control accumulation of specific gene products.

ACKNOWLEDGMENTS

We would like to acknowledge the skilled technical assistance of Susan Leary, Cecilia Liu, and Peter Kuhn. The KAP peptide was synthesized and provided to us by Dr. Jean Rivier and the coupling reaction and antibody production were carried out by Dr. Rosemarie

Thau. The authors also thank Jean Schweis for preparation of the manuscript. This work was supported by NIH Grant HD-13541.

REFERENCES

Aiman, J., and Griffin, J. E. (1982). *J. Clin. Endocrinol. Metab.* **54,** 725.
Bardin, C. W., and Catterall, J. F. (1981). *Science* **211,** 1285.
Bardin, C. W., Bullock, L. P., Sherins, R. J., Mowszowicz, I., and Blackburn, W. R. (1973). *Recent Prog. Horm. Res.* **29,** 65.
Bardin, C. W., Brown, T. R., Mills, N. C., Gupta, C., and Bullock, L. P. (1978). *Biol. Reprod.* **18,** 74.
Berger, F. S., Gross, K. W., and Watson, G. (1981). *J. Biol. Chem.* **256,** 7006.
Berger, F., Szymanski, P., Read, E., and Watson, G. (1984). *J. Biol. Chem.* **259,** 7941.
Berger, F. G., Loose, D. S., and Meisner, H. M. (1986). *Biochemistry* (in press).
Brandt, J. T., Pierce, D. A., and Fausto, N. (1972). *Biochim. Biophys. Acta* **279,** 184.
Brock, M. L., and Shapiro, D. J. (1983a). *J. Biol. Chem.* **258,** 5449.
Brock, M. L., and Shapiro, D. J. (1983b). *Cell* **34,** 207.
Bullock, L. P., and Bardin, C. W. (1970). *J. Clin. Endocrinol. Metab.* **31,** 113.
Bullock, L. P., and Bardin, C. W. (1972). *J. Clin. Endocrinol. Metabl.* **35,** 935.
Bullock, L. P., Bardin, C. W., and Ohno, S. (1971). *Biochem. Biophys. Res. Commun.* **44,** 1537.
Catterall, J. F., and Jänne, O. A. (1986). *In* "Modern Cell Biology" (B. Satir, ed.), Liss, New York (in press).
Catterall, J. F., and Leary, S. L. (1983). *Biochemistry* **22,** 6049.
Catterall, J. F., Watson, C. S., Kontula, K. K., Jänne, O. A., and Bardin, C. W. (1985). *In* "Molecular Mechanism of Steroid Hormone Action" (V. K. Moudgil, ed.), p. 587. de Gruyter, New York.
Chang, C. H., Rowley, D. R., and Tindall, D. J. (1983). *Biochemistry* **22,** 6170.
Choi, J. H., and Scheffler, I. E. (1983). *J. Biol. Chem.* **258,** 12601.
Compere, S. J., McKnight, G. S., and Palmiter, R. D. (1981). *J. Biol. Chem.* **256,** 6341.
Compton, J. G., Schrader, W. T., and O'Malley, B. W. (1983). *Proc. Natl. Acad. Sci. U.S.A.* **80,** 16.
Derman, E. (1981). *Proc. Natl. Acad. Sci. U.S.A.* **75,** 5425.
Fitzgerald, M., and Shenk, T. (1981). *Cell* **24,** 251.
French, F. S., VanWyck, J. J., Baggett, B., Easterling, W. E., Talbert, L. M., and Johnston, F. R. (1966). *J. Clin. Endocrinol.* **26,** 493.
Ganschow, R. E., and Bunker, B. G. (1970). *Biochem. Genet.* **4,** 127.
Griffin, J. E., and Wilson, J. D. (1980). *N. Engl. J. Med.* **302,** 198.
Griffin, J. E., Leshin, M., and Wilson, J. D. (1982). *Am. J. Physiol.* **243,** E81.
Groner, H., Kennedy, N., Skroch, P., Hynes, N. E., and Ponta, H. (1984). *Biochim. Biophys. Acta* **781,** 1.
Gubler, U., and Hoffman, B. J. (1983). *Gene* **25,** 263.
Hall, C. W., Cantz, M., and Neufeld, E. F. (1973). *Arch. Biochem. Biophys.* **155,** 32.
Henningsson, A., Persson, L., and Rosengren, E. (1978). *Acta Physiol. Scand.* **102,** 385.
Herrup, K., and Mullen, R. J. (1977). *Biochem. Genet.* **15,** 641.
Hickok, N. J., Seppänen, P. J., Kontula, K. K., Jänne, P. A., Bardin, C. W., and Jänne, O. A. (1986). *Proc. Natl. Acad. Sci. U.S.A.* **83,** 594.
Higgins, S. J., and Burchell, J. M. (1978). *Biochem. J.* **174,** 543.

Isomaa, V. V., Pajunen, A. E. I., Bardin, C. W., and Jänne, O. A. (1982). *Endocrinology* **111**, 833.
Isomaa, V. V., Pajunen, A. E. I., Bardin, C. W., and Jänne, O. A. (1983). *J. Biol. Chem.* **258**, 6735.
Jänne, J., Pösö, H., and Raina, A. (1978). *Biochim. Biophys. Acta* **473**, 241.
Jänne, O. A., Kontula, K. K., Isomaa, V. V., and Bardin, C. W. (1984). *Ann. N.Y. Acad. Sci.* **438**, 72.
Kahana, C., and Nathans, D. (1985). *Proc. Natl. Acad. Sci. U.S.A.* **82**, 1673.
Katzenellenbogen, B. (1980). *Annu. Rev. Physiol.* **42**, 17.
Keenan, B. S., Meyer, W. J., Hadjian, A. J., Jones, H. W., and Migeon, C. J. (1974). *J. Clin. Endocrinol. Metab.* **38**, 1143.
King, W. J., and Greene, G. L. (1984). *Nature (London)* **307**, 745.
Kochakian, C. D., and Harrison, D. G. (1962). *Endocrinology* **70**, 99.
Kontula, K. K., Torkkeli, T. K., Bardin, C. W., and Jänne, O. A. (1984). *Proc. Natl. Acad. Sci. U.S.A.* **81**, 731.
Kontula, K. K., Seppänen, P. J., Van Duyne, P., Bardin, C. W., and Jänne, O. A. (1985). *Endocrinology* **116**, 226.
Lalley, P. A., and Shows, T. B. (1974). *Science* **185**, 442.
Loeb, D., Houben, P. W., and Bullock, L. P. (1984). *Mol. Cell. Endocrinol.* **38**, 67.
Lusis, A. J., and Paigen, K. (1977). *In* "Isozymes: Current Topics in Biological and Medical Research" (M. C. Ratazzi, J. G. Scandalios, and G. S. Whitt, eds.), Vol. 2, p. 63. Liss, New York.
Lusis, A. J., Chapman, V. M., Wagenstein, R. W., and Paigen, K. (1983). *Proc. Natl. Acad. Sci. U.S.A.* **80**, 4398.
McConlogue, L., and Coffino, P. (1983). *J. Biol. Chem.* **258**, 8384.
McConlogue, L., Gupta, M., Wu, L., and Coffino, P. (1984). *Proc. Natl. Acad. Sci. U.S.A.* **81**, 540.
McKnight, G. S., and Palmiter, R. D. (1979). *J. Biol. Chem.* **254**, 9050.
Melanitou, E., Cohn, D. A., Bardin, C. W., and Jänne, O. A. (1986). *J. Biol. Chem.* (submitted for publication).
Meredith, S. A., and Ganschow, R. E. (1978). *Genetics* **90**, 725.
Mitchell, J. L. A., and Wilson, J. M. (1983). *Biochem. J.* **214**, 345.
Mulvihill, E. R., Lepennec, J. P., and Chambon, P. (1982). *Cell* **24**, 621.
Nicholson, W. E., Levine, J. A., and Orth, D. N. (1976). *Endocrinology* **98**, 123.
O'Malley, B. W., and Means, A. R. (1974). *Science* **183**, 610.
O'Malley, B. W., Roop, D. R., Lai, E. C., Nordstrom, J. L., Catterall, J. F., Swaneck, G. E., Colbert, D. A., Tsai, M. J., Dugaiczyk, A., and Woo, S. L. C. (1979). *Recent Prog. Horm. Res.* **35**, 1.
Okayama, H., and Berg, P. (1982). *Mol. Cell. Biol.* **2**, 161.
Page, M. J., and Parker, M. G. (1982). *Mol. Cell. Endocrinol.* **27**, 343.
Paigen, K. (1979). *Annu. Rev. Genet.* **13**, 417.
Paigen, K., Labarca, C., and Watson, G. (1979). *Science* **203**, 554.
Pajunen, A. E. I., Isomaa, V. V., Jänne, O. A., and Bardin, C. W. (1982). *J. Biol. Chem.* **257**, 8190.
Palmer, R., Gallagher, P. M., Bozko, W. L., and Ganschow, R. E. (1983). *Proc. Natl. Acad. Sci. U.S.A.* **80**, 7596.
Palmiter, R. D., Mulvihill, E. R., Shephard, J. H., and McKnight, G. S. (1981). *J. Biol. Chem.* **256**, 7910.
Parker, M. G., Scrace, G. T., and Mainwaring, W. I. P. (1978). *Biochem. J.* **170**, 115.
Payvar, F., Wrange, Ö., Carlstedt-Duke, J., Okret, S., Gustafsson, J.-Å., and Yamamoto, K. R. (1981). *Proc. Natl. Acad. Sci. U.S.A.* **78**, 6628.

Payvar, F., DeFranco, D., Firestone, G. L., Edgar, B., Wrange, Ö., Okret, S., Gustafsson, J.-Å., and Yamamoto, K. R. (1983). *Cell* **35**, 381.
Pegg, A. E. and McCann, P. P. (1982). *Am. J. Physiol.* **243**, C212.
Pfister, K., Chapman, V., Watson, G., and Paigen, K. (1985). *J. Biol. Chem.* **260**, 11588.
Renkawitz, R., Schutz, G., von der Ahe, D., and Beato, M. (1984). *Cell* **37**, 503.
Ringold, G. M., Yamamoto, K. R., Bishop, J. M., and Varmus, H. E. (1977). *Proc. Natl. Acad. Sci. U.S.A.* **74**, 2879.
Schrader, W. T. (1984). *Nature (London)* **308**, 17.
Seely, J. E., and Pegg, A. E. (1983). *J. Biol. Chem.* **258**, 2498.
Slater, E. P., Rabenau, O., Karin, M., Baxter, J. D., and Beato, M. (1985). *Mol. Cell. Biol.* **5**, 2984.
Sly, W. S., Quinton, B. A., McAlister, W. H., and Rimoin, D. L. (1973). *J. Pediat.* **82**, 249.
Snider, L. D., King, D., and Lingrel, J. B. (1985). *J. Biol. Chem.* **260**, 9884.
Southern, E. M. (1975). *J. Mol. Biol.* **98**, 503.
Spelsberg, T. C., Gosse, B. J., Littlefield, B. A., Toyoda, H., and Seelke, R. (1984). *Biochemistry* **23**, 5103.
Stanley, A. J., and Gumbreck, L. G. (1964). *Prog. Annu. Meet. Endocrine Soc. 46th*, p. 40.
Stanley, A. J., Gumbreck, L. G., Allison, J. E., and Easley, R. B. (1973). *Recent Prog. Horm. Res.* **29**, 43.
Swaneck, G. E., Nordstrom, J. L., Kreuzaler, F., Tsai, M. J., and O'Malley, B. W. (1979). *Proc. Natl. Acad. Sci. U.S.A.* **76**, 1049.
Swank, R. T., Paigen, K., and Ganschow, R. E. (1973). *J. Mol. Biol.* **81**, 225.
Swank, R. T., Paigen, K., Davey, R., Chapman, V., Labarca, C., Watson, G., Ganschow, R., Brandt, E. J., and Novak, E. (1978). *Recent Prog. Horm. Res.* **34**, 401.
Tabor, C. W., and Tabor, M. (1984). *Annu. Rev. Biochem.* **33**, 749.
Thomas, P. (1980). *Proc. Natl. Acad. Sci. U.S.A.* **77**, 5201.
Tomino, S., Paigen, K., Tulsiani, D. R. P., and Touster, O. (1975). *J. Biol. Chem.* **250**, 8503.
Toole, J. J., Hastie, N. D., and Held, W. A. (1979). *Cell* **17**, 441.
Tulsiani, D. R. P., and Keller, R. K. (1975). *J. Biol. Chem.* **250**, 4770.
Vanin, E. F. (1984). *Biochim. Biophys. Acta* **782**, 231.
Verhoeven, G. F. M., and Wilson, J. D. (1979). *Metabolism* **28**, 253.
Warne, G. L., Gyorki, S., Risbridger, G. P., Khalid, B. A. K., and Funder, J. W. (1983). *J. Steroid Biochem.* **19**, 583.
Watson, C. S., and Catterall, J. F. (1986). *Endocrinology* (in press).
Watson, C. S., Salomon, D., and Catterall, J. F. (1984). *Ann. N.Y. Acad. Sci.* **438**, 101.
Welshons, W. V., Lieberman, M. E., and Gorski, J. (1984). *Nature (London)* **307**, 747.
White, B. A., Preston, G. M., Lufkin, J. C., and Bancroft, C. (1985). *Mol. Cell. Biol.* **5**, 2967.
Wilson, C. M., Erdos, E. G., and Dunn, J. F. (1977). *Proc. Natl. Acad. Sci. U.S.A.* **74**, 1185.
Yamamoto, K. R., Payvar, F., Firestone, G. L., Maler, B., Wrange, Ö., Carlstedt-Duke, J., Gustafsson, J.-Å., and Chandler, V. L. (1983). *Cold Spring Harbor Symp. Quant. Biol.* **47**, 977.

DISCUSSION

W. D. Odell. First, can you tell us the relative affinities of the androgen receptors for T, DHT, and androstanediol, and whether these latter two compounds stimulate gene expression? Second, is there in fact a cytosol androgen receptor in the kidney?

J. F. Catterall. To answer the first question, yes, testosterone and DHT do affect gene expression in this tissue. DHT can be used interchangeably with testosterone to induce these genes. The relative affinites of testosterone and DHT for the androgen receptor are in the range of 1–5 nM. We have not studied the induction of gene expression in the mouse kidney by androstanediol. As for your second question, my feeling about a cytosolic receptor is that it is an artifact of the way the receptor is prepared. During the isolation procedure in the absence of hormone, the receptor will actually leak out of the nucleus and appear in the cytosol fraction. This does not indicate, however, that there is a cytoplasmic receptor.

M. Melner. I was curious; you made a comment about the *Gus-r* cis regulatory gene that you expected to find the regulatory elements in the 5' region of the gene, and I wonder if why, a priori, you would expect to find those regions in that area of the gene, since interesting cis regulatory elements have now been found in introns of genes such as the glucocorticoid regulatory element in the first intron of the growth hormone gene.

J. F. Catterall. I meant that only as a working model. In the absence of any physical means for identifying *Gus-r*, we think that our best approach is to identify the transcriptional start site of the *Gus-s* structural gene and then compare the sequences of different alleles in the 5' flanking region. However, eventually, it will be necessary to reconstitute *Gus-r* phenotypic expression *in vitro* (for instance, in cultured cells) in order to show what sequences are responsible for this activity. The only data available on the position of *Gus-r* come from genetic studies which are consistent with a location near or within the structural gene.

T. Horton. First, have you given any thought to examining your model system in lower vertebrates such as turtles, specifically mud turtles and pythons, which lack sex chromosomes, but show environmental sex determination based on the incubation temperature of the eggs?

J. F. Catterall. No, we haven't considered that, but it would be a very interesting system, I suppose, if similar gene probes were available that would allow us to assay the phenotypes.

J. E. Rall. Have you examined a nuclear run-off experiment so that you can have any idea as to whether the increase in message level is due to an increase in transcription or message stabilization, either cytoplasmic or intranuclear?

J. F. Catterall. Yes, we have done those experiments but, as you might guess, they are preliminary since I did not present any of the data. Berger *et al.* (*Biochemistry*, 1986, in press) showed that there was only a 2- to 3-fold increase in transcription of the ODC gene(s) which is insufficient to account for the extent of enzyme induction by testosterone. We can infer from this that mRNA stabilization must play a role, but the mechanism by which this occurs is unknown. Our own unpublished data indicate that increases in the initiation of transcription (as assayed in the nuclear run-off experiments) cannot explain the effects on the expression of these genes by androgens.

J. E. Rall. The question was whether there is perhaps some common effect of androgen on stabilizing message, if it does not do anything on increasing transcriptionally.

J. F. Catterall. Yes, that's possible. In that regard we have to keep in mind that the genetic studies of Paigen show that *Gus-r* is a cis-acting regulator. Therefore, at least in the case of the β-glucuronidase gene, the mechanism for increasing steady-state levels of mRNA must be consistent with this cis action.

S. Y. Ying. I would like to know if you have ever tried other types of substances or steroids on the possible regulation of these genes; and have you ever used tissues other than kidney to see if there would be differential regulation of these genes by androgen?

J. F. Catterall. Yes, we've looked at some other steroids particularly in relation to KAP because of the unusual induction of KAP in the Tfm/y mutant. We tested 5β-DHT and found no effect on KAP gene expression which indicates that the β-androgen receptor is not responsible for the induction of KAP gene expression. We have also tested estradiol, progesterone, and dexamethasone for their ability to induce KAP gene expression. Only progesterone had any effect. Other tissues that express KAP antigenic activity are the liver and submaxillary glands. We have not yet tested the effect of testosterone on KAP expression in these tissues. ODC is induced by testosterone in prostate and seminal vesicles, but β-glucuronidase is regulated by androgens only in the kidney.

D. K. Granner. I'm interested in the remarkable difference in induction of ODC in different strains of mice, and seem to recall that the family tree of these different strains has been worked out reasonably well. Are the 3 or 4 strains that are good inducers on the same branch of that tree?

J. F. Catterall. Actually I haven't looked at it in that way. We have determined the testosterone levels and the levels of receptor in these strains, but we have not tried to correlate their ability to respond with their evolutionary relationships.

D. K. Granner. I was going to ask whether there might be restriction fragment heterogeneity between inducing and noninducing strains at the 5' end, but if there is no transcriptional control such an experiment wouldn't be reasonable. Is the message restriction analysis the same with these strains, or do they induce the same mRNA?

J. F. Catterall. We haven't looked at the heterogeneity but the message levels do change similar to the enzyme levels.

D. K. Granner. Is it the same mRNA in these different strains of mice?

J. F. Catterall. That hasn't been done yet.

M. Saffran. Two questions about KAP. First of all, have you looked for it in the gut?

J. F. Catterall. No, we haven't.

M. Saffran. The gut is a gold mine for many peptides, and the colon has been described as just a big kidney. My second question about KAP is whether it is possible that KAP is itself an internal message in the kidney cell, turning on the other genes?

J. F. Catterall. It's certainly a possibility for β-glucuronidase. Since there is such a long lag, we've been wondering for some time if there are other proteins involved that must be produced first before β-glucuronidase can respond. I think it's unlikely in the case of ornithine decarboxylase, however.

G. B. Cutler. It looked as if there might be a substantial proportion of KAP synthesized independently of androgens. Have you examined mice in which you totally got rid of androgens by castration and adrenalectomy to see if there's a proportion of gene expression which is independent of androgen?

J. F. Catterall. There is a constitutive level of KAP mRNA after castration. The level is lower in castrates than in normal females.

G. B. Cutler. Might not your interpretation of sensitivity be altered if there were a substantial proportion of KAP which is androgen independent? For example, what if you redid your sensitivity analysis examining just the proportion of androgen-inducible protein? One could postulate that there are 2 KAP genes, one that is independent of androgen and one that is dependent on androgen, and that the androgen-dependent gene only produced about 25% of total KAP synthesis. If you then gave flutamide, for example, and if that gene making 25% of KAP were really normally sensitive to inhibition by flutamide, the fact that you get KAP levels down to 75% of control doesn't necessarily mean that the sensitivity to androgen is different from the other androgen-dependent proteins that are inhibited nearly 100%.

J. F. Catterall. We have no evidence for there being two genes. However, the assays that we have done could not distinguish the expression of two similar or identical genes as you suggest, so this remains a formal possibility until we can directly show that there is a single KAP gene.

C. S. Nicoll. Am I correct in my understanding that all of your treatments were done *in vivo*?

J. F. Catterall. That's right.

C. S. Nicoll. Have you been able to induce transcription of these genes in any *in vitro* experiments?

J. F. Catterall. *In vitro* transcription experiments? No, that has not been tried yet.

C. S. Nicoll. In that case is it possible that the differential gene expression that you observed is due to a metabolite of androgens or something that is being produced elsewhere in the body in response to androgens?

J. F. Catterall. Yes, as I said we have tested several things in that regard. While no single experiment will eliminate all the possibilities we have tried other hormones (estradiol, progesterone, and dexamethasone) and we have looked at an alternative receptor (the β receptor) thinking that the β metabolites might be responsible for this particular gene. All of those have turned out to be negative with the exception of a slight increase in the presence of progesterone.

C. S. Nicoll. I understand there has been difficulty in stimulating or inducing transcription of these renal androgen-responsive genes *in vitro*. Do you have any comments on why this difficulty might exist?

J. F. Catterall. I assume that by *in vitro* systems you mean expressing these genes in cultured cells. I know of no published report of this type of experiment. However, the fact that it hasn't been done yet may indicate that it has proven difficult for those who have tried it. Similar experiments using androgen-regulated genes in the rat ventral prostate have proven to be very difficult [Page, M. J., and Parker, M. G., *Cell* **32**, 495 (1983)].

J. M. Hershman. In the mouse, have you correlated any of the changes in gene expression that you find induced by testosterone with alterations of renal physiology or anatomy, particularly in regard to ornithine decarboxylase?

J. F. Catterall. No, we have done no anatomy studies on the kidney.

J. Geller. I gather that the androgen-mediated gene expression here is a very specific system that intimately involves the androgen receptor. There are other systems such as the prostate in which, for instance, Dr. Cunha has shown that androgen-mediated expression can be induced in the Tfm/Y mouse which lacks an androgen receptor with stromal extracts or just the presence of stroma. I wonder if you could comment on this since I thought the androgen receptor was a key part of your hypothesis.

J. F. Catterall. Yes, the androgen receptor we feel is the key part. β-Glucuronidase and ODC are made in the tubule epithelial cells of the proximal convoluted tubule. Androgens have been shown to increase β-glucuronidase activity in dispersed cells, so other cell types do not appear to be involved.

E. Knobil. I wonder if you could reassure me a little more about the specificity of this effect of androgen. What happens to your 3 proteins in the course of kidney hypertrophy which follows unilateral nephrectomy. In the same vein what is the relationship between the androgen sensitive β-glucuronidase in the kidney and the estrogen-sensitive β-glucuronidase in the accessory organs of reproduction in the female?

J. F. Catterall. As to your first question I haven't done any nephrectomy experiments. Could you repeat your second question.

E. Knobil. What is the relationship between the system which contains the androgen-sensitive β-glucuronidase in the kidney and whatever regulates the estrogen-sensitive β-glucuronidase system in accessory organs of reproduction in the female?

J. F. Catterall. We presume that there are kidney-specific factors that interact with the

Gus-r regulatory gene which make it specific for the kidney and that these are not present in these other tissues.

E. Knobil. It has been conjectured at one time that the major increases that one sees in β-glucuronidase activity in hormone-sensitive organs is not the primary consequence of reaction of the hormone but is secondary to the induction of rapid growth in these hormone-sensitive tissues. Is this still a possibility?

J. F. Catterall. No. Dr. Paigen and co-workers have shown in single cells by immunocytochemistry that single cells are indeed responding to the androgen stimulation.

W. J. Bremner. For a variety of experimental and clinical purposes it would be interesting to have a quantitative *in vivo* monitor of androgen effects. You mentioned that KAP was measurable in urine. I wondered whether you had any data or thoughts on the usefulness of any of the three proteins you have discovered either in serum or in urine as an *in vivo* quantitative monitor of androgen effect.

J. F. Catterall. β-Glucuronidase has certainly been used by scientists as a quantitative marker of androgen action. We have no evidence that these genes are expressed under androgenic control in humans. In principle, however, the excretion of the products of differentially regulated genes into the urine could form the basis of a clinical assay.

S. A. Atlas. The tissue distribution of KAP reminded me somewhat of renin which, in the mouse, is known to exist not only in kidney but also in the submaxillary gland, at least in certain inbred strains. Carol Wilson and her colleagues at Southwestern have shown that mouse submaxillary renin is androgen inducible in these strains. Ken Gross and his colleagues at Roswell Park have shown that there are two renin genes in the mouse, one of which is androgen regulated in the submaxillary gland and the other of which, in the kidney, is not. I wondered, therefore, if you had evidence that among the family of ornithine decarboxylase genes there are those which are and are not androgen regulated, and, although you showed evidence for only a single KAP message, whether you have definitively ruled out that there may be more than one KAP gene which is expressed differentially in the kidney and submaxillary gland.

J. F. Catterall. We don't have any direct evidence for the lack of expression of particular ODC genes. However, we believe that there are two being expressed using restriction site polymorphisms, and specific probes have been developed on the basis of differential DNA sequence between two cDNA groups. We have native genes for ODC now as well, and they fall into at least eight groups by restriction mapping. So by difference, since there are obviously more than two genes present then there must be genes in the kidney which are not responding to androgens but we have no direct assays for them. In response to your second question, the KAP gene has been shown to be unique by solution hybridization experiments, so to the extent that such an assay can distinguish between 1 and 2 or maybe 1 and 3 genes we would have to say that there is a single gene, but I would say at this point that there is a formal possibility of more than one gene for KAP.

I. Callard. I have a question and a comment related to the question asked by Dr. Rall on stabilization. David Shapiro, working with estrogen, i.e., induction of vitellogenesis, feels that the attainment of very high levels of expression of vitellogenesis in RNA and of vitellogenin production is due to the stabilization of mRNA by estrogen. Apparently your system is different as the product itself is stabilized. What is meant by "stabilization" and how does the steroid affect stabilization, the product?

J. F. Catterall. One might have expected, based on recent work with other steroids, that transcriptional control would play an important role in the regulation of these genes by androgens. In addition, the genetic data for the androgenic regulation of *Gus-s* by *Gus-r* are consistent with a transcriptional mechanism. However, recent data on KAP and ODC genes (Berger *et al., Biochemistry*, in press) and our own preliminary results from transcriptional run-off experiments suggest that changes in the initiation of transcription cannot account for increases in enzyme activity, enzyme synthesis, or steady-state mRNA levels.

I. Callard. I also have a comment regarding the previous comment by Dr. Horton in which she mentioned the usefulness of certain turtles as models because of the temperature sensitivity of sex differentiation in these species. I think reptiles would be useful models because the squamate reptiles (lizards and snakes) have an androgen-dependent "sexual" segment of the kidney which is morphologically distinct and produces hormone-inducible secretory products. This system would be a useful one as a model for androgen induction of renal proteins.

M. Walters. I was interested in the question of the cellular localization of these proteins, in particular KAP. I have two questions. One is whether there anything is known about the cell type localization of the androgen receptor among the kidney cells. The second is whether you have considered using *in situ* hybridization with your cDNA probe to localize the message for KAP protein.

J. F. Catterall. Yes, the androgen receptor is mainly present in the same cells which are producing β-glucuronidase and ornithine decarboxylase, the tubule epithelium of the proximal convoluted tubule. Approximately 70 to 80% can be accounted for in those cells. In response to your second question, yes we have considered this approach but we haven't done it yet. We are also considering doing immunocytochemical studies with the KAP antibody.

H. Kronenberg. The fact that the *Gus-r* gene acts in cis suggests that androgens might have a direct effect on transcription. An alternative possibility would be that if in fact message stabilization or half-life is the major regulating factor, *Gus-r* alleles could in fact reflect structural changes in the messenger RNA itself. These sequence changes in the mRNA could be responsible for half-life changes. The prediction would be that cDNA made from mRNA from different *Gus-r* alleles would have different sequences or S-1 analysis. Using the s clone against the r message might well produce differences or similarities that could eliminate that sort of problem.

J. F. Catterall. *Gur-r* is a cis-acting regulator and probably does not make a messenger RNA.

H. Kronenberg. No, I understand that, but couldn't *Gus-r* reflect, say, a couple of point mutations in the messenger RNA for glucuronidase that would make the message more or less susceptible to turnover by testosterone?

J. F. Catterall. *Gus-r* could indeed be included in the *Gus-s* gene and be part of the *Gus-s* messenger RNA, that's true.

H. Kronenberg. But an easy test of that would be to compare the transcripts.

J. F. Catterall. Compare the different alleles of the (*Gus*) complex?

H. Kronenberg. And see for instance if the messenger RNAs have the same sequences no matter what the *Gus-r* alleles. If the mRNAs of different *Gus-r* alleles are identical, it is unlikely that *Gus-r* is working through a posttranscriptional or processing effect.

J. F. Catterall. Yes, our structural studies so far have involved only the a allele. We have isolated the gene in the b allele but prior to our comparative studies we wanted to complete the characterization of *Gus-s*[a] and then use that as a basis for comparison of the others. We just haven't gotten to that yet.

P. Singh. I agree with the comment of Dr. Walters, but from our past experience all target glands of steroids have been shown to have cytosol receptors under the *in vitro* conditions they are measured in. So the liklihood of finding a nuclear receptor in the absence of a cytosol receptor, which appears in the cytosol due to possible leakage from the nucleus during preparation is remote. However, I agree that we need to look for nuclear receptors before we understand the mechanism.

J. F. Catterall. Thank you.

T. Hökfelt. Just a brief question with regard to your negative finding on the presence of KAP in the brain. Did you analyze the whole brain or did you attempt to make a regional analysis?

J. F. Catterall. No, we took whole brain and made an extract. We haven't done anything further than this. We hope to be more specific in that regard when we get to the immunocytochemical studies.

M. Walters. I have a comment relative to Dr. Singh's. Now that the steroid receptors are recognized as likely residing in the nuclei *in vivo*, one must be cautious in interpreting the absence of cytosolic receptors as indicative of their absence in the tissue. This caveat is especially important in tissues where an apparent lack of cytosol receptor contrasts with demonstrable steroid effects.

Insulin Regulates Expression of the Phosphoenolpyruvate Carboxykinase Gene

DARYL K. GRANNER, KAZUYUKI SASAKI, TERESA ANDREONE, AND ELMUS BEALE

Department of Molecular Physiology and Biophysics, Vanderbilt University School of Medicine, Nashville, Tennessee

I. Introduction

Insulin regulates metabolic processes in two general ways. It influences the rate of entry of substrates into cells and it affects, through alterations of enzyme activity, the intracellular modification of these substrates. Changes of enzyme activity result from covalent modification of a fixed amount of protein, generally by a phosphorylation–dephosphorylation mechanism, or from changes of the absolute amount of the protein. Both mechanisms are involved in the more than 50 proteins reported to be affected by insulin, but, while the processes involved in covalent modification have been studied extensively, little is known about how insulin regulates the synthesis of specific proteins.

General protein synthesis is reduced in adipose tissue (Crofford *et al.*, 1970; Crofford and Reynold, 1965), mixed fiber skeletal muscle (Pain and Garlick, 1974; Jefferson *et al.*, 1972, 1974, 1977), and heart muscle (Morgan *et al.*, 1971) of diabetic rats, and insulin supplementation restores protein synthesis to near normal values in these tissues. An effect of insulin on the phosphorylation of ribosomal proteins (Avruch *et al.*, 1982), and on the activation–inactivation of translation factors (Pilkis and Park, 1974), might explain how insulin regulates general protein synthesis in a tissue. It is difficult to imagine how these mechanisms could affect the synthesis of specific proteins, however, since stimulatory and inhibitory responses occur in the same tissue. Until recently the prevailing notion was that insulin regulated the synthesis of specific proteins by altering the rate of translation of a fixed amount of a given mRNA. Indeed, the most recent review of insulin action on protein synthesis made only one, brief mention of an effect of the hormone on a specific mRNA (Jefferson, 1980).

Several years ago we decided to define a system with which we could examine the actions of insulin on the synthesis of a specific protein. After checking several enzymes in a number of tissues and cultured cells we decided to study the gluconeogenic enzyme phosphoenolpyruvate carboxykinase (GTP) (EC 4.1.1.32; PEPCK). PEPCK is present at relatively high specific activities in liver, kidney cortex, and white adipose tissue, and at lower specific activities in lung, jejunum, and brain (Hanson and Garber, 1972). The enzyme catalyzes the conversion of oxalacetate to phosphoenolpyruvate (see Fig. 1), which is the rate-limiting reaction in hepatic and renal gluconeogenesis (Utter, 1969), and the reaction is essential for the synthesis of α-glycerol phosphate in adipose tissue (Hanson and Garber, 1973).

Although all species studied contain immunologically distinct cytoplasmic and mitochondrial forms of PEPCK (Ballard and Hanson, 1969), only the cytoplasmic form responds to hormonal and dietary stimuli (Shrago et al., 1963; Nordlie et al., 1965). PEPCK activity is altered in vivo by the administration of glucagon, glucocorticoids, insulin, norepinephrine, epinephrine, ACTH, and thyroxine. The response to many of these hormones is tissue specific. For example, insulin decreases PEPCK enzyme activity in liver (Shrago et al., 1967) and adipose tissue (Reshef and Shapiro, 1970) but has no effect on the kidney enzyme (Kamm et al., 1974). Glucocorticoids increase hepatic (Gunn et al., 1975) and renal (Longshaw and Pogson, 1972) PEPCK synthesis but decrease the synthesis of the adipose tissue enzyme (Meyuhas et al., 1976). Cyclic AMP increases

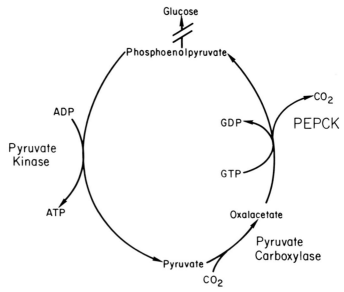

FIG. 1. The reaction catalyzed by PEPCK.

PEPCK synthesis in liver (Hanson et al., 1975) and adipose tissue (Reshef and Hanson, 1972) but has no effect on the renal enzyme (Iynedjian et al., 1975).

One can thus analyze the mechanism of action of insulin and the multihormonal regulation of the synthesis of a protein using PEPCK as a model. However, studies of insulin effects on protein synthesis in intact animals are hard to interpret. Insulin-induced hypoglycemia results in changes in availability of various substrates and in the release of many counterregulatory hormones which themselves regulate many of the enzymes in question. It is therefore difficult to determine whether the insulin effect is direct or indirect. It is also difficult to determine whether the effects of insulin in intact animals are mediated through the insulin receptor or through growth peptide receptors. Insulin and several serum growth factors, including the insulin-like growth factors IGF-I and IGF-II (Van Wyk and Underwood, 1975; King et al., 1980), have an overlapping spectrum of biologic activity. Insulin is structurally related to these peptides and, although each apparently has unique cell surface receptors, at the high concentrations of insulin generally used in intact animal studies there may be significant binding of this hormone to the growth peptide receptor.

One can avoid many of these difficulties by employing cultured cell systems that respond to insulin. For example, H4IIE cells are a clonal line of rat hepatoma cells (Pitot et al., 1964) that were adapted to culture from the Reuber H35 transplantable hepatoma (Reuber, 1961). H4IIE cells have normal insulin receptors (Iwamoto et al., 1981) and the synthesis of cytoplasmic PEPCK in these cells is induced by cAMP analogs or glucocorticoids (Barnett and Wicks, 1971) and inhibited by insulin (Tilghman et al., 1975). This article is a summary of our analysis of the action of insulin on PEPCK synthesis in H4IIE cells.

II. Results

A. STRATEGY USED TO STUDY HOW INSULIN REGULATES PEPCK SYNTHESIS IN H4IIE CELLS

Assays that quantitate the rate of synthesis of PEPCK and of $mRNA^{PEPCK}$ activity and amount allow one to determine whether insulin affects the rate of translation of a fixed amount of mRNA, the intrinsic activity of the mRNA, or the amount of the mRNA. A change of the amount of $mRNA^{PEPCK}$ in the cytoplasm could result from an increased rate of degradation, from changes of the transport of this mRNA from the nucleus, or from a change of the production or stability of nuclear mRNA^{PEPCK}. We developed the assays required for the quantitation of each of these processes, then addressed a series of questions related to mechanism(s) involved in the regulatory action of insulin on PEPCK.

B. INSULIN IS A DIRECT, PHYSIOLOGIC REGULATOR OF PEPCK SYNTHESIS

H4IIE cells were placed in serum-free medium 24 hours before the addition of insulin to exclude the possibility that other hormones might be involved in the action of insulin on PEPCK synthesis. Since an inhibitory effect of insulin was expected, PEPCK synthesis and mRNA^{PEPCK} activity were first increased by an analog of cAMP. The dibutyryl (Bt$_2$) derivative of cAMP was initially used but analogs more resistant to metabolism, such as 8-bromo-cAMP and 8-(4-chlorophenylthio)cAMP (8-CPT-cAMP), were employed later as these afforded a more reproducible level of induction.

Bt$_2$cAMP quadruples the rate of synthesis of PEPCK and causes a proportional increase of the activity of mRNA^{PEPCK} (see Table I). In these H4IIE cells, synthesizing PEPCK at a high rate, 1 nM insulin is a potent and specific inhibitor. Insulin (1 nM), a postprandial concentration in the portal circulation, decreases the rate of synthesis of PEPCK to or below control levels within 3 hours and this is accompanied by a parallel diminution of mRNA^{PEPCK} activity (see Table I).

The observation that 1 nM insulin causes a maximal repression of both PEPCK synthesis and mRNA^{PEPCK} activity indicates that these responses

TABLE I
Effect of Insulin on P-Enolpyruvate Carboxykinase Synthesis and mRNA^{PEPCK} Activity in H4IIE Cells[a]

Treatment	PEPCK synthesis (% of total protein)	mRNA^{PEPCK} activity (% of total mRNA)
Control	0.06 ± 0.01	0.06 ± 0.01
Bt$_2$cAMP	0.27 ± 0.01	0.26 ± 0.01
Bt$_2$cAMP + insulin	0.04 ± 0.01	0.04 ± 0.00

[a] Confluent H4IIE cells were treated with Bt$_2$cAMP (0.5 mM), theophylline (1 mM), and insulin (1 nM) for 3 hours then the rate of PEPCK synthesis and mRNA^{PEPCK} activity were determined. PEPCK synthesis is expressed as the mean ± SE of aliquots of cells taken from 9 flasks. The remaining cells were pooled in groups of three, and poly(A)$^+$ RNA was isolated. Therefore, mRNA^{PEPCK} values represent the mean ± SE of three samples, each of which consists of cells from three flasks. Reprinted, with permission, from Andreone et al. (1982).

REGULATION OF METABOLIC PROCESSES BY INSULIN

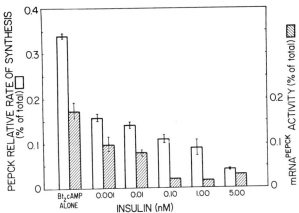

FIG. 2. Dependence of PEPCK synthesis and mRNA^{PEPCK} activity on insulin concentration. Confluent H4IIE cells were treated with 0.5 mM Bt$_2$cAMP and 1.0 mM theophylline in serum-free medium for 24 hours prior to harvest. Insulin, in fresh serum-free medium which also contained 0.5 mM Bt$_2$cAMP and 1.0 mM theophylline, was added 3 hours prior to cell harvest. PEPCK synthesis and mRNA^{PEPCK} activity are expressed as described in the footnote to Table I. Reprinted, with permission, from Andreone et al. (1982).

are extremely sensitive to insulin. Indeed, both decrease in concert when H4IIE cells are exposed to insulin concentrations in the picomolar to nanomolar range for 3 hours (Fig. 2). The half maximal effect occurs at 1–2 pM insulin and 1 nM insulin reduces mRNA^{PEPCK} activity to the level measured in control cells. This response of H4IIE cells to insulin is not unique; the half maximal concentration of insulin required for tyrosine aminotransferase induction (Iwamoto et al., 1981) or the stimulation of glucose oxidation (Hoffman et al., 1980) in these cells is 10 pM.

C. THE INSULIN RECEPTOR IS INVOLVED IN THE REPRESSION OF mRNA^{PEPCK} ACTIVITY

The potency hierarchy of porcine insulin >> porcine proinsulin >> guinea pig insulation is a predictable characteristic of effects mediated by the insulin receptor (Freychet et al., 1971). Porcine proinsulin and guinea pig insulin are 50 and 1000 times less effective than porcine insulin at decreasing mRNA^{PEPCK} activity (Fig. 3A). These relative biologic potencies correlate very well with differences in the affinity of binding each ligand has for the insulin receptor, as illustrated in Fig. 3B. Conversely, porcine and guinea pig insulins bind to insulin-like growth peptide receptors with nearly equal affinity and porcine proinsulin binds to these receptors with equal or greater affinity than insulin (King et al., 1980; Megyesi

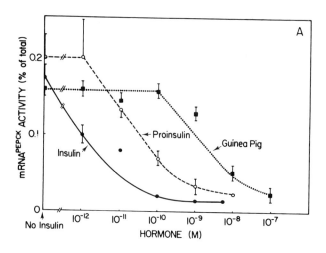

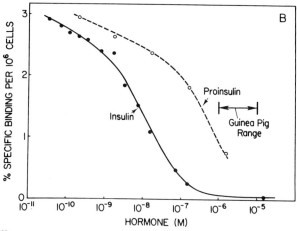

FIG. 3. Effects of insulin, proinsulin, and guinea pig insulin on mRNA^{PEPCK} activity and ^{125}I-labeled insulin binding. The effects of insulin, proinsulin, and guinea pig insulin on mRNA^{PEPCK} activity are shown in A. Each point represents the mean ± SE of assays performed on three samples of poly(A)$^+$ mRNA, each of which included cells from three flasks. The value for "No Insulin" was from flasks treated with Bt$_2$cAMP and theophylline only. Arrows designate the half-maximal effective concentrations. The binding of ^{125}I-labeled insulin in the presence of various concentrations of porcine insulin, porcine proinsulin, and guinea pig insulin is shown in B. Reprinted, with permission, from Andreone *et al.* (1982).

et al., 1974). Since very different concentrations of these insulins are required to depress mRNA^{PEPCK}, and since proinsulin is only 2–5% as potent as insulin, the possibility that the effect of insulin on mRNA^{PEPCK} is mediated through an insulin-like growth peptide receptor is unlikely.

In H4IIE cells, as in many other cells, biologic effects of insulin are elicited when only a small fraction of the insulin receptors is occupied (Andreone *et al.*, 1982; Ginsberg, 1985). When receptor occupancy and biologic activity are expressed as a function of ambient insulin concentration it is apparent that a maximal effect of insulin on mRNA^{PEPCK} activity is obtained when less than 1% of the receptors are occupied by the hormone (see Fig. 4). The same phenomenon is observed when proinsulin or guinea pig insulin is used. Hence occupancy of very few receptors is required for PEPCK repression and there are "spare receptors" in this response of insulin (Andreone *et al.*, 1982).

The low concentration of insulin required, and the relative order of potency of various insulins, makes it extremely likely that the regulation of mRNA^{PEPCK} involves the insulin receptor. Since these effects are also obtained in cells maintained for 24 hours in serum-free medium, the possibility that other hormones or counterregulatory substances are involved in this action of insulin is excluded.

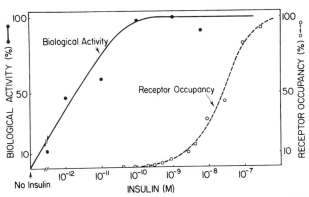

FIG. 4. Correlation of biologic activity and receptor occupancy. Biologic activity (mRNA^{PEPCK} deinduction) was calculated using the value from cells treated with Bt$_2$cAMP alone as 0% biologic activity and the value from cells treated with 10^{-9} M insulin as 100% biologic activity (see Fig. 3A) Receptor occupancy ($R/R_0 \times 100$) was determined by computer calculation of R_0 (total receptor number) from a computer-fitted Scatchard plot of the insulin data in Fig. 3B. R (number of receptors occupied by insulin) was calculated at each point by specific binding (^{125}I-labeled insulin bound/^{125}I-labeled insulin added to the flask) × total insulin concentration. Reprinted, with permission, from Andreone *et al.* (1982).

D. EFFECTS OF INSULIN ON CYTOPLASMIC mRNA^{PEPCK} METABOLISM IN H4IIE CELLS

1. Comparison of mRNA^{PEPCK} Activity and Amount

Having demonstrated that insulin decreases PEPCK synthesis and mRNA activity in parallel, we next compared the effect of the hormone on mRNA activity and amount. The addition of 5 nM insulin to H4IIE cells results in a rapid decrease of mRNA^{PEPCK} activity and amount (see Fig. 5A). Various concentrations of insulin also decrease mRNA^{PEPCK}

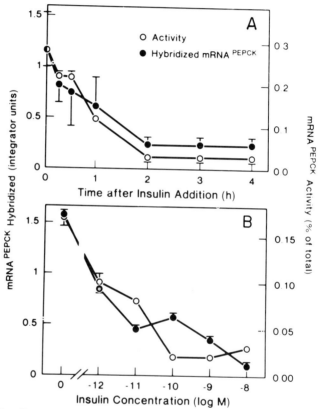

FIG. 5. Insulin repression of mRNA^{PEPCK} activity and amount in H4IIE cells. Messenger mRNA^{PEPCK} activity was determined using reticulocyte lysate translation and mRNA^{PEPCK} was quantified using the dot blot hybridization assay. The change of mRNA^{PEPCK} activity (○) and amount (●) with time after the addition of 1 nM insulin is illustrated in A and at various concentrations of insulin in B. Each point represents the mean ± SE of three samples. From Granner et al. (1983). Reprinted by permission from Nature. Copyright © 1983 Macmillan Journals Limited.

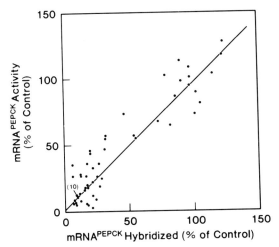

FIG. 6. Summary of 15 experiments in which mRNA^{PEPCK} activity and amount were determined in H4IIEC cells in the absence and presence of insulin. Confluent cells were treated with no hormones, 1 nM insulin, or 0.5 mM Bt$_2$cAMP plus (1 pM to 1 nM) insulin. Each dot represents a single RNA sample isolated from cells from three flasks. Messenger RNA^{PEPCK} activity and amount were determined on each sample and are expressed as a normalized value, which is the percentage of the "no hormone" control. Reprinted, with permission, from Granner and Andreone (1985).

activity and amount in parallel (Fig. 5B). A plot of mRNA activity versus amount, using poly(A)$^+$ RNA samples from a number of experiments, reveals a straight line association with a correlation coefficient of 0.94 (see Fig. 6), and a similar relationship is noted when the rate of synthesis of PEPCK is plotted against either mRNA^{PEPCK} activity or amount. This observation establishes the direct relationship between PEPCK synthesis and cytoplasmic mRNA^{PEPCK}. Insulin therefore has little, if any, effect on the translational efficiency of this mRNA in H4IIE cells.

2. Reversibility of This Inhibitory Effect of Insulin

The repression of mRNA^{PEPCK} by insulin occurs quickly and, given the rapid response of mRNA^{PEPCK} to various nutritional and hormonal regimens, it should be readily reversible. In H4IIE cells treated with insulin to maximally suppress mRNA^{PEPCK}, then washed at pH 6.0 to remove receptor-bound insulin (Kosmakos and Roth, 1980), mRNA^{PEPCK} returns to the level found in untreated cells with 3 hours (see Table II). The fact that such cells also respond to Bt$_2$cAMP indicates that the wash effective-

TABLE II
Reversibility of the Insulin Repression of mRNA^{PEPCK}[a]

Addition during 3 hour preincubation	pH 6 wash	Addition during 3 hour incubation	mRNA^{PEPCK} hybridized (integrator units)
None	–	None	0.24 ± 0.04
None	–	Insulin	Not detectable
None	+	None	0.22 ± 0.06
None	+	Insulin	Not detectable
None	+	Bt$_2$cAMP	1.76 ± 0.10
Insulin	+	None	0.18 ± 0.04
Insulin	+	Bt$_2$cAMP	1.40 ± 0.36

[a] An initial incubation of confluent cultures of H4IIE cells was done with or without insulin for 3 hours. At the end of this preincubation the cells were washed at pH 6.0 to remove bound insulin (Kosmakos and Roth, 1980). Fresh medium was added and the incubation was continued for 3 hours in the presence of 0.1 nM insulin or Bt$_2$cAMP and theophylline at 0.5 and 1 mM, respectively. Integrator units are expressed as the mean ± SE of triplicate RNA samples. Reprinted, with permission, from Sasaki et al. (1984).

ly removes insulin (see Table II). Given the 30–40 minute $t_{1/2}$ of mRNA^{PEPCK}, the return to the normal level in less than 3 hours indicates that insulin repression is reversed as soon as the hormone is removed (Sasaki et al., 1984).

3. Effect of Insulin on Cytoplasmic mRNA^{PEPCK} Turnover

mRNA^{PEPCK} decreases with a $t_{1/2}$ of about 30 minutes after the addition of insulin (see Fig. 5). This rate of deinduction is in good agreement with the half-life of mRNA^{PEPCK} estimated under a variety of conditions in rat liver (Cimbala et al., 1980; Beale et al., 1982; Kioussis et al., 1978; Nelson et al., 1980; Granner and Andreone, 1985), hence it is unlikely that insulin decreases mRNA^{PEPCK} by enhancing its degradation.

E. EFFECT OF INSULIN ON NUCLEAR mRNA^{PEPCK} METABOLISM

1. Size, Abundance, and Kinetics of Nuclear RNA^{PEPCK}

The decrease of cytoplasmic mRNA^{PEPCK} and the lack of an effect of insulin on mRNA^{PEPCK} turnover suggest an action of the hormone on nuclear mRNA^{PEPCK} production or egress. These possibilities were tested by blot transfer of nuclear and cytoplasmic RNA from H4IIE cells which were first treated with 8-CPT-cAMP then exposed to insulin for various times. H4IIE cell nuclei contain at least 11 RNA species that hybridize specifically to a genomic DNA probe that represents more than 80% of the rat liver PEPCK gene sequence (Fig. 7A, left lane). Five of these species are larger than the 3.2-kilobase (kb) form that represents the only RNA^{PEPCK} found in the cytoplasm (Fig. 7A), and five are smaller.

The relative abundance of each of these forms, and the change of each with time after the addition of insulin, was determined using scanning densitometry. Scans of the lanes representing samples isolated at the 0, 0.5, and 2.0 hour time points are shown in Fig. 7B (inset). To simplify the presentation only the data representing the largest (6.8 kb), the most abundant (4.0 kb), and the mature 3.2 kb nuclear and cytoplasmic forms are illustrated in Fig. 7B. The 6.8 kb transcript decreases most rapidly, followed in turn by the 4.0 kb form, the mature 3.2 kb nuclear form, and at least by cytoplasmic mRNA^{PEPCK}. This observation is consistent with a precursor–product relationship, although pulse-chase experiments are required to firmly establish this point. Effects similar to these are seen in cells exposed only to insulin but, since initial levels of nuclear PEPCK transcripts are quite low in such cells, quantification is more difficult.

2. Insulin Has No Effect on Nuclear RNA^{PEPCK} Turnover

The rate of disappearance of the various forms of RNA^{PEPCK} can be estimated by applying a semilog plot analysis to the data presented in Fig. 7B. The 6.8 kb presumptive primary transcript (open triangles) decreases with a $t_{1/2}$ of 22 minutes, the nuclear 3.2 kb transcript (open circles) declines with $t_{1/2}$ of 28 minutes, and, after a 25 minute lag, the cytoplasmic transcript (closed circles) exhibits a half time of disappearance of 24 minutes. These values are in excellent agreement with the estimates of mRNA^{PEPCK} turnover cited above (Cimbala *et al.*, 1980; Beale *et al.*, 1982; Kioussis *et al.*, 1978; Nelson *et al.*, 1980; Granner and Andreone, 1985), so it is apparent from this study that an effect of insulin on nuclear RNA turnover must be minimal, if it exists at all. The nuclear transcripts smaller than 3.2 kb disappear as rapidly as the 4.0–5.0 kb species, but the relationship of these molecules to the production of mature mRNA^{PEPCK} is obscure.

3. Insulin Does Not Affect Nuclear RNA^{PEPCK} Egress

An effect of insulin on the egress of mRNA from the nucleus has been proposed (Purello *et al.*, 1982; Schumm and Webb, 1981). To affect mRNA^{PEPCK} in this manner insulin would have to selectively decrease the transport of this RNA out of the nucleus which, if production is not inhibited to the same extent, should result in the accumulation of nuclear

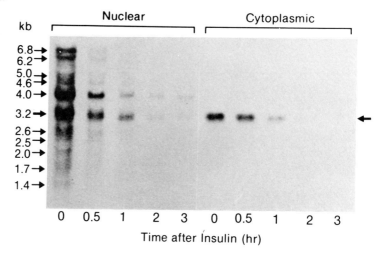
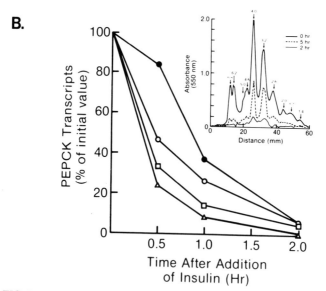

FIG. 7. Northern blot transfer analysis of nuclear and cytoplasmic RNA following insulin treatment. Insulin (5 nM) was added to H4IIE cells after 3 hours of preincubation with 0.5 mM dibutyryl cyclic AMP and the cells were collected at the indicated times. Nuclear and cytoplasmic RNAs were isolated, subjected to electrophoresis, transferred to nitrocellulose, and hybridized to a cloned 5.8-kb EcoRI fragment of PEPCK genomic DNA carried in recombinant plasmid pλPC112.R3 (Beale et al., 1985). (A) Autoradiograph of the Northern transfer. (B, inset) Densitometry scan of the 0, 0.5, and 2 hour lanes of the nuclear transcripts shown in A. (B) Analysis of the changes in the 6.8 (△), 4.0 (□), and 3.2 (○) kb nuclear and the 3.2 kb cytoplasmic transcript (●) shown in A. Data represent the percentage of the zero time values obtained from the scan shown in the inset in B. From Granner et al. (1983). Reprinted by permission from Nature. Copyright © 1983 Macmillan Journals Limited.

mRNA^{PEPCK}. This is clearly not the case, as illustrated in Fig. 7A, hence any role of insulin on transport of mRNA^{PEPCK} from nucleus to cytoplasm must be of minor importance.

F. EFFECT OF INSULIN ON PEPCK GENE TRANSCRIPTION

1. Insulin Inhibits cAMP-Induced PEPCK Gene Transcription

The decrease of nuclear PEPCK mRNA could result from accelerated degradation of nuclear mRNA^{PEPCK} or its precursors, from decreased processing of RNA^{PEPCK} into the mature form, or from decreased gene transcription. We had acquired presumptive evidence that degradation was unaffected (Fig. 7B) so, since processing is difficult to assess, the next step evaluated was the effect of insulin on PEPCK gene transcription. This was done by quantifying the labeling of nascent RNA transcripts in nuclei isolated from hormone-treated cells (McKnight and Palmiter, 1979). Treatment of H4IIE cells with dibutyryl cyclic AMP results in a prompt quadrupling of PEPCK gene transcription (see Expt. 1, Table III) and the addition of insulin results in a significant inhibition within 10 minutes and a reduction to the basal level within 30 minutes.

In a second experiment, in which the analog 8-CPT-cAMP was used as the inducer, 5 nM insulin reduced transcription from 832 to 53 ppm which is less than the uninduced rate of 150 ppm. In both experiments the amount of isotope incorporated into nascent mRNA^{PEPCK} in the presence of insulin was equivalent to or less than that obtained with a concentration of α-amanitin which selectively inhibits class II gene transcription. Insulin therefore completely abolishes the transcription of the PEPCK gene.

2. Concentration–Response Relationship

PEPCK gene transcription is progressively repressed through an insulin concentration range from 0.5 pM to 1 nM and the half maximal effective concentration is approximately 10 pM (Fig. 8). This observation is in excellent agreement with the concentration range required for suppression of nuclear and cytoplasmic mRNA^{PEPCK}, and for the inhibition of PEPCK synthesis. Porcine proinsulin is 50 times less effective than porcine insulin in the repression of PEPCK gene transcription.

3. Time Course of Inhibition of Transcription by Insulin

Since nuclear and cytoplasmic P-enolpyruvate carboxykinase RNA decrease with a $t_{1/2}$ of 20–30 minutes after H4IIE cells are exposed to insulin, transcription should be inhibited with a $t_{1/2}$ of less than 20 minutes if this is the primary effect of the hormone. H4IIE cells were first exposed to 8-CPT-cAMP, which increased PEPCK gene transcription 9-fold

TABLE III
Effect of Insulin on Transcription of the PEPCK Gene[a]

Expt.	Treatment	mRNA[PEPCK] synthesis (ppm)	Total RNA synthesis (cpm × 10^{-6})
1	None	136 ± 72	8.6 ± 0.1
	Bt$_2$cAMP for 40 minutes	454 ± 151	8.5 ± 0.1
	+ insulin for last 10 minutes	324 ± 72	9.9 ± 0.2
	Bt$_2$cAMP for 60 minutes	592 ± 65	10.0 ± 0.2
	+ insulin for 30 minutes	142 ± 31	11.0 ± 0.3
2	None	150 ± 33	5.2 ± 0.1
	8-CPT-cAMP for 90 minutes	832 ± 95	9.1 ± 0.9
	+ insulin for last 60 minutes	53 ± 31	8.8 ± 1.4

[a] Confluent H4IIE cells were placed in serum-free medium and 21 hours later 0.5 mM Bt$_2$cAMP, 0.1 mM 8-CPT-cAMP, and 5 nM insulin were added for the indicated times. Nuclei were isolated and the transcription assay was conducted (McKnight and Palmiter, 1979) using pBR322 or pλPC112.R3 (Beale et al., 1985) as control and PEPCK-specific probes respectively. [^{3}H]cRNA, prepared using pλPC112.R3 as a template, was used to determine hybridization efficiency which was typically 20–40%. α-Amanitin (1 μg/ml) completely inhibited the incorporation of [^{32}P]UTP into PEPCK RNA in nuclei isolated from control, dibutyryl cyclic AMP-treated, or insulin-treated cells. Results are expressed as PEPCK transcription in parts per million (ppm) calculated as follows: ppm = [(cpm on pλPC112.R3 filter − cpm on pBR322 filter) × 6800/5400 × 10^{-6}]/efficiency × total cpm in hybridization mixture (Granner et al., 1983). Values are the mean ± SD from duplicate (Experiment 2) or triplicate cultures (Experiment 1). Reprinted, with permission, from Granner and Andreone (1985).

within 30 min (Fig. 9). Transcription continues at the maximal rate of about 1500 ppm for an additional 60 min then decreases to 800 ppm by 150 min after the addition of 8-CPT-cAMP if no other additions are made. The addition of insulin 30 minutes after the cyclic nucleotide, at a time when the transcription rate is at its peak, results in a rapid decrease. Inhibition is apparent within 5 minutes and by 60 minutes the rate of PEPCK gene transcription is at the basal level. The $t_{1/2}$ of this inhibition is thus approximately 12–15 minutes which is notably faster than the decrease of cytoplasmic mRNA[PEPCK] (Granner et al., 1983; Beale et al., 1982; Nelson et al., 1980; and see Fig. 6B), as would be expected if transcription is the step that is primarily affected.

4. Insulin Inhibits Transcription in the Absence of cAMP

In all of the studies on the effects of insulin on PEPCK gene transcription discussed so far, H4IIE cells were first exposed to a cyclic AMP

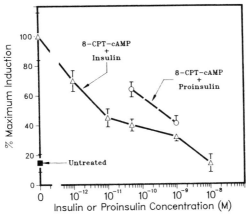

FIG. 8. Effects of 8-CPT-cAMP, insulin, and proinsulin on mRNA^{PEPCK} transcription as a function of concentration. H4IIE cells were exposed to 0.1 mM 8-CPT-cAMP for 90 minutes and to various concentrations of insulin or proinsulin for the final 60 minutes. Each point represents the mean of three transcription assays, expressed as the percentage of maximum induction. Reprinted from Sasaki et al. (1984).

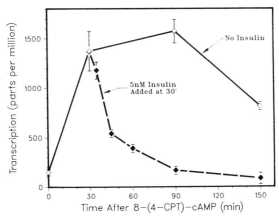

FIG. 9. Effects of 8-CPT-cAMP and insulin on mRNA^{PEPCK} transcription as a function of time. Nuclei were isolated from H4IIE cells that had been exposed to 0.1 mM 8-CPT-cAMP for the times indicated, or from cells to which 5 nM insulin was added 30 minutes after the cyclic nucleotide. The results represent the average ± SD of triplicate transcription assays. Reprinted from Sasaki et al. (1984).

analog. This was deemed necessary for technical reasons, that is, an enhanced rate of transcription made it easier to detect the postulated inhibitory effect of insulin. As we gained experience with the transcription assay it became possible to test whether insulin exerted an effect in the absence of cAMP. The results illustrated in Table IV indicate that insulin concentrations of 10 pM and 1 nM decrease PEPCK gene transcription from the basal rate of 177 to 71 and 25 ppm, respectively. In this experiment the rate of incorporation of labeled UTP into nascent RNA^{PEPCK} chains in the presence of α-amanitin, a general inhibitor of class II gene transcription, was also 25 ppm. Therefore insulin completely abolishes PEPCK gene transcription in H4IIE cells.

5. Specificity of This Effect of Insulin

The amount of total RNA transcribed is determined as a part of each transcription experiment, and in no case does insulin (or any other hormone tested to date) affect total genomic activity in H4IIE cells (see Tables III and V). This effect of insulin is thus exerted on a very small fraction of the active genes in the H4IIE cell, so the effect is quite selective.

6. Evidence That This Is a Direct Effect of Insulin

The action of some hormones on the transcription of specific genes requires on-going protein synthesis and thus presumably the intervention of an intermediary gene product (McKnight, 1979; Ray, 1979). Yamamoto and Alberts proposed a general test of whether or not a hormone exerts a direct action on a gene (Yamamoto and Alberts, 1976). In this test protein synthesis is inhibited before the hormone is added. If the effect persists protein synthesis is obviously not required and the action is presumably exerted directly on the gene in question. In H4IIE cells the 10-fold de-

TABLE IV
Effect of Insulin on mRNA^{PEPCK} Transcription[a]

Insulin concentration	mRNA^{PEPCK} synthesis (ppm)
0	177 ± 12
10 pM	71 ± 3
1 nM	25 ± 14

[a] Cells were incubated with and without insulin for 60 minutes then the transcription assay was performed. The results indicate the average ± SD determined from three independent assays. Reprinted, with permission, from Sasaki et al. (1984).

TABLE V
Cycloxheximide Does Not Influence the Effects of 8-CPT-cAMP and Insulin
on P-Enolpyruvate Carboxykinase Gene Transcription[a]

Treatment	mRNA^{PEPCK} synthesis (ppm)	Total RNA synthesis (cpm × 10^{-6})
None	186 ± 30	1.44 ± 0.16
+ Cycloheximide	195 ± 22	1.55 ± 0.14
8-CPT-cAMP	1597 ± 88	1.14 ± 0.15
+ cycloheximide	1196 ± 112	1.63 ± 0.18
8-CPT-cAMP + insulin	447 ± 28	1.39 ± 0.26
+ cycloheximide	383 ± 36	1.34 ± 0.17

[a] Confluent H4IIE cells were induced with 0.1 mM 8-CPT-cAMP for 30 minutes, then the incubation was continued for an additional 60 minutes in the presence or absence of 10 μM cycloheximide. Insulin (5 nM) was added 30 minutes before the cells were harvested. Results are expressed as the mean ± SD of triplicate assays. Reprinted, with permission, from Sasaki et al. (1984).

crease of mRNA^{PEPCK} caused by insulin is not affected by cycloheximide, emetine, or puromycin at concentrations (10 μM, 10 μM, and 1 mM, respectively) that inhibit protein synthesis by greater than 95%, and that have no effect on total RNA concentration (Granner and Andreone, 1985).

The specific inhibition of PEPCK gene transcription by insulin also does not require protein synthesis. Cycloheximide has no effect on the rate of transcription of this gene in either untreated or insulin plus 8-CPT-cAMP treated H4IIE cells, nor does it affect total RNA transcription under these conditions (Table V). These observations indicate that the regulation of PEPCK gene transcription by cAMP and insulin does not require the synthesis of a protein with a rapid turnover time.

G. SUMMARY OF THE ACTION OF INSULIN ON mRNA^{PEPCK} IN H4IIE CELLS

The data shown in Figs. 5, 7, and 9, coupled with previous observations of insulin repression of mRNA^{PEPCK} (Andreone et al., 1982; Granner et al., 1983; Sasaki et al., 1984; Cimbala et al., 1982), indicate that the primary action of insulin is exerted at the level of transcription. To test this hypothesis we plotted changes of the transcription rate, the amount of the putative primary transcript, the amount of the mature mRNA^{PEPCK} in the nucleus, and the amount of cytoplasmic mRNA^{PEPCK} as a function of time after insulin addition (see Fig. 10). PEPCK gene transcription decreases first, followed shortly thereafter by a decrease of the primary transcript. The decrease of the mature nuclear mRNA^{PEPCK} is further

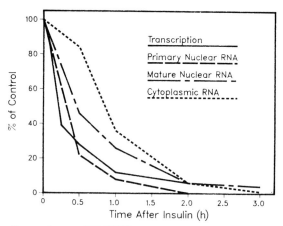

FIG. 10. Insulin regulates mRNA^{PEPCK} amount by inhibiting transcription. The data shown in Fig. 7 and 9 were plotted together in this representation. The data are expressed as the percentage of the maximum rate of transcription or RNA sequence abundance. Reprinted, with permission, from Sasaki et al. (1984).

delayed, as is the decline of the cytoplasmic mRNA^{PEPCK}. The rate and the magnitude of these changes are consistent with the hypothesis that the primary effect of insulin is the inhibition of transcription of the PEPCK gene.

H. MULTIHORMONAL REGULATION OF PEPCK GENE TRANSCRIPTION

The Dominant Role of Insulin

A major goal in our laboratory is to determine how several hormones act in concert to regulate the synthesis of a specific gene product. As cited in the Introduction, glucagon (cAMP) is a more potent stimulator of gluconeogenesis than are glucocorticoids, and insulin is a potent inhibitor of gluconeogenesis. It was known that these hormones caused similar changes in the rate of synthesis of PEPCK (Gunn et al., 1976) and mRNA^{PEPCK} (Chrapkiewicz et al., 1982), but the effect combinations of these hormones have on PEPCK gene transcription had not been tested. The results of such an experiment are illustrated in Table VI. Dexamethasone, a nonmetabolizable glucocorticoid, results in a 6-fold increase of the rate of transcription of this gene and 8-CPT-cAMP results in a 13-fold increase, in keeping with the greater ability of glucagon (cAMP) to promote gluconeogenesis. The combination of dexamethasone and 8-CPT-

TABLE VI
Insulin Exerts the Dominant Effect in Cells Treated with 8-CPT-cAMP, Dexamethasone, and Insulin[a]

Treatment	mRNA^{PEPCK} synthesis (ppm)	
	No Insulin	Insulin
Control	108 ± 3	70 ± 6
Dexamethasone	606 ± 3	58 ± 23
8-CPT-cAMP	1355 ± 43	365 ± 6
Dexamethasone + 8-CPT-cAMP	1619 ± 131	520 ± 10

[a] H4IIE cells were treated for 30 minutes with 0.1 mM 8-CPT-cAMP, 0.5 μM dexamethasone, or the combination of these two, in the absence or presence of 5 nM insulin. Results represent the mean ± SD of triplicate assays. Reprinted, with permission, from Sasaki *et al.* (1984).

cAMP results in an additive response, which indicates that these effectors act by independent mechanisms.

The other half of the cells in this experiment received identical treatment except that 5 nM insulin was also added to the culture medium. Insulin abolishes transcription in basal and dexamethasone-treated cells (the amount of isotope incorporated in the presence of insulin was similar to that noted when α-amanitin was added) and it significantly inhibits transcription in cells treated with 8-CPT-cAMP, or the combination of dexamethasone plus 8-CPT-cAMP. The maximal effect of insulin in the presence of 8-CPT-cAMP was not achieved since a 30 minute incubation was employed in this experiment (see Fig. 9). It should be reemphasized that the amounts of dexamethasone and 8-CPT-cAMP used in the experiment shown in Table VI afford maximal induction by themselves and neither of these compounds is metabolized under these experimental conditions. Thus, in this combination of hormones, insulin exerts the dominant effect at the level of PEPCK gene transcription.

I. MECHANISM OF ACTION OF INSULIN ON PEPCK GENE TRANSCRIPTION

The processes involved in the transcription of class II genes (mRNA encoding) in higher eukaryotic cells are poorly understood. The general steps involve initiation, elongation, and termination of the transcript. Several factors are probably involved in each of these processes, thus there are several possible sites at which a hormone could exert a regulatory

TABLE VII
Insulin Does Not Attenuate PEPCK Gene Transcription[a]

Treatment	(E-E)	(E-K)	(S-S)	(S-C)	(C-E)
Control	133 ± 18	119 ± 33	215 ± 4	112 ± 17	57 ± 37
8-CPT-cAMP	1164 ± 67	1420 ± 125	1385 ± 143	1108 ± 184	1097 ± 68
8-CPT-cAMP +insulin	184 ± 17	121 ± 23	185 ± 46	103 ± 12	138 ± 25

[a] H4IIE cells were incubated with 0.1 mM 8-CPT-cAMP for 30 minutes, then 5 nM insulin was added to the culture medium. Cells were harvested 30 minutes later and transcription was quantitated utilizing the genomic DNA fragments shown in Fig. 11. Results are expressed in ppm corrected for variations of fragment length.

influence. Prokaryotic cells regulate transcription by an additional process called attenuation in which the nascent transcript is blocked at a certain point on the gene, thus impeding elongation of the transcripts. A precise analysis of all these mechanisms requires a cell-free system in which transcription initiates at the proper site, proceeds at the *in vivo* rate, and is regulated by hormones. Such a cell-free system does not exist for the PEPCK gene but the nuclear run-off assay can be used to analyze whether insulin causes attenuation or retards transcript elongation.

1. Insulin Does Not Attenuate Transcription

We conducted the experiment illustrated in Table VII and Fig. 11 to address the question of whether transcription is blocked at a certain region of the PEPCK gene in insulin-treated H4IIE cells. The standard nuclear "run-off" assay was performed except that, in addition to the full size .R3 DNA probe, genomic probes representing four regions of the PEPCK gene (Fig. 11) were bound to nitrocellulose filters and used to

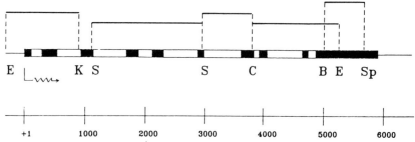

FIG. 11. Structure of PEPCK gene. Solid and open areas represent exon and intron regions, respectively. Solid lines show DNA fragments used for the run-off transcription assays shown in Table VII and Fig. 12. Restriction enzyme sites: E, *Eco*RI; K, *Kpn*I; S, *Sma*I; C, *Cla*I; B, *Bgl*II; SP, *Sph*I.

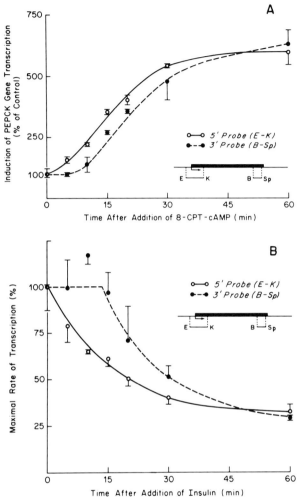

FIG. 12. Effects of 8-CPT-cAMP and insulin on PEPCK gene transcript elongation. Cells were incubated for 3 hours at 32°C (A) 8-CPT-cAMP was added to the medium and cells were harvested at the indicated times. (B) Insulin was added to the medium after 1 hour of pretreatment with 8-CPT-cAMP and cells were harvested at the indicated times. Transcription of 5' and 3' regions were measured with fragments (E-K) and (B-Sp), respectively. Refer to Fig. 11.

select labeled PEPCK RNA isolated from nuclei from control, 8-CPT-cAMP, and 8-CPT-cAMP plus insulin-treated cells. The radioactivity incorporated into PEPCK RNA measured using a 5' region probe (E-K or S-S) should exceed that detected by 3' region probes (S-C or C-E) if insulin causes transcript attenuation. This is not the case as transcription is uniformly decreased across the entire gene in insulin-treated cells (see Table VII).

2. *Insulin Retards Transcription Elongation*

A similar approach was used to assess the effect insulin treatment has on transcript elongation. Probes representing 5' (E-K) and 3' (B-Sp) regions of the PEPCK gene were employed (Fig. 11) and nuclei were pulsed with [^{32}P]UTP for short durations. As shown in Fig. 12A, incorporation at the 5' end of the gene increases immediately after the addition of 8-CPT-cAMP and a similar change is seen at the 3' end about 5 minutes later. In H4IIE cells exposed to insulin, incorporation of the isotope at the 5' end decreased immediately but the affect was not detected at the 3' end until more than 10 minutes had elapsed (see Fig. 12B). Insulin therefore retards PEPCK RNA transcript elongation, in comparison to 8-CPT-cAMP at least, but this effect cannot account for the dramatic action the hormone has on the transcription of this gene. We conclude that insulin must be inhibiting the initiation of PEPCK RNA transcription.

III. Discussion

A major focus of our recent research has been the mechanism of action of insulin on gene transcription. The inhibitory effect of insulin on PEPCK gene transcription is (1) achieved at physiologic concentrations of the hormone, (2) mediated through the insulin receptor, (3) specific, (4) seen in the absence of on-going protein synthesis, (5) readily reversible, (6) dominant over the action of inducing hormones, and (7) probably exerted at the step of transcript initiation.

Studies of the regulation of mRNA^{PEPCK} synthesis provided the first evidence of an effect of insulin on gene transcription and would be of interest even if this were the only example of this action of the hormone. In recent years several additional examples of regulation of specific mRNA metabolism by insulin have been reported and it appears that the regulation of PEPCK synthesis represents a prototype of a general regulatory role of insulin. The mRNAs known to be regulated by insulin, many at the level of transcriptional control, are listed in Table VIII, and this list is expanding at a rapid rate (Granner and Andreone, 1985). Indeed this list now has more entries than one which shows proteins whose activity is

TABLE VIII
Messenger RNAs Regulated by Insulin

Intracellular enzymes
Phosphoenolpyruvate carboxykinase[a]
Fatty acid synthetase
Pyruvate kinase[a]
Glucokinase
Glyceraldehyde phosphate dehydrogenase[a]
Glycerol phosphate dehydrogenase[a]
Glucose-6-phosphate dehydrogenase
Secreted proteins/enzymes
Albumin[a]
Amylase
α_{2u}-Globulin
Growth hormone
Proteins involved in reproduction
Ovalbumin[a]
Casein[a]
Structural proteins
δ-Crystallin

[a] Insulin regulates the rate of transcription.

modulated by insulin-mediated changes in phosphorylation or dephosphorylation.

Several of these mRNAs direct the synthesis of enzymes which have a well-established metabolic connection to insulin (i.e., PEPCK, glucokinase, fatty acid synthetase, pyruvate kinase, glycerol phosphate dehydrogenase, glyceraldehyde phosphate dehydrogenase, glucose-6-phosphate dehydrogenase) while others represent major secretory proteins (albumin, amylase, growth hormone, α_{2u}-globulin) in which the metabolic role of insulin is obscure. The production of ovalbumin and casein is involved in reproductive function in birds and mammals, and δ-crystallin is a structural protein. Insulin therefore regulates mRNAs that represent a spectrum of different actions in a number of tissues including liver, pancreas, oviduct, mammary gland, and lens.

The diversity of the functions of the proteins encoded by insulin-regulated mRNAs, and the variety of tissues affected, suggests that this is a major mechanism by which insulin regulates the amount of protein. If so, other examples of effects of insulin on gene transcription should be forthcoming. It will be interesting to see whether transcriptional control by insulin can be implicated in control of "total" protein synthesis in liver, muscle, and adipose tissue, and on differentiation and organogenesis.

Exactly how a hormone regulates the transcription of a specific gene is a question that perplexes all investigators involved in this aspect of hormone action. Unfortunately, the general mechanism of class II gene (mRNA) transcription in mammalian cells is poorly understood, but this is an extremely active area of investigation and basic principles should be elucidated in the near future. There are probably many factors that control the rate and fidelity of transcript initiation, just as there are probably several factors involved in elongation, termination, and release of the mRNA molecules. The rapidity of the effect, and the fact that protein synthesis is not required, suggests that this action of insulin, like so many others, may involve the covalent modification of a protein that somehow selectively inhibits the transcription of this gene. The fact that insulin inhibits the basal rate of transcription from the PEPCK gene, as well as that stimulated by two effectors with different mechanisms of action

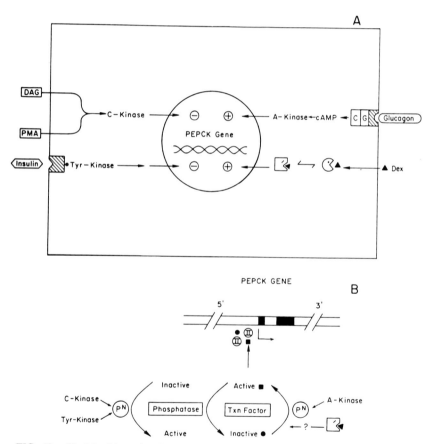

FIG. 13. Model of how insulin, cyclic AMP, glucocorticoids, PMA, and DAG may regulate PEPCK gene transcription.

(cAMP and glucocorticoids), suggests an effect at or near the site of RNA polymerase II initiation; some evidence for this exists (Table VII and Fig. 12). The action of insulin must not be exerted on RNA polymerase II itself, nor on any other general transcription factor, since the effect is very specific and total gene transcription is not affected.

A model which incorporates these ideas is illustrated in Fig. 13. The effectors discussed in this article, and their presumed mediators, are shown in Fig. 13A. This figure includes a representation of phorbol myristate acetate (PMA) and diacylglycerol (DAG) which, like insulin, inhibit PEPCK gene transcription (Chu and Granner, unpublished observations). PMA and DAG activate protein kinase C, insulin may act through the tyrosine kinase that is an integral part of the insulin receptor, cAMP presumably acts through protein kinase A, and glucocorticoids act by combining with specific intracellular receptors. The combined action of these four unique classes of effectors is illustrated, as a working hypothesis, in Fig. 13B. We postulate that a protein(s) involved in PEPCK gene transcription exists in multiple phospho and dephospho forms and that its activity is related to the extent of phosphorylation. The phosphoprotein formed by the action of cAMP would facilitate the action of polymerase II, perhaps with an assist from glucocorticoids. The dephospho protein would be formed by the action of a phosphatase, which in turn is activated by insulin (through tyrosine kinase) and phorbol esters (through C-kinase). The model accounts for the individual action of each effector and can explain how insulin (or phorbol esters) exerts a dominant effect on transcription of the PEPCK gene.

IV. Conclusion

Considerable progress has been made toward understanding how insulin regulates the rate at which a specific protein is synthesized in a target tissue. There are now many examples in which insulin affects mRNA metabolism, and in several of these instances this hormone influences the rate of transcription of specific genes. Based on the rate at which additional examples are appearing it is obvious that this is an important manifestation of insulin action.

ACKNOWLEDGMENTS

We would like to recognize colleagues who played an integral role in the research described in this article: Robert Bar, Cathy Caldwell, Nancy Chrapkiewicz, Timothy Cripe, Mark Granner, James Hartley, Stephen Koch, Michael Peacock, and Dan Peterson. This research was supported by NIH Grant AM 35107, by the Iowa (AM 25295) and Vanderbilt

(AM 20593) Diabetes Research Centers, and by the Veterans Administration. We thank Patsy Raymer and Maylene Long for help in preparing this manuscript.

REFERENCES

Andreone, T. L., Beale, E., Bar, R., and Granner, D. K. (1982). *J. Biol. Chem.* **257**, 35.
Avruch, J., Alexander, M. C., Palmer, J. L., Pierce, M. W., Nemenoff, R. A., Blackshear, P. J., Tipper, J. P., and Witters, L. A. (1982). *Fed. Proc., Fed. Am. Soc. Exp. Biol.* **41**, 2629.
Ballard, F. J., and Hanson, R. W. (1969). *J. Biol. Chem.* **245**, 5625.
Barnett, C. A., and Wicks, W. D. (1971). *J. Biol. Chem.* **246**, 7201.
Beale, E. G., Hartley, J. L., and Granner, D. K. (1982). *J. Biol. Chem.* **257**, 2022.
Beale, E. G., Chrapkiewicz, N. B., Scoble, H., Metz, R. J., Noble, R. L., Quick, D. P., Donelson, J. E., Biemann, K., and Granner, D. K. (1985). *J. Biol. Chem.*, in press.
Chrapkiewicz, N., Beale, E., and Granner, D. (1982). *J. Biol. Chem.* **257**, 14428.
Cimbala, M. A., Van Lelyveld, P., and Hanson, R. W. (1980). *Adv. Enzyme Regul.* **19**, 205–214.
Cimbala, M. A., Lamers, W. H., Nelson, K., Monahan, J. E., Yoo-Warren, H., and Hanson, R. W. (1982). *J. Biol. Chem.* **257**, 7629.
Crofford, O. B., and Reynold, A. E. (1965). *J. Biol. Chem.* **240**, 3237.
Crofford, O. B., Minemura, T., and Kono, T. (1970). *Adv. Enzyme Regul.* **8**, 219–238.
Freychet, P., Roth, J., and Neville, Jr., D. M. (1971). *Proc. Natl. Acad. Sci. U.S.A.* **68**, 1833.
Ginsberg, B. H. (1985). *Biol. Regul. Dev.* **3B**, 59–67.
Granner, D. K., and Andreone, T. L. (1985). "Diabetes Metabolism Reviews," Vol. 1. Wiley, New York.
Granner, D. K., Andreone, T., Sasaki, K., and Beale, E. (1983). *Nature (London)* **305**, 5934.
Gunn, J. M., Hanson, R. W., Meyuhas, O., Reshef, L., and Ballard, F. J. (1975). *Biochem. J.* **150**, 195.
Gunn, G. M., Ballard, F. J., and Hanson, R. W. (1976). *J. Biol. Chem.* **251**, 3586.
Hanson, R. W., and Garber, A. J. (1972). *Am. J. Clin. Nutr.* **25**, 1010.
Hanson, R. W., and Garber, A. J. (1973). *Am. J. Clin. Nutr.* **26**, 55.
Hanson, R. W., Fisher, L., Ballard, F. J., and Reshef, L. (1975). *Enzyme* **15**, 97.
Hoffman, C., Marsh, J. N., Miller, B., and Steiner, D. F. (1980). *Diabetes* **29**, 865.
Iwamoto, Y., Wong, K. Y., and Goldine, I. D. (1981). *Endocrinology* **108**, 44.
Iynedjian, P. B., Ballard, F. J., and Hanson, R. W. (1975). *J. Biol. Chem.* **250**, 5596.
Jefferson, L. S. (1980). *Diabetes* **29**, 487.
Jefferson, L. S., Rannels, D. E., Munger, B. L., and Morgan, H. E. (1974). *Fed. Proc., Fed. Am. Soc. Exp. Biol.* **33**, 1098.
Jefferson, L. S., Koehler, J. O., and Morgan, H. E. (1972). *Proc. Natl. Acad. Sci. U.S.A.* **69**, 816.
Jefferson, L. S., Li, J. B., and Rannels, S. R. (1977). *J. Biol. Chem.* **252**, 1476.
Kamm, D. E., Strope, G. L., and Kuchmy, B. L. (1974). *Metabolism* **23**, 1073.
King, G. L., Kahn, C. R., Rechler, M. M., and Nissley, S. P. (1980). *J. Clin. Invest.* **66**, 130.
Kioussis, D. M., Reshef, L., Cohen, H., Tilghmanm, S. M., Iynedjian, P. B., Ballard, F. J., and Hanson, R. W. (1978). *J. Biol. Chem.* **253**, 4327.
Kosmakos, F. C., and Roth, J. (1980). *J. Biol. Chem.* **255**, 9860.
Lamers, W. H., Hanson, R. W., and Meisner, H. M. (1982). *Proc. Natl. Acad. Sci. U.S.A.* **79**, 5137.

Longshaw, I. D., and Pogson, C. I. (1972). *J. Clin. Invest.* **51**, 2277.
McKnight, G. S. (1979). *Cell* **14**, 403.
McKnight, G. S., and Palmiter, R. (1979). *J. Biol. Chem.* **254**, 9040.
Megyesi, K., Kahn, C. R., Roth, J., Froesch, E. R., Humbel, R. E., Zapf, J., and Neville, Jr. D. M. (1974). *Biochem. Biophys. Res. Commun.* **57**, 307.
Meyuhas, O., Reshef, L., Gunn, J. M., Hanson, R. W., and Ballard, F. J. (1976). *Biochem. J.* **158**, 1.
Morgan, H. E., Earl, D. C. N., Broadus, A., Wolpert, E. B., Giger, K. E., and Jefferson, L. S. (1971). *J. Biol. Chem.* **246**, 2152.
Nelson, K., Cimbala, M., and Hanson, R. (1980). *J. Biol. Chem.* **255**, 8509.
Nordlie, R. C., Varricchio, F. E., and Holten, D. D. (1965). *Biochim. Biophys. Acta* **97**, 214.
Pain, V. M., and Garlick, P. J. (1974). *J. Biol. Chem.* **249**, 4510.
Pilkis, S. J., and Park, C. R. (1974). *Annu. Rev. Pharmacol.* **14**, 368.
Pitot, H. C., Peraino, C., Morse, P. A., and Potter, V. A. (1964). *Natl. Cancer Inst. Monogr.* **13**, 229.
Purrello, F., Vigneri, R. M., Clawson, G. A., and Goldfine, I. D. (1982). *Science* **216**, 1005.
Ray, A. K. (1979). "Biochemical Actions of Hormones," Vol. 6, pp. 482–517. Academic Press, New York.
Reshef, L., and Hanson, R. W. (1972). *Biochem. J.* **127**, 809.
Reshef, L., and Shapiro, B. (1970). "Adipose Tissue," p. 136. Academic Press, New York.
Reuber, M. D. (1961). *J. Natl. Cancer Inst.* **26**, 891.
Sasaki, K., Cripe, T. R., Koch, S. R., Andreone, T. L., Petersen, D. D., Beale, E. G., and Granner, D. K. (1984). *J. Biol. Chem.* **259**, 15242.
Schumm, D. E., and Webb, T. E. (1981). *Arch. Biochem. Biophys.* **210**, 275.
Shrago, E., Lardy, H. A., Nordlie, R. C., and Foster, D. O. (1963). *J. Biol. Chem.* **238**, 3188.
Shrago, E., Young, J. W., and Lardy, H. A. (1967). *Science* **158**, 1572.
Tilghman, S. M., Gunn, J. M., Fisher, L. M., Hanson, R. W., Reshef, L., and Ballard, F. J. (1975). *J. Biol. Chem.* **250**, 3322.
Utter, M. F. (1969). "Citric Acid Cycle" (J. M. Lowenstein, ed.), pp. 249–296. Academic Press, New York.
Van Wyk, J. J., and Underwood, L. E. (1975). *Annu. Rev. Med.* **26**, 427.
Yamamoto, K. R., and Alberts, B. M. (1976). *Annu. Rev. Biochem.* **45**, 721.

DISCUSSION

J. H. Clark. That was an excellent talk. You mentioned that maximal response could be obtained by occupancy of 1% of the insulin receptors. This seems like a very small number. Can you reassure us that the number of receptors measured *in vitro* is quantitatively similar to those present *in vivo–in situ*? Is it possible that the apparent excess receptors are due to a preparation artifact and that the physiological level is actually much lower?

D. K. Granner. Jim, one of your problems is that you live in Texas so you are inclined to think big. You have to learn to think small also. Indeed, this response occurs when even less than 1% of the receptors are occupied but I always hesitate to show the slide because it confuses people who do not work with peptide hormones. It is very hard to do these studies under conditions that exactly mimic those used to obtain physiological responses, because insulin receptor assays are done at a different temperature than the biologic assay. That is because there are degradation and on-off rate problems at 37°C that are minimized at the lower temperature used to measure receptor binding. I don't pretend to be a receptorologist. I have always told people I had but three goals in my life. One was to never measure a receptor, one was to never own a station wagon, and one was never to carry a pocket beeper. A friend who studies peptide hormones has told me to "buy the station wagon." In any case, it takes a very small amount of insulin to shut off the PEPCK gene and if you just go through the calculations using the assays as performed one must conclude that less than 1% of the receptors need be occupied, and the rest are "spare."

J. H. Clark. Would you describe the conditions under which the receptor measurements were made?

D. K. Granner. It's done under the same conditions that we do the biologic assays (PEPCK mRNA repression) except that the temperature is different.

J. H. Clark. The question really is: are those cells growing *in vitro* (regardless of whether they are tumor cells or not) manifesting receptors that would not be seen *in vivo*? Is there any way to settle the question of artifactual receptor numbers?

D. K. Granner. The spare receptor concept with respect to insulin was not just derived from studies such as PEPCK regulation in hepatoma cells, or from other experiments using cultured cells. It is also based on a number of responses of adipose, liver, and muscle tissue in intact animals. The number of receptors that must be occupied to elicit a given response varies from tissue to tissue, and can even vary within a tissue (adipose tissue) depending on what response you measure. But, in every instance that I know of with insulin, only a small fraction of the receptors has to be occupied to get a maximal biological response. That's just the way it is, and it is the same for most peptide hormones.

Y. A. Lefevre. You mentioned toward the beginning of your talk that you did not mean to imply that there are any direct actions of insulin at the nucleus, and certainly your model does not propose that as a possibility. Can you exclude with time courses or by any other means direct actions of insulin at the nucleus?

D. K. Granner. No. Receptor internalization happens within the time frame of what I have shown you, although it's stretching it a bit.

Y. A. Lefevre. With inhibitors of internalization?

D. K. Granner. We have not done that experiment, but others have studied the process of internalization of the insulin–receptor complex and this occurs in a few minutes. I suspect that this action of insulin I described is actually exerted very rapidly, perhaps within 1 minute, and that the limitation on showing is that our assay just isn't sensitive enough. If we extrapolate the data it looks as though PEPCK gene transcription is shut off almost instantaneously, which would argue for a catalytic process and against a process which requires the synthesis of something. We cannot exclude insulin action per se inside the cell.

S.-Y. Ying. I just wonder if you can tell us more about this transcription factor, how big is it, and does it bind to any part of the PEPCK?

D. K. Granner. All I can tell you is that it is square.

M. Melner. In relation to the problems you mentioned in establishing precursor–product relationships with your RNA, had you considered using regions of genomic DNA to probe your northern blots which code exclusively for the introns?

D. K. Granner. That's an interesting question. Dick Hanson and co-workers have recently reported that these multiple forms are processing intermediates. We have not done detailed studies, but in a couple of early experiments some of these bands disappeared when we used different probes.

L. Murphy. In your characterization of the genomic PEPCK gene, have you been able to identify any 5′ sequences that are absolutely necessary for the action of insulin, or any of the other hormones that affect the gene.

D. K. Granner. No, but we are in the process of doing those studies. Again, Hanson and co-workers have identified a region, very near the transcription initiation site, that confers partial response of the PEPCK gene to cyclic AMP, but not to glucocorticoids or insulin. These studies represent the next step.

P. A. Kelly. You indicated that phorbol esters could completely duplicate all of the actions of insulin on PEPCK. Is this true for the other enzyme systems that you discussed in the slide at the end of your presentation? In other words, do you think that tyrosine kinase is the mechanism by which insulin acts for all of those systems?

D. K. Granner. I do not think there is any compelling evidence to suggest that the tyrosine kinase activity that is an integral part of the insulin receptor, and that is activated when insulin binds to the receptor, has anything to do with insulin action. It is the most recent suggestion for mediating insulin action so people are interested in exploring this possibility. This potential mechanism fits the phosphorylation–dephosphorylation concept that has been part of insulin action for 15 or 20 years. There is interesting literature about the interaction of C kinase activated by phorbol esters, and the phosphorylation of other receptors. How this all correlates is still not clear. At this point we have the observation that insulin and phorbol esters result in the same biologic response, presumably by different mechanisms, but they are converging. Our problem is to find out where this occurs.

I. Callard. Do you have any indication that insulin influences either the amount of glucocorticoid receptor or its association with the nucleus?

D. K. Granner. No.

I. Callard. We have some evidence from another multihormonal system, the regulation of vitellogenin in reptiles, that growth hormone is necessary for the maintainance of the receptor for estrogen, and that estrogen doesn't exert its full effect in the absence of growth hormone. This is certainly a longer time course experiment than you're discussing, but I wonder whether insulin might be important for glucocorticoid agents in your system?

D. K. Granner. One would have to imagine a situation wherein the receptor was rendered unable to bind the steroid because the response occurs too rapidly to be accounted for by a change of the concentration of the receptor. Or you could imagine a possibility in which the accessibility of the receptor to the gene was decreased. We have not done anything related to your question.

J. E. Rall. If I understand correctly, you could have a stronger statement about the effects of cyclic AMP on the rate of initiation. Using the 5' and 3' probe you showed about a 15 minute delay in the decrease after insulin. I don't know how many nucleotides there are between these probes, but I am assuming about 10 kb between them.

D. K. Granner. The PEPCK RNA transcript is about 6.8 kb, which is still a large number.

J. E. Rall. Other data show that from 12 to 15 nucleotides form per second with polymerase II in systems like this, so that would just about give you exactly what you want, i.e., in about 15 minutes you would synthesize about 10 kb before insulin inhibited initiation.

D. K. Granner. We have done the same calculations. The one missing part is that we cannot do this experiment in cells that haven't seen any hormone because we can't get enough cpm incorporated into PEPCK RNA. Therefore, we don't know whether the insulin result reflects the normal situation and cyclic AMP enchances the transit time or whether it is the other way around. All we can say is that in comparing the two there is a difference, but it isn't sufficient to account for the insulin effect.

J. E. Rall. You had the different messages which are regulated by insulin but doesn't fatty acid synthase go up with insulin?

D. K. Granner. Yes.

J. E. Rall. So what you had, your little arrow with insulin in some instances it has to make squares and in some instances it has to make circles; so you are going to have a little problem with that.

D. K. Granner. In that case we turn the model over. I did not get into that, but any model will obviously have to account for the fact that insulin stimulates and inhibits transcription. In fact, in every other instance, with the possible exception of growth hormone mRNA in pituitary cells, the actions of insulin on mRNA are stimulatory. At this point we are just working with an inhibition effect. Later, we will have to worry about how to make the squares into circles or vice versa.

H. M. Kronenberg. A few related questions about the beautiful biological models you are using. I wonder whether there are any substrate effects on gluconeogenesis in these cells and whether they are reflected at all in regulation of transcription.

D. K. Granner. Like glucose? This is an excellent question because it again illustrates the problem of using intact animals. There is a suggestion in the literature that insulin affects PEPCK synthesis by promoting increased glucose metabolism, and that this somehow is the inhibitor. We addressed this using the H4IIE cell system and indeed we thought we had an effect because as we added increasing concentrations of glucose to the cells we decreased PEPCK mRNA. But it took about 400 mg/dl to do this so I thought there might be an osmotic effect. We did the same experiment with mannitol and it mimicked the "glucose" effect. This may be an interesting side project, i.e., the osmotic effect on gene transcription. No we haven't been able to see an effect of glucose per se in the H4IIE cell system.

H. M. Kronenberg. I also wondered whether you have had the opportunity to look at the different hepatoma lines that you alluded to. Are any of them selective in responding to insulin but not to glucocorticoids or the cyclic AMP-mediated hormones but not to insulin and so forth? If so, use of those lines would be a lovely way to dissect out complicated interrelationships.

D. K. Granner. There are some mutant CHO cell lines that do not respond to cyclic AMP and we are testing these with transfected genes. Unfortunately there are no hepatoma cell lines that would allow us to do the experiments you propose.

H. M. Kronenberg. The other tissue that undergoes gluconeogenesis is the kidney. Do any of these hormones have a similar effect on renal gluconeogenesis?

D. K. Granner. Insulin and cyclic AMP do not have an effect on the kidney enzyme; glucocorticoids do.

T. Chen. Considering insulin can activate so many proteins and enzymes are you surprised to find that the total RNA was not changed?

D. K. Granner. No. Consider that there are 10,000 genes actively transcribed in the liver. If 100 genes are affected by a hormone and some are stimulated or inhibited, this will tend to cancel out in a total RNA assay. I think it would be more surprising to find a change of total RNA transcription. Most actions of hormones, glucocorticoids, cyclic AMP, and insulin included, are exerted on a very, very small number of genes, generally less than 1% of the total.

C. S. Nicoll. In view of the rapidity of the response to insulin, have you considered the possibility that changes in intracellular Ca^{2+} concentration might be involved?

D. K. Granner. Yes, that is how we ended up examining the effects of phorbol esters. We looked for changes in intracellular calcium using the Quin 2 assay, and couldn't see that any of these hormones were changing intracellular calcium, nor could we effect any changes using calcium ionophores.

J. Gorski. I wonder whether you had any thoughts about the complexity of the response that you see in your system and how the box in your model functions. It would seem that if you examine the sequences in enough of these genes that we ought to find some kind of repetitive sequence present in blocks of genes that are turned on or turned off. I wondered whether you actually looked into that or have discussed it with others who may have done so?

D. K. Granner. We discuss this all the time. There is no other gene to look at with respect to insulin, since the PEPCK gene is the only one that has been extensively characterized and sequenced. The 5' regions of several others will have to be sequenced before we can pick out a unique sequence that confers insulin responsiveness, if such regions even

REGULATION OF METABOLIC PROCESSES BY INSULIN 141

exist. It is conceivable, and I think you have alluded to this very important point, that a model based by analogy to how the steroid hormone and its receptor sits on certain unique regions of DNA, and thus confers the steroid response, may not apply to all effectors. It is possible that things happen in the vicinity of a gene but not right on it. We don't know what the transcription complex is, but it undoubtedly consists of several proteins. The one (or more) affected by insulin may not come close to the DNA itself, therefore I think we have to consider that there may not be a region of DNA that we can find using the techniques that have been so successful with steroid receptors. The same may be true with the cyclic AMP effect. Evidence is beginning to accumulate that the cyclic AMP responsive regions map very close to the promotor region so that when one deletes DNA in that region one also loses basal promoter activity. It therefore may be that there are no unique regions necessary for the cAMP response. That implies that the protein isn't binding directly to the DNA, rather it is interacting with another protein which is, such as polymerase or another general transcription factor. Do you follow that?

J. Gorski. I understand what you are saying but I think we often get tangled up with the same problem. I have reservations about steroid receptor binding to DNA, but the problem remains that in some way or other a very specific recognition is necessary to explain the fact that just a small number of genes are turned on. On the other hand it is obvious that far more than one gene is affected. There is a fair amount of evidence just as you have shown in your system that the hormone itself isn't directly interacting with the DNA. I think we need to sit back and come up with newer models or newer thoughts on this problem rather than trying to make this fit the Lac operon model.

D. H. Granner. Yes. What I have suggested is a little different from the steroid hormone action. The insulin effect is seen in cells that haven't been exposed to any other hormone, in cells exposed to either glucocorticoids or cyclic AMP which must have unique mechanisms of action and in cells exposed to both. It seems to me that this pinpoints the insulin effect to somewhere near the promoter. Steroid binding sites are found as far away as 2000 bases upstream, so one way to block their effect is to put some impedence near the general area of transcription initiation. This could be done by preventing polymerase II from binding or by preventing the interactions of some other proteins with the polymerase. We don't really have any way of knowing what is happening at this point, but it is likely to be different from what is known about steroid action. I, too, am more than a little skeptical that one model is going to answer all these questions.

A. Means. The DNA sensitivity experiments, however, would suggest that like several steroid hormone-inducible systems there may be an accessible region of flanking DNA. By mapping this region you may be able to determine whether this inhibitory case exposes huge regions of open DNA that are affected by DNase as is the case in the steroid-induced genes or whether this is a much more restricted case. The hypersensitive site you mentioned could constitute a large portion of a small DNA domain affected by insulin.

D. K. Granner. Exactly, and we are very interested in pursuing that observation.

K. Lederis. Returning to the earlier discussion between Dr. Clark and yourself as to whether you are dealing with *in vivo* or an *in vitro* system, and accepting that semantically both views may be correct, it would be of some importance to know whether the phenomena that you describe, i.e., the stimulation or the inhibition of the PEPCK enzyme, has been determined in a cell or in a system *in vitro* or *in vivo* that might be considered more normal than the hepatoma cell?

D. K. Granner. Yes, in hepatocytes, if you will accept the isolated hepatocyte as being a more normal cell. We can see these effects in liver, but I didn't want to get into that since the animals must be diabetic and special manipulations are required. Under proper conditions we can see suppression of PEPCK mRNA by insulin in liver and stimulation by glucagon.

Molecular Characterization of Fibroblast Growth Factor: Distribution and Biological Activities in Various Tissues

ANDREW BAIRD, FREDERICK ESCH, PIERRE MORMÈDE, NAOTO UENO, NICHOLAS LING, PETER BÖHLEN, SHAO-YAO YING, WILLIAM B. WEHRENBERG, AND ROGER GUILLEMIN

Laboratories for Neuroendocrinology, The Salk Institute, San Diego, California

I. Introduction

From the late 1960s to the early 1970s several laboratories described the presence of mitogenic activity in pituitary extracts and in partially purified hormone preparations (Holley and Kiernan, 1968; Clark *et al.*, 1972; Corvol *et al.*, 1972; Armelin, 1973). It was only in 1974, however, that Gospodarowicz established the preliminary characteristics of a fibroblast growth factor (FGF) found in both pituitary and brain (Gospodarowicz, 1974). Within a short period of time, Gospodarowicz (1975) had described the biochemical characteristics of the molecule present in these acid extracts as being a cationic polypeptide (pI > 9.6) with a molecular weight of 14,000–16,000. Furthermore, he was also able to predict the presence of distinct activities characterized by neutral or acidic pI, suggesting the existence of structurally distinct classes of mitogens in the brain (Gospodarowicz *et al.*, 1975). Within a few years, however, it became obvious that the capacity of FGF to stimulate mitosis in fibroblasts was only one of several biological activities of the growth factor. Preparations of both pituitary and brain FGF were found to be potent stimulators of the proliferation of several cell types including chondrocytes, adrenocortical cells, vascular smooth muscle cells, and vascular endothelial cells (reviewed by Gospodarowicz and Moran, 1976; Gospodarowicz, 1979). Therefore, while the name FGF implied its activity on fibroblasts, it became inherently obvious that its activity could be expanded to include several cell types closely, but not exclusively, connected by their derivation from the primary and secondary mesenchyme.

While the characterization of FGF was hampered by the inability to obtain homogeneous preparations of the growth factor, the available preparations of FGF were studied for their *in vivo* and *in vitro* activities. The effects of pituitary and brain FGF on cultured cells and the mechanisms through which they stimulate cell proliferation were extensively studied (reviewed by Gospodarowicz, 1984). Their capacity to stabilize the phenotypic expression of cultured cells (Vlodavsky *et al.*, 1979a,b) and eventually delay their senescence *in vitro* (Gospodarowicz and Bialecki, 1978; Simmonian *et al.*, 1979; Simmonian and Gill, 1979) made it an invaluable tool for the maintenance of primary cell lines and the study of cell growth and differentiation. Yet throughout this time, and indeed to this day, the *in vivo* physiological function of FGF has remained unclear. Early studies by Gospodarowicz *et al.* (1979) were able to establish that FGF could induce neovascularization suggesting at least functional similarities with the tumour angiogenic factor described by Folkman (1975,1984) and the corpus luteum angiogenic factor described by Gospodarowicz and Thakral (1978). Further studies indicated that the same preparations of the growth factor could enhance wound healing and promote limb regeneration (reviewed by Gospodarowicz and Mescher 1980; Gospodarowicz, 1984). It remained unclear, however, how each of these activities might correlate with the ability of the "pituitary" and "brain" FGFs to stimulate the proliferation of adrenocortical and granulosa cells *in vitro*. Were these fortuitous activities or did they represent a physiological function for the pituitary- and brain-derived growth factor?

Aware of the importance of identifying the structure responsible for FGF activity, Westall *et al.* (1978) reported the identification of brain FGF with fragments of myelin basic protein. While this finding would soon be questioned (Thomas *et al.*, 1980; Lemmon *et al.*, 1982) and eventually disproved (Gospodarowicz *et al.*, 1984), it raised several questions as to the relationship of brain FGF to other growth factors and in particular to pituitary FGF. Several hypotheses were presented to suggest that the generation of FGF occurred as a physiological response to injury from the autolytic digestion of myelin. But these paradigms could not account for the identity of pituitary FGF and the FGF activities detected in other tissues.

During the course of the *in vitro* and *in vivo* studies with pituitary and brain preparations of FGF, several other laboratories, including our own (Jones *et al.*, 1980), were also examining the presence of mitogenic activities in various tissue extracts. Thomas *et al.* (1980,1984) characterized by amino acid analysis an acidic form of FGF activity that was biologically related to FGF and very similar both to the second form of FGF activity detected by Gospodarowicz in 1975 (Gospodarowicz et al., 1975) and to the endothelial cell growth factor described by Maciag et al. (1978). In brain (Kellet et al., 1981), as well as in other tissues such as the corpus luteum (Gospodarowicz and Thakral, 1978; Jakob et al., 1977), the ovary (Koos and Le Maire, 1983; Makris et al., 1983,1984), the kidney (Preminger et al., 1980; Ekblom, 1981), the adrenal (Kiss, 1975), the retina (Glaser et al., 1980; Barritault et al., 1981), macrophages (Leibovich and Ross, 1976), and tumors (Folkman, 1975), it had been possible to detect activities that were similar to those of FGF. Extracts of each of these tissues showed potent abilities to stimulate the proliferation of mesoderm-derived cells and the activities appeared to be linked with either acidic, basic, or neutral polypeptides that were very similar to FGF. Because FGF activity had been a priori reported to be contained mostly in brain and pituitary (Gospodarowicz, 1974), it was unclear if these other factors were structurally related to FGF. Moreover, the lack of a known chemical structure to FGF precluded a possible comparison with the corpus luteum angiogenic factor, ovarian growth factor, retina-derived growth factors, and others. It was with the goal of identifying the characteristics of these growth factors that we set out to isolate and characterize pituitary fibroblast growth factor.

II. Isolation of Fibroblast Growth Factor

A. BIOASSAY

Growth factor purification was monitored by testing chromatography fractions for their ability to stimulate the proliferation of adult bovine aortic arch endothelial (BAAE) cells or in some cases that of brain capillary endothelial cells or adrenocortical cells (Fig. 1). Stock cultures, maintained in the presence of Dulbecco's modified Eagle's medium H-16 (GIBCO) supplemented with 10% calf serum (HyClone, Logan, UT), and antibiotics [gentamycin (Schering) at 50 μg/ml and Fungizone (Squibb) at 0.25 μg/ml] and pituitary FGF were passaged weekly at a split ratio of 1:64. Cell monolayers from stock plates (at passage 3–10) were dissociated by exposure (2–3 minutes, 24°C) to a trypsin solution (GIBCO; 0.9% NaCl/0.01% EDTA) and were then plated at an initial density of 2 × 10^4 cells per 35-mm plastic culture dish (Falcon) containing 2 ml of Dulbecco's modified Eagle's medium H-16/10% calf serum and antibiotics. Diluted test samples (10 μl) were added to triplicate dishes every 48 hours and after 4 days of culture, plates were treated with trypsin and the detached cells were counted with a Coulter particle counter.

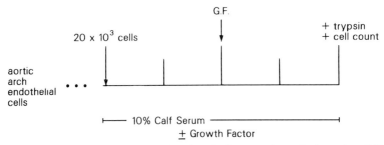

FIG. 1. Bioassay for mitogenic activity. Standard preparations of mitogenic activity or aliquots of column fractions were added in 10–20 µl to cells 6 to 8 hours after their attachment to 35-mm tissue culture dishes or miniwell plates. Forty-eight hours later the treatment was repeated, and on day 5 of the experiment the cells were harvested after trypsin digestion and counted with a Coulter particle counter. The same protocol was used for bovine aortic arch endothelial cells, brain or adrenal capillary endothelial cells, adrenocortical cells, granulosa cells, and chondrocytes.

B. CLASSICAL PROCEDURE

The original protocol devised for the isolation of FGF required a 2-fold approach since our initial studies had clearly indicated the presence of both acid-labile and acid-stable mitogenic activity in the pituitary extracts. Since this observation was not compatible with the reported acid lability of FGF (Gospodarowicz and Moran, 1976) we had naturally assumed that there were at least two distinct factors in the extract. Therefore, while routine reverse-phase HPLC was used to isolate the mitogen that remained active under denaturing conditions, a second protocol using ion exchange chromatography was devised to isolate the mitogen active only under nondenaturing conditions (Böhlen et al., 1984). The purification procedure using either method resulted in the isolation of a homogeneous peptide preparation that is shown in Fig. 2A and B. Amino acid analyses of these proteins clearly demonstrated that they were very similar and probably related. Amino-terminal sequence analyses of each of these peptides using a gas-phase sequenator identified the amino terminus as being Pro-Ala-Leu-Pro-Glu-Asp-Gly-Gly-Ser.

As was eventually shown (Böhlen et al., 1984), both procedures resulted in the isolation of an identical mitogen. In fact, the loss of biological activity of FGF under denaturing conditions was readily noticed when the potencies of each preparation were compared (Fig. 3). The ED_{50} for the unmodified growth factor (30 pg/ml) was indeed 20–25 times lower than that of its denatured counterpart (600 pg/ml), though both had full intrinsic activity. The results explained the loss of 95% of mitogenic activity upon acid treatment by the generation of the less potent form of fibroblast

MOLECULAR CHARACTERIZATION OF FGF 147

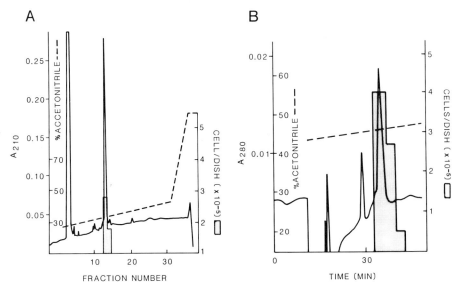

FIG. 2. (A) The pituitary growth factor isolated by nondenaturing conditions was concentrated for chemical analysis using reverse-phase HPLC on a C_3 column (7.5 × 0.46 cm, 5 μm particle size, 300 Å pore size, Altex, Berkeley, CA). Elution was with a 90-minute gradient from 23 to 33% (v/v) acetonitrile in 0.1% trifluoroacetic acid at room temperature at a flow rate of 0.6 ml/minute. (B) Final purification of the pituitary growth factor isolated by denaturing conditions using reverse-phase HPLC on a C_8 column (Aquapore RP-300, 25 × 0.46 cm, 10 μm particle size, 300 Å pore size, Brownlee, Santa, Clara, CA). Chromatography was performed at room temperature with a 90-minute gradient from 42 to 52% acetonitrile in 0.2% heptafluorobutyric acid at a flow rate of 1 ml/minute. After a run the column was washed with 99.8% acetonitrile/0.2% heptafluorobutyric acid. (Taken from Böhlen et al., 1984.)

growth factor. It is also important to note that, even denatured, FGF has a high intrinsic activity well in the 10^{-10} M range.

Although the specific activity of the proteins sequenced was quite high (ED_{50} = 30–300 pg), we were still concerned that the mitogenic activity might not be attributable to the protein identified. This was particularly important since the wrong identification of brain FGF with fragments of myelin basic protein (Westall et al., 1978; see Section I) had generated considerable confusion in the field. In an effort to avoid a similar error, we synthesized the synthetic peptide [Tyr^{10}]FGF(1–10) and conjugated it to BSA. Antibodies raised in rabbits with this conjugate (Baird et al., 1985a) were used to immunoneutralize the effect of FGF on vascular endothelial cells (Fig. 4), thereby demonstrating the identity of the sequenced protein with the mitogenic activity.

The availability of sequence-specific antibodies enabled us to conduct

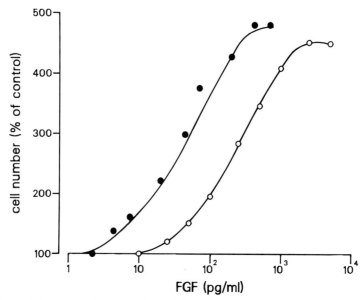

FIG. 3. Mitogenic activity of FGF on adult bovine aortic arch endothelial cells. (○) FGF purified by ion-exchange/reverse-phase HPLC (denatured); (●) FGF purified by gel filtration/ion-exchange chromatography (not denatured). Replicates were within 6% of mean values. The two preparations were tested in different assays; maximal cell densities were reached at 7.2×10^5 and 5×10^5 cells per dish for FGFs prepared by procedures I and II, respectively. (From Böhlen et al., 1984.)

tests to determine their suitability for use in a radioimmunoassay for FGF (Fig. 5). For this purpose the cross-reactivity of the antibodies to the synthetic peptide was tested with known amounts of FGF as determined by amino acid analysis. The results demonstrated that these antibodies exhibited a high specificity for a free amino terminal of FGF (inset of Fig. 5) and established a functional RIA for FGF that could be applied to its measurement, at least in early studies, in column fractions. The high specificity of the antibodies and their equimolar cross-reactivity with native FGF clearly meant that the turnaround time for the purification procedure could be reduced from 5 days to less than 24 hours. Moreover, the antibodies could be used for the large-scale isolation of FGF by means of immunoaffinity column chromatography. These methods had been extensively used by our laboratory to identify corticotropin- and growth hormone-releasing factors from many species (Esch et al., 1984; Ling et al., 1984, review 1985).

MOLECULAR CHARACTERIZATION OF FGF   149

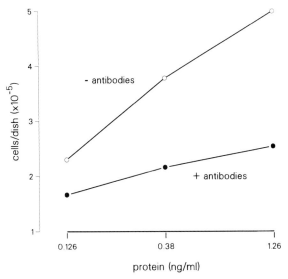

FIG. 4. Immunoneutralization of FGF proliferative activity on adult bovine aortic endothelial cells in culture. Cells were incubated in the presence (●) or absence (○) of 20μl of anti-FGF with various amounts of FGF. Ammonium sulfate-precipitated antibodies were used instead of antiserum to exclude the possibility of artifactual stimulation of cells by growth factors present in the rabbit serum. Antibodies and growth factor (in 10 μl) were added together on days 1 and 3. Cells were counted on day 4. Individual points are averages from duplicate determinations. Replicates were within 6% of mean values.

C. IMMUNOAFFINITY COLUMN CHROMATOGRAPHY

Pools of antibody generated against [Tyr^{10}]FGF(1–10) were processed by ammonium sulfate precipitation and the IgG linked to Affi-Gel as described by Ling et al. (1984). A column of anti-FGF was built and its capacity to retain FGF in tissue extracts was tested. The columns were capable of retaining 1–2 nmol of FGF in a single pass and were going to provide a way to bypass the long and tedious procedure used initially. The FGF bound to the column was eluted with acid (1 M HOAc) and after chromatography on Sephadex G-100 the growth factor was to be purified to homogeneity by one or two steps of reverse-phase HPLC. The methodology was ideal for the rapid and large-scale purification required for structural analysis by classical protein chemistry.

While the immunoaffinity chromatography was being validated and scaled up, a most interesting manuscript appeared in *Science* (Shing et al., 1984) describing the use of heparin-Sepharose affinity chromatogra-

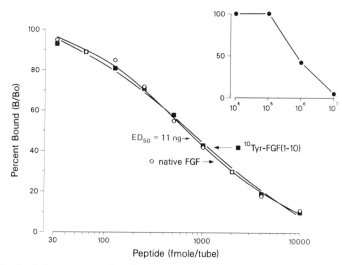

FIG. 5. Radioimmunoassay for FGF. Antisera were diluted to a final dilution of 1:10,000 and incubated with the radioiodinated hapten in the presence or absence of the cold peptide or native FGF. The intact mitogen was purified to homogeneity (as demonstrated by sequence analysis) and quantitated by amino acid analysis, assuming a molecular mass of 15,800 Da. Inset shows the cross-reactivity (0.01%) of N-acetyl-[Tyr^{10}]FGF(1–10) in the RIA. (From Baird et al., 1985a.)

phy to purify an angiogenic peptide from an extract of chrondrosarcoma. In view of the similarities between the activities of FGF and those of the chondrosarcoma, we investigated the possible application of this method to the purification of FGF (Gospodarowicz et al., 1984). During this time sequence analyses of the FGF isolated from the initial procedure (Section II,B) were continuing at a steady pace.

D. HEPARIN-SEPHAROSE AFFINITY CHROMATOGRAPHY

Bovine pituitaries were extracted in 0.15 M NH$_4$(SO$_4$)$_2$ and proteins in the supernatants were precipitated with ammonium sulfate (Gospodarowicz et al., 1984). Carboxymethyl Sephadex C-50 chromatography retained the biological activity in the extract, with a 60-fold purification factor. The active fractions were then pumped onto a bed of heparin-Sepharose and the mitotic activity was strongly retained but could be eluted with 1.5 M NaCl (Fig. 6A). The final purification factor of 350,000 obtained after heparin-Sepharose affinity chromatography resulted in a preparation of homogeneous mitogen, as assessed by reverse-phase HPLC (Fig. 6B). The biologically active fractions, which corresponded to

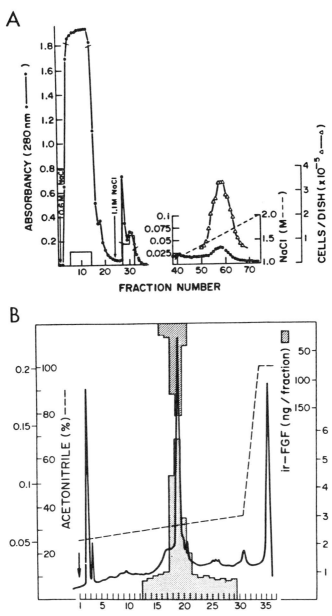

FIG. 6. Purification of pituitary FGF by heparin-Sepharose affinity chromatography. (A) The mitogenic activity that was retained by the carboxymethyl Sephadex column and eluted with 0.6 M NaCl was pumped directly onto a bed of heparin-Sepharose at a flow rate of 35 ml/hour. The column was washed with 0.6 M NaCl in 10 mM Tris buffer, pH 7, until the absorbency at 280 nm became negligible and then washed again with 1.1 M NaCl. A linear gradient of 1.1 to 2 M NaCl was then used to elute the major part of the mitogenic activity,

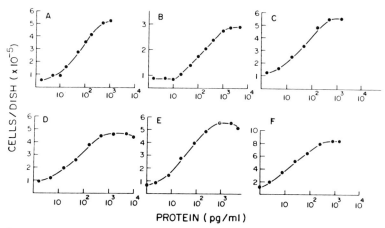

FIG. 7. Potency of pituitary fibroblast growth factor on various mesodermal cell types. The growth factor was added to (A) vascular smooth muscle cells, (B) human umbilical endothelial cells, (C) bovine brain capillary endothelial cells, (D) granulosa cells, (E) adrenocortical cells, or (F) chondrocytes in various doses and cell number determined 4 to 5 days later using a Coulter particle counter. (From Gospodarowicz et al., 1985b.)

the major peak of activity, also had full FGF immunoreactivity, thereby demonstrating the identity of the growth factor isolated by this methodology with FGF identified by classical methods.

Amino acid analysis of the protein showed distinct similarities with the FGF isolated by Böhlen et al. (1984). Gas-phase sequence analysis identified the amino-terminus Pro-Ala-Leu-Pro-Glu-Asp-Gly, which was identical to the amino-terminus established for pituitary FGF (Böhlen et al., 1985a). The protein yield was very high (6 nmol/kg), allowing for a rapid and reliable method of providing sufficient material for both chemical analysis and biological studies.

The final protocol, which does not necessarily have to include the dialysis step, permits the efficient purification of FGF in less than 24 hours. At the step after heparin-Sepharose affinity chromatography, the mitogen can be considered pure, as assayed by HPLC, amino acid analysis, etc. Furthermore, because denaturation steps have not been performed, the

---

which was further purified by reverse-phase HPLC. Note the change in absorbency units before the salt gradient. (B) Purity of the heparin-Sepharose retained FGF was established by reverse-phase HPLC using a 0.46 × 25 cm, 5 μm C_4 Vydac column and a linear gradient of 25–32% acetonitrile over a 90-minute period. Mitogenic activity was detected using bovine aortic arch endothelial cells and immunoreactive FGF measured by radioimmunoassay (Gospodarowicz et al., 1984).

growth factor has high specific activity (ED_{50} = 20–60 pg/ml) and is a potent stimulator of the proliferation of several distinct cell types (Fig. 7). Structural characterization was performed on the FGFs isolated from each of these procedures and has demonstrated that it is the same protein that is isolated by each method. By virtue of its efficiency, the application of heparin-Sepharose affinity chromatography has become routine for the isolation of large amounts of mitogen for biological studies.

III. Structural Characterization of Pituitary Fibroblast Growth Factor

The structural characterization of pituitary FGF was described by Esch *et al.* (1985) using growth factor isolated from each of the methodologies described earlier.

A. AMINO ACID ANALYSIS

Peptides were hydrolyzed in sealed evacuated ignition tubes containing 5 µl 6 N HCl and 7% thioglycollic acid for 24 hours at 110°C. Amino acid analyses of 5–30 pmol of peptide hydrolysate were performed with a Liquimat III amino acid analyzer (Kontron, Zurich, Switzerland) equipped with a proline conversion, fluorescence detection system (Böhlen and Mellet, 1979).

B. STRUCTURAL CHARACTERIZATION

Gas-phase microsequencing of intact FGF and its digestion fragments was carried out by methods previously described (Esch, 1984). >PhNCS-[^{14}C]carboxyamidomethylcysteine was identified during sequence analysis by liquid-scintillation counting of the residues from the sequencer. The identification of cysteic acid in a given cycle was accomplished by comparison of the amino acid composition of the peptide and the remainder of its sequence as determined by Edman degradation. Carboxypeptidase Y was obtained from Pierce Chemicals and utilized according to the manufacturer's recommendations. Prior to any proteolytic or chemical digestion the FGF (with or without modified cysteines) was dried in a polypropylene microfuge tube in a Speed Vac vacuum centrifuge to remove volatile HPLC solvents. The dried FGF (1–5 nmol) was dissolved in 0.01 ml 0.5 *M* Tris–HCl, pH 7.7, 10 m*M* EDTA, 6 *M* guanidine–HCl and then diluted to 1 ml with 1% NH_4HCO_3. Submaxillaris protease or chymotrypsin was added in a 1/50 (w/w) ratio while digestions with *Staphylococcus aureus* V8 employed a 1:35 (mol:mol) ratio of enzyme to substrate. Incubations were allowed to proceed overnight at 37°C and peptides were purified by HPLC. For cyanogen bromide digestions, the dried, alkylated FGF (5–6 nmol) was first dissolved in 0.05 ml 70% formic acid and treated with a solution of 2.9 M N-methylmercaptoacetamide in 7% formic acid for 24 hours at 37°C. The alkylated, reduced FGF was purified by HPLC, dried in a Speed Vac vacuum centrifuge, and redissolved in 0.1 ml deoxygenated 70% formic acid. A 100-fold excess of cyanogen bromide was added and the incubation continued overnight at room temperature in the dark. Purifications of modified FGF and its digestion fragments were accomplished with a Brownlee RP-300 reverse-phase column (0.46 × 25 cm) and a 0.1% TFA/acetonitrile or a 0.1% heptafluorobutyric acid (HFBA)/acetonitrile solvent system (Esch *et al.*, 1983).

Cysteine residues were either reduced and alkylated with [^{14}C]iodoacetamide (New England Nuclear) or oxidized with performic acid as indicated below. In either case the FGF in 0.1% TFA/acetonitrile was dried in a 1.5 ml polypropylene microfuge tube in a Speed Vac vacuum centrifuge (Savant, Inc.) just prior to modification. The dried FGF was dissolved in 0.1 ml deoxygenated 0.5 M Tris–HCl, pH 7.7, 10 mM EDTA, 6 M guanidine–HCl. Dithiothreitol was added to a final concentration of 5–50 mM and the reduction was allowed to proceed at 37°C for 30 minutes. A 0.5-fold molar excess of [^{14}C]iodoacetamide (24 mCi/mmol) over total sulfhydryl groups was added and the incubation continued at 37°C for 60 minutes in the dark. The alkylation was terminated by addition of a large excess of dithiothreitol over iodoacetamide and the alkylated FGF was purified by reverse-phase liquid chromatography.

The results from these studies are presented in Fig. 8. Basic pituitary FGF is a 146-amino acid protein with several particularly interesting features. For one thing, there are two clusters of basic residues at positions FGF(18–22) and FGF(107–110), a feature which was found in other molecules with heparin-binding activity (Schwarzbauer *et al.*, 1983). Computer-assisted analysis of the protein sequence failed to identify any proteins homologous with FGF. There is, however, the presence of an inverted repeat in the sequence at residues 32–39 and 46–53 that make the structure unusual. A close examination of the primary sequence also identified three related sequences that are remarkably similar to the cell-recognition site of fibronectin (Pierschbacher and Ruoslahti, 1984a,b). The sequences, illustrated in Fig. 9, show the presence of residues Asp-Gly-Arg-Val FGF(37–40) and the sequence Asp-Gly-Arg-Leu at FGF(79–82). Each of these sequences is the inverse of the X-Arg-Gly-Asp-X sequence of fibronectin that is required for cell recognition. A third peptidic fragment corresponding to Lys-Ser-Asp-X at FGF(46–49) is a conservative change from the direct fibronectin-cell recognition sequence mentioned above. All of these sequences are detected in the more hydrophilic

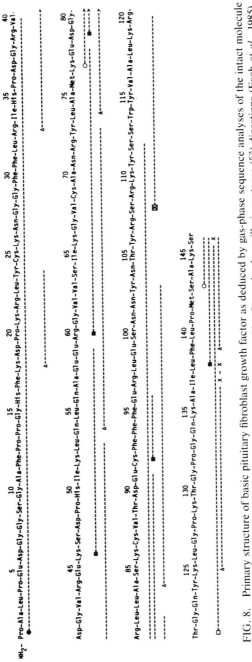

FIG. 8. Primary structure of basic pituitary fibroblast growth factor as deduced by gas-phase sequence analyses of the intact molecule (●), chymotrypsin (▲), S. aureus V8 protease (■), cyanogen bromide (○), and submaxillaris protease (⊠) digestions (Esch et al., 1985).

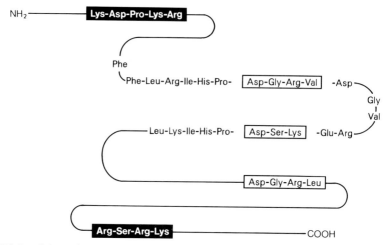

FIG. 9. Schematic representation of the primary structure of pituitary FGF including the basic clusters of amino acids at the amino and carboxyl-termini of the molecule (dark boxes). Sequences related to the fibronectin cell recognition site (open boxes) and the inverted repeat sequence of FGF(32–53) that contains the active core of antagonist activity are also presented.

regions of FGF (see below), which would make them compatible with being on the exposed surface of the molecule (Hopp *et al.*, 1983).

All the evidence obtained so far indicates that none of the four cysteine residues identified in the primary sequence of FGF participates in sulfhydryl bridging (Esch *et al.*, 1985). Although highly speculative, it is possible that this may account for the inherent lability of FGF to denaturing conditions. The significance of this finding remains to be established.

C. FUNCTIONAL DOMAINS IN THE STRUCTURE OF FGF

Several biological and chemical characteristics of FGF make it interesting to attempt to map the functional domains in the molecule. These domains include the receptor-binding site and the locus that interacts with heparin. Moreover, the role of the free cysteine resides in the molecule and the possible involvement of cofactors in its biological activity are not known. The approach we have used to map the functional domains of FGF has been to synthesize key peptidic fragments and test them for biological activity.

A computer-assisted analysis for antigenic determinants in a protein was developed by Hopp and Woods (1983) and used to map the relative hydrophilicity of the FGF sequence (Fig. 10). Based on this analysis, various peptide fragments were synthesized, including [Tyr^{87}]FGF(73–

MOLECULAR CHARACTERIZATION OF FGF 157

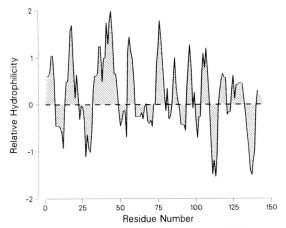

FIG. 10. Hydrophilicity analyses of the primary structure of FGF by the method of Hopp and Wood (1983) revealed the presence of a large zone of hydrophilic amino acids in the region of the inverted repeat sequence of FGF(32–53). Peptides corresponding to each wide area of hydrophilicity were synthesized for antibody generation and antagonist testing on the premise that increased hydrophilicity correlated with the presence of that sequence on the outside surface of the molecule.

87)NH_2 and [$Tyr^{69,87}$]FGF(69–87)NH_2. Antibodies were generated against these peptides and used to immunoblot the FGF-like proteins isolated from various tissues (see Section IV). They have also provided a useful tool for establishing the routine RIA for FGF.

In an effort to make a complete analysis of the molecule, several distinct peptides were synthesized for testing (see Fig. 11). They were designed either to contain a hydrophilic region or to represent a possible

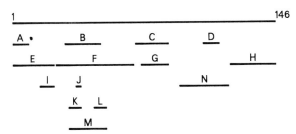

FIG. 11. Synthetic peptides made from knowledge of the primary structure of FGF were prepared by solid-phase methodology and used to test for the functional domains of the molecule. Peptides correspond to the sequences of (A) [Tyr^{10}]FGF(1–10)OH, (B) [Tyr^{50}]FGF(30–50)NH_2, (C) [$Tyr^{69,87}$]FGF(69–87)NH_2, (D) FGF(106–115)NH_2, (E) FGF(1–24)NH_2, (F) FGF(24–68)NH_2, (G) [Tyr^{87}]FGF(73–87)NH_2, (H) FGF (120–146)NH_2, (I) FGF(16–24)NH_2, (J) FGF(36–39)NH_2, (K) FGF (32–39)NH_2, (L) FGF(46–53)NH_2, (M) FGF(32–53)NH_2, (N) FGF(93–120)NH_2.

functional core of FGF. As such, [Tyr^{50}]FGF(30–50)NH$_2$ was synthesized to encompass a wide region of hydrophilicity shown in Fig. 3 and [Tyr^{10}]FGF(1–10)OH represents a hydrophilic area at the amino-terminus. [Tyr^{69,87}]FGF(69–87)NH$_2$ represents a similar area in the mid-portion of the molecule. Other fragments such as [Tyr^{25}]FGF(25–68)NH$_2$ were synthesized to contain the complete inverted repeat sequence in FGF and FGF(106–115) was made as a possible heparin-binding site.

Each of the synthetic peptides was tested for its ability to modulate endothelial cell growth (Table I). It became obvious that one set of fragments, all related to [Tyr^{50}]FGF(30–50)NH$_2$, could inhibit both basal and FGF-stimulated cell proliferation. In view of the identification of FGF(36–39) as an inverted sequence of the fibronectin-cell recognition site (Pierschbacher and Ruoslahti, 1984a,b), this fragment was tested and shown to contain the active core of antagonistic activity. An amino- and carboxyl-terminal extended analog, synthesized to contain the inverted repeats detected between FGF(32–39) and FGF(46–53) (see Fig. 9), was found to be 200 times more potent than the small-molecular-weight peptide (Fig. 12). It therefore seemed clear that a relatively small peptidic

TABLE I

Biological Activity of Synthetic Fragments of FGF[a]

Peptide	Sequence	Basal cell number ($\times 10^{-4}$)[b]	Stimulated cell number ($\times 10^{-4}$)[c]
Control	—	4.3	15.1
A	[Tyr^{10}]FGF(1–10)OH	4.6	14.8
B	[Tyr^{50}]FGF(30–50)NH$_2$	3.5[d]	12.8[d]
C	[Tyr^{69,87}]FGF(69–87)NH$_2$	4.8	17.7
D	FGF(106–115)NH$_2$	4.3	16.4
E	FGF(1–24)NH$_2$	4.0	15.6
F	FGF(24–68)NH$_2$	0.7[d]	1.2[d]
F^1	[Tyr^{25}]FGF(25–68)NH$_2$	0.6[d]	0.7[d]
G	[Tyr^{87}]FGF(73–87)NH$_2$	4.4	15.9
H	FGF(121–146)NH$_2$	3.9	16.7
I	FGF(16–24)NH$_2$	4.7	17.9
J	FGF(36–39)NH$_2$	2.7[d]	7.0[d]
K	FGF(32–39)NH$_2$	—	—
L	FGF(46–53)NH$_2$	4.7	—
M	FGF(32–53)NH$_2$	4.0	16.4

[a] All peptides were tested at a final concentration of 100 μg/ml.

[b] Basal cell growth was increased from ~1 × 10^4 cells/well seeding density to 4 × 10^4 cells/well at the end of the experiment.

[c] Stimulated cell growth was achieved by the simultaneous addition of 2 ng/ml bovine pituitary FGF prepared as described (Gospodarowicz *et al.*, 1984).

[d] Significantly different from control.

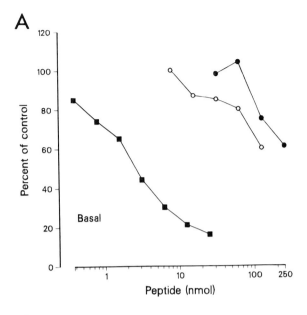

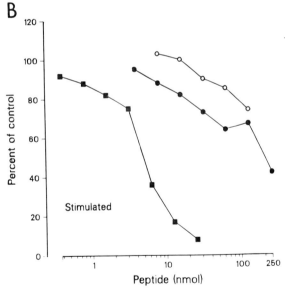

FIG. 12. Antagonist activity of synthetic peptides was tested in the absence (A) or presence (B) of 2 ng of pituitary FGF. The results are expressed as the percentage of control to facilitate comparisons. Bovine aortic arch endothelial cells were incubated with the peptide for 96 hours with two treatments of FGF. On the fifth day, cells were trypsinized and counted.

fragment could be used as an antagonist for FGF activity. It also suggested that the sequence of FGF responsible for receptor activation may be in the core sequence of FGF(25–68).

Because endothelial cells grow well under the basal conditions of the bioassay, the active peptides also show full antagonist activity to basal cell growth. These results in themselves, while precluding the possibility of detecting agonistic activity per se, indicate that basal cell growth *in vitro* is dependent on FGF, possibly derived from the serum present in the medium or from the autocrine release of FGF by the endothelial cells themselves. The peptides offer the unique opportunity for determining the physiological role of FGF *in vivo*. The antagonists will permit the specific dissection of the relative contributions of FGF as compared to other growth factors in normal growth and development.

IV. Distribution of Fibroblast Growth Factor in Tissues Other Than Pituitary

A. BRAIN

In view of the historical inaccuracies as to the identity of FGF with fragments of myelin basic protein (see Section I), it was of particular importance to establish whether the pituitary and brain growth factors showed any homologies. Using the heparin-Sepharose methodologies described in Section II, FGF was isolated to homogeneity from the bovine brain (Gospodarowicz *et al.*, 1984) and later the human brain (Böhlen *et al.*, 1985a). As shown in Fig. 13, the affinity chromatography step gave a very efficient purification (300,000-fold) of the mitogenic activity in the extract. Further purification of the brain FGF-like growth factor that was recovered in the 2 M NaCl eluate is shown in Fig. 14B. The HPLC and RIA results compared well with those obtained for pituitary FGF (Fig. 2A). Although some microheterogeneity is observed, all peaks of bioactivity clearly show radioimmunological similarities with FGF. Amino acid analyses of brain FGF showed these peaks to be statistically indistinguishable from those of pituitary FGF (Table II) and gas-phase sequence analysis identified the amino-terminal residues of brain FGF as Pro-Ala-Leu-Pro-Glu-Asp-Gly-Gly-Ser, which are identical to those of the amino-terminus of pituitary FGF.

Unlike extracts of pituitary, however, it was clear that there was significant biological activity in the 1 M NaCl eluate of the heparin-Sepharose affinity column. When this mitogenic activity was examined more closely, it was found to be devoid of any detectable immunoreactive FGF, using the amino-terminal-directed antibodies (Böhlen *et al.*, 1985b). Reverse-

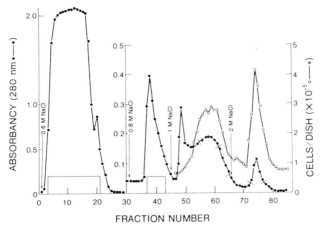

FIG. 13. Heparin-Sepharose affinity chromatography of bovine brain FGF. Fractions containing mitogenic activity from the carboxymethyl Sephadex C-50 chromatography were loaded onto a heparin-Sepharose column that had been equilibrated at room temperature with 0.6 M NaCl in 10 mM Tris–HCl, pH 7.0. Chromatography was performed at room temperature by washing the column, after sample loading, with the equilibration buffer until the absorbance of the eluate at 280 nm became negligible, followed by stepwise elution with 0.8, 1.0, and 2.0 M NaCl in 10 mM Tris–HCl, pH 7.0. The flow rate was 35 ml/hour. Fractions of 9 ml were collected during loading and washing of the column; afterwards the fraction size was 2 ml. (From Böhlen et al., 1985b.)

phase HPLC of this material (Fig. 15), however, established a homogeneity which, upon amino acid analysis and amino-terminal sequence analysis, identified a totally new growth factor (Table III) with characteristics that appear indistinguishable from those of acidic FGF (Thomas et al., 1984). This material was referred to as acidic FGF. Sequence analyses of this protein demonstrated that there are striking homologies between FGF and acidic FGF (Esch et al., 1985). Our studies have concentrated on the biological and chemical characteristics of FGF, although full primary sequence determination of acidic FGF is in progress. It seems likely that each of these growth factors (FGF and acidic FGF) represents the two subclasses of endothelial cell growth factors that have been described by several groups under different names. These would include eye-derived growth factor, retina-derived growth factor, macrophage-derived growth factor, endothelial cell growth factor, hypothalamus-derived growth factor, chondrosarcoma-derived growth factor, ovarian growth factor, and others. Evidence for this is presented in a later section.

The possible important clinical interests in FGF and the potential implications of using a radioimmunoassay that recognizes human FGF for

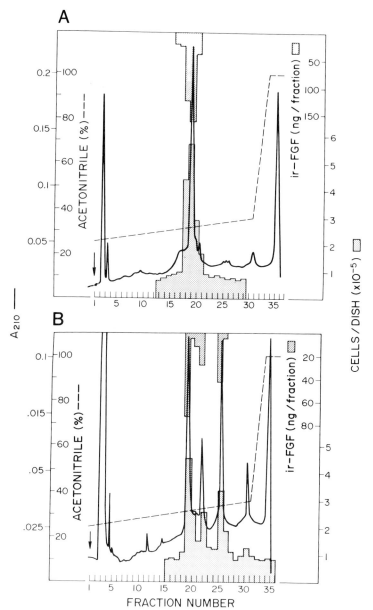

FIG. 14. Reverse-phase HPLC of heparin-purified pituitary and brain FGF. (A) Pituitary FGF was diluted with 930 µl of 0.2 N acetic acid and 1 ml was injected into a Vydac C_4 column (0.46 × 25 cm) equilibrated in 0.1% (v/v) trifluoroacetic acid. Protein was eluted with a linear 90-minute acetonitrile gradient in 0.1% trifluoroacetic acid [26–36% (v/v) acetonitrile]. Flow rate was 0.6 ml/minute and fractions of 1.8 ml were collected. Aliquots of each fraction were bioassayed and 20µl aliquots were subjected to RIA. (B) Brain FGF detected

TABLE II
Amino Acid Composition of Brain and Pituitary FGF[a,b]

	Purified brain FGF	Purified pituitary FGF	Pituitary FGF[c]
Asx	12.0	11.7	10.5
Thr	4.6	4.6	4.2
Ser	9.3	9.2	9.5
Glx	11.9	13.5	12.6
Gly	16.9	16.1	16.8
Ala	9.8	9.5	9.5
Val	5.5	5.9	5.3
Met	2.1	2.4	2.1
Ile	3.2	3.5	3.4
Leu	12.7	12.7	11.6
Tyr	6.1	6.2	7.3
Phe	7.6	6.8	7.3
His	1.9	2.4	3.2
Trp	1.1	2.1	1.1
Lys	14.9	14.7	13.7
Arg	10.8	10.5	11.6
Cys	4.8	4.2	6.3
Pro	9.7	9.7	6.3

[a] Values represent residues per molecule determined from 24-hour hydrolysates of 10–30 pmol of protein. Compositions are calculated for a 146-amino acid protein.

[b] Growth factors were purified by heparin-Sepharose affinity chromatography. Values are means of duplicate determinations.

[c] Amino acid composition of pituitary FGF purified as described by Böhlen et al. (1984) and corrected for 146 residues.

diagnostic purposes led us to obtain a partial structure of human brain FGF (Böhlen et al., 1985a). In these studies, brains obtained at the time of autopsy were collected and extracted exactly as described for bovine pituitary and brain FGF. Heparin-Sepharose affinity chromatography of the extract (Fig. 16) led to the isolation of a protein that, upon amino-terminal sequence analysis, had the same amino-terminus as bovine FGF: Pro-Ala-Leu-Pro-Glu-Asp-Gly-Gly-Ser-Gly-Ala-Phe-Pro (Böhlen et al., 1985a). This was in agreement with the ability of the RIA to recognize

---

in the 2 M eluate from the heparin affinity column was chromatographed as described for A. Bioassay was carried out as described, except that aliquots of each fraction were diluted 1:10 with DMEM/0.5% bovine serum albumin for assay. RIA was carried out as described above. Material in fraction 30, corresponding to the last UV-absorbing peak, was immunoreactive when assayed at a higher dose. (From Gospodarowicz et al., 1984.)

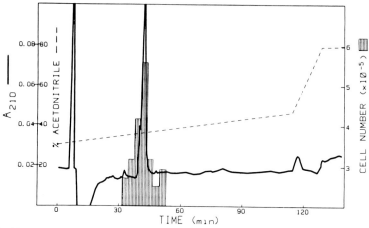

FIG. 15. Reverse-phase HPLC of bovine acidic FGF. Batches of heparin-Sepharose-purified FGF activity detected in the 1 M eluate were chromatographed on a Vydac C_4 column (0.46 × 25 cm, 300 Å pore size, 5 µm particle size), using a linear 2-hour gradient of 30–45% (v/v) acetonitrile in 0.1% (v/v) trifluoroaetic acid as the mobile phase. The flow rate was 0.6 ml/minute and fractions of 3 ml were collected. Bioassays were performed as described in Figs. 1 and 14. (From Böhlen et al., 1985b.)

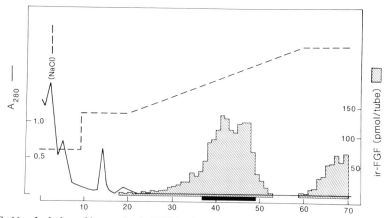

FIG. 16. Isolation of human brain FGF by heparin-Sepharose affinity chromatography. The active fraction from carboxymethyl-Sephadex C-50 was directly loaded onto a heparin-Sepharose column (1.6 × 5 cm) that had been equilibrated with 0.6 M sodium chloride in 0.01 M Tris–Cl, pH 7.0. The column was first washed with the same buffer (flow rate 90 ml/hour) until the absorbance at 280 nm returned to the baseline value, and then with 1.1 M sodium chloride in 0.01 M Tris–Cl, pH 7). Mitogenic activity was eluted with a 60-minute linear salt gradient of 1.1–2.0 M sodium chloride in 10 mM Tris–Cl, pH 7. The flow rate was 1.5 ml/minute and fractions of 1.5 ml were collected. Aliquots of 20 ml of each fraction were used for RIA. ir-FGF, Immunoreactive FGF. (From Böhlen et al., 1985a.)

TABLE III
Amino Acid Composition of Bovine Acidic FGF[a]

	Brain acidic FGF	Data from Thomas et al. (1984)
Asx	15.6	14.1
Thr	8.6	8.6
Ser	9.9	9.8
Glx	15.5	16.4
Pro	6.0	6.8
Gly	13.3	14.2
Ala	4.7	4.6
Cys	3.6	4.1
Val	4.6	4.7
Met	1.0	0.9
Ile	5.4	5.6
Leu	20.0	18.8
Tyr	7.1	7.4
Phe	8.1	7.0
His	5.7	5.4
Lys	12.9	13.0
Trp	1.1	1.2
Arg	5.4	5.8

[a] Values represent residues/molecule determined from 24-hour hydrolysates of 10 pmol of protein. The composition is calculated for a 149-residue protein to facilitate comparison with the results of Thomas et al. (1984) and taken from Böhlen et al. (1985).

FGF in extracts of human tissues. It was clear, therefore, that pituitary and brain fibroblast growth factors were indistinguishable. To date all antibodies generated against synthetic portions of pituitary FGF have been able to recognize brain FGF.

B. CORPUS LUTEUM

Angiogenesis plays an extremely important role in the development of the functional corpus luteum, mainly because of the vascular changes that take place in the capillary wreath surrounding the follicle at the time of ovulation (Bassett, 1943). Capillary sprouts grow into the granulosa cell layer and develop into a complex network of sinusoidal vessels which invade avascular cell layers, induce luteinization, and later nourish the parenchyma of the corpus luteum.

Gospodarowicz and Thakral (1978) had previously demonstrated that granulosa and luteal cells produced a diffusible substance that could trigger the early vascular changes that occur during the development of the corpus luteum. Similar activities were described by Jakob et al. (1977), who demonstrated that bovine corpora lutea grafted on the chick chorioallantoic membrane could elicit a strong angiogenic response. Similarly Koos and Le Maire (1983) have described gonadotropin-responsive angiogenesis in rat follicles and corpora lutea. Frederick et al. (1984), using human follicular fluid, also established the presence of an angiogenic factor that they associated with the perifollicular neovascularization that occurs during folliculogenesis. Taken together, the results suggested that granulosa and luteal cells could produce an angiogenic factor that triggers capillary invasion in the avascular granulosa cell layer. This possibility was also supported by the observation that crude extracts of both porcine and bovine corpus luteum is mitogenic for vascular endothelial cells *in vitro* (Gospodarowicz and Thakral, 1978). It seemed logical, therefore, that the activity present in this tissue was due to the local presence of an FGF-like growth factor. The RIA for FGF in our laboratory had already demonstrated the presence of an immunoreactive FGF in the ovary (see Section V), suggesting a wide distribution of the growth factor.

In a series of further studies on this topic, it became apparent that indeed the methodology of heparin-Sepharose-affinity chromatography could be used to purify the corpus luteum angiogenic protein (Gospodarowicz et al., 1985a). In these experiments, 2.3 kg of bovine corpora lutea was able to yield 67 µg of pure FGF-like growth factor (Fig. 17). Unlike in other tissues, however, the protein isolated lacked the intact amino-terminus characteristic of FGF (Pro-Ala-Leu-Pro-Glu-Asp) but was present as the des(1–15) form of the FGF molecule as characterized from pituitary sources. Gas-phase sequence analysis identified the amino-terminal sequence as His-Phe-Lys-Asp-Pro-Lys-Arg-Leu-Tyr, which corresponds to FGF(16–24). Therefore, except for the amino-terminal, 15-amino-acid truncation, the corpus luteum molecule appeared identical to the pituitary form of FGF. Our major concern that amino-terminal truncation of the corpus luteum-FGF occurred as a function of the extraction procedure appeared well founded since, in these extracts, while it is possible to detect and purify the truncated 15-amino acid molecule, altering the conditions of extraction prevents loss of the amino-terminal (Mormède et al., 1985a).

The growth factor isolated from the corpus luteum is a potent mitogen for endothelial cells and has the same target cell population as pituitary

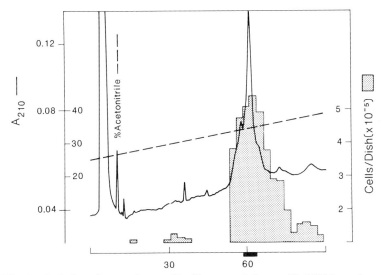

FIG. 17. Isolation of corpus luteum fibroblast growth factor. CL-FGF from the pooled fractions of the 2 M NaCl eluate of a heparin-Sepharose column was diluted with 3 ml of 0.2 M acetic acid. The sample was loaded by pumping it onto a Vydac C_4 column (25 × 0.46 cm, 5 μ particle size, 300 Å pore size). Protein was eluted with a linear gradient of 25–39% acetonitrile in 0.1% TFA (v/v) over 90 minutes. Flow rate was 0.6 ml/minute and fractions of 1.8 ml were collected. Aliquots of each fractions were diluted 1:10 with DMEM/0.5% BSA and bioassayed as described. (From Gospodarowicz et al., 1985a.)

FGF. It has also been tested for its ability to induce angiogenesis on the chick chorioallantoic membrane, where it is a potent inducer of neovascularization (Gospodarowicz et al., 1985a). In a later section (Section VI) its possible biological significance in the corpus luteum is discussed in further detail and its presence in the nonluteinized ovary is demonstrated. Clearly, however, it was immediately established that the angiogenic activity described by many groups, including Gospodarowicz and Thakral (1978), Jakob et al. (1977), Koos and Le Maire (1983), and even Makris et al. (1983,1984), could be due to the local presence of FGF in the ovary. This was also the first demonstration that the presence of FGF as a characterized molecule was not restricted to brain and pituitary.

C. KIDNEY

One of the first extracts shown to be capable of stimulating thymidine incorporation into 3T3 cells was made from kidney tissue (Holley and Kiernan, 1968). Since fibrosis and proliferative forms of glomerulonephritis are important parameters involved in the progression of chronic and acute renal disease, it was of particular importance to identify the agent(s) responsible. The initial studies of Mormède et al. (1985a) had established a high concentration of immunoreactive FGF in extracts of the rat kidney. Because urinary and renal angiogenic activity was associated with both the normal (Preminger et al., 1980) and metanephric (Sariola et al., 1983) kidney, we investigated the possibility that these activities may be due to the local presence of an FGF-like molecule, possibly FGF itself, in the kidney.

Using the same methods as described for brain and corpus luteum, we established a partial sequence for the first 17 residues of the kidney-derived mitogen and showed that they were identical to those of basic pituitary FGF(16–32) (Baird et al., 1985b). The growth factor is a potent stimulator of endothelial cell growth in vitro (ED_{50} = 50 pg/ml). Several other lines of evidence support the hypothesis that it is related, and perhaps identical to an amino-truncated form of pituitary FGF. The kidney FGF is retained on heparin-Sepharose columns with the same characteristics of FGF, has similar retention behavior on HPLC, and has the same target cell population as basic pituitary FGF. It also shows positive staining in immunoblots using antisera generated against several synthetic fragments of FGF.

As in the extracts of corpus luteum, there appears to be one major difference between mitogens of pituitary and kidney origin. Using the methodology developed for the isolation of pituitary FGF, we isolated from kidney several microheterogeneous forms of FGF that can only be distinguished by HPLC (Baird et al., 1985b). This microheterogeneity was not detected in pituitary extracts and may possible be due, in part, to amino- or even carboxyl-terminal truncation. This is supported by our detection of a small-molecular-weight peptide (2000) in the extract that cross-reacts with anti-[Tyr^{10}]FGF(1–10)OH antibodies in the RIA. Kidneys extracted under neutral conditions (Mormède et al., 1985a) fail to show the existence of this small-molecular-weight amino-terminal fragment and instead show an immunoreactive form of high molecular weight, thereby suggesting that the amino truncation (and possibly the extensive microheterogeneity) is generated by the extraction conditions. Similar conclusions to these, as discussed earlier, can be drawn from other tissues that contain the amino-truncated form, including the corpus luteum (Gospodarowicz et al., 1985a).

The biological activity of the preparation seems unaffected by the amino-terminal truncation. The potency of the kidney FGF (ED_{50} = 35 pg/ml) is comparable to that of pituitary and brain FGF (Fig. 18). Moreover, the kidney-derived mitogen has all of the same target cell types as basic

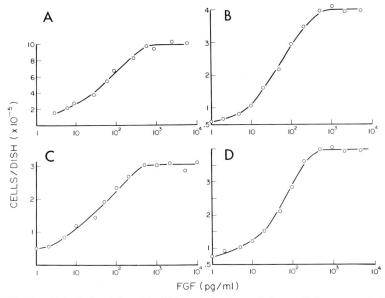

FIG. 18. Biological activity of the kidney-derived growth factor. Cells were seeded at an initial density of 2×10^4 in 35-mm dishes as described in the text. (A) Capillary endothelial cells were derived from the bovine adrenal cortex and maintained in DMEM supplemented with 10% calf serum. (B) Bovine granulosa cells were maintained in DMEM supplemented with 5% calf serum. (C) Bovine adrenocortical cells were maintained in F-12 medium supplemented with 10% calf serum. (D) Bovine vascular smooth muscle cells were maintained in DMEM supplemented with 5% defibrinated bovine plasma. Increasing amounts of the kidney-derived growth factor were added to the cells every other day in 10-μl aliquots. Six or seven days after seeding, triplicate cultures were trypsinized and counted with a Coulter counter. The final densities in control cultures were 1.65×10^5, 4.7×10^4, 4.9×10^4, and 4.7×10^4 for each cell types, respectively. (From Baird *et al.*, 1985b.)

pituitary FGF. Clearly, on this basis alone the kidney may be an inexpensive source for large amounts of the mitogen for biological studies.

The detection, isolation, and identification of FGF in the kidney have several implications in the understanding of the mechanisms that underlie both normal processes and proliferative diseases in the kidney. In view of the potent activity of FGF on many cell types, including endothelial cells, it is apparent that this molecule may be responsible in part or in toto for the angiogenic effects of kidney explants (Preminger *et al.*, 1980). Moreover, the wide distribution of FGF in several tissues suggests a paracrine, if not an autocrine, function for FGF in the kidney. Whether this mitogen and other growth factors that have been described in the kidney (Rall *et al.*, 1985) are involved in the etiology of proliferative forms of glomerulonephritis remains to be established. The availability of large quantities of pure growth factor, a sensitive radioimmunoassay for basic FGF, and the ability to measure changes in FGF content under various experimental conditions should clarify its *in situ* physiological and pathophysiological function.

D. ADRENAL

Of all endocrine organs, the adrenal gland has the greatest capacity for regeneration. Following incomplete ablations, the adrenal cortex will readily regenerate (Skelton, 1959). When adrenocortical fragments are implanted under the kidney capsule, the grafts become vascularized by the host capillaries and within a short time the adrenal cortex becomes functional. Kiss and Krompecher (1962; and Kiss and Szabo, 1969) have reported that adrenal extracts are similarly effective in stimulating vascularization. In 1985 we reported the isolation and characterization of a potent angiogenic factor from the adrenal which shares all the properties of FGF (Gospodarowicz *et al.*, 1986) and may account for the angiogenic properties of the adrenal cortex. In these studies we identified, by gas-phase sequence analysis of their amino-terminus, two FGF-related polypeptides that corresponded to intact FGF and the amino-truncated FGF. The purified mitogens stimulated the proliferation of a large number of cell types, all having the same characteristics as pituitary FGF. Since the spectrum of cell types sensitive to the adrenal-derived mitogen is not restricted to vascular endothelial cells but encompasses cells derived from the primary and secondary mesenchyne (vascular smooth muscle cells, granulosa and adrenal cortex cells, chondrocytes and myoblasts, as well as neuroectoderm-derived cells such as corneal endothelial cells), the growth factor, like pituitary FGF, is therefore specific not so much to a particular cell type as to the germ layer from which the cells originate. The observation that chondrocyte growth is stimulated by the mitogen may also explain previous reports which have described enhanced *in vivo* ossification by crude extracts of the adrenal gland (Kiss, 1975).

Although the specific significance of the presence *in situ* of an adrenal FGF is not clear, several investigators (Simmonian *et al.*, 1979; Simmonian and Gill, 1979; Gospodarowicz *et al.*, 1977; Hornsby and Gill, 1978) have made extensive studies on its proliferative and steroidogenic effects *in vitro*. As with the pituitary (see Section VI), the presence of FGF in the adrenal may not necessarily reflect a mitogenic activity per se, but might be involved in the maintenance of differentiated function. These effects and the possible biological function in the adrenal *in situ* are discussed further in Section VI.

E. RETINA

Over 30 years ago Michaelson (1948) proposed the existence of a vasculogenic factor produced by the ischemic retina that could induce neovascularization. Although 20 years would be required for the direct *in vitro* and *in vivo* evidence demonstrating angiogenic and cell-stimulating activity of the retina (Arruti and Courtois, 1978; D'Amore *et al.*, 1981; Glaser *et al.*, 1980; Barritault *et al.*, 1981), this work was pursued by Ashton *et al.* (1953, 1954), who predicted the involvement of a vasculogenic factor in the etiology of diabetic retinopathy, retinal fibroplasia, and retinal vein occlusion.

Heparin-Sepharose affinity chromatography made possible the detection of two distinct mitogens in extracts of bovine retina (Fig. 19) that were highly reminiscent of the results found in the brain (see Section

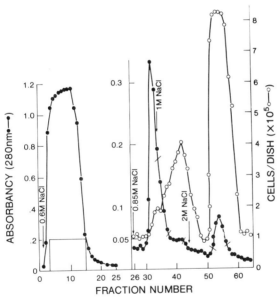

FIG. 19. Purification of the retina-derived growth factors by heparin-Sepharose affinity chromatography. The partially purified preparation obtained from the 0.6 M NaCl eluate of a carboxymethyl-Sephadex C-50 column was chromatographed on a heparin-Sepharose column. Fractions of 9 ml were collected during sample loading and column washing. Aliquots of pooled fractions 3–19 were diluted 10-fold with DMEM/0.5% BSA and 10-μl aliquots containing 5 μg protein were added to low-density cell cultures. The column was eluted stepwise with 0.85, 1, and 2 M NaCl in 10 mM Tris–HCl, pH 7.0. Fractions of 1.4 ml were collected and aliquots of the 0.85 and 1 M eluates diluted 25-fold in DMEM/0.5% BSA. Fractions collected in the 2 M NaCl elution were diluted 200-fold prior to the bioassay and 10-μl aliquots were added to the cells. (From Baird *et al.*, 1985c.)

IV,A). Purification of each of these activities led to the isolation of acidic and basic eye FGF (Baird et al., 1985c) and identified a structure for the biological activities described by D'Amore and Klagsbrun (1984, retina-derived growth factor) and Courty and colleagues (1985, eye-derived growth factor). Gas-phase sequence analysis identified the amino-terminal sequence of the major form of the mitogen as being identical to residues 1–35 of bovine basic fibroblast growth factor (FGF). Sequence analysis of the second form identified residues 1–28 that are indistinguishable from acidic FGF(1–28). The possibility that these retina-derived endothelial cell growth factors are related, if not identical, to basic and acidic FGF is supported by observations that they have similar molecular weights (15K–16K), similar retention behavior on all steps of chromatography (ion-exchange, heparin), and similar amino acid compositions. They cross-react with antibodies to the amino-terminus of basic and acidic FGF and stain by immunoblotting with antibodies raised against synthetic fragments of basic and acidic FGF. Moreover, the eye-derived growth factors, like FGF, are potent stimulators of the proliferation of several cell types, including capillary endothelial cells.

With the identification of FGF as the possible vasculogenic factor described by Michaelson (1948), it is interesting to speculate on the physiological and pathophysiological implications of this finding. The effects of FGF on fibroblasts as well as on endothelial cells may explain the association of fibrosis with the neovascularization that is seen in conditions of retrolental fibroplasia and diabetic retinopathy. The possibility that FGF is involved in the etiology of vasoproliferative diseases of the eye is an attractive hypothesis that emphasizes the importance of expanding these results. With the availability of large amounts of the pure growth factor and a defined chemical structure that can be associated with the agent responsible for vasoproliferative diseases of the eye, it has already been possible to apply the sequence-specific RIA for its measurement in ocular fluid (Baird et al., 1985d). Knowledge of its structure may even lead to the development of antagonists for use in clinical studies. Each of these approaches is currently being worked on in our laboratory and the significance of the findings is discussed in further detail in Section V.

F. MONOCYTES, MACROPHAGES, AND CELL LINES

One of the first studies demonstrating that macrophages were involved in the proliferation of nonlymphoid mesenchymal cells was that of Leibovich and Ross (1976), who demonstrated the required presence of monocytes and macrophages for fibroplasia and wound repair. On this basis, Martin et al. (1981) demonstrated the synthesis of a macrophage-derived growth factor capable of stimulating cell proliferation *in vitro*. Recent data have also clearly established the capacity of macrophages to induce neovascularization in several pathophysiological processes (Clark et al., 1976; Polverini et al., 1977; Thakral et al., 1979). Since the macrophage-derived growth factor also had *in vitro* activities resembling those of FGF (Greenberg and Hunt, 1978; Wall et al., 1978; Martin et al., 1981; Knighton et al., 1983), it seemed logical to propose that the growth factors were related, if not identical (Baird et al., 1985e).

In these studies FGF was detected in extracts of macrophages obtained from thioglycollate-stimulated mice and purified by heparin-Sepharose affinity chromatography and HPLC (Fig. 20). The purified protein appeared indistinguishable from pituitary FGF although no sequence data were obtained. Recent studies in collaboration with the laboratories of Russell Ross have confirmed the presence of FGF in macrophages and suggest that, together with a PDGF-like mitogen (Shimokado et al., 1986), the macrophage FGF provides a significant amount of the growth factor activity in this cell type.

The concentration of FGF in macrophage-derived cell lines is also

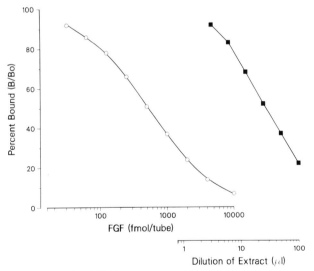

FIG. 20. Immunoreactive FGF in an extract of mouse peritoneal macrophages. Cells obtained from peritoneal lavage of thioglycollate-stimulated mice were extracted for FGF and the presence of the growth factor was established by parallel displacement in the FGF RIA and later retention on heparin-Sepharose and reverse-phase HPLC. (From Baird et al., 1985e.)

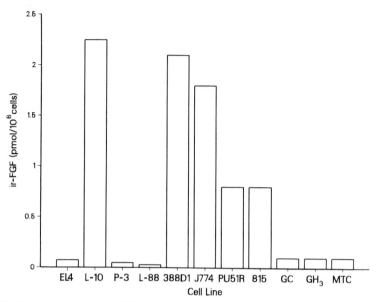

FIG. 21. Immunoreactive FGF was measured in extracts of various cell lines. Cells were extracted at densities of 10^7–10^9 cells and the amounts of ir-FGF were established by RIA.

higher than in lines derived from any other cell types (Fig. 21). Cells grown under standard conditions in the presence or absence of serum were washed and extracted with 0.15 M $(NH_4)_2SO_4$. As is clearly evident in the figure, cells derived from monocyte lineage, including the J774, P-388D1, P-815 and Pu51R, contained high amounts of ir-FGF. The L-10 cells, derived from a β cell lymphoma, also contained remarkable amounts of the growth factor. All other cell types, including pituitary-derived GH_3 or GC, thyroid-derived MTC, or the thymus-derived EL_4, had low to undetectable levels of FGF.

The question as to whether or not FGF is the classically described macrophage-derived growth factor remains to be answered, although at this time it is clear that the presence of FGF in macrophages establishes its identity with at least some MDGF activity. While it may be simplistic to propose FGF as being responsible for the macrophage-mediated pathogenesis of atherosclerosis and other diseases, it appears likely that interactions between FGF, PDGF, and possibly other mitogens (TGF-β), are responsible for these processes. Once again, the availability of sequence-specific antibodies, putative antagonists, and large amounts of the growth factor will permit a dissection of these complex relationships between growth factors to establish the specific function of each.

G. FGF IN SERUM

One of the first applications of the RIA for FGF was to establish whether there was a detectable amount of the growth factor in serum and plasma. In a study of the effects of hypophysectomy on serum FGF, Mormède et al. (1985a) established that an immunoreactive FGF circulated as a high-molecular-weight protein (150,000) and that its concentration in serum actually increased after hypophysectomy. The radioimmunoassay for circulating FGF was validated by demonstrating parallelism in bovine, human, and rat serum (Fig. 22). With these results in hand, it was possible to establish that in each of these species, ir-FGF circulated as a 150,000-Da molecule (Fig. 23). The possibility that FGF was present as a bound molecule was investigated by Mormède et al. (1985b), who demonstrated that it was possible to partially dissociate the high-molecular-weight ir-FGF to a form with lower molecular weight. Since the form generated under these conditions had the same behavior as the 70,000-Da form of ir-FGF present in neutral extracts of tissues (see Section V) and was distinct from the behavior of the radioiodinated 16,000-Da FGF, it was also concluded that the form of ir-FGF found in serum was the 70,000-Da molecule. Its purification is currently in process in our laboratory.

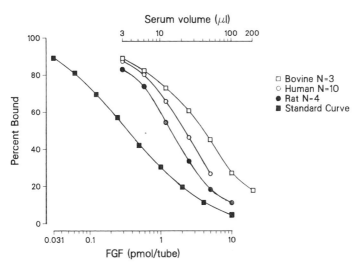

FIG. 22. Parallelism of the immunoreactive FGF in bovine human and rat serum. Bovine serum was collected at a local abattoir and rat serum was collected from normal male Sprague-Dawley rats. Human serum was collected from normal volunteers. The samples were serially diluted and the amounts of ir-FGF were measured by RIA using a second antibody precipitation.

Normal serum

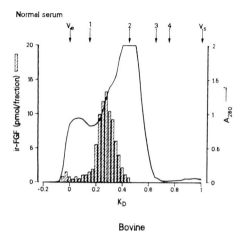

Bovine

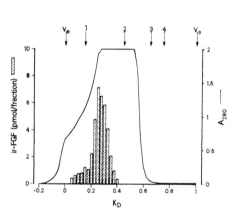

Rat

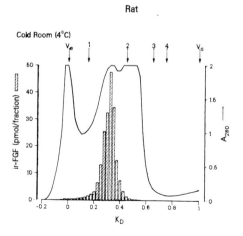

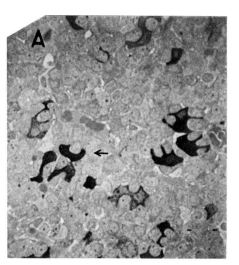
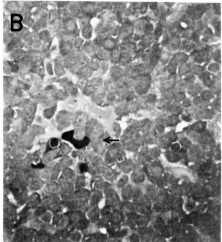

FIG. 24. Cellular localization of pituitary FGF. Antibodies raised against [Tyr[10]]FGF(1–10) or $ACTH_{1-39}$ were incubated with adjacent thin slices (10 μm) of rat pituitary and staining was visualized with horseradish-peroxidase-labeled anti-rabbit IgG. (A) Staining of corticotropes with anti-ACTH; (B) staining of adjacent section with anti-FGF.

H. CELLULAR LOCALIZATION OF FGF

In spite of the availability of several distinct antibodies that cross-react with FGF, there have been no reports of its cellular location in any of the tissues so far examined. Although preliminary studies have indicated the possible existence of a specific subpopulation of cells in the anterior pituitary that stain for FGF, their identity with other hormone-containing cells is currently being examined. The early results obtained so far indicate that an ir-FGF may be localized in a subpopulation of corticotropes (Fig. 24). Their number appears unaffected by adrenalectomy, indicating that they may not be functional ACTH-containing cells that respond to CRF (Smith et al., in preparation).

FIG. 23. Gel permeation of human, bovine, and rat sera. Aliquots (2 ml) of serum were loaded onto Sephacryl S-300 Superfine columns (100 × 1.75 cm) and eluted with phosphate buffer (0.067 M, pH 7.4). Aliquots of the column fractions were tested for their ir-FGF content by RIA. Molecular weight markers included (1) blue dextran (M_r = 2,000,000), (2) fibrinogen (M_r = 400,000), (3) bovine serum albumin (M_r = 76,000), (4) chymotrypsinogen (M_r = 25,000), (5) cytochrome c (M_r = 12,400). The void volume (V_e) and salt volume (V_s) are indicated.

V. Molecular Forms of Fibroblast Growth Factor

Mormède et al. (1985a) identified the presence of high-molecular-weight forms of ir-FGF in several tissues including those from which FGF was isolated and characterized (Fig. 25). It also became apparent that the protocol used for extraction could select for the molecular form of ir-FGF in these extracts. A simple extract at pH 7 revealed the almost exclusive

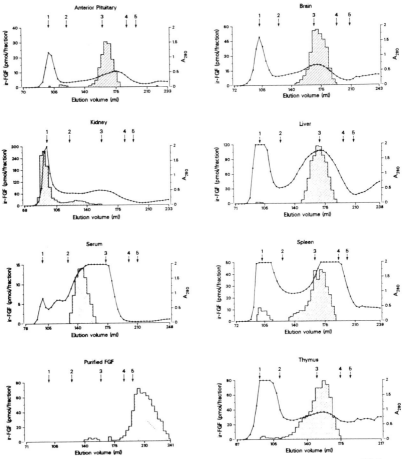

FIG. 25. Gel filtration column chromatography of serum and tissue extracts. High-speed supernatants of tissue extract were loaded onto (100 × 1.75 cm) Sephacryl S-300 columns and eluted with phosphate buffer (0.067 M, pH 7.4). Aliquots of the column fractions were tested for their ir-FGF content. Molecular weight markers: (1) blue dextran (M_r = 2,000,000), (2) fibrinogen (M_r = 400,000), (3) bovine serum albumin (M_r = 67,000), (4) chymotrypsinogen (M_r = 25,000), (5) cytochrome c (M_r = 12,400). (From Mormède et al., 1985a.)

presence of a 70,000-Da molecule, while extraction procedures performed to isolate the 16,000-Da biologically active growth factor failed to show the presence of a high-molecular-weight FGF. This observation may well account for the development of the extraction procedure at pH 4.5, which was optimized for the extraction of growth factor activity rather than immunoreactivity. The protocol, which includes a 2-hour mixing of the crude extracts at 4°C (Gospodarowicz and Bialecki, 1978), would allow for the generation of the 16,000-Da molecule from a high-molecular-weight molecule and the detection of increased biological activity. Maciag and colleagues (1982) had previously demonstrated a precursor–product relationship with endothelial cell growth factor (ECGF), the anionic brain homolog of FGF (see Section II). It therefore seems likely that, in tissues, FGF exists as a 70,000-Da molecule which (upon release) generates a low-molecular-weight 16,000-Da protein with full intrinsic and high specific activity.

It is interesting to speculate that the presence of a relatively inactive high-molecular-weight form of FGF in several tissues may regulate the bioavailability of the 16,000-Da FGF. Under this paradigm, the availability of the mitogen to its surrounding target cells would be totally dependent on the generation of a bioactive 16,000-Da FGF from a high-molecular-weight complex. Enzymatic cleavage of this molecule would represent a highly regulated step of growth factor bioavailability and would explain the presence of immunoreactive but not bioactive FGF in tissues that are not undergoing growth or angiogenic responses. Isolation and characterization of the 70,000-Da form of ir-FGF should answer this possibility.

VI. Biological Activities of Fibroblast Growth Factor

With the isolation and characterization of FGF from several distinct and seemingly unrelated tissues, it became inherently obvious that it was necessary to reconsider the possible physiological functions of "pituitary" or "brain" FGF action on adrenal, ovarian, or chondrocyte cell growth. It seemed more likely that FGF, locally present in each of these tissues, played a paracrine, if not autocrine, function in regulating cell growth and possibly cell differentiation, survival, and function. To this end, we investigated each of the tissues where we had isolated and identified FGF to determine its possible *in situ* biological significance. Particular emphasis was placed on such endocrine tissues as the pituitary and the ovary, although several lines of investigation have suggested a function for FGF in the adrenal, retina, macrophages, and brain.

A. PITUITARY

Initial experiments were designed to investigate the possibility that the anterior pituitary could elaborate an FGF-like molecule *in vitro*. In these studies, bovine anterior pituitaries were maintained in culture for 3 days, and on the third day cells were incubated with 50 mM KCl for 4 hours. The conditioned medium was collected, mixed with heparin-Sepharose, and the FGF eluted with 2 M NaCl (Baird *et al.*, 1985a). As shown in Fig. 26, the conditioned medium contained an immunoreactive FGF that appeared indistinguishable from authentic FGF.

It was of particular interest, however, to determine whether the amounts of FGF measured in conditioned medium could be modulated by factors added to the cells. It was already observed that basal FGF levels in a 4-hour incubation medium were low, although they could be increased by 50 mM KCl. Consistent with the hypothesis that FGF fulfils a paracrine intrapituitary function, none of the hypothalamic releasing factors was able to increase the release of FGF. Indeed, only preincubation with estradiol had any significant effect on the amount of ir-FGF released into the conditioned medium evoked by 50 mM KCl (Fig. 27). Basal FGF appeared unaffected by estradiol, with the only detectable changes occur-

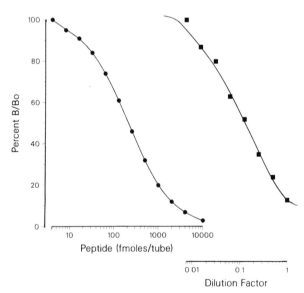

FIG. 26. Cross-reactivity of ir-FGF in conditioned medium. The supernatants obtained from bovine anterior pituitary cells in culture were passed through a heparin-Sepharose affinity column and the 2–3 M NaCl eluate was desalted and tested for ir-FGF content by RIA. (From Baird *et al.*, 1985a.)

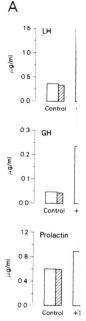

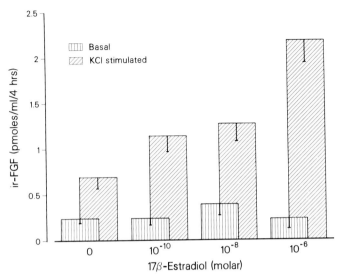

FIG. 27. The effect of 17β-estradiol on KCl-induced release of ir-FGF. Bovine anterior pituitary cells were maintained in culture for 3 days a transferred to serum-free medium supplemented with 0.1% bovine serum albumin and various concentrations of 17β-estradiol. The cells were washed 48 hours later and incubated with 50 mM KCl for 4 hours. The amounts of ir-FGF were measured by RIA. (From Baird et al., 1985a.)

ring after KCl-mediated release. In these experiments, cells had to be preincubated with estradiol for 48 hours prior to incubation with KCl to demonstrate an effect of the steroid.

With the first indication that estrogen treatment could modulate pituitary FGF came the hypothesis that FGF could be involved in the development of estrogen-dependent pituitary tumors. To approach this problem, we examined the development of prolactin-secreting tumors in estrogen-treated Fischer 344 rats. In these experiments we demonstrated that indeed the pituitaries of these animals contain more FGF than those of untreated animals (Fig. 28). Whether this change reflects a causal relationship is not clear, particularly since, at the concentrations tested, antibodies to FGF failed to prevent the onset of these tumors. Remarkably, estrogens have no effect on the amounts of FGF measured in the pituitaries of Sprague–Dawley rats (Mormède et al., 1985b), indicating that the species specificity of these estrogen-dependent tumors may also be linked to the inducibility of FGF. Similar results were obtained in vitro, where it was not possible to demonstrate KCl-evoked FGF release by the rat anterior pituitary. This occurs presumably because of the low levels of ir-FGF in the rat as compared to bovine pituitary.

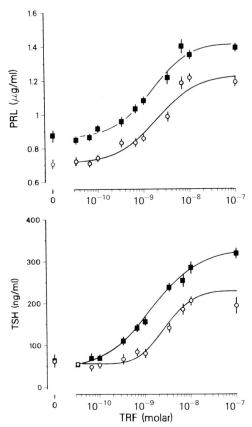

FIG. 30. TRF dose–response curve in FGF-treated cells. On day 3 of culture, pituitary cells were incubated with 2 ng of FGF for two successive periods of 24 hours. On day 5 of culture, the cells were incubated with various doses of TRF. (○) Results from control cells; (■) results from FGF-treated cells. (From Baird et al., 1985f.)

the effect of FGF on estradiol by their simultaneous addition to cells *in vitro*. While these results could be confirmed on normal pituitary cells, we turned to the GH_3 cell line developed by Tashjian et al. (1968) to examine the interactions between FGF and estradiol. Schonbrunn et al. (1980) had previously described a similar biological activity for FGF and EGF in this cell line and established the ability of FGF to increase prolactin and to decrease growth hormone and cell growth.

In our laboratory, Mormède et al. (1985b) used this model to dissect the mechanism of action between FGF and estradiol. In these studies they demonstrated a synergistic action of these treatments on the release of

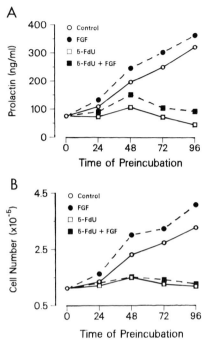

FIG. 31. Effect of cell growth on the response to FGF. (A) Prolactin release by rat anterior pituitary cells was measured over a period of 96 hours under conditions where growth was either inhibited by 5 fluorodeoxyuridine (5 FdU) or stimulated by fibroblast growth factor (FGF). In each instance FGF was capable of increasing the release of prolactin by these cells. (B) Cell growth in this same preparation was determined by trypsinization and counting on a Coulter particle counter.

prolactin and established the ability of FGF to modulate differentiated cell function (Fig. 32). Preliminary studies have indicated that, unlike estradiol, FGF has little or no effect on the number of TRF receptors. Moreover, growth stimulators such as PMA failed to elicit the potentiated response of FGF and estradiol when added with either one.

There are several implications in the finding that FGF is involved in the release of prolactin and TSH in the normal pituitary. Perhaps the major observation is that the results establish that the presence of FGF in the pituitary need not reflect its mitogenic activity per se, but instead may reflect its presence as a differentiating factor. This possibility would be consistent with a role for FGF in the development of estrogen-dependent prolactin-secreting tumors and in instances of abnormal or accentuated prolactin release such as during lactation.

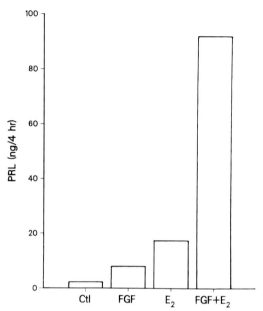

FIG. 32. The effect of FGF and estradiol on prolactin release by GH_3 cells. Cells were incubated with FGF (2 ng/ml), estradiol (100 pg/ml), or both (FGF + E_2) for 24 hours and the amount of prolactin released over a 4-hour period after incubation was measured by radioimmunoassay.

B. OVARY AND CORPUS LUTEUM

The possible biological activities of FGF in the ovary and corpus luteum appear, at first glance, to be related to their high content of angiogenic activity (Jakob et al., 1977; Gospodarowicz and Thakral, 1978; Koos and Lemaire, 1983). On the one hand, as an angiogenic factor, ovarian FGF becomes a likely participant in the development of the functional corpus luteum, but on the other, as in the pituitary, we considered it necessary to determine whether FGF plays a role in the modulation of cell function. It was to this end that we examined ovarian FGF in two ways. First, we looked at the nonluteinized ovary to determine (1) whether FGF is present, and (2) whether, like other growth factors (Hsueh et al., 1984), it might fulfil a functional role. Second, we examined the amounts of ir-FGF while the ovary was undergoing luteinization and attempted to make a correlation with the onset of angiogenic activity.

The presence of FGF in the nonluteinized rat ovary was established using the RIA for FGF. Neutral extracts of this tissue revealed the presence of an immunoreactive FGF that appears indistinguishable from pituitary FGF (Fig. 33). In fact, gel filtration on Sephacryl S-300 columns

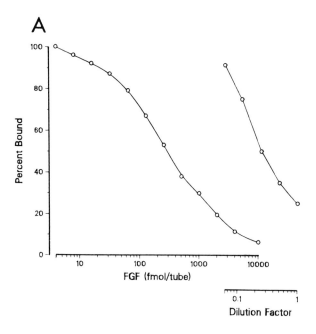

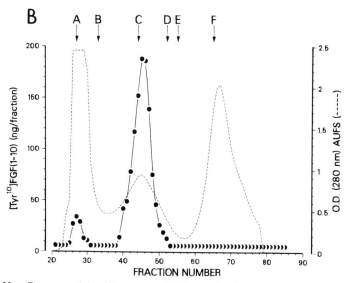

FIG. 33. Cross-reactivity of rat ovarian FGF. (A) Ovaries were collected from immature female Sprague–Dawley rats and extracted in 0.15 M $(NH_4)_2SO_4$, pH 7, as described by Mormède et al. (1985a). Parallelism was tested in the RIA for FGF using amino-terminal-directed antibodies raised against synthetic [Tyr[10]]FGF(1–10). (B) The extract was loaded onto a column of Sephacryl S-300 and eluted with phosphate buffer (0.067 M, pH 7). Molecular weight markers include (A) blue dextran (M_r = 2,000,000), (B) fibronigen (M_r = 400,000), (C) bovine serum albumin (M_r = 67,000), (D) chymotrypsinogen (M_r = 25,000), (E) cytochrome c (M_r = 12,400), and (F) NaCl (M_r = 58).

established that the molecular form of FGF is identical to that in the pituitary and demonstrated that altered conditions of extraction could avoid the amino-terminal truncation observed by Gospodarowicz et al. (1985a) during its isolation and sequence analysis. When corpus luteum development was stimulated by pregnant mare's serum gonadotropin and hCG, there was only a small increase in intraovarian FGF from 3.4 to 5.1 fmol/mg protein. However, since both angiogenic activity and FGF can be detected in follicles prior to their lutealization (Koos and LeMaire, 1983), the possibility remained that intraovarian FGF might be playing a more functional role, independent of its capacity to induce neovascularization. In these experiments (performed in collaboration with Aaron Hseuh of the University of California at San Diego) granulosa cells were prepared from rat ovaries and the effect of FGF was tested on their ability to form estradiol from exogenous androstenedione when stimulated with FSH. This *in vitro* model had served Hsueh *et al.* (1984) and others (Erickson *et al.*, 1985) as an invaluable tool in dissecting intraovarian mechanisms of hormone regulation. When FGF was introduced into the bioassay, it was found to be a potent inhibitor of FSH-stimulated aromatase activity (Fig. 34). Doses that were capable of eliciting a half-maximal mitogenic response in several cell types (30 pg/ml, see Section II) were capable of inhibiting the conversion of androstenedione to estradiol. At concentrations of 300 and 3000 pg/ml, the effects of FSH were completely inhibited. Whether FGF is the same molecule whose activity has been described by Di Zerega *et al.* (1982; Di Zerega and Hodgen, 1980) remains to be established. It seems clear, however, that the same factor has two unusually opposite actions. On the one hand, FGF is a potent angiogenic factor, thereby promoting follicular development and corpus luteum formation. On the other, by virtue of its action on aromatase activity, the same molecule can inhibit estradiol formation and effectively mediate follicular atresia. How these two results, remarkably divergent in activity, might explain a role for FGF in ovarian function remains to be established.

C. ADRENAL AND KIDNEY

The physiological role of FGF in the adrenal is not known. Simmonian *et al.* (1979; Simmonian and Gill, 1979) have made extensive studies of the role of FGF in stimulating adrenocortical cell growth and promoting steroidogenic function *in vitro*. Whether FGF fulfils a similar function *in vivo* remains to be determined. It is interesting to hypothesize that, as in the

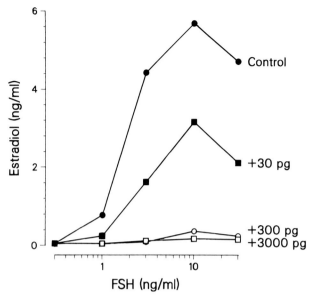

FIG. 34. Effect of FGF on aromatase activity in rat granulosa cell cultures. Cells were prepared as described by Hsueh et al. (1982) and incubated with androstenedione and varying concentrations of follicle-stimulating hormone (FSH). The formation of estradiol was measured by RIA and the effect, 0 (control), 30 pg/ml (+30), 300 pg/ml (+300 pg) and 3000 pg/ml (+3000 pg) of FGF on the stimulated conversion of androstenedione to estradiol was examined.

pituitary and ovary, FGF might participate in the maintenance of differentiated function. In vitro studies to date have not examined the effects of FGF on adrenocortical function when growth has been halted. In particular, the possibility that FGF may represent an intraadrenal mechanism to mediate the steroidogenic changes that characterize adrenarche, for example, is an interesting hypothesis. These current studies, ongoing in our laboratory, will exploit the availability of both the antibodies and the antagonists to FGF.

The presence of FGF as a distinct entity in extracts of kidney agrees well with the detection of angiogenic and fibroblast-stimulating activities in this tissue. The results also extend those obtained by Ekblom and colleagues (Ekblom, 1981; Ekblom et al., 1982), which demonstrated similar activities in embryonic tissues. Further work will be required, however, to determine whether FGF in the kidney is acting as an angiogenic factor or whether its presence reflects a more functional role above and beyond tissue maintenance and repair.

D. MACROPHAGES, MONOCYTES, AND TUMORS

Perhaps the major implication of the detection and identification of FGF in macrophages is with its identification as a macrophage-derived angiogenic factor. Its presence in this cell type allows for an understanding of why "pituitary" and "brain" FGF would have both angiogenic (Gospodarowicz *et al.*, 1979) and wound-healing (Gospodarowicz *et al.*, 1976) activities. Obviously, synthesis and delivery of FGF by macrophages invading the sites of wounds allow for an efficient, paracrine mechanism that can interact with the other growth-promoting substances coreleased from the same cell type (i.e., PDGF, TGF-α, TGF-β, etc).

There is also a good argument to support the hypothesis that FGF may be related to the tumor angiogenic factor described by Folkman and colleagues (Folkman, 1975, 1984; Shing *et al.*, 1984). Its identification in extracts of chondrosarcoma (Baird *et al.*, 1985g) suggested its identity with the chondrosarcoma-derived growth factor purified by Shing *et al.* (1984) in the first description of heparin-Sepharose affinity chromatography.

E. RETINA

The biological significance of the identification of FGF as the angiogenic factor in the retina lies in understanding the pathogenesis of vasoproliferative diseases of the eye. To this end, we have used our current knowledge about FGF to develop the methods to aid in its measurement in human vitreous fluid. As shown in Fig. 35, human ocular fluid contains an immunoreactive and biologically active FGF. The growth factor can be retained on heparin-Sepharose columns and eluted with 2 M salt (Fig. 36). Both bioactive and immunoreactive forms appear to coelute and show identical behavior. All the immunoreactive FGF detected in the ocular fluid is retained by the heparin column, a result in marked contrast to ir-FGF in serum or in neutral extracts.

Using the RIA for FGF, we have begun to measure the amounts of FGF in vitreous and ocular fluid. In 6 samples from 4 patients, ir-FGF levels in ocular fluid have been found to be in the range of 0.6–1.8 pmol/ml (Baird *et al.*, 1985d). So far we have obtained one sample of ocular fluid from a patient with diabetes mellitus complicated by proliferative retinopathy. In this sample, the amount of ir-FGF was 8.9 pmol/ml. The marked increase under this particular pathological condition emphasizes the necessity of obtaining more samples and expanding these studies. It is also highly suggestive that FGF may play an important role in mediating the pathogenesis of the disease. The capacity to regulate FGF content in the eye

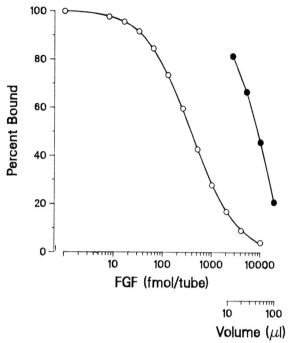

FIG. 35. Fibroblast growth factor in human ocular fluid. A sample of human ocular fluid was obtained at the time of gas-fluid exchange procedures for retinal detachment. Parallelism of the ir-FGF detected was established in the RIA for FGF using antibodies directed against [Tyr^{10}]FGF(1–10).

with either antibodies or synthetic antagonists may provide significant therapeutic value to the prognosis of this and other proliferative diseases of the eye.

F. BRAIN

The *in vivo* role of FGF in the brain is not known. Recent data have suggested that, at least *in vitro*, FGF can act as a survival agent, preventing the death of hippocampal neurons in tissue culture (Walicke *et al.*, 1986). In view of the multifunctional role that FGF can play in the various peripheral tissues, it is likely that the growth factor plays several distinct roles in the CNS. These roles may involve its capacity to stimulate cell proliferation but may also involve its ability to maintain cell-differentiated function and survival (Morrison and de Veillis, 1981; Eccleston and Silverberg, 1985).

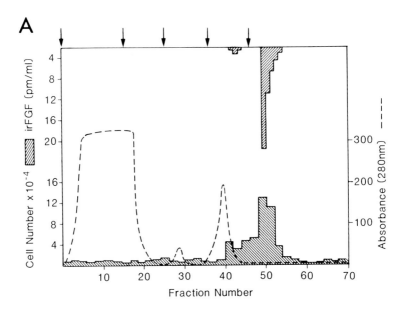

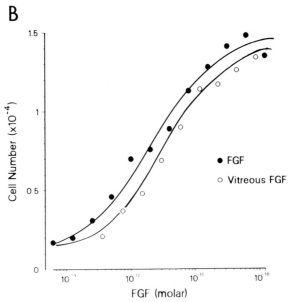

FIG. 36. (A) Heparin-Sepharose affinity chromatography of human vitreous was performed with a sample of vitreous obtained from a patient undergoing vitrectomy. The solution was loaded directly onto the heparin-Sepharose column and the sample eluted with step gradients of 0.15, 0.6, 1, and 2 M NaCl. Aliquots of column fractions were taken for the determination of bioactive and immunoreactive FGF content. Both bioactive and immunoreactive fractions elute at the predicted location of FGF. (B) The biological activity of the heparin-purified sample was compared with the FGF standard using bovine aortic arch endothelial cells as described in the text. The amount of ir-FGF added to the cells was determined by amino acid analysis for the standard and RIA for the ocular sample.

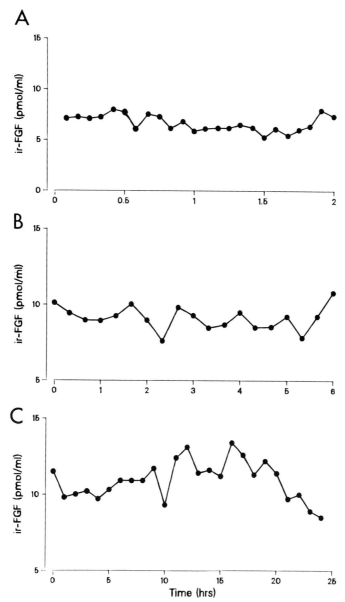

FIG. 37. Variations in the serum levels of ir-FGF in rat. Chronically catheterized male Sprague–Dawley rats (250 g) were sampled (200 µl) every 5 minutes for 2 hours (A), every 20 minutes for 6 hours (B), or every hour for 24 hours (C). Aliquots of serum were measured for ir-FGF content by RIA and each point is the mean of 4–7 samples.

G. SERUM

The significance of the detection of an immunoreactive FGF in serum is also not known. The possibility that it may contribute to the proliferative activities of serum *in vitro* is supported by the observation that the synthetic antagonists to FGF can inhibit serum-induced proliferation of endothelial cells (see Section X). As with other growth factors, however, it is clear that ir-FGF in serum is present as a high-molecular-weight component in either a bound or an aggregate form. This limitation of the bioavailability of circulating FGF may reflect the mechanisms that underlie its physiological regulation but need not preclude the possibility that a serum-derived FGF can modulate the cell response in various distinct target tissues. The identification of the tissue source of serum FGF and of pathophysiological conditions where it is elevated or decreased in serum should provide insight into its possible physiological significance in normal cellular homeostasis.

We have used the RIA for FGF to determine whether there exists a dirunal rhythm in the circulating levels of ir-FGF. In these studies, chronically catheterized male rats were sampled every 5 minutes for 60 minutes (Fig. 37A), every 20 minutes for 6 hours (Fig. 37B) or every hour for 24 hours (Fig. 37C). There was no evidence for a pulsatile release of ir-FGF, and the mean circulating levels of ir-FGF in serum appeared remarkably constant in the 24-hour period, except for a small but consistent nocturnal increase. Similar results were obtained in human serum samples provided

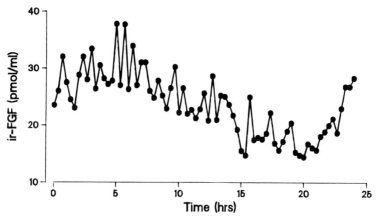

FIG. 38. Variations in the circulating levels of ir-FGF in human serum. Samples were collected every 20 minutes from four normal women volunteers over a period of 24 hours. Aliquots of serum were measured for ir-FGF content by RIA.

by Drs. Yen and Liu of the University of California at San Diego. In these studies, the circulating serum levels of ir-FGF in normal women were increased during waking hours (Fig. 38). Although the significance of this finding is not clear, the detection of a change in circulating ir-FGF suggests a homeostatic balance in the maintenance of its levels in serum.

VII. Conclusions

From the results presented here it has been possible to establish the identity of a new class of growth factor, characterized in part by their high affinity for heparin and their capacity to stimulate the proliferation of cells that are derived mainly from the primary or secondary mesenchyme. The growth factors are all related to either the acidic or the basic FGFs that were initially purified by conventional isolation procedures (Thomas *et al.*, 1984; Böhlen *et al.*, 1984). The review presented here was mainly concerned with our studies of one form of this growth factor, basic FGF.

The identification of FGF has enabled the development of sequence-specific antibodies that established the widespread distribution of the growth factor in several tissues. This finding is in contrast with the initial observations of Gospodarowicz (1974) that suggested a pituitary and brain specificity for FGF but agrees with the subsequent detection of FGF-like biological activities in several tissue sources (see Section III). The results explain the capacity of a "pituitary" FGF to have biological effects on several cell types (chondrocytes, granulosa, and adrenocortical cells) when it is, a priori, inaccessible to these target tissues. The local presence of FGF in each tissue would account for their responsiveness to pituitary and brain preparations of FGF. As such, while brain, pituitary, adrenal, ovarian, retinal, and macrophage FGFs may be identical in structure, their specific function in each tissue relates to the *in situ* environment. Thus FGF in the pituitary plays a nonmitogenic role in the maintenance of prolactin and TSH responsiveness, while the same molecule in macrophages may promote wound healing and in the corpus luteum angiogenesis.

The language of polypeptides is a relatively new concept which proposes that the *in situ* localization of a polypeptide predetermines its *in fine* function (Guillemin, 1985). It is based on the diverse biological activities of somatostatin in brain, hypothalamus, stomach, and pancreas (reviewed in Raptis *et al.*, 1984) and the wide distribution of numerous neuropeptides in both the central nervous system and the gastrointestinal tract (Said, 1980). The concept is equally applicable to FGF and perhaps other growth factors that show a wide distribution in many tissues. The hypothesis also emphasizes the importance of understanding the local mechanisms that regulate the bioavailability of the growth factor to the target cells.

The results presented here also suggest that heretofore unidentified biologic activities may be related if not identical to acidic and basic FGFs. These would include ovarian growth factor (Makris *et al.*, 1982, 1983), macrophage-derived growth factor (Polverini and Leibovich, 1985), cartilage growth factor (Sullivan and Klagsbrun, 1985), tumor angiogenic factor (Shing *et al.*, 1984), eye-derived growth factor (Glaser *et al.*, 1980), retina-derived growth factor (Arruti *et al.*, 1978; Barritault *et al.*, 1981), corpus luteum angiogenic peptide (Gospodarowicz and Bialecki, 1978), follicular angiogenic factor (Koos and LeMaire, 1983), adrenal vascularizing factor (Kiss, 1975), and endothelial cell growth factors (Maciag *et al.*, 1978, 1982, 1984). Whether their biological significance is related to their *in vitro* activities will, as mentioned above, be strictly related to their local environment. It is likely that future studies will clarify the importance of the FGFs in each of these tissues and provide the foundations for the determination of their physiological role *in vivo*.

ACKNOWLEDGMENTS

Research in the Laboratories for Neuroendocrinology was supported by program Grants HD-09690 and AM18811 from the National Institutes of Health and by the Robert J. Kleberg, Jr. and Helen C. Kleberg Foundation. Research in the Laboratories of Dr. Denis Gospodarowicz was supported by Grant EV-02186 from the National Institutes of Health. The authors wish to acknowledge the active participation of Dr. Denis Gospodarowicz in many of these studies and thank Drs. Floyd Culler, Luc Denoroy, Aaron Hseuh, Ken Jones, James Liu, Philip Smith, Toratane Munegumi, and Samuel Yen for their contributions and helpful suggestions. The authors also wish to thank the secretarial staff of the Laboratories for Neuroendocrinology for typing the manuscript and Darlene Gore for preparation of the figures. The dedication of Fred Castillo, Terri Durkin, Fred Hill, Robert Klepper, Roland Schroeder, Karen Von Dessonneck, and the entire technical staff of the Laboratories for Neuroendocrinology is greatly appreciated.

REFERENCES

Alessandri, G., Raju, K., and Gullino, P. M. (1983). *Cancer Res.* **43**, 1790–1797.
Armelin, H. A. (1973). *Proc. Natl. Acad. Sci. U.S.A.* **70**, 2702–2706.
Arruti C., and Courtois, Y. (1978). *Exp. Cell Res.* **117**, 283–292.
Ashton, N., Ward, B., and Serpell, G. (1953). *Br. J. Ophthalmol.* **37**, 513–520.
Ashton, N., Ward, B., and Serpell, G. (1954). *Br. J. Ophthalmol.* **38**, 397–432.
Baird, A., Böhlen, P., Ling, N., and Guillemin, R. (1985a). *Regul. Peptides* **10**, 309–317.
Baird, A., Esch, F., Böhlen, P., Ling, N., and Gospodarowicz, D. (1985b). *Regul. Peptides* **12**, 201–213.
Baird, A., Esch, F., Gospodarowicz, D., and Guillemin, R. (1985c). *Biochemistry* **24**, 7855–7860.

Baird, A., Culler, F., Jones, K. L., and Guillemin, R. (1985d). *Lancet* **2**, 563.
Baird, A., Mormède, P., and Böhlen, P. (1985e). *Biochem. Biophys. Res. Commun.* **126**, 358–364.
Baird, A., Mormède, P., Ying, S.-Y., Wehrenberg, W. B., Ueno, N., Ling, N., and Guillemin, R. (1985f). *Proc. Natl. Acad. Sci. U.S.A.* **82**, 5545–5549.
Baird, A., Mormède, P., and Böhlen, P. (1985g). *J. Cell Biochem.* Suppl. **9A**, 351.
Baird, A., *et al.* (1985h).
Barritault, D., Arruti, C., Courtois, Y. (1981). Differentiation **18**, 29–42.
Bassett, D. L. (1943). *Am. J. Anat.* (1973). 251–291.
Böhlen, P., and Mellet, M. (1979). *Anal. Biochem.* **94**, 313–321.
Böhlen, P., Baird, A., Esch, F., Ling, N., and Gospodarowicz, D. (1984). *Proc. Natl. Acad. Sci. U.S.A.* **81**, 5364–5368.
Böhlen, P., Esch, F., Baird, A., Jones, K. L., and Gospodarowicz, D. (1985a). *FEBS Lett.* **185**, 177–181.
Böhlen, P., Esch, F., Baird, A., and Gospodarowicz, D. (1985b). *EMBO J.* **4**, 1951–1956.
Clark, J. L., Jones, K. L., Gospodarowicz, D., and Sato, G. H. (1972). *Nature (London)* **236**, 180–181.
Clark, R. A., Stone, R. D., Leung, D. Y. K., Silver, I., Hohn, D. C., and Hunt, T. K. (1976). *Surg. Forum* **27**, 16–23.
Clemmons, D. R., Isley, W. L., and Brown, M. T. (1983). *Proc. Natl. Acad. Sci. U.S.A.* **80**, 1641–1645.
Corvol, M.-T., Malemud, C. J., and Sokoloff, L. (1972). *Endocrinology* **90**, 262–271.
Courty, J., Loret, C., Moenner, M., Chevallier, B., Lagente, O., Courtois, Y., and Barritault, D. (1985). *Biochimie* **67**, 265–269.
D'Amore, P., and Klagsbrun, M. (1984). *J. Cell Biol.* **99**, 1545–1549.
D'Amore, P. A., Glaser, B. M., Brunson, S. K., and Fenselau, A. H. (1981). *Proc. Natl. Acad. Sci. U.S.A.* **78**, 3068–3072.
Di Zerega, G. S., and Hodgen, G. D. (1980). *J. Clin. Endocrinol. Metab.* **50**, 819–825.
Di Zerega, G. S., Goebelsmann, U., and Nakamura, R. M. (1982). *J. Clin. Endocrinol. Metab.* **54**, 1091–1096.
Eccleston, P. A., and Silberberg, D. H. (1985). *Dev. Brain Res.* **21**, 315–318.
Ekblom, P. (1981). *J. Cell Biol.* **91**, 1–10.
Ekblom, P., Sariola, H., Karkinen-Jaaskelainen, M., and Saxen, L. (1982). *Cell Differ.* **11**, 35–39.
Erickson, G. F., Magoffin, D. A., Dyer, C. A., and Hofeditz, C. (1985). *Endocrine Rev.* **6**, 371–399.
Esch, F. (1984). *Anal. Biochem.* **136**, 39–47.
Esch, F., Ling, L., and Böhlen, P. (1983). In "Methods in Enzymology: Hormone Action Part II, Neuroendocrine Peptides" (P. Michael Conn, ed.), Vol. 103, pp. 72–89. Academic Press, New York.
Esch, F., Ling, N., Böhlen, P., Baird, A., Benoit, R., and Guillemin, R. (1984). *Biochem. Biophys. Res. Commun.* **122**, 899–905.
Esch, F. E., Baird, A., Ling, N., Ueno, N., Hill, F., Denoroy, L., Klepper, R., Gospodarowicz, D., Böhlen, P., and Guillemin, R. (1985). *Proc. Natl. Acad. Sci. U.S.A.* **82**, 6507–6511.
Folkman, J. (1975). *Am. J. Intern. Med.* **82**, 96–100.
Folkman, J. (1984). *Symp. Fundam. Cancer Res.* **36**, 201–208.
Frederick, J. L., Shimanuki, T., and diZerega, G. S. (1984). *Science* **224**, 389–390.
Glaser, B. M., D'Amore, P. A., Michels, R. G., Patz, A., and Fenselau, A. J. (1980). *J. Cell Biol.* **84**, 298–304.

Gospodarowicz, D. (1974). *Nature (London)* **249**, 123–127.
Gospodarowicz, D. (1975). *J. Biol. Chem.* **250**, 2515–2520.
Gospodarowicz, D. (1976). *In* "Membranes and Neoplasia: New Approaches and Strategies" (V. T. Marchesi, ed.), pp. 1–19. Liss, New York.
Gospodarowicz, D. (1979). *Mol. Cell. Biochem.* **25**, 79–110.
Gospodarowicz, D. (1984). *Horm. Proteins and Peptides* **12**, 205–230.
Gospodarowicz, D. (1985). *In* "Mediators in Cell Growth and Differentiation" (R. Ford and A. Maizel, eds.). Raven, New York.
Gospodarowicz, D., and Bialecki, H. (1978). *Endocrinology* **103**, 854–865.
Gospodarowicz, D., and Moran, J. S. (1976). *Annual Rev. Biochem.* **45**, 531–558.
Gospodarowicz, D., and Mescher, A. L. (1980). *Ann. N.Y. Acad. Sci.* **339**, 151–174.
Gospodarowicz, D., and Thakral, T. K. (1978). *Proc. Natl. Acad. Sci. U.S.A.* **75**, 847–851.
Gospodarowicz, D., Weseman, J., and Moran, J. (1975). *Nature (London)* **256**, 216–219.
Gospodarowicz, D., Moran, J., Braun, D., and Birdwell, C. (1976). *Proc. Natl. Acad. Sci. U.S.A.* **73**, 4120–4124.
Gospodarowicz, D., III, C. R., Hornsby, P. J., and Gill, G. N. (1977). *Endocrinology* **100**, 1080–1089.
Gospodarowicz, D., Bialecki, H., and Thakral, T. K. (1979). *Exp. Eye Res.* **28**, 501–514.
Gospodarowicz, D., Vlodavsky, I., and Savion, N. (1980). *J. Supramol. Struct.* **13**, 339–372.
Gospodarowicz, D., Cheng, J., and Lirette, M. (1983). *J. Cell Biol.* **97**, 1677–1685.
Gospodarowicz, D., Cheng, J., Lui, G. M., Baird, A., and Böhlen, P. (1984). *Proc. Natl. Acad. Sci. U.S.A.* **81**, 6963–6967.
Gospodarowicz, D., Cheng, J., Lui, G. M., Baird, A., Esch, F., and Böhlen, P. (1985a). *Endocrinology* **117**, 2383–2391.
Gospodarowicz, D., Massoglia, S., Cheng, J., Lui, G. M., and Böhlen, P. (1985b). *J. Cell Biol.* **122**, 323–333.
Gospodarowicz, D., Baird, A., Cheng, J., Lui, G. M., Esch, F., and Böhlen, P. (1986). *Endocrinology* **118**, 82–90.
Greenberg, G. B., and Hunt, T. K. (1978). *J. Cell. Physiol.* **97**, 353–360.
Guillemin, R. (1985). *Physiologist* **28**, 391–396.
Heder, G., Jakob, W., Halle, W., Maisersberger, B., Kambach, G., and Jentzsch, K. D. (1979). *Exp. Pathol.* **17**, 143–150.
Holley, R. W., and Kiernan, J. A. (1968). *Proc. Natl. Acad. Sci. U.S.A.* **60**, 300–304.
Hopp, T. P., and Woods, K. R. (1983). *Mol. Immunol.* **20**, 483–489.
Hornsby, P. J., and Gill, G. N. (1978). *Endocrinology* **102**, 926–936.
Hsueh, A. J. W., Adashi, E. Y., Jones, P. B. C., and Welsh, T. H. (1984). *Endocr. Rev.* **5**, 76–127.
Jakob, W., Jentzsch, B., Bauerberger, and Oehme, P. (1977). *Exp. Pathol.* **13**, 231–239.
Jones, K. J., Juilliard, J., and Guillemin, R. (1980). *Pediatr. Res.* **14**, 958.
Kellett, J. G., Tanaka, T., Rowe, J. M., Shiu, R. P. C., and Friesen, H. G. (1981). *J. Biol. Chem.* **256**, 54–58.
Kiss, F. A. (1975). *Stud. Biol. Acad. Sci. Hung.* 41–75, 153–154.
Kiss, F. A., and Szabo, S. (1969). *Acta Morphol. Acad. Sci. Hung.* **17**, 351–352.
Kiss, F. A., and Krompecher, I. (1962). *Acta Biol. Acad. Sci. Hung. Suppl.* **4**, 39–40.
Klagsbrun, M., and Smith, S. (1980). *J. Biol. Chem.* **75**, 10859–10866.
Knighton, D. R., Hunt, T. K., Scheuenstuhl, H., Halliday, B. J., Werb, Z., and Banda, M. J. (1983). *Science* **221**, 1283–1285.

Koos, R. D., and LeMaire, W. J. (1983). In "Factors Regulating Ovarian Function" (G. S. Greenwald and P. F. Terranova, eds.), pp. 191-195. Raven, New York.
Leibovich, S. J., and Ross, R. (1976). *Am. J. Pathol.* **78**, 71-77.
Lemmon, S. K., Riley, M. C., Thomas, K. A., Hoover, G. A., Maciag, T., and Bradshaw, R. A. (1982). *J. Cell Biol.* **95**, 162-169.
Linder, M. C., Bryant, R. R., Lim, S., Scott, L. E., and Moore, J. E. (1979). *Enzyme* **24**, 85-90.
Ling, N., Esch, F., Böhlen, P., Baird, A., and Guillemin, R. (1984). *Biochem. Biophys. Res. Commun.* **122**, 1218-1224.
Ling, N., Zeytin, F., Böhlen, P., Esch, F., Brazeau, P., Wehrenberg, W., Baird, A., and Guillemin, R. (1985). *Annu. Rev. Biochem.* **54**, 403-423.
McAuslan, B. R. (1980). In "Control Mechanisms in Animal Cells" (A. Jemenez, R. Levi-Montalcini, and R. Shields, eds.), pp. 285-292. Raven, New York.
Maciag, T., Cerundolo, J., Isley, S., Kelley, P. R., and Forand, R. (1978). *Proc. Natl. Acad. Sci. U.S.A.* **76**, 5674-5678.
Maciag, T., Hoover, G. A., and Weinstein, R. (1982). *J. Biol. Chem.* **257**, 5333-5336.
Maciag, T., Mehlman, T., Friesel, R., and Schreiber, A. B. (1984). *Science* **225**, 932-935.
Makris, A., Klagsbrun, M. A., Yasumizu, T., and Ryan, K. J. (1983). *Biol. Reprod.* **29**, 1135-1141.
Makris, A., Ryan, K. J., Yasumizu, T., Hill, C. L., and Zetter, B. R. (1984). *Endocrinology* **115**, 1672-1677.
Martin, B. M., Gimbrone, M. A., Unanue, E. R., and Cotran, R. S. (1981). *J. Immunol.* **126**, 1510-1521.
Michaelson, I. C. (1948). *Trans. Ophthalmol. Soc. UK* **68**, 137-180.
Mormède, P., Baird, A., and Pigeon, P. (1985a). *Biochem. Biophys. Res. Commun.* **120**, 1108-1113.
Mormède, P., Baird, A., Wehrenberg, W. B., and Guillemin, R. (1985b). Proceedings of the Fifth Meeting of the French Endocrine Society, Bordeaux, France, October 3-5, 1985.
Morrison, R. S., and De Veillis, J. (1981). *Proc. Natl. Acad. Sci. U.S.A.* **78**, 7205-7209.
Neill, J. D., and Frawley, L. S. (1983). *Endocrinology* **112**, 1135-1137.
Pierschbacher, M. D., and Ruoslahti, E. (1984a). *Nature (London)* **309**, 30-33.
Pierschbacher, M. D., and Ruoslahti, E. (1984b). *Proc. Natl. Acad. Sci. U.S.A.* **81**, 5985-5988.
Polverini, P. J., and Leibovich, S. J. (1985). *J. Leukocyte Biol.* **37**, 279-288.
Polverini, P. J., Cotran, R. S., Gimbrone, M. A., and Unanue, E. R. (1977). *Nature (London)* **269**, 804-806.
Preminger, G. M., Koch, W. E., Fried, F. A., and Mandell, J. (1980). *Am. J. Anat.* **159**, 17-24.
Pruss, R. M., Bartlett, P. F., Gavrilovic, J., Lisak, R. P., and Rattray, S. (1982). *Dev. Brain Res.* **2**, 19-35.
Raju, K. S., Allessandri, G., Ziche, M., and Gillino, P. M. (1982). *J. Natl. Cancer Inst.* **5**, 1183-1188.
Raju, K. S., Allesandri, G., and Gillino, P. M. (1984). *Cancer Res.* **44**, 1579-1584.
Rall, L. B., Scott, J., Bell, G. I., Crawford, R. J., Penschow, J. D., Niall, H. D., and Coghlan, J. P. (1985). *Nature (London)* **313**, 288-231.
Raptis, S., Rosenthal, J., and Gerich, J. (1984). *Int. Symp. Somatostatin, 2nd, Greece, 1981.*
Said, S. I. (1980). In "Frontiers in Neuroendocrinology" (L. Martini and W. F. Ganong, eds.), Vol. 6. Raven, New York.

Sariola, H., Ekblom, P., Lehtonen, E., and Saxen, L. (1983). *Dev. Biol.* **96**, 427–435.
Schonbrunn, A., Krasnoff, M., Westerdorf, J. M., and Tashjian, A. H., Jr. (1980). *J. Cell Biol.* **85**, 786–797.
Schwarzbauer, J. E., Tamkun, J. W., Lemishka, I. R., and Hynes, R. O. (1983). *Cell* **35**, 421–431.
Shimokado, K., Raines, E., Madtes, D. K., and Ross, R. (1986). *J. Cell Biol.* (in press).
Shing, Y., Folkman, J., Sullivan, R., Butterfield, C., Murray, J., and Klagsbrun M. (1984). *Science* **223**, 1296–1299.
Simmonian, M. H., and Gill, G. N. (1979). *Endocrinology* **104**, 588–595.
Simmonian, M. H., Hornsby, P. J., Ill, C. R., O'Hare, M. J., and Gill, G. N. (1979). *Endocrinology* **105**, 99–108.
Skelton, F. R. (1959). *Physiol. Rev.* **39**, 162–182.
Sullivan, R., and Klagsbrun, M. (1985). *J. Biol. Chem.* **260**, 2399–2403.
Tashjian, A. H., Jr., Yasumura, Y., Levine, L., Sato, G. H., and Parker, M. L. (1968). *Endocrinology* **82**, 342–351.
Thakral, T. K., Goodson, W. H., and Hunt, T. K. (1979). *J. Surg. Res.* **26**, 430–442.
Thomas, K. A., Riley, M. C., Lemmon, S. K., Baglan, N. C., and Bradshaw, R. A. (1980). *J. Biol. Chem.* **255**, 5517–5520.
Thomas, K. A., Rios-Candelore, M., and Fitzpatrick, S. (1984). *Proc. Natl. Acad. Sci. U.S.A.* **81**, 357–361.
Unger-Waron, H., Gluckman, A., Spira, E., Waron, M., and Trainin, Z. (1978). *Cancer Res.* **38**, 1296–1299.
Vlodavsky, I., and Gospodarowicz, D. (1979a). *J. Supramol. Struct.* **12**, 73–114.
Vlodavsky, I., Johnson, L. K., Greenburg, G., and Gospodarowicz, D. (1979b). *J. Cell Biol.* **83**, 468–486.
Volm, M., Wayss, K., Wesch, H., and Zimmerer, J. (1972). *Arch. Geschwlstforsch.* **40**, 248–258.
Walicke, P., Cowan, M., Ueno, N., Baird, A., and Guillemin, R. (1986). *Proc. Natl. Acad. Sci. U.S.A.* (in press).
Wall, R. T., Harker, L. A., Quadracci, L. J., and Striker, G. E. (1978). *J. Cell. Physiol.* **96**, 203–214.
Wehrenberg, W. B., Baird, A., Ying, S.-Y., Rivier, C., Ling, N., and Guillemin, R. (1984). *Endocrinology* **114**, 1995–2001.
Westall, F. C., Lennon, V. A., and Gospodarowicz, D. (1978). *Proc. Natl. Acad. Sci. U.S.A.* **75**, 4675–4678.
Ziche, M., Jones, J., and Guillino, P. M. (1982). *J. Natl. Cancer Inst.* **69**, 475–482.

DISCUSSION

R. Ryan. In the pituitary the molecular form of FGF has been reported as 12,000 Da, while in brain there is a slightly larger species that elutes heparin agarose at a high salt concentration. You said they have the same N-terminal sequence; what accounts for the size difference?

A. Baird. There is indeed in brain (Fig. 13) and in retina (Fig. 19) a second form. It is actually a bit smaller than FGF, putting it at about 15,500 Da. We purified and sequenced this form. There is quite a bit of activity eluting with 1 M NaCl, prior to the elution of basic FGF. In fact, this molecule is a molecule distinct from basic FGF. We've identified it as being acidic FGF [Esch *et al.*, *Biochem. Biophys. Res. Commun.*, **133**, 554–562 (1985)] and, as shown it contains about 50–75% structural homology with basic FGF. Its presence

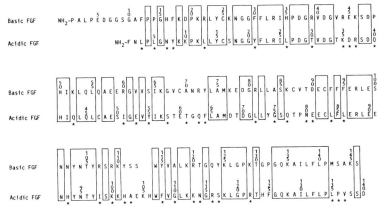

FIG. 39. Structural homology between bovine pituitary basic FGF and brain acidic FGF.

accounts for the detection of acidic endothelial cell growth factors in bovine brain, and I think it has the same activity that was described by Maciag et al. (1984) as being endothelial cell growth factor (ECGF) and is identical to the structure characterized by Thomas et al. [Science, 230, 1385–1388 (1985)] as being acidic FGF (Fig. 39).

R. Ryan. There is another activity peak of ir-FGF that comes off around 2.5–3 M sodium chloride in extracts of human brain, while your last elution of bovine FGF was with 2 M.

A. Baird. That was a second immunoreactive form of basic FGF (Fig. 16). We've partially characterized the amino-terminus of this molecule. It appears identical by all criteria to pituitary basic FGF, so it would be another microheterogeneous form. We are not sure why this form has a greater affinity for heparin than the form we have completely characterized.

M. R. Sherman. I have two questions related to possible clinical applications of your results, particularly of your antibody to FGF. First, have you used your radioimmunoassay to compare the levels of FGF in human ovaries that are found to be benign or malignant at the time of ovariectomy?

A. Baird. No, we have only begun to look at FGF in humans.

M. R. Sherman. If you did find significantly higher levels of FGF in malignant ovaries than in benign ovaries or serum, your antibody might be useful in the development of a noninvasive technique for detecting ovarian cancer at an early stage. What I envision is that the antibody would be labeled with a gamma-emitting isotope such as ^{125}I and injected into the patient with suspected ovarian cancer. The body could then be scanned for excessive localization of the radioactivity labeled anti-FGF antibodies in the ovaries, as an adjunct to present noninvasive techniques such as ultrasound.

A. Baird. The problem we've had using passive immunization is that the concentration of FGF in serum was very high.

J. D. Veldhuis. Two questions are pertinent to the possible actions of FGF in the ovary. First, you have shown interesting data from your group and Aaron Hsueh that FGF inhibits FSH stimulated aromatase in the rat system. In contrast, the distinct but somewhat related growth factor IGF-I curiously enhances aromatase activity both in the rat and in the pig. Can you speculate on a subcellular mechanism by which these two peptide hormones would exert divergent effects on the end point aromatase?

A. Baird. It's certainly an interesting observation, although the subcellular mechanism, at this point, is anyone's conjecture. It will inevitably entail relating bioavailability of each growth factor. In other words, the effects of IGF and the opposite reactions to FGF will be a strict function of when they're present as a biologically active form to exert an action in that tissue.

J. D. Veldhuis. My second question relates to your observation in the pituitary in which estradiol appears to increase the production of FGF in that organ. Is this the case in the ovary, and, if so, would increasing estradiol levels enhance FGF production which would in turn inhibit estradiol synthesis? It would be a curious autonegative feedback effect within the ovary.

A. Baird. We haven't looked at the effects of estrogen on FGF in the ovary.

C. S. Nicoll. I am curious to know how you could establish the specificity of your radioimmunoassay for measuring plasma or serum levels of the FGF. Generally, when one develops an RIA for a peptide hormone, such as growth hormone, one must at least remove the source of the hormone to check that its levels go down essentially to zero. I don't see how you can do that with FGF, so how can you rule out the possibility of nonspecific cross-reacting materials?

A. Baird. You're right. It's a major question that has to be addressed and I'm not really sure how it's approachable. You must remember that the serum FGF results are very preliminary and we emphasize that the measurement that is made is that of an immunoreactive form. Future experiments will certainly determine whether the measurement really reflects the distinct biological entity which is FGF. It's also important to remember, however, that even in the measurements of "known" and "defined" hormones such as prolactin and growth hormone it is still unclear as to what form is being measured. Therefore, it's really difficult to get a fully validated serum RIA for any hormone.

C. S. Nicoll. In view of the fact that this material has numerous other effects besides simulating fibroblast proliferation, has anyone suggested a more appropriate name for the material?

A. Baird. Right. Moreover, the material measured in serum may not even reflect biological activity since it appears at such a high molecular weight.

C. S. Nicoll. In view of the fact that this material has numerous other effects besides simulating fibroblast proliferation, has anyone suggested a more appropriate name for the material?

A. Baird. Yes. In fact it appears, as I mentioned, that this one protein has at least half a dozen different names. We didn't change the name so that it would be accurate in a historical context.

R. Ryan. Could I follow up on Dr. Nicoll's question. Can you dissociate material in the serum or examine it under dissociating conditions, and show that the immunoactivity moves as a smaller molecule, or does it always behave at 150,000?

A. Baird. If you use dissociated conditions you find the circulating form of FGF immunoreactivity to be a 70,000 molecule which is the form that we found in tissues.

T. T. Chen. In a mutant mouse strain called Blind/Sterile in which the animal cannot see and cannot reproduce, I wondered if anybody has looked at the FGF levels?

A. Baird. No.

T. T. Chen. The reason I ask is that blindness is one of the consequences of diabetes and since FGF can inhibit steroid synthesis, I wondered if there is any connection in which FGF is the common denominator?

A. Baird. It's certainly possible.

T. T. Chen. In the ovarian cell treated with FGF, do you know if the cells go into cell cycle or not?

A. Baird. The only way I can answer that would be to say that the granulosa cells under these conditions do not proliferate. We have not studied whether they go into G_1 or S. The cell numbers under these conditions did not change, and it's a 72-hour incubation.

M. V. Nekola. Would you care to speculate on what blocks FGF activity *in vivo*? Does this imply an FGF IF?

A. Baird. In fact, several tissues including follicular fluid have the unusual capacity to inhibit the effect of FGF, suggesting that there is something in these tissues that can modulate its activity. Determining how these factors modify the bioavailability of FGF to various target cells is probably the most exciting challenge of the next few years.

M. V. Nekola. When you measured FGF and prolactin release from pituitary cells, did you measure any other peptide hormones to see how specific the response was?

A. Baird. It appears only to be for prolactin and TSH.

M. V. Nekola. But that was in a tumor line.

A. Baird. No. The *in vitro* studies were done with cultures of normal rat pituitaries.

H. G. Friesen. Could you give us some information about the stability and recovery of FGF in serum?

A. Baird. The stability of FGF in serum is truly amazing. You can take a sample of serum ir-FGF, put it in the cold room for up to a month, go back and measure it, and you'll still measure the same amount there and always as this 150,000-Da molecule.

H. G. Friesen. Could you speculate on the interaction of heparin and FGF. How do you envisage this might take place? Is the binding of heparin and FGF fortuitous or do you believe heparin might have a physiologic role interacting with FGF?

A. Baird. I think that's probably one of the most interesting aspects of FGF. There's no question that a lot of proteins will bind heparin, but very few proteins will actually bind heparin with the affinities that FGF seems to exhibit. I think it's going to be inherently related to its biological activity *in vivo*. The binding and angiogenic activities were actually the first clue that gave Folkman's group the idea of testing the use of heparin affinity chromatography. The fact that several cells have GAG proteins as an intrinsic part of their cellular membrane and that they are inherent parts of the extracellular matrix suggest that indeed interactions with FGF and GAG proteins represent an important physiological role in the modulation of FGF activity.

H. G. Friesen. I'd just like to mention that Drs. John Rowe and Steven Henry in my laboratory have examined the interaction of heparin with human pituitary FGF in a number of situations in which it acts as a mitogen. In these cases there seems to be a synergistic effect of heparin on FGF in promoting growth. I wondered if you had an opportunity to study this.

A. Baird. Which FGF?

H. G. Friesen. Pituitary FGF.

A. Baird. So it would be the basic FGF. We haven't studied this, however, I understand Denis Gospadorowicz is studying that in detail.

H. Papkoff. I recall your saying that the concentration of FGF was higher in liver and kidney than in pituitary, and it seems to me I once heard that you were getting as much as 30 mg/kg of pituitary. Could you give some hard numbers as to how much FGF might be present in these tissues?

A. Baird. Not 30 mg. First of all we have to consider two different things. If we're talking about the amount of bioactive 16,000 Da basic FGF that can be isolated from a kilogram of bovine pituitary, you're talking about 10-30 nanomoles which would be of the order of 160–500 µg. In the studies I mentioned in which we looked at the difference between tissues, we were looking at neutral extracts. Under those conditions, where we didn't go to final purification, the final yield is actually anybody's guess. What I was alluding to in these studies was that there is more immunoreactive FGF in the neutral extracts of liver and kidneys, than there is in similar extracts of pituitary. To date, however, the pituitary is by far the best source for the growth factor if you want it homogeneous at the step of heparin-Sepharose affinity chromatography.

H. Papkoff. On a weight basis?

A. Baird. Yes, on a weight basis. However the rat pituitary contains less FGF than the bovine pituitary, so we're making comparisons not only in terms of methods of extraction here, we're making comparisons in terms of different species.

D. Goltzman. In view of the role that you have postulated for this material in diabetic retinopathy, have you had a chance to see if insulin affects either biosynthesis or release of FGF from any of the tissues in which it originates? In particular, does it inhibit FGF biosynthesis or release?

A. Baird. We haven't looked at that, but, indeed, I think this possibility is going to be of major interest in the next few years.

L. A. Meserve. Is FGF an important angiogenic substance in goitrogenic proliferation of capillaries in the thyroid gland? Does anyone know?

A. Baird. I don't think that anyone knows. My guess at this point would be that since FGF is so widespread, the activity in thyroid is probably due to at least either the basic or the acidic forms of FGF.

L. A. Meserve. I just thought it fascinating that in your pituitary studies TSH was another hormone the release of which was potentiated by FGF, and wonder if there is any sort of communication that might be established between pituitary FGF and thyroid FGF if they are indeed involved in that process?

A. Baird. I think its certainly possible. At this point though I would consider it more likely that the growth factor in the thyroid would be in the same hormonal milieu, which may than have the same tropic effect in the thyroid.

D. L. Keefe. I have a question about another possible clinical implication of FGF. Endometrosis in women is associated with local macrophage accumulation fibrosis formation, as well as neovasculization. I wonder if it would be worthwhile to look at immunoreactive FGF in peritoneal fluid of women with endometriosis?

A. Baird. Yes. In fact that is one of the reasons that we are very excited about the possibility of developing antagonists to FGF molecules that could be used to see if, indeed, we could go in irrespective of the molecular forms of FGF and interfere with the biologically active molecules responsible for neovascularization. Once again, however, it is going to be of critical importance to determine what regulates the pathophysiological expression of the growth factor under this condition.

R. F. Laherty. I was very happy to hear your elegant studies on FGF, and the fact that brain FGF now actually is the same as pituitary FGF. I was particularly interested in your studies in the ovary, and I was wondering if you have done any studies in the ovaries, or in other tissues, which would reveal what might modulate the control of the levels of FGF found in these tissues?

A. Baird. That is of major concern right now. The fact that FGF is widely distributed in various tissues, what indeed is modulating the expression? The fact that in these tissues we have been able to detect this high-molecular-weight seemingly less active molecule would suggest to me that one of the regulatory control mechanisms may be the generation of the truly biologically active molecule from the 70,000-Da molecule.

R. F. Laherty. I would be particularly interested in knowing if it played a role in atresia?

A. Baird. That would be the underlying correlate to its inhibiting aromatase activity.

M. H. Melner. I wondered if you have had a chance to search for the presence of and measure the levels of FGF in certain tumors which display an extensive amount of angiogenesis such as angiosarcomas?

A. Baird. We haven't looked in the angiosarcomas per se but we have looked in several tumors, all of which contain significant amounts of immunoreactive FGF, but again nothing particularly outstanding compared to normal tissues. One thing I would like to emphasize is that the immunoassay may be measuring the inactive and the active forms, i.e., the 70,000- and 16,000-Da molecules. I think the important thing at this point is going to be the development of a radioreceptor assay which will specifically measure the active forms. With these tools it will be possible to do these studies and distinguish total FGF versus biologically active FGF in tissues.

D. K. Pomerantz. You indicated that you have an antibody of reasonable quality that you use for an RIA and you also have fibroblast growth factor. Have you used this antibody to neutralize endogenous growth factor to show that in fact tissues respond to a loss of the growth factor?

A. Baird. Actually, that is a very good question. The problem we have had is that the serum levels of FGF are very high. As an example, when we noticed the increased FGF in these prolactin-secreting tumors, we tested whether administration of estrogen plus antibody to FGF could inhibit development of these tumors. These experiments were unsuccessful. We think it is because of our inability to overcome the amounts of circulating FGF and get a critical neutralizing titer of what would be endogenous pituitary FGF. In other studies we have used a transplantable mouse chondrosarcoma which grows to a gram in less than 15 days to show that the antibodies indeed were effective in slowing down the growth of these tumors. That is an indirect answer to your question. The only tissue for which we have been able to establish a requirement needs the factor for growth not function.

D. K. Pomerantz. Have you attempted to use that antibody in any *in vitro* situation to show that you can alter tissue function?

A. Baird. No.

J. E. Rall. Is fibroblast growth factor the same as the angiogenesis factor that Folkman has been talking about for so many years?

A. Baird. Yes, I would say so, although there is some good evidence that, at least in tumors, it is aminoterminally extended.

Atrial Natriuretic Factor: A New Hormone of Cardiac Origin

STEVEN A. ATLAS

Cardiovascular Center and Department of Medicine Cornell University Medical College, New York, New York

A considerable body of clinical observation and experimental evidence has led to the long-held suspicion that atrial stretch receptors may be involved in fluid volume regulation (1–5). Early work suggested that the rapid diuretic response to mechanical distension of the left atrium was mediated in part by vagal afferent nerves (6,7), leading to inhibition of vasopressin release (1,6,8) and/or to decreased renal sympathetic nerve traffic (3,9). Other responses to atrial stretch, including the natriuresis and decreases in renin and aldosterone release that are particularly prominent following distension of the right atrium (10,11), have been less convincingly ascribed to neural reflexes alone. In keeping with earlier suggestions (3), evidence has accumulated recently that a humoral mediator likely contributes to the multiform effects of atrial stretch, based upon the demonstration by deBold and co-workers in 1981 that extracts of atrial muscle cells contain a potent and rapidly acting natriuretic and diuretic factor (12).

The notion that the cardiac atria might have a secretory function in fact dates back nearly 30 years when it was discovered that atrial muscle cells contain cytoplasmic granular inclusions not present in mammalian ventricular muscle (13,14). Although an endocrine function was not explicitly postulated for these inclusions, they were characterized as membrane-bound, electron-dense granules that bore an intimate relationship to an abundant Golgi complex in atrial muscle (14), and thus reminiscent of secretory granules found in hormone-producing cells. Later studies suggested that the degree of atrial granularity is altered somewhat following perturbations of fluid volume balance (15,16), but it was the direct demonstration of an "atrial natriuretic factor" (12) that provided the first convincing evidence linking the responses to atrial distension with the existence of specific atrial granules.

The idea of a hormonal substance that promotes sodium excretion is not, of course, new. Since the initial work of deWardener and colleagues in 1961 (17), numerous investigators have suggested the existence of potential natriuretic substances ("third factor") in blood or urine of volume-expanded animals (18–22). More recent work has focused on linking these putative natriuretic factors to circulating ouabain-like inhibitors of Na^+,K^+-ATPase (23–25), perhaps of hypothalamic origin (26). To date, however, investigators have been unable to fully characterize these substances or to establish their physiological significance.

In contrast, investigation regarding the atrial natriuretic factor (ANF) has progressed at an impressive pace. This is in large part attributable to the establishment of an unequivocal tissue source for this factor. Further, it became apparent from early studies that atrial extracts, in addition to producing marked natriuresis, induced major alterations in renal hemodynamics (27) and also caused relaxation of precontracted vascular smooth muscle (28). In addition to providing fundamental insights into the nature and potential actions of ANF, these findings provided the basis for alternative bioassays which facilitated its subsequent isolation and purification.

I. Structure of ANF and the ANF Precursor

Early studies revealed that ANF is comprised of heat-stable, protease-degradable polypeptides existing in both high and low-molecular-weight forms (29–32). Low-molecular-weight fractions of rat atrial extracts were resolved into several biologically active forms by cation-exchange chromatography or reversed-phase HPLC (33–38). These findings are illustrated in Fig. 1, which shows that both the large-molecular-weight and multiple low-molecular-weight forms of ANF possess both natriuretic and vasorelaxant activities. Further isolation studies were initially reported by deBold and Flynn (33) and Grammer et al. (35), who reported purification to apparent homogeneity of 5000- and 3800-Da rat atrial peptides, respectively, and provided amino acid composition data.

A. AMINO ACID SEQUENCE OF ANF PEPTIDES

Subsequently, a series of structurally related smaller peptides, ranging between 21 and 35 amino acid residues, were isolated and sequenced in several laboratories (34, 37–45). The reasons for the discrepancy in molecular size and in amino acid composition between these peptides and the larger ones mentioned above are not clear, although difficulties in precise molecular weight estimation and possible impurities in the earlier preparations (33,35) may have contributed.

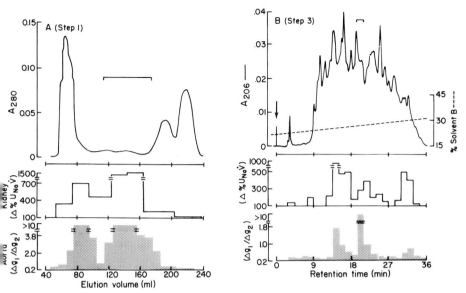

FIG. 1. Isolation of natriuretic and vasorelaxant peptides from rat atria. Atria from 200 rats were homogenized in 12 volumes of 1 N acetic acid containing protease inhibitors and, following centrifugation, the supernatant was lyophilized and reconstituted in 6 ml of 1 N acetic acid. (A) Gel filtration on a 2.5×45 cm column of Sephadex G-50 (fine). Aliquots of each fraction were lyophilized, reconstituted in phosphate-buffered saline, and assayed for natriuretic activity in the isolated perfused kidney and for vasorelaxant activity in angiotensin II-contracted rabbit aortic rings. Two broad regions of natriuretic and vasorelaxant activity were identified, corresponding to MW >10,000 and 2000–5000 (bracket). The latter fractions were pooled, lyophilized, and then further purified by reversed-phase HPLC on a 0.39×30 cm μBondapak C_{18} column. A single broad region of activity was found (Step 2, not shown). (B) Rechromatography of the low-molecular-weight material by reversed-phase HPLC. Using a gradient of 22 to 34% acetonitrile (solvent B) in 0.1% trifluoroacetic acid over 48 minutes, the active material was resolved into 3 principal components of coincident natriuretic and vasorelaxant activity. Peaks were subsequently purified to homogeneity (using a total of 1400 rat hearts) by additional steps of reversed-phase HPLC. Taken from Atlas et al. (37). Reprinted by permission from Nature. Copyright © 1984 Macmillan Journals Limited.

Some of the peptides sequenced to date are shown in Fig. 2, where they are compared to the amino acid sequence of the C-terminal portion of the ANF precursor (to be discussed below). A number of generic terms other than ANF have been proposed for the atrial peptides, including cardionatrin (33,34), atriopeptin (38), and auriculin (37). These related peptides have in common a core sequence including a 17-member ring, formed by an intra-chain cystine disulfide bridge, and vary in the lengths of their N- and C-terminal extensions. The human and rat sequences in this region of

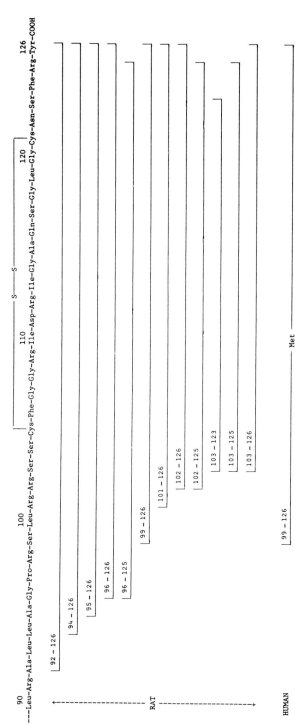

FIG. 2. Amino acid sequences of ANF peptides isolated from rat and human atria. The peptides isolated are compared to the amino acid sequence of the rat ANF precursor (pro-ANF). These peptide sequences were identified in studies by several investigators: Ala^{92}-Tyr^{126} (41,43), Leu^{94}-Tyr^{126} (41,44,45), Ala^{95}-Tyr^{126} (41,44,45), Gly^{96}-Tyr^{126} (41,43–45), Gly^{96}-Arg^{125} (43), Ser^{99}-Tyr^{126} (34,41), Arg^{101}-Tyr^{126} (45), Arg^{102}-Tyr^{126} (37,41–43), Arg^{102}-Arg^{125} (37), Ser^{103}-Ser^{123} (38), Ser^{103}-Arg^{125} (38), Ser^{103}-Tyr^{126} (55), and human Ser^{99}-Tyr^{126} (40). The latter sequence is identical to that of the rat peptides except for substitution of Met for Ile at position 110.

the precursor are identical save for a single amino acid (Met substituted for Ile in position 110 of the rat precursor). Of note, only a single small peptide, termed α-human ANP, has been identified in human atrium (40), corresponding to the 28-residue rat peptide (34); in addition, an unusual 56-residue peptide has been isolated from human atrial extracts ("β-hANP") which corresponds to an antiparallel dimer of α-hANP formed by two intermolecular disulfide bridges (46).

Since very large-molecular-weight (>10,000) forms of ANF, which are now known to correspond to the precursor or to fragments thereof (see below), are a major storage form in atrial tissue (29,30,37; see Fig. 1), it is likely that the multiple peptides isolated have arisen as artifacts due to proteolysis during extraction and purification. The possibility that the multiple peptides could reflect polymodal processing of the precursor *in vivo* has not, however, been entirely excluded. A variety of extraction methods and means of inhibiting proteolysis have been used in the reports published to date. Several peptides have been isolated by most of the laboratories concerned, though their exact sequences have differed from laboratory to laboratory, and variations in extraction conditions have clearly been shown to alter the species of peptides obtained (41,47). Although the issue is far from resolved, there is very recent evidence that the principal low-molecular-weight species of immunoreactive ANF present in blood may correspond to the 28-residue peptide termed cardionatrin I in the rat (34,47–49).

B. STRUCTURES OF THE ANF PRECURSOR AND GENE

The availability of amino acid sequence data enabled several groups of investigators, in nearly simultaneous reports, to identify and sequence the complementary DNA corresponding to the ANF precursor in cDNA clones established from rat or human atrial poly(A)$^+$ (messenger) RNA (50–53). It was thus predicted early on that the rat atrial peptides are derived from a 152-amino acid precursor (prepro-ANF) containing a hydrophobic signal peptide (approximately 24 residues in the rat), typical of secretory proteins, and the precursor proper (126 amino acids). The C-terminal portion of this sequence is shown in Fig. 2 and the entire precursor sequence is provided in Fig. 3. The C-terminal basic dipeptide (Arg-Arg) predicted from rat cDNA would not necessarily be expected in the isolated precursor since it would be readily subject to hydrolysis during processing. The remainder of this predicted sequence has, in fact, been validated since the intact 126-residue precursor (46,47,54) and 48-, 73-, 78-, 106-, and 111-residue fragments of the precursor (47,55–58) have been purified from rat and human atria and sequenced. The predicted human precursor has a similar general structure (Fig. 3), although it con-

	-25					-20										-10									-1
Human	Met	Ser	Ser	Phe	Ser	Thr	Thr	Thr	Val	Ser	Phe	Leu	Leu	Leu	Leu	Ala	Phe	Gln	Leu	Leu	Gly	Gln	Thr	Arg	Ala
Rat		Gly				Ile	---		Lys	Gly		Phe		Phe				Trp		Pro		His	Ile	Gly	
Mouse		Gly				Ile	---		Leu	Gly		Phe		Val				Trp		Pro		His	Ile	Gly	

	1									10										20						
Human	Asn	Pro	Met	Tyr	Asn	Ala	Val	Ser	Asn	Ala	Asp	Leu	Met	Asp	Phe	Lys *	Asn	Leu	Leu	Asp	His	Leu	Glu	Glu	Lys	Met
Rat			Val		Ser					Thr																
Mouse			Val		Ser					Thr																

				30										40										50	
Human	Pro	Leu	Glu	Asp	Glu	Val	Val	Pro	Pro	Gln	Val	Leu	Ser	Glu	Pro	Asn	Glu	Glu	Ala	Gly	Ala	Ala	Leu	Ser	Pro
Rat		Val					Met				Ala				Gln	Thr	Asp								Ser
Mouse		Val					Met				Ala				Gln	Thr									Ser

									60										70							
Human	Leu	Pro	Glu	Val	Pro	Pro	Trp	Thr	Gly	Glu	Val	Ser	Pro	Ala	Gln	Arg	Asp	Gly	Gly	Ala	Leu	Gly	Arg	Gly	Pro	Trp
Rat		Ser										Asn		Ser												
Mouse												Asn		Pro	Leu				Ser		Ser	Arg		Ser		

			80										90										100		
Human	Asp	Ser	Ser	Asp	Arg	Ser	Ala	Leu	Leu	Lys	Ser	Lys	Leu	Arg	Ala	Leu	Leu	Thr	Ala	Pro	Arg	Ser	Leu	Arg	Arg
Rat		Pro																Ala	Gly						
Mouse		Pro																Ala	Gly						

								110										120						126		
Human	Ser	Ser	Cys	Phe	Gly	Gly	Arg	Met	Asp	Arg	Ile	Gly	Ala	Gln	Ser	Gly	Leu	Gly	Cys	Asn	Ser	Phe	Arg *	Tyr		
Rat								Ile																	Arg	Arg
Mouse								Ile																	Arg	Arg

FIG. 3. Deduced amino acid sequences of the human, rat, and mouse ANF precursors (prepro-ANF). Amino acid residues are numbered according to the predicted (51,53) and subsequently proven (46,47,54) cleavage point of the signal peptide [i.e., pro-ANF = residues 1 to 126 and the signal peptide = residues −1 to −25 (human) or −24 (rat,mouse)]. The complete human sequence is shown and only residues which differ in the rat and mouse sequences are indicated. The human sequence was deduced from both cDNA (51) and genomic DNA (59,60), the rat sequence from cDNA (53), and the mouse from genomic DNA (60). The asterisks between positions Lys[16]-Asn[17] and Arg[125]-Tyr[126] indicate the points at which

tains only 151 residues, having a longer (25-residue) signal peptide and lacking the C-terminal basic dipeptide (51); the latter finding is accounted for by a single base substitution, resulting in replacement of the first Arg codon in the rat by a termination signal in the human. Thus both the rat and human precursor proper (i.e., pro-ANF) contain 126 residues. There is nearly complete amino acid homology between the rat and human precursors in the C-terminal portion from which the biologically active peptide(s) (ANF) are derived, and the N-terminal 30 residues are rather well conserved, but considerable differences occur in other portions of the precursor sequence, particularly in the signal peptide (Fig. 3).

More recently, the nucleotide sequence of the entire human and mouse genes coding for the ANF precursor has been determined, using labeled cloned cDNA as a probe to identify the gene in cloned genomic DNA (59,60). It has thus been established that the ANF precursor is derived from three coding sequences, with two intervening sequences (introns) that are spliced out during the processing of messenger RNA (illustrated schematically in Fig. 4). There is a large intron separating the second coding sequence (which contains nearly the entire sequence coding for the biologically active ANF peptides) and the third sequence, which codes only for the C-terminal tyrosinyl residue (Tyr-Arg-Arg in the rat and mouse). This rather unusual finding has been interpreted as indicating either that the tyrosine is critical for biologic expression (60) or that it is relatively unimportant (59). The latter interpretation is more consistent with the known structure–activity relationships to be discussed below. Since it has been observed that introns frequently separate functional domains of expressed proteins, and since the C-terminal Arg-Arg dipeptide (which is a common proteolytic processing site in polypeptide hormone precursors) has been lost in the evolution to humans, the preservation of this large intron and third coding sequence has led to speculation that the ANF gene may have evolved from a larger, multifunctional ancestral gene (59). Of interest, a putative glucocorticoid recognition sequence has been identified in this intron in the human gene (59), and there is a preliminary report indicating that glucocorticoids may regulate ANF gene expression in certain tissues (61).

C. HOMOLOGY WITH OTHER PROTEINS

Sequence analysis of the cDNA and genomic DNA coding for ANF has revealed no substantial homology with any other known protein. It should be noted, in this regard, that some investigators have postulated that the ANF precursor encompasses two distinct biologically active peptides, ANF itself at the C-terminus and "cardiodilatin" (62) at the N-terminus (i.e., beginning at position 25 of the cDNA-predicted rat precursor). Thus

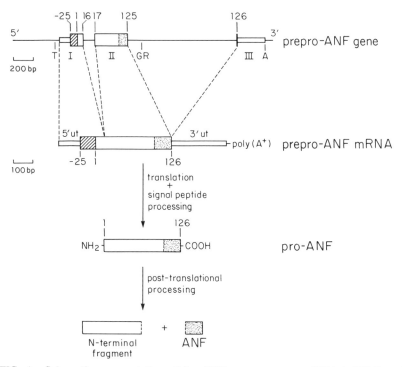

FIG. 4. Schematic representation of the ANF gene, messenger RNA (mRNA), and precursor (pro-ANF). In the gene structure, the horizontal lines represent the 5' and 3' flanking regions and the intervening sequences (introns) that separate the three coding blocks (exons I, II, and III). Features typical of eukaryotic genes include a "TATA" box (T) in the 5' flanking region and a polyadenylation signal (A) near the 3' end; in addition, a sequence homologous to the putative glucocorticoid receptor binding site (GR) occurs in the second intron. Roman numerals correspond to the amino acid sequence of the translated product, indicating the locations of the respective codons in the gene and mRNA. The narrow and wide horizontal bars indicate, respectively, the untranslated (ut) and translated portions of the mRNA. The signal peptide (cross-hatched portion, residues −1 to −25) is presumably cleaved co-translationally. Post-translational processing of pro-ANF leads to formation of the mature ANF peptide(s) at the C terminus (shaded portion); whether this occurs intracellularly or extracellularly is uncertain.

have arisen the terms "cardiodilatin-ANF precursor" or "pronatriodilatin" (57,63,64). This speculation is based on the substantial homology between the N-terminus of the rat ANF precursor and the partial N-terminal sequence analysis reported by Forssmann et al. (62) for a vasorelaxant polypeptide (cardiodilatin) isolated from pig atria, which was estimated to have a molecular weight of approximately 8000 by gel filtration. The notion of a distinct biological activity at the N-terminus was thus based on an imprecise estimate of molecular size and on an unsubstantiated statement that the vasorelaxant pig polypeptide lacked detectable natriuretic activity (62). In fact, it is quite likely that porcine cardiodilatin could represent the intact ANF precursor, which is known to have weak biological activity (46,47,54). Thus, until better documentation is obtained, there is no basis for assuming that the ANF precursor codes for more than a single class of biologically active peptides.

II. Biological Effects of ANF

Many of the established actions of ANF were first described using crude atrial extracts or partially purified preparations, and these have been confirmed with pure synthetic peptides. Given the large variety of peptides isolated and studied, a brief word about structure–activity relationships is in order.

A. STRUCTURE–ACTIVITY REQUIREMENTS

Although it must be emphasized that current understanding of this issue is based upon rather imprecise and at best semiquantitative bioassays, it is reasonably certain that the minimum requirement for nearly full biological activity in the rat resides in the 23-residue peptide Ser^{103}-Arg^{125} ("atriopeptin II," numbered according to the position of residues in the 126 amino acid pro-ANF; see Fig. 2). Thus, the presence of the C-terminal tyrosine (Tyr^{126}) has little appreciable effect on either natriuretic or vasorelaxant activity. Absence of the next two amino acids (Phe^{124}–Arg^{125}), as in "atriopeptin I," was originally thought to result in loss of vasorelaxant activity but retention of natriuretic activity and the ability to relax the carbachol-contracted chick rectum (38,39). Further work has indicated, however, that this peptide retains all activities but exhibits a marked reduction in biological potency (70 to 150-fold) of similar magnitude for both vasorelaxation and natriuresis (65–67). Thus, although further study is clearly needed, there is no evidence at present for a dissociation of these two fundamental properties of ANF. Extensions at the N-terminus, resulting in peptides of up to 33 amino acid residues (Fig. 2), appear to have little discernible effect on potency in rat tissues (67), but peptides extended by a single N-terminal arginyl residue or more, such as Arg^{102}-Tyr^{126} ["auriculin B" (37,42)], Arg^{101}-Tyr^{126} ["ANF 8-33" (45)], and Ser^{99}-Tyr^{126} ["cardionatrin I" (34)], exhibit greater activity in other species (66,68,69). It must be considered, however, that, depending on the experimental model used, a greater activity may reflect not only biological potency (i.e., receptor recognition) but also metabolism and clearance. On the other hand, the biological activities of some of the even larger forms that have been isolated (73 to 126 amino acids) appear to be reduced (46,47,54,58,67).

B. EFFECTS ON ISOLATED SMOOTH MUSCLE

The original observation of Deth et al. (28), who reported that atrial extracts cause relaxation of norepinephrine-contracted rabbit aortic strips, has been confirmed in many laboratories (35,37,70–73). ANF also causes relaxation of the isolated chick rectum precontracted with carbachol (70). The vasorelaxant effect of ANF does not depend upon an intact endothelial lining (73,74). ANF has generic actions on vascular smooth muscle, as noted by Kleinert et al. (72), since it relaxes isolated vessels precontracted with a variety of hormonal and nonhormonal vasoconstrictor substances. Despite this lack of specificity, the effect of ANF on angiotensin II-induced contraction appears to be especially marked (72), since angiotensin is uniquely unable to overcome the inhibitory effect (Fig. 5). Recent work has indicated that ANF affects mainly arteries or the artery-like rabbit facial vein (73), and appears to have lesser effects on venous tone *in vitro* (75). As will be discussed in a subsequent section, the vasorelaxant effect of ANF is associated with activation of particulate guanylate cyclase (see below).

C. RENAL HEMODYNAMIC EFFECTS

A rather marked renal hemodynamic effect of ANF was first observed in the isolated perfused rat kidney (27,76,77). In kidneys perfused with

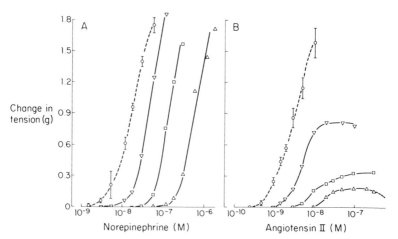

FIG. 5. Inhibition of norepinephrine- (A) and angiotensin II- (B) induced contraction of isolated rabbit thoracic aortic rings by increasing concentrations of ANF. Each curve depicts the increase in tension caused by increasing concentrations of vasoconstrictor (i.e., norepinephrine or angiotensin II). (○), Control rings; (▽), (□), (△), increasing concentrations of partially purified ANF added to the bath. Modified from Kleinert et al. (72).

artificial medium in the absence of innervation or of circulating vasoconstrictors, ANF induces a marked increase in glomerular filtration rate (GFR) associated with slight *increases* in renal vascular resistance and an increase in filtration fraction (77). Thus, in the absence of circulating vasoconstrictors, ANF leads to net renal vasoconstriction (a phenomenon that has not to date been observed in isolated vessels); this vasoconstrictive effect is likely to be exerted mainly at the efferent arteriole of the glomerulus (77). ANF causes net renal vasodilation, on the other hand, when the renal vasculature is first precontracted with hormonal or nonhormonal agonists (77), which probably explains the potent renal vasodilation induced by ANF in the *in situ* blood-perfused kidney (78), a highly vasoconstricted model. Although ANF has been reported to increase renal blood flow under nonsteady-state conditions or in certain experimental preparations (78–80), there is evidence that constant infusion of synthetic ANF does not lead to sustained renal vasodilation under basal conditions *in vivo* (81,82); in contrast, significant increases in GFR have been observed in intact animals in most (81-87), though not all (88), studies. Either systemic (81) or intrarenal artery (82) infusion of ANF in intact dogs leads to a transient increase in total renal blood flow lasting less than 2 minutes. Thereafter, renal blood flow returns to or below control levels, whereas increases in GFR (and natriuresis) are sustained for up to 1 hour (81,82). Thus, although ANF clearly has greater potential effects on the renal vasculature compared to other vascular beds (78–80) and may thus be an important renal vasodilator in states characterized by intense renal vasoconstriction, its central observable renal hemodynamic effect is, according to most investigators, to increase GFR. This effect appears to result from an increase in glomerular capillary hydrostatic pressure brought about by a combination of efferent vasoconstriction and afferent vasodilation (77,81,89a,89b), and there is no evidence at present that ANF directly increases glomerular permeability. Although there is some evidence that ANF may induce redistribution of flow to deeper nephrons (83), this is probably of little importance to the effect on GFR per se, since single nephron and total GFR increase to comparable degrees (87). A more comprehensive discussion of these issues can be found in a recent review (90).

D. MECHANISM OF NATRIURESIS

Early studies (12,77,83,91–94), which have been confirmed with synthetic peptides (37,81,82,87,88), indicated that ANF acts directly on the kidney (77,94), independent of renal prostaglandin synthesis (92), to produce striking increases in absolute and fractional excretion of sodium, chloride, other solutes, and water, with lesser, presumably secondary effects on potassium excretion. Some investigators thus considered it likely that ANF might directly inhibit tubular reabsorption of sodium chloride, and early micropuncture studies were interpreted to support that view (91,93). Subsequently, however, numerous studies have failed to demonstrate an effect of ANF on ion transport in micropuncture experiments, in isolated perfused nephron segments, or in isolated epithelia (see reference 90 for a more detailed review of this subject). A direct tubular effect is, naturally, difficult to rule out, and further work in this area is needed. At present, however, the available data suggest that the rather unusual renal hemodynamic effects of ANF are likely to fully explain its natriuretic action (90). Although the exact way in which this occurs remains to be proved, the major increase in filtered load of sodium, consequent to the increase in GFR, is likely to contribute. The degree of natriuresis is, however, larger than that expected if glomerulotubular balance were maintained. Maintenance of the latter depends on several conditions, including preservation of adequate solute concentration in the inner medullary interstitium for passive sodium reabsorption in the ascending limb of Henle's loop. Since ANF consistently produces a washout of the inner medulla (81,83,95), this condition is not fulfilled. Therefore, it is possible that the ANF-induced natriuresis can be accounted for by an increase in luminal fluid load into a washed-out inner medulla (77,81). There are other conceivable consequences of the renal hemodynamic actions of ANF (e.g., an increase in interstitial pressure) which could also indirectly influence tubular reabsorption. Recent evidence from our laboratories supports these concepts, since we find that partial clamping of the renal artery, which prevents the renal hemodynamic effects of systemically administered ANF, also abolishes its natriuretic action (96).

E. EFFECTS ON BLOOD PRESSURE AND SYSTEMIC HEMODYNAMICS

deBold and co-workers found that bolus infusion of crude atrial extract, in addition to inducing marked diuresis and natriuresis, caused a significant reduction in arterial pressure (12). This effect is clearly due to ANF, since it is mimicked by continuous infusion of synthetic peptides in conscious or anesthetized normotensive dogs (81,88) and rats (97,100). At an infusion rate of 40 pmol/kg body weight per minute, ANF induces a sustained 10% reduction in mean blood pressure in dogs. The consequent reduction in renal perfusion pressure makes it all the more remarkable that GFR is simultaneously increased by as much as 30% (81).

Not surprisingly, ANF causes significantly greater falls in arterial pressure in hypertensive animal models. Antihypertensive effects have been noted in rats with experimental renovascular hypertension (97–100) or hypertension induced by deoxycorticosterone plus high salt intake (DOC-salt hypertension) (100,101), and in the Dahl salt-sensitive (102) and spontaneously hypertensive (103–105) rat strains. A striking aspect of these studies is the lack of significant reflex tachycardia during continuous infusions despite often profound falls in arterial pressure (97,100,104). There is controversy regarding the relative sensitivity of these models to ANF. Volpe et al. (97) have shown that conscious rats with renin-dependent two-kidney, one-clip renovascular hypertension exhibit greater ANF-induced falls in blood pressure than do rats with one-kidney, one-clip renovascular hypertension, a model which is not renin dependent; and Garcia et al. (99) have reported that continuous ANF infusion for up to 1 week can actually normalize blood pressure in two-kidney, one-clip rats. It has thus been suggested (97) that ANF may preferentially lower arterial pressure in hypertension induced by angiotensin II or other vasoconstrictor substances (i.e., in states characterized by heightened arteriolar tone). On the other hand, Seymour et al. (100) have reported greater antihypertensive effects in DOC-salt rats, a model of volume-expanded hypertension.

This apparent contradiction may be resolved by two important considerations: (1) the dose of ANF infused, and (2) the mechanism of the blood pressure-lowering effect. The doses employed by Seymour et al. to demonstrate a greater effect in DOC-salt hypertension (100) were 5-fold higher than those used by Volpe et al. in renovascular models (97). With regard to mechanism, although it has been widely assumed that the effect of ANF on blood pressure may be due to a relaxant effect on resistance vessels, as has been demonstrated in isolated larger vessels (see above), this is not necessarily the case. Bolus administration of crude atrial extracts has been shown to reduce cardiac output in rats (106) and recent studies with constant infusion of synthetic ANF have shown that, in normotensive dogs and hypertensive rats, the lowering of arterial pressure may be associated with falls in cardiac output and, if anything, a tendency for total peripheral resistance to increase (101,104,107). It has also been reported that significant sustained reductions in systemic and regional vascular resistance are mainly observed in normotensive animals only after baroreceptor denervation (108). Further work is needed to determine whether the ANF-induced fall in cardiac output results from venodilation, direct or indirect effects on myocardial contractility, or contraction of blood volume (e.g., due to extravascular fluid shifts). The latter mechanism is suggested by the marked increase in hematocrit induced by ANF infusion (12,81) and by the prominence of binding sites on vascular endothelium (see below). From presently available data it would appear that when ANF infusion reduces cardiac output, the fall in arterial pressure is blunted, in the steady state, by reflex changes in vascular resistance. This may also contribute to the failure of ANF to induce a sustained increase in renal blood flow *in vivo*, although it should be emphasized, as noted above, that ANF can increase renal vascular resistance in the denervated isolated kidney (77). The absence of reflex tachycardia with ANF does not rule out reflex sympathetic activation, since there is some evidence that ANF might have vagomimetic effects as well (106). Further studies are needed, however, to define more clearly the mechanisms of the complex hemodynamic effects of administered ANF.

It remains possible that ANF can act as a systemic vasodilator in certain circumstances. Recent work from our laboratories shows that at low rates of infusion (24 and 120 pmol/kg/minute), ANF induces significant falls in blood pressure in renin-dependent two-kidney, one-clip rats (but not in DOC-salt rats), due primarily to a fall in peripheral resistance (101). Higher doses (e.g., 600 pmol/kg/minute) are required to demonstrate a reduction in cardiac output in this model; at such doses, significant falls in arterial pressure are also observed in DOC-salt rats, but this is due exclusively to a dose-dependent fall in stroke volume and cardiac output (101). These findings are illustrated in Table I.

In summary, ANF clearly has complex effects on systemic (and regional) hemodynamics which are dose dependent and some of which are likely to be indirect. It appears that at low rates of infusion ANF preferentially lowers arterial pressure (via a reduction in vascular resistance) in hypertensive states induced by vasoconstrictors such as angiotensin II. In the absence, however, of data on circulating levels of ANF in various hypertensive models, the significance of these possibly pharmacological effects of ANF in the pathophysiology of hypertension must remain speculative.

F. EFFECTS ON RENIN SECRETION

As shown in Fig. 6, constant infusion of ANF in normal dogs causes a prompt and profound reduction in renin secretion rate (81,82) together with a more gradual fall in peripheral plasma renin levels (81). A fall in plasma renin has also been demonstrated in human subjects (109). This effect may be a consequence of the renal hemodynamic actions of ANF, reflecting either GFR-dependent increases in the filtered load of sodium chloride reaching the macula densa or, possibly, increased hydrostatic pressure at the juxtaglomerular cells lining the afferent arteriole. Given that ANF infusion reduces renal perfusion pressure, the latter mechanism would be unlikely unless ANF were to cause major afferent vasodilation.

TABLE 1

Hemodynamic Responses to Graded Infusion of Synthetic ANF in Two-Kidney, One-Clip (2K, 1C) and Deoxycorticosterone (DOC)-Salt Hypertensive Rats[a]

	2K,1C rats			DOC-salt rats		
	24[b]	120[b]	600[b]	24[b]	120[b]	600[b]
Change in						
Mean blood pressure (mm Hg)	−10 ± 2[c]	−20 ± 2[c]	−31 ± 4[c]	+6 ± 5	−10 ± 5	−22 ± 2[c]
Cardiac output (ml/minute)	−5 ± 5	−7 ± 5	−19 ± 4[c]	−10 ± 4[c]	−35 ± 9[c]	−49 ± 11[c]
Total peripheral resistance [(mm Hg · minute/ml) × 100]	−5 ± 3	−14 ± 4[c]	−9 ± 3	+11 ± 3	±25 ± 9[c]	+30 ± 10[c]

[a] Data taken from Volpe et al. (101).
[b] ANF infusion rate (pmol/kg/minute).
[c] $p < 0.05$ vs control.

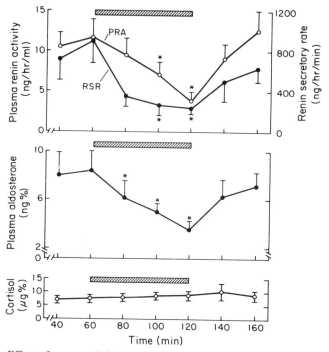

FIG. 6. Effect of constant infusion of synthetic ANF on renin secretory rate (RSR), plasma renin activity (PRA), plasma aldosterone, and plasma cortisol in anesthetized dogs. Synthetic auriculin A (37) was administered as a 1 µg/kg bolus followed by infusion of 0.1 µg/kg/min for 1 hour (horizontal hatched bar). Reproduced from Maack et al. (81).

A recent preliminary report indicates that ANF does not lower renin secretion in the nonfiltering kidney (110). In addition, this renin-suppressing effect is not observed in the hemodynamically compromised kidney, at least acutely. In rats with chronic two-kidney, one-clip hypertension, ANF administration for 1 hour actually increases peripheral plasma renin activity slightly but significantly (97,98), probably because ANF is unable to alter intrarenal hemodynamics in the clipped, renin-secreting kidney so that the renin-stimulating effect of a fall in systemic (i.e., renal perfusion) pressure predominates. These findings are illustrated in Fig. 7 (see below). This concept has been validated in dogs with acute unilateral renal artery constriction; in this model, ANF infusion fails to decrease renin secretion but instead tends to increase it (111). On the other hand, ANF is able to suppress renin levels in sodium-depleted one-kidney, one-clip rats (97), possibly reflecting heightened sensitivity of the macula densa mechanism following sodium depletion, so that even subtle increases in distal delivery of NaCl could turn off secretion. Such a mechanism could also explain why plasma renin can be suppressed in two-kidney, one-clip rats following continuous ANF administration for 1 week (99). It remains possible that ANF could also have direct effects on the juxtaglomerular cells themselves, and there is a preliminary report suggesting that high ANF concentrations may inhibit isoproterenol-induced renin release in kidney slices (112). The preceding observations suggest, however, that a direct inhibitory effect, if present, is easily counterbalanced by decreased renal perfusion.

At present, there are no data indicating to what degree the effects of ANF on renin secretion contribute to its other observable actions. Nonetheless, in view of the now widely accepted importance of the renin–angiotensin system in the regulation of arterial pressure, the effects of ANF on renin secretion must be considered an interaction of potential importance in its short- and long-term actions.

G. EFFECTS ON ALDOSTERONE

ANF infusion in normotensive dogs and humans causes a significant and reversible fall in plasma aldosterone concentration (81,109) (Fig. 6). While this might reflect, in part, the ANF-induced fall in circulating renin (and, presumably, angiotensin II), a direct effect of ANF on the adrenal cortex is likely to be involved. Independent reports from several laboratories have established that both crude atrial extracts (113) and synthetic ANF (114–117) inhibit aldosterone production by isolated adrenal zona glomerulosa cells. ANF inhibits "basal" aldosterone production from such cells and also antagonizes the stimulation of aldosterone by agonists such as angiotensin II (113–117), ACTH (113,117), dibutyryl cAMP (116), and potassium (116). Its effect is specific for the outer zone of the cortex in some species, since there is no effect on corticosterone production by isolated rat fasciculata-reticularis cells (113) or on plasma cortisol in intact dogs (81). Inhibition of cortisol production by ANF in bovine adrenal has, however, been reported (115).

Studies by Goodfriend *et al.* (116) have shown that the inhibitory action of ANF on aldosterone production by bovine adrenal cells is principally exerted at the early portion of the steroidogenic pathway. Thus ANF also inhibits basal and angiotensin II-stimulated pregnenolone production by bovine glomerulosa (116). It does not, however, block the stimulation of steroidogenesis by exogenous 25-hydroxycholesterol (116), a polar derivative which passes through the mitochondrial membrane more easily than cholesterol itself, thus implying that ANF does not inhibit cholesterol side-chain cleavage. Taken together, these results suggest that ANF acts largely at a point prior to mitochondrial uptake and metabolism of cholesterol (116). An inhibitory effect on the late pathway (i.e., conversion of progesterone to aldosterone) is not detectable in the basal state (116), but is reported to occur in angiotensin II-stimulated cells (118).

Although ANF inhibits the effect of a variety of agonists on the adrenal glomerulosa, as it does on vascular smooth muscle (see above), there is some evidence for a preferential effect on angiotensin II-stimulated steroidogenesis in some species. Thus in isolated rat glomerulosa cells, angiotensin II is, unlike other agonists, not able to overcome the inhibitory effect of ANF on aldosterone production (119). This finding is reminiscent of our observations on agonist-induced contraction of the isolated aorta (see Fig. 5 above).

Evidence for a direct adrenal effect of ANF *in vivo* is provided by the observation that ANF induces at times profound reductions in plasma aldosterone in renin-dependent two-kidney, one-clip rats despite concurrent increases in plasma renin activity (97,98) (Fig. 7) ANF also inhibits the aldosterone-stimulating effect of angiotensin II infusion *in vivo* (120). These findings also emphasize the potential importance of antagonism by ANF of the actions of angiotensin II. Although it is unlikely that ANF-

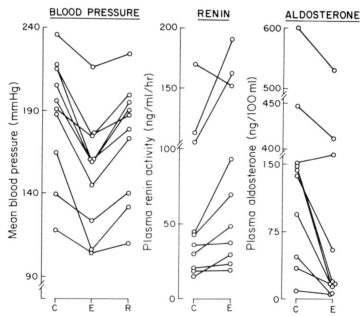

FIG. 7. Effect of synthetic ANF on blood pressure, plasma renin activity, and plasma aldosterone in rats with chronic, saralasin-responsive 2-kidney, 1-clip renovascular hypertension. Synthetic auriculin A (37) was administered as a 2 μg/kg bolus followed by infusion of 0.3 μg/kg/minute for 1 hour. C, Control; E, experimental; and R, recovery periods. Reproduced from Volpe *et al.* (97).

induced inhibition of aldosterone secretion contributes to its acute natriuretic action, the potential impact of this effect in terms of long-term fluid volume homeostasis is obviously of major importance.

H. EFFECTS ON VASOPRESSIN

Administration of ANF has been shown to inhibit dehydration- and hemorrhage-induced vasopressin release in intact dogs (121). Preliminary reports suggest that ANF may directly affect vasopressin release by cultured pituitary cells, but the results are somewhat contradictory in that they indicate a stimulation of basal release (122) but an inhibition of potassium-stimulated release (123). Confirmation and extension of these observations will be quite important, since they raise the possibility of yet another level at which ANF could interact in a coordinated regulation of fluid volume.

III. Tissue Distribution of ANF

Immunohistochemical studies have confirmed that ANF is localized within the specific granules of atrial muscle cells of a variety of mammalian species (123). Cross-reacting immunoreactivity has also been detected in both atria and ventricles of several nonmammalian vertebrate species (124), consistent with the previously established distribution of secretory granules in nonmammals (125). Granularity has been found to be most intense in the subpericardial region and is greater in the right than left atrium (123), similar to the reported concentrations of extractable ANF in the two atria (126–128). As far as is known at present, the heart appears to be the principal, if not exclusive, source of ANF in blood.

ANF-like immunoreactivity has also been identified in extracts of pituitary and hypothalamus (129) and has been localized in a variety of specific brain regions by immunocytochemical techniques (130,131). There is also one unconfirmed report that immunoreactive ANF is present in parenchymal cells of the rat salivary gland (123).

Sites of synthesis of ANF have been elucidated by measurement of tissue mRNA levels; Northern blot analysis has generally been performed to confirm that radiolabeled cDNA hybridizes to poly(A)$^+$ RNA of a molecular size corresponding to the ANF precursor. ANF mRNA is highly abundant in atrial muscle, accounting for 1–2% of total cellular mRNA (53,59,60,63). Far lower levels of mRNA expression have been identified in pituitary, hypothalamus, lung, and cardiac ventricles of the rat (61), although immunoreactive ANF has not been reported to date in the latter two tissues.

The presence of ANF in the brain raises the possibility of its having a neurotransmitter function, as has been postulated for a number of other polypeptide hormones and autacoids. The distribution of brain ANF deserves some comment. The greatest density of immunoreactive ANF has been identified in neurons of the hypothalamus, particularly in the anteroventral portion of the third ventricle (130,131). This region has been implicated in the central actions of angiotensin II and is thought to play a role in the regulation of both blood pressure and fluid and electrolyte balance. This finding provides further indication of the potential significance of the actions of systemically administered ANF described above.

IV. ANF Receptors and Target Organs

High-affinity membrane binding sites for ANF have been characterized in homogenates of rabbit aorta (132), rat kidney cortex (132), and bovine adrenal glomerulosa (133) as well as cultured vascular smooth muscle (134,135) and endothelial (135) cells. Scatchard analysis has revealed a single class of high-affinity sites ($K_d = 10^{-10}$ to 10^{-11} M in fresh tissue preparations, and approximately 10^{-9} M in cultured cells) in most studies, although there has been suggestion of a second, lower affinity class of binding sites in some tissues (133,135). Little information is available on the cofactor requirements for binding, although monovalent and divalent cations appear to enhance binding in the adrenal (133). In recently reported studies, bifunctional cross-linking reagents and photoaffinity labeling have been used to assess the subunit molecular weight of the ANF receptor (136a–138). Discrepant estimates have been reported (60–160 kDa), possibly owing to the use of intact membranes (rather than solubilized receptor preparations) in some studies.

The distribution of binding sites in various tissues has been evaluated by autoradiographic techniques and by stimulation of second messenger levels. Following intraaortic injection, radiolabeled ANF binding in the kidney has been found to be most intense in the cortex, principally due to dense labeling of glomeruli (139). Binding has also been localized to endothelial cells of the vasa recta in the medulla and papilla (139). A similar dominance of glomerular binding sites is suggested by the ability of ANF to stimulate cGMP levels in various kidney subfractions *in vitro*. In these studies, ANF elicited a 50-fold rise in cGMP in isolated glomeruli but only a 2- to 3-fold rise in fractions enriched in collecting duct or thick ascending limb fragments, and no effect in proximal tubules (140,141).

By *in vivo* autoradiography, ANF binding sites are also prominent on endothelium and vascular smooth muscle of adrenal cortex, liver, and lung as well as the endothelium of the heart and the smooth muscle of the colon (139). ANF infusion in rats has also been shown to activate particulate guanylate cyclase (see below) in lung, liver, and intestine, as well as adrenal and kidney (142). In addition, guanylate cyclase activation was detected in testis (142) and a recent report indicates that ANF can stimulate testosterone production by interstitial cells of the mouse (143).

Using an *in vitro* technique, Quirion et al. (144) found that binding sites are also widely distributed in the brain, including areas accessible to the circulation (area postrema and subfornical organ) and a variety of regions protected by the blood-brain barrier, most prominently the median eminence, nucleus tractus solitarius, olfactory bulb, choroid plexus, and ependyma. Although this distribution of binding sites is quite unique, striking overlap with the distribution of angiotensin II binding sites was noted (144). At present, the physiological significance of ANF and ANF receptors in the central nervous system remains speculative; it is of interest, however, that in recent studies intracerebral ventricular administration of ANF has been shown to block the dipsogenic effects of angiotensin II (145,145a) and dehydration (145a).

V. Cellular Actions of ANF

Hamet et al. (140) reported that injection of purified or synthetic ANF into rats causes marked increases in plasma and urine cGMP levels. This finding suggested that ANF might have a mechanism of action similar to that of certain vasodilators, such as sodium nitroprusside and the endothelium-dependent vasodilators, which are known to cause cGMP accumulation by direct or indirect activation of soluble guanylate cyclase. It was subsequently shown that ANF has a unique action, namely to activate the particulate form of guanylate cyclase (142). This effect has been demonstrated in vascular tissues (142,146), and in a variety of other potential target tissues (141,142), as discussed above. It seems likely, though unproved, that stimulation of cGMP may be directly involved in the mechanism of ANF-induced vasorelaxation; one postulated consequence of increased intracellular cGMP levels is activation of the sarcolemmal Ca^{2+}-ATPase extrusion pump, which would lead to decreased cytosolic-free calcium (147). It should be noted, however, that the ED_{50} of ANF for cGMP stimulation is considerably higher than that for binding in cultured smooth muscle cells (135,148). A recent report indicates that ANF can also stimulate cGMP in adrenal cortical cells (149), but there is little evidence at present of a role for cGMP in the regulation of steroidogenesis.

Another action of ANF that has been described in several tissues is a modest (approximately 25%) but highly significant inhibition of adenylate cyclase (150,151). This effect could conceivably contribute to the inhibitory action of ANF on steroidogenesis, but it is unlikely to represent the major mechanism since angiotensin II-induced aldosterone production, which is markedly inhibited by ANF (97,120), is not dependent on intracellular generation of cAMP.

The renal hemodynamic and natriuretic effects of ANF in the isolated perfused rat kidney have been shown to depend on the presence of extracellular calcium (76,77). On the other hand, the vasorelaxation induced by ANF in precontracted vascular smooth muscle is, as implied above, likely to depend ultimately on a decrease in cytosolic calcium, and such an effect could also be a plausible explanation for ANF-induced inhibition of aldosterone synthesis. To date, however, there is no evidence that ANF alters either membrane calcium fluxes or intracellular calcium concentration in its target tissues. While ANF opposes the actions of a variety of agonists on vascular smooth muscle and adrenal cortex, certain observations suggest a particularly prominent antagonism of angiotensin II-mediated responses (72,97,119,120). This apparent selectivity, which requires further documentation, is not due to antagonism at a receptor level (116) and cannot be explained easily by any of the cellular actions of ANF described thus far.

VI. Biosynthesis, Secretion, and Metabolism

Reasonably convincing evidence has begun to emerge which suggests that ANF is present in blood and that its release is stimulated by distension of the atria. It will, perhaps, be easier to attempt to place in perspective the rather fragmentary information available concerning the regulation of biosynthesis, post-translational processing, and storage after first considering what is known about circulating ANF.

A. ANF IN BLOOD

Immunoreactive ANF has been detected in rat (129,152,153), dog (154), and human (155–159) plasma. For the most part, these assays have employed antisera directed against low-molecular-weight synthetic ANF peptides. There has, however, been marked discrepancy in reported plasma levels, ranging from as low as 1 pmol/liter in human plasma (155,159) to as high as 2 nmol/liter in plasma from anesthetized rats (152). These differences are explained, at least in part, by methodologic considerations and by the stimulatory effects of anesthesia. Higher values have generally been found by direct radioimmunoassay of unextracted plasma, likely due to nonspecific cross-reacting substances and/or to other artifacts leading to increased apparent displacement of radiolabeled ANF. The most commonly employed extraction procedure has involved application to octadecylsilane (C_{18}) cartridges, which when used under proper conditions are known to bind all forms of ANF present in atrial extracts (47). Assays using this or similar procedures have found "basal" levels of approximately 20 pmol/liter in conscious rats (153) and in the range of 1–20 pmol/liter in normal human subjects on unrestricted sodium intake (155,159). Even these values must be accepted only tentatively, however, until better data are available on the validity of these extraction procedures, on the cross-reactivity of the antibodies used with all the forms of ANF potentially present in plasma, and on the influences of environmental and physiological factors on sampling conditions.

Available evidence suggests that the major form of immunoreactive ANF in rat plasma (129,153) or of immunoreactive or biologically active ANF released by the isolated perfused rat or rabbit heart (153,160,161) corresponds to one or more of the low-molecular-weight peptides. These studies do not necessarily rule out secretion of high-molecular-weight forms (e.g., the precursor), since their cross-reactivity with the antisera used is unknown and their detection by bioassays is likely to be insensitive; furthermore, it is conceivable that the precursor, if released, might be rapidly hydrolyzed in the circulation. Two groups of investigators have recently reported purification of low-molecular-weight ANF from rat plasma (48,49) and have found, by amino acid sequence analysis, that the principal form corresponds to the 28-residue peptide (Ser^{99}-Tyr^{126}) termed cardionatrin I by deBold (34,47). Although accomplishing this task required unphysiological procedures to sufficiently stimulate circulating levels of ANF, the results obtained (48,49) are in agreement with the chromatographic properties of immunoreactive ANF in rat plasma collected under more physiological conditions (153). Data of comparable specificity are not available for human plasma ANF, but, as shown in Fig. 8, the major low-molecular-weight immunoreactive form that we have identified in human plasma (159) also coelutes with the 28-residue peptide ("α-hANP") isolated from human atrium by Kangawa and Matsuo (40). We have consistently observed a smaller, earlier-eluting peak in human plasma (Fig. 8) whose presence is not affected by collection conditions (including the use of various protease inhibitors) and which is not formed when synthetic ANF is added to plasma *in vitro,* thus suggesting that it is an endogenously produced form. It remains to be determined whether or not this represents a distinct low-molecular-weight peptide, since it could conceivably arise from sulfoxidation of the methionyl residue (present in human, but not rat, ANF), which might alter retention time on reversed-phase HPLC.

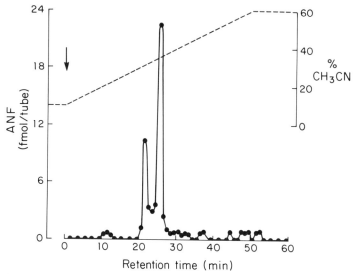

FIG. 8. Reversed-phase HPLC of an extract of pooled human plasma. Plasma was collected in EDTA and extracted on C_{18}-silica cartridges. The extract was injected (arrow) onto a 0.39 × 30 cm μBondapack C_{18} column and eluted with a linear gradient of 10 to 60% acetonitrile (CH_3CN) over 50 minutes. Immunoreactive ANF was assayed in each fraction. The major form of immunoreactivity had a retention time (26 minutes) identical to that of synthetic Ser^{99}-Tyr^{126} human ANF. The earlier eluting peak (retention time 21 minutes) was found consistently under a variety of sampling conditions (see text). Reproduced from Epstein et al. (159).

B. ANF RELEASE

Immunoreactive ANF concentration has been found to be markedly elevated in coronary sinus, compared to peripheral, blood, thus suggesting that ANF secreted by either the left or right atrium may be released into the right atrial chamber. Other routes of release have not, however, been excluded. A variety of maneuvers has been shown to increase peripheral plasma ANF levels acutely in intact animals and humans; it should be mentioned that qualitatively similar findings have been reported with assays employing either extracted or unextracted plasma. The stimuli used include acute blood volume expansion in rats (153), mechanical distension of the left atrium in dogs (154), infusion of a variety of pressor agents, leading to increases in right and left atrial pressure (162), and head-out water immersion in humans (159), a maneuver known to increase cardiopulmonary blood volume (5). The latter effect is illustrated in Fig. 9, in which it can be seen that the increases in plasma immunoreactive ANF are associated with reciprocal changes in plasma renin and

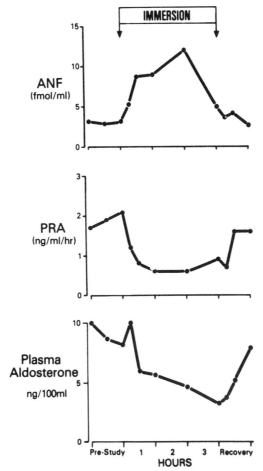

FIG. 9. Effect of water immersion to the neck on plasma immunoreactive ANF, plasma renin activity (PRA), and plasma aldosterone in a normal male subject. The subject remained seated throughout the 3-hour immersion period and the 1-hour control (pre-study) and recovery periods. Reproduced from Epstein *et al.* (159).

aldosterone (159). A common denominator of all these maneuvers is an increase in atrial wall tension, a mechanism for release that has also been demonstrated in the isolated heart (153,161). There is evidence that the response to mechanical distension of the atrium *in vivo* is not dependent on cardiac innervation (154). Thus, taken together, these observations permit the tentative conclusion that atrial stretch is an important mechanism subserving ANF release *in vivo*. These findings may thus explain the observations that plasma ANF is increased transiently during atrial tachyarrhythmias (163) and increased chronically in patients with decompensated congestive heart failure (155,158).

Other potential factors regulating ANF secretion have been less clearly defined, and there is no information available on the mechanisms of stimulus-secretion coupling. Sonnenberg and colleagues have reported that vasopressin, acetylcholine, and epinephrine (but not isoproterenol) induce release of natriuretic activity from isolated atrial fragments *in vitro* (164,165), and have postulated on this basis that stimulation of the phosphatidylinositol pathway may be involved in secretion. Confirmation and extension of these observations have yet to be reported.

Finally, the evidence that chronic perturbations of extracellular volume alter plasma ANF concentration is rather limited at present. Oral sodium loading has been reported to increase immunoreactive ANF slightly in anesthetized rats (129) and there is a weak, albeit significant, relationship between urinary sodium excretion rate and plasma ANF levels in humans during changes in sodium intake (155). The changes described, however, are at best modest. It is unclear at present whether plasma ANF changes only transiently or in a sustained fashion during chronic changes in sodium balance. There are observations which suggest changes in ANF synthesis during alterations in fluid and electrolyte balance (see below), but further study of these issues with recently developed immunoassays for plasma ANF is needed.

C. BIOSYNTHESIS AND METABOLISM

Current understanding of the regulation of ANF biosynthesis and storage is based on rather fragmentary and, in some cases contradictory, observations, so that only very tentative conclusions and/or speculations can be offered. Using highly stringent extraction conditions, deBold and co-workers have found that the intact 126-residue precursor is by far the major form present in atrial tissue, with only small amounts of low-molecular-weight peptide (Ser^{99}-Tyr^{126}) being isolated (47). It is not clear whether this finding can be attributed to a high basal level of gene transcription or whether the precursor is actually stored in the atrial granules. Based on the findings in plasma (see above) it is assumed that the low-molecular-weight form is mainly released, so that if the precursor is actually the major storage form one must invoke a rapid proteolytic cleavage in some way linked to the process of secretion; alternatively, it is possible that the precursor is released and rapidly processed in the extracellular space, or that the mature peptide hormone is the major form in the granules per se. The nature of the processing mechanism is unknown. It has been reported that trypsin and kallikrein promote the generation of low-molecular-weight active peptide(s) from crude preparations of large-molecular-weight ANF (166,167), but the specificity of these findings is open to question given the evidence that the precursor is readily (and nonspecifically) hydrolyzed during purification procedures.

The abundance of ANF-specific mRNA in atrial tissue (53,59,60,63) does, indeed, suggest a high basal rate of precursor synthesis, and there is evidence that mRNA levels are decreased by dehydration (63) and increased during volume expansion induced by deoxycorticosterone (61); in the latter study, mRNA levels were found to peak at the same time that the animals were escaping from the sodium-retaining effects of deoxycorticosterone. These preliminary reports thus raise the possibilities that altered rates of biosynthesis might contribute to changes in circulating ANF levels, that increased levels might be sustained during the course of alterations in fluid balance, and that ANF might contribute to the phenomenon of mineralocorticoid escape; however, measurements of plasma ANF levels will be needed to confirm these suspicions. It is noteworthy that these changes in tissue mRNA levels are directionally opposite to those previously reported to occur in the degree of atrial granularity, namely that dehydration tends to increase, and volume expansion tends to decrease, the number of granules (15,16). And yet, the changes reported in tissue levels of extractable natriuretic activity tend to parallel (31,168) those suggested for mRNA. These findings suggest a rather complex interaction between synthesis, storage, and release of ANF, and pose an interesting area for future research. In view of these complexities, it is not possible at present to interpret the findings of differences in tissue ANF levels between the right and left atria (126–128) or of altered tissue levels in certain animal models of pathologic states (95,128,169).

Little is known at present about the metabolism of circulating ANF, other than the fact that infused synthetic ANF has a very short half-life, on the order of 3 minutes (170), consistent with the rapid decay of its biological effects when a constant infusion is stopped (81). The relative importance of tissue uptake and degradation by peptidases has yet to be assessed. We had previously suggested the possibility that ANF might be metabolized by angiotensin converting enzyme (37) and a recently reported study suggests that the natriuretic response to crude atrial extracts is increased by pretreatment with converting enzyme inhibitor (171). Aside from the difficulties in interpreting this sort of experiment, we have been unable to confirm this finding with synthetic ANF (unpublished observations); moreover, we find that plasma levels of immunoreactive ANF are not altered acutely by converting enzyme inhibition, at least in subjects with normal plasma angiotensin II levels (unpublished data). Although a significant role for angiotensin-converting enzyme thus appears questionable, it remains to be tested more definitively. Of interest, another dipeptidylcarboxypeptidase, distinct from angiotensin-converting enzyme, was recently isolated from bovine atrium and shown to metabolize synthetic ANF (172). The role of this and other hydrolases in both the processing (activation) and metabolic degradation of ANF remains to be established.

VII. Current and Future Perspectives

In the relatively short span of time since its discovery in 1981, there has been remarkable progress in elucidating fundamental aspects of the biology of ANF. In this overview, we have attempted to highlight the advances made in several intensively studied areas. Given the explosive growth of the literature on this subject, unintentional oversights are inevitable. Moreover, it can be anticipated that by the time this work is published, new insights may well have emerged. Although there are several major issues which await resolution, a general concept of the possible role of this new hormone is beginning to take shape.

Aside from the enhanced understanding of fluid and circulatory regulation that is likely to result from further study of ANF, some unusual properties of this substance indicate its potential usefulness as a tool for investigation in a number of biological processes. Areas that can be readily identified include (1) the control of the renal circulation and of glomerular hemodynamics, (2) the regulation of particulate guanylate cyclase (of which ANF appears to be the only known mammalian activator) and its role in vascular contractility and other cellular processes, and (3) the regulation of polypeptide hormone biosynthesis and posttranslational processing.

The information available at present does not permit any definitive conclusions about the role of ANF in physiological or pathological processes. It remains to be demonstrated which (if any) of the documented effects elicited by ANF infusion actually occur at the levels which circulate in normal or disease states. Clearly, development of specific antagonists (or possibly of high affinity antibodies) will be essential in this regard, since a convincing model of the classical endocrine ablation experiment is not possible in this case. Nonetheless, some tentative statements can reasonably be made. The actions of infused ANF, together with its distribution in the central nervous system, suggest a fundamental interaction with the hormonal and neural control of fluid and electrolyte balance and of arterial pressure and the circulation. There is convincing evidence that ANF is released from the heart in response to atrial stretch, and the peptide clearly has actions which might explain some of the known effects of atrial distension. It remains to be determined what other factors may influence ANF secretion. For the moment, it appears that ANF might work, in concert with neural reflexes and other secondary hormonal effects, to modulate extracellular volume and cardiovascular function in response to acute volume overload sensed by the atria. In addition to its direct and indirect effects on renal excretory function, there is circumstantial evidence that ANF might regulate vascular volume by increasing capillary permeability. As indicated by the scheme in Fig. 10, the similarities of the potential actions of ANF and of the proposed neurally mediated consequences of atrial distension are rather striking. It must be emphasized that this scheme is intended to summarize succinctly these potential interactions and not to imply an established regulatory pathway. There continues to be nearly as much controversy concerning the neural effects of atrial distension (1–3) as there is uncertainty regarding the physiological actions of ANF. Further investigation will be required to unravel the relative importance of each of these postulated actions.

There is also evidence that increased levels of plasma ANF can be sustained chronically in pathologic states, such as congestive heart fail-

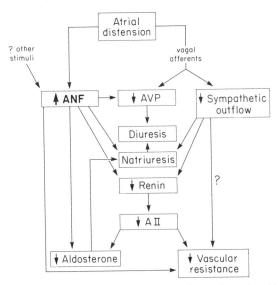

FIG. 10. Potential interactions of ANF with the renin–angiotensin–aldosterone system and other neurohumoral cardiovascular control systems. The scheme serves only to indicate the considerable overlap between the documented responses to ANF infusion and the neurally mediated effects that have been described during atrial distention in experimental animals. It must be emphasized that most of the effects indicated by the arrows have yet to be proven of physiological significance (see text).

ure, characterized by marked volume overload and chronic increases in atrial wall tension, but there is little indication at present that ANF is involved in the long-term regulation of extracellular volume, renal function or arterial pressure under normal conditions. Although there has been considerable excitement over its potential as a therapeutic agent, it will be necessary first to have a clearer understanding of the role of endogenous ANF in health and disease. Yet the evidence at hand for an endocrine function of the heart suggests many exciting avenues for future clinical and basic investigation.

ACKNOWLEDGMENTS

Many of the studies described as part of this review represent collaborative efforts with the laboratory of Dr. Thomas Maack in the Department of Physiology at Cornell and would not have been possible without the major contributions of our colleagues Dr. Hollis D. Kleinert, Dr. Maria J. F. Camargo, Dr. Massimo Volpe, Dr. R. Ernest Sosa, Dr. E. Darracott Vaughan, Jr., Mr. Donald Marion, and Ms. Rose M. Aceto. Our work has been benefitted invaluably by the efforts of Dr. John Lewicki and his colleagues at California Biotechnology, Inc. and has been influenced by important collaborations with Dr. Theodore Goodfriend of the University of Wisconsin and Dr. Murray Epstein of the University of Miami School of Medicine. The author is also indebted to Dr. John H. Laragh and Dr. Jean E. Sealey for their constant support and advice and for many invaluable discussions, and to Ms. Linda Stackhouse for her expert assistance in the preparation of this manuscript. Finally, as should be readily apparent to the reader, the current state of knowledge on this subject has resulted from the work of numerous investigators throughout the world, particularly in Canada, the United States, and Japan.

REFERENCES

1. Gauer, O. H., and Henry, J. P. (1963). *Physiol. Rev.* **43**, 423–481.
2. Goetz, K. L., Bond, G. C., and Bloxham, D. D. (1975). *Physiol. Rev.* **55**, 157–205.
3. Linden, R. J. (1979). *Am. J. Cardiol.* **44**, 879–883.
4. Wood, P. (1963). *Br. Heart. J.* **25**, 273–282.
5. Epstein, M. (1978). *Physiol. Rev.* **58**, 529–581.
6. Henry, J. P., and Pearce, J. W. (1956). *J. Physiol. (London)* **131**, 572–585.
7. Ledsome, J. R., and Linden, R. J. (1968). *J. Physiol. (London)* **198**, 487–503.
8. Share, L. (1968). *Am. J. Physiol.* **215**, 1384–1389.
9. Kappagoda, C. T., Linden, R. J., Mary, D. A. S. G., and Weatherill, D. (1978). *J. Physiol. (London)* **280**, 61P–62P.
10. Anderson, C. H., McCarry, M., and Farrell, G. C. (1959). *Endocrinology* **64**, 202.
11. Brennan, L. A., Malvin, R. L., Jochim, K. E., and Roberts, T. F. (1971). *Am. J. Physiol.* **221**, 273.
12. deBold, A. J., Borenstein, H. B., Veress, A. T., and Sonnenberg, H. (1981). *Life Sci.* **28**, 89–94.
13. Kisch, B. (1956). *Exp. Med. Surg.* **114**, 99–112.
14. Jamieson, J. D., and Palade, G. E. (1964). *J. Cell Biol.* **23**, 151–172.
15. Marie, J. P., Guillemot, H., and Hatt, P. Y. (1976). *Pathol. Biol. (Paris)* **24**, 549–554.
16. deBold, A. J. (1979). *Proc. Soc. Exp. Biol. Med.* **161**, 508–511.

17. DeWardener, H. E., Mills, I. H., Clapham, W. F., and Hayter, C. J. (1961). *Clin. Sci.* **21**, 249–258.
18. Sealey, J. E., Kirshman, J. D., and Laragh, J. H. (1969). *J. Clin. Invest.* **48**, 2210.
19. Bourgoignie, J., Klahr, S., and Bricker, N. S. (1971). *J. Clin. Invest.* **50**, 303–311.
20. Sonnenberg, H., Veress, A. T., and Pearce, J. W. (1972). *J. Clin. Invest.* **51**, 2631–2644.
21. Buckalew, V. M., and Nelson, D. B. (1974). *Kidney Int.* **5**, 12–22.
22. deWardener, H. E. (1977). *Clin. Sci. Mol. Med.* **53**, 1–8.
23. Haddy, F. J., and Overbeck, H. W. (1976). *Life Sci.* **19**, 935–948.
24. Blaustein, M. (1977). *Am. J. Physiol.* **232**, C165–173.
25. Gruber, K. A., Whitaker, J. M., and Buckalew, V. M., Jr. (1980). *Nature (London)* **287**, 743–745.
26. Haupert, G. T., and Sancho, J. (1979). *Proc. Natl. Acad. Sci. U.S.A.* **76**, 4658–4660.
27. Kleinert, H. D., Camargo, M. J. F., Sealey, J. E., Laragh, J. H., and Maack, T. (1982). *Physiologist* **25**, 298.
28. Deth, R. C., Wong, K., Fukozawa, S., Rocco, R., Smart, J. L., Lynch, J., and Awad, R. (1982). *Fed. Proc.* **41**, 983.
29. deBold, A. J. (1982). *Proc. Soc. Exp. Biol. Med.* **170**, 133–138.
30. Trippodo, N. C., MacPhee, A. A., Cole, F. E., and Blakesley, H. L. (1982). *Proc. Soc. Exp. Biol. Med.* **170**, 502–508.
31. Thibault, G., Garcia, R., Cantin, M., and Genest, J. (1983). *Hypertension* **5**, I-75-I-80.
32. Trippodo, N. C., MacPhee, A. A., and Cole, F. E. (1983). *Hypertension* **5**, I-81–I-88.
33. deBold, A. J., and Flynn, T. G. (1983). *Life Sci.* **33**, 297–302.
34. Flynn, T. G., deBold, M. L., and deBold, A. J. (1983). *Biochem. Biophys. Res. Commun.* **117**, 859–865.
35. Grammer, R. T., Fukumi, H., Inagami, T., and Misono, K. S. (1983). *Biochem. Biophys. Res. Commun.* **116**, 696–703.
36. Thibault, G., Garcia, R., Seidah, N. G., Lazure, C., Cantin, M., Chrétien, M., and Genest, J. (1983). *FEBS Lett.* **164** (2), 286–290.
37. Atlas, S. A., Kleinert, H. D., Camargo, M. J., Januszewicz, A., Sealey, J. E., Laragh, J. H., Schilling, J. W., Lewicki, J. A., Johnson, L. K., and Maack, T. (1984). *Nature (London)* **309**, 717–720.
38. Currie, M. G., Geller, D. M., Cole, B. R., Siegel, N. R., Fok, K. F., Adams, S. P., Eubanks, S. R., Galluppi, G. R., and Needleman, P. (1984). *Science* **223**, 67–69.
39. Geller, D. M., Currie, M. G., Wakitani, K., Cole, B. R., Adams, S. P., Fok, K. F., Siegel, N. R., Eubanks, S. R., Galluppi, G. R., and Needleman, P. (1984). *Biochem. Biophys. Res. Commun.* **120**, 333–338.
40. Kangawa, K., and Matsuo, H. (1984). *Biochem. Biophys. Res. Commun.* **118**, 131–139.
41. Kangawa, K., Fukuda, A., Kubota, I., Hayashi, Y., and Matsuo, H. (1984). *Biochem. Biophys. Res. Commun.* **121**, 585–591.
42. Misono, K. S., Fukumi, H., Grammer, R. T., and Inagami, T. (1984). *Biochem. Biophys. Res. Commun.* **119**, 524–529.
43. Misono, K. S., Grammer, R. T., Fukumi, H., and Inagami, T. (1984). *Biochem. Biophys. Res. Commun.* **123**, 444–451.
44. Napier, M. A., Dewey, R. S., Albers-Schonberg, G., Bennett, C. D., Rodkey, J. A., Marsh, E. A., Whinnery, M., Seymour, A. A., and Blaine, E. H. (1984). *Biochem. Biophys. Res. Commun.* **120**, 981–988.

45. Seidah, N. G., Lazure, C., Chrétien, H., Thibault, G., Garcia, R., Cantin, M., Genest, J., Nutt, R. F., Brady, S. F., Lyle, T. A., Paleveda, W. J., Colton, C. D., Ciccarone, T. M., and Veber, D. F. (1984). *Proc. Natl. Acad. Sci. U.S.A.* **81**, 2640–2644.
46. Kangawa, K., Fukuda, A., and Matsuo, H. (1985). *Nature (London)* **313**, 397–400.
47. Flynn, T. G., Davies, P. L., Kennedy, B. P., deBold, M. L., and deBold, A. J. (1985). *Science* **228**, 323–325.
48. Schwartz, D., Geller, D. M., Manning, P. T., Siegel, N. R., Fok, K. F., Smith, C. E., and Needleman, P. (1985). *Science* **229**, 397–400.
49. Thibault, G., Lazure, C., Schiffrin, E. L., Jutkowska, J., Chartier, L., Garcia, R., Seidah, N. G., Chrétien, M., Genest, J., and Cantin, M. (1985). *Biochem. Biophys. Res. Commun.* **130**, 981–986.
50. Maki, M., Takayanagi, R., Misona, K. S., Pandey, K. N., Tibbetts, C., and Inagami, T. (1984). *Nature (London)* **309**, 722–724.
51. Oikawa, S., Imai, M., Veno, A., Tanaka, S., Nogushi, T., Nakazato, H., Kangawa, K., Fukuda, A., and Matsuo, H. (1984). *Nature (London)* **309**, 724–726.
52. Seidman, C. E., Duby, A. D., Choi, E., Graham, R. M., Haber, E., Homcy, C., Smith, J. A., and Seidman, J. G. (1984). *Science* **225**, 324–326.
53. Yamanaka, M., Greenberg, B., Johnson, L., Seilhamer, J., Brewer, M., Friedemann, T., Miller, J., Atlas, S., Laragh, J., Lewicki, J., and Fiddes, J. (1984). *Nature (London)* **309**, 719–722.
54. Kangawa, K., Tawaragi, Y., Oikawa, S., Mizuno, A., Sakuragawa, Y., Nakazato, H., Fukuda, A., Minamino, and N., Matsuo, H. (1984). *Nature (London)* **312**, 152–155.
55. Geller, D. M., Currie, M. G., Siegel, N. R., Fok, K. F., Adams, S. P., and Needleman, P. (1984). *Biochem. Biophys. Res. Commun.* **121**, 802–807.
56. Kangawa, K., Fukuda, A., Minamino, N., and Matsuo, H. (1984). *Biochem. Biophys. Res. Commun.* **119**, 933–940.
57. Lazure, C., Seidah, N. G., Chrétien, M., Thibault, G., Garcia, R., Cantin, M., and Genest, J. (1984). *FEBS Lett.* **172**, 80–86.
58. Thibault, G., Garcia, R., Cantin, M., Genest, J., Lazure, C., Seidah, N. G., and Chrétien, M. (1984). *FEBS Lett.* **167**, 352–356.
59. Greenberg, B. D., Bencen, G. H., Seilhamer, J. J., Lewicki, J. A., and Fiddes, J. C. (1984). *Nature (London)* **312**, 656–658.
60. Seidman, C. E., Bloch, K. D., Klein, K. A., Smith, J. A., and Seidman, J. G. (1984). *Science* **226**, 1206–1209.
61. Gardner, D. G., Lewicki, J. A., Fiddes, J. C., Metzler, C. H., Ramsay, D. J., Trachewsky, D., Hane, S., and Baxter, J. D. (1985). *Clin. Res.* (Abstr.) **33**, 553A.
62. Forssmann, W. G., Hok, D., Lottspeich, F., Henschen, A., Kreye, V., Christmann, M., Reinecke, M., Metz, J., Carlquist, M., and Mutt, V. (1983). *Anat. Embryol.* **168**, 307–313.
63. Nakayama, K., Ohkubo, H., Hirose, T., Inayama, S., and Nakanishi, S. (1984). *Nature (London)* **310**, 699–701.
64. Nemer, M., Chamberland, M., Sirois, D., Argentin, S., Drouin, J., Dixon, R. A. F., Zivin, R. A., and Condra, J. H. (1984). *Nature (London)* **312**, 654–656.
65. Sugiyama, M., Fukumi, H., Grammer, R. T., Misono, K. S., Yabe, Y., Morisowa, Y., and Inagami, T. (1984). *Biochem. Biophys. Res. Commun.* **123**, 338–344.
66. Garcia, R., Thibault, G., Seidah, N. G., Lazure, C., Cantin, M., Genest, J., and Chrétien, M. (1985). *Biochem. Biophys. Res. Commun.* **126**, 178–184.

67. Thibault, G., Garcia, R., Carrier, F., Seidah, N. G., Lazure, C., Chrétien, M., Cantin, M., and Genest, J. (1984). *Biochem. Biophys. Res. Commun.* **125**, 938–946.
68. Katsube, N., Wakitani, K., Fok, K. F., Adams, S. P., Tjoeng, F. S., Zupec, M. E., Eubanks, S. R., and Needleman, P. (1985). *Biochem. Biophys. Res. Commun.* **128**, 325–330.
69. Wakitani, K., Oshima, T., Loewy, A. D., Cole, B. R., Adams, S. P., Fok, K. F., Currie, M. G., and Needleman, P. (1985). *Circ. Res.* **56**, 621–627.
70. Currie, M. G., Geller, D. M., Cole, B. R., Boylan, J. G., YuSheng, W., Holmberg, S. W., and Needleman, P. (1983). *Science* **221**, 71–73.
71. Garcia, R., Thibault, G., Cantin, M., and Genest, J. (1984). *Am. J. Physiol.* **247**, R34–R39.
72. Kleinert, H. D., Maack, T., Atlas, S. A., Januszewicz, A., Sealey, J. E., and Laragh, J. H. (1984). *Hypertension* **6** (Suppl. I), I-143–I-147.
73. Winquist, R. J., Faison, E. P., and Nutt, R. F. (1984). *Eur. J. Pharmacol.* **102**, 169–173.
74. Scivoletto, R., and Carvalho, M. H. C. (1984). *Eur. J. Pharmacol.* **101**, 143–146.
75. Kleinert, H. D., Volpe, M., Atlas, S. A., Camargo, M. J. F., Laragh, J. H., and Maack, T. (1984). *Circulation* (Abstr.) **70**, II-333.
76. Camargo, M. J. F., Kleinert, H. D., Sealey, J. E., Laragh, J. H., and Maack, T. (1983). *Kidney Int.* (Abstr.) **23**, 251.
77. Camargo, M. J. F., Kleinert, H. D., Atlas, S. A., Sealey, J. E., Laragh, J. H., and Maack, T. (1984). *Am. J. Physiol.* **264**, F447–F456.
78. Oshima, T., Currie, M. G., Geller, D. M., and Needleman, P. (1984). *Circ. Res.* **54**, 612–616.
79. Koike, H., Sada, T., Miyamoto, M., Oizumi, K., Sugiyama, M., and Inagami, T. (1984). *Eur. J. Pharmacol.* **104**, 391–392.
80. Hintze, T. H., Currie, M. G., and Needleman, P. (1985). *Am. J. Physiol.* **248**, H587–H591.
81. Maack, T., Marion, D. N., Camargo, M. J. F., Kleinert, H. D., Laragh, J. E., Vaughan, E. D., Jr., and Atlas, S. A. (1984). *Am. J. Med.* **77**, 1069–1075.
82. Burnett, J. C., Jr., Granger, J. P., and Opgenorth, T. S. (1984). *Am. J. Physiol.* **247**, F863–F866.
83. Borenstein, H. B., Cupples, W. A., Sonnenberg, H., and Veress, A. T. (1983). *J. Physiol.* **334**, 133–140.
84. Keeler, A. U., and Azzarolo, R. (1983). *Can. J. Physiol. Pharmacol.* **61**, 996–1002.
85. Vaughan, E. D., Jr., Marion, D. N., Sealey, J. E., Camargo, M. J. F., Kleinert, H. D., Maack, T., and Laragh, J. H. (1983). *Surg. Forum* **34**, 690–692.
86. Beasley, D., and Malvin, R. L. (1985). *Am. J. Physiol.* **248**, F24–F30.
87. Huang, C. L., Lewicki, J., Johnson, L. K., and Cogan, M. G. (1985). *J. Clin. Invest.* **75**, 769–773.
88. Seymour, A. A., Blaine, E. H., Mazack, E. K., Smith, S. G., Stabilito, I. I., Haley, A. B., Napier, M. A., Whinnery, M. A., and Nutt, R. F. (1985). *Life Sci.* **36**, 33–44.
89a. Ichikawa, I., Dunn, B. R., Tray, J. L., Maack, T., and Brenner, B. M. (1985). *Clin. Res.* **33**, 487A.
89b. Fried, T. A., McCoy, R. N., Osgood, R. W., Reineck, H. J., and Stein, J. H. (1985). *Clin. Res.* **33**, 584A.
90. Maack, T., Camargo, M. J. F., Kleinert, H. D., Laragh, J. H., and Atlas, S. A. (1985). *Kidney Int.* **27**, 607–615.
91. Briggs, J. P., Steipe, B., Schubert, G., and Schnermann, J. (1982). *Pflüger's Arch.* **395**, 271–276.

92. Keeler, T. (1982). *Can. J. Physiol. Pharmacol.* **60,** 1078–1082.
93. Sonnenberg, H., Cupples, W. A., deBold, A. J., and Veress, A. T. (1982). *Can. J. Physiol. Pharmacol.* **60,** 1149–1152.
94. Baines, A. D., deBold, A. J., and Sonnenberg, H. (1983). *Can. J. Physiol. Pharmacol.* **61,** 1462–1466.
95. Hirata, Y., Ganguli, M., Tobian, L., and Iwai, J. (1984). *Hypertension* **6,** 1148–1155.
96. Sosa, R. E., Volpe, M., Marion, D. N., Atlas, S. A., Laragh, J. H., Vaughan, E. D., Jr., and Maack, T. (1986). *Am. J. Physiol.* **250** (*Renal Fluids* **19**), F520–F524.
97. Volpe, M., Odell, G., Kleinert, H. D., Müller, F., Camargo, M. J. F., Laragh, J. H., Maack, T., Vaughan, E. D., Jr., and Atlas, S. A. (1985). *Hypertension* **7** (Suppl. I), I-43–I-48.
98. Volpe, M., Odell, G., Kleinert, H. D., Camargo, M. J. F., Laragh, J. H., Lewicki, J. A., Maack, T., Vaughan, E. D., Jr., and Atlas, S. A. (1984). *J. Hypertension* **2** (Suppl. 3), 313–315.
99. Garcia, R., Thibault, G., Gutkowska, J., Hamet, P., Cantin, M., and Genest, J. (1985). *Proc. Soc. Exp. Biol. Med.* **178,** 155–159.
100. Seymour, A. A., Marsh, E. A., Mazack, E. K., Stabilito, I. I., and Blaine, E. H. (1985). *Hypertension* **7** (Suppl. I), I-35–I-42.
101. Volpe, M., Sosa, R. E., Muller, F. B., Camargo, M. J. F., Glorioso, N., Laragh, J. H., Maack, T., and Atlas, S. A. (1986). *Am. J. Physiol.* **250** (*Heart Circ. Physiol.* **19**), H871–H878.
102. Snajdar, R. M., and Rapp, J. P. (1985). *Hypertension* **7,** 775–782.
103. Garcia, R., Thibault, G., Gutkowska, J., Horky, K., Hamet, P., Cantin, M., and Genest, J. (1985). *Proc. Soc. Exp. Biol. Med.* **179,** 396–401.
104. Lappe, R. W., Smits, J. F. M., Todt, J. A., Debets, J. J. M., and Wendt, R. L. (1985). *Circ. Res.* **56,** 606–612.
105. Marsh, E. A., Seymour, A. A., Haley, A. B., Whinnery, M. A., Napier, M. A., Nutt, R. F., and Blaine, E. H. (1985). *Hypertension* **7,** 386–391.
106. Ackermann, U., Irizawa, T. G., Milojevic, S., and Sonnenberg, H. (1984). *Can. J. Physiol. Pharmacol.* **62,** 819–826.
107. Kleinert, H. D., Volpe, M., Camargo, M. J. F., Atlas, S. A., Laragh, J. H., and Maack, T. (1985). *Fed. Proc.* **44,** 1729.
108. Brody, M. J., O'Neill, T. P., Porter, J. P., Bonham, A. C., and Needleman, P. (1984). *Hypertension* (Abstr.) **6,** 783.
109. Cody, R. J., Covit, A. B., Laragh, J. H., and Atlas, S. A. (1985). *Hypertension* (Abstr.) **7,** 845.
110. Opgenorth, T. J., Burnett, J. C., Jr., Granger, J. P., and Scriven, T. A. (1986). *Am. J. Physiol.,* in press.
111. Sosa, R. E., Volpe, M., Marion, D. N., Glorioso, N., Laragh, J. H., Vaughan, E. D., Jr., Maack, T., and Atlas, S. A. (1985). *J. Hypertension* **3** (Suppl. 3), S299–S302.
112. Henrich, W., McAllister, L., Smith, P., Needleman, P., and Campbell, W. (1985). *Clin. Res.* (Abstr.) **33,** 528A.
113. Atarashi, K., Mulrow, P. P., Franco-Saenz, R., Snajdar, R., and Rapp, J. (1984). *Science* **224,** 992–994.
114. Chartier, L., Schiffrin, E., and Thibault, G. (1984). *Biochem. Biophys. Res. Commun.* **122,** 171–174.
115. De Léan, A., Racz, K., Gutkowska, J., Nguyen, T.-T., Cantin, M., and Genest, J. (1984). *Endocrinology* **115,** 1636–1638.
116. Goodfriend, T. L., Elliott, M., and Atlas, S. A. (1984). *Life Sci.* **35,** 1675–1682.
117. Kudo, T., and Baird, A. (1984). *Letters to Nature* **312,** 756–757.
118. Campbell, W. B., Currie, M. G., and Needleman, P. (1985). *Circ. Res.* **57,** 113–118.

119. G. Aguilera, personal communication.
120. Chartier, L., Schiffrin, E., Thibault, G., and Garcia, R. (1984). *Endocrinology* **115**, 2026–2028.
121. Samson, W. K. (1985). *Neuroendocrinology* **40**, 277–279.
122. Januszewicz, P., Gutkowska, J., DeLéan, A., Thibault, G., Garcia, R., Genest, J., and Cantin, M. (1985). *Proc. Soc. Exp. Biol. Med.* **178**, 321–325.
123. Cantin, M., Gutkowska, J., and Thibault, G. (1984). *Histochemistry* **80**, 113–127.
124. Cantin, M., and Genest, J. (1985). *Endocrine Rev.* **6**, 107–127.
125. Bencosme, S. A., and Berger, J. M. (1971). *In* "Methods of Achievement of Experimental Pathology" (E. Bajusz and G. Jasmin, eds.), Vol. 5, p. 173. Karger, Basel.
126. deBold, A. J. (1982). *Can. J. Physiol. Pharmacol.* **60**, 324–330.
127. Garcia, R., Cantin, M., Thibault, G., Ong, H., and Genest, J. (1982). *Experientia* **38**, 1071–1073.
128. Sonnenberg, H., Milojevic, S., Chong, C. K., and Veress, A. T. (1983). *Hypertension* **5**, 672–675.
129. Tanaka, I., Misona, K. S., and Inagami, T. (1984). *Biochem. Biophys. Res. Commun.* **124**, 663–668.
130. Jacobowitz, D. M., Skofitsch, G., Keiser, H. R., Eskay, R. L., and Zamir, N. (1985). *Neuroendocrinology* **40**, 92–94.
131. Saper, C. B., Standaert, D. G., Currie, M. G., Schwartz, D., Geller, D. M., and Needleman, P. (1985). *Science* **227**, 1047–1049.
132. Napier, M. A., Vandlen, R. L., Albers-Schonberg, G., Nutt, R. F., Brady, S., Lyle, T., Winquist, R., Faison, E. P., Heinel, L. A., and Blaine, E. H. (1984). *Proc. Natl. Acad. Sci. U.S.A.* **81**, 5946–5950.
133. De Léan, A., Gutkowska, J., McNicoll, N., Schiller, P. W., Cantin, M., and Genest, J. (1984). *Life Sci.* **35**, 2311–2318.
134. Hirata, Y., Tomita, M., Yoshima, H., and Ikeda, M. (1984). *Biochem. Biophys. Res. Commun.* **125**, 562–568.
135. Schenck, D. B., Johnson, L. K., Schwartz, K., Sista, H., Scarborough, R. M., and Lewicki, J. A. (1985). *Biochem. Biophys. Res. Commun.* **127**, 433–442.
136a. Yip, C. C., Laing, L. P., and Flynn, T. G. (1985). *J. Biol. Chem.* **260**, 8229–8232.
136b. Hirose, S., Akiyama, F., Shinjo, M., Ohno, H., and Murakami, K. (1985). *Biochem. Biophys. Res. Commun.* **130**, 574–579.
137. Misono, K. S., Grammer, R. T., Rigby, J. W., and Inagami, T. (1985). *Biochem. Biophys. Res. Commun.* **130**, 994–1001.
138. Vandlen, R. L., Arcuri, K. E., and Napier, M. A. (1985). *J. Biol. Chem.* **260**, 10889–10892.
139. Bianchi, C., Gutkowska, J., Thibault, G., Garcia, R., Genest, J., and Cantin, M. (1985). *Histochemistry* **82**, 441–452.
140. Hamet, P., Tremblay, J., Pang, S. C., Garcia, R., Thibault, G., Gutkowska, J., Cantin, M., and Genest, J. (1984). *Biochem. Biophys. Res. Commun.* **123**, 515–527.
141. Tremblay, J., Gerzer, R., Vinay, P., Pang, S. C., Beliveau, R., and Hamet, P. (1985). *FEBS Lett.* **181**, 17–22.
142. Waldman, S., Rapoport, R. M., and Murad, F. (1984). *J. Biol. Chem.* **259**, 14322–14334.
143. Bex, F., and Corbin, A. (1985). *Eur. J. Pharmacol.* **115**, 125–126.
144. Quirion, R., Delpé, M., deLean, A., Gutkowska, J., Cantin, M., and Genest, J. (1984). *Peptides* **5**, 1167–1172.
145. Nakamura, M., Katsuura, G., Nakao, K., and Imura, H. (1985). *Neurosci. Lett.* **58**, 1–6.

145a. Antunes-Rodrigues, J., McCann, S. M., Rogers, L. C., and Samson, W. K. (1985). *Proc. Natl. Acad. Sci. U.S.A.* **82**, 8720–8723.
146. Winquist, R. J., Faison, E. P., Waldman, S. A., Schwartz, K., Murad, F., and Rapoport, R. M. (1984). *Proc. Natl. Acad. Sci. U.S.A.* **81**, 7661–7664.
147. Popescu, L. M., Panoiu, C., Hinescu, M., and Nuto, O. (1985). *Eur. J. Pharmacol.* **107**, 393–394.
148. Hirata, Y., Tomita, M., Tanaka, S., and Yoshimi, H. (1985). *Biochem. Biophys. Res. Commun.* **128**, 538.
149. Matsuoka, H., Ishii, M., Sugimoto, T., Hirata, Y., Sugimoto, T., Kangawa, K., and Matsuo, H. (1985). *Biochem. Biophys. Res. Commun.* **127**, 1052–1056.
150. Anand-Srivastava, M. B., Franks, D. J., Cantin, M., and Genest, J. (1984). *Biochem. Biophys. Res. Commun.* **121**, 855–862.
151. Anand-Srivastava, M. B., Genest, J., and Cantin, M. (1985). *FEBS Lett.* **181**, 199–202.
152. Gutkowska, J., Horky, K., Thibault, G., Januszewicz, P., Gantin, M., and Genest, J. (1984). *Biochem. Biophys. Res. Commun.* **125**, 315–323.
153. Lang, R. E., Thoelken, H., Ganten, D., Luft, F. C., Ruskoaho, H., and Unger, T. H. (1985). *Nature (London)* **314**, 264–266.
154. Ledsome, J. R., Wilson, N., Rankin, A. J., and Courneya, C. A. (1985). *Con. J. Physiol. Pharmacol.* **63**, 739–742.
155. Shenker, Y., Sider, R. S., Ostafin, E. A., and Grekin, R. J. (1985). *J. Clin. Invest.* **76**, 1684–1698.
156. Gutkowska, J., Bourassa, M., Roy, D., Thibault, G., Garcia, R., Cantin, M., and Genest, J. (1985). *Biochem. Biophys. Res. Commun.* **128**, 1281–1287.
157. Sugawara, A., Nakao, K., Morii, N., Sakamoto, M., Suda, M., Shimokura, M., Kiso, Y., Kihara, M., Yamori, Y., and Nishimura, K. (1985). *Biochem. Biophys. Res. Commun.* **129**, 439–446.
158. Tikkanen, I., Fyhrquist, R., Metsarinne, K., and Leidenius, R. (1985). *Lancet* **2**(8446), 66–69.
159. Epstein, M., Loutzenhiser, R., Friedland, E., Aceto, R., Camargo, M. J. F., and Atlas, S. A. (1986). *J. Hypertension* **4** (Suppl. 2), in press.
160. Currie, M. G., Sukin, D., Geller, D. M., Cole, B. R., and Needleman, P. (1984). *Biochem. Biophys. Res. Commun.* **124**, 711–717.
161. Dietz, J. R. (1984). *Am. J. Physiol.* **247**, R1093–R1096.
162. Manning, P. T., Schwartz, D., Katsube, N. C., Holmberg, S. W., and Needleman, P. (1985). *Science* **229**, 395–397.
163. Schiffrin, E. L., Gutkowska, J., Kuchel, O., Cantin, M., and Genest, J. (1985). *New Engl. J. Med.* **312**, 1196–1197.
164. Sonnenberg, H., Krebs, R. F., and Veress, A. T. (1984). *IRCS Med. Sci.* **12**, 783.
165. Sonnenberg, H., and Veress, A. T. (1984). *Biochem. Biophys. Res. Commun.* **124**, 443–449.
166. Currie, M. G., Geller, D. M., Cole, B. R., and Needleman, P. (1984). *Proc. Natl. Acad. Sci. U.S.A.* **81**, 1230–1233.
167. Currie, M. G., Geller, D. M., Chao, J., Margolius, H. S., and Needleman, P. (1984). *Biochem. Biophys. Res. Commun.* **120**, 461.
168. Ackermann, U., and Irizawa, T. G. (1984). *Am. J. Physiol.* **247**, R750.
169. Chimoskey, J. E., Spielman, W. S., Brandt, M. A., and Heidemann, S. R. (1984). *Science* **223**, 820–822.
170. Atlas, S. A., unpublished observations.
171. Wang, S. L., and Gilmore, J. P. (1985). *Can. J. Physiol. Pharmacol.* **63**, 220–223.
172. Harris, R. B., and Wilson, I. B. (1984). *Arch. Biochem. Biophys.* **233**, 667.

DISCUSSION

R. Rittmaster. Has anyone tried to immunoneutralize ANF to look at its effects physiologically?

S. A. Atlas. If you mean to administer antibodies to animals, no one has published such a study. Lack of success in this regard may reflect insufficiently high affinity of available antibodies. In addition, with antibodies or even Fab fragments, there is the theoretical problem of limitation of diffusion. Obviously, the classical endocrine ablation experiment cannot be done in this case, although investigators have attempted to approximate it by removing the atrial appendages. There are reports from the laboratories of Sonnenberg and of Trippodo that removal of both atrial appendages will blunt the natriuretic response to acute volume expansion; however ANF exists in other parts of the atrium, so that this approach is not necessarily definitive. I know that people are looking for analogs of the peptide that might be receptor antagonists, but there are none as yet that I know of.

R. M. Carey. That was a lovely description of the biochemistry and physiology of ANF. My question concerns the relationship of the physiologic action of ANF to suppression of intrarenal angiotensin formation via suppression of renin release. If you suppress angiotensin II intrarenally, you certainly obtain a profound natriuretic and diuretic response. I wonder if you could clarify the issue of whether the majority of the effects of ANF in the kidney are related to suppression of intrarenal angiotensin II or whether there is an additional direct action of ANF aside from that on suppression of intrarenal angiotensin II.

S. A. Atlas. First I guess we don't really know what ANF does to intrarenal angiotensin generation since the effects on renin released into the periphery and the effects on renin within the kidney may be different. Our studies on the isolated rat kidney clearly show that ANF has potent natriuretic actions in the absence of angiotensin II, since angiotensinogen was not present in the perfusate. There is a recent paper, I believe by Gilmore and his associates, in which they have reported the effects of angiotensin converting enzyme inhibition on the natriuretic action of ANF. What they find is that the activity is, if anything, slightly enhanced which they interpret as evidence that enzyme inhibition may block catabolism of the peptide. Whatever the mechanism, these studies at least provide further evidence that the natriuretic action of ANF does not depend on an ability to suppress intrarenal angiotensin formation. In fact, the opposite may be true. There is clear evidence that the action of systemically administered atrial natriuretic factor can be sustained by simultaneously infusing angiotensin II. Thus maintenance of intrarenal hemodynamics by angiotensin II may facilitate the expression of the ANF-induced renal hemodynamic changes and thereby actually potentiate its natriuretic effect.

R. M. Carey. It certainly is a curious parallel that both angiotensin II and ANF seem to have preferential effects on efferent arteriolar tone and have profound effects on filtration fraction. Since ANF suppresses renin release, I would expect that intrarenal angiotensin would be reduced. Studies are needed to sort out whether suppression of intrarenal angiotensin II is involved in the renal action of ANF. A second question is related to the effects of ANF on renin secretion per se. Has anyone in your laboratory or elsewhere done studies on renal slice preparations or isolated renal juxtaglomerular cell preparations to determine whether there is a direct effect of ANF on renin release and whether there are ANF receptors on the membranes of these cells?

S. A. Atlas. As far as receptors are concerned I don't know the answer. There are two preliminary reports in abstract form on inhibition of renin release from rat kidney slices. In one study a very large concentration, I believe 10^{-7} M, of ANF was found to inhibit isoproterenol-stimulated renin release from rat kidney slices. So I think it is certainly possible that direct effects on the JG cell might contribute to ANF-induced suppression of renin release. I think the fact that we are able to eliminate this effect by constricting the renal arteries suggests that direct effects on the JG cell may be relatively unimportant at the doses that we are infusing *in vivo*, which result in circulating concentrations of 10^{-10} M. or at the very most 10^{-9} M.

R. M. Carey. I would like to ask whether you think your enhanced antihypertensive effect of ANF may be due to a reduction of either angiotensin II or PRA in the one-clip, two-kidney model?

S. A. Atlas. The answer is no, because in the two-kidney one-clip model, ANF actually stimulates peripheral renin probably because systemic infusion of the peptide lowers perfusion pressure to a hemodynamically compromised kidney in which ANF may be unable to increase distal NaCl delivery. In this case, the renin-stimulating effect of renal hypoperfusion may dominate. On the other hand, Garcia and his co-workers in Montreal have shown that when you chronically administer ANF to two-kidney, one-clip Goldblatt rats, using mini osmopumps, you will lower plasma renin activity. Perhaps you can still increase the load of NaCl to the distal tubule when you give the peptide chronically, but certainly the acute antihypertensive effect can't be explained by reduction of renin in this model.

G. Anogianakis. I was looking at the distribution of the different sites for ANF in the brain. Do you consider that apart from the central mediation, i.e., neuronal mediation of the effects you described, it is also possible that you may have some direct effects on the extrachoroidal production of CSF?

S. A. Atlas. I suppose I was focusing on the areas that reflect my own interest, namely the binding of ANF to brain regions that are important in salt and water metabolism and cardiovascular homeostasis, but I believe the work of Quirion *et al.* demonstrated a high density of binding sites in the choroid plexus as well, and so certainly there could be a role of ANF in the regulation of CSF production. There was also considerable binding in the retina, and that might suggest some role in fluid reabsorption in the eye. What the mechanism of those potential effects would be is unclear to me because of the rather disappointing lack of an effect of these peptides on epithelial ion transport, at least in mammals, as far as I know.

G. Anogianakis. But you haven't measured, for example, the *in situ* pressure of the CSF under administration of ANF have you?

S. A. Atlas. No. we have not done that.

G. T. Campbell. I would like to ask a couple of questions regarding the regulation of release of ANF. In the dog you consistently can find the natriuretic and diuretic reflex if you blow up a balloon catheter in the atrium, or if you volume expand isotonically. In the rat and subhuman primates, if you blow up the balloon catheter you do not consistently find this reflex, but if you volume expand you do find the associated diuresis and natruresis. I wondered if there is a dual regulation of ANF in which you would have a stimulatory effect by the atrial mechanical receptor but yet an inhibitory influence by the efferent cardiac sympathetics. Maybe, in some species, ANF cannot be released unless you remove the inhibition by the sympathetics. Second, in the absence of evidence regarding peripheral osmoreceptors, what sort of system do you invision which may cause the release of ANF if you were to have a hypernatremic and normovolemic state?

S. A. Atlas. What kind of state?

G. T. Campbell. Elevated salt but normal volume.

S. A. Atlas. I think first we assume that atrial distension is a stimulus for the release of ANF, and the existing data strongly suggest this. Whether this results from simple mechanical distension of stretch-responsive cells or whether it involves central neural reflexes is unclear. I should point out that the known atrial stretch receptors are at the junction of the great veins with the atrium, and this is an area in which ANF itself is very poorly localized, the greatest abundance being in the auricular appendages. Sonnenberg and his colleagues have shown that muscarinic cholinergic agonists will stimulate ANF from isolated atria as will adrenaline. I don't know how the latter observation will tie into your theory about the inhibition by cardiac sympathetic efferents, but certainly all of the small inconclusive bits of information that we have suggests that sympathetic stimulation tends to increase ANF rather than decrease it. As far as what the other signals for release are, I feel that if ANF is important in long-term volume regulation, I have difficulty with the concept that atrial distension can be the only signal for release. I think it is likely that in addition to sympathetic control we will find some interaction with other hormonal systems. I think we have to explore what the effects of angiotensin II and aldosterone are on release, since the renin–angiotensin system is such an important reflector of extracellular volume. Sonnenberg and his associates also reported that vasopressin will stimulate release from the isolated atria, and this could be the sort of signal that you would like to see in the setting of hypernatremia without volume expansion. More recently, Needleman and his colleagues infused vasopressin *in vivo* and also showed a marked increase in plasma ANF levels. However, in their study they markedly increased both right and left atrial pressure during systemic infusion of vasopressin, and it is impossible to dissociate a direct effect on the atrial cardiocytes from an effect of atrial distention. I think that this is an area that needs to be explored quite vigorously.

M. R. Sherman. Has anyone undertaken reciprocal competitive binding studies between angiotensin II and ANF? In other words, have they measured the effectiveness of ANF as a competitor for the classical angiotensin II binding sites, and, conversely, have they measured effectiveness of angiotensin II as a competitor for ANF binding?

S. A. Atlas. Concerning the second, I don't know, but the first has been studied. Goodfriend and his associates in Madison have shown that there is no effect of ANF on angiotensin II binding to bovine glomerulosa. Indira Sen in our institution has examined the effects of ANF on membrane-bound, solubilized and purified angiotensin II receptors and has found that ANF has no effect on the angiotensin binding constants.

M. I. New. How do you envision ANF works to stimulate steroidogenesis, solely on the glomerulosa the way angiotensin II does? If that's so did I understand that it also acts to stimulate steroidogenesis in the testes?

S. A. Atlas. No, I don't know what it does in the testes. It stimulates particulate guanylate cyclase and cyclic GMP in the testes, but a functional counterpart to that effect has yet to be reported. Regarding the adrenal effects, in the rat ANF appears to be specific for the glomerulosa in that corticosterone production by isolated fasciculata cells is not affected by ANF. Presumably this relfects the specificity of binding sites, present in the glomerulosa and absent from the fasciculata and reticularis. There is one report, that of deLéan and co-workers, which has shown binding sites in the bovine fasciculata and a decrease in cortisol production in isolated cells from that species. Of interest, in the bovine adrenal angiotensin is known to have effects on both the glomerulosa and fasciculata, suggesting an interesting parallelism between ANF and angiotensin II. As I pointed out, in the dog we see no effect on plasma cortisol. In man we find that ANF may cause slight suppression of plasma cortisol. So I think it's possible that there may be species variability as to whether the effects are specific for aldosterone or whether they embrace both glucocorticoid and mineralocorticoid production, and it appears, at least from the available evidence, that that specificity resides in where the receptors are, not in the mechanism of its effects on steroidogenesis. All I can say about the relative specificity for which agonists ANF opposes is that there are some inconsistencies between *in vitro* data and *in vivo* observations. Clearly, the most profound inhibition of aldosterone production that we've seen *in vivo* has been in the setting of angiotensin II-stimulated steroidogenesis, either in the case of the Goldblatt rat or by infusing angiotensin II. Lesser effects have been observed when ANF is administered to ACTH-stimulated rats. I have difficulty translating what is observed in isolated cells to the intact animal, and it certainly remains possible that the effects of ANF on angiotensin II-stimulated steroidogenesis may be the most important.

B. A. Scoggins. We've been interested in the relationship between natriuretic and the circulatory effects of ANP, and our studies have been done in the conscious sheep. Our studies show intrarenal infusion of human (1-28) ANP produces potent natriuretic effects. What we then did was to infuse ANP intravenously into sheep. a 60-minute infusion, 100 μg/hr, into conscious sheep had a very small natriuretic effect. However, it produced a striking fall in cardiac output, which occurred very rapidly. This was due entirely to a fall in stroke volume; peripheral resistance increased. This suggested to us that ANP is in fact a venodilator and does not in any way act in the normal animal as a vasodilator. I think it is also clear from this study that this is a very good maneuver to reduce preload on the heart. However, the levels of ANP produced in these experiments would never been seen in a normal animal. What we would suggest is that if you looked at the effect in the volume-expanded animal, perhaps you would see similar cardiovascular effects but a dramatic increase in the renal sensitivity to the hormone. Our preliminary data would suggest that this is exactly what happens. In other words the relationship between the natriuretic and circulatory effects of ANP is closely related to the volume status of the animal. A different relationship may also be seen in hypertension. In hypertensive animals the balance may be changed the other way, and you may see an exaggerated hemodynamic effect with no change in renal sensitivity.

S. A. Atlas. I think that's a very nice demonstration of the relationship between the hemodynamic and natriuretic effects. Ed Blaine and his co-workers at Merck have shown a bell-shaped dose–response curve of natriuresis to ANF and what they find is that sodium excretion increases as you increase the dose to a certain point, but beyond that point there is a fall in sodium excretion with higher doses. This is probably a reflection of the fact that with these very large, supraphysiological doses you are reducing cardiac output, as I showed in the DOCA-salt volume expanded rat; we've also seen this in normotensive dogs with large doses. In man, preliminary work by Bob Cody in our department shows that at a dose of 40 pmol/kg/minute, that is, 0.1 μg/kg/minute, which is a dose that would increase plasma levels in man to about 800 pmol/liter, ANF produces a significant decrease in right atrial pressure and in pulmonary capillary wedge pressure, suggesting either a venodilating effect or intravascular volume contraction, perhaps due to an increase in capillary permeability (as is suggested by the profound hemoconcentrating effect of ANF infusion). However, in these supine normotensive subjects cardiac output did not fall. Obviously the net effect on cardiac output probably depends on the concurrent effect of ANF on arterial resistance.

C. S. Nicoll. Growth hormone and prolactin are frequently reported to have effects on renal function. Sometimes it is claimed that they have direct effects on water and/or electrolyte excretion, and others have reported that they modulate the effects of ADH or aldosterone. I wondered if you or anyone else has looked for possible effects of ANF on the secretion of either of these two hormones.

S. A. Atlas. That's a very good question. It hasn't been studied in our laboratories, and I don't know that it has been studied by others.

J. Geller. I enjoyed your talk very much. I wonder if you could explain the seeming paradox about the role of ANF in congestive heart failure. It seems as if everything ANF does (decreasing aldosterone and renin, and increasing glomerular filtration rate and sodium excretion) is found to be the opposite of what is found in congestive heart failure and yet ANF levels are high. Is ANF simply not biologically important or can you give us another explanation for this paradox?

S. A. Atlas. I think that ANF does not operate in a vacuum, and clearly the example that you cite indicates that there are other factors in cardiovascular homeostasis that may override its effects. In some of the earlier studies in which a decreased atrial content of ANF was found in hamsters with spontaneous congestive failure, it was hypothesized that decreased circulating levels may be responsible for some of the manifestations of heart failure. Based on the observations made in several institutions in humans, this does not seem to be the case because plasma ANF levels have been elevated uniformly in patients with heart failure. I view that as simply a homeostatic response, perhaps to increased distension of the atrium with ventricular dysfunction or due to volume retention. I think the other effects of cardiac failure on systemic hemodynamics, on sympathetic stimulation, and on stimulation of the renin–angiotensin system are apparently able to override the effects of ANF. Whether there are specific defects in renal or other tissue responsiveness to ANF or changes in receptor density with prolonged sitmulation by increased circulating levels of the hormone is an important question that needs to be answered. On the other hand, it remains possible that ANF may actually mediate some of the manifestations of heart failure. For instance, ANF might conceivably promote edema formation as a mechanism to protect the heart against volume overload.

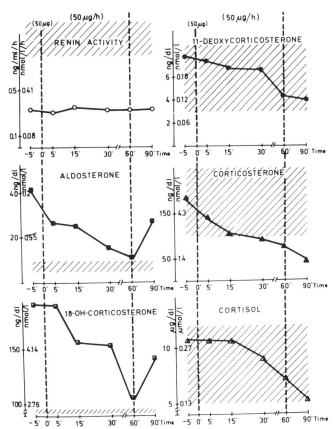

FIG. 11. Hormonal effect of human atrial natriuretic peptide infusion in a patient with aldosteronoma. After E. Glàz, I. Varga, R. Kiss, E. Rácz, and P. Vecsei (*Acta Endocrinologica Congress*, Helsinki, 1985).

P. Vecsei. We have given ANF in a case of primary aldosteronism, and have also studied the effect of ANF on the adrenal hormone patterns. Results are shown in Fig. 11. As you can see the plasma renin activity remained unchanged; the aldosterone was decreased within a short period of time after the administration of ANF, more so than was cortisol and other corticosteroids. This shows that the alteration of the plasma renin is not necessary for the depression of the plasma aldosterone activity, and that the plasma cortisol and other corticosteroids are influenced by ANF, at least in human beings, and primary aldosteronism. This is only one of the interesting observations in the relationship of renin and aldosterone.

S. A. Atlas. Yes, this would go along with our observations in dogs in which we prevented renin from falling by constricting the renal artery; we still saw a decrease in aldosterone. Regarding the effect on cortisol, during our studies on ANF infusion in man we also observed a decrease in cortisol that was greater than the diurnal variation that occurred in placebo studies. However the difference was not statistically significant. Over what time period was your study?

P. Vecsei. Of course experiments were done to control the circadian alterations; the depression of cortisol was significant in comparison to the control patterns. However, I would like to emphasize that this occurred only in one case. The ANF suppresses aldosterone, cortisol, corticosterone, 19-OH-corticosterone, and DOC when given to isolated aldosteroma cells.

K. Lederis. A number of years ago we administered chronically a peptide, urotensin I, a homolog of CRF, into spontaneously hypertensive rats and measured the systemic blood pressure and various parameters of renal function, including GFR, urine flow, and sodium excretion (Medakonic *et al., Proc. West. Pharm. Soc.* **18,** 384–388; 1975). During the first 24 hours of the experiment, renal function, including urine flow and Na^+ excretion, was inhibited significantly. In spite of continued and prolonged low systemic blood pressure, renal function returned to normal on the second day and remained normal. Has any possible interaction (antogonistic) between CRF and ANF been considered in your studies?

S. A. Atlas. I do not know of any, but I think that is a very provocative observation that needs to be followed up on.

B. F. Rice. Is it your feeling that the atrial factor has a direct effect without having to exaggerate all the other physiologic situations? The second question is general: are there any studies on other excretory products such as calcium or phosphate to define the specificity of this phenomenon?

S. A. Atlas. The answer to the last question is that calcium, phosphorus, and magnesium excretion are all increased in parallel with the increase in sodium and chloride excretion. Regarding the functional properties of ANF, perhaps my emphasis was a bit misleading. What I was referring to in the various experimental models was specifically the renal vasodilator and blood pressure lowering effects of ANF. The tentative conclusion I came to was that under normal circumstances I did not see very impressive evidence that ANF is likely to play a major role in the control of vascular resistance, but that where it seems to have very important actions is in antagonizing other hormonal and neural vasoconstrictor mechanisms. This is completely independent of its effects on sodium excretion which, while apparently dependent on the rather unique renal hemodynamic effects of ANF, are potent and very impressive even under basal conditions.

M. Katz. Thank you for a very nice discussion. Could you comment on ANF effects on renovascular adenylate cyclase and whether these effects might play any role in the vasodilation induced by ANF?

S. A. Atlas. Yes, I neglected to mention it, but studies by Anand-Srivastava *et al.* in Montreal have shown about a 25 to 30% decrease in adenylate cyclase activity in vascular tissue I believe. It's a rather mild decrease in activity of the cyclase but evidently the effect has an ED_{50} of about 10^{-10} M so it could be biologically important. How it fits into any of the biological phenomena that have been observed is not clear to me. Certainly it is unlikely to contribute to the effects on angiotensin-stimulated aldosteronogenesis.

G. Callard. Could you comment on possible biological effects of ANF on salt and water transporting tissues other than the kidney, for example, the gills of salt glands? And could these tissues possibly be used to reveal effects of the peptide on vasculature versus sodium per se?

S. A. Atlas. Yes, there have been studies reported on the shark rectal gland which it is felt represents an inverted thick ascending limb, and clearly ANF promotes chloride secretion by the shark rectal gland. In these studies, by Epstein, Solomon, and Silva in Boston and Providence, ANF appears to act indirectly. There is no evidence for a direct effect of the peptide on chloride secretion in slices of the shark rectal gland, but there is an impressive effect in the isolated perfused gland and in the intact shark. These workers have proposed that the effect of ANF may be mediated by the release of VIP from nerve terminals in the shark rectal gland. I have heard of an effect of ANF on transport in the colon of the winter flounder, but I am not really familiar with the details of these experiments. I think we still await direct demonstration of an effect of ANF on epithelial ion transport.

H. Friesen. Are there any data on ANF levels in pregnancy in fetal life or in the newborn?

S. A. Atlas. None that I know of as yet.

H. M. Kronenberg. My superficial understanding of the functional derangements in hepatorenal syndrome is that because of a renal hemodynamic disorder, there is decreased GFR and sodium retention. This sounds like a disease in which atrial natriuretic factor might be of help. I wondered whether it had been given to any patients with the syndrome and whether the level of hormone has been measured in their blood.

S. A. Atlas. Elevated levels of ANF have been found in patients with chronic renal insufficiency, and there is a recent report on a marked decrease in plasma levels following hemodialysis. It has not been studied in hepatorenal syndrome to my knowledge. We have only had the chance to study a few cirrhotic patients taken off all medication and found mild increases in circulating ANF levels which increased markedly during water immersion, but in hepatorenal syndrome per se a deficiency of ANF has not been shown.

R. M. Carey. Many of the effects of ANF are similar to those of dopamine. Has anyone looked at the effects of dopamine antagonists, particularly the DA-1 receptor antagonists, on the renal effects of ANF?

S. A. Atlas. Yes, there is one report, I believe by Briggs, Schnermann, and their colleagues, and unconfirmed to my knowledge, that the DA-1 receptor antagonists will blunt the natriuretic response to ANF. There also is a report by Sowers *et al.*, however, that ANF infusion will not alter dopamine production or clearance by the kidney.

Recent Progress in the Control of Aldosterone Secretion

ROBERT M. CAREY AND SUBHA SEN

Division of Endocrinology and Metabolism, Department of Internal Medicine, University of Virginia Medical Center, Charlottesville, Virginia, and The Research Division, Cleveland Clinic Foundation, Cleveland, Ohio

I. Introduction

The renin–angiotensin–aldosterone system is a coordinated hormonal cascade which regulates the homeostasis of peripheral vascular resistance and the volume and composition of body fluids. The proteolytic enzyme, renin, the octapeptide, angiotensin II, and the mineralocorticoid, aldosterone, have been at the center of interest in investigation of the physiologic and/or pathophysiologic importance of this intricate system. During the past 20 years, our knowledge of the control of the renin–angiotensin–aldosterone system has increased progressively, due in large part to the availability of assays for measurement of each component of the system. A vast literature has accumulated on the regulation of renin and aldosterone secretion. Recently, evidence has indicated that aldosterone secretion may be controlled by neuroendocrine determinants. The principal purpose of this review is to examine new evidence for neuroendocrine control of aldosterone secretion by aldosterone-stimulating factor and dopamine.

Aldosterone secretion is regulated physiologically by several known humoral factors. Angiotensin II, potassium ion, and ACTH all stimulate aldosterone secretion directly at the adrenal cortex. In addition, large decreases in serum sodium concentration can stimulate aldosterone secretion *in vivo* but the physiological role of hyponatremia is uncertain. These four trophic stimuli are universally accepted as stimulating aldosterone biosynthesis and secretion. The majority of the evidence suggests that the renin–angiotensin system is predominant, but the other established factors may also be important in certain situations (Davis, 1975; Fraser *et al.*, 1979; Coghlan *et al.*, 1979; Reid and Ganong, 1979). However, there is considerable evidence that other factors besides those which are well established are involved. Two of these factors are an aldosterone-stimulating factor (ASF) secreted by the anterior pituitary gland and dopamine.

II. Established Factors

ACTH is an important factor in the control of aldosterone secretion (Davis, 1975; Fraser *et al.*, 1979; Coghlan *et al.*, 1979; Reid and Ganong, 1979). ACTH stimulates aldosterone biosynthesis proximally by stimulating the conversion of cholesterol to pregnenolone. ACTH stimulates steroidogenesis predominantly by activating adenylate cyclase with production of intracellular cyclic AMP (Fujita *et al.*, 1979), but calcium is required for ACTH-stimulated cyclic AMP production and the aldosterone steroidogenic response (Fakunding *et al.*, 1979). Although ACTH stimulates aldosterone secretion acutely, it plays little or no role in the chronic maintenance of aldosterone secretion under basal conditions or in response to other specific stimuli. Variations in aldosterone secretion are not mediated by ACTH because suppression of ACTH secretion with dexamethasone does not alter the circadian rhythm of aldosterone (Catts *et al.*, 1975; Fraser *et al.*, 1979). Infusion of ACTH increases plasma aldosterone concentration in sustained fashion, suggesting that ACTH may mediate acute increases in aldosterone secretion physiologically (Nichol *et al.*, 1975; Kem *et al.*, 1975). Chronic regulation of aldosterone secretion is not dependent upon ACTH, however, because circulating aldosterone is normal in hypophysectomized animals and ACTH-deficient man (Himathongham *et al.*, 1975; McKenna *et al.*, 1978). Sodium depletion enhances the aldosterone response to ACTH, but chronic ACTH deficiency reduces the aldosterone response to sodium depletion (Kem *et al.*, 1975; Fraser *et al.*, 1979). Thus, although physiological levels of ACTH may regulate acute fluctuations of circulating aldosterone, ACTH is not important in the chronic maintenance of aldosterone secretion.

Potassium stimulates aldosterone secretion and potassium depletion inhibits aldosterone secretion (Davis, 1975; Himathongham *et al.*, 1975; Fraser *et al.*, 1979; Coghlan *et al.*, 1979; Reid and Ganong, 1979). Potassium stimulates both the early and late steroid metabolic pathway in the adrenal zona glomerulosa (McKenna *et al.*, 1978). The early phase is stimulated by calcium-dependent mechanisms (Fakunding and Catt, 1980), but the mechanism of the late phase stimulation by potassium is unknown. In man and experimental animals, relatively small alterations in potassium balance as well as larger acute increments in serum potassium concentration can stimulate aldosterone production. In man, an increase of as little as 0.1 mEq/liter in plasma potassium concentration increases plasma aldosterone concentration significantly (Himathongham *et al.*, 1975). The importance of potassium ion in the control of aldosterone secretion can be demonstrated by the observation that responses to other stimuli are subnormal in the presence of potassium depletion. In the rat, low potassium diet increases and high potassium enhances the aldosterone reponse to angiotensin II (Campbell and Schmitz, 1978; Douglas, 1980). During sodium depletion, potassium may assume an enhanced role, of equal importance to that of angiotensin, in the acute regulation of aldosterone secretion (Dulhy *et al.*, 1977). Since aldosterone increases urinary potassium excretion, inhibition of its secretion by potassium depletion may be regarded as a protective feedback mechanism.

Hyponatremia can increase aldosterone secretion. However, a very large reduction in serum sodium concentration (10–20 mEq/liter) is necessary to stimulate aldosterone secretion. Thus, the physiological role of hyponatremia is minimal (Davis, 1975; Fraser *et al.*, 1979; Coghlan *et al.*, 1979; Reid and Ganong, 1979).

The renin–angiotensin system plays an important primary role in the regulation of aldosterone secretion (Kaplan, 1965). Angiotensin II stimulates both the early and late phases of aldosterone biosynthesis. Angiotensin stimulates the early phase by a calcium-dependent mechanism (Fakunding and Catt, 1980). The mechanism(s) by which angiotensin stimulates the late phase of aldosterone biosynthesis is unknown. Several studies have demonstrated parallel changes in plasma renin activity and aldosterone under a variety of physiological conditions, suggesting that the renin–angiotensin system is the principal regulator of aldosterone secretion (Davis, 1975; Fraser *et al.*, 1979; Coghlan *et al.*, 1979; Reid and Ganong, 1979; Swartz *et al.*, 1980). Intravenously administered renin, angiotensin II, and angiotensin III[(des-aspartyl^1)-angiotensin II] stimulate aldosterone secretion (Ganong *et al.*, 1962, 1966; Laragh *et al.*, 1960; Carey *et al.*, 1978). The ability of angiotensin to stimulate aldosterone secretion has been shown *in vivo* and *in vitro*. Administration of angiotensin II into the adrenal arteries of sheep and dogs stimulates aldosterone production (Blair-West *et al.*, 1970; Ganong *et al.*, 1962). At low infusion rates in man, intravenously administered angiotensin II results in an increase of aldosterone production (Ames *et al.*, 1965; Ganong *et al.*, 1966). Incubation of glomerulosa cells with physiological concentrations of angiotensin II results in an increase of aldosterone secretion in all species studied (Bravo *et al.*, 1975a,b; Douglas *et al.*, 1978; Marieb and Mulrow, 1965). Plasma aldosterone concentration decreases in response to blockade of the renin–angiotensin system with angiotensin II antagonists or converting enzyme inhibitors (Williams *et al.*, 1978; Fraser *et al.*, 1979). Angiotensin III also stimulates aldosterone secretion, but is not as efficacious as angiotensin II, predominantly because of the shorter half-life of angiotensin III in the circulation (Carey *et al.*, 1978; Mendelsohn and

Kachel, 1980). In summary, present evidence indicates that the renin–angiotensin system is the prime regulator of aldosterone secretion under physiologic conditions in man and experimental animals.

III. Other Potential Regulators

Experimental evidence has suggested that other as yet unidentified factors control aldosterone secretion. Although most studies support the primary role of the renin–angiotensin system in the aldosterone response to sodium depletion, several studies indicate the existence of the process which may act conjointly with angiotensin II in the regulation of aldosterone production. Aldosterone secretion is stimulated by sodium depletion in nephrectomized rats, without a functioning renin–angiotensin system, in the absence of a change in circulating ACTH or electrolytes (Palmore *et al.*, 1969). In sodium-deficient sheep, rapid sodium intake suppresses aldosterone secretion within 15–30 minutes without change in plasma sodium, potassium, angiotensin II, or ACTH (Denton *et al.*, 1977). The animals were rendered sodium deficient by withdrawal of dietary sodium and the presence of a unilateral parotid fistula. Rapid ingestion of 900 mmol of sodium bicarbonate suppressed aldosterone secretion in the presence of a fixed plasma angiotensin II concentration by intravenous angiotensin infusion. Circulating electrolytes and ACTH, as reflected by cortisol secretion, did not account for this change. In dexamethasone-suppressed, nephrectomized dogs, acute reduction in plasma sodium increases aldosterone secretion (McCaa *et al.*, 1974). Acute sodium depletion was accomplished by hemodialysis; plasma sodium concentration was lowered by 20 mEq/liter, while plasma potassium and total body fluid volume were held constant. In the absence of a functioning renin–angiotensin system, aldosterone was markedly stimulated by sodium depletion. This response was blocked by hypophysectomy, suggesting that a pituitary factor other than ACTH may stimulate aldosterone secretion under these circumstances (McCaa *et al.*, 1972, 1974). The failure of aldosterone to respond to sodium depletion in hypophysectomized animals was not related to ACTH, as nephrectomized animals with intact pituitary glands during dexamethasone suppression responded to sodium depletion with an increase in aldosterone secretion. Taken altogether, this evidence suggests an unidentified factor or factors controlling aldosterone secretion, possibly a pituitary factor other than ACTH.

Experimental evidence for a new factor stimulating aldosterone secretion has been demonstrated in man as well. In 1959, Mulrow and co-workers reported the presence of a factor in human urine which stimulated steroid production from rat adrenal glands (Mulrow *et al.*, 1959).

Because of the absence of a dose–response relationship, these investigators discounted the physiological significance of these observations. In 1962, Petersen and Mueller provided additional evidence for the presence of an aldosterone-stimulating factor in human urine (Petersen and Mueller, 1962). Partially purified fractions of human urine during low sodium intake stimulated aldosterone production in rat adrenocortical tissue *in vitro*.

In 1979, Brown *et al.* reported a fraction of human urine which stimulated aldosterone secretion in rats. Urine was processed with kaolin followed by gel filtration with Sephadex G-50. A single peak of biological activity was found; the active material was reported to have molecular weight of approximately 30,000 (Brown *et al.*, 1979).

In man, a dissociation of plasma aldosterone and its known stimuli has been demonstrated. Systemic administration of angiotensin II in sodium-replete man which produces circulating angiotensin II concentrations similar to those observed during sodium deficiency does not elevate plasma aldosterone to levels observed in sodium-deficient subjects (Boyd *et al.*, 1969). A change in posture from recumbent to sitting increases plasma aldosterone concentration without altering the metabolic clearance rate of aldosterone or blood angiotensin II concentrations (Carey, 1982). Suppression of plasma renin activity by sodium loading in hypertensive patients does not result in suppression of aldosterone secretion (Collins *et al.*, 1970). Chronic blockade of angiotensin II generation by inhibition of angiotensin-converting enzyme activity does not lead to subnormal plasma aldosterone concentration and does not prevent the aldosterone response to upright posture (Bravo and Tarazi, 1979). Aldosterone secretion is stimulated by hyponatremia in bilaterally nephrectomized humans maintained on renal hemodialysis without an endogenous renin–angiotensin system (McCaa *et al.*, 1972).

Studies from our laboratory have shown that the dopamine antagonist, metoclopramide, stimulates aldosterone secretion in man and experimental animals without alterations of any of the known factors which affect aldosterone secretion (Carey *et al.*, 1979). As shown in Fig. 1, bolus administration of metoclopramide acutely increased aldosterone secretion in normal human subjects. There was no change in aldosterone metabolic clearance engendered by metoclopramide. No increase in aldosterone secretion accompanied vehicle administration. As shown in Fig. 2, metoclopramide administration was not accompanied by any change in known regulatory factors which control aldosterone secretion. Serum potassium concentration, plasma 11-hydroxycorticoids, reflecting ACTH secretion, and plasma renin activity were not altered by metoclopramide. In response to metoclopramide, no hemodynamic changes which could

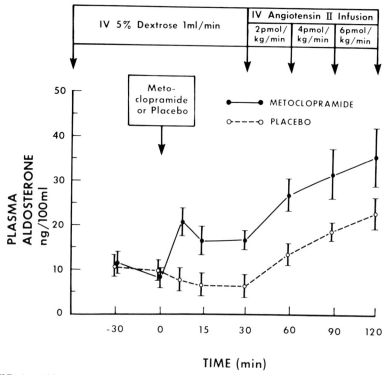

FIG. 1. Aldosterone responses to metoclopramide and to metoclopramide placebo of normal subjects ($n=6$). Responses to three cumulative doses of intravenous angiotensin II also are shown. (From Carey et al., J. Clin. Invest. 63, 727–735, 1979, with permission.)

have altered aldosterone secretion were demonstrated in this study. Blood pressure was not affected by administration of the dopamine antagonist.

In summary, a wide variety of evidence indicates that as yet unidentified factors regulate aldosterone secretion. These factors may operate independently of the primary control mechanism, the renin–angiotensin system, or may interact with angiotensin II by modulating the action of the octapeptide at the adrenal zona glomerulosa.

IV. Discovery of Aldosterone-Stimulating Factor

In 1975, Sen and Bumpus described a protein fraction of normal human urine which produced hypertension in rats (Sen and Bumpus, 1975). Since this urine fraction stimulated aldosterone secretion, and the hypertension was abolished by adrenalectomy, the compound was termed "aldoster-

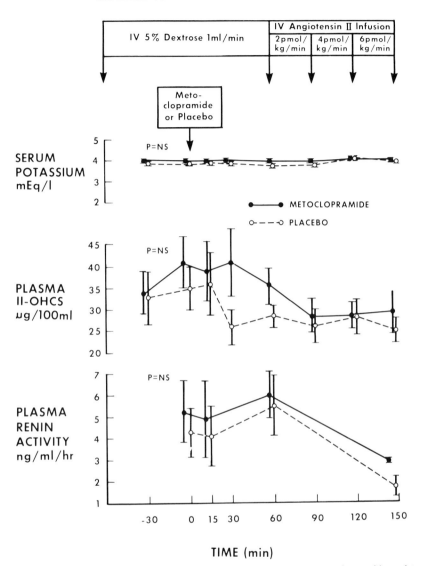

FIG. 2. Potassium, 11-OHCS, and renin activity responses to metoclopramide and to metoclopramide placebo of normal subjects ($n=6$). Responses to three cumulative doses of intravenous angiotensin II also are shown. (From Carey et al., J. Clin. Invest. **63,** 727–735, 1979, with permission.)

one-stimulating factor," or ASF. ASF was discovered during the investigation of the cause of hypertension in spontaneously hypertensive rats. Sen and Bumpus discovered that in spontaneously hypertensive rats, normalization of arterial blood pressure with antihypertensive drug treatment was associated with correction of erythrocytosis. These investigators then proceeded to isolate erythropoietic-like factors from human urine for injection into rats (Sen et al., 1977). In the process of this investigation, they found a fraction which was devoid of erythropoietic activity but produced a delayed rise in arterial blood pressure during its prolonged administration. Subsequently, this compound has been purified to homogeneity and determined to be a glycoprotein with a molecular weight of 26,000. This compound was shown to stimulate aldosterone production *in vivo* and *in vitro*.

V. Isolation and Purification of ASF

Aldosterone-stimulating factor was isolated from normal human urine by benzoic acid absorption followed by ion-exchange and molecular sieve chromatography (Table I) (Sen et al., 1977). Purification was achieved by ion-exchange chromatography, affinity chromatography, and high-performance liquid chromatography (HPLC). The molecular weight has been determined by both equilibrium ultracentrifugation and by HPLC using a Protein Pak column.

The homogeneity of the purified material was determined by DEAE cellulose chromatography (Fig. 3). Four peaks were identified and only one, eluted from the column by acetate buffer, was biologically active (Sen et al., 1977). Figure 4 shows a typical elution pattern from gel filtration on Sephadex G-100. Filtration through Sephadex using phosphate buffer, pH 7, yielded one active peak.

A considerable degree of purification was achieved by DEAE-cellulose

TABLE I

Aldosterone-Stimulating Factor: Isolation and Purification[a]

---
1. Adsorption of protein on benzoic acid
2. Removal of benzoic acid with Tris-ethanol and ethanol
3. DEAE—cellulose chromatography
4. Gel filtration on Sephadex G-100
5. Affinity chromatography on Con A- Sepharose 4B
6. Gel filtration
7. Dialysis
---

[a] Starting material = normal human urine.

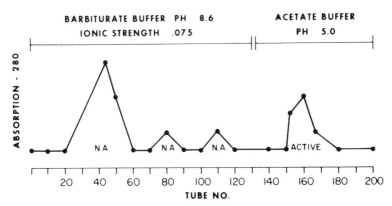

FIG. 3. A typical elution pattern from DEAE-cellulose chromatography. NA, Nonbiologically active. (From Sen et al., Circ. Res. (Suppl. I) **40,** I-5–I-10, 1977, with permission.)

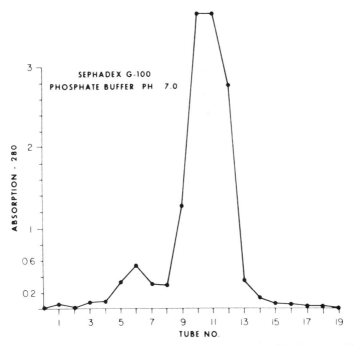

FIG. 4. A typical elution pattern from gel filtration. Sephadex G-100 was used in phosphate buffer (0.1 M), pH 7.0. Fraction containing pooled material from tubes 7–15 was biologically active. (From Sen et al., Circ. Res. (Suppl. I) **40,** I-5–I-10, 1977, with permission.)

chromatography and gel filtration. Figure 5 shows the degree of purification as assessed by paper electrophoresis before purification by chromatography in A on the left and at the end of purification in B on the right, in which the active compound was homogeneous. Figure 6 shows a polyacrylamide gel electrophoresis pattern of the hypertension-producing compound after purification (left). In this system, the active compound was homogeneous. As shown on the right side of Fig. 6, there was a shift in the R_f value of the compound after treatment with neuraminidase, suggesting that the compound is a glycoprotein.

Three lines of evidence indicate that this material is a glycoprotein. As

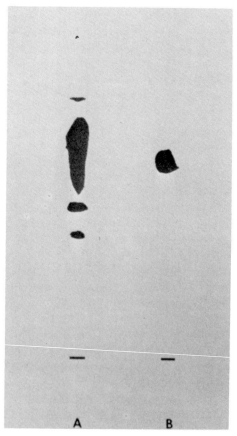

FIG. 5. Electrophoretic mobility of hypertension-producing compound. Strips were stained with ninhydrin and Pauly's reagent. (A) Before purification by chromatography. (B) At the end of purification. (From Sen et al., Circ. Res. (Suppl. I) **40**, I-5–I-10, 1977, with permission.)

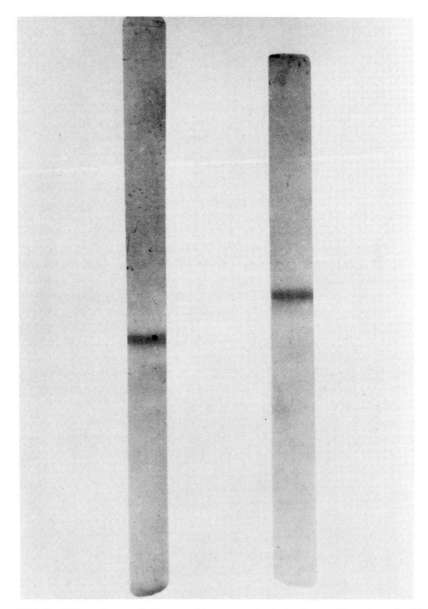

FIG. 6. Effect of neuraminidase on aldosterone-stimulating factor. The polyacrylamide gel on the left represents aldosterone-stimulating factor before and after neuraminidase treatment: the gel on the right shows the effect of neuraminidase treatment. A 7% polyacrylamide gel preparation was used and the protein was stained with Coomassie blue. (From Sen et al., Hypertension **3**, 4–10, 1981, with permission.)

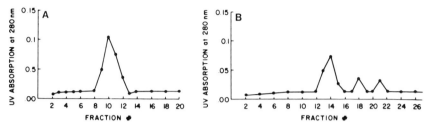

FIG. 7. A typical elution pattern of aldosterone-stimulating factor before and after neuraminidase treatment. (A) Before treatment. (B) After treatment. Sephadex G-75 was used for gel filtration and phosphate buffer (0.01 M), pH 7.0, was used for elution. (From Sen *et al.*, *Hypertension* **3**, 4-10, 1981, with permission.)

shown in Fig. 7, neuraminidase treatment caused major changes in the electrophoretic mobility of ASF. In Fig. 7A, a typical elution pattern of ASF before neuraminidase treatment on a Sephadex G-75 column is shown (Sen *et al.*, 1981). In Fig. 7B, after treatment with neuraminidase, three ultraviolet absorbing materials were obtained, a major component and two minor components. These fractions were pooled separately and their biological activities determined by their effects on aldosterone production *in vitro* (Sen *et al.*, 1981). A 2-fold increase in aldosterone production was found in fraction A after neuraminidase; the two other fractions were biologically inactive (Table II).

Two other lines of evidence that the material is a glycoprotein are the binding of the protein to Con A-Sepharose and qualitative tests of sugar content. According to these tests, the sugar content of the material is approximately 15-20%. When subjected to radioimmunoassay for growth hormone, prolactin, ACTH, and renin activity, none of these substances was detected in the purified ASF preparation.

TABLE II

Aldosterone Production by Aldosterone-Stimulating Factor (ASF) after Treatment with Neuraminidase

	Aldosterone production (ng/ml/2 hours)
Cells alone (100,000)	7.5 ± 3
ASH (10^{-7} M)	23.5 ± 4
ASF after neuraminidase treatment	
Fraction A (10^{-7} M)	47 ± 5.6
Fraction B	4.5 ± 3
Fraction C	6.2 ± 2

CONTROL OF ALDOSTERONE SECRETION 263

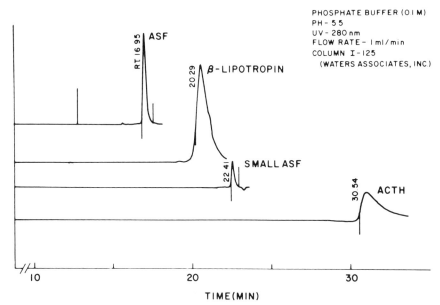

FIG. 8. A typical HPLC chromatograph of aldosterone secretogogues, ASF, β-lipotropin, small ASF, and ACTH. The retention times are considerably different: ASF (16.95), β-lipotropin (20.29), small ASF (22.44), ACTH (30.54). (From Sen *et al., Hypertension* (Suppl. I) **3**, I-81–I-86, 1981, with permission.)

Figure 8 shows a typical HPLC chromatogram of ASF. When ASF is prepared and purified at room temperature without the addition of the usual proteolytic enzyme inhibitors (PMSF, NEM, and EDTA), it is hydrolyzed to four smaller fragments. Using HPLC, one small fragment with a retention time of 22.5 minutes was separated. This small fragment of ASF is estimated to be approximately 4000 Da.

To compare ASF with the small fragment, when large ASF and small ASF were added in equamolar quantity to isolated adrenal zona glomerulosa cells, they both stimulated aldosterone production (Fig. 9). The small fragment was 34% less potent than the larger molecule. The structural requirement for steroidogenesis of ASF, thus, may reside in the smaller fragment.

VI. Steroidogenic Characteristics of ASF

In order to establish the requirement of ASF for adrenal steroidogenic activity, and to assess the site and mode of action of ASF, experiments have been conducted *in vitro* using adrenal zona glomerulosa cells in

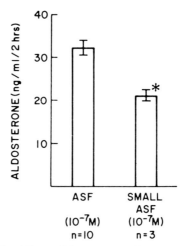

FIG. 9. Effect of ASF and its small fragment on aldosterone stimulation *in vitro*. Each tube contained 100,000 cells in 0.9 ml Medium 199: 0.1 ml of either diluent or stimulant was added (final volume 1 ml) and incubated for 2 hours at 37°C under 95% O_2 and 5% CO. When added in an equal molar basis (10^{-7}) small ASF produced 34% less aldosterone compared to ASF at the end of 2 hour incubation. (From Sen *et al.*, *Hypertension* (Suppl. I) **3**, I-81–I-86, 1981, with permission.)

rabbit and rat (Saito *et al.*, 1981). Figure 10 depicts the dose–response curves for the purified glycoprotein in collagenase-dispersed rat adrenal zona glomerulosa cells *in vitro*, as compared with that of angiotensin II. ASF stimulated aldosterone production in a dose-dependent manner from 10^{-9} to 10^{-4} M. The maximum adrenocortical response was 4-fold at 10^{-7} M after 2 hours incubation. As shown in Fig. 11, the steroidogenic potency of ASF was compared with that of two other known secretagogues, ACTH and angiotensin II. The threshold and ED_{50} responses to ACTH were greater than those to angiotensin II, which were greater than those of ASF in the rabbit. In other studies, not shown in Fig. 11, the potencies of angiotensin II and ASF were equivalent when rat adrenocortical cells were employed. Furthermore, the maximum aldosterone stimulatory effect of ASF was similar to that of ACTH and greater than that of angiotensin II in these *in vitro* systems. ACTH was a more potent stimulator of corticosterone production than either angiotensin II or ASF (Fig. 11).

An extensive evaluation of the steroidogenic mechanisms of ASF also has been conducted. ASF stimulates the late pathway of steroidogenesis by a noncyclic AMP-dependent mechanism. Figure 12 shows the relationship between aldosterone and cyclic AMP production in response to ASF. Significant increases in aldosterone were demonstrated starting at concentrations of 10^{-9} M. ASF failed to increase cyclic AMP production in

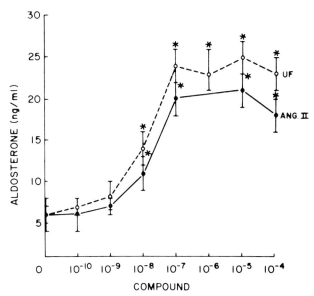

FIG. 10. Effect of urinary aldosterone stimulating factor and aldosterone on collagenase-dispersed rat adrenal zona glomerulosa cells *in vitro*. Each point represents mean ± SEM from 15 tubes. Asterisks, statistically significantly different from control. (From Sen *et al.*, Hypertension **3**, 4–10, 1981, with permission.)

concentrations which significantly stimulated aldosterone biosynthesis. The cyclic AMP response to angiotensin II was similar to that of ASF.

Similar to angiotensin II and ACTH, ASF appears highly dependent upon extracellular potassium for its steroidogenic activity. Figure 13 shows the effect of increasing concentrations of potassium upon aldosterone production in response to angiotensin II (10^{-7} M), ACTH (10^{-9} M), and ASF (10^{-7} M). In the complete absence of potassium, none of these secretogogues stimulated aldosterone production. Potassium itself was a potent stimulus to aldosterone production. At a concentration of 8 mM, the increase in aldosterone achieved was not different from that produced when combined with maximal doses of either angiotensin II, ACTH, or ASF. ACTH and ASF appeared to be less dependent on potassium than angiotensin II. The response to angiotensin II did not reach a maximum until the potassium concentration was at least 4.0 mM, but both ACTH and ASF produced maximal responses with potassium concentrations of 2.0 mM.

The aldosterone response to ASF is completely blocked by ouabain (1 mM), an inhibitor of membrane-bound Na^+, K^+-ATPase activity. Ouabain completely inhibited the aldosterone response to ASF without induc-

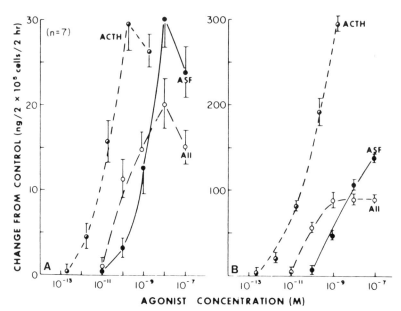

FIG. 11. Aldosterone (A) and corticosterone (B) production by rabbit adrenocortical cells stimulated by ACTH, ASF, and angiotensin II (AII). Each point indicates mean ± SEM for 9–16 vials from 7 different experiments. (From Saito et al., Hypertension 3, 300–305, 1981, with permission.)

ing any changes in basal aldosterone production. Similar results were obtained with ACTH and angiotensin II. Thus, for all of these agonists, intracellular transport of potassium is required for steroidogenesis.

Using the *in vitro* system with adrenal glomerulosa cells, all hormonal agonists of aldosterone production—angiotensin II, ASF, and ACTH—significantly increased the conversion of corticosterone to aldosterone. The percentage increases of steroidogenic conversion for ASF, angiotensin II, and ACTH were 17.4, 19.1, and 31.5, respectively. Thus, ASF and angiotensin II stimulated the late pathway of aldosterone production to a similar degree, whereas ACTH was twice as potent as angiotensin II and ASF in stimulating the conversion of corticosterone to aldosterone (Saito et al., 1981).

A series of studies has been carried out to determine if ASF acts at the same adrenocortical receptor site as angiotensin II or ACTH (Fig. 14). Equimolar concentrations of [Sar^1,Thr^8]-angiotensin II and [Ile^9]-ACTH were employed in the presence of angiotensin II (10^{-7} M), ACTH (10^{-9} M), and ASF (10^{-7} M). The data have been normalized so that the aldosterone production stimulated by each agonist in the absence of analog was taken as 100%. Neither antagonist affected ASF-induced aldosterone se-

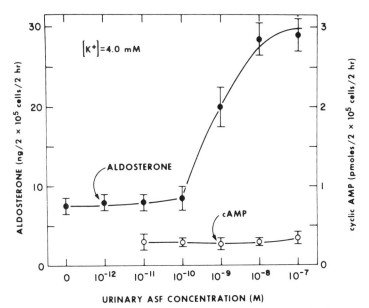

FIG. 12. Relationship between aldosterone and cyclic AMP production in response to urinary ASF. Each point represents mean ± SEM of triplicate determinations from three different experiments. Significant increases in aldosterone were demonstrated starting at concentrations of 10^{-9} M, which were associated with unchanged cyclic AMP production. (From Saito et al., Hypertension **3**, 300–305, 1981, with permission.)

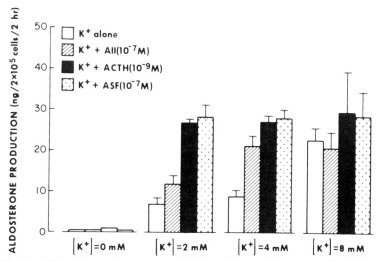

FIG. 13. Effect of potassium concentration in media upon aldosterone-stimulating activity of maximal doses of ACTH, urinary ASF, and angiotensin II (AII). Height of bars ± SEM are mean values of 4–12 separate determinations from 5 different experiments. (Fram Saito et al., Hypertension **3**, 300–305, 1981, with permission.)

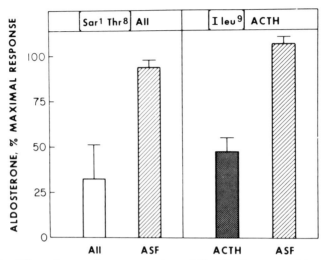

FIG. 14. Effect of equamolar concentrations of [Sar^1,Thr^8]-AII and [Ile^9]-ACTH upon aldosterone production in response to AII (10^{-7} M), urinary ASF (10^{-7} M), and ACTH (3.5 × 10^{-9} M). The data have been normalized so that the aldosterone production stimulated by each agonist in the absence of the analog was taken as 100%. Neither antagonist effected ASF-induced aldosterone production. (From Saito et al., Hypertension **3**, 300–305, 1981, with permission.)

cretion. However, the aldosterone response to angiotensin II was inhibited by approximately 70%. Similarly, in the presence of [Ile^9]-ACTH, the specific competitive antagonist of ACTH, the aldosterone response to ACTH was significantly inhibited (52%), whereas the response to ASF was unchanged. A similar result was obtained when [Ile^9]-ACTH was used with β-lipotropin, demonstrating that ASF acts at different adrenocortical receptor sites than angiotensin II, ACTH, or β-lipotropin.

The *in vitro* studies described above demonstrate that ASF is a potent stimulator of aldosterone biosynthesis in isolated rabbit and rat adrenal zona glomerulosa cells. These studies confirm earlier results obtained in the cat (Bravo et al., 1980). The maximal aldosterone stimulatory effect of ASF is similar to that of ACTH and greater than that of angiotensin II in these *in vitro* systems. However, the threshold dose and the ED$_{50}$ for ASF were higher than those of either ACTH or angiotensin II.

Thus, ASF shares the following similarities with angiotensin II: (1) both are relatively weak agonists of aldosterone production; (2) both stimulate aldosterone biosynthesis without increasing cyclic AMP production; and (3) both stimulate the late pathway of steroidogenesis. However, the inability of [Sar^1,Thr^8]-angiotensin II to inhibit steroidogenic activity indicates separate receptor sites.

On the other hand, ASF can readily be distinguished from ACTH by ASF's failure to stimulate cyclic AMP and by lack of antagonism by the ACTH analog.

VII. Physiological Studies

When ASF is injected chronically (10–20 days) into rats at a dose level of 5 μg/day IP (in 0.1 ml saline), ASF produces a rise in blood pressure that is sustained for the duration of the injection, as demonstrated in Fig. 15. Blood pressure begins to rise after 5 days of treatment and achieves a hypertensive plateau after 10 days at approximately 165 mm Hg. When treatment is stopped, blood pressure returns to control values. Saline alone (vehicle) or human albumin administered in the same volume does not alter blood pressure, indicating that volume expansion by the vehicle or protein does not contribute to hypertension.

Figure 16 depicts the results of ASF administration to bilaterally adrenalectomized rats. The bar on the left shows the rise in blood pressure from control values (crosshatched area) and after two weeks of administration of ASF (clear area). On the right, when 12 bilaterally adrenalectomized rats were treated similarly, they remained normotensive.

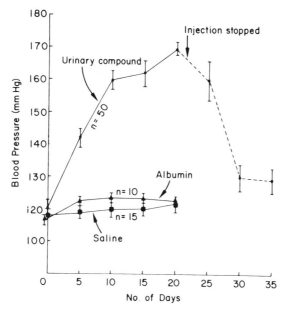

FIG. 15. The course of development of hypertension by urinary compound in normal rats. (From Sen *et al.*, *Circ. Res.* (Suppl. I) **40**, I-5–I-10, 1977, with permission.)

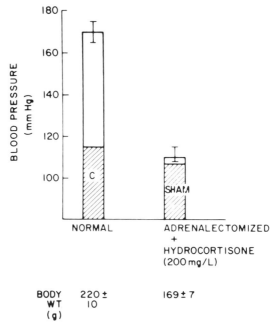

FIG. 16. Effect of ASF on adrenalectomized rats. Hatched column, control period; open column, after 2 weeks administration of ASF (6 μg/rat/day). There were 12 rats in each group. (From Sen et al., *Hypertension* (Suppl. I) **3**, I-81–I-86, 1981, with permission.)

The rise in blood pressure in normal rats with ASF administration was accompanied by a reduction in urinary sodium excretion from 1931 ± 138 to 1186 ± 197 μEq per total volume ($p < 0.05$), volume expansion from 8.6 to 10.4 ml ($p < 0.001$), an increase in total body weight from 230 ± 8 to 242 ± 4 g, an increase in heart weight from 0.710 ± 0.038 to 0.795 ± 0.048 g ($p < 0.05$), an increase in adrenal weight from 0.092 ± 0.04 to 0.102 ± 0.03 g ($p < 0.05$), and a reduction in urinary sodium:potassium ratio from 0.698 ± 0.03 to 0.47 ± 0.09 ($p < 0.01$). The plasma renin activity decreased from 21.9 ± 3.2 to 16.5 ± 1.5 ng/0.1 ml per 15 hours ($p < 0.05$). These changes suggested that the hypertension was mediated by increased adrenal mineralocorticoid secretion.

Figure 17 shows blood pressure and plasma corticosteroid concentrations in rats treated for 5 and 10 days with ASF at a dose of 5 μg/day. A significant increase in plasma aldosterone concentration was observed in rats administered ASF for both 5 and 10 days. However, plasma corticosterone was unchanged as a result of ASF treatment. After 10 days of ASF administration, the rats were hypertensive, but blood pressure levels were not increased after 5 days of administration. Thus the increase in

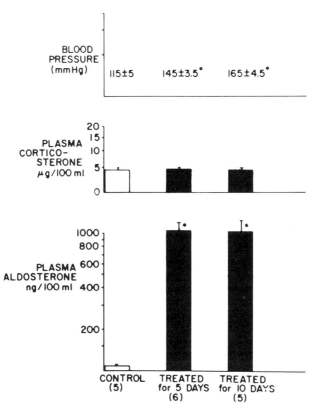

FIG. 17. Effect of urinary compound on plasma aldosterone, plasma corticosterone, and blood pressure of normal rats. Open bars represent the control rats and solid bars represent 5- and 10-day treated rats. Figures in parentheses represent the number of rats. (From Sen et al., Circ. Res. (Suppl. I) **40**, I-5–I-10, 1977, with permission.)

circulating aldosterone preceded the rise in blood pressure in these animals. This observation constitutes evidence that the increase in aldosterone secretion was responsible for the hypertension.

Chronic administration of ASF did not alter renal renin activity, plasma angiotensin II, or angiotensinogen concentrations in the rat.

The above studies have demonstrated the existence of a previously unidentified factor in human urine which stimulates aldosterone production from the adrenocortex *in vivo* and *in vitro* and upon chronic administration increases blood pressure significantly. Several lines of evidence indicate that this compound does not possess any intrinsic pressor activity. ASF does not stimulate the release of any known pressor substance and does not generate substances which are acutely pressor. This hypothesis is supported by the observation that the rise in blood pressure with ASF is not acute but is delayed in onset and occurs only after aldosterone secretion has risen. The inability of ASF to raise arterial pressure acutely either in the intact or the nephrectomized rat constitutes further evidence against a direct effect on vascular smooth muscle and direct stimulation of catecholamine release. To date, the available evidence suggests that the mechanism by which ASF stimulates aldosterone production and produces hypertension is mediated via the adrenal cortex. In adrenalectomized animals, the compound fails to increase blood pressure. Taken altogether, the changes observed with ASF administration, including adrenocortical hyperplasia, plasma volume expansion, sodium retention, and urinary potassium loss implicate the adrenal cortex as the target organ for the glycoprotein.

VIII. Localization of ASF

Antibodies to human ASF have been raised in the goat using purified ASF antigen (Fig. 20). On immunodiffusion studies, a single precipitin line between ASF and its antibody was obtained, suggesting immunological homogeneity. There was no cross-reactivity of these ASF antibodies with other known aldosterone secretogogues, such as ACTH, angiotensin II, and β-lipotropin (Sen et al., 1981). ASF isolated from dog urine has been shown to cross-react with the antibody to purified human ASF. After using preimmunization plasma from the same goat, no precipitin line was obtained. Thus, ASF appears immunochemically distinct from other known aldosterone secretogogues.

Direct immunofluorescence with fluorescein isothiocyonate-labeled goat anti-human ASF, generated using the purified ASF antigen, showed specific focal, bright apple-green stain in the cells of the anterior pituitary gland (Fig. 19). In addition, direct immunofluorescent studies with ASF antibody were negative in the posterior pituitary gland, brain, peripheral nerves, parasympathetic and sympathetic ganglia, skeletal muscle, heart, lungs, liver, kidneys, gastrointestinal tract, spleen, thyroid, and adrenal glands. High power magnification of the immunofluorescent studies show cytoplasmic localization in some but not all of the anterior pituitary cells. Control studies with fluorescein-labeled goat nonimmune serum were essentially negative.

Another method of demonstrating the pituitary origin of ASF was affinity chromotography/HPLC assay of the glycoprotein, which is described in detail below. These measurements demonstrated the existence of ASF in the anterior pituitary gland of the pig and the dog, but not in dog heart,

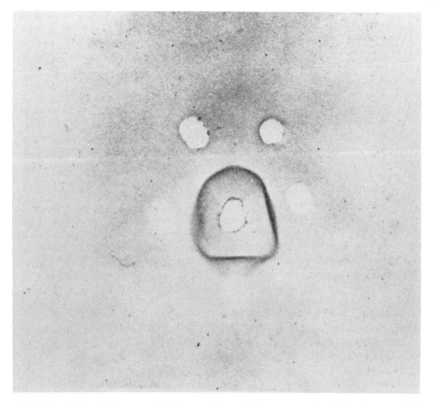

FIG. 18. A typical immunodiffusion pattern of aldosterone-stimulating factor and its antibody. Center well, antigen (ASF); outer wells, antibody, clockwise from the right. Dilutions 1=1:16, 1:10, 1:4, 1:5. Stain, Coomassie blue. (From Sen et al., Hypertension 3, 4–10, 1981, with permission.)

muscle, or adrenal glands. These tissues were saline washed prior to ASF assay.

IX. Measurement of Plasma and Urinary ASF

Sen et al. (1983) have developed methods for measuring human plasma and urinary ASF. The assay for ASF is based on affinity chromatography and HPLC. ASF is isolated from plasma and urine by affinity chromatography. Antibodies raised against ASF, coupled to Affi-Gel 10, are used as the affinity column. ASF adsorbed from plasma or urine onto the column is eluted with 4 M urea. ASF, thus isolated, is quantified by HPLC on the

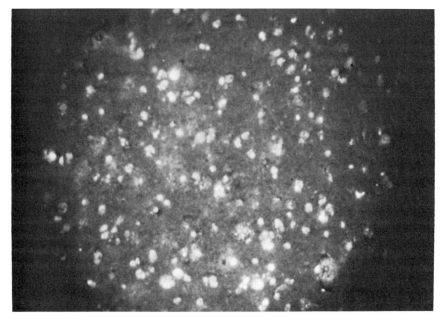

FIG. 19. Photomicrograph of anterior pituitary showing focal specific apple-green stain. Fluorescein isothiocyanate-labeled goat antihuman ASF. (From Sen et al., *Hypertension* (Suppl. I) **3**, 181–186, 1981, with permission.)

basis of the area of the peak emerging at the retention time of ASF in comparison with the area of the peak of pure ASF standards.

As mentioned previously, the antibody used in the affinity column is specific for ASF and does not cross-react with ACTH, angiotensin II, renin, or β-lipotropin. In the assay, virtually all of the ASF binds immediately to the ligand and recovery of ASF is 93 to 97%. Assay sensitivity is 2.5 ng of ASF. In both the plasma and urinary assays, inter- and intraassay variability is less than 3%. The specificity of the chromatography-HPLC assay for ASF is based on the different retention time of ASF on HPLC compared to ACTH, angiotensin II, or β-lipotropin, and by the antibody specificity.

X. Physiological and Pathophysiological Significance of ASF in Man

Since ASF administration to rats was associated with hypertension, sodium retention, volume expansion, decreased urinary sodium/potassium ratio, increased adrenal weight, and circulating aldosterone, we hypothesized that aldosterone-stimulating factor may be involved in the pathogenesis of human primary aldosteronism. We reasoned that the most likely form of primary aldosteronism which might be related to increased ASF secretion would be idiopathic hyperaldosteronism with bilateral adrenocortical hyperplasia. The syndrome of primary hyperaldosteronism due to bilateral adrenocortical hyperplasia (idiopathic hyperaldosteronism) is characterized by hypertension, sodium retention, volume expansion, hypokalemia, inappropriate kaliuresis, lack of response of plasma renin activity to stimulation, and inability to suppress aldosterone secretion. Although some of these traits may be absent, inappropriate secretion of aldosterone is a consistent finding (Winberger et al., 1979; Streeten et al., 1979; Herf et al., 1979 and Bravo et al., 1983). It has also been demonstrated that patients with idiopathic aldosteronism have enhanced aldosterone responsiveness to angiotensin and ACTH (Brown et al., 1979; Fraser et al., 1981; Carey et al., 1979; Kem et al., 1978). It has been unclear, however, whether adrenal sensitivity to these known secretogogues is sensitivity in the presence of an expanded mass of adrenocortical cells. The pathogenesis of idiopathic hyperaldosteronism heretofore has been unknown.

Another form of primary aldosteronism, aldosterone-producing adenoma, also is characterized by hypertension, hypokalemia, nonsuppressible hyperaldosteronism with sodium loading, and low plasma renin activity. However, idiopathic hyperaldosteronism differs from aldosterone-producing adenoma by the presence of bilateral adrenal hyperplasia, increased aldosterone responsiveness to angiotensin II, and lack of correction of the hypertension by the mineralocorticoid antagonist spironalactone in the case of idiopathic hyperaldosteronism.

Figure 20 shows the aldosterone response to stimulation by endogenous angiotensin III evoked by upright posture in patients with aldosterone-producing adenoma and idiopathic hyperaldosteronism. Patients with idiopathic hyperaldosteronism have greater than normal aldosterone responses to angiotensin III, whereas patients with aldosterone-producing adenoma have no response or an actual decrease in aldosterone production (Carey et al., 1979). In these same patients, renin activity was suppressed and was attenuated in response to upright posture in both patients with aldosterone-producing adenoma and those with idiopathic hyperaldosteronism. Thus, the enhanced aldosterone responsiveness to upright position depicted in Fig. 17 appears either to be hypersensitivity to circulating angiotensin II or stimulation by another factor which may interact with angiotensin II.

Figure 21 shows aldosterone responses to exogenous angiotensin for normal subjects (Group I), patients with idiopathic hyperaldosteronism (Group II), and aldosterone-producing adenoma (Group III). The hepta-

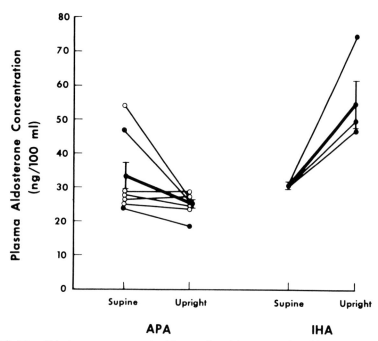

FIG. 20. Aldosterone responses to 4 hours of upright posture in subjects with primary aldosteronism due to aldosterone-producing adenoma (APA) and idiopathic hyperaldosteronism (IHA). Mean values are plotted with the heavy bars. (From Carey et al., J. Clin. Invest. 63, 718–726, 1979, with permission.)

peptide, or angiotensin III, was used for these studies. Patients with idiopathic hyperaldosteronism have enhanced adrenal responsiveness to angiotensin III, whereas patients with aldosterone-producing adenoma have decreased responsiveness (Carey et al., 1979). When adrenocortical responsiveness to angiotensin II is compared among normal subjects and patients with various disease states, marked differences in responsiveness are observed (Brown et al., 1979). Patients with aldosterone-producing adenoma have lower than normal responsiveness. Those with low-renin essential hypertension and idiopathic hyperaldosteronism have supraphysiologic responses. Thus, idiopathic hyperaldosteronism has been considered by many investigators as the tail end of a continuum of essential hypertension, the disorder being manifested by inappropriately high aldosterone secretion, volume expansion, and suppression of plasma renin activity.

To investigate the possible role of aldosterone-stimulating factor in idiopathic hyperaldosteronism, we compared concentrations of the factor

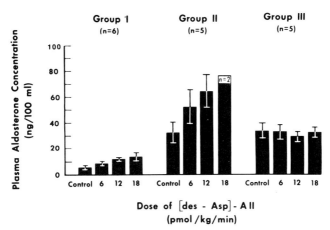

FIG. 21. Aldosterone responses with subjects in three study groups to angiotensin III ([Des-Asp]-angiotensin II) infusion. Subjects in Group I constitute normal control subjects. Subjects in Group II constitute patients with idiopathic hyperaldosteronism. Subjects in Group III constitute patients with aldosterone-producing adenoma. (From Carey et al., J. Clin. Invest. 63, 718–726, 1979, with permission.)

in urine and plasma specimens from seven patients with idiopathic hyperaldosteronism with those in specimens from four patients after surgical removal of an aldosterone-producing adenoma and from 15 normal control subjects.

As shown in Fig. 22, plasma aldosterone stimulating factor was 81 ng/dl in normal subjects and was elevated at 185 ng/dl ($p < 0.001$) in patients with idiopathic hyperaldosteronism. Values for aldosterone-producing adenoma were not significantly different from those of normal subjects. The observation that patients who had undergone unilateral adrenalectomy for aldosterone-producing adenoma had normal circulating levels of ASF indicates that it is unlikely that increased pituitary secretion of ASF plays a part in the pathogenesis of aldosterone-producing adenoma. The marked differences in the control of aldosterone secretion between primary aldosteronism due to idiopathic adrenal hyperplasia and that due to aldosterone-producing adenoma, as well as the aforementioned clinical differences between these two disease forms, are consistent with this hypothesis.

In our study (Carey et al., 1984), plasma renin activity in patients with idiopathic hyperaldosteronism in the supine position was lower than normal and did not increase normally in response to upright posture. However, in the supine position, plasma aldosterone concentrations were uniformly elevated and in response to upright posture plasma aldosterone

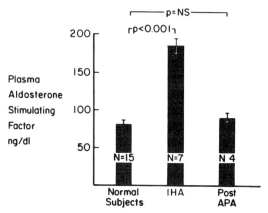

FIG. 22. Plasma ASF of normal subjects and patients with idiopathic hyperaldosteronism (IHA) or post-aldosterone-producing adenoma (APA) in the supine position.

increased from a mean of 38 to 78 mg/dl ($p < 0.01$)(Fig. 23). ASF (Fig. 23) increased from a mean of 185 to 290 ng/dl ($p < 0.05$) with upright position. Thus, in idiopathic hyperaldosteronism, upright position is associated with parallel increases in the concentrations of plasma aldosterone and ASF in the absence of a significant increase in renin activity. Our two patients with the smallest responses of circulating ASF and aldosterone to

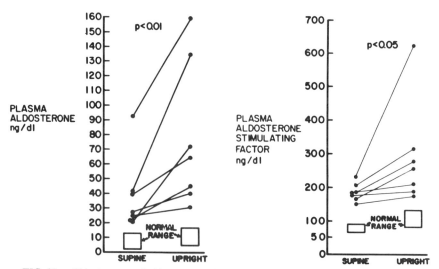

FIG. 23. Aldosterone and aldosterone-stimulating factor responses to upright posture in idiopathic hyperaldosteronism.

upright posture were limited to the sitting position, and the patient with the largest response of aldosterone-stimulating factor also had the largest response of aldosterone to upright posture. Thus, in patients with idiopathic hyperaldosteronism, ASF may stimulate aldosterone responses to upright posture in the absence of increased adrenal sensitivity to angiotensin II. Alternatively, ASF may interact with angiotensin II to facilitate aldosterone responses to upright position.

Figure 24 shows responses of plasma ASF and cortisol to dexamethasone for the individual patients with idiopathic hyperaldosteronism. Dexamethasone, at a dose which suppressed plasma cortisol to < 2 mg/dl, did not suppress circulating ASF in any patient. Plasma ACTH was 40 pg/ml after dexamethasone in all instances. Figure 25 depicts urinary excretion of ASF measured by affinity chromatography and HPLC. In normal subjects, urinary ASF excretion was 145 ng/24 hours. In patients with idiopathic hyperaldosteronism, urinary ASF was 463 ng/24 hours ($p < 0.01$). In response to dexamethasone, no change in urinary ASF was observed.

Several investigators have reported that certain derivatives of proopiomelanocortin, such as β-lipotropin, β-endorphin, α-MSH, γ-MSH and an amino-terminal segment of proopiomelanocortin, stimulate aldosterone secretion, whereas other investigators have suggested that these compounds are not effective at physiological concentrations (Matsuoka *et al.*, 1981; Gullner and Gill, 1983; Pinsin *et al.*, 1980; Lis *et al.*, 1981; Washburn *et al.*, 1982). Several lines of evidence suggest that the concentrations of ASF reported for our study do not constitute measurement of proopiomelanocortin derivatives. First of all, the antibody employed for

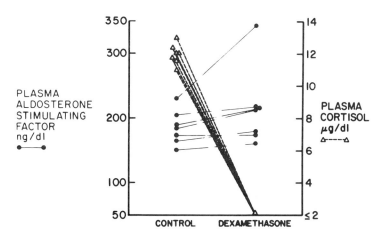

FIG. 24. Effect of dexamethasone on plasma cortisol and aldosterone-stimulating factor in idiopathic hyperaldosteronism.

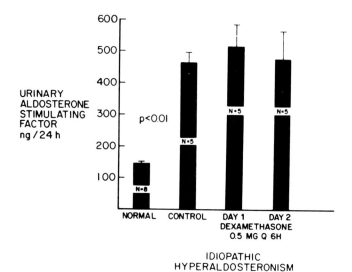

FIG. 25. Urinary excretion of ASF in normal subjects and patients with idiopathic hyperaldosteronism.

the assay of aldosterone-stimulating factor does not cross-react with ACTH, β-lipotropin, or γ_3-MSH. Second, after affinity chromatography has been used to extract aldosterone-stimulating factor, HPLC yields a single peak with a retention time of 16.95 minutes. In contrast, the retention times for derivatives of proopiomelanocortin are uniformly longer (e.g., ACTH, 30.54 minutes; β-lipotropin, 20.29 minutes). In addition, dexamethasone does not suppress plasma or urinary ASF in idiopathic hyperaldosteronism but it does suppress ACTH in all other biologically active proopiomelanocortin derivatives in human beings and laboratory animals (Tinaka et al., 1978; Oki et al., 1982). Dexamethasone also suppresses the amino-terminal fragment of proopiomelanocortin (sequence 1–76) by 80% in normal human subjects (Gaspar et al., 1985). This fragment contains the structure of γ-MSH. Since the proopiomelanocortin derivatives are released in equimolar concentrations, it would be highly unlikely for dexamethasone not to suppress all of these derivatives in coordinated fashion (Hope et al., 1981). Also, previous studies have demonstrated that the proopiomelanocortin derivatives, β-endorphin, β-lipotropin, α-MSH, and ACTH were not elevated in the circulation of two patients with idiopathic hyperaldosteronism (Gullner et al., 1983). Although a recent study has described an elevation of plasma immunoreactive γ-MSH in patients with idiopathic hyperaldosteronism, for the above reasons we do not believe that ASF represents γ-MSH or any other proopiomelanocortin derivative (Griffing et al., 1985).

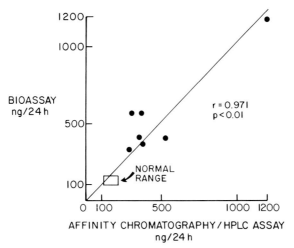

FIG. 26. Urinary aldosterone-stimulating factor. Comparison of affinity chromatography/HPLC assay with bioassay in patients with idiopathic hyperaldosteronism.

Figure 26 illustrates a comparison of ASF measured by affinity chromatography/HPLC assay and bioassay for the patients with idiopathic hyperaldosteronism. The range of values for eight normal subjects is shown in the box. HPLC and bioassay results were closely correlated with an R value of 0.971 and a p value of 0.01. Figure 27 shows the responses of aldosterone-stimulating factor to dexamethasone and dietary sodium de-

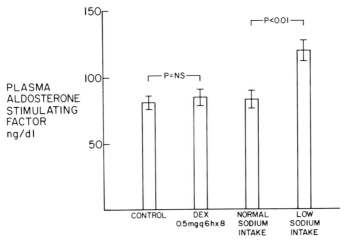

FIG. 27. Effect of dexamethasone and dietary sodium depletion on plasma ASF in normal subjects ($N=6$).

pletion in normal subjects. Dexamethasone did not suppress ASF in normal subjects. On the seventh day of dietary sodium restriction, plasma ASF was increased in comparison with values on normal sodium intake. These studies suggest that sodium and/or volume mechanism may participate in the feedback regulation of ASF secretion, since both ASF and aldosterone rise in response to sodium depletion.

XI. Dopaminergic Mechanisms in the Control of Aldosterone Secretion

During the past 6 years, studies have suggested that aldosterone secretion is inhibited by dopaminergic mechanisms. The initial evidence that dopaminergic mechanisms inhibit aldosterone secretion derived from *in vivo* studies employing the dopamine antagonist, metoclopramide. Metoclopramide is a competitive antagonist of dopamine in the central nervous system and the periphery (Jenner *et al.*, 1976; Valenzuela, 1976; Day and Blower, 1975). As stated in the initial section of this review, metoclopramide administration in man increased plasma aldosterone concentration and urinary aldosterone excretion with no change in aldosterone metabolic clearance (Carey *et al.*, 1979, 1980; Brown *et al.*, 1981). Thus, in response to antagonism of dopamine receptors by metoclopramide, an increase in aldosterone secretion occurs.

In all of the studies with metoclopramide, none of the known stimuli for aldosterone secretion such as the renin–angiotensin system, ACTH, or potassium was altered. Metoclopramide-induced aldosterone secretion is not influence by bilateral nephrectomy (absence of the renin–angiotensin system), angiotensin blockers (saralasin and captopril), or dexamethasone (ACTH suppression)(Sowers *et al.*, 1981). In addition, ganglionic blockade with trimethaphan in doses which cause postural hypotension and reduction of plasma norephinephrine by 75% do not alter metoclopramide-induced aldosterone secretion (Wilson *et al.*, 1983). Taken altogether, these results make a strong case that metoclopramide stimulates aldosterone secretion independently of aldosterone-regulating factors.

A large amount of experimental evidence now demonstrates that dopamine itself inhibits aldosterone secretion. As shown in Fig. 28, we have found that the ability of metoclopramide to stimulate aldosterone secretion can be blocked by dopamine in dose-dependent fashion. These studies were conducted in normal human subjects in metabolic balance on 150 mEq sodium, 60 mEq potassium constant intake. An intravenous infusion of dopamine at 4 μg/kg/minute or vehicle was initiated at 60 minutes and continued for 2 hours. At time 0, a bolus dose of metoclopramide 10 mg iv was administered. In subjects receiving vehicle for dopamine metoclo-

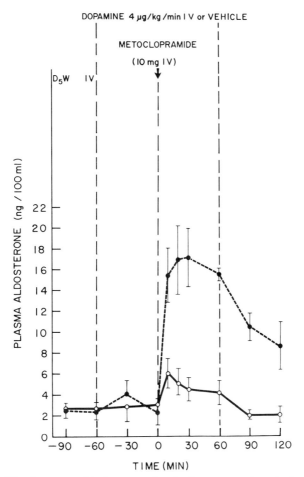

FIG. 28. Blood pressure, aldosterone, and prolactin responses to metoclopramide 10 mg iv in five normal subjects (mean ± SE). Dashed lines and solid circles represent data for Study Day 1 (vehicle infusion). Solid lines and open circles represent data for Study 2 (dopamine 4 μg/kg/min infusion). Metoclopramide-induced increases in plasma aldosterone and serum prolactin concentrations were inhibited significantly by dopamine 4 μg/kg/min. (From Carey et al., J. Clin. Invest. **66**, 10–18, 1980, with permission.)

pramide engendered a large increment in plasma aldosterone concentration within 5 minutes. Dopamine at doses which suppressed basal serum prolactin to undectable levels did not alter basal plasma aldosterone concentration. However, dopamine inhibited metoclopramide-induced aldosterone responses by approximately 75%. The ability of dopamine to block metoclopramide-induced aldosterone secretion has been shown to be dose dependent (Carey et al., 1980). The aldosterone response to metoclopramide thus appears to be mediated by an antagonist activity at dopamine receptors. We have taken these data to indicate that aldosterone secretion in under maximum tonic dopaminergic inhibition in man. Maximum dopaminergic inhibition can be unmasked by administration of a competitive antagonist of dopamine.

XII. Interactions of Dopamine and the Renin–Angiotensin–Aldosterone System

Further evidence that dopamine inhibits aldosterone secretion has been derived from studies of the effects of dopamine on angiotensin-induced aldosterone secretion. During normal or high sodium intake, aldosterone responses to angiotensin II and to ACTH are not altered by dopamine (Drake et al., 1984; Carey, 1982). However, at low sodium intake, dopamine reduces the slope of the angiotensin II–aldosterone dose–response curve to a level indistinguishable from that observed during normal or high sodium intake (Drake et al., 1984) (Fig. 29). Thus, dopamine inhibits angiotensin II-induced aldosterone secretion only during sodium depletion. Recent studies also have shown the converse of this phenomenon. In the presence of high sodium intake, angiotensin II does not stimu-

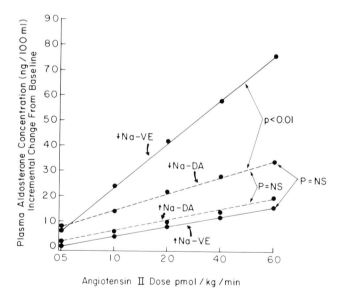

FIG. 29. Comparison of angiotensin aldosterone dose–response relationships during high and low sodium balance.

late aldosterone secretion appreciably. However, when metoclopramide is added, angiotensin II becomes a potent stimulator of aldosterone secretion (Gordon *et al.*, 1983). The ability of dopamine to inhibit angiotensin II-induced aldosterone secretion is selective for angiotensin II, as dopamine does not inhibit ACTH-induced aldosterone secretion during low sodium intake. Taken altogether, the evidence indicates that alterations of angiotensin-induced adrenocortical responsiveness with different sodium balance states may be related to dopaminergic modulation. The specificity of the effect of dopamine on angiotensin II-induced aldosterone secretion has been studied in normal subjects on low sodium intake. The data indicate that the significant inhibition of the angiotensin II effect by dopamine is not present when ACTH is substituted for angiotensin II. That is, the inhibitory effect of dopamine appears selective for angiotensin II.

XIII. Physiologic Significance of Dopaminergic Control of Aldosterone Secretion

In addition to the ability of dopamine to inhibit angiotensin II-induced aldosterone secretion, dopamine also inhibits the aldosterone response to upright posture (Hughes *et al.*, 1985). In normal subjects during normal sodium balance, dopamine inhibits the aldosterone response to standing position by approximately 70%. The expected increase in plasma renin activity with upright posture was similar in the presence and absence of dopamine. Thus, it seems possible that aldosterone responsiveness to upright position may be facilitated largely by reduction of dopaminergic inhibition of aldosterone secretion, which may normally be present in the supine position. This suggestion is strengthened by the observation that aldosterone responsiveness to upright posture is inhibited during normal sodium intake when dopamine does not influence the angiotensin-induced aldosterone response. Thus, a reduction of dopaminergic activity may operate independently of angiotensin II to allow aldosterone secretion to increase during upright position.

We have further demonstrated that dopaminergic activity is directly proportional to sodium intake or balance status (Carey *et al.*, 1981). In normal subjects placed initially on a normal sodium intake, and then subjected to sodium deprivation, urinary dopamine excretion was reduced significantly on the first day of sodium restriction and remained low throughout the period of dietary sodium deprivation. As renal dopaminergic activity declined, aldosterone secretion rose. We have hypothesized that a reduction in renal dopaminergic activity in response to sodium deprivation reduces dopamine and allows aldosterone secretion to rise, stimulated at least in part by the renin–angiotensin system as well.

XIV. Site of Dopaminergic Inhibition of Aldosterone Secretion

The majority of the evidence favors a direct adrenocortical site for dopamine action. The adrenal cortex contains a high concentration of dopamine (McCarty et al., 1984). However, the origin of adrenocortical dopamine is unknown. Adrenocortical dopamine is not derived from primary adrenocortical dopamine synthesis or by secretion of dopaminergic neurons (McCarty et al., 1984). Possibilities include uptake of sulfo-conjugated dopamine from the circulation by the adrenal cortex with subsequent deconjugation to free dopamine or, alternatively, uptake from the adrenal medulla with subsequent storage in the cortex. Since circulating quantities of dopamine are low, and orders of magnitude higher concentrations of circulating dopamine are required to inhibit metoclopramide-induced aldosterone secretion, it is unlikely that plasma-free dopamine contributes to aldosterone regulation. Recently, dopamine receptors have been identified in the rat and calf adrenal zona glomerulosa (Dunn and Bosmann, 1981; Bevilacqua et al., 1982).

XV. Summary and Conclusions

Aldosterone secretion is regulated by several known mechanisms, including the renin–angiotensin system, ACTH, and potassium. Additional new mechanisms participate in the control of aldosterone secretion including aldosterone-stimulating factor (ASF) and dopamine.

ASF has been purified to homogeneity and determined to be a glycoprotein with a molecular weight of 26,000. A biologically active fragment with a molecular weight of approximately 4000 also has been identified. The purified glycoprotein, ASF, has a different retention time on HPLC than ACTH, angiotensin II, or β-lipotropin and stimulates aldosterone secretion by a non-cyclic AMP-dependent mechanism. ASF-stimulated aldosterone secretion is not blocked by specific competitive antagonists of ACTH or angiotensin II. Fluorescein-labeled antibody in direct measurement has demonstrated ASF in the pituitary gland and in no other tissue. To date, experimental evidence suggests that the source of ASF is the anterior pituitary gland and that the adrenal cortex is the target organ for its biological activity.

ASF circulates in man. Plasma ASF increases with dietary sodium intake and is not dexamethasone suppressible. Basal plasma and urinary ASF is elevated in idiopathic hyperaldosteronism. Circulating ASF does not change with upright posture in normal subjects, but increases in parallel with aldosterone in idiopathic hyperaldosteronism. Plasma and urinary

ASF do not suppress with dexamethasone in patients with idipathic hyperaldosteronism.

Aldosterone secretion is under maximum tonic dopaminergic inhibition. Dopaminergic mechanisms selectively inhibit aldosterone responses to angiotensin II. Dopaminergic suppression of angiotensin II-induced aldosterone secretion is dependent on sodium balance state. Dopaminergic activity is directly related to sodium balance status. The adrenal cortex contains a large quantity of dopamine which is not neuronally derived, and dopamine receptors exist in the adrenal cortex. Dopamine suppresses posture-induced aldosterone responses independent of known aldosterone-regulating mechanisms.

Aldosterone-stimulating factor and dopamine represent previously unrecognized mechanisms regulating aldosterone secretion in man. Much work still needs to be done on characterization of the chemical structure, precise cellular localization, and control of secretion of ASF. As suggested by our data, increased secretion of ASF may be responsible for the pathogenesis of idiopathic hyperaldosteronism, and potentially may contribute to other clinical states of aldosterone excess. The pathophysiological significance of the modulatory effect of dopamine still needs to be explored.

REFERENCES

Bevilacqua, M., Vago, T., Sorza, D. et al. (1982). *Biochem. Biophys. Res. Commun.* **108**, 1669.
Blair-West, J. R., Coghlan, J. P., and Denton, D. A. (1965). *Circ. Res.* **17**, 386.
Boyd, G. W., Adamson, A. R., and James, V. H. T. (1969). *Proc. R. Soc. Med.* **62**, 1253.
Bravo, E. L., and Tarazi, R. C. (1979). *Hypertension* **1**, 39–46.
Bravo, E. L., Khosla, M. C., and Bumpus, F. M. (1975a). *J. Clin. Endocrinol.* **40**, 530–533.
Bravo, E. L., Khosla, M. C., and Bumpus, F. M. (1975b). *Circ. Res.* **38**, 104–107.
Bravo, E. L., Saito, I., Zanella, T., Sen, S., and Bumpus, F. M. (1980). *J. Clin. Endocrinol. Metab.* **51**, 176–178.
Bravo, E. L., Tarazi, R. C., Dustan, H. P., et al. (1983). *Am. J. Med.* **74**, 641–651.
Brown, R. D., Wisgerhof, M., Carpenter, S., Brown, G., and Hegstad, R. (1979). *Gen. Steroid Biochem.* **11**, 1043–1050.
Brown, R. D., Wisgerhoff, M., Jiang, N. S., Kao, P., and Hagstad, R. (1981). *J. Clin. Endocrinol. Metab.* **52**, 1014–1018.
Campbell, W. B., and Schmitz, J. M. (1978). *Endocrinology* **103**, 2098–2104.
Carey, R. M. (1982a). *J. Clin. Endocrinol. Metab.* **54**, 463–469.
Carey, R. M. (1982b). *In* "Neuroendocrine Perspectives" (R. M. MacLeod and E. E. Muller, eds.), pp 253–303. Elsevier-North Holland, Amsterdam.
Carey, R. M., Vaughan, E. D., Jr., Peach, M. J. and Ayers, C. R. (1978). *J. Clin. Invest.* **61**, 20–31.
Carey, R. M., Ayers, C. R., Vaughan, E. D., Jr., Peach, M. J., and Herf, S. M. (1979). *J. Clin. Invest.* **63**, 718–726.

Carey, R. M., Thorner, M. O., and Ortt, E. M. (1979b). *J. Clin. Invest.* **63,** 727–735.
Carey, R. M., Thorner, M. O., and Ortt, E. M. (1980). *J. Clin. Invest.* **66,** 10–18.
Carey, R. M., Van Loon, G. R., Baines, A. D., and Ortt, E. M. (1981). *J. Clin. Endocrinol. Metab.* **52,** 903–909.
Coghlan, J. P., Blair-West, J. R., Butkus, A., Denton, D. A., Hardy, K. J., Leksell, L., McDougall, G. C., McKinley, M. J., Scoggins, B. A., Tarjan, E., Weisinger, R. S., and Wright, R. D. (1980). *In* "Endocrinology 1980" (I. A. Cumming, J. W. Funder, and F. A. O. Mendelsohn, eds), pp. 385–388. Elsevier/North-Holland Biomedical Press, Amsterdam.
Collins, R. D., Weinberger, M. H., Dowdy, A. J., Nokes, G. W., Gonzoles, C. M., and Luetscher, J. A. (1970). *J. Clin. Invest.* **49,** 1415–1426.
Davis, J. O. (1975). *In* "Handbook of Physiology: Adrenal physiology" (H. Blascko, G. Sayers, and A. D. Smith, eds.), pp. 77–106. American Physiological Society, Washington, D. C.
Day, M. D., and Blower, P. R. (1975). *J. Pharm. Pharmacol.* **27,** 276–278.
Douglas, J. G. (1980). *Endocrinology* **106,** 983–990.
Douglas, J., Agulera, G., Kondo, T., and Catt, K. J. (1978). *Endocrinology* **102,** 685–696.
Drake, C. R., Jr., Ragsdale, N. V., Kaiser, D. L., and Carey, R. M. (1984). *Metabolism* **33,** 696–701.
Dunn, M. G., and Bosmann, H. B. (1981). *Biochem. Biophys. Res. Commun.* **99,** 1081.
Fakunding, J. L., Chow, R., and Catt, K. J. (1979). *Endocrinology* **105,** 327–333.
Fraser, R., Brown, J. J., Lever, A. F., Mason, P. A., and Robertson, J. I. S. (1979). *Clin. Sci. Mol. Med.* **56,** 389–399.
Fraser, R., Beretta-Piccoli, G., Brown, J. J., *et al.* (1981). *Hypertension* **3**(3), Suppl, I-87–I-92.
Fujita, K., Aguilera, G., and Catt, K. J. (1979). *J. Biol. Chem.* **254,** 8567–8574.
Ganong, W. F., Mulrow, P. J., Boryczka, A., and Cera, G. (1962). *Proc. Soc. Exp. Biol. Med.* **109,** 381–384.
Ganong, W. F., Biglieri, E. G., and Mulrow, P. J. (1969). *Recent Prog. Horm. Res.* **22,** 381–430.
Gasper, L., Chan, J. S. D., Seidah, N. J., and Chretien, M. (1985). *J. Clin. Endocrinol. Metab.,* in press.
Gordon, M. B., Moore, T. J., Dluhy, R. G., and Williams, G. H. (1983). *J. Clin. Endocrinol. Metab.* **56,** 340–345.
Gullner, H-G, and Gill, J. R., Jr. (1983). *J. Clin. Invest.* **71,** 124–128.
Gullner, H-G, Nicholson, W. E., Gill, J. R., Jr., and Orth, D. N. (1983). *J. Clin. Endocrinol. Metab.* **56,** 853–855.
Herf, S. M., Teates, D. C., Tegtmeyer, C. J., Vaughan, E. D., Jr., Ayers, C. R., and Carey, R. M. (1979). *Am. J. Med.* **67,** 397–402.
Himathongham, T., Dluhy, R. J., and Williams, G. H. (1975). *J. Clin. Endocrinol. Metab.* **41,** 153–159.
Hope, J., Ratter, S. J., Estivariz, F. E., McLaughlin, L., and Lowry, P. J. (1981). *Clin. Endocrinol.* **15,** 221–227.
Hughes, J., Malchoff, C. D., Sen, S., and Carey, R. M. (1985). *Clin. Res.* **32,** 554A.
Kaplan, N. M. (1965). *J. Clin. Invest.* **44,** 2029–2039.
Katz, F. H., Romfh, P., and Smith, J. A. (1975). *J. Clin. Endocrinol. Metab.* **40,** 125–134.
Kem, D. C., Gomez-Sanchez, C., Kramer, N. J., Holland, O. B., and Higgins, J. R. (1975). *J. Clin. Endocrinol. Metab.* **40,** 116–124.
Kem, D. C., Weinberger, M. H., Higgins, J. R., Kramer, N. J., Gomez-Sanchez, C., and Holland, O. B. (1978). *J. Clin. Endocrinol. Metab.* **46,** 552–560.

Laragh, J. H., Angers, M., Kelly, W. G., and Lieverman, S. (1960). *J. Am. Med. Assoc.* **174**, 234–240.
Lis, M., Hamet, P., Gutkowska, J., *et al.* (1981). *J. Clin. Endocrinol. Metab.* **52**, 1053–1056.
McCaa, R. E., McCaa, C. S., and Reid, D. G. (1972). *Circ. Res.* **31**, 473–480.
McCaa, R. E., Young, D. B., and Guyton, A. C. (1974). *Circ. Res.* **34-35**(Suppl. I), 115–125.
McCarty, R., Kirby, R. F., and Carey, R. M. (1984). *Am. J. Physiol.* **247**, E709–E713.
McKenna, T. J., Island, D. P., Nicholson, W. P., and Liddle, G. W. (1978). *Endocrinology* **103**, 1411–1416.
Marieb, N. J., and Bulrow, P. J. (1965). *Endocrinology* **76**, 657–664.
Matsuoka, H., Mulrow, P. J., Franco-Saenz R., and Li, C. H. (1981). *J. Clin. Invest.* **68**, 752–759.
Mendelsohn, F. A. O., and Kachel, C. D. (1980). *Endocrinology* **106**, 1760–1768.
Murow, P. J., Shmagranoff, G. L., Lieberman, A. H., Slade, C. L., and Luetscher, J. A., Jr. (1959). *Endocrinology* **64**, 631–637.
Nicholls, M. G., Espiner, E. A., and Donald, R. A. (1975). *J. Clin. Endocrinol. Metab.* **41**, 186–188.
Oki, S., Nakao, K., Tanaka, I., Kinoshita, F., Nakai, Y., and Immura, H. (1982). *Endocrinology* **111**, 418–424.
Reid, I. A., and Ganong, W. F. (1979). *In* "Hypertension" (J. Genest, E. Koiw, and O. Kuchel, eds.), pp. 265–292.
Sowers, J. R., Brickman, A. S., Sowers, D. K., and Berg, G. (1983). *J. Clin. Endocrinol. Metab.* **52**, 1078–1084.
Streeten, D. H. P., Tomycz, N., and Anderson, G. H., Jr. (1979). *Am. J. Med.* **67**, 403–413.
Tanaka, K., Nicholson, W. E., and Orth, D. N. (1978). *J. Clin. Invest.* **62**, 94–104.
Valenzuela, J. E. (1976). *Gastroenterology* **71**, 1019–1022.
Vinson, G. P., Whitehouse, B. J., Dell, A., Etienne, T., and Morris, H. (1980). *Nature (London)* **284**, 464–467. Washburn, D. D., Kem, D. C., Orth, D. N., Nicholson, W. E., Chretien, M., and Mount C. D. (1982). *J. Clin. Endocrinol. Metab.* **54**, 613–618.
Weinberger, M. H., Grim, C. E., Hollifield, J. W., *et al.* (1979). *Ann. Intern. Med.* **90**, 386–395.
Williams, G. H., Hollenberg, N. K., Brown, C., and Mersey, J. H. (1978). *J. Clin. Endocrinol. Metab.* **47**, 725–731.
Wilson, T. A., Kaiser, D. L., and Carey, R. M. (1983). *J. Clin. Endocrinol. Metab.* **57**, 200–203.

DISCUSSION

J. R. Gill. It is important to note that peptides which may or may not be related to aldosterone-stimulating factor have been implicated as physiological or pathophysiological stimulators of aldosterone secretion. We have observed that infusion of β-endorphin, 3 pmol/minute into the adrenal glands of nephroctomized, hypophysectomized dogs significantly increased the secretion of aldosterone but did not affect cortisol (*J. Clin. Invest.* **71**, 124, 1983). Other investigators have observed supranormal values for plasma immunoreactive β-endorphin in patients with idiopathic hyperaldosteronism (*J. Clin. Endocrinol. Metab.* **60**, 315, 1983). More recently these investigators have also reported that plasma immunoreactive γ-melanotropin may also be increased in patients with idiopathic hyperaldosteronism (*J. Clin. Invest.* **76**, 163, 1985). These findings suggest the importance of determining whether aldosterone-stimulating factor is in any way related to proopiomelanocortin.

R. M. Carey. Thank you for that comment, Dr. Gill. We share your thesis that ASF cannot be a derivative of proopiomelanocortin (POMC). Several of the POMC derivatives, such as β-lipotropin and MSH and the NH_2 terminal fragment, have been demonstrated to stimulate aldosterone secretion *in vitro*, but all at supraphysiological doses. Melby et al recently have demonstrated that γ_3-MSH is elevated in idiopathic hyperaldosteronism. However, he did not use dexamethasone suppression in his studies. Since all of the POMC derivatives are released quantitatively, all are suppressible with dexamethasone. ASF was not suppressible with dexamethasone.

M. Raj. Is there any evidence that ASF might be prolactin? Have you determined whether your antibody to ASF cross-reacts with prolactin? Also, do you have any information about the precise pituitary cell of origin of ASF?

R. M. Carey. First, purified ASF does not cross-react with prolactin or growth hormone antibody and is undetectable in immunoassays for these hormones. Also, there is a lot of evidence that prolactin does not stimulate aldosterone secretion. Maneuvers which alter prolactin do not affect aldosterone *in vivo* in man. We have not had an opportunity as yet to study the precise pituitary cell of origin of ASF. We are certainly planning to go ahead with some immunocytochemical studies with double antibody techniques to look at the cell of origin of ASF. The immunofluorescent studies which I showed indicate a focal deposition of the stain, so that I would anticipate localization to a particular cell type and it seems to be a fairly plentiful cell type within the pituitary. However, the distribution does not seem to be consistent with the lactotroph because it appears much less frequent. The double antibody studies will be important. Also, we hope to be able to perform studies with the reverse hemolytic plaque assay with the isolated pituitary cell and the ASF antibody to determine which specific cells are involved.

C. S. Nicoll. I have three questions. First, have you identified the pituitary cell type that contains the ASF? Second, is the responsiveness of the adrenal to ASF altered in hypophysectomized animals? Third, can you interrelate ASF with prolactin and the high concentration of prolactin binding sites in the adrenal cortex?

R. M. Carey. Those studies have not yet been done. What has been done is studies of the circulating and urinary excretory levels of ASF in hypophysectomized subjects. In these studies, ASF is undectable. We don't know about adrenal sensitivity to ASF though this is an important issue.

B. D. Albertson. We have been interested in a putative stimulator of adrenal androgen in children which has been hypothesized by us and some other groups to be of pituitary origin. I am interested in your pituitary immunologic studies and in particular what types of cells you believe are producing ASF. And second, is there a prehormone or a prohormone of this molecule or fragment of ASF that may have some other actions, notably at the level of the zona reticularis in the adrenal?

R. M. Carey. I don't have any information on this issue. It's an interesting idea, but there is no basis for speculating at this point.

B. D. Albertson. One more question. In reference to ASF, adrenal mechanisms of action, and the independence of cyclic AMP, have you looked at cyclic GMP levels?

R. M. Carey. No, we have not.

J. Geller. Congratulations on an excellent paper. It sounds as if it is now possible to distinguish idiopathic hyperaldosterism from an aldosterone-producing adenoma. In aldosterone-stimulating factor, comparing idiopathic hyperaldosteronism to the tumor, you showed only tumor plasma values of ASF after treatment. Do you have any data on what plasma levels of aldosterone-stimulating factor are in aldosteronomas prior to treatment and compared to the idiopathic hypoaldosteronisms?

R. M. Carey. Yes, we do. The reason for the description of the postadrenal excision data is our original hypothesis that ASF participates in or is responsible for the pathogenesis of the aldosterone-producing, adenoma. In the case where the adenoma has been removed, elevated ASF levels would be anticipated. This was not the case. With a limited number of patients with aldosterone-producing adenomas *in situ* the ASF values are normal. Interestingly, ASF is not suppressed in the patients, suggesting that aldosterone may not be the major feedback regulator of ASF secretion.

J. Geller. One last question. Has anybody demonstrated a pituitary adenoma in patients with idiopathic hyperaldosteronism?

R. M. Carey. Yes. There are a couple of interesting patients. One is a patient with idiopathic adrenal hyperplasia reported by Dr. Mulrow in which a postmortem examination showed a pituitary intermediate lobe tumor. No assays for ASF were performed because that study predated our work. We presently are studying a very interesting patient who has anterior pituitary adenoma and who has marked hyperaldosteronism with idiopathic adrenal hyperplasia. We don't have any ASF data yet, but I think this patient has a good possibility of ASF hypersecretion from the pituitary adenoma. However, we have CT scanned a number of patients with idiopathic hyperaldosteronism and are not able to demonstrate a pituitary tumor in any of these patients. Thus the patient with the anterior pituitary adenoma under study now is unique. The vast majority of patients with idiopathic hyperaldosteronism have no demonstrable pituitary tumors on CT scan. These results do not signify that pituitary tumors are not present, but merely indicates that we can't find them on head CT scan.

M. I. New. Thank you very much for this wonderful review. In order to study the separation of aldosterone-stimulating factor from other pituitary hormones, we were lucky enough to get Dr. Sen to measure ASF in a prepubertal boy who had absolutely normal levels of growth hormone as well as all other pituitary hormones. He had normal cortisol diurnal variation and secretion rates but he had hyperaldosteronism with bilateral adrenal hyperplasia. It appeared that in this child, who did not have elevated prolactin or any other pituitary hormones, the elevation of ASF appeared to explain (or at least it was a desirable explanation for the fact) the hyperaldosteronism. Thus this child who had low renin and hyporeninemia had none of the factors that are known to stimulate aldosterone, but had measurable increased ASF to very high levels which might stimulate aldosterone and cause adrenal hyperplasia.

R. M. Carey. Thank you Dr. New.

S. A. Atlas. Thanks very much for that lovely review. I have three related questions concerning the possible role of ASF in idiopathic hyperaldosteronism. First, you told us that the effect of ASF on isolated adrenal cells is less dependent on potassium than that of angiotensin II or of ACTH, but you didn't say anything about whether ASF might modulate the response of those cells to angiotensin, since it is well known that the effect of angiotensin II is exaggerated, as you told us, in patients with IHA. A related question concerns the findings of Melby and co-workers who have reported that prolonged angiotensin-converting enzyme inhibition will normalize hyperaldosteronism and hypertension in at least a subset of patients with idiopathic hyperaldosteronism. I was wondering if there is any information available on the effect either of converting enzyme inhibitors or angiotension II on plasma levels of ASF?

R. M. Carey. There is no information available on ASF levels after administration of angiotensin II or converting enzyme inhibitors at present. There does appear to be a potentiating response of ASF *in vitro*, particularly in the rat cell suspension, on angiotensin II-induced aldosterone secretion. Therefore, it is tempting to speculate that in those associated responses of ASF and aldosterone to upright posture there is an interaction of angiotensin II with ASF in these patients. The patients with idiopathic hyperaldosteronism appear biochemically similar to normal subjects who have been sodium depleted. That is, they have superphysiological adrenal sensitivity to angiotensin II. We are interested in continuing studies on a potential interaction of these two aldosterone secretagogues.

S. A. Atlas. The third part of my question concerns one of the most dramatic findings that has been reported in patients with idiopathic hyperaldosteronism, namely that there is a marked suppression of plasma aldosterone produced by the serotonergic antagonist cyproheptadine in dexamethasone-treated patients, as reported by Grekin and associates. Evidently there is also evidence from Grekin's group that 5-hydroxytryptophan can stimulate adlosterone when administered to man and that this effect is not blocked by peripheral decarboxylase inhibitors. Is there any information on possible central serotonerigc effects on ASF secretion?

R. M. Carey. No, but it is tempting again to speculate because the effect of serotonin does appear to be central and because the data of Dr. Roger Grekin at the University of Michigan appear to suggest a possible pathophysiological role of serotonin in idiopathic hyperaldosteronism. I would just like to extend that comment to the observation of Dr. Brian Holland of the University of Texas/Galveston that patients with idiopathic hyperaldosteronism normalize their blood pressure and hyperaldosteronism with administration of dopamine or L-dopa. I would like to link these observations with our preliminary data that ASF secretion is increased with metochopramide and speculate that perhaps dopaminergic mechanisms and possibly also central serotoninergic mechanisms may participate in the control of secretion of ASF.

R. Osathanondh. I have two questions. Could you discuss a bit the stability of these ASF glycoproteins in biological fluids, especially plasma or serum, and how long they can be kept in a freezer? Do you have any data on the ASF levels in pregnant women?

R. M. Carey. To my knowledge, plasma ASF levels are not available for pregnant women. ASF appears quite stable *in vitro* under freezer conditions for long periods of time. ASF secretory responses seem to be fairly stable. That is pulsatile release of this material appears not to occur in the same fashion as other anterior pituitary glycoproteins such as LH. Circulating ASF levels measured time after time appear to be fairly consistent in the same individual.

R. Osathanondh. Is the level the same when measured in the plasma and in the serum?

R. M. Carey. We have not done any serum ASF measurements. All our measurements have been conducted in plasma.

G. Segre. The first question is directed toward the concept that small ASF and big ASF are related. I would like to know how rigorous was the data that those two entities are related. Have you been able to generate by some chemical manipulation the small form from the big form?

R. M. Carey. We incubate the large form of the glycoprotein at room temperature without the addition of proteolytic enzyme inhibitors. The 4000 molecular weight material is formed by hydrolysis.

G. Segre. Is there anything known about the peptide structure of the big and the little forms?

R. M. Carey. No, the structure of small ASF is not yet known, but the amino acid composition is known. We hope to be able to obtain the sequence of this material since that material retains full biologic activity.

G. Segre. The second question related to some of the *in vivo* studies you presented. Have you tried to block the *in vivo* effect of ASF and any of the other factors that stimulate aldosterone secretion by use of the antibodies you developed to ASF?

CONTROL OF ALDOSTERONE SECRETION 293

R. M. Carey. No. This would be possible in experimental animals and is a good suggestion.

J. M. Hershman. In regard to IHA, spironolactone apparently will normalize the sodium and potassium abnormalities, but not control the hypertension, and, as you said, removal of the adrenals also does not result in a long-term cure of the hypertension. Does this have any implications in regard to the role of ASF in the pathogenesis of IHA?

R. M. Carey. Yes, I could speculate that the aldosteroidism occurs first. Then a process within the vasculature itself may occur independently of the aldosteronism after chronic hypertension and hyperaldosteronism. At that point adrenalectomy either biochemically or surgically does not cure the hypertension. Studies with bilateral one and four-fifths adrenalectomy in these patients were performed quite a long time ago. I believe these studies were done fairly well, but I still question whether a total bilateral adrenalectomy should be done to cure the hypertension. Of course, the possibility exists that the hyperplastic gland may regrow with time unless all of the adrenal mass is removed.

J. M. Hershman. In regard to your response to Dr. Raj concerning cross-reaction of pituitary hormones in the ASF assay, the pituitary glycoproteins are TSH and the gonadotropins. I assume that you tested those hormones and their subunits for cross-section in the ASF assay, and they did not cross-react?

R. M. Carey. We tested prolactin and growth hormone, and the reason for that is that these are the two anterior pituitary hormones which have been demonstrated either to have permissive effects on aldosterone secretion or controversial effects acutely. We have not tested LH, FSH, or TSH because these hormones have not been demonstrated to affect aldosterone secretion.

J. M. Nolin. With Nicoll, and perhaps a few others, I'm not yet totally convinced that this hormone is not good old-fashioned prolactin. However, I'm compelled to tell you that insome studies we did with the adrenal where we could clearly isolate glomerulosa from fasciculata and reticularis, in terms of the findings defining the three zones as a target for prolactin, we found no evidence that the glomerulosa is a target, whereas, in parallel, we obtained strong evidence that the fasciculata and reticularis are clearly targets for this hormone.

R. M. Carey. Let me extend these observations. We've performed extensive studies including prolactin administration to man and we cannot demonstrate (in spite of huge increases in circulating prolactin) any change in aldosterone secretion. We are quite sure that prolactin does not affect aldosterone secretion in man.

B. A. Scoggins. You said that in normal man the aldosterone secretion is under maximum tonic inhibition by dopamine. If that is true, how do you envision that additional dopamine produces the effects that you see on aldosterone if it's already under maximum tonic inhibition.

R. M. Carey. The reason for stating that aldosterone secretion is under maximum tonic dopaminergic inhibition is that we don't see any reduction in basal aldosterone secretion with exogenous dopamine. Under all experimental circumstances so far, aldosterone secretion has to be stimulated first; then it can be inhibited. In that regard, dopamine is similar to atrial natriuretic factor, which blocks angiotensin-induced aldosterone secretion much more readily than basal aldosterone secretion.

D. J. Morris. I wonder if you could comment further on the fact that the steroid spironolactone doesn't inhibit the hypertension in idiopathic hypertension. And, once one infuses ASF have you looked in the urine for either hypertensinogenic agents or other steroids other than aldosterone?

R. M. Carey. I can't really say more than I've already stated regarding the pathogenesis of the hypertension per se. You can give adrenal mineralocorticoid receptor blocking doses of spironolactone and normalize the hyperaldosteronism, and yet the patients do remain hypertensive. That's a fact which is poorly understood.

D. J. Morris. In your experiments in which you infuse or administer ASF and cause hypertension in rats, have you looked for other hypertensinogenic agents in the urine of those rats?

R. M. Carey. No, we have not. Would you like to suggest steroids that we might study?

D. J. Morris. Figure 30 shows chemically synthesized derivatives of aldosterone; one is 19-hydroxyaldosterone and the other is 19-noraldosterone. We do not know if they are present in man; they could be quite important molecules. 19-Hydroxyaldosterone is equally potent with aldosterone in the isolated toad bladder system and 19-noraldosterone is equally potent in the rat mineralocorticoid bioassay as well as in the toad bladder system. 19-Hydroxyaldosterone has the possibility of being synthesized by the mitochondria of the adrenal glomerulosa cells, and 19-noraldosterone has the possibility of being synthesized in the kidney, since 19-nor-DOC and 19-nortestosterone have both been shown to be synthesized there. In Fig. 31 (the right panel) infusion of 19-noraldosterone at dosages of 0.1 and 0.5 µg/day into adrenalectomized SHR cause marked increases in blood pressure of this rat model. And in fact, if you compare the results with the left panel, the hypertensinorgenic activity of 19-noraldosterone may even be a little more potent than the effects of 0.5 and 0.1 µg of aldosterone.

R. W. Carey. Could you clarify the metabolic pathway by which the 19-nor compound is derived?

D. J. Morris. According to the work of Melby and Gomez-Sanchez, the adrenal-secreted 19-hydroxy substance is converted to a 19-oxo and a 19-carboxy derivative, which is than converted in a target tissue, the kidney, and is decarboxylated to the nor-substance. It's not thoroughly understood yet.

R. M. Carey. That's an excellent idea; thank you, Dr. Morris.

J. F. Tait. Dr. Sylvia Tait and I feel we have been privileged to hear these excellent two presentations indicating that at long last we may be beginning the end of the hunt for the missing factors controlling aldosterone secretion, which perhaps started some 32 years ago after the first Laurentian Hormone Conference. As you know I have been interested in the observation that Sen's substance may not increase cyclic AMP in the glomerulosa. We have not tested the atrial natriuretic factors but the angiotensins do not increase cyclic AMP, so Sen's substance would be fairly exceptional in this regard. It's quite difficult to obtain results showing that cAMP is not increased. What alternative molecular messenger are you postulating? Could it be phospholipase C, associated with changes in calcium transport or through nonphospholipase C calcium effects. These systems with cyclic AMP I believe are now established as two glomerulosa mechanisms. Which one are you selecting for your substance?

R. M. Carey. I'm electing, if I have that privilege, a calcium-dependent mechanism. Dr.

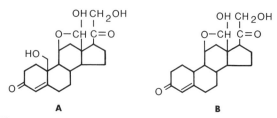

FIG. 30. (A) 19-hydroxyaldosterone and (B) 19-noraldosterone.

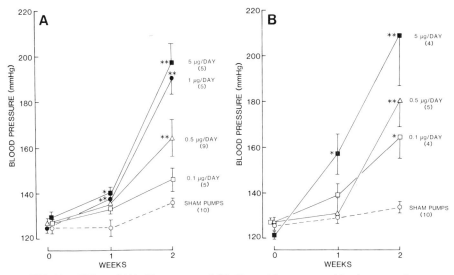

FIG. 31. Effect of (A) aldosterone and (B) 19-noraldosterone on blood pressure in a rat model.

Sen has conducted some preliminary studies indicating that there is a calcium-dependent mechanism which is responsible for ASF-induced steroidogenesis.

J. F. Tait. But, you dont' get any effects on phospholipase C.

R. M. Carey. That has not been investigated adequately to make a comment.

R. Rittmaster. Have you attempted inferior petrosal sinus sampling in any of your patients with idiopathic hyperaldosteronism?

R. M. Carey. We are doing that now.

R. Rittmaster. And have you found either lateralization or a central to peripheral gradient?

R. M. Carey. We don't have enough data to make a comment. We are studying that now.

M. Raj. If I understood your presentation correctly you mentioned that ASF increased the side chain cleavage of cholesterol. Am I correct?

R. M. Carey. No, I don't think I said that. What I did say was that in cells incubated with corticosterone you can get an increased in conversion to aldosterone, and that ASF does in fact stimulate the late pathways of steroidogenesis. There is some preliminary evidence also, but its just preliminary at this point, that in the early pathway the conversion of cholesterol to pregnenalone also is stimulated.

M. Raj. Do you think that the effects of ACTH and ASF would be additive?

R. M. Carey. Well I don't think there is any additive effect in aldosterone production when you just look at the production rates in the *in vitro* preparation.

P. Vecsei. You said that you treated rats with aldosterone-stimulating factor. What was the morphology of the adrenals after the treatment?

R. M. Carey. The adrenal weight is increased and the morphology is consistent with hyperplasia but, as you know, it is difficult to distinguish hyperplastic from normal glands on the basis of ordinary histology.

P. Vecsei. And what about the zona glomerulosa?

R. M. Carey. The zona glomerulosa appears to be wider than normal.

P. Vecsei. What about the 18-hydroxycorticosterone?

R. M. Carey. We haven't measured that.

J. F. Tait. I think I should just mention that from what we have seen here there is some implication that if you take rat capsular cells (and these are capsular cells I think not prezona glomerulosa cells) and add ACTH, the steroid would originate from the glomerulosa cells. In our studies with the usual 5% contamination with zona fasciculata cells, I am afraid this would not necessarily be the case and any alteration could be from zona fasciculata cells. So I think one has to be a bit cautious. In addition, recently, Mr. Chen and Dr. Hyett in our laboratory have I think good evidence (although not completely certain perhaps because morphologically they are indistinguishable from glomerulosa cells) that there is another population of cells in these capsular cells, which make aldosterone but are highly responsive to ACTH. This the the population we have been looking at theoretically as the source. You could say they are a subclass of glomerulosa cells and we are still talking about glomerulosa cells. Also I think the evidence as to population will have to be confirmed. However I think the evidence is good because one can separate them from the glomerulosa cell with the McDougall-Williams column. They have certain surface properties, which indicates that they are absorbed on the column, whereas zona glomerulosa cells are eluted.

R. M. Carey. Is that a density difference?

J. F. Tait. No, it is not a density difference and it is not a radius difference, but they are separated on unit gravity sedimentation studies. Dr. Hyett has proposed that this is a clumping effect and is based on the same kind of effect which results in their adsorption on the column. Hence after the column they don't reclump. It is a definite new population which is being observed.

R. M. Carey. Thank you.

Evolution of a Model of Estrogen Action

JACK GORSKI, WADE V. WELSHONS,[1] DENNIS SAKAI,[2]
JEFFREY HANSEN, JANE WALENT, JUDY KASSIS,[3] JAMES SHULL,[4]
GARY STACK,[5] AND CAROLYN CAMPEN[6]

Departments of Biochemistry and Animal Sciences, College of Agricultural and Life Sciences, University of Wisconsin, Madison

I. Preface

Our current conceptualization of how estrogens initiate target cell response is illustrated in Fig. 1. The unoccupied estrogen receptor (no estrogen ligand) is thought to be a nuclear protein bound to nuclear components by low affinity interactions. Cytoplasmic exclusion may also influence the nuclear localization. Estrogens are lipophilic and therefore can diffuse through cell membranes, cytoplasm, and nuclear envelope to interact with the nuclear receptor. As a result of this interaction rapid changes occur in the conformation of the receptor protein. These conformational changes result in new physical properties, including a higher affinity for nuclear components which prevents low salt extraction of the transformed estrogen–receptor complex. The nature of the interaction between estrogen–receptor and nucleus is still unknown but nuclear components involved could include chromatin proteins, the nuclear matrix, DNA, or various combinations of these. It is clear, however, that estrogen binding to the receptor causes increased rates of transcription of a variety of genes, depending upon the respective target cell.

---

[1] Present address: Department of Human Oncology, University of Wisconsin Clinical Cancer Center, Madison, Wisconsin.
[2] Present address: Department of Molecular and Microbiology, Case Western Reserve University School of Medicine, Cleveland, Ohio.
[3] Present address: Department of Biochemistry and Biophysics, University of California Medical Center, San Francisco, California.
[4] Present address: Department of Oncology, McArdle Labs, University of Wisconsin, Madison, Wisconsin.
[5] Present address: Institut de Chimie Biologique, Faculte de Medecine, Strasbourg, France.
[6] Present address: The Salk Institute, P.O. Box 85800, San Diego, California.

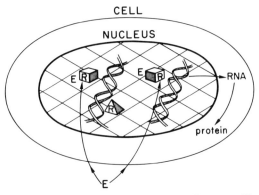

FIG. 1. "New" model of estrogen receptor. R, Receptor; E, estrogen.

II. Historical Perspective

The roots of this model can be found in the long history of investigation of steroid hormone biology and chemistry. However, most current thought on this subject of estrogen action has evolved from studies in the late 1950s carried out in the laboratories of Gerald Mueller and Elwood Jensen. Figure 2 shows a model presented by Mueller at the 1957 Laurentian Hormone Conference (Mueller et al., 1958) which summarized studies indicating that estrogen played a role in regulating gene expression. It also suggested that estrogen coupled to a protein might directly interact

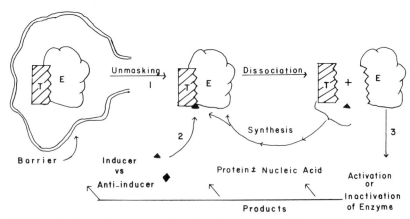

FIG. 2. A scheme depicting possible sites of hormonal regulation of induced biosynthesis. T, Template; E, enzyme; 1, 2, or 3, possible sites of hormone action. (From Mueller et al., 1958.)

with the genetic apparatus of a target cell. Jensen's group was concurrently establishing the groundwork for the isolation and characterization of estrogen and other steroid hormone receptors (Jensen and Jacobson, 1960).

A variety of studies from numerous labs led to the model independently proposed in 1968 by Jensen et al. (1968) and our laboratory (Gorski et al., 1968). In this model the receptor was assumed to be a cytoplasmic protein, which, when complexed with estrogen, translocated to the nucleus where it modified nuclear function. A major difference between the current model that is presented in the preface and the 1968 model is the subcellular localization of the unoccupied receptor. This in turn has led to concerns regarding the cause of nuclear localization and the nature of estrogen-induced changes in the receptor.

III. Evidence for the Old Model

The cytoplasmic location of steroid hormone receptors was based on the finding that unoccupied receptor was present in cytosols prepared by homogenizing estrogen target tissues such as uterus, pituitary, oviduct, etc., in hypotonic buffers (Gorski et al., 1968; Jensen et al., 1968). While there were some variations in the distribution of receptor between particulate and cytosolic fractions, the general finding was that the majority of unoccupied receptor was present in cytosols. In cases in which a larger fraction of receptor was found in nuclear preparations (Zava and McGuire, 1977), more extensive washing of the nuclei generally extracted the receptor (Edwards et al., 1980). In contrast, after in vivo injections of estrogen into whole animals or incubation of intact cells or tissues with estrogen, a marked increase in nuclear bound receptor was noted, as shown in Fig. 3 (Shull and Gorski, 1985). In vitro studies indicated that addition of estrogen to preparations of cytosolic receptor resulted after mild heating in a form of estrogen–receptor complex which bound to nuclei in a cell-free system (Brecher et al., 1967). This fit the model of ligand-induced translocation. Autoradiographic studies reported low levels of steroid binding in the cytoplasm but not in the nucleus when cells were incubated with labeled estrogen at low temperatures. These results were interpreted as indicating cytoplasmic localization of unoccupied receptor (Jensen et al., 1968). Finally, early immunocytochemistry studies also indicated that receptor was present in the cytoplasm of estrogen-free tissues, but after exposure to estrogen, immunocytochemical staining was found primarily in the nucleus. Essentially the same data were seen for all the steroid hormones, supporting a unified model of cytoplasmic–nuclear translocation in steroid hormone action.

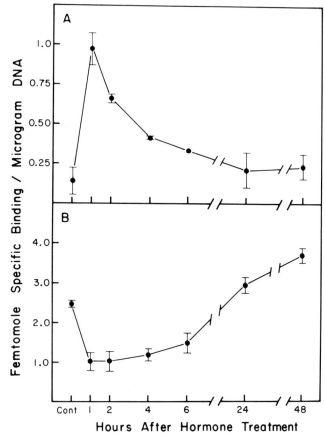

FIG. 3. Time course of the levels of the nuclear and cytosol forms of the pituitary estrogen receptor following a single injection (10 g, ip) of 17β-estradiol. Nuclear (A) and cytosol (B) receptors were assayed. Each data point represents the mean and SEM of specific binding measurements in individual anterior pituitaries ($n=3$) (Shull and Gorski, 1985).

IV. Evidence for the New Model

While this unified model became the standard dogma of textbooks and lectures, there were still some data which did not support it. In particular, autoradiographic studies by Sheridan et al. (1979) led them to conclude that unoccupied steroid receptors were distributed in equilibrium between cytoplasmic and nuclear compartments. McCormack and Glasser (1980) reported high nuclear concentrations of unoccupied estrogen receptor in uterine cell cultures. Perhaps of greater impact was the finding that thyroxine and 1,25-dihydroxyvitamin D receptors were nuclear proteins, with or without ligand (Samuels and Tsai, 1973; Walters *et al.*, 1980).

In another line of investigation, Siiteri *et al.* (1973) showed that 4 S (nontransformed-occupied) receptor was found in nuclei along with 5 S (transformed-occupied) receptor. Thus, the relationship of changes in receptor form to translocation or subcellular localization was questioned. All of these studies demonstrated the need for new approaches to the study of the site of action and physical nature of the steroid receptor.

A. RECEPTOR CONFORMATIONAL CHANGES INDUCED BY LIGAND

In the old model there was a tendency to think of the receptors *in vivo* as having two general states, cytoplasmic and nuclear, and *in vitro* as having cytosolic (4 S or 8 S) and nuclear (5 S) forms. In reality there are at least three states of solubilized receptor: unoccupied, nontransformed (occupied), and transformed. In the new model all forms are assumed to be present in the nucleus. The nontransformed state may be a transitory intermediate in the intact cell (maintained at physiological temperatures) and probably accounts for certain observations reported in the literature. Cell-free studies using the cytosolic receptor have distinguished all three receptor forms in very few cases. In most studies, only the nontransformed and the transformed receptors have been investigated. However, it has been reported that the unoccupied receptors have different properties from the nontransformed in terms of heat inactivation (Baulieu, 1973; Katzenellenbogen *et al.*, 1973) and light inactivation (Katzenellenbogen *et al.*, 1975). Most other physical characteristics of the receptor, such as sedimentation velocity, are not affected by ligand binding. The transformed receptor, however, shows marked differences in sedimentation velocity, DNA binding, and other characteristics as compared to the unoccupied and nontransformed forms.

We have recently turned to aqueous two-phase partitioning (ATPP) to reexamine changes due to ligand binding (Hansen and Gorski, 1985a) and have found this to be a useful tool for quantitation of receptor modifications. Figure 4 shows the partition values (K_{obs}) for unoccupied and nontransformed (occupied) receptors in an ATPP system using dextran-poly (ethylene glycol) phase systems. It is apparent that a marked difference in partitioning properties exists between the nontransformed and unoccupied receptors. Figure 5 shows the thermodynamic equation which describes how ionic (Z) and conformational (K_0) characteristics influence partitioning. In principle, varying salt and pH differentiates between conformational and charge contributions to the K_{obs}. Figure 6 shows that

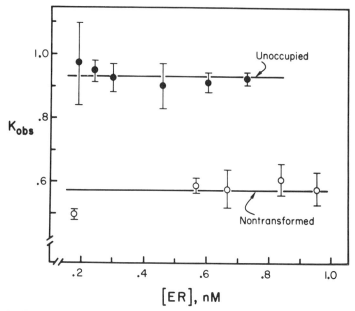

FIG. 4. Partition coefficients of unoccupied (●) and nontransformed/occupied (○) ER as a function of receptor concentration in TED/0.1 M Li$_2$SO$_4$ systems. Tubes containing 5.4% (w/w) of both dextran (M_r 510,000) and polyethylene glycol, 0.1 M Li$_2$SO$_4$, and varying amounts of rat uterine cytosol were partitioned at pH 8.2 (●) or 7.8 (○) and assayed. For both ER forms, the final ER concentration was manipulated by diluting concentrated cytosol with buffer such that all other cytosolic components were being equally altered. ER concentrations are expressed as the total dpm of ER recovered after phase partitioning per total phase H$_2$O. Lines represent the mean of all K_{obs} values for each ER form. (From Hansen and Gorski, 1985a.)

varying salts and pH markedly affects receptor partitioning, which indicates that these systems are sensitive to the electrostatic properties of the receptor. Both unoccupied and nontransformed receptors have the same pI as judged by pH at the points of intersection. In contrast, the partitioning properties of unoccupied and nontransformed receptors in the absence of molecular charge are quite different, leading to the conclusion that a conformational change occurs upon ligand binding. This conformational change makes the receptor less hydrophobic, but does not involve net changes in surface ionic properties. Calculations of the thermodynamics of such changes indicate that addition of ligand results in changes of magnitude comparable to those from heat-induced transformation. The importance of these observations will become apparent in later discussion concerning receptor interaction with nuclear components. Ligand interaction with receptor by itself seems to be able to bring about large changes

At partitioning equilibrium, the free energy necessary to transfer one mole of receptor from the top phase to the bottom phase under standard conditions is:

$$\Delta G°_{TR} = RT \ln \frac{(\gamma C)_{TOP}}{(\gamma C)_{BOTTOM}} + ZF \Delta \Psi$$

Experimentally, one measures the partition coefficient (K_{obs}):

$$K_{OBS} = \frac{[ER]_{TOP}}{[ER]_{BOTTOM}}$$

Neglecting activity coefficients (γ), K_{obs} is also a measure of $\Delta G°_{tr}$, which leads to the general equation describing partitioning behavior:

$$\ln K_{OBS} = \ln K_0 + \frac{ZF}{RT} \Delta \Psi$$

where: Z = net receptor charge (pH sensitive)

$\Delta \Psi$ = interfacial potential difference across the phase boundary (salt sensitive).

K_0 = partition coefficient when either Z or $\Delta \Psi$ is absent. This value reflects solute/solvent interactions.

FIG. 5. General thermodynamic theory that governs phase partitioning.

in the protein, which could lead to modified nuclear function. On the other hand this conformational change does not appear to involve the differences in net charge that are assumed to be the basis for the high affinity of transformed estrogen–receptor complex for the nucleus.

Figure 7 shows the time course of attaining receptor partitioning equilibrium. Equilibrium is reached within 5 seconds of vortexing and a reliable, quantitative estimate of both conformational and electrostatic receptor properties is obtained in a very short time, as compared to the techniques of sucrose gradients and column chromatography. The speed of ATPP and its sensitivity to modifications in protein structure should make it a valuable tool in receptor studies. For example, we have used ATPP to discriminate between estrogen and antiestrogen occupied forms of the estrogen receptor (Hansen and Gorski, 1985b). A limit of the one-step partitioning is that it does not separate the different forms and only averages of mixtures are obtained. ATPP was developed by Albertsson (1971) and has been used in steroid receptor studies previously by Andreasen (Andreasen, 1978, 1981, 1982; Andreasen and Mainwaring, 1980; Andreasen and Gehring, 1981) and Baulieu (Alberga et al., 1976).

There is an increasing awareness and interest in different states of the steroid receptors and in particular phosphorylation–dephosphorylation-

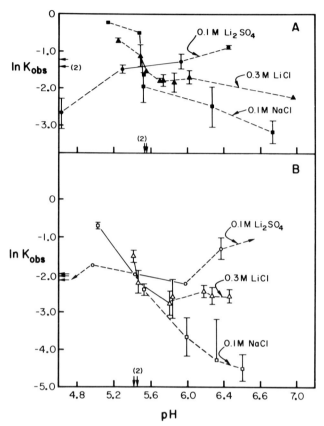

FIG. 6. "Cross-point" analysis of unoccupied (A) and nontransformed (B) rat uterine estrogen receptor. Tubes containing either unoccupied (A) or nontransformed (B) ER, 5.4% of both dextran (M_r 510,000) and poly(ethylene glycol), and either 0.1 M NaCl (▲, △), 0.3 M LiCl (■, □), or 0.1 M Li$_2$SO$_4$ (●, ○) were partitioned at the indicated pH. All systems were buffered with TED. Each point represents the mean ± the standard error of one K_{obs} derived from duplicate partitionings. If no error bars are shown, the standard error is smaller than the data point. Arrows indicate the points of intersection of each salt pair on each axis. The K_0 of the unoccupied ER under these conditions is 0.25 ± 0.03 and that of the nontransformed ER is 0.12 ± 0.02. Both K_0 values represent the mean ± standard error of the three points of intersection on the K_{obs} axis. The pI of the unoccupied ER is 5.55 ± 0.15 and that of the nontransformed ER is 5.45 ± 0.15. These values represent the mean of the three points of intersection on the pH axis ± the combined standard errors of the points of intersection and the pH determinations (Hansen and Gorski, 1985a).

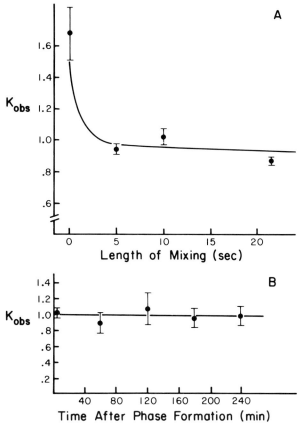

FIG. 7. Determination (A) and stability (B) of the equilibrium value of the unoccupied estrogen receptor partition coefficient is TED 0.1 M Li_2SO_4 systems. Tubes containing unlabeled rat uterine cytosol, 5.4% (w/w) of both dextran (M_r 461,000) and poly(ethylene glycol), 0.1 M Li_2SO_4, and TED buffer, pH 8.0, were vortexed for the indicated times (A) or for 10 seconds (B), centrifuged for 5 minutes at 6000 g and placed at 0°C. Either immediately (A) or at the times indicated (B), aliquots of each phase were removed, assayed, and used to calculate the corresponding partition coefficients. In (A), the amount of ER recovered at 0, 5, 10, and 20 seconds of vortexing was 69, 66, 68, and 75%, respectively. In (B), the amount of ER recovered at 1, 2, 3, and 4 hours after phase formation was 74, 74, 67, and 73%, respectively. Each data point represents the mean ± standard error of one K_{obs} derived from triplicate partitioning (A) or the mean ± standard error of two to six K_{obs} values determined under similar conditions (B) (Hansen and Gorski, 1985a).

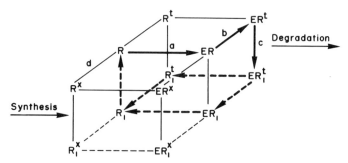

FIG. 8. Model of potential receptor forms. E, Estrogen; R, unoccupied receptor; ER, nontransformed/occupied receptor; R^t, transformed receptor; R^x, modified receptor.

induced changes in the affinity of receptor–steroid interactions. Rather than just the three "binding" forms we have discussed, there may be a complex array of forms as illustrated in Fig. 8. While a preferred equilibrium may exist in the intact cell in which only two or three forms predominate, such an equilibrium might change due to cell type, metabolic state, etc. Possibly the biological activities of different forms of the receptor are also different. If a large receptor protein has several distinct functional domains, as has been postulated recently for other proteins, then a change in just one domain might exert considerable influence on the functional activity of a receptor. Different forms of the receptor might reflect modifications of a single protein domain. Whether such a complexity exists remains to be seen, but it would be compatible with our ATPP results as well as data of others. This also would provide the theoretical basis for a single gene to code for a multifunctional receptor protein.

B. NUCLEAR LOCALIZATION

A paper on subcellular distribution of DNA polymerase α (Herrick et al., 1976) had a marked influence on our thinking about receptor localization. They reported that DNA polymerase α, which is generally found in cytosols, was not found in enucleated cells (cytoplasts) prepared by a cytochalasin enucleation procedure. Instead, the polymerase was present in the nucleoplasts which contain nuclei with a rim of cytoplasm. Recently Welshons et al. (1984) have used enucleation to reexamine the distribution of estrogen receptor in cellular compartments.

Figure 9 shows the distribution of unoccupied receptors in the cytoplast and nucleoplast fractions prepared from GH_3 cells. The pituitary tumor-derived GH_3 cells contain a typical receptor concentration of approximately 25,000 receptor sites per cell and increase synthesis of prolactin in

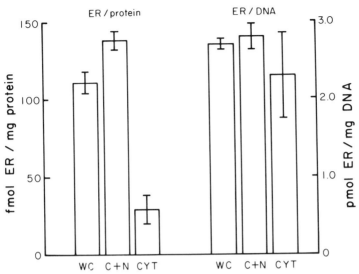

FIG. 9. Estrogen receptor (ER) concentration in whole cells (WC), in cells + nucleoplasts (C + N) and in cytoplasts (Cyt) (Welshons et al., 1984).

response to estrogen. When GH_3 cells are treated with cytochalasin and then centrifuged on a Percoll gradient, the more dense nuclei are centrifuged away from the less dense cytoplasm. Cytoplasm "buds off" to yield an enucleated cell (or cytoplast) which is still surrounded by an intact plasma cell membrane. Nuclei are retained in a nucleoplast which also contains a rim of cytoplasm and an intact plasma cell membrane. It is clear that unoccupied receptors are present only in very low concentrations in cytoplasts. Since the ratio of cytoplasm to nucleus decreases with increasing density, the nucleoplasts can be further fractionated on Percoll gradients.

As seen in Fig. 10, the concentration of unoccupied estrogen receptor per unit of protein increases in the more dense nucleoplasts, which contain decreasing amounts of cytoplasm. Per unit of DNA, the receptor concentration is relatively constant over all fractions, suggesting but not proving nuclear localization of receptor. The cytoplasts obtained by this method are intact in the sense that they exclude dyes, as shown in Table I. More importantly, they still synthesize prolactin, one of the principal physiological functions of the pituitary-derived GH_3 cells. Table II shows total protein synthesis and prolactin synthesis (measured by immunoprecipitation of labeled protein) in cytoplasts and intact cells. While total protein synthesis is about 50% less than that of intact cells, prolactin synthesis in cytoplasts is comparable to the intact cells. Variation in

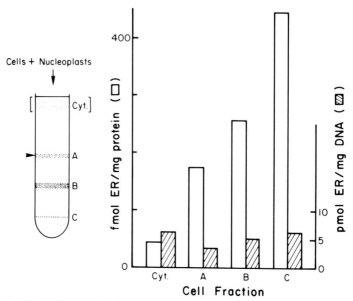

FIG. 10. The cell + nucleoplast fraction was further fractionated on a density step gradient. The (intact) cells at higher density steps have had more cytoplasm removed. Arrowhead indicates density of cells before enucleation, and the position cytoplasts would have occupied is indicated in brackets (Welshons et al., 1984).

TABLE I

Dye Exclusion of Cells and Fractions[a]

	Excluding dye (%)	
	Initial	Final
Whole cells	97 ± 1	95 ± 1
Cells + nucleoplasts	98 ± 1	93 ± 1
Cytoplasts	98 ± 2	73 ± 4
Means ± standard error, $n = 6$ to 9		

[a] Trypan blue exclusion just after enucleation (initial) and after measuring the receptor content by whole cell uptake at 37°C (final) (Welshons et al., unpublished, in Shull et al., 1985).

TABLE II

Prolactin Synthesis by GH_3 Cells and Fractions[a]

	Leucine incorporation: $10^{-6} \times$ dpm per 10^6 cells or equivalent	Prolactin synthesis: % of total protein synthesis
Whole cells	6.2	1.2 ± 0.02
Cells + nucleoplasts	5.7	1.3 ± 0.06
Cytoplasts	3.0[b]	3.4 ± 0.2

[a] Cells, cells plus nucleoplasts, and cytoplasts were incubated with [^{3}H]leucine to measure general protein synthesis by TCA-precipitable leucine incorporation, and to measure prolactin synthesis using precipitation with antiprolactin antibody.

[b] Assuming 5 cytoplasts per whole cell equivalent (Welshons et al., unpublished, in Shull et al., 1985).

mRNA turnover could account for much of this difference. Since protein synthesis is a complex process requiring energy, amino acid transport, etc., we feel these data are a good index of cell or cytoplast integrity and indicate that the receptor-free cytoplasts were viable. Enucleation of GH_3 cells can also be carried out without cytochalasin treatment by increasing centrifugation speeds. The results of enucleation with or without cytochalasin are similar; little or no estrogen receptor is present in the cytoplasts (Welshons et al., 1985).

Figure 11 shows the distribution of unoccupied glucocorticoid receptor in GH_3 cells enucleated without cytochalasin (Welshons et al., 1985). As with the estrogen receptors, unoccupied glucocorticoid receptors are not present in cytoplasts but can be recovered in the nucleoplast fraction. Figure 11 also shows the distribution of a typical cytoplasmic enzyme marker, lactate dehydrogenase. As expected, a high concentration of the enzyme was observed in the cytoplasts. The decrease in concentration of the nucleoplast fraction, while not large, is appropriate for the amount of cytoplasm lost in enucleation.

When similar studies were carried out on unoccupied progesterone receptor, we observed a 50% loss of these receptors by the end of the enucleation process (Welshons et al., 1985). Unoccupied progesterone receptor that was recovered was present in the nucleoplasts and not in the cytoplasts, suggesting a similar distribution within the cell of all these steroid hormone receptors. Recently Gravanis and Gurpide (1985) have used a different enucleation procedure with cells derived from human endometrium and observed nuclear localization of unoccupied estrogen receptor.

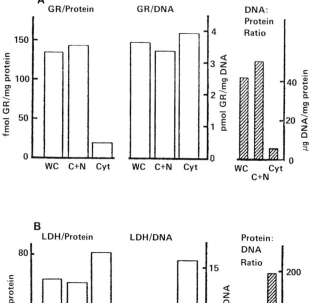
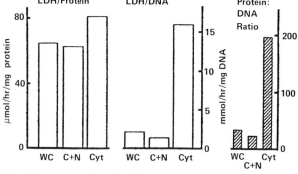

FIG. 11. GH₃ cells were enucleated without cytochalasin. Glucocorticoid receptor (GR), protein, DNA, and cell number were measured in the untreated, density-selected whole cells (WC), in the cell + nucleoplast fraction (C + N), and in the cytoplast fraction (Cyt) (A). The cytosolic marker lactate dehydrogenase (LDH) was measured instead of glucocorticoid receptor in a separate experiment (B). The DNA/protein ratio is shown by the glucocorticoid receptor graphs (A), while the protein/DNA ratio is shown by the lactate dehydrogenase graphs (B), to facilitate comparisons. (From Welshons et al., 1985.)

Although the enucleation studies indicate that the unoccupied estrogen receptor is not present in the cytosol, they do not absolutely prove a nuclear localization. It might be possible that a cytoplasmic fraction closely associated with the nuclei would behave in the manner observed in our studies. However, immunocytochemical studies also conclude that unoccupied as well as occupied estrogen receptors are located in cell nuclei. In contrast to their earlier reports, King and Greene (1984) used modified procedures which showed that a monoclonal antibody to estrogen receptor interacted only with nuclei of target cells. Nuclear staining was not localized to any one site, such as the nuclear envelope. A subsequent study by McClellan et al. (1984) using the same antibodies also showed nuclear localization of both unoccupied and occupied estrogen receptors. Using immunocytochemical methods, they noted that the unoccupied receptor appeared to be more easily extracted from tissue sections than the occupied (transformed) receptor. Studies by Baulieu (Gasc et al., 1984a,b) and Milgrom (Perrot-Applanat et al., 1985) on immunocytochemical localization of the progesterone receptor also concluded that both unoccupied and ligand-occupied forms are present in the nucleus.

On the other hand, Raam et al. (1982, 1983) report immunocytochemical studies, using a polyclonal antibody to estrogen receptors, which indicate cytoplasmic localization of unoccupied and nuclear localization of occupied estrogen receptors. Similarly Robertson et al. (1985), using a monoclonal antibody to the glucocorticoid receptor, observed two-compartment localization that fits the translocation model. Further studies will be necessary to confirm the validity of our model and its general application to all cells and all steroid hormones.

Other observations in the literature are of interest in relationship to the current model. Jordan et al. (1985) have observed that certain low affinity estrogens and antiestrogens have biological effects, although no nuclear receptor forms have ever been observed. One can explain such data with the single-compartment nuclear model by the ready extraction of nuclear receptor upon ligand dissociation.

Another area of some controversy has been the proposed membrane location of the estrogen receptor, reviewed extensively by Szego (Szego and Pietras, 1984). Both our enucleation studies (Welshons et al., 1984, 1985) and the immunocytochemical studies (King and Greene, 1984) indicate little or no membrane localization. Cytoplasts contain an outer cell membrane that excludes dyes and transports amino acids and energy-yielding substrates into cells, yet the cytoplasts do not contain steroid receptors. None of the immunocytochemical studies indicates localization of receptor to the periphery of the cell. Work by a number of investigators (Kilvik et al., 1985; Müller and Wotiz, 1979) indicates that steroid uptake is a nonsaturable process that is not limiting to specific receptor binding at physiological concentrations of steroid hormones. Thus it does not appear that cellular transport of steroid is a factor, regardless of subcellular localization of receptor. The steroid affinity for lipid membranes may permit a two- rather than a three-dimensional diffusion that might speed intracellular movement. This limited type of diffusion has been proposed to speed the movement of DNA-binding proteins along the genome (Von Hippel et al., 1984).

V. What Keeps the Receptors in the Nucleus?

Feldherr and Pomerantz (1978) have shown that most nuclear proteins do not leave the nucleus even when the nuclear envelope is disrupted in a manner which should permit equilibrium between nuclear and cytoplasmic compartments. These investigators concluded that nuclear proteins are bound to nuclear components, shifting the equilibrium in favor of the nucleus. Thus one mechanism which would account for the nuclear localization of the unoccupied estrogen receptor is its binding to nuclear components. The validity of such a mechanism is supported by data that contrast *in vivo* versus *in vitro* equilibrium binding of estrogen to its receptor.

Notides *et al.* (1981) have made a series of elegant observations concerning the interaction of estrogen receptor proteins and have presented evidence that, *in vitro*, the soluble receptor shows positive cooperativity in binding estrogen. These important studies show that the cooperative binding of estrogen occurs at physiological concentrations of receptor (10^{-9} to 10^{-8} M in target cells). Notides shows that, at the lower concentrations of receptor generally used in previous equilibrium binding studies, binding of estrogen is not cooperative. These observations have been confirmed by Sakai and Gorski (1984) and by Thompson *et al.* (1985).

In contrast to these *in vitro* studies, older (Williams and Gorski, 1972) as well as more recent investigations (Walent and Gorski, 1985; Traish *et al.*, 1979) have shown that estrogen binding in intact cells or tissues is not cooperative and behaves in a manner which suggests that there is no interaction between estrogen receptor subunits (Fig. 12). The validity of the intact cell observations can be questioned based on the inability to directly measure the intracellular concentration of free estrogen (Notides *et al.*, 1985). In order to validate this system, it has been argued that the intracellular pool of free steroid must be in equilibrium with the extracel-

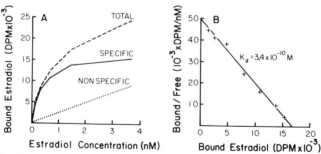

FIG. 12. Saturation analysis of the whole cell uptake of [^{3}H]estradiol at 37°C by GH$_3$ cells (Welshons *et al.*, unpublished, in Shull *et al.*, 1985).

lular pool and the nonspecific binding components inside and outside the cell. In reality the problem in estimating free steroid is the same in both intact cells and in cell-free systems, where the free steroid is in equilibrium with specific and nonspecific binding entities. This is illustrated by the studies recently reported by Thompson et al. (1985) using centrifugal dialysis to estimate the free steroid in cytosols. The actual level of free steroid was about 1% of the total nonspecific estrogen binding. Similarly, in vivo, complex equilibria occur between free steroid, serum proteins, nonspecific tissue binding, and the receptors. Thus in all cases, with the possible exception of studies using 100% purified receptor, one deals with complex binding equilibria. Pardridge (1981) has discussed this problem extensively. We believe, if care is taken to accurately measure the free extracellular concentration of steroid at equilibrium, that this estimate is directly proportional to the available intracellular pool of free steroid. Experimentally this is confirmed by a direct proportionality (linearity) in the nonspecific binding of estrogen to cells with increasing concentrations of estrogens in the medium incubated with cultured cells (Fig. 12). Thus we believe that the straight line Scatchard plots and Hill coefficients, $n=1$, seen in binding studies with intact cells are valid and reflect the noncooperative binding of steroid to receptor in the intact cell (Scatchard, 1949).

A possible explanation for the difference between cell-free and intact cell systems that we favor is based on studies by Sakai and Gorski (1984) which report that receptor immobilized by prebinding to hydroxylapatite shows estrogen binding characteristics that are similar to those seen in intact cells. Figure 13 shows the equilibrium binding of estradiol to cytosolic estrogen receptor either in solution or when receptors are prebound to hydroxylapatite. Both the Scatchard and Hill plots indicate that native receptor prebound to hydroxylapatite at a concentration of 0.75 nM shows cooperative binding characteristics, as first shown by Notides et al. (1981). However, as shown in Fig. 14, if the receptor is treated with 0.4 M KCl in order to disrupt subunit interaction prior to binding to the hydroxylapatite, one observes straight line Scatchard and Hill plots with slopes that indicate no cooperativity. Furthermore, in Fig. 15 it can be seen that, when dissociation kinetics were compared under these differing conditions, the dissociation of estrogen from the receptor shows the same kinetic characteristics, whether in the monomeric form bound to hydroxylapatite (Fig. 15E) or the polymeric form in solution (Fig. 15A). Müller et al. (1985) have reported similar observations. The two stage dissociation of estrogen from receptor has been reported to be due to the transformation of the occupied receptor. The slower dissociation rate of the transformed receptor is similar to that of the nuclear receptor that results from

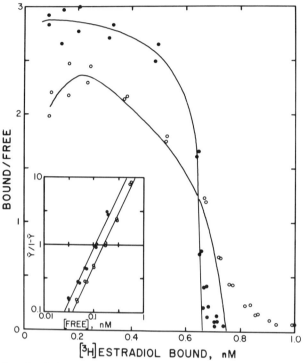

FIG. 13. Scatchard plot of equilibrium binding at 0°C of [³H]estradiol to estrogen receptors in solution or adsorbed to hydroxylapatite. Lines represent receptor binding after correction of the data points for the contribution of type II sites. Inset: Hill plots of corrected data. Lines represent linear least-squares fit to the data. (●) Binding to receptor in solution; receptor concentration = 0.66 nM $S_{0.5}$ = 0.11 nM, n_H = 1.60. (○) Binding to receptor adsorbed to HAP; receptor concentration = 0.75 nM, $S_{0.5}$ = 0.18 nM, n_H = 1.59. (From Sakai and Gorski, 1984.)

in vivo binding of estrogen. Thus the monomeric cytosolic receptor, when immobilized on hydroxylapatite, has similar equilibrium binding characteristics as the receptor in the intact cell. One interpretation of these data is that the unoccupied estrogen receptor in the intact cell is bound to nuclear components in a monomeric form. This would explain both localization of the receptor in the nuclear compartment and the equilibrium binding data in intact cells versus cytosols. The equilibrium between receptor binding to nuclear components and binding to another receptor subunit would be the product of the receptor's affinity for nuclear components and the concentration of such components, versus the affinity for another receptor subunit and the concentration of the receptor subunit. The large mass of nuclear components such as matrix, DNA, histones,

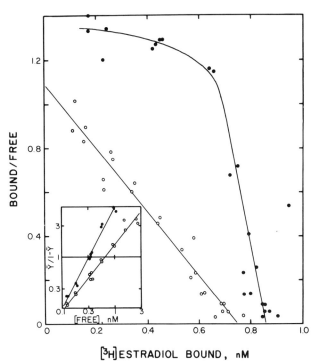

FIG. 14. Scatchard plot of equilibrium binding at 0°C of [³H]estradiol to native and monomeric estrogen receptors. Inset: Hill plots. (●) Binding to receptor in solution; receptor concentration = 0.85 nM, $S_{0.5}$ = 0.32 nM, n_H = 1.54. (○) Binding to monomeric receptor generated by treatment with 0.4 M KCl and adsorption onto HAP; receptor concentration = 0.73 nM, $S_{0.5}$ = 0.64 nM, n_H = 1.01. (From Sakai and Gorski, 1984.)

etc., would favor binding to nonreceptor components of the nucleus. This conclusion is based on extrapolation from limited data but represents a hypothesis that is worth further testing.

What is the nature of the nuclear components to which the unoccupied receptor is bound? Because of the well-documented effects of estrogen on gene transcription, binding of estrogen receptor to the genetic apparatus, and DNA in particular, is a popular area of investigation. The unoccupied cytosolic receptor binds to nonspecific DNA with relatively high affinity, apparently by ionic interactions (Yamamoto and Alberts, 1972). In recently reported experiments, Skafar and Notides (1985) show that 7.6 ionic interactions are involved in receptor-DNA binding, in the absence of estrogen, and 12.8 with estrogen. The implication is that the unoccupied estrogen receptor might interact directly with DNA in the nucleus. However, unlike the occupied estrogen receptor, the unoccupied estrogen

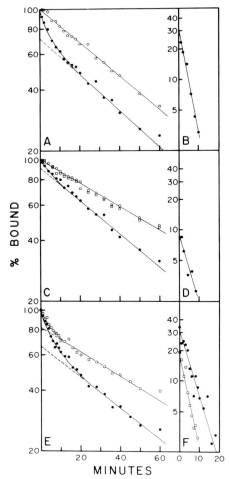

FIG. 15. Dissociation of [^{3}H]estradiol–receptor complexes in solution or adsorbed to hydroxylapatite. (A) Receptor in solution. [^{3}H]Estradiol–receptor complexes were equilibrated at 0 (●) or 28°C (○) for 0.5 hour and dissociation of the complexes at 28°C was monitored by isotopic dilution with a large excess of DES. Receptor concentration = 1.3 nM. Calculated dissociation rate constants: (●) $k_f = 3.8 \times 10^{-3}$ sec^{-1}, $k_s = 3.3 \times 10^{-4}$ sec^{-1}; (○) $k_s = 3.2 \times 10^{-4}$ sec^{-1}. (C) Receptor adsorbed to HAP. [^{3}H]Estradiol–receptor complexes were equilibrated at 0 (●,□) or 28°C (○) for 0.5 hour and adsorbed onto HAP. (□) Complexes heated at 28°C for 0.5 hour after adsorption onto HAP. After the samples were washed with TD buffer, dissociation rates at 28°C were measured. Receptor concentration = 0.7 nM. Calculated dissociation rate constants: (●) $k_f = 2.7 \times 10^{-3}$ sec^{-1}, $k_s = 3.2 \times 10^{-4}$ sec^{-1}; (○) $k_s = 2.2 \times 10^{-4}$ sec^{-1}; (□) $k_s = 2.3 \times 10^{-4}$ sec^{-1}. (E) Monomeric receptor. Receptor monomers were generated by treatment with 0.4 M KCl and adsorption onto HAP. After samples were washed with TD buffer plus 0.4 M KCl, they were reequilibrated with [^{3}H]estradiol in TD buffer at 0 (●) or 28°C (□) for 1.5 hours. Dissociation was measured at

receptor is readily extracted from its native state in the cell into cytosols with dilute salt buffer, suggesting that it is not participating in electrostatic interactions with DNA. There are few other data that address the association between the unoccupied receptor and the nucleus.

Circumstantial evidence is consistent with the hypothesis that unoccupied steroid receptors interact with chromatin via low affinity, hydrophobic interactions. This proposal is based on two lines of evidence. In our studies with ATPP (Hansen and Gorski, 1985a), we found that the estrogen receptor is a hydrophilic protein but that the unoccupied receptor possesses a hydrophobic domain that is less apparent on the nontransformed receptor. Second, the work of Shanbhag (Axelsson and Shanbhag, 1976; V. P. Shanbhag, personal communication), using hydrophobic affinity partitioning, has established that all histone subtypes possess considerable hydrophobicity. Thus, it seems feasible that the unoccupied receptor is bound weakly to chromatin, leaving virtually no free receptor because of the almost infinite capacity of the chromatin surface to adsorb unoccupied receptor. In this hypothesis, ligand binding to the unoccupied receptor initially causes a conformational change that disrupts the random, hydrophobic interactions with chromatin, while subsequently inducing an additional transition (transformation) that shifts the nuclear binding to that of a salt-dependent mode. Perhaps the most appealing aspect of this hypothesis is that it predicts the classically observed extraction of unoccupied steroid receptors upon homogenization in low salt aqueous buffers and dilution of the volume-excluding cytoplasmic environment.

As mentioned above, one other major influence on nuclear localization is probably physicochemical exclusion of receptor from the cytoplasmic compartment. Gannon *et al.* (1976) and Sheridan *et al.* (1979, 1981) have suggested that cytoplasmic exclusion might play a role in subcellular partitioning of receptors. Sheridan used this concept to explain the equilibrium distribution he observed in studies discussed earlier. The fact that the vitamin D receptor (Walters *et al.*, 1980) and the dioxin receptor (Whitlock and Galeazzi, 1984) are more readily extracted into cytosols as the volume of homogenization buffer is increased fits this concept. It is

---

28°C. Receptor concentration = 0.5 nM. Calculated dissociation rate constants: (●) $k_{f,monomer}$ = 2.5 × 10^{-3} sec^{-1}, $k_{s,monomer}$ = 3.0 × 10^{-4} sec^{-1}; (□) $k_{f,monomer}$ = 3.5 × 10^{-3} sec^{-1}, $k_{s,monomer}$ = 2.3 × 10^{-4} sec^{-1}. B, D, and F show replots of fast-dissociating components of biphasic curves from A, C, and E, respectively, and were calculated by subtracting values from the extrapolated curves representing the slow dissociating components (dashed lines) from the experimental values. The intercepts at time = 0 represent the percentage contributions of the fast-dissociating components. (From Sakai and Gorski, 1984.)

likely that cytoplasmic exclusion plays at least some part in the nuclear localization of estrogen receptors, possibly by facilitating diffusion of the unoccupied receptor into the nucleus after translation in the cytoplasm.

VI. The Transformed Receptor in the Nucleus

Compared to several aspects of our model discussed above, the localization of the estrogen–receptor complex in the nucleus seems the least controversial. In all manner of studies, the steroid–receptor complex resides in the nucleus. This is closely correlated with the increased affinity of the transformed receptor for DNA and also for other polyanions. As noted by Skafar and Notides (1985), transforming cytosolic receptor by heat or salt in the presence of steroid increases the affinity of receptors for nonspecific DNA by about 10-fold. In the last several years, the availability of specific cloned sequences of DNA has led to the study of receptor binding to such DNA sequences. The general finding is that purified or crude cytosolic receptor binds to specific DNA regions associated with target genes. For example, the glucocorticoid receptor binds to mouse mammary tumor virus DNA (Payvar *et al.*, 1981; Pfahl, 1982), to the growth hormone gene (Moore *et al.*, 1985), and to metallothionein DNA (Karin *et al.*, 1984). Estrogen receptor binds to flanking regions of the prolactin gene (Maurer, 1985) and the egg white protein genes (Renkawitz *et al.*, 1984). Concensus sequences have been derived and the question of nuclear binding and response might appear to be resolved. Certain questions have arisen, however, that need to be dealt with. There is little evidence that binding of steroid to the receptor affects the binding of receptors to the "so called" concensus sequences. Furthermore, the affinity of the receptors for specific versus nonspecific sequences seems to be only 10-fold.

The relative affinity of a protein for a specific DNA sequence versus nonspecific sequences is crucial. This has been the subject of a number of discussions, often citing Lin and Riggs (1975) for a theoretical basis. An interesting exchange of views on this subject has been published (Travers, 1983, 1984; Ptashne, 1984). In general these discussions conclude that a minimal ratio of specific to nonspecific binding of at least 1000-fold (or 3 orders of magnitude) is required to permit discrimination of a specific DNA sequence from a nonspecific one. In the case of a globin gene binding protein recently reported by Felsenfeld's laboratory (Emerson *et al.*, 1985), the ratio of specific to nonspecific binding constants is 3.5×10^4. This appears to be significantly greater than what has been observed with steroid hormone receptor binding to specific sequences of DNA but close to theoretical predictions.

It should be noted that in the nucleus, DNA is associated with a complex of proteins that we call chromatin. The structure and composition of chromatin at regulatory regions of a gene are known to change with different states of cellular activity, and it is unlikely that DNA *in vivo* is ever completely free of chromatin proteins. Thus renaturing of whole chromatin with the associated steroid receptors would be required to give a complete picture of how and where the receptor interacts with the genome. It must be noted that the above approach assumes a cis relationship between receptor binding sites and regulated genes, with both proximal to each other on the same chromosome. Trans regulation, where receptor binding site and gene are located on different chromosomes, is also possible and will require more work to decipher. It also must be mentioned that transformed steroid–receptor complexes bind with high affinity to many other nuclear components, such as RNA, histones, etc. (Gorski and Gannon, 1976), and that these could be the actual site of action (e.g., mRNA stability, etc.).

Related to this problem is the vast array of genes seemingly regulated at the transcriptional level by estrogens. For example, in the pituitary, estrogen increases lactotroph transcription of prolactin gene, but decreases FSH and LH production in gonadotrophs. Genes for the egg yolk proteins in avian liver are stimulated by estrogen, while genes for the egg white proteins are stimulated in avian oviduct. A common denominator, such as a repetitive DNA sequence or a common chromatin protein, could explain these observations.

Another aspect of receptor binding to nuclear components concerns the nuclear matrix or scaffold. This is a controversial area of study because many investigators are skeptical of the concept of a matrix. The matrix is thought to represent a fibrillar network in the nucleus comparable to the microtubular and microfilamental organization present in the cytoplasm. It is difficult to precisely define the elements of such a matrix, and the presence of potential artifacts compromises these studies. A number of observations about chromosome localization in the nucleus, thalassemic mutants in globin genes, etc. (Vanin *et al.*, 1983), have been explained by the matrix concept. Barrack and Coffey (1983) have shown that occupied estrogen receptors are associated with the nuclear matrix. Clark and Markaverich (1982) have confirmed these findings. Evidence has also been presented that promoter regions and the initiation of transcription of the ovalbumin gene are associated with nuclear matrix in estrogen-stimulated oviduct cells (Robinson *et al.*, 1982; Ciejek *et al.*, 1983). Finally, recent reports have linked eukaryotic topoisomerases to the nuclear matrix (Berrios *et al.*, 1985). The nuclear matrix could provide a strategic location in the chromatin for receptor binding, and could explain the divergent effects of steroid hormones on cellular function, as well as binding to specific DNA sequences.

VII. Conclusions

Study of the mechanisms of steroid hormone action has come a long way since the 1958 review in these proceedings by Mueller *et al.* (1958). It has not yet come to a final conclusion, but rather is in an exciting state of flux with more questions than answers. We have speculated a great deal in this discussion with the hope of encouraging divergent points of view and new approaches to investigate these questions. The next decade of research should lead to the evolution of a more detailed, and hopefully more conclusive model of steroid hormone action.

ACKNOWLEDGMENTS

Supported by the College of Agricultural and Life Sciences, University of Wisconsin, Madison, NIH Grants HD08192, 5T3207259, CA18110, and the National Foundation for Cancer Research. Supported in part by NRS Fellowship 5F32 HD-06008.

REFERENCES

Alberga, A., Ferrez, M., and Baulieu, E.-E. (1976). *FEBS Lett.* **61**, 223.
Albertsson, P. A. (1971). "Partition of Cell Particles and Macromolecules." Wiley (Interscience), New York.
Andreasen, P. A. (1978). *Biochim. Biophys. Acta* **540**, 484.
Andreasen, P. A. (1981). *Biochim. Biophys. Acta* **676**, 205.
Andreasen, P. A. (1982). *Mol. Cell. Endocrinol.* **28**, 563.
Andreasen, P. A., and Gehring, U. (1981). *Eur. J. Biochem.* **120**, 443.
Andreassen, P. A., and Mainwaring, W. I. P. (1980). *Biochim. Biophys. Acta* **631**, 334.
Axelsson, C.-G., and Shanbhag, V. P. (1976). *Eur. J. Biochem.* **71**, 419.
Barrack, E. R., and Coffey, D. S. (1983). *In* "Biochemical Actions of Hormones" (G. Litwack, ed.), pp. 23–90. Academic Press, New York.
Baulieu, E.-E. (1973). *Adv. Exp. Med. Biol.* **36**, 80.
Berrios, M., Osheroff, N., and Fisher, P. A. (1985). *Proc. Natl. Acad. Sci. U.S.A.* **82**, 4142.
Brecher, P., Vigersky, R., Wotiz, H. S., and Wotiz, H. H. (1967). *Steriods* **10**, 635.
Ciejek, E. M., Tsai, M.-J., and O'Malley, B. W. (1983). *Nature (London)* **306**, 607.
Clark, J. H., and Markaverich, B. M. (1982). *In* "The Nuclear Envelope and the Nuclear Matrix" (G. Maul, ed.), pp. 259–269. Liss, New York.
Edwards, D. P., Martin, P. M., Horwitz, K. B., Chamness, G. C., and McGuire, W. L. (1980). *Exp. Cell Res.* **127**, 197.
Emerson, B. M., Lewis, C. D., and Felsenfeld, G. (1985). *Cell* **41**, 21.
Feldherr, C. M., and Pomerantz, J. (1978). *J. Cell Biol.* **78**, 168.
Gannon, F., Katzenellenbogen, B., Stancel, G., and Gorski, J. (1976). *In* "The Molecular Biology of Hormone Action" (J. Papaconstantinou, ed.), pp. 137–149. Academic Press, New York.

Gasc, J.-M., Renoir, J.-M., Radanyi, C., Joab, I., and Baulieu, E.-E. (1984a). *C. R. Acad. Sci. Paris* **295**, 707.
Gasc, J.-M., Renoir, J.-M., Radanyi, C., Joab, I., Tuohimaa, P., and Baulieu, E.-E. (1984b). *J. Cell Biol.* **99**, 1193.
Gorski, J., and Gannon, F. (1976). *Annu. Rev. Physiol.* **38**, 425.
Gorski, J., Toft, D., Shyamala, G., Smith, D., and Notides, A. (1968). *Recent Prog. Hormo. Res.* **24**, 45.
Gravanis, A., and Gurpide, E. (1985). *Steroid Biochem.* (in press).
Hansen, J. C., and Gorski, J. (1985a). *Biochemistry* **24**, 6078.
Hansen, J. C., and Gorski, J. (1985b). *Endocr. Soc. Prog. Abstr., 67th Annu. Meet* p. 3.
Herrick, G., Spear, B., and Veomett, G. (1976). *Proc. Natl. Acad. Sci. U.S.A.* **73**, 1136.
Jensen, E. V., and Jacobson, H. I. (1960). *In* "Biological Activities of Steroids in Relation to Cancer" (G. Pincus and E. P. Vollmer, eds.), pp. 161–178. Academic Press, New York.
Jensen, E. V., Suzuki, T., Kawashima, T., Stumpf, W. E., Jungblut, P. W., and DeSombre, E. R. (1968). *Proc. Natl. Acad. Sci. U.S.A.* **59**, 632.
Jordan, V. C., Tate, A. C., Lyman, S. D., Gosden, B., Wolf, M. F., Bain, R. R., and Welshons, W. V. (1985). *Endocrinology* **116**, 1845.
Karin, M., Haslinger, A., Holtgreve, H., Richards, R. I., Krauter, P., Westphal, H. M., and Beato, M. (1984). *Nature (London)* **308**, 513.
Katzenellenbogen, J. A., Johnson, H. J., Jr., and Carlson, K. E. (1973). *Biochemistry* **12**, 4092.
Katzenellenbogen, J. A., Ruh, T. S., Carlson, K. E., Iwamoto, H. S., and Gorski, J. (1975). *Biochemistry* **14**, 2310.
Kilvik, K., Furu, K., Haug, E., and Gautvik, K. M. (1985). *Endocrinology* **117**, 967.
King, W. J., and Greene, G. L. (1984). *Nature (London)* **307**, 745.
Lin, S.-Y., and Riggs, A. D. (1975). *Cell* **4**, 107.
McClellan, M., West, N., Tacha, D., Greene, G., and Brenner, R. (1984). *Endocrinology* **114**, 2002.
McCormack, S. A., and Glasser, S. R. (1980). *Endocrinology* **106**, 1634.
Maurer, R. A. (1985). *DNA* **4**, 1.
Moore, D. D., Marks, A. R., Buckley, D. I., Kapler, G., Payvar, F., and Goodman, H. M. (1985). *Proc. Natl. Acad. Sci. U.S.A.* **82**, 699.
Mueller, G. C., Herranen, A. M., and Jervell, K. F. (1958). *Recent Prog. Hormo. Res.* **14**, 95.
Müller, R. E., and Wotiz, H. H. (1979). *Endocrinology* **105**, 1107.
Müller, R. E., Traish, A. M., Hirota, T., Bercel, E., and Wotiz, H. H. (1985). *Endocrinology* **116**, 337.
Notides, A. C., Lerner, N., and Hamilton, D. E. (1981). *Proc. Natl. Acad. Sci. U.S.A.* **78**, 4926.
Notides, A. C., Sasson, S., and Callison, S. (1985). *In* "Molecular Mechanism of Steroid Hormone Action" (V. K. Moudgil, ed.), pp. 173–197. De Gruyter, Berlin.
Pardridge, W. M. (1981). *Endocrin. Rev.* **2**, 103.
Payvar, F., Wrange, O., Carlsstedt-Duke, J., Okret, S., Gustafsson, J.-A., and Yamamoto, K. R. (1981). *Proc. Natl. Acad. Sci. U.S.A.* **78**, 6628.
Perrot-Applanat, M., Logeat, F., Groyer-Picard, M. T., and Milgrom, E. (1985). *Endocrinology* **116**, 1473.
Pfahl, M. (1982). *Cell* **31**, 475.
Ptashne, M. (1984). *Nature (London)* **308**, 753.

Raam, S., Nemeth, E., Tamura, H., O'Brian, D. S., and Cohen, J. L. (1982). *Eur. J. Cancer Clin. Oncol.* **18**, 1.
Raam, S., Richardson, G. S., Bradley, F., MacLaughlin, D., Sun, L., Frankel, F., and Cohen, J. L. (1983). *Breast Cancer Res. Treat.* **3**, 179.
Renkawitz, R., Schutz, G., von der Ahe, D., and Beato, M. (1984). *Cell* **37**, 503.
Robertson, N. M., Kusmik, W. F., Grove, B., Miller-Diener, A., Webb, M. L., and Litwack, G. (1985). *Endocr. Soc. Prog. Abstr., 67th Annu. Meet* p. 20.
Robinson, S. I., Nelkin, B. D., and Vogelstein, B. (1982). *Cell* **28**, 99.
Sakai, D., and Gorski, J. (1984). *Biochemistry* **23**, 3541.
Samuels, H., and Tsai, J. (1973). *Proc. Natl. Acad. Sci. U.S.A.* **70**, 3488.
Sasson, S., and Notides, A. C. (1983). *J. Biol. Chem.* **258**, 8113.
Scatchard, G. (1949). *Ann. N.Y. Acad. Sci.* **51**, 660.
Sheridan, P. J., Buchanan, J. M., and Anselmo, V. C. (1979). *Nature (London)* **282**, 579.
Sheridan, P. J., Buchanan, J. M., Anselmo, V. C., and Martin, P. M. (1981). *Endocrinology* **108**, 1533.
Shull, J. D., and Gorski, J. (1985). *Endocrinology* **116**, 2456.
Shull, J. D., Welshons, W. V., Lieberman, M. E., and Gorski, J. (1985). In "Molecular Mechanism of Steroid Hormone Action" (V. K. Moudgil, ed.,), pp. 539–562. De Gruyter, Berlin.
Siiteri, P. K., Schwarz, B. E., Moriyama, I., Ashby, R., Linkie, D., and MacDonald, P. C. (1973). *Adv. Exp. Med. Biol.* **36**, 97.
Skafar, D. F., and Notides, A. C. (1985). *J. Biol. Chem.* **260**, 12208.
Szego, C. M., and Pietras, R. J. (1984). *Int. Rev. Cytol.* **88**, 1.
Thompson, M. A., Murai, J. T., Spady, R., and Siiteri, P. K. (1985). *Endocr. Soc. Prog. Abstr., 67th Annu. Meet.* p. 83.
Traish, A. M., Müller, R. E., and Wotiz, H. H. (1979). *J. Biol. Chem.* **254**, 6560.
Travers, A. (1983). *Nature (London)* **303**, 755.
Travers, A. (1984). *Nature (London)* **308**, 754.
Vanin, E. F., Henthorn, P. S., Kioussis, D., Grosveld, R., and Smithies, O. (1983). *Cell* **35**, 701.
Von Hippel, P. H., Bear, D. G., Morgan, W. D., and McSwiggen, J. A. (1984). *Annu. Rev. Biochem.* **53**, 389.
Walent, J. H., and Gorski, J. (1985). *Endocr. Soc. Prog. Abstr., 67th Annu. Meet.* p. 83.
Walters, M. R., Hunziker, W., and Norman, A. W. (1980). *J. Biol. Chem.* **255**, 6799.
Welshons, W. V., Lieberman, M. E., and Gorski, J. (1984). *Nature (London)* **307**, 747.
Welshons, W. V., Krummel, B. M., and Gorski, J. (1985). *Endocrinology* **117**, 2140.
Whitlock, J. P., Jr., and Galeazzi, D. R. (1984). *J. Biol. Chem.* **259**, 980.
Williams, D., and Gorski, J. (1972). *Proc. Natl. Acad. Sci. U.S.A.* **69**, 3464.
Yamamoto, K. R., and Alberts, B. (1972). *Proc. Natl. Acad. Sci. U.S.A.* **69**, 2105.
Zava, D. T., and McGuire, W. L. (1977). *J. Biol. Chem.* **252**, 3703.

DISCUSSION

G. Anogianakis. Concerning the diagram of the ratio of bound over free estradiol, I am reminded of the work in another field, i.e., the calculation of the ionic and hydration ratio for a host of different compounds. When you described the partition coefficient behavior of the estrogen–receptor complex in the cytosol solutions, you suggested that surface conformational changes rather than surface charge changes were responsible for the phenomena you described. It would seem to me that one way to explain these rather large free energy changes, without invoking the surface changes, would be to assume a multiunit receptor where the units come together or exhibit cooperative effects only when they interact with the estrogen molecule. Is it possible that some units may be cytoplasmic and some may be nuclear? That could explain why some people find the receptors in the nucleus. There are of course others who think the receptors also exist in the cytoplasm.

J. Gorski. I am not sure I follow your question on cytoplasmic distribution. What we have done in the experiments with aqueous two-phase partitioning is to use a range of concentrations which Notides suggests results in cooperative binding. We don't see any effect on the partitioning behavior. Thus we are skeptical about the polymeric model. However, when you are working with crude cytosols you have a lot of potential problems. One problem, I imagine, is that we are getting some counteracting influences. In other words, dilution may bring in one particular aspect and concentration another aspect which compensates so that the partition coefficient stays the same. We don't see any evidence that the concentration of receptor has any effect on partitioning but that is as far as we can go at the present time.

G. Anogianakis. What I am asking is a little different. Is it possible that there are two types of receptors?

J. Gorski. As I showed in Fig. 8, we think there could be a whole range of receptor forms, but we can't identify them in any of the studies that we have done so far. Actually more could be done to separate these forms by countercurrent distribution or some other kind of separation such as Mary Sherman's columns and gradients. I don't find all of the data convincing that many different forms exist, but certainly there is a growing literature on phosphorylation–dephosphorylation and other kinds of modifications. There is no way I could rule out such modified forms of the receptor.

J. H. Clark. I would like to comment on some of these points. I am not totally convinced that true cooperativity exists, and all the discussion of subunit interactions may be invalid. It is well known that concentrated solutions of binding sites (approximating or exceeding the K_d of the binding interaction) will bind ligand in a linear fashion until saturation is reached. Such a condition will yield a Scatchard plot with a linear component parallel to the Y axis which, at the point of saturation, curves downward. The Scatchard plots that you showed as well as those of Notides have these characteristics. Such curves can be obtained without subunit interactions or cooperativety.

J. Gorski. Yes, that raises an interesting problem because of the nonspecific binding in the concentrated cytosols. I favor the idea that polymeric interactions are an *in vitro* artifact which is seen only in cytosol. But I think it is of interest because it is different than what we observed in the intact cell. On the other hand, if there isn't such a phenomenon, one of our arguments for there being a monomer is weakened.

H. L. Bradlow. Accepting for the moment that the estrogen receptors are actually in the nucleus, isn't it about time to consider abandoning thermotransformation as a true phenomenon of receptors, since it is really only observed if we maintain the receptors at 4°C during the binding process which is manifestly an unnatural state for a mammalian cell, or even a cytosol. At 37°C the receptor has already passed the point where it would immediately transform as soon as it met the estrogen molecule, if indeed it is not already transformed. Essentially is this whole study of thermal transformation not simply an artifact of our experimental convenience?

J. Gorski. I think that's really a good point, and we are focusing on estrogen binding to the unoccupied receptor. If it turns out we are really correct on the nuclear localization then a nuclear receptor in a cytosol preparation faces an environment it never ordinarily would encounter. There are a number of nuclear proteins that are extracted into cytosols. For example, cell-free transcription systems require a cytosolic extract which probably contains factors of nuclear origin. However, you also have all the cytoplasmic proteins, and, as mentioned, ribonucleoproteins as well as all kinds of other things. This puts the receptor in a very strange environment, so when cytosol is heated to transform receptor you can get a variety of artifacts. On the other hand, how can we unravel what's going on in the intact cell without taking it apart and trying to work out the component parts. I think we are going to continue to work on cell extracts in order to understand the functions of the receptor. Probably one of the biggest problems in the field is that we don't really have purified receptors to work with. Most of the work on receptors is being done with crude preparations or purified preparations in which receptor is still only one part in 100 or 1000.

H. L. Bradlow. There are ways of stabilizing the receptor at 30°C or greater by charcoal treatment, and at this point at least you're dealing with the receptor at something close to the real environment. This might be better than staying at 4°C.

J. Gorski. We are trying to work with receptors at physiological temperatures and then study estrogen binding with ATPP. Because we can partition so rapidly we can see some things that other methods might not permit.

M. Walters. I have a comment, and then a question. My comment pertains to the range of steroid receptors for which nuclear localization has been described. Several years ago, Walters, Hunziken, and Norman used a different set of criteria to describe the predominant nuclear association of unoccupied receptors for the steroid hormone 1,25-dihydroxyvitamin D with crude nuclei on chromatin and with purified nuclei. These results were interpreted as suggesting that the receptor might be an inherent nuclear protein. With respect to the question, we're all left with the burden of looking at function. Earlier someone made the comment that no one has been able to establish function for a receptor within isolated nuclei. I wonder if you've thought of using the nucleoplast (formed in the absence of cytochalasin) to look at estrogen function in the nuclei, for example, by looking at induction of prolactin nessenger RNA?

J. Gorski. Welshons is continuing studies with the nucleoplasts. However, you must remember that these nucleoplast fractions are not optimal because they always have some cytoplasm. If nucleoplasts can be prepared relatively free of cytoplasm, some interesting studies could be done. I'm sorry I didn't mention other studies that have suggested nuclear localization of steroid receptors. I'd like to mention particularly Peter Sheridan's work as he was really one of the first people to raise questions about the translocation models.

M. M. Grumbach. I have three questions. They come from a nonreceptorologist. First, where is this receptor synthesized? Is it a yo-yo phenomenon between the nucleus and the cytoplasm? Second, this is a very important controversy and I know it's critical to resolve. In terms of our thinking physiologically, how do you think this will influence a shift from the emphasis on the cytoplasm to the nucleus. Third, as the estrogen steroid receptor interactions go, do all of the other steroid receptor interactions go?

J. Gorski. In answer to your last question some of the other hormone receptors were identified as nuclear proteins before the estrogen receptor. The first, as Marian Walters brought out, were thyroxin and vitamin D. I think that nuclear localization is going to turn out to be universal for all the steroid hormones, but that's a pretty big extrapolation from our data. That would be my guess because the receptors in general have such similar characteristics.

M. M. Grumbach. You favor a parsinonian's hypothesis.

J. Gorski. Yes, but we just have to wait and see what the data will show. As to your question about the physiological implications, I think for people outside of the steroid action area that it may not be so important. However, if you are working on steroid action, nuclear localization has a lot of implications. You have to be more concerned about where the unoccupied receptor is initially. Perhaps the unoccupied receptor is bound to the target site already. That would be very important in understanding the mechanisms of steroid hormone action. Our data suggest that the receptor is not moving around, but we can't be certain of that at present. One might imagine that the unoccupied receptor is bound to DNA or other nuclear components with a low affinity and is never completely soluble.

M. M. Grumbach. The first question is where do you think it's synthesized?

J. Gorski. I don't know of any evidence for synthesis of proteins in the nucleus, although this has been discussed in the literature. I think most people in the field feel that nuclear proteins are synthesized in the cytoplasm. They all have certain characteristics that make them bind to nuclear components so that they stay there, or the nuclear protein has components that facilitate moving the protein through the nuclear pores into the nucleus. Only a few nuclear proteins have been studied thoroughly, for example, the lamins. They have very special characteristics that make them nuclear proteins, and I would assume that when the structure of the receptor proteins is known we will see such structures in the receptors. We will be able to modify the genes at some regions and make a protein that will no longer get into the nucleus.

Y. A. Lefevre. I wonder if part of the problem could be explained by a population of receptors on the outer surface of the nuclei? Have you subjected nucleoplasts to any form of treatment to see if you could get off a population of receptors, for instance, with Triton treatment?

J. Gorski. We haven't done that, but I'm sure if you treated nucleoplasts in that way the cytosol and nuclear proteins would be solubilized.

Y. A. Lefevre. I think if you used a low concentration of Triton, you would selectively remove outer nuclear membranes. I don't know if, in your case, you'd be able to remove the plasma membrane.

J. Gorski. I think it may be more important to consider the studies using autoradiography from King and Greene as well as Brenner's group. They do not observe any indication of localization at the periphery of the nucleus. Rather, the only thing that I've seen so far in any kind of subnuclear localization is that the receptor is not in the nucleolus and that there appears to be less in the dense chromatin and more in the less dense chromatin. Otherwise the receptor is spread pretty uniformly throughout the nucleus, which would not be consistent with its being associated with the nuclear envelope.

Y. A. Lefevre. You showed the steroid moving through the cytoplasm. Do you have any comments on the work of Puca and Sica which shows that there might be interactions of steroid with cytoplasmic microtubules?

J. Gorski. In our data, as well as King and Greene's and Brenner's data, there is no evidence for any high affinity binding of estrogen in the cytoplasm. We assume that a compound as lipophylic as the steroid when diffusing through the cytoplasm is associated with proteins and lipid that are present there. The estrogen is not soluble in the free water space of the cytoplasm to any great extent. Therefore it is probably bound with low affinity to cytoplasm, but I don't know of any data that suggest it is strongly associated with any particular component of cytoplasm.

B. F. Rice. I have been impressed with how difficult it is to get gonadotropin off the receptor once it got stuck. There are thousands of estrogen receptor assays done on breast cancer specimens around the world and many decisions about therapy, prognosis, protocols, etc., are based on the results of those estrogen receptor assays. Now based on what you have told us tonight, if you were in that position of having to make decisions about therapy would you utilize estrogen receptor assays as they are currently being done today?

J. Gorski. The unoccupied receptor protein is still a cytosolic protein. Therefore when you are using cytosolic protein assays you can measure the unoccupied receptor. If you are concerned about how many receptors are occupied, you may have to use some other assays, such as Jim Clark's exchange assays, to look at the occupied nuclear forms of the receptor.

However if you grind up a tissue in almost any kind of medium you are going to extract the unoccupied receptor from the nucleus into the extract.

B. F. Rice. Concerning receptor occupancy, I have never really been satisfied with some of the answers people have given to the results one gets based on *in vivo* concentration. In your system, for example, has an experiment been done in which the cells were grown in an animal that was ovariectomized and adrenalectomized versus an animal that still had its ovaries and adrenals, to try to get some idea of what an *in vivo* situation does to the *in vitro* experiment.

J. Gorski. I am not sure I understand. Do you mean lower endogenous steroid and then measure receptors?

B. F. Rice. That's right. In other words, the *in vivo* situation in terms of what the amount of estradiol in circulation of an experimental animal or a patient would be, for example.

J. Gorski. We have done some studies with certain dwarf mice strains that have very little pituitary function and very underdeveloped reproductive tracts. These dwarf mice still have normal estrogen receptor concentration in the uterus. I can't recall offhand if ovariectomized and adrenalectomized mice have been studied. I know Jim Clark ovariectomized newborn rats in our lab and then measured receptors and found that receptor development was perfectly normal.

J. H. Clark. I think I understand exactly what you mean. If you were to make the measurements in intact animals, there would be more occupied receptors, and they would be in the nucleus. So one would have to measure the unoccupied as well as the occupied receptors in the nucleoplast, and more receptors would be occupied.

B. F. Rice. Just one point of clarification. Premenopausal women most commonly are estrogen receptor negative and the explanation that has been given is that those are not occupied receptors.

J. H. Clark. That is because the receptors are in the nucleus where they are not being measured.

B. F. Rice. I think that explains many of the complicated problems that our clinicians have been faced with.

J. H. Clark. I totally agree, and a lot of us have realized for a long time that in order to get any accuracy out of that system one would have to measure both compartments and ideally would want to know occupied versus unoccupied. Of course that turns out to be such a horrendous effort that I don't think very many people could do it routinely.

J. Gorski. I think the problem is that most people are measuring only the unoccupied receptors. However the number of unoccupied receptors depends very much on the physiological state of the tissue being assayed. The exact stage of the cycle and a variety of other things can complicate these assays.

G. Callard. For the record I would like to mention that there are a number of systems in which estrogen receptors and other steroid receptors remain associated with nuclei during homogenization procedures which ordinarily extract ligand-free receptors into the cytosol. These systems are the estrogen receptor of adult hen and *Xenopus* liver, the ecdysone receptor of insects, and the system we have been working with, the estrogen receptor of dogfish testis (Callard, G., and Mak, P., *Proc. Natl. Acad. Sci. U.S.A.* **82**, 1336,1985). An interesting question is what properties of these particular receptors cause them to remain associated with nuclei irrespective of the presence of ligand? These exceptions also raise the issue of whether there might be some dysfunctional states in which even the well-studied mammalian receptors might remain adherent to nuclei and not be solubilized in the predictable manner. I would like to suggest that the dogfish testis, and perhaps these other exceptional tissues, may be good systems in which to look at interactions between the different receptor forms and the nucleus. We have preliminary evidence using the dogfish system that

A MODEL OF ESTROGEN ACTION   327

both occupied and unoccupied receptors may in fact be bound in some way to the nucleus since both forms require quite high salt concentrations to extract them into the cytosol.

J. Gorski. That sounds like an interesting system. I can't comment other than to note that not only the receptor but also the environment of a particular nucleus could be different. I think you had mentioned earlier that in some of these species localization varies from one tissue to another. Unless there are different genetic forms of the receptor or the receptor has undergone some modification, you would have to assume something in that nuclear environment prevents the receptor from being extracted. That might even be an artifact since you still aren't 100% sure whether this is what the receptor was associated with in the intact cell.

H. Kronenberg. I was confused about one of the minor points in your discussion of the cytoplasts. You showed us that the cytoplasts are rather vigorous in protein synthesis, and I wasn't sure whether you were using that as a criterion to determine if the cytoplast was generally healthy or intact. The reason I'm surprised that protein synthesis would be a criterion for intactness is that cell-free extracts are often vigorous in protein synthesis with no membranes at all.

J. Gorski. We would have a difficult time getting a lot of prolactin synthesis with a broken cell preparation. We have also studied dye uptake and the cytoplasts act very much like intact cells. In other words, they seem to have a good intact membrane around them. These intact membranes around the cytoplasts do not contain receptors. This contradicts certain reports of receptors associated with the outer plasma membrane.

S. Cohen. My question concerns the condition of pregnancy. I assume that there's nothing about pregnancy that invalidates your suggestion?

J. Gorski. As far as we know there may be some changes due to the other hormones in a pregnant state that would have some influence, but we have not studied this.

S. Cohen. If we follow the concentration of estradiol, it should be the same in cytoplasm as it is in the blood stream?

J. Gorski. We believe that the free steroid present in the plasma, which is a small part of the total, is going to be in equilibrium with that steroid present in the cell. The exact levels may vary some, but we believe that they're quite similar.

S. Cohen. We have done some assays on a rat nuclear fraction of human myometrium obtained at Caesarian section. We found that the estradiol concentration ran around 4 ng/g of tissue. In the blood stream, the similar free-estradiol runs around 30 ng/ml. Others have reported similar figures. Betra has reported 3 ng of estradiol in myometrium, whereas a number of people have shown that the blood stream runs around 30 ng/ml of blood. Can you explain this discrepancy?

J. Gorski. You probably have the problem of plasma proteins. In other words, the absolute free steroid present in the plasma is a small fraction of the total steroid.

S. Cohen. No, I'm talking about the free component.

J. Gorski. How is that measured?

S. Cohen. Free estrogen is that portion not conjugated and not bound to protein and has been measured by a number of different methods.

J. Gorski. About the only way you can really assay free steroid is to use some kind of a dialysis procedure to differentiate free and bound steroids. I don't know of very many studies that have measured the dialyzable steroid in the plasma versus the free dialyzable steroid present in the tissue.

S. Cohen. If you extract blood with ether that will extract very little conjugate of protein bound tissue.

J. Gorski. I think that's not true because the actual amount of steroid that is protein bound with a low affinity is rather large. Probably a very small percentage of what you would assay as free steroid is truly solubilized in the water compartment.

S. Cohen. I have never done the blood assays. I was just quoting assays in the literature. We do tissue assays using alcohol that will denature the receptor and free the estrogen and make it alcohol soluble. Also, the free estriol:estradiol ratio in the blood stream is about 5 to 1, but in the myometrium it is about 1 to 2. How can you explain that?

J. Gorski. I couldn't, but maybe someone else here could.

J. H. Clark. I think it has more to do with the relative binding affinities with which E_3 and E_2 are bound in the blood. The actual free level is difficult to determine. E_3 is generally not bound as tightly as E_2.

S. Cohen. Loriaux has worked out the method for assay. We have found a difference in estrogen assays depending on when the Caesearian was performed; those obtained from elective Caesarians versus those obtained from emergency Caesarians after labor has started. During the latter condition the ratio of E_2/E_3 is reversed and is present at about 2 to 1 over estriol whereas the same ratio was 1 to 2 for the strips obtained from elective Caesarians.

J. Gorski. I would not be able to answer. I have no comments about that.

S. Cohen. One of the people showed estradiol stimulating growth and I think it was a fibroblast growth stimulating factor that also stimulated growth, but the two of them gave a much greater response than the summation. How can that be explained on a molecular biology basis?

J. Gorski. There are some interesting reports of very dramatic effects of estrogen on epidermal growth factor receptors. I don't know of any particular studies on the fibroblast-stimulating factor, but I can imagine it would be worth studying. The estrogen can stimulate the production of another receptor, and then you have the estrogen plus another regulator to bring about a much larger effect. We used to call this synergism but it's really just the additive effect of a complex array of responses.

G. B. Cutler. I would like to pursue your curvilinear Scatchard plots a bit further. Positive cooperativity is one thing that can produce this sort of plot. Other potential ways in which they can arise is failure to reach equilibrium at the low ligand concentrations, instability of the receptor at the low concentrations, and overestimation of the nonspecific and a corresponding underestimation of the specific binding by failing to saturate all the specific sites at the low concentrations. The latter possibility can happen particularly if one is using a 100-fold excess. I wondered how carefully these possibilities have been ruled out?

J. Gorski. I think they are all possibilities. Siitteri has reported that he has observed cooperativity under conditions in which he had a better estimate of the free steroids using centrifical dialysis to get an estimate of the truly free steroid. He saw positive cooperativity and Hill plots that suggest it is really cooperative and not just a little hook at one end of the plot. He was working with free steroid levels of less than $10-12\ M$. This is similar to what we see in intact cells incubated without serum. The absolute amount of free steroid is very low, and measurement may be a problem.

G. B. Cutler. A number of years ago we had the experience with a rather concentrated preparation of cytosol from liver, while looking at glucocorticoid receptor and using the standard 100-fold excess to estimate nonspecific binding, of getting a curvilinear Scatchard plot. It then disappeared when we changed to $10^{-6}\ M$ unlabeled steroid throughout all the nonspecific tubes. In that particular case we attributed the curvilinear Scatchard plot to failure to saturate the specific binding sites at the low ligand concentrations.

J. Gorski. This can be a problem, but as far as I know Notides has been very careful with all this. In our case we used a constant, saturating level of unlabeled steroids.

J. M. Nolin. I just have one general comment that relates to 23 years of estrogen receptor research as it pertains to those of us who are responsible for the endocrine section of a first year medical curriculum in which an hour lecture time has been allotted exclusively to the estrogen receptor. I personally recommend that everybody simply put a sign up saying "Cancelled."

J. Gorski. It might be a good idea!

J. M. Lakoski. Both intracellular and extracellular recording studies have identified effects of estrogens on active and passive properties of neuronal membranes as CSN. You have been focusing on estrogen receptors, both in the cytoplasm and the nucleus, and reacting to the discussion of estrogen receptors on the plasma membrane. How would your model account for these data or do you think they are artifacts?

J. Gorski. I don't really know enough about these studies. I know there have been reports in the literature of effects of estrogen that are very rapid and which are difficult to explain in terms of nuclear function. Some of these I think are artifacts but others I really am not sure of. A lot depends on how the hormone is administered. I know of artifacts that have risen by the injection of the hormone that induces another kind of regulator. There are some other reports that I have seen for which I don't have a good explanation. I believe nuclear function accounts for the bulk of the responses, but there may be estrogen functions that we don't know much about as yet.

J. M. Lakoski. So you think there could be tissue-specific differences in estrogen receptors that may account for these effects?

J. Gorski. That is a possibility I cannot rule out.

Determinants of Puberty in a Seasonal Breeder[1]

DOUGLAS L. FOSTER,[2] FRED J. KARSCH, DEBORAH H. OLSTER,[3]
KATHLEEN D. RYAN,[4] AND STEVEN M. YELLON[5]

Reproductive Endocrinology Program, Consortium for Research in Developmental and Reproductive Biology, and Departments of Obstetrics and Gynecology and Physiology, and Division of Biological Sciences, The University of Michigan, Ann Arbor, Michigan

I. Introduction

Most female mammals in temperate climates are seasonally rather than continuously fertile. With the exception of the human being and some species that man protects from the pressures of annual environmental changes, conception is restricted to a certain time of the year—the *breeding season*. At other times, females remain anovulatory and do not exhibit mating behavior—the *anestrous season*. This reproductive stategy forces births and early postnatal development to occur when food is plentiful. As pointed out in our earlier contribution to this series (Karsch *et al.*, 1984), the season of conception varies with the length of pregnancy. Typically species in which gestation is very short (weeks) or long (ca. 1 year) ovulations occur during spring and summer, whereas those species with intermediate length pregnancies (several months) mate in autumn. In addition to this consideration of seasonal breeding is the type of the environmental cue that times annual periods of fertility and infertility. Several are known: food availability, social cues, photoperiod, temperature, humidity, and rainfall. Which cues are used depends upon a variety of factors including the physical size of the species and the seasonal variation of the local environment (Bronson, 1985).

---

[1] This contribution is dedicated to Professor Andrew V. Nalbandov, reproductive biologist and mentor (D.L.F. and F.J.K.).
[2] Correspondence address: Room 1101, 300 North Ingalls Building, The University of Michigan, Ann Arbor, Michigan 48109.
[3] Present address: Department of Obstetrics and Gynecology, Columbia University College of Physicians and Surgeons, New York, New York.
[4] Present address: Department of Physiology, University of Pittsburgh School of Medicine, Pittsburgh, Pennsylvania.
[5] Present address: Department of Physiology, Division of Perinatal Biology, Loma Linda University School of Medicine, Loma Linda, California.

The evolution of seasonal breeding, "nature's contraceptive" (Lincoln and Short, 1980), influences the time when fertility is first attained in the young female. Puberty occurs only during the breeding season. Thus, not only must the developing seasonal breeder be able to determine when she is sufficiently mature to begin reproductive cycles, she must also be able to determine when during the year onset of fertility will produce young during the spring and summer.

This article will focus on the development of reproductive endocrine function in the female domestic sheep (*Ovis aries*), a seasonally breeding species that exhibits rapid growth (Fig. 1). At birth the body weight is similar to that of the human, however, the female lamb matures much more rapidly than the child. Whereas by 30 weeks the weight of the human infant has only doubled, that of the lamb has increased 10-fold (Fig. 1, inset). Furthermore, in relative terms, the degree of sexual maturation achieved in the lamb during this 30 weeks is not attained until age 13–15 years in girls when puberty occurs. Despite the distinct differences in absolute time of sexual maturity, reproductive function is ultimately controlled by the same pathway—the hypothalamus and its modulation of anterior pituitary function. Bearing this in mind, we will summarize our understanding of the general timing of puberty by photoperiod and growth in the lamb and then develop concepts about how these external and internal factors become integrated to modulate the tempo of sexual maturation.

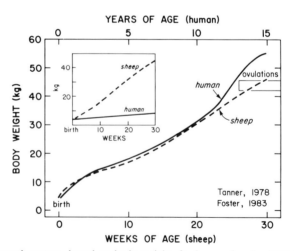

FIG. 1. Growth curves, based on body weight, for human females (fiftieth percentile) and female sheep (mean) from birth through the initiation of ovulation. Inset shows absolute growth during the first 30 weeks. From Foster *et al.* (1985a), which was modified from Tanner and Foster in a previous publication.

II. Seasonal Breeding in Adult Sheep

Although the sheep has been closely associated with man for several thousand years, it retains a seasonal pattern of reproduction. Some primitive breeds, such as the Soay, have a restricted mating season lasting only a few months in late autumn and winter (Lincoln and Short, 1980). The duration of the sheep breeding season has been expanded by domestication and artificial selection (Hafez, 1952), and certain breeds are capable of mating for as many as 8–10 months of the year (Fletcher and Geytenbeek, 1970; Mallampati *et al.*, 1971). Furthermore, individuals of particular breeds will exhibit estrous cycles throughout the year (Watson and Gamble, 1961; Wiggins *et al.*, 1970) indicating that the potential for continuous reproduction exists in the sheep.

The Suffolk breed, which we use in our research, was chosen because it has a distinct breeding season lasting about 6 months (Fig. 2). Ovulation and estrus begin in late summer, when day length is decreasing, and occur at 16-day intervals until pregnancy (148 days duration) is established. As

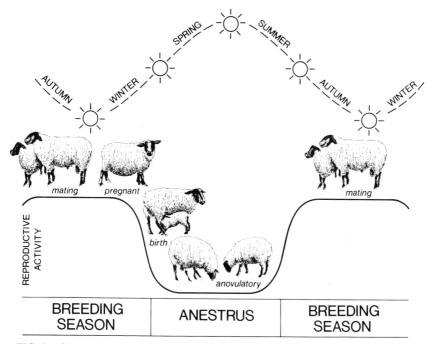

FIG. 2. Seasonal reproduction in the Suffolk ewe. Broken line represents annual photoperiod curve. During the breeding season when reproductive activity is high, ovulations occur at 16-day intervals, and mating is possible; during anestrus breeding does not occur because the female is anovulatory.

day length increases in late winter, estrous cycles cease, and a 6-month period of anestrus ensues. This annual reproductive pattern concentrates births in late winter and early spring.

The role of photoperiod in timing the annual reproductive cycle of the mature female has been extensively reviewed recently (Karsch *et al.*, 1984). It is possible to reverse the timing of the breeding season by reversing the annual changes in photoperiod by means of artificial lighting. Moreover, exposure of ewes to alternating 90-day photoperiods of long days and short days can produce two breeding seasons and two anestrous seasons each year. Such results, obtained by simply altering photoperiod without changing other environmental variables, provide strong evidence for a deterministic role for day length in the timing of breeding activity in the mature female sheep.

III. Multiple Factors Time Puberty in the Lamb

Lambs, usually born in late winter and spring, develop rapidly during the summer nonbreeding season when adult females are anovulatory. Puberty (first ovulation, first estrus) then occurs in autumn between 25 and 35 weeks of age, well after the onset of reproductive cycles in the adult female (Foster and Ryan, 1979a). There is long-standing evidence that the month of birth of the lamb will subsequently influence the age of puberty (Fig. 3). Hammond (1944) observed that regardless of the season of birth, lambs would exhibit first estrus only during the breeding season.

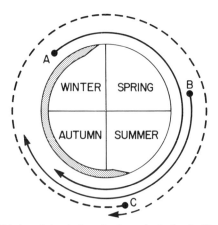

FIG. 3. Season of birth and the timing of puberty in the lamb. Birth is denoted by closed circles and puberty (first estrus) occurs at the end of the arrow. Shaded area designates the adult breeding season. Note, lambs reach puberty only during the breeding season. Schematic based primarily on the data of Hammond (1944).

Furthermore, winter-born lambs and spring-born lambs attained puberty at about the same time of year, but because the latter were born several months later, they were younger at first estrus. Births later in the year lead to an even younger age at puberty (Mallampati et al., 1971; Fitzgerald and Butler, 1982; Foster and Ryan, 1986). There is a point, however, when being born late in the year no longer reduces the age of first mating. When lambs are born in late summer and autumn, as a consequence of being conceived later in the breeding season, they often fail to achieve sexual maturity during the first year; rather, puberty is delayed until they are beyond 1 year of age (Fig. 3C) (Hammond, 1944; Mallampati et al., 1971; Foster, 1981). Clearly, some environmental factor strongly influences the timing of puberty in the female sheep.

The conclusion that multiple factors time puberty may be reached when growth rate, rather than season of birth, is varied. Even when lambs are born in the same season, there can be great individual differences in the age of first ovulation when the rate of growth varies. This is illustrated in Fig. 4 for spring-born lambs in which the rate of growth was experimentally altered by a period of restricted nutrition followed by *ad libitum* feeding. Some lambs were well fed from birth, and puberty occurred normally (30 weeks, Group A). Other females, born at the same time, failed to initiate reproductive cycles at the normal age when growth was retarded by restricting level of nutrition from weaning (Groups B, C, and D). When such lambs were induced to grow rapidly during the breeding season of the autumn and early winter, in response to unlimited quantities of food, ovulations began within a few weeks (Groups B and C). However, when the phase of rapid growth was delayed until late winter and spring during the adult anestrous season, puberty failed to occur, and the females grew well beyond the normal size required to initiate reproductive cycles (Group D). They remained anovulatory throughout most of the summer. Moreover, when ovulations began the subsequent autumn, they did so in the absence of any further increase in growth; the onset of cycles was coincident with the onset of the breeding season.

The foregoing observations about the effects of season of birth and growth rate on timing of puberty can be interpreted meaningfully when viewed against the background of seasonal reproduction in the adult ewe (Fig. 5). The female lamb must be born sufficiently early in the year to attain the appropriate physiologic size for puberty during the breeding season. If such growth does not occur by virtue of being born late in the year or through inadequate nutrition, then puberty is delayed until the next breeding season. This strategy ensures continued synchrony of breeding among females of all ages, because the first matings are clustered during the times of year when adult reproductive activity is high. More-

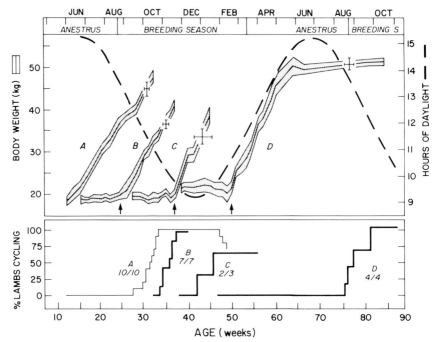

FIG. 4. Influence of season and growth on the initiation of reproductive cycles in the lamb (top, mean ± SEM; bottom, individuals, histograms). The lambs were born in spring (March) and were raised outdoors (dashed line, photoperiod). They were either fed *ad libitum* after weaning at 10 weeks of age or were placed on a restricted diet of similar composition; at various ages (arrows), *ad libitum* feeding was begun in food-restricted lambs. The general limits of the anestrus and breeding seasons of the adult female are indicated at the top. From Foster *et al.* (1985b).

over, it prevents the establishment of pregnancy in a female that is too small to carry an offspring to term without jeopardizing her own survival.

Recognizing that multiple factors relating to growth and season serve as determinants for puberty in the female sheep, several questions arise. For example, what environmental cue does the lamb use for its seasonal timing of puberty? Is it photoperiod—the same factor used by the adult to shape its breeding season? How does the lamb monitor its growth to determine when it has attained some genetically predetermined physiologic size to support a successful pregnancy? How are these internal and external signals integrated to time the sequence of endocrine events that underlie the transition into adulthood? Answers to some of these questions are emerging.

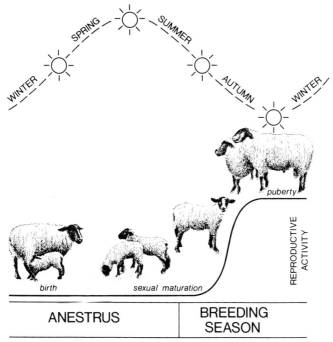

FIG. 5. Time of puberty in the spring-born lamb. Note, growth and development occur during the long days of the spring and summer anestrous season, and reproductive cycles including matings begin during the decreasing day lengths of the autumn breeding season. See Fig. 2 for greater details of seasonal reproduction in the adult female.

IV. Photoperiodic Timing of Puberty

The seasonal factor responsible for modifying the time of puberty in the female lamb is day length. This conclusion arises from numerous demonstrations that artificial photoperiod can markedly alter the age of onset of reproductive cycles. The first example is shown in Fig. 6 in which the age at puberty in natural photoperiod (A) is compared between spring-born (B) and autumn-born lambs (C); in addition, the time of puberty is shown for autumn-born females (E) raised in an annually reversed artificial photoperiod simulating the one they would have experienced had they been born in the spring (D). Spring-born lambs began reproductive cycles at 25–35 weeks of age, but as indicated earlier (Fig. 3), lambs born sufficiently late in the year delay puberty until the next year. In the example illustrated, the autumn-born lambs attained the appropriate age for puberty (ca. 25–35 weeks) during the anestrous season, but they remained

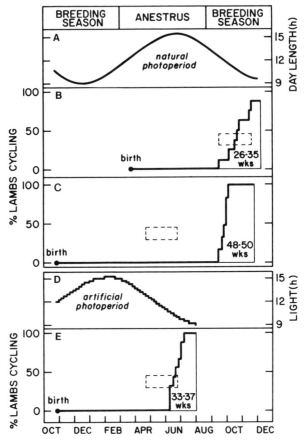

FIG. 6. Month and age at initiation of reproductive cycles in (A) natural photoperiod in (B) spring-born (March) and (C) autumn-born (October) lambs and in (D) an artificial, seasonally reversed photoperiod in (E) autumn-born lambs. The general limits of the anestrous and breeding seasons of the adult female sheep are indicated at the top. First ovulation was based upon the appearance of luteal phase levels of circulating progesterone. In (C) and (E) the dashed rectangle identifies the age range (26–35 weeks) during which ovulation was initiated in control spring-born lambs in (B). Note, reversal of annual photoperiod in autumn-born lambs prevents delay in puberty that occurs in autumn-born lambs in natural photoperiod. From Foster and Ryan (1981).

anovulatory. Such lambs eventually became sexually mature, albeit some 20 weeks later with the onset of the breeding season. Most importantly, this delay in onset of reproductive cycles was almost entirely prevented by rearing the autumn-born females in a seasonally reversed artificial photoperiod. Thus, the delay in puberty in late-born lambs is mediated through some mechanism involving day length. As discussed below and in the next sections, additional evidence has accumulated to support the view that photoperiod alters the timing of puberty and that it does so by changing the activity of the hypothalamo–hypophyseal mechanisms regulating hormone secretion.

A. THE TIMING OF PUBERTY BY ALTERNATE PHOTOPERIODS

In view of the importance of photoperiod in the expression of sexual maturation, the question arises as to which aspect of the annual photocycle times puberty in the female sheep? Because the first reproductive cycles begin during decreasing day lengths, regardless of season of birth (Figs. 4 and 6), perhaps the simplest hypothesis is that the lamb monitors day length, and when it is sufficiently short, ovulations begin. The corollary is that long days inhibit the initiation of reproductive cycles. The onset of puberty well after the onset of the adult breeding season (Fig. 5) could be explained by a requirement of the lamb for shorter day lengths than the adult.

To determine if long day lengths prevent puberty, spring-born lambs were reared from birth under continuous artificial long days (15L:9D) (Fig. 7B). During the period when repetitive ovulatory cycles began in young females raised outdoors (ca. 30 weeks, Fig. 7A), long-day reared lambs remained anovulatory or exhibited isolated reproductive cycles with luteal phase defects. Therefore, long photoperiods can indeed prevent the onset of reproductive cycles at the usual age supporting the general notion that the long days of summer serve to inhibit puberty. Interestingly, however, continued exposure to long photoperiods cannot block puberty indefinitely. Eventually, long-day reared females will exhibit reproductive cycles (after 1 year of age: Yellon and Foster, 1985). This latter finding indicates that the lamb is able to reject photoperiod information (become photorefractory) and attain puberty despite its living in a light environment which normally does not facilitate sexual maturation.

In view of the demonstration that long days delay puberty, it could be predicted that short days would facilitate attainment of sexual maturity. Moreover, raising lambs from an early age under short days should reveal the youngest age at which puberty is possible. However, the results of such a light treatment do not support these predictions (Fig. 8). When spring-born lambs were exposed only to artificial short days (9L:15D), surprisingly, the time of first ovulation was not advanced; rather, it was delayed much the same as for females that experienced entirely long days. Most short-day reared lambs remained anovulatory during the first year

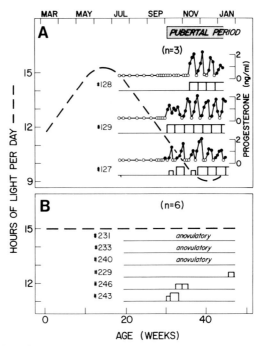

FIG. 7. Initiation of reproductive cycles in long days. Spring-born lambs were raised in (A) natural photoperiod or (B) entirely in artificial long days (15L:9D). Broken lines indicate photoperiods. In (A) the blocks below each progesterone cycle are used for coding purposes with normal luteal phase cycles represented by large blocks and short luteal phase cycles (<10 days) by small blocks. Each horizontal line shows data for an individual lamb with the beginning of each line representing the start of the period of blood sample collection. Note, long days block puberty during the first year. From Foster et al. (1985b).

(Fig. 8B), and it was not until the second year that repetitive reproductive cycles occurred (Fig. 8C). During prolonged treatment with artificial short days, cycles began and persisted throughout the anestrous season when females outdoors remained anovulatory (March–August). Clearly, photoperiodic timing of puberty in the female lamb is much more complex than envisioned at the outset when we thought that merely the experience of sufficiently short days would initiate reproductive cycles.

The failure of short days alone to induce puberty in the female sheep prompted an alternative hypothesis. It was reasoned that because in natural conditions the lamb normally experiences long day lengths during the summer before the shorter day lengths of autumn, perhaps a combination of long days and short days is required for puberty. Indeed, it was initially determined that a 10-week block of long days (12–22 weeks of age) would initiate repetitive cycles in lambs otherwise raised in short days (Fig. 9B).

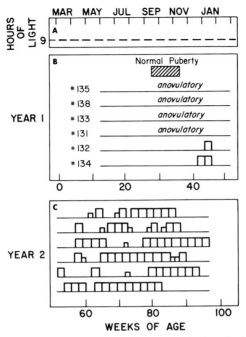

FIG. 8. Initiation of reproductive cycles in spring-born lambs raised for 2 years entirely in artificial short days (9L:15D, broken line). Large blocks designate reproductive cycles with normal luteal phases and small blocks cycles with short luteal phases (see Fig. 7 for examples of progesterone profiles). Note, short days delay puberty until second year. Redrawn from Yellon and Foster (1985).

Reducing the block of long days to only 5 weeks also induced reproductive cycles (Fig. 9C). Remarkably, as little as 1 week of long days (week 21) was effective in establishing repetitive cycles, but puberty was not entirely normal. The 1-week block of long days produced a large number of short luteal phase cycles during subsequent exposure to short days. The finding of a long day requirement for initiation of ovulations in this young, short-day breeder raises the possibility that the lamb maintains a "photoperiod history" and that it uses the long days of spring and summer as a reference to time puberty to the short days of autumn. As will be noted below, the age when long days are experienced is of importance.

B. TRANSMISSION OF PHOTIC CUES

To begin to appreciate how photic cues regulate the time of first ovulation via the hypothalamic–pituitary system, it is necessary to consider the mechanism that codes day length (Fig. 10). The pathway for transmission

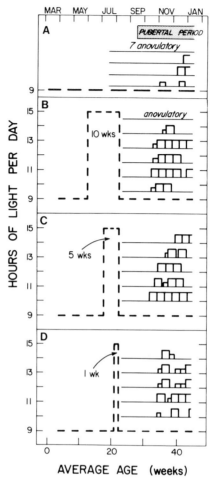

FIG. 9. Alternate photoperiods initiate reproductive cycles. Spring-born lambs were raised in artificial short days (9L:15D) (A) entirely without any long days or with (B) 10 weeks, (C) 5 weeks, (D) 1 week of long days (15L:9D). Broken lines indicate photoperiod. Large blocks designate reproductive cycles with normal luteal phases and small blocks cycles with short luteal phases; each horizontal line shows data for an individual lamb except for (A) which indicates 7 of 10 lambs remained anovulatory (see Fig. 7 for examples of progesterone profiles). Note, only 1 week of long days induces repeated reproductive cycles in short-day reared females. Redrawn from Yellon and Foster (1985).

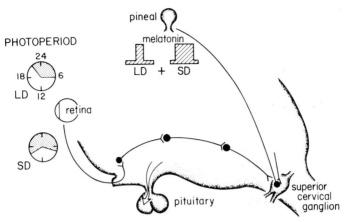

FIG. 10. Proposed pathway for transmission of photoperiod information in the sheep. Shaded portions of 24-hour clocks represent the dark phase of the photoperiods; widths of the melatonin bars indicate relative durations of its nighttime rises. LD and SD, long day and short day, respectively. See text for details. Redrawn from Foster et al. (1985b).

of photoperiod information in the adult sheep has been reviewed earlier (Karsch et al., 1984). Key anatomical links in the photoneuroendocrine pathway are (1) the *retinal photoreceptor*, demonstrated by loss of photoperiod control of reproduction in blinded ewes (Legan and Karsch, 1983), (2) a monosynaptic *retinohypothalamic tract*, as shown by intraneuronal transport of tritiated proline from the retina to the suprachiasmatic nuclei (Legan and Winans, 1981), (3) the *superior cervical ganglia*, demonstrated by loss of photoperiod control of reproduction after their removal in the ram (Lincoln and Short, 1980, for review), (4) the *pineal gland*, as evidenced by the loss of photoperiod control of reproduction after its removal in the ewe, and (5) *melatonin*, as demonstrated by its ability to drive reproductive responses attributed to photoperiod (Karsch et al., 1984). Studies of this pathway have led to the conclusion that the pineal gland of the adult mediates photoperiod information through the secretion of melatonin. The circadian-based melatonin rhythm codes for day length, and the duration of the daily rise in circulating melatonin is proportional to the hours of darkness.

Bearing in mind the foregoing pathway for transmission of photoperiodic information in the mature female sheep, we can question whether this pathway exists in the lamb. Is there a circadian pattern of melatonin that reflects the prevailing photoperiod? If so, when does it develop? Does disruption of the pathway alter the timing of puberty? The results of our initial efforts have determined that the prepubertal lamb produces a melatonin rhythm which is responsive to the light/dark cycle. This is exemplified in Fig. 11 for females raised in a sequence of artificial photoperiods in which the continuous short days (9L:15D) were interrupted by 5 weeks of long days (15L:9D). During short days (15 hours of darkness), the duration of the nocturnal melatonin rise is greater than during the period of long days (9 hours of darkness). Thus, melatonin has the potential to serve as a timekeeping hormone in the lamb, much like in the adult.

Disruption of the melatonin rhythm of the lamb early in life and its subsequent effects on the timing of puberty have been studied. Bilateral removal of the superior cervical ganglia to ablate the sympathetic innervation to the pineal gland of 6-week-old lambs raised outdoors abolished the melatonin rhythm (Fig. 12B). Melatonin was still present in the circulation, albeit at low levels, but modest increases were not consistently related to any particular time of day. Ganglionectomy prevented initiation of reproductive cycles during the first year indicating that normal pineal function is necessary for timing puberty to the decreasing daylengths of autumn. Interestingly, surgical removal of the pineal of the young crossbred Merino sheep was initially reported not to influence sexual maturation, a finding attributed to the relative insensitivity of the Merino breed to photoperiod (Matthews *et al.*, 1981). More recently, a study from the

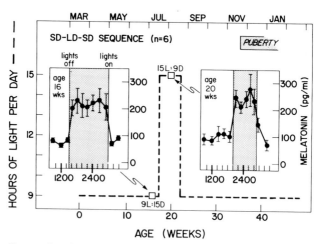

FIG. 11. Twenty-four hour patterns of circulating melatonin (means ± SEM, insets) in spring-born female lambs in artificial short days (9L:15D) at 16 weeks of age and then in artificial long days (15L:9D) at 20 weeks of age. Broken line indicates photoperiod, shaded area in insets delineates dark phase of photoperiod, and horizontal bar at top denotes time of puberty based upon circulating progesterone (Fig. 9C for onset of cycles). SD and LD, short days and long days, respectively. Note, duration of melatonin rise reflects the dark period. From Foster *et al.* (1985b).

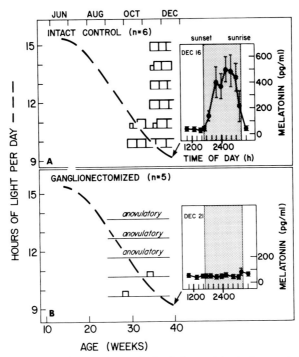

FIG. 12. Twenty-four hour patterns of circulating melatonin (means ± SEM, insets) under natural photoperiod in 40-week-old spring-born lambs with (A) normal or (B) sympathetically denervated pineal glands. Broken line indicates photoperiod, shaded area in insets delineates dark phase of photoperiod, and the large and small boxes on the horizontal lines normal and short luteal phases, respectively (see Fig. 7 for details of coding of progesterone profiles). Denervation of pineal gland was accomplished by bilateral removal of the superior cervical ganglia at 6 weeks of age. Note, ganglionectomy prevented onset of repetitive reproductive cycles at the normal age. From Foster et al. (1985b).

same laboratory (Kennaway et al., 1985) has determined that pinealectomy prevented the onset of reproductive cycles at the usual age in crossbred Merino lambs (× Border Leicester × Dorset). As is the case for inappropriate photoperiods (uninterrupted long days or short days), puberty does occur when pineal function is impaired (ganglionectomy) or absent (pinealectomy), but it is markedly delayed. In the ganglionectomized lambs of Fig. 12B, the first normal luteal phase did not occur until 65 weeks of age, and at that time of the year (May–June) day length was increasing (data not shown, S. M. Yellon and D. L. Foster, unpublished). This offers additional evidence that photoperiod is used to modify the timing of sexual maturation, but that it cannot advance or forestall puberty outside of certain limits.

Replacement of the melatonin rhythm would be expected to restore the time of puberty to normal in the absence of a functional pineal gland. To test this postulate, we infused melatonin by means of a portable backpack infusion system (Bittman *et al.*, 1983) into lambs with denervated pineal glands (Fig. 13C). Untreated ganglionectomized lambs (Fig. 13B) were exposed to a photoperiod of continuous short days interrupted by a 5-week block of long days, a sequence effective in initiating reproductive cycles in unoperated females (Fig. 13A). Ganglionectomized lambs without melatonin replacement generally remained anovulatory despite their exposure to an appropriate photoperiod sequence for induction of puberty (Fig. 13B). The melatonin-treated lambs were infused initially with a short-day melatonin pattern (15 hour melatonin) immediately after gangli-

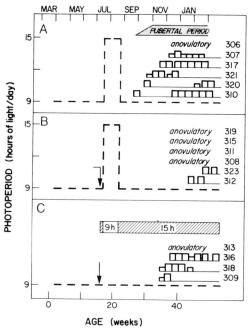

FIG. 13. Melatonin mediates photoperiod induction of puberty. Initiation of reproductive cycles was determined in (A) normal lambs and in lambs with denervated pineal glands that (B) received no further treatment and (C) received a nightly infusion of melatonin. Broken lines indicate artificial photoperiod treatment, and each line shows the data for an individual female with large blocks representing normal luteal phases and small blocks short luteal phases. Vertical arrow denotes denervation of pineal gland by bilateral superior cervical ganglionectomy; the nightly duration of melatonin infusion, either 9 hours to mimic long days (15L:9D) or 15 hours to mimic short days (9L:15D) is shown in the horizontal bar in C. Note, the delay in puberty in lambs with denervated pineal glands was prevented by exogenous melatonin. Redrawn from Yellon and Foster (1986).

onectomy in artifical short days. The infusion was then changed to the long-day pattern (9 hour melatonin) for 5 weeks, after which the short-day infusion patterns were continued (Fig. 13C). Replacement of the long-day, short-day melatonin sequence prevented the delay in puberty in 3 of 4 lambs that should have occurred in continuous short days or in the absence of normal pineal function. Thus, as in the adult, melatonin codes photoperiod for timing of reproductive phenomena.

Melatonin has also been administered chronically by means of implants to pineal-intact lambs in natural photoperiod. This simpler approach to providing exogenous melatonin is pharmacologic because of the circadian nature of melatonin secretion which is not mimicked by a continuous release device. On the other hand, Lincoln and Ebling (1985) have reported that continuous administration of melatonin produces the same effects as short days in both young and fully mature rams. Female lambs implanted with melatonin (subcutaneously) at 3–4 weeks of age exhibit delayed puberty (Kennaway and Gilmore, 1984). If chronic administration of melatonin mimics short days, then the previous 3–4 weeks of natural long days during spring before treatment began should have satisfied the long-day requirement for induction of puberty under short days. Obviously they do not in view of the delayed puberty in lambs with such a melatonin treatment. This agrees with our own findings that lambs raised in long days during the first 4 weeks after birth exhibit delayed puberty in subsequent short days (Fig. 14C), despite the fact that only 1 week of long days is necessary when they are older (Fig. 14B). Even as many as 10 weeks of long days immediately after birth did not consistently induce puberty in short-day reared lambs (Fig. 14D). An explanation for this finding is provided in the next section, namely that the lamb may not be photoperiodic during the neonatal period. Finally, it is noteworthy that chronic administration of melatonin (intravaginal implant) beginning at 19 weeks of age advanced puberty several weeks in pineal-intact female lambs raised outdoors (Nowak and Rodway, 1985). Clearly, a great deal remains to be learned from such approaches about the age-dependent response to melatonin and the timing of puberty.

C. ONTOGENY OF MELATONIN SECRETION

The foregoing results of melatonin administration to the young female sheep indicate that we must identify periods of differential sensitivity to melatonin during development, with regard to timing sexual maturation. Certainly, there is no evidence to support the notion that melatonin is an inhibitor of gonadotropin secretion in the lamb as has been proposed for children. This hypothesis derives largely from an observed decrease in

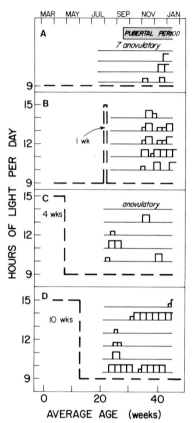

FIG. 14. Effects of early exposure to long days on the initiation of reproductive cycles. Spring-born lambs were raised in (A) artificial short days (9L:15D) entirely or were treated with long days (15L:9D) during the (B) middle of postnatal development and (C and D) during the neonatal period. Broken lines indicate photoperiod. Large blocks designate reproductive cycles with normal luteal phases and small blocks cycles with short luteal phases; each horizontal line shows data for an individual lamb except for (A) which indicates 7 of 10 lambs remained anovulatory (see Fig. 7 for examples of progesterone profiles). Note, exposure to long days shortly after birth is largely ineffective in inducing puberty at the normal age during subsequent short days. Redrawn from Yellon and Foster (1985).

nighttime melatonin levels in children, and thus, an inverse relationship between melatonin and the well-documented pubertal increase in nocturnal LH concentrations in the human (Waldhauser et al., 1984). However, other workers have pointed out that the decline in nighttime melatonin concentrations is not exclusively related to puberty; in fact, there appears to be a continuous decline in nighttime melatonin levels throughout life, from infancy to senescence (Attanasio et al., 1985). Finally, a puberty-associated increase in amplitudes of daytime melatonin secretory episodes has been reported along with the suggestion that melatonin may be stimulated by the higher levels of sex steroids that occur during sexual maturation (Penny, 1985). The discordance among these findings, as well as others reviewed by the aforementioned studies, lead to the conclusion that consensus has not been achieved on the role, if any, for melatonin as a regulator of human puberty. Penny (1985) points out a variety of possible causes for differences between the results of these studies, including the episodic nature of melatonin secretion, age, and method of sampling; an additional consideration may relate to assay methodology.

In the lamb, melatonin has been detected in the pineal gland and circulation shortly before birth (Kennaway *et al.*, 1977). Further, as indicated earlier, the prepubertal lamb can adjust its pattern of daily melatonin secretion to conform to the photoperiod (16–20 weeks of age, Fig. 11). To examine the ontogeny of the melatonin rhythm, we have monitored at early ages 24-hour patterns of circulating melatonin in young females raised in artificial or natural photoperiods. In lambs only experiencing artificial short days (9L:15D), an adult-like pattern of melatonin secretion was evident from 10 weeks of age onward; the melatonin rise was restricted to the dark period (Fig. 15). We have some evidence that before 10 weeks melatonin concentrations can be increased periodically during the day, as well as night (data not shown; Yellon and Foster, 1986b), suggesting that at very young ages, the melatonin rhythm may not be entrained to the light/dark cycle. Alternatively, the pattern of melatonin secretion may not be temporally organized in a circadian rhythm. The failure of the very young lamb to confine strictly its increased melatonin secretion to the dark period may provide an explanation for the ineffectiveness of the early placement of long days (first 5–10 weeks) to induce puberty in short-day reared lambs (see Fig. 14C and D). At very young ages, the sheep simply may not be able to transduce photoperiod cues into an appropriate melatonin rhythm; hence, the requisite long-day melatonin pattern has not been experienced before the short-day pattern. Finally, it must be noted that no decrease with age in amplitude of the nocturnal melatonin rise has been observed in our studies conducted in either natural or artificial photoperiods. Rodway *et al.* (1985) also failed to find a decrease in the nighttime level of melatonin in developing lambs as has been reported for the human (see earlier discussion).

The above findings lead us to a much different view for how melatonin times puberty in the lamb relative to that held by some workers for its function in regulating human sexual maturation (see above discussion). Rather than assigning a progonadal or antigonadal role to the pineal and its secretion of melatonin during sexual maturation, we consider that the

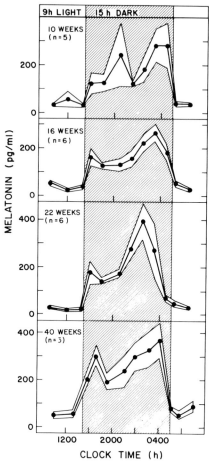

FIG. 15. Ontogeny of melatonin secretion. Twenty-four hour patterns of circulating melatonin (means ± SEM) were determined at various ages in spring-born female lambs raised in artificial short days (9L:15D). The shaded area designates the dark phase of the photoperiod. Note, after 10 weeks of age duration of the melatonin rise is proportional to the dark phase of the photoperiod. From S. M. Yellon and D. L. Foster (unpublished).

pineal transduces photic cues into a hormonal signal, melatonin, and that it is the duration of the melatonin rise which provides information about the length of the dark phase of the day. Moreover, there is no evidence that the melatonin rhythm is perturbed in lambs with photoperiodically delayed puberty. The nocturnal melatonin rise in postpubertal females at 40 weeks in natural photoperiod was found to be remarkably similar to that in short-day females at 40 weeks of age which had not yet begun reproductive cycles (Foster et al., 1985a). This suggests that merely the production of a short-day melatonin pattern does not necessarily ensure that puberty will occur at the normal age. Moreover, it raises the possibility that the target tissue of melatonin action may not be responsive to the short-day melatonin pattern until it has experienced a minimal duration (1 week or less) of the long-day pattern. These and other hypotheses relating to the identity of the target issue and its programming by melatonin remain to be tested in the young lamb.

V. Hypothalamic–Pituitary Mechanisms Regulating Puberty

The information accumulated thus far indicates that, with the exception of the early postnatal period, the sheep has developed pathways to transduce photic cues into hormonal signals long before puberty, and it uses day length as an important environmental cue in the timing of the first reproductive cycle. To better understand how external and internal cues might govern the transition into adulthood, we now consider the neuroendocrinology of puberty (Fig. 16).

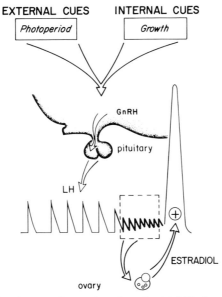

FIG. 16. Internal and external cues time the pubertal follicular phase initiated by the increase in LH pulse frequency (within dashed square). The ovary and the LH surge mechanism are capable of function before puberty (see text for evidence). Redrawn from Foster et al. (1985b).

The simplest view of puberty in the female is that it is the completion of the first follicular phase that successfully results in ovulation. Whether fertility will result is a more complex question, for as will be discussed below, initiation of ovulation may be associated with luteal phase defects and may not be accompanied by behavioral sexual receptivity. The sequence of endocrine events that lead to ovulation during the 2–3 day follicular phase of the mature sheep are (1) an increase in frequency of LH pulses to hourly or faster (ca. 30 minutes), (2) a sustained rise in estradiol, and (3) a preovulatory gonadotropin surge (Fig. 16) (Karsch et al., 1984, for review).

Which of the follicular phase events cannot occur in the immature female? She remains anovulatory before puberty because of a failure of the preovulatory gonadotropin surge mechanism to function. This could be due to insensitivity to the stimulatory feedback action of estradiol. On the other hand, the ovary may not be able to produce the requisite estradiol rise to trigger the LH surge. Maybe the immature female cannot produce the high frequency LH pulses to drive the ovarian follicle to the preovulatory stage and increase estradiol production. As discussed below, our studies indicate that the system limiting onset of reproductive cyclicity in the developing lamb is the one responsible for generating high frequency LH pulses. Representative evidence for this hypothesis will be presented (see other reviews for more complete documentation—Foster and Ryan, 1979b, 1981, 1986; Foster et al., 1985a,b).

A. PREOVULATORY GONADOTROPIN SURGE SYSTEM

Preovulatory gonadotropin surges do not occur spontaneously in the immature lamb. This is not due to the inability of the surge mechanism to respond to the stimulatory feedback action of estradiol. Exogenous estradiol can readily induce LH surges long before puberty (Land et al., 1970; Squires et al., 1972; Foster and Karsch, 1975; Tran et al., 1979). As shown in Fig. 17, administration of estradiol produced LH surges of progressively increasing magnitude throughout sexual maturation. By 12–20 weeks of age, the magnitude of the LH surge was in the range of that capable of producing ovulation (K. D. Ryan and D. L. Foster, unpublished; also see Fig. 23 for LH surges during puberty). Moreover, when the response to estradiol positive feedback was determined at 20 weeks of age (ca. 10 weeks before expected puberty), the sensitivity was equal to that of the adult, and as little as a 2 pg/ml increment in circulating estradiol produced an LH surge (Foster, 1984). This represents an important species difference for in the rhesus monkey (Dierschke et al., 1974a; Terasawa et al., 1983) or human (Reiter et al., 1974; Kelch et al., 1973),

PUBERTY IN A SEASONAL BREEDER 353

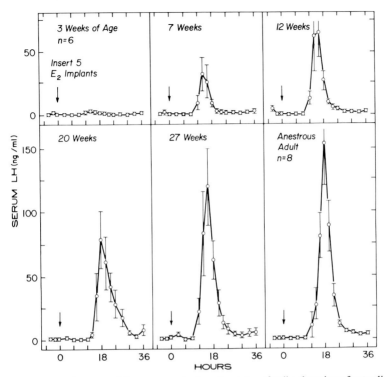

FIG. 17. Development of the response to the stimulatory feedback action of estradiol on LH secretion. Estradiol was administered by sc Silastic capsules for 96 hours at each age. Note, after 3 weeks of age estradiol induced an LH surge during each treatment period. From Foster and Karsch (1975).

estradiol alone cannot produce LH surges until rather late stages of development. Therefore, in primates, the immaturity of the preovulatory surge system limits onset of ovulation throughout much of prepubertal development. This may be due to a deficit in GnRH production (see Foster *et al.*, 1983, for discussion).

B. OVARY

If the preovulatory gonadotropin surge system of the lamb is exquisitely sensitive to the stimulatory feedback action of estradiol then it could be reasoned that there is a deficit in estradiol secretion. This raises questions about the maturity of the ovary. Certainly, the failure of ovulation to occur in response to the estradiol-induced gonadotropin surge (Fig. 17) is more of an example of neuroendocrine–ovarian asynchrony, rather than ovarian immaturity, because a preovulatory follicle had not been developed in preparation for the induced gonadotropin surge. When provided with exogenous gonadotropins, the ovary of the lamb exhibits a demonstrably advanced stage of maturity before puberty. It has the potential to produce large amounts of estradiol as evidenced by the sustained rise in circulating concentrations of the steroid to preovulatory levels after injection of pregnant mare serum gonadotropin at 25 weeks of age (Foster and Ryan, 1981). The ability to develop a preovulatory follicle in response to a massive gonadotropic stimulus has been used to practical advantage. Pharmacologic treatments involving the use of exogenous gonadotropins for induction of ovulation are used successfully to facilitate the breeding of young females before the age of natural puberty (Quirke *et al.*, 1983, for review).

Can the ovary of the immature lamb respond to physiologic concentrations of gonadotropin? More specifically, can the lamb ovary produce the concentrations of estradiol required to induce the preovulatory gonadotropin surge if stimulated with a pattern and level of LH characteristic of the follicular phase? To answer this critical question about ovarian competence, 18-week-old lambs were injected (iv) with purified ovine LH each hour for 48 hours, and patterns of circulating LH and estradiol were monitored (Foster *et al.*, 1984). The hourly frequency of injection produced correspondingly rapid increments in circulating LH, and the dose used resulted in peak LH concentrations indistinguishable from those occurring during endogenous secretory episodes (Fig. 18). This treatment stimulated a sustained production of estradiol, which in turn, resulted in the induction of a preovulatory gonadotropin surge and ovulation in many females. Such findings lead to the inference that the anovulatory condition of the prepubertal lamb, at least during the final one-third sexual maturation, is due to a deficiency in pulsatile LH secretion, rather than to inadequate development of the ovary or preovulatory surge system.

C. PULSATILE LH SECRETION

It could be predicted on the basis of the foregoing studies that high-frequency LH pulses for initiation of a follicular phase and induction of the sequence of endocrine events leading to ovulation would first occur during puberty. Certainly, slow LH pulses are characteristic of the immature lamb, and the interval between pulses is 2–3 hours or greater (Bindon and Turner, 1974; Foster *et al.*, 1975a,b; Echternkamp and Laster, 1976). Some evidence has been obtained for a marked increase in LH pulse frequency, but not amplitude, shortly before first ovulation (Huffman and Goodman, 1985; D. L. Foster and K. D. Ryan, unpublished).

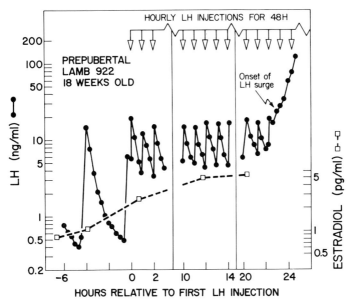

FIG. 18. Patterns of circulating estradiol and LH at various times before and during hourly injections (arrows) of purified ovine LH in a prepubertal female sheep. The dose per injection was 15.5 μg NIH-LH-S1 equivalents; blood samples were collected each 20 minutes. Note, a 5-fold rise in circulating estradiol occurred in response to exogenous LH and an LH surge beginning during hour 23 was initiated in response to the increase in estradiol. From Foster et al. (1985b).

Such an increase in pulsatile LH secretion is exemplified in Fig. 19. In this female, we were able to identify the first LH surge as it fortuitously occurred during a period of rapid blood sample collection. By at least 7 days before the first gonadotropin surge, the LH pulse frequency approached an hourly discharge rate. No increase in LH pulse amplitude was evident, and, in fact, in the preliminary report by Huffman and Goodman (1985) the amplitude of LH pulses was noted to decrease during the final week before first ovulation.

Perhaps puberty occurs when the neuroendocrine system of the lamb first becomes competent to produce hourly LH pulses. This is clearly not the case because such high frequency LH secretion is possible very early in life. Removal of the ovaries of the 2-week-old lamb results in the production of hourly LH pulses within a few weeks (Fig. 20C), much the same as those which occur during the follicular phase (Fig. 20A). Therefore, the developing hypothalamo–hypophyseal system is endowed with the *continuous* ability to produce high-frequency LH pulses from a very young age. Again, this represents an important species difference with

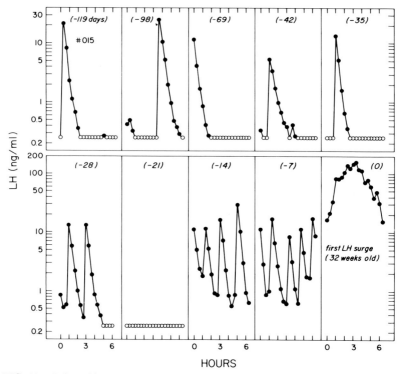

FIG. 19. Pubertal increase in LH pulse frequency in the sheep. Patterns of circulating LH were measured at various ages in blood samples collected every 20 minutes during a 6-hour period. Time 0 is the day of the first LH surge, and the day of each sample collection series relative to this time is indicated in parentheses. Values for LH concentrations are plotted on a logarithmic scale. Note an increase in frequency of LH pulses at 7 and 14 days before the LH surge. From Foster and Ryan (1986).

regard to the female primate as evidenced by the patterns and levels of circulating LH in females in which the ovary is nonfunctional or has been surgically removed (Winter and Faiman, 1972; Dierschke et al., 1974b; Conte et al., 1975; Foster et al., 1983; Terasawa et al., 1983). After a period of hyperactivity during the neonatal period, the system governing tonic LH secretion in the primate female becomes quiescent. The hiatus in gonadotropin secretion extends throughout much of the prepubertal period, and it is not until relatively late in maturation does the potential for high LH secretion return in the developing child or rhesus monkey.

If the ovary of the young lamb is capable of responding to high-frequency LH pulses by developing a preovulatory follicle, and if the neuroendocrine system is capable of producing such a pattern of LH early in

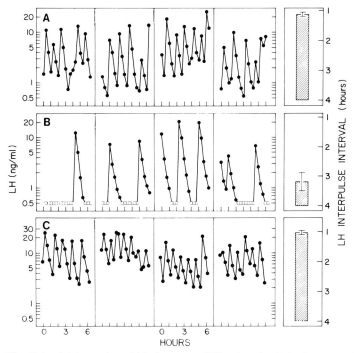

FIG. 20. Potential to produce high frequency LH pulses is attained at an early age. Patterns of circulating LH were determined in individual lambs (A) 30–40 weeks of age during a postpubertal follicular phase, (B) 25 weeks of age before puberty, and (C) 9 weeks of age after ovariectomy at 2 weeks of age. Right, mean (± SEM) interpulse interval for the 4 lambs of each group. Blood samples were collected at 20 minute intervals during each 6-hour period. Values for LH concentrations are plotted on a logarithmic scale. Note, removal of the ovaries before puberty increases LH pulse rate to a high frequency. Redrawn from Foster and Ryan (1981).

life, then how are LH pulses maintained at the slow frequency before puberty? A clue to the answer arises from earlier studies in the rat in which the *"Gonadostat Hypothesis"* emerged as a conceptual explanation for the pubertal increase in tonic LH secretion. The origins of this hypothesis date back more than 50 years, and in the interim, it has been refined by many investigators (Andrews *et al.*, 1981, for brief review). According to this hypothesis, low concentrations of circulating gonadotropins are maintained in the immature female because the hypothalamic system governing tonic gonadotropin secretion, the "gonadostat," is exquisitely sensitive to the inhibitory feedback action of ovarian steroids. As sexual maturation advances, the sensitivity to steroid negative feedback decreases, and gonadotropin secretion increases to stimulate ovarian follicular development in an adult manner. This concept of "resetting the gonadostat" may be applied to the developing female lamb because levels of steroid that are able to suppress LH secretion before puberty are no longer effective in this regard afterward. In the experimental model used, the ovariectomized lamb, estradiol was maintained chronically within physiologic limits for several months by a Silastic implant (Fig. 21B). This steroid replacement treatment maintained circulating LH at low concentrations in agonadal females until the time when ovulations began in ovarian-intact lambs (Fig. 21A). In more detailed studies of this phenomenon, we have determined that low LH concentrations in estradiol-treated ovariectomized lambs during the prepubertal period reflect slow frequency pulses, whereas during the pubertal period the high LH concentrations reflect high-frequency pulses (Foster and Olster, 1985; D. L. Foster, unpublished).

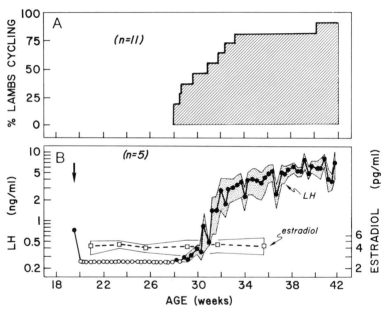

FIG. 21. Response to estradiol inhibition of tonic LH secretion decreases during puberty in the spring-born lamb. (A) Onset of reproductive cycles in ovarian-intact lambs based on the appearance of luteal phase concentrations of progesterone. (B) Mean (± SEM) concentrations of circulating LH and estradiol in ovariectomized lambs bearing sc Silastic implants of estradiol. Hormone concentrations are plotted in a logarithmic scale. Arrow indicates time of ovariectomy and the beginning of estradiol treatment. Note, inhibitory action of estradiol is much less after the age when reproductive cycles begin. Redrawn from Foster and Ryan (1979a).

D. HYPOTHESIS FOR INITIATION OF REPRODUCTIVE CYCLES

Many of the foregoing observations and considerations of the pubertal process in the female sheep can be integrated and summarized in the form of a working hypothesis for the initiation of reproductive cycles (Fig. 22). The lamb is endowed with the ability to produce high-frequency LH pulses from an early age, but they are maintained at a slow frequency before puberty. Rapid LH pulses are not normally produced because the system governing tonic LH secretion is hypersensitive to the small amounts of LH-stimulated estradiol secretion from the ovaries. Thus, the preovulatory endocrine sequence cannot be initiated, and the gonadotropin surge system remains inactive. The transition into adulthood occurs

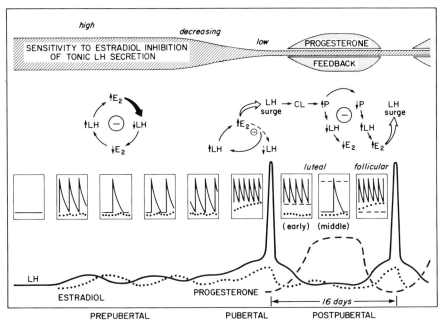

FIG. 22. Schematic hypothesis for endocrine events during the transition into adulthood in the female lamb. The hypothesis is based upon a decrease in sensitivity to estradiol inhibition of tonic LH secretion. Width of crosshatched area at top depicts degree of sensitivity to estradiol negative feedback; after the first ovulation, progesterone assumes the role of a feedback regulator of LH secretion. Insets illustrate detailed patterns of hormone secretion during a 6-hour period during the various stages of the pubertal transition. LH, solid line; E_2, estradiol, dotted line; P, progesterone, broken line;—, negative feedback; CL, corpus luteum. See text for details. From Foster and Ryan (1986).

when sensitivity to estradiol feedback inhibition decreases sufficiently to allow the expression of the requisite high-frequency LH pulses. In concert with the reduction in sensitivity to negative feedback, estradiol becomes able to accelerate LH pulses to frequencies beyond those possible in the absence of steroids (Foster and Olster, 1985; Foster et al., 1985b, for discussion). These high-frequency LH pulses then develop the ovarian follicle to the preovulatory stage, and induce the sustained rise in estradiol production. The dormant gonadotropin surge system is activated, and the first ovulation occurs.

E. EVENTS FOLLOWING THE FIRST LH SURGE

Although not discussed above, steroid feedback control of LH secretion shifts from estradiol to progesterone during the pubertal transition after the initiation of ovulation (Fig. 22). Progesterone serves as a potent inhibitor of LH secretion in the lamb (Foster and Karsch, 1976), and LH pulses slow in response to this steroid emanating from the newly formed corpus luteum (Foster et al., 1975b). During the luteal phase of the cycle, such infrequent pulses do not provide a sufficient stimulus to drive the follicle to the preovulatory stage, and hence for its secretion of estradiol to levels that trigger the gonadotropin surge mechanism. Following luteal regression and the decline in progesterone secretion, high-frequency LH pulses once again occur to initiate the next 2–3 day follicular phase. Another postpubertal reproductive cycle begins. (See Karsch et al., 1984; and Martin, 1984, for details of neuroendocrine regulation of the estrous cycle of the ewe.)

Finally, as shown in Fig. 23, it is worthy of note that the first gonadotropin surge often produces a short luteal phase (Ryan and Foster, 1978; Berardinelli et al., 1980; Keisler et al., 1983), much the same as occurs in the adolescent rhesus monkey (Dierschke et al., 1974a; Foster, 1977). Whereas in primates repetitive short luteal phases may occur, the lamb normally exhibits fewer defective luteal phases during the pubertal period. Some never produce a short luteal phase under natural conditions (K. D. Ryan and D. L. Foster, unpublished). However, certain artificial photoperiod treatments yield multiple short luteal phases (Figs. 8 and 14), and therefore, extend the period of adolescent infertility. Whether the short luteal phase reflects luteinization of a follicle or a defective corpus luteum is not yet known (Berardinelli et al., 1980). We have suggested, based upon pretreatment of prepubertal lambs with progesterone before exogenous gonadotropin, that progesterone may play a key role in timing the LH surge relative to follicular development (Ryan and Foster, 1978); a

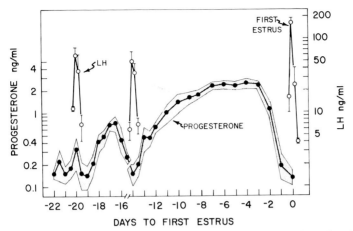

FIG. 23. Three LH surges occur during the transition into adulthood in the female lamb. Mean (± SEM) concentrations of LH, during each preovulatory surge, and progesterone are shown for the 22 days preceding first estrus for 4 females. Note, the overt sign of puberty, first estrus, does not typically occur until the time of the third LH surge. From K.D. Ryan and D.L. Foster (unpublished).

similar conclusion has been reached recently by Pearce et al. (1985) from studies conducted in the mature ewe. In the setting of the pubertal transition, we propose that the LH surge mechanism is exquisitely responsive to the positive feedback action of estradiol such that the first LH surge is triggered by early increases in circulating estradiol from the developing preovulatory follicle. The premature ovulation or follicular luteinization that would occur would result in only a meager production of progesterone (short luteal phase). The brief rise in circulating progesterone, by acting on the LH surge mechanism, would then delay the second LH surge until a later phase of estradiol secretion, and therefore, a more appropriate stage of follicular development for ovulation. A normal luteal phase would follow.

Regardless of whether a normal or short luteal phase occurs after the first gonadotropin surge, mating does not occur during the initial reproductive cycles because sexual receptivity is not manifest (Fig. 23). Progesterone priming of behavioral centers is necessary for estradiol to induce estrous behavior in the sheep (Dutt, 1953; Robinson, 1954), and until sufficient quantities of progesterone are produced during the pubertal luteal phases, estrus cannot occur in response to the sustained rise in estradiol during the first one or two follicular phases. In this regard, there are typically two "silent" ovulations before the overt sign of puberty, first estrus, is expressed at the time of the third ovulation.

VI. The GnRH Pulse Generator and the Integration of Internal and External Cues Timing Puberty

The production of high-frequency LH pulses is clearly the pivotal force driving the transition into adulthood in the lamb. Therefore, further understanding of the timing of sexual maturation in this seasonal breeder depends upon our ability to unravel how developmental and environmental signals modify the activity of the system generating pulsatile LH secretion by the pituitary. Again, many fundamental questions must be answered. Where is the oscillator regulating pulsatile LH secretion? What are the neuroanatomical characteristics of this system? Is this oscillator influenced directly by internal and external cues or are steroids required to modify its activity? What are the internal and external cues?

A THE GnRH PULSE GENERATOR

In recent years interest in the mechanism governing pulsatile LH secretion has heightened. Not only has it been implicated in regulating the tempo of sexual maturation, it is also considered a key component in the regulation of the estrous or menstrual cycle, spermatogenesis, and seasonal breeding (Knobil, 1980; Lincoln and Short, 1980; Karsch et al., 1984). A rather simple principle is emerging, namely that rapid LH pulses initiate or enhance reproductive activity, and conversely, gonadal function is reduced or fails entirely in the presence of only slow LH pulses. Several lines of evidence support the hypothesis that the varying patterns of LH (and FSH) secretion are produced in response to the varying activity of the GnRH "pulse generator," a hypothalamic oscillator. In the mature female sheep, each LH pulse is associated with a GnRH pulse in hypothalamic portal blood (Clarke and Cummins, 1982) or in perfusate from the median eminence (Levine et al., 1982). Administration of antisera against GnRH to mature male sheep prevents pulsatile LH secretion (Lincoln and Short, 1980). Finally, LH pulses occur simultaneously with volleys of electrical activity recorded from multiple cells in the arcuate region of the rat (Kawakami et al., 1982) and in the medial basal hypothalamus of the monkey (Wilson et al., 1984). Additional evidence for a GnRH pulse generator and further consideration of its properties have been reviewed by Lincoln et al. (1985).

Based upon the work conducted in the adult and the patterns of LH observed in the lamb, there is little reason to suspect that the pubertal increase in LH pulse frequency in the young female sheep reflects anything other than an increase in frequency of the GnRH pulse generator. Whether this inference holds for other species is unclear, for in the female rat, the amplitude of LH pulses, rather than frequency, appears to increase during puberty (Urbanski and Ojeda, 1985). Moreover, the release of GnRH *in vitro* by the hypothalamus of the rat during the final stages of development remains relatively constant, about one discharge per 30–40 minutes which is similar to the LH pulse frequency *in vivo* during that period of life (Bourguignon and Franchimont, 1984). This fascinating observation of an invariant hypothalamic GnRH rhythm and its implications for regulation of the timing of sexual maturity in the rat are discussed by Ojeda *et al.* (this volume).

The pubertal increase in LH pulses is occasioned by a lessening of negative feedback control of the GnRH pulse generator. This tenet is well founded because a dose of estradiol that completely inhibits tonic LH secretion before the time of puberty has little, if any, suppressive action in the postpubertal lamb (Fig. 21). The observed differential responsiveness to estradiol inhibition of LH secretion, however, does not constitute a mechanistic explanation for the decrease in sensitivity to the negative feedback of estrogen. At least two can be readily proposed to account for the gonadostat hypothesis in greater detail. The simplest is that the GnRH pulse generator increases its inherent activity during puberty and that more estrogen is required to decrease its pulsatile secretion. An increase in synaptogenesis (Matsumoto, 1976; Ruf, 1982) or alteration in GnRH cytoarchitecture (Wray and Hoffman-Small, 1984) could provide a morphologic basis for this fundamental change. A more complicated explanation for the gonadostat hypothesis has been put forth by Bhanot and Wilkins (1983) on the basis of studies conducted with opiate antagonists in the developing rat. They propose that the inhibitory effects of opiate peptides on LH secretion gradually lessens with age. In the lamb, a limited study has provided suggestive evidence that opiates may be involved in the control of LH secretion (Mathews and Murdoch, 1984); however, whether sensitivity to opiate inhibition of the GnRH pulse generator decreases with sexual maturation remains to be determined for the sheep. Lincoln *et al.* (1985) in these proceedings last year considered that, indeed, the various principles learned about opiate inhibition of LH secretion might serve to further greatly our understanding of set-point of the GnRH pulse generator to steroid negative feedback.

B. MODULATION OF THE GnRH PULSE GENERATOR BY PHOTOPERIOD

The finding that photoperiod is a primary *seasonal* determinant timing puberty may now be considered in the context of the GnRH pulse generator (Fig. 24). Our current working hypothesis is that photoperiod modu-

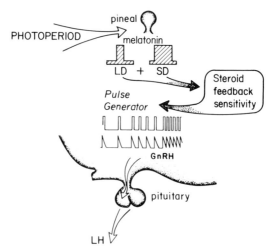

FIG. 24. Model for photoperiod regulation of the GnRH pulse generator in the developing female lamb. Widths of the melatonin bars represent the durations of the nocturnal rises in melatonin during long days and short days (LD and SD, respectively). See text for details. Redrawn from Foster *et al.* (1985b).

lates GnRH pulse generator activity through changing its sensitivity to inhibitory steroid feedback. There is no direct photoperiodic drive to the GnRH pulse generator in the absence of steroids. Support for this hypothesis is presented below.

Photoperiod alters sensitivity to steroid feedback inhibition as evidenced by the age of the decrease in response to steroid negative feedback in lambs born during different times of the natural photocycle (Fig. 25). As discussed earlier, spring-born lambs attain puberty during decreasing daylength at about 30 weeks of age; however, autumn-born lambs at that age are experiencing increasing daylength, and they delay puberty for an additional 20 weeks until day length has decreased (Fig. 25A). This prolongation of the prepubertal period is due to a delay in the decrease in response to estrogen feedback inhibition. The autumn-born lamb remains hypersensitive to estradiol for 20 weeks longer than the spring-born lamb, and therefore, the pubertal increase in circulating LH is delayed (Fig. 25B). When autumn-born lambs are raised under artificial photoperiod simulating the annual day length changes they would have experienced had they been born in spring, the timing of the decrease in estradiol feedback response is returned to normal (Foster, 1981). These findings prompt the suggestion that long days maintain high sensitivity of the GnRH pulse generator to estradiol inhibition. Furthermore, after the

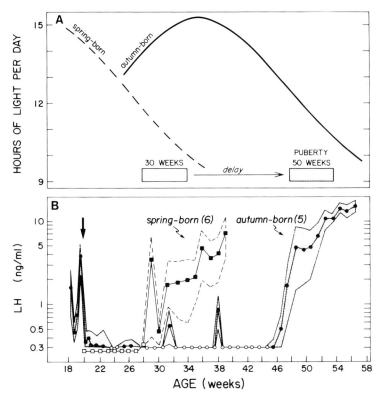

FIG. 25. Photoperiod modulates response to estradiol inhibition of tonic LH secretion in the lamb. (A) Onset of reproductive cycles in spring-born and autumn-born lambs (puberty, rectangle). (B) Mean ± SEM concentrations of circulating LH in spring-born and autumn-born lambs that were ovariectomized (arrow) and treated with low levels of estradiol by sc Silastic implant. Note, prolonged period of high responsiveness to estradiol inhibition of LH secretion in autumn-born lamb during long days delays initiation of reproductive cycles. See Fig. 6 for evidence that photoperiod alters the timing of puberty in the autumn-born lamb. Redrawn from Foster and Ryan (1986).

appropriate exposure to long days, short days lead to a decrease in steroid feedback sensitivity. Indeed, when lambs experience long days only for a brief period after birth (ca. 10 weeks), the system responsive to steroid feedback inhibition of LH secretion is perturbed during subsequent short days (Foster, 1983). Periods of hypersensitivity are interspersed with those of hyposensitivity leading to irregular cycles and an inordinate number of short luteal phases. Furthermore, some reproductive cycles may even occur during increasing day lengths when sensitivity to negative feedback is reduced transiently during an inappropriate time of the year (anestrous season). The negative feedback control system does not seem to be permanently impaired after experiencing inappropriate photoperiods during development because exposure to a natural photocycle during the second year after birth results in the predicted reduction in estradiol inhibition of LH secretion and normal onset of the breeding season (Foster, 1983).

Steroid-independent control of the GnRH pulse generator of the developing lamb has received little attention. Our limited studies have not yet revealed any direct drive of photoperiod on tonic LH secretion in the absence of gonadal steroids. In this regard, the prolonged period of hypersensitivity to estradiol in the autumn-born lamb (Fig. 25) does not appear to result from a reduced potential to secrete LH. In the absence of steroid feedback, LH concentrations (determined biweekly) in ovariectomized autumn-born lambs are not reduced relative to those of spring-born females without ovaries (see Fig. 7 in Foster *et al.*, 1985b). Even when LH pulses, rather than infrequent measurements of circulating LH are made, there is no evidence for a direct photoperiod drive to the GnRH pulse generator. This is shown in Fig. 26B where the LH pulse frequency was determined in lambs without ovaries during an artificial photoperiod treatment that is known to induce puberty at the normal age (Fig. 26A). No additional thrust to pulsatile LH secretion occurred during the pubertal period and LH pulse frequency remained constant. These composite findings force us to contend at present that photoperiod regulation of the GnRH pulse generator of the maturing female sheep is solely through modification of its sensitivity to estradiol inhibitory feedback. If this is the case, then the photoneuroendocrine control of the lamb remains incompletely developed even during the early postpubertal period. In the fully mature ewe without ovaries, the frequency of LH pulses is inversely proportional to day length; furthermore, LH pulse frequency can be set by artificial light with long photoperiods producing slow pulses and short photoperiods resulting in rapid pulses (Robinson *et al.*, 1985b).

Based upon studies conducted in the adult female sheep (Karsch *et al.*, 1984), we predict that melatonin mediates the photoperiod-induced decrease in sensitivity to estradiol inhibition of LH secretion in the lamb. In the pinealectomized ewe, perhaps the most convincing experiments have been those in which the replacement of melatonin and length of day were in the opposite direction. In one such experiment, short-day melatonin patterns were infused during artificial long days (Yellon *et al.*, 1985). Despite the experimental mismatch of the inductive melatonin pattern with the inhibitory photoperiod, sensitivity to estradiol negative feedback

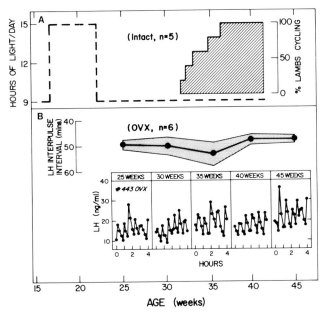

FIG. 26. Failure of photoperiod to regulate LH secretion in the absence of steroid negative feedback in the lamb. (A) Initiation of reproductive cycles, based upon progesterone concentrations, in ovarian intact lambs raised in a sequence of short days, long days and short days (broken line shows photoperiod). (B) Mean (± SEM) LH interpulse interval in ovariectomized lambs raised in the same photoperiod as intact lambs; inset illustrates patterns of LH in an individual female. Samples were collected at 12 minute intervals for 4 hours. Note, LH pulse frequency remains constant in ovariectomized females during photoperiod-induced onset of reproductive cycles in intact females. Redrawn from Foster and Yellon (1985b).

decreased. Our unraveling of the development of this photoneuroendocrine mechanism will remain incompletely documented until similar information is obtained for the lamb. Finally, we do not have convincing evidence for a site of action of melatonin, even in the adult. Because the acute LH response to GnRH is not influenced by melatonin in the prepubertal lamb (Rodway and Swift, 1983) or in the mature ewe (Symons and Arendt, 1982; Robinson *et al.*, 1985a), its regulatory action is most likely within the central nervous system. Interestingly, a pituitary site of action for melatonin has been found in the developing rat during the early postnatal period (Martin and Sattler, 1979; Martin *et al.*, 1977, 1980). Clearly, identification of the system coupling melatonin and the GnRH pulse generator, including its ontogeny, represents the next chapter in the story of the timing of puberty by photoperiod.

C. MODULATION OF THE GnRH PULSE GENERATOR BY GROWTH AND NUTRITION

The emphasis of the discussion thus far has been placed on photoperiodic cues governing the initiation of reproductive cycles, but in a seasonal breeder, environmental cues are not the sole determinants timing puberty. There is also a growth requirement, much the same as in nonseasonal breeders. Our current working hypothesis is that growth-related cues modify the activity of the GnRH pulse generator both directly and through a steroid feedback mechanism (Fig. 27). Support for this hypothesis is presented below.

As shown earlier (Fig. 4), lambs raised on low nutrition remain anovulatory during the first breeding season because growth is retarded. Their failure to achieve puberty at the normal age is because they remain hypersensitive to estradiol inhibition. According to our hypothesis (Fig. 22), these growth-retarded females cannot express the high-frequency LH pulses required for development of a preovulatory follicle. Experimentally, it can be demonstrated that in the absence of the ovaries, small amounts of exogenous estradiol exert a potent negative feedback action, and LH remains suppressed during the time when puberty would occur had the lambs not been growth retarded (Fig. 28B). When such food-restricted females are fed *ad libitum,* sensitivity to estradiol inhibition is reduced and LH increases (Fig. 28B), reflecting the production of high-frequency pulses (Foster and Olster, 1985). This initiates puberty as evidenced by a similar time course in the onset of reproductive cycles in

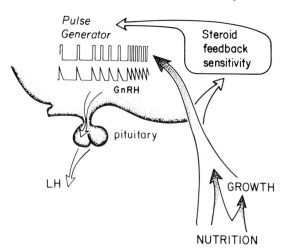

FIG. 27. Model for regulation of the GnRH pulse generator by nutrition and growth in the developing female lamb. Redrawn from Foster *et al.* (1985b).

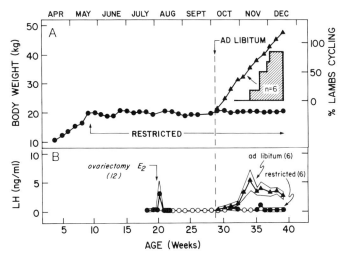

FIG. 28. Nutrition modulates response to estradiol inhibition of tonic LH secretion in the lamb. (A) Growth curves for lambs placed on a restricted diet at weaning (10 weeks of age); some remained on restricted nutrition until 28 weeks of age after which they were fed *ad libitum*. Histogram indicates time of onset of cycles during *ad libitum* feeding. (B) Mean (± SEM) concentrations of circulating LH in ovariectomized lambs treated with low levels of estradiol (E_2) by Silastic implant; their growth curves are shown in (A). Note, high response to estradiol inhibition during food restriction; the onset of reproductive cycles during *ad libitum* feeding occurs as a consequence of a reduction in response to estradiol negative feedback. Redrawn from Foster and Olster (1985).

ovarian-intact lambs that were first maintained on a restricted diet and then fed *ad libitum* (Fig. 28A). Therefore, nutritionally influenced growth cues alter sensitivity to estradiol inhibition of LH secretion.

Why is the undernourished lamb so responsive to estrogen feedback inhibition of tonic LH secretion? Perhaps it is because the inherent ability to produce LH pulses is reduced. Removal of steroid negative feedback by ovariectomy revealed that the postcastration LH rise was attenuated and that with prolonged undernutrition, the LH pulse frequency remained slow in the absence of the ovaries (Foster and Olster, 1985). This is exemplified in Fig. 29A for ovariectomized lambs raised on a food-restricted diet. Furthermore, such females respond quickly to an increased level of nutrition by increasing LH pulse frequency. In many lambs, an increase in rate of LH discharges can be detected within 48 hours after the start of feeding *ad libitum* (Fig. 29B). In contrast to its dramatic effects on LH pulse frequency, level of nutrition has no influence on the amplitude of LH pulses in the ovariectomized lamb (Foster and Olster, 1985). Moreover, administration of a low physiologic dose of GnRH (2 ng/kg) to these undernourished females with hypogonadotropism rapidly produces a

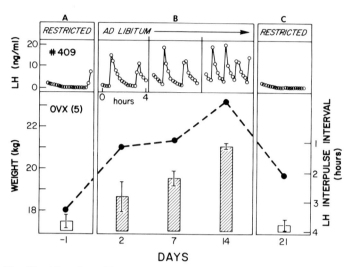

FIG. 29. Nutrition alters LH secretion in the absence of the ovaries in the lamb. LH pulse frequency was determined in chronically undernourished, ovariectomized lambs during (A) food restriction, (B) a 14-day period of *ad libitum* feeding, and (C) 7 days after return to the food-restricted diet. Top, individual. Bottom, mean body weight (broken line) and mean (± SEM, bars) LH interpulse intervals. The females were 37 weeks of age at the beginning of the study; they had been on low nutrition from 8 weeks of age and were ovariectomized at 31 weeks. Samples were obtained at 12-minute intervals during each sampling period. Note, slow LH pulse frequency during food restriction and rapid increase during *ad libitum* feeding. D.L. Foster and A.F. Micka (unpublished).

well-defined LH discharge of high amplitude (A. F. Micka and D. L. Foster, unpublished), implying that the deficiency in LH secretion is at the level of frequency modulation of the GnRH pulse generator.

It could be proposed that the effects of nutrition on the GnRH pulse generator are indirect and that they merely alter its response to photoperiod. According to this argument the undernourished lamb may not be able to distinguish between photoperiods. This is not the case as determined from a study of steroid negative feedback control of LH secretion in nutritionally growth-retarded lambs subjected to different photoperiods (Fig. 30). One group of undernourished, estradiol-treated young females without ovaries was exposed to a block of long days before they were fed *ad libitum* under short days; the other group did not experience long days, but was otherwise treated similarly. During the period of catch-up growth, the ability of estradiol to suppress LH depended upon the photoperiod that was experienced when the females had been undernourished. Those maintained in only short days continued to be hyperresponsive to exogenous estradiol, and LH remained suppressed. However, females

PUBERTY IN A SEASONAL BREEDER

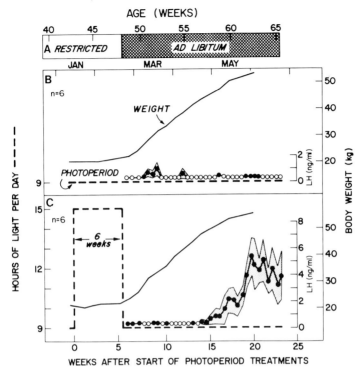

FIG. 30. Demonstration that undernourished lambs remain photoperiodic. Mean (± SEM) concentrations of circulating LH were monitored during (A) *ad libitum* feeding in ovariectomized, estradiol-treated females, (B) exposed to continuous short days (9L:15D) and (C) exposed to a 6-week block of long days (15L:9D) when undernourished and then to short days when level of nutrition was increased. Lambs were 42 weeks of age at the beginning of the study; they were ovariectomized at 20 weeks of age, and for this experiment, a small estradiol implant was inserted 5 days before *ad libitum* feeding began. Continuous line is weight; broken line indicates photoperiod. Note in (C) the increase in LH reflecting the decrease in response to estradiol inhibition induced by exposure to long days when they were undernourished. From Foster and Yellon (1985a).

that experienced the long-day block while undernourished decreased their sensitivity to estradiol inhibition during their growth spurt. This provides compelling evidence that the malnourished female lamb can keep track of day length; an appropriate response will be manifest only when she achieves the appropriate physiologic size for reduction of sensitivity to negative feedback.

Whether another mechanism(s), other than an inherently slow GnRH pulse generator, contributes to estradiol hypersensitivity in the undernourished condition is not yet known. It is extremely difficult to assess mechanisms of steroid feedback control when there is such a pronounced steroid-independent regulation of LH secretion. Regardless, it can now be concluded that of the two major influences on LH that have been studied, photoperiod and nutrition, the effects of the latter are more direct because they can alter the pattern of secretion in the absence of steroid feedback whereas photoperiod requires its presence (compare Figs. 26 and 29). A further difference is related to the time course required. Nutritional modulation of puberty is more rapid than photoperiod induction of reproductive cycles which requires several weeks for its manifestation (compare Figs. 9, 28, and 30).

Relatively few hypotheses have been put forth to explain how the developing hypothalamus monitors physiologic size and well being. Frisch (1984, 1985), largely on the basis of body composition estimates, considers that both absolute and relative amounts of fat may be used to govern the tempo of sexual maturation; the mechanisms proposed for this include adipose tissue regulation of sex steroid hormone concentrations and their metabolism. The observation of an increase in LH pulse frequency within 48 hours after increasing level of nutrition (Fig. 29) in the ovariectomized lamb would not support such a mechanism for the sheep. First, significant quantities of sex steroids would not be present in the absence of the ovaries, and second, significant fat deposition would not be expected to occur within such a short time period. Rather than having a deterministic role, we might consider that "fatness" per se merely reflects a state of metabolic well being or maturity. In the sheep, however, the percentage or absolute amount of carcass fat, protein, or water has not been found to be a good index of degree of sexual maturation (Moore *et al.*, 1985).

Some type of growth-related cue(s) must be used to time the increase in activity of the GnRH pulse generator. Such cues, for example, may be metabolic hormones or precursors or products of metabolic pathways they regulate. Steiner *et al.* (1983) and Cameron *et al.* (1985) have suggested that developmental changes in levels of insulin and certain amino acid precursors for neurotransmitter synthesis in the brain may influence the GnRH pulse generator. Furthermore, they raise the possibility that such cues may be used to reduce its activity during undernutrition. Alterations in thyroid function have recently been found to change LH pulse frequency in ram lambs (Chandrasekhar *et al.*, 1985), but no information in this regard is available for the developing ewe lamb. Finally, it should be pointed out that endogenous opioids may be part of the mechanism by which level of nutrition or growth modulates the GnRH pulse generator. This is based upon studies of underfed adult female rats with naloxone (Dyer *et al.*, 1985). Interestingly, the effects of opioids were on amplitude, rather than frequency regulation of LH secretion, although this type of regulation may be a species difference between the rat and sheep as pointed out in several places earlier in this paper. Nevertheless, the conclusion of Dyer et al. (1985) that fasting activates a pathway involving opioid inhibition of LH secretion raises a cautionary note about the validity of using undernutrition to study growth-related signals timing normal puberty. Perhaps, this approach mobilizes pathophysiologic mechanisms, ones that are used only during abnormal conditions associated with inadequate food intake. On the other hand, level of nutrition may well participate in the timing of sexual maturity in many seasonal breeders (Bronson, 1985), including the sheep. Some lambs are raised under inadequate nutrition in certain management conditions where competition is keen for limited food resources. Many of these females may not reach puberty during the first year after birth because they are undernourished and have not yet developed completely (Hammond, 1944; Hafez, 1952; Dyrmundsson and Lees, 1972).

D. INTEGRATED MODEL

The rather simple observations at the outset that to achieve puberty the developing lamb must attain an appropriate weight during a fixed period of the year (Fig. 4) have led to numerous questions about how this coordination of size and season is accomplished. In seeking solutions to these questions, we offer the scheme presented in Fig. 31. Central to this working hypothesis is the GnRH pulse generator which integrates information about the developing internal environment and the changing external environment. A sustained increase in GnRH pulse generator activity occurs when growth-related signals indicate that a sufficient physiologic size has been achieved and when the history of melatonin patterns reflects that the long-day, short-day photoperiod sequence has been experienced. Failure of either set of signals to occur does not allow GnRH pulse generator activity to increase, and the slow LH pulse frequency characteristic of the sexually immature condition is maintained. However, if both size and photoperiod history are perceived to be appropriate, the increase in GnRH activity increases the frequency of LH pulses to which the ovary is exposed. This drives the follicle to the preovulatory stage and initiates the neuroendocrine sequence of events culminating in first ovulation.

It is evident that we have only begun to put into place the pieces of a very complicated puzzle of how photoperiod and growth information is channeled into the brain for central processing in the developing sheep. Clearly, in other seasonal breeding species, nonphotoperiodic signals such as pheromones, plant hormones, and behavioral cues including

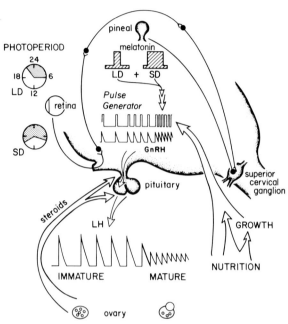

FIG. 31. Model for regulation of timing of puberty in the female lamb through modulation of the GnRH pulse generator by internal cues (growth, metabolic signals, melatonin) and external cues (photoperiod, nutrition). Shaded portions of 24-hour clocks represent the dark phases of the photoperiods; widths of the melatonin bars indicate relative durations of the nighttime melatonin rises. LD and SD, long day and short day, respectively. See text for details. From Foster et al. (1985b).

stress, can be used to provide a detailed picture about the physical and social environments. It is indeed intriguing to consider how this continuous barrage of detailed information is transformed first into neural signals and then into endocrine signals to synchronize the initiation of fertility in the young female with the seasonal breeding of the adult. Are the signals equal in their influence or are they hierarchical? If we can free ourselves of the notion that a specific "trigger" exists for puberty then we may proceed more rapidly in our conceptual understanding of the transition into adulthood. Finally, we continue to be intrigued with the question often raised about whether or not a seasonal breeder undergoes puberty annually (Foster and Ryan, 1979b; Karsch and Foster, 1981). Certainly, the common neuroendocrine feature underlying the onset of puberty and the onset of the breeding season is the increase in GnRH pulse generator activity. However, the set of signals timing this increase differs between the two transitional states. In the adult, external signals are used to phase the biologic rhythms of reproduction, whereas in the young, developmental signals must be monitored as well to initiate the rhythms.

ACKNOWLEDGMENTS

We are grateful to a number of colleagues, staff, and friends for their contributions to this research. Eric L. Bittman, Robert L. Goodman, Sandra J. Legan, James A. Lemons, and Jane E. Robinson are recognized for their conceptual and technical help. Douglas D. Doop expertly cared for the experimental lambs and assisted in surgery. Laboratory services were provided by Tovaghgol E. Adel, Janice A. Clayton, Erica K. Paslick, and Robert A. Drongowski; analytical reagents were prepared by Marjorie R. Hepburn and Mark D. Byrne. Hormone preparations and antisera were most generously provided by Gordon D. Niswender, Mark D. Rollag, Leo E. Reichert, Jr., and Harold Papkoff. We especially acknowledge the administrative and secretarial staff including Diane E. Belleba, Ruth M. Lum, Susan A. Bareis, and Marilyn D. Katz-Pek for their supporting services. The Reproductive Endocrinology Program, under the inspired leadership of A. Rees Midgley, Jr. provided the setting for critical evaluation of experimental design and interpretation. This work was made possible by funds from The University of Michigan through the Department of Obstetrics and Gynecology, the Rackham School of Graduate Studies, and Biomedical Research Council, from the Ford Foundation, and from the National Institutes of Child Health and Human Development.

REFERENCES

Andrews, W. W., Advis, J. P., and Ojeda, S. R. (1981). *Endocrinology* **109**, 2022.
Attanasio, A., Borrelli, P., and Gupta, D. (1985). *J. Clin. Endocrinol. Metab.* **61**, 388.
Berardinelli, J. G., Dailey, R. A., Butcher, R. L., and Inskeep, E. K. (1980). *Biol. Reprod.* **22**, 233.
Bhanot, R., and Wilkins, M. (1983). *Endocrinology* **113**, 596.
Bindon, B. M., and Turner, H. N. (1974). *J. Reprod. Fertil.* **39**, 85.
Bittman, E., Dempsey, R. J., and Karsch, F. J. (1983). *Endocrinology* **113**, 2276.
Bourguignon, J.-P., and Franchimont, P. (1984). *Endocrinology* **114**, 1941.
Bronson, F. H. (1985). *Biol. Reprod.* **32**, 1.
Cameron, J. L., Hanson, P. D., McNeill, T. H., Koerker, D. J., Clifton, D. K., Rogers, K. V., Bremner, W. J., and Steiner, R. A. (1985). *In* "Adolescence in Females" (C. Flamigni, S. Venturoli, and J. R. Givens, eds.), p. 59. Year Book Medical Publishers, Chicago.
Chandrasekhar, Y., D'Occhio, M. J., Holland, M. K., and Setchell, B. P. (1985). *Endocrinology* **117**, 1645.
Clarke, I. J., and Cummins, J. T. (1982). *Endocrinology* **111**, 1737.
Conte, F. A., Grumbach, M. M., and Kaplan, S. L. (1975). *J. Clin. Endocrinol. Metab.* **40**, 670.
Dierschke, D. J., Weiss, G., and Knobil, E. (1974a). *Endocrinology* **94**, 198.
Dierschke, D. J., Karsch, F. J., Weick, R. F., Weiss, G., Hotchkiss, J., and Knobil, E. (1974b). *In* "Control of the Onset of Puberty" (M. M. Grumbach, G. D. Grave, and F. E. Mayer, eds.), p. 104. Wiley, New York.
Dyrmundsson, O. R., and Lees, J. L. (1972). *J. Agric. Sci. Camb.* **78**, 39.
Dutt, R. H. (1953). *J. Anim. Sci.* **12**, 515.
Dyer, R. G., Mansfield, S., Cobet, H., and Dean, A. D. P. (1985). *J. Endocrinol.* **105**, 91.
Echternkamp, S. E., and Laster, D. B. (1976). *J. Anim. Sci.* **42**, 444.

Fitzgerald, J., and Butler, W. R. (1982). *Biol. Reprod.* **27**, 853.
Fletcher, I. C., and Geytenbeek, P. E. (1970). *Aust, J. Exp. Agric. Anim. Husb.* **10**, 267.
Foster, D. L. (1977). *Biol. Reprod.* **17**, 584.
Foster, D. L. (1981). *Biol. Reprod.* **25**, 85.
Foster, D. L. (1983). *Endocrinology* **112**, 111.
Foster, D. L. (1984). *Endocrinology* **115**, 1186.
Foster, D. L., and Karsch, F. J. (1975). *Endocrinology* **97**, 1205.
Foster, D. L., and Karsch, F. J. (1976). *Endocrinology* **99**, 1.
Foster, D. L., and Olster, D. H. (1985). *Endocrinology* **116**, 375.
Foster, D. L., and Ryan, K. D. (1979a). *Endocrinology* **105**, 896.
Foster, D., and Ryan, K. (1979b). *Ann. Biol. Anim. Biochem. Biophys.* **19**, 1369.
Foster, D. L., and Ryan, K. D. (1981). *J. Reprod. Fertil. Suppl.* **30**, 75.
Foster, D. L., and Ryan, K. D. (1986). *In* "Control of the Onset of Puberty II" (P. C. Sizonenko, M. L. Aubert, and M. M. Grumbach, eds.). Williams & Wilkins, Baltimore (in press).
Foster, D. L., and Yellon, S. M. (1985a). *J. Reprod. Fertil.* **75**, 203.
Foster, D. L., and Yellon, S. M. (1985b). *Biol. Reprod.* **32** (Suppl. 1), Abstr. 279.
Foster, D. L., Jaffe, R. B., and Niswender, G. D. (1975a). *Endocrinology* **96**, 15.
Foster, D. L., Lemons, J. A., Jaffe, R. B., and Niswender, G. D. (1975b). *Endocrinology* **97**, 985.
Foster, D. L., Rapisarda, J. J., Bergman, K. S., Lemons, J. A., Jaffe, R. B., Steiner, R. A., and Wolf, R. C. (1983). *In* "Neuroendocrine Aspects of Reproduction" (R. L. Norman, ed.), p. 149. Academic Press, New York.
Foster, D. L., Ryan, K. D., and Papkoff, H. (1984). *Endocrinology* **115**, 1179.
Foster, D. L., Olster, D. H., and Yellon, S. M. (1985a). *In* "Adolescence in Females" (C. Flamigni, S. Venturoli, and J. R. Givens, eds.), p. 1. Year Book Medical Publishers, Chicago.
Foster, D. L., Yellon, S. M., and Olster, D. H. (1985b). *J. Reprod. Fertil.* **75**, 327.
Frisch, R. E. (1984). *Biol. Rev.* **59**, 161.
Frisch, R. E. (1985). *In* "Adolescence in Females", (C. Flamigni, S. Venturoli, and J. R. Givens, eds.), p. 131. Year Book Medical Publishers, Chicago.
Hafez, E. S. E. (1952). *J. Agric. Sci.* **42**, 189.
Hammond, J. Jr. (1944). *J. Agric. Sci.* **34**, 97.
Huffman, L. J., and Goodman, R. L. (1985). *Biol. Reprod.* **32** (Suppl. 1), Abstr. 342.
Karsch, F. J., and Foster, D. L. (1981). *In* "Environmental Factors in Mammal Reproduction" (D. Gilmore and B. Cook, eds.), p. 30. Macmillan, London.
Karsch, F. J., Bittman, E. L., Foster, D. L., Goodman, R. L., Legan, S. J., and Robinson, J. E. (1984). *Recent Prog. Horm. Res.* **40**, 185.
Kawakami, M., Uemura T., and Hayashi, R. (1982). *Neuroendocrinology* **35**, 63.
Keisler, D. H., Inskeep, E. K., and Dailey, R. A. (1983). *J. Anim. Sci.* **57**, 150.
Kelch R. P., Kaplan, S. L., and Grumbach, M. M. (1973). *J. Clin. Endocrinol.* **36**, 424.
Kennaway, D. J., and Gilmore, T. A. (1984). *J. Reprod. Fertil.* **70**, 39.
Kennaway, D. J., Matthews, C. D., Seamark, R. F., Phillipou, G., and Schilthuis, M. (1977). *J. Steroid Biochem.* **8**, 559.
Kennaway, D. J., Gilmore, T. A., and Dunstan, E. A. (1985). *J. Reprod. Fertil.* **74**, 119.
Knobil, E. (1980). *Recent Prog. Horm. Res.* **36**, 53.
Land, R. B., Thimonier, J., and Pelletier, J. (1970). *C. R. Acad. Sci. [D] (Paris)* **271**, 1549.
Legan, S. J., and Karsch, F. J. (1983). *Biol. Reprod.* **29**, 316.
Legan, S. J., and Winans, S. S. (1981). *Gen. Comp. Endocrinol.* **45**, 316.

Levine, J. E., Pau K.-Y. F., Ramirez, V. D., and Jackson, G. L. (1982). *Endocrinology* **111**, 1449.
Lincoln, D. W., Fraser, H. M., Lincoln, G. A., Martin, G. B., and McNeilly, A. S. (1985). *Recent Prog. Horm. Res.* **41**, 369.
Lincoln, G. A., and Ebling, F. J. P. (1985). *J. Reprod. Fertil.* **73**, 241.
Lincoln, G. A., and Short, R. V. (1980). *Recent Prog. Horm. Res.* **36**, 1.
Mallampati, R. S., Pope, A. L., and Casida, L. E. (1971). *J. Anim. Sci.* **33**, 1278.
Martin, G. B. (1984). *Biol. Rev.* **59**, 1.
Martin, J. E., and Sattler, C. (1979). *Endocrinology* **105**, 1007.
Martin, J. E., Engel, J. N., and Klein, D. C. (1977). *Endocrinology* **100**, 675.
Martin, J. E., McKellar, S., and Klein, D. C. (1980). *Neuroendocrinology* **31**, 13.
Mathews, M. V., and Murdoch, W. J. (1984). *Domest. Anim. Endocrinol.* **1**, 167.
Matthews, C. D., Kennaway, D. J., and Seamark, R. F. (1981). *In* "Pineal Function" (C. D. Matthews and R. F. Seamark, eds.), p. 137. Elsevier, Amsterdam.
Matsumoto, A. (1976). *Cell Tissue Res.* **169**, 143.
Moore, R. W., Bass, J. J., Winn, G. W., and Hockey, H.-U. P. (1985). *J. Reprod. Fertil.* **74**, 433.
Nowak, R., and Rodway, R. G. (1985). *J. Reprod. Fertil.* **74**, 287.
Pearce, D. T., Martin, G. B., and Oldham, C. M. (1985). *J. Reprod. Fertil.* **75**, 79.
Penny, R. (1985). *J. Clin. Endocrinol. Metab.* **560**, 751.
Quirke, J. F., Adams, T. E., and Hanrahan, J. P. (1983). *In* "Sheep Production" (W. Haresign, ed.), p. 409. Butterworths, London.
Reiter, E. O., Kulin, H. E., and Hamwood, S. M. (1974). *Pediatr. Res.* **8**, 740.
Robinson, J. E., Kaynard, A. H., and Karsch, F. J. (1985a). *Annu. Meet., 67th Endocr. Soc.* Abstr. No. 1039.
Robinson, J. E., Radford, H. M., and Karsch, F. J. (1985b). *Biol. Reprod.* **33**, 324.
Robinson, T. J. (1954). *Endocrinology* **55**, 403.
Rodway, R. G., and Swift, A. D. (1983). *Horm. Metab. Res.* **15**, 349.
Rodway, R. G., Swift, A. D., Nowak, R., Smith, J. A., and Padwick, D. (1985). *Anim. Reprod. Sci.* **8**, 241.
Ruf, K. B. (1982). *Front. Horm. Res.* **10**, 65.
Ryan, K. D., and Foster, D. L. (1978). *Biol. Reprod.* **18** (Suppl. 1), Abstr. 118.
Squires, E. L., Scaramuzzi, R. J., Caldwell, B. V., and Inskeep, E. K. (1972). *J. Anim. Sci.* **34**, 614.
Steiner, R. A., Cameron, J. L., McNeill, T. H., Clifton, D. K., and Bremner, W. J. (1983). *In* "Neuroendocrine Aspects of Reproduction" (R. L. Norman, ed.). p. 183. Academic Press, New York.
Symons, A. M., and Arendt, A. (1982). *J. Reprod. Fertil.* **64**, 103.
Terasawa, E., Nass T. E., Yeoman, R. R., Loose, M. D., and Schultz, N. J. (1983). *In* "Neuroendocrine Aspects of Reproduction" (R. L. Norman, ed.), p. 149. Academic Press, New York.
Tran, C. T., Edey, T. N., and Findlay, J. K. (1979). *Aust. J. Biol. Sci.* **32**, 463.
Urbanski, H. F., and Ojeda, S. R. (1985). *Endocrinology* **117**, 644.
Waldhauser, F., Frisch, H., Waldhauser, M., Weiszenbacher, G., Zeithuber, U., and Wurtman, R. J. (1984). *Lancet* **1**, 362.
Watson, R. H., and Gamble, L. C. (1961). *Aust. J. Agric. Res.* **12**, 124.
Wiggins, E. L., Barker, H. B., and Miller, W. W. (1970). *J. Anim. Sci.* **30**, 405.
Wilson, R. C., Kesner, J. S., Kaufman, J-M., Uemura, T., and Knobil, E. (1984). *Neuroendocrinology* **39**, 256.

Winter, J. S. D., and Faiman, C. (1972). *J. Clin. Endocrinol. Metab.* **35,** 561.
Wray, S., and Hoffman-Small, G. E. (1984). *J. Steroid Biochem.* **20,** Abstr. B15.
Yellon, S. M., and Foster, D. L. (1985). *Endocrinology* **116,** 2090.
Yellon, S. M., and Foster, D. L. (1986). *Endocrinology* (in press).
Yellon, S. M., Bittman, E. L., Lehman, M. N., Olster, D. H., Robinson, J. E., and Karsch, F. J. (1985). *Biol. Reprod.* **32,** 523.

DISCUSSION

H. Kulin. I should like to provide some comparative data in man which bear on the escape from negative feedback that you discussed in sheep. In order to compare the level of feedback interaction between gonadal steroids and gonadotropins in children and adults we must assess responses in subjects who have similar levels of gonadotropins (and, presumably, similar central GnRH drive). Results in patients with gonadal disgenesis provide an open loop feedback system similar to your own model and do allow comparative studies at different ages.

Shown in Fig. 32 are data obtained from 4 girls between the ages of 12 and 16 with bone ages of 10 to 13. These patients were given 0.3 mg of Premarin daily to initiate puberty artificially and the drug was continued throughout the study period of 24–30 months. Measurements of urine gonadotropins as assessed by radioimmunoassay decreased from adult castrate levels into the prepubertal range over several months and then escape occurred just as you found in the sheep. The point I wish to emphasize by means of these data is that the

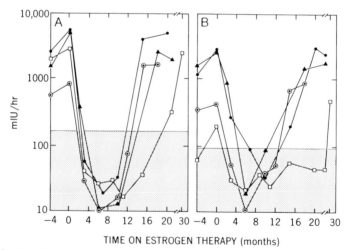

FIG. 32. Time course of suppression and escape of urinary gonadotropins in 4 patients with gonadal dysgenesis given 0.3 mg/day of conjugated estrogen continuously to initiate the physical changes of puberty. (A) FSH; (B) LH. Shaded areas indicate the limits of the normal prepubertal range. Note the log scale on the ordinate. From Kulin, H.E., in "Adolescence in Females:Endocrinological Development and Implications on Reproductive Function" (J.R. Givens, S. Venturoli, and E. Porcu, eds.), pp. 179–196, Year Book, Chicago, 1985.

events take place over many months, more than 2 years for some individuals. Even as a proportion of the pubertal period this is a far longer period of time than is required by the sheep. Changes in feedback sensitivity in man, therefore, may be a very protracted phenomenon.

In Fig. 33 are shown similar studies in three additional patients given only 0.15 mg of Premarin daily on a continuous basis. These girls were somewhat younger, ages 11–13, with a bone age between 8 and 10. Nonetheless, they still excreted castrate levels of gonadotropins when the study began, and the measurements fell promptly into the prepubertal range; escape than took place as noted previously. What I wish to point out here is not only the long duration of time involved but also the marked degree of feedback sensitivity displayed by these children as compared to adults. The castrate, sexually mature woman would require more than 10 times as much Premarin to effect similar suppressive changes. So the quantitative differences in set point in maturing man may be very marked indeed.

D. L. Foster. Have the patterns of pulsatile LH secretion been determined in these experimental patients?

H. Kulin. Yes, they have but perhaps Dr. Cuttler could provide the answer. His group has studied the subject and, to my understanding, changes in pulse frequency do not occur with age. Is that correct?

G. B. Cutler. We studied about 10 girls with q. 20 minute samples around the clock, and looked at LH, and FSH, pulse frequency. I'm not sure these methods are adequate to define pulse frequency rigorously in the open loop situation in humans. I think we might need to sample more frequently to make a definitive statement, but in the study as it was done, the prepubertal aged girls had an overall gonadotropin pulse frequency that didn't differ from the adults. The transition at the pubertal age was primarily an increase in amplitude, and this is different from what you see in the lamb.

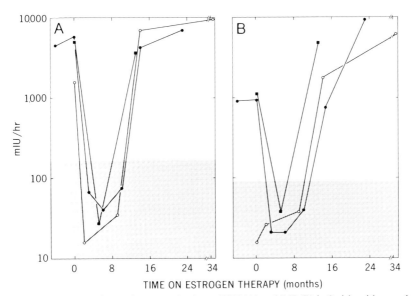

FIG. 33. Suppression and escape of urinary FSH (A) and LH (B) in 3 girls with gonadal dysgenesis given 0.15 mg/day of conjugated estrogen (see legend to Fig. 32).

N. B. Schwartz. You have presented us with a very tight scheme between estrogens and LH. As you know the evidence in other species both with puberty and with transitions from nonstimulatory to stimulatory periods suggests that FSH secretion has to precede the rise in LH, not only because the ovary needs to see it first, but because there are apparently signals which trigger FSH secretion first. Jill Milegge, Fred Turek, and I had presented an abstract on the Hungarian hamster at the Endocrine meetings. When the hamster is moved from the short photoperiod to a stimulatory photoperiod, serum FSH goes up much faster than LH over a 5-day period. We blocked the FSH rise with follicular fluid, and were able to block testicular recrudescence. Have you been able to do any FSH measurements?

D. L. Foster. We have no FSH measurements under photoperiodically induced puberty. However, during the transition to adulthood under natural photoperiod, we find an increase in circulating FSH, but it's very slight. Furthermore, we have not become extremely interested in FSH because, as I have shown, administration of LH alone to immature sheep initiates the entire sequence of events that lead to first ovulation. These observations lead us to conclude that the prepubertal sheep really has sufficient FSH and that it is the increase in LH that is the hormonal change timing puberty.

N. B. Schwartz. Have there been any studies of ovarian development in the sheep under the relative influence of the FSH and LH? I know you haven't done them, but I'm not familiar with sheep being used as a model for FSH–LH studies.

D. L. Foster. I don't recall any such studies.

G. B. Cutler. I'd like to get your further thoughts on the nutritional control of puberty. It appears from your data that the rapidity with which the oscillator begins to speed up when you improve nutrition is too fast to be accounted for by body weight changes. I wonder if you have looked at what's going on with, for example, blood glucose levels and other intermediary metabolic signals that might be being read by the LHRH oscillator, or have tried different forms of nutrition that were isocaloric, such as different mixtures of fat, protein, and carbohydrate?

D. L. Foster. This is clearly a fruitful area for further studies. We have not done more than I have shown you, largely because we've been trying to work out a model system in which we can repetitively show which changes occur. I think those are excellent ideas, and perhaps, the food restricted lamb may be a model system that we can explore for elements of nutrition or growth that are responsible for changing LH pulse frequency. It could be one of the major metabolic hormones, it could be a substrate in a metabolic pathway, or it could be a product in the pathway.

G. B. Cutler. It appears that the lamb is quite different from the human in regard to the open loop state in the prepubertal age period. The human with Turner's syndrome and also castrate primates have quite low gonadotropin secretion. Would you agree with that assessment?

D. L. Foster. Yes, I would agree with that. You must remember that the lamb is a precocial species, that is, it is very well developed at birth. It has a relatively active GnRH pulse generator throughout most of postnatal development. Before puberty its activity is not fast enough to induce cycles, but it is sufficient to keep the pituitary stimulated. If there is an analogous stage to the human or to the rhesus monkey, it may occur *in utero* in the sheep, as I have noted in one of our earlier reviews [D. L. Foster, "Problems in Pediatric Endocrinology." eds. (C. LaCausa and A. W. Root, eds.), pp. 193–210. Academic Press, London, 1980].

M. Walters. I have two questions which are pretty much unrelated. The first is relative to the superior cervical ganglionectomy-associated delay in puberty. I wondered whether you have information on how the puberty came about in those animals, and whether it was synchronized?

D. L. Foster. Puberty in those animals, as you recall from the data presented, was delayed beyond the normal time. Puberty did occur though; it was extremely late, and it was asynchronous among individuals. Apparently, interference with the photic loops may well alter the timing of puberty but it can't entirely prevent puberty. I think that's a very important principle that we are beginning to realize.

M. Walters. Do you think that some sort of degeneration of the negative feedback mechanism ultimately results in the puberty in those animals?

D. L. Foster. In lambs that experience inappropriate photoperiods there's clearly a primary pretubation in the negative feedback loop which then affects the timing of the cycles (D. L. Foster, *Endocrinology* **116**, 2090, 1985). Although we have not conducted steroid feedback studies in lambs with denervated pineal glands, I would predict that they have a delay in the time of their decreases in sensitivity to steroid feedback inhibition. The irregularity of cycles after puberty would also suggest that they oscillate between periods of hypo- and hypersensitivity to estradiol negative feedback.

M. Walters. Do you have any information on the mechanism of the steroid negative feedback? For example, is it a classic steroid receptor-mediated gene repression?

D. L. Foster. We have no information on that very important question.

R. A. Hoffman. In light of your promise to talk about north and south, I wonder if you could tell us when an equatorial lamb would go through puberty?

D. L. Foster. We are not familiar to any great extent with the equatorial breeds, but what I understand is that the adults show a lesser degree of seasonal reproduction. I am not aware of any rigorously conducted studies on the timing of puberty in such breeds. If one could extrapolate from other species, such as the deer, it is likely that the lambs may be born and reach puberty throughout the entire year.

R. E. Peter. You have shown that the after birth photoperiodic history is very important for determining the onset of puberty. In the experiments that you showed us I believe that the lambs were all born in March, probably early March, before they would have experienced any *in utero* exposure to long photoperiods. We know that animals *in utero* can respond to light. If lambs are born 4 or 6 weeks later, in April or May when long photoperiods already occur, do they show an earlier age of onset of puberty?

D. L. Foster. The answer to that question is derived from a number of studies in the literature. If they are born late in the year, such as July or August, they will reach puberty late during the same year but at a very young age (ca. 20–25 weeks). On the other hand, if animals are born very early in the year, such as January, they reach puberty during the early part of the breeding season. These females are then much older at first ovulation (ca. 35–40 weeks).

R. E. Peter. The age of puberty remains at 30 weeks of age?

D. L. Foster. It is pretty well locked at 30 to 35 weeks unless you get extreme times of birth such as January or July.

R. E. Peter. Are there differences in the growth rates of animals that are born later in the year versus those born in the spring?

D. L. Foster. There may well be since long photoperiods favor more rapid growth. Lambs that are born, for example, in June and July have early growth that is more rapid than those born earlier in the year.

S. Y. Ying. In your model, GnRH plays a very important role influenced by the photic factors as well as the internal cue. I am quite sure that you are aware of the latest paper published by Nikolies *et al.* in *Nature* (**316**, 511, 1985) regarding the C-terminal of the procusor of LRF. They demonstrated that this C-terminal peptide acts as FRF and PIF. Have you ever measured the levels of prolactin in your photoperiod-induced puberty?

D. L. Foster. No, we haven't but we are very interested in measuring prolactin in the neonatal period, largely from another standpoint that is to determine when the lamb becomes photoperiodic. Based upon the melatonin pattern, very young lambs are not able to distinguish between photoperiods very well. This raises the question of whether in those animals prolactin secretion is normal, since prolactin secretion is sensitive to photoperiod. Therefore, as another test for photosensitivity, we would be interested in examining prolactin under various artificial photoperiods in newborn lambs.

S. Y. Ying. In your model you also emphasized the negative feedback of estrogen. As you know in other species such as in the rat, low doses of estrogens can induce puberty. Have you also looked at this positive feedback of estrogen in your model?

D. L. Foster. As I indicated, the surge system can respond to the positive feedback action of estrogen throughout most of the development, so that system is competent to respond, but it remains dormant. Even if we do activate the system with exogenous estradiol, we don't induce ovulation because the prepubertal lamb doesn't have a preovulatory follicle waiting for the induced gonadotropin surge.

S. Y. Ying. Is that true also with your restricted diet? Is it also possible that these lambs didn't have mature follicles to be ovulated. Have you ever tried to use GH or GRF to see if you could compensate for this?

D. L. Foster. No, we haven't but we would predict in the ovarian intact, anovulatory, undernourished lamb that they can't ovulate because they can't develop a preovulatory follicle. They are so sensitive to estradiol negative feedback they can't induce a rapid train of LH pulses.

G. Anogianakis. Did you observe the high-frequency pulses of LH when you ablated the superior cervical ganglion?

D. L. Foster. No, we have not looked at that.

G. Anogianakis. The reason I am asking you is that other than the pineal gland, the anterior hypothalamic area and the suprachiasmatic nucleus are also considered to be a day to day "clock" in animals and I wondered whether they are the ones that may be pacing the LH pulses?

D. L. Foster. We don't have any anatomical information on the system that times the LH (GnRH) oscillators. We recognize that doing the superior cervical ganglionectomy is a rather dirty operation because you denervate more than the pineal gland.

D. L. Keefe. Your model designates separate pathways for the effects of nutrition and photoperiod on the GnRH pulse generators. I wondered if there is a direct effect of melatonin on nutrition as reported in other animals in which melatonin has nonreproductive effects such as on metabolism and appetite.

D. L. Foster. I am not aware of any studies in that regard, but I would like to address the question of the separate pathways. One of the potential criticisms of the studies of undernourished animals is that we may be rendering the photoperiod-monitoring device inoperative, and this could explain delayed puberty. We have examined this in a recent experiment and we can demonstrate that these very small, undernourished lambs can readily read photoperiods. Therefore, the absence of puberty would be due to the nutritional deficit we have imposed upon them. Not a photoperiod problem.

H. Urbanski. My question relates to the critical photoperiod in the sheep. In your experiments you used two photoperiods, short days and long days, and you got a very clearcut response, almost all-or-none. I wonder, if you have done experiments involving intermediate photoperiods such as 12L:12D and would you expect to get intermediate responses?

D. L. Foster. We have not done those experiments in the developing lamb, but my colleague Dr. Robinson has in the adult. I wonder if she would like to comment on that?

J. Robinson. In the adult ewe the time course for steroid-independent changes in gonadotrophin secretion is very similar to that for the steroid-dependent changes. When ovary-intact, ovariectomized or ovariectomized ewes bearing Silastic capsules of estradiol are placed on a stimulatory day length there is a lag period of approximately 50 days before any increases in LH secretion are observed. The reasons for this relatively long lag period to a photoperiodic response are unclear at present. However, we do know that the lag does not lie at the level of the pineal gland. Thus the diurnal pattern of melatonin secretion becomes appropriate to day length rapidly following a photoperiodic shift. And so the 50 day lag for a photoperiodic response in the ewe must reside somewhere between the generation of the melatonin signal and the LH pulse generator.

H. Urbanski. In relation to puberty then, would you expect lambs that have been reared on a 12L:12D photoperiod to go through puberty say at 40 weeks of age rather than 30 or 50?

D. L. Foster. I could make really no prediction if they would be reading it as a long photoperiod or a short photoperiod. However, judging the inability to induce puberty at the normal age with a constant photoperiod, either long days or short days, I would expect that lambs raised on 12L:12D would delay puberty.

H. Urbanski. How long does it take for gonadotropins secretion to show a response following the change in photoperiod? I mean how long is it before you see a detectable change in the pulse frequency pattern?

D. L. Foster. In the open loop system?

H. Urbanski. No, in the closed loop system.

D. L. Foster. In the closed loop system we do not have that information as well characterized as we would like. The latent period would be about the same as if it is for onset of cycles, based upon the limited information we have. So the lag period would be perhaps 10 weeks.

H. Urbanski. The reason I ask this question is because in other photoperiodic species, such as Japanese quail and the golden hamster, if you transfer the animal from a nonstimulatory photoperiod to one that is stimulatory you can usually detect an increase in gonadotropin secretion as early as 1 or 2 days after the switch. Can you comment on why it takes the sheep so long for its neuroendocrine system to make that change?

D. L. Foster. I think that the sheep may use its neuroendocrine system as a predictive system and be less perhaps opportunistic than the bird to photoperiod changes. I don't know why other than we continually note that this difference exists between the two types of animals—the very short interval in birds and very long intervals in sheep.

H. Urbanski. I wonder if this could be related to the fact that in the quail and in the hamster the photoperiodic mechanism appears to be primarily direct, or steroid independent, whereas your data suggest that in the sheep the mechanism is primarily steroid dependent and for some reason the steroid-dependent changes take considerably longer to come about?

D. L. Foster. That is possible but the hypothesis wouldn't be quite that simple based upon work that has been done in the adult sheep. In the mature ewe we do see a direct photoperiodic modulation of LH pulse frequency in the absence of the ovaries, much the same as in the bird. I wonder if Dr. Robinson could comment upon the time course of the changes in LH pulse frequency following light shifts in the ovariectomized ewe?

J. Robinson. We have recently published the results of an experiment which was designed to test whether photoperiod can set the level of activity of the LH pulse generator in the adult ewe (*Biol. Reprod.* **33**, 324–334, 1985). Three groups of ovariectomized ewes were transferred from a nonstimulatory day length of 16L:8D to a highly stimulatory day length of 8L:16D or two intermediate day lengths of 11.5L:12.5D or 13.5L:10.5D. Controls remained on 16L:8D. After 91 days on these day lengths LH pulse frequencies in the control group remained low while those in the animals transferred to 8L:16D were relatively fast. The two intermediate day lengths produced intermediate pulse frequencies. The correlation between day length and LH pulse frequency on day 91 of the experiment was highly significant suggesting that day length can set, in a graded fashion, the level of activity of the LH pulse generator in the ewe. This conclusion is similar to those derived from studies in the Japanese quail and the hamster.

D. L. Foster. I think we should also recall that this discussion we are now having is oscillating back and forth between adult and immature animals, and I caution that the timing of many of the neuroendocrine responses may not be the same between the two ages.

The Onset of Female Puberty: Studies in the Rat

SERGIO R. OJEDA, HENRYK F. URBANSKI, AND CAROL E. AHMED

Department of Physiology, University of Texas Health Science Center at Dallas, Dallas, Texas

I. Introduction

One of the most fascinating aspects of mammalian neuroendocrinology resides in the extraordinarily complex series of events that leads to acquisition of reproductive maturity. Generations of scientists have been intrigued by the perfection of the process, the dynamics of the neuroendocrine interrelationships, and, most of all, by the marvelous functional synchronization of the various components. During the initial years of research the main approach to an understanding of the pubertal process was merely the examination of morphological changes. New impetus was given to the field first by the isolation and purification of various peptidergic hormones that participate in the process, and, subsequently, by the development of specific assay methods for them. The 1970s witnessed a marked increase in research activity with substantial progress made on several fronts. At present, the ontogeny of hormone production is well defined in several species, and with the advent of a variety of new physiological techniques the intricacies of the different control mechanisms underlying these ontogenic changes are being unravelled even further (for recent reviews see Reiter and Grumbach, 1982; Foster *et al.*, 1985; Ojeda *et al.*, 1984a).

In trying to more clearly comprehend the basis of female pubertal development it became evident to us that to make further progress our studies should comprise both *in vivo* and *in vitro* approaches. These approaches need to be diverse because of the complexity and number of the issues under scrutiny. Examination of the recent literature indeed reveals a growing tendency to employ more techniques originating in fields as diverse as those of immunocytochemistry, cell biology, neurobiology, and molecular biology to study the neuroendocrine development of reproductive function.

This article presents an account of our efforts to understand the genesis of female puberty. In pursuing this goal we have used an interdisciplinary approach and have investigated, both *in vivo* and *in vitro*, the developmental regulation of the neuroendocrine reproductive axis at the level of its three basic components: the hypothalamus, the anterior pituitary gland, and the ovaries. The information to be presented derives exclusively from experiments performed in the laboratory rat. We hope that the implications such observations may have for the understanding of puberty as a whole will be clarified by this article and also by those of Foster *et al.* and Cutler *et al.* (this volume) which deal with the onset of puberty in sheep and humans, respectively.

II. The Animal Model

The rat is born very immature, at an age equivalent to 150 days of human gestational life (Tanner, 1974). The gestational period in the rat lasts, on the average 22 days and the first ovulation in most laboratory stocks occurs 35–45 days after birth. Externally, the only signal that puberty has occurred is canalization of the vagina which normally is imperforated during immature days and later becomes patent as a consequence of estrogenic stimulation. Vaginal opening usually occurs on the day after the first preovulatory surge of gonadotropins has occurred (Critchlow and Bar-Sela, 1967; Meijs-Roelofs *et al.*, 1975; Ojeda *et al.*, 1976b). In most cases, the cytology of vaginal lavages at opening shows the majority of cells as being cornified (estrus), a condition that is followed within 1–2 days by the appearance of leukocytes which soon become the predominant cell type (first diestrous phase of puberty).

The postnatal developmental events implicated in the initiation of reproductive cyclicity are, therefore, compressed within a period of about 5 weeks. This makes definition of such events a particularly difficult endeavor as some of them may last for only a few hours. The urgency of maturation in the rat is exemplified by the animal's impressive rate of growth. Between birth and first ovulation body weight increases 15-fold and body length 3-fold (Fig. 1).

Developmental phases in the female rat have been defined mainly in relation to the maturation of the ovary (Hisaw, 1947; Dawson and McCabe, 1951; Critchlow and Bar-Sela, 1967). Ramirez (1973) was the first to propose a classification primarily based on physiological parameters such as the changes in circulating gonadotropin levels and the alterations in steroid feedback mechanisms occurring at different postnatal ages. We have proposed a classification which considers both morphological and physiological parameters (Ojeda *et al.*, 1980). According to this classifica-

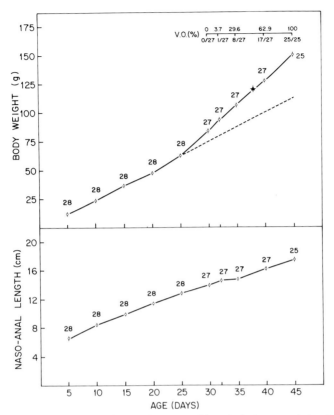

FIG. 1. Changes in body weight and naso-anal length during sexual development of female rats. V. O., Vagina open. (▲) Mean body weight (121.2 ± 2.7 g) and mean age (38.4 ± 0.58 days) at vaginal opening with their corresponding SEM (n = 25–28). The dashed line is an extrapolation of the curve of bodily growth increase between days 5 and 25, whose slope was 2.48 g/day as compared with the actual slope of 4.35 g/day between days 25 and 45. From Ojeda and Jameson (1977a), reprinted with permission.

tion[1] postnatal development of the female rat can be divided into four phases: a neonatal period that is initiated at birth and ends on postnatal day 7, an infantile period which extends from day 8 to 21, a juvenile or prepubertal period which ends around day 30-32,[2] and a peripubertal period which has a variable duration, and is highlighted by the occurrence of first ovulation.

[1] Inclusion of a fetal period seems appropriate as several developmental events initiated at this time appear to have significant repercussions on subsequent maturational steps leading to the onset of puberty. Since LHRH can be first detected in the fetal brain as early as day 12 of gestation (Aubert et al., 1985) and since it appears to influence the differentiation of gonado-

In order to offer a more orderly view of the sequence of events that precedes the activation of the gonadotropin surge mechanism we will discuss each developmental period separately. Also, each component of the system will be considered individually, but interrelated regulatory mechanisms controlling their function will be emphasized.

III. The Developmental Periods

A. THE FETAL PERIOD

Perhaps one of the most significant events that occurs during the fetal period is the initiation of LHRH production on gestational day 12 (Aubert et al., 1985). Indeed, the peptide may even be transported to the pituitary anlage via vascular connections since such connections also appear to become established around day 12 (Szabo and Csanyi, 1982). Once LHRH gains access to the primordial gland it may play a decisive role in determining the onset of gonadotropin secretion as in vitro experiments have clearly established the ability of LHRH to initiate the functional and morphological differentiation of pituitary gonadotrophs (Begeot et al., 1984).

Circulating gonadotropins are not observed until after day 17 (Chowdhury and Steinberger, 1976; Salisbury et al., 1982) and levels remain very low until the day of birth (Chiappa and Fink, 1977). In spite of this, both the reproductive hypothalamus and the ovary appear engaged in activities which strongly suggest the presence of functional communications between them. It is remarkable that the hypothalamic–preoptic area of the rat fetus acquires noticeable aromatase activity in a period of no more than 48 hours (between day 14 and 16 of gestation) (Fig. 2), and becomes the only fetal tissue demonstrating such activity at this time (George and Ojeda, 1982). Aromatase activity decreases very rapidly after gestational day 20, becoming minimal by postnatal day 20. While it is tempting to assume that such an abrupt increase in hypothalamic capacity to produce estrogens is intimately linked to sexual differentiation of the brain this would not appear to be the only, or most important, function. The capacity of estradiol to induce neuronal growth and differentiation (Toran-Allerand, 1976; Naftolin and Brawer, 1977; Matsumoto and Arai,

---

trophs (Begeot et al., 1984), we will define the fetal period as that extending from gestational day 12 to day 22.

[2] Viewed in the light of recent information (Urbanski and Ojeda, 1985a) the end of juvenile development can be considered to be signaled by the establishment of morning–afternoon differences in pulsatile LH release.

MECHANISM OF FEMALE PUBERTY   389

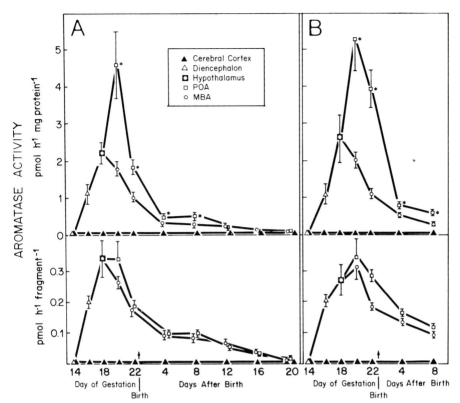

FIG. 2. Developmental pattern of aromatase activity in the brains of embryonic, neonatal, and infantile rats. (A) Female; (B) male. Brain fragments (0.05–0.25 mg protein) were incubated in 0.2 ml Eagle's minimal essential medium, pH 7.4, containing 0.1 μM [1β-^{3}H]testosterone for 1 hour at 37°C. The tubes were gassed with 95% O_2–5% CO_2. The reactions were stopped with 1 ml chloroform, and the ^3H_2O formed during the incubation was purified and measured as described (George and Ojeda, 1982). Each point represents the mean ± SEM of three to six determinations. The MBH and POA could not be separated from each other before day 20 of embryonic development. The hypothalamus could not be individualized before day 18. *Activity significantly higher than that in the MBH of the same animals. From George and Ojeda (1982), reprinted with permission.

1976) provides a rather compelling argument in favor of a role for locally formed estrogens in hypothalamic growth and in the establishment of neuronal circuitry. Little has been done, however, to evaluate this possibility.

The fetal ovary, on the other hand, does not possess follicles and when examined *in vitro* fails to respond to gonadotropin stimulation with increased aromatase activity (George and Ojeda, 1984). The ovary, however, does have the intracellular machinery to produce the enzyme and the transmembrane signaling system which can activate aromatase gene expression. This conclusion derives from the observation that, in contrast to their lack of response to gonadotropins, 19-day-old fetal ovaries in culture respond to cyclic AMP and to forskolin or cholera toxin (two activators of adenylate cyclase) with marked increases in aromatase activity (George and Ojeda, 1984). The effect of cyclic AMP appears to involve synthesis of new enzyme and not activation of preexisting aromatase, as both cycloheximide and actinomycin D blocked the increase in enzyme activity induced by forskolin without diminishing the increase in cyclic AMP formation (George and Ojeda, 1986).

The capacity of the fetal ovary to respond to forskolin, but not to gonadotropins, suggests that an extracellular messenger different from LH or FSH may control early ovarian function. This notion was considerably strengthened by the finding that vasoactive intestinal peptide (VIP), which in older rats is found in ovarian nerves (Larsson *et al.*, 1977; Ahmed *et al.* 1985) and is already present in 2-day-old neonate ovaries (Ahmed *et al.*, unpublished), increased both cyclic AMP production and aromatase activity from cultured 19-day-old fetal ovaries (George and Ojeda, 1986). This finding raises the exciting possibility that at the onset of ovarian development the central nervous system controls the maturation of the gland via direct neural connections rather than through circulating gonadotropins. While much work needs to be done to test such a possibility it is important to bear in mind that early ovarian development does occur independently of gonadotropin control (Lamprecht *et al.*, 1973; Kraiem *et al.*, 1976; Hunzicker-Dunn and Birnbaumer, 1976; for a review see Peters, 1979).

B. THE NEONATAL PERIOD

One of the most striking features of the neonatal period is the initiation of increased secretion of gonadotropins. However, the hormonal link between the hypothalamic–adenohypophysial unit and the ovaries is not yet fully operative. The ovary is relatively insensitive to gonadotropins for at least the first 4–5 postnatal days (Ben-Or, 1963; Funkenstein *et al.*, 1980; Peters *et al.*, 1973) and (E_2) negative feedback is not functional as demonstrated by the inability of neonatal ovariectomy to activate gonadotropin release (Goldman *et al.*, 1971). It is likely that ovarian unresponsiveness to gonadotropins in newborn rats is due to the lack of gonadotropin receptors. Both LH and FSH receptor contents are very low at postnatal day 4 (Siebers *et al.*, 1977; Peluso *et al.*, 1976; Smith-White and Ojeda, 1981b). In contrast, E_2 negative feedback fails to operate, not because of a lack of specific hypothalamic–pituitary estrogen receptors, but rather because the serum and tissues contain extremely high levels of α-fetoprotein (AFP), which binds estrogen avidly (Nunez *et al.*, 1971; Raynaud *et al.*, 1971). The consequences that such amounts of AFP may have for subsequent reproductive development are not completely understood. While it appears clear that AFP protects the brain from exposure to excessive amounts of E_2 (Plapinger and McEwen, 1975; Vannier and Raynaud, 1975), the intracellular occurrence of AFP (Benno and Williams, 1978) suggests that the protein may, in fact, play a modulatory role which, as in other cells (Soto and Sonnenschein, 1980), may regulate the amount of E_2 available to developing estrogen-sensitive neuronal systems.

Acknowledging the risk of oversimplification, the neonatal period may be seen as the developmental phase during which the brain begins to strengthen its grip over ovarian function. Thus, FSH acquires the capacity to facilitate the ovarian conversion of testosterone (T) to E_2 by postnatal day 4 (Funkenstein *et al.*, 1980). Moreover, we have recently observed that suppression of gonadotropin release by injection of dihydrotestosterone propionate (DHTP), a nonaromatizable androgen, during postnatal days 1–5 markedly decreased ovarian FSH receptor content measured on day 12 (Smith and Ojeda, 1986). Since FSH receptor number increases more noticeably between neonatal day 4 and the second half of the infantile period (day 16), we concluded that neonatal release of gonadotropins, and in particular FSH, plays an essential role in the subsequent acquisition of FSH receptors by the developing ovary. This conclusion was further strengthened by the demonstration that FSH replacement therapy during the first 5 postnatal days effectively reversed the suppressive effect of DHTP-induced gonadotropin deficiency on FSH receptor content.

Neonatal gonadotropins may become essential for the maintenance of follicular development sometime after postnatal day 2 since primary follicles, which are absent at this time, become recognizable at day 4 (Funkenstein and Nimrod, 1982; George and Ojeda, 1986).

As predicted by the foregoing observations, gonadotropin receptors become detectable by the end of the neonatal period (Peluso *et al.*, 1976; Smith-White and Ojeda, 1981b) and steroid responsiveness to gonadotropins becomes evident (Lamprecht *et al.*, 1976; Funkenstein *et al.*, 1980). Interestingly, the operation of a dual direct neural control of the gland also appears established by this time as judged by the presence of both VIP (Ahmed and Ojeda, unpublished) and norepinephrine (NE) (Ben-Jonathan *et al.*, 1984) in the neonatal ovary. The roles that these substances may play in subsequent ovarian development will be discussed later in this article.

An additional, and unexpected, control mechanism appears to become established shortly after birth. Most remarkably it does not originate in the neonate, but is rather provided by the mother. Rat milk, like that from several other species, contains an LHRH-like substance which, as judged from its chromatographic behavior in Sephadex G-25 and HPLC, appears indistinguishable from hypothalamic LHRH (Amarant *et al.*, 1982; Smith-White and Ojeda, 1984). We have observed that after suckling, LHRH immunoreactivity can be readily detected in the stomach content of the pups and its concentration increases in their plasma. Available LHRH receptors in the pup's ovaries decrease (Fig. 3), an effect that can be prevented by prior intravenous administration of an antiserum to LHRH. That the decrease in available receptors is not the consequence of suckling per se or due to stomach distension was demonstrated by the finding that intragastric administration of milk, but not saline, reproduced the decrease in ovarian LHRH receptors associated with suckling. These observations led us to the conclusion that LHRH of maternal origin is transferred to the pup via the milk; it crosses the gastrointestinal epithelium and reaches the ovary via the blood stream where it binds to specific receptors.

Milk LHRH behaves like hypothalamic LHRH in that it is able to both stimulate gonadotropin release from the anterior pituitary *in vitro* and to

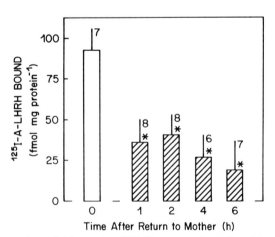

FIG. 3. Changes in available ovarian LHRH receptors in 10-day-old rats after suckling periods of 1, 2, 4, and 6 hours subsequent to a 6-hour fast (0). Vertical lines represent SEM and numbers above bars are number of animals per group. *$p < 0.01$ vs 0 hour. From Smith-White and Ojeda (1984), reprinted with permission.

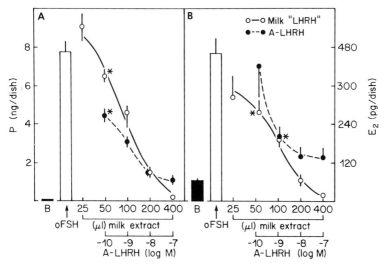

FIG. 4. Effect of milk LHRH on FSH-induced estradiol (E_2, B) and progesterone (P, A) release from granulosa cells in culture (48 hours). The effect of an LHRH agonist (A-LHRH) was also studied for comparative purposes. Each point represents the mean of 4–5 dishes ± SEM (vertical lines). *The lowest dose producing a significant ($p \leq .05$) inhibition of the FSH effect. oFSH, Ovine FSH. From Smith-White and Ojeda (1984), reprinted with permission.

inhibit gonadotropin-induced E_2 and progesterone (P) reiease from granulosa cells in culture (Fig. 4). Since the rat pup suckles frequently throughout the entire day it would be expected that milk LHRH is almost continuously available for binding to the infant ovary. Chronic exposure to continuous levels of LHRH are known to depress ovarian function (see Hsueh and Jones, 1981, for a review), and thus the suggestion may be made that milk LHRH plays a physiological role in restraining neonatal–infantile development of the pup ovary. Such a remarkable phenomenon may represent an evolutionary mechanism by which the mother rat regulates gonadal development of her offspring beyond intrauterine life. Indeed, available ovarian LHRH receptor content increases after postnatal days 15–20 (Dalkin *et al.,* 1981; Smith-White and Ojeda, 1984), i.e., at the time when pups begin to eat regular food and nursing episodes become less frequent.

It would thus appear that very soon after birth sources other than those residing within the immature hypothalamic–pituitary–ovarian axis might play a fundamental role in regulating the development of both hypothalamic (AFP) and ovarian (milk LHRH) functions. We suspect, however, that milk LHRH is involved in other undisclosed functions that are linked to more general growth processes rather than specifically to reproduction.

If milk LHRH is involved in modulating neonatal reproductive functions, what is the contribution of hypothalamic LHRH and how developed is the LHRH neuronal releasing system at this time? While there is no doubt that the primary role of hypothalamic LHRH in the neonates, as in older animals, is to positively influence gonadotropin release, little is known regarding the functional development of the LHRH releasing system. Because of the low content of LHRH found in the hypothalamus of neonatal rats (Araki et al., 1975; Chiappa and Fink, 1977; Aubert et al., 1985), and the already elevated serum FSH levels seen at this time we assumed that LHRH release would not occur in a fashion similar to that of older rats. Although we expected that LHRH release would be pulsatile, we presumed that the frequency of the secretory episodes would differ substantially from that observed later in life. To our surprise when preoptic area-medial basal hypothalamic (POA-MBH) fragments excluding the lateral hypothalamus and the mammillary bodies were analyzed for pulsatile LHRH release in a perifusion system,[3] it was found that LHRH secretory episodes occurred regularly with an interpulse frequency of approximately 30 minute (Fig. 5). Such a pattern of release is similar to that observed in juvenile animals (vide infra) in which the frequency of LH pulses in plasma and the low FSH circulating levels make entirely predictable a frequency of LHRH dischanges corresponding to that of LH.

Based on these observations, we have tentatively concluded that the neonatal LHRH neuronal system located in the POA-MBH has already the intrinsic capacity of generating LHRH secretory episodes with a rhythmicity similar to that observed in pre- and peripubertal rats. It would thus appear that developmental modulation of this basic LHRH activity depends primarily on inputs originating outside the POA-MBH island. Of particular note is a recent report by Melrose (1985) which demonstrated that isolated LHRH neurons in culture release LHRH in a pulsatile fashion in the absence of any neural connections, thus suggesting that pulsatility is an intrinsic property of the LHRH neuron itself.

Figure 6 summarizes some of the events that may occur during fetal–neonatal development in the female rat.

---

[3] Our own experiments as well as those of Bourguignon and Franchimont (1984) and Ramirez et al. (1985) have shown that single POA-MBH fragments incubated *in vitro* in either a perifusion system or for short-term periods with frequent removal of the medium exhibit pulsatile release of LHRH.

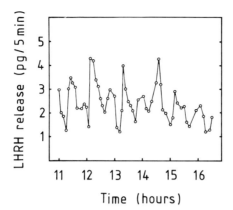

FIG. 5. Demonstration of pulsatile release of LHRH from the preoptic area-medial basal hypothalamus (POA-MBH) of a neonatal, 6-day-old female rat. The LHRH profile depicted corresponds to a single POA-MBH which was perifused in vitro for 5 hours with Krebs-Ringer bicarbonate buffer, pH 7.4, containing both glucose (0.1%) and BSA (0.01%). The flow rate was 75 μl/minute and LHRH was measured by RIA in 5-minute fractions. From Urbanski and Ojeda, unpublished.

C. THE INFANTILE PERIOD

Between postnatal day 7 and 21 the CNS–pituitary axis undergoes changes that, we believe, represent the first developmental events having a direct impact on the timing of puberty onset. Serum FSH levels increase rapidly to reach peak levels around day 12 (Ojeda and Ramirez, 1972; Kragt and Dahlgren, 1972; Meijs-Roelofs et al., 1973). Thereafter, levels decline steadily to reach their lowest value shortly before the first proestrus (Meijs-Roelofs et al., 1975; Ojeda et al., 1976b). The elevated FSH not only behaves chromatographically similar to serum FSH of juvenile or adult rats, but is also biologically active (Ojeda and Jameson, 1977b).

As previously indicated there is substantial evidence that development of ovarian follicles at this time comes under strong gonadotropin control (Eshkol et al., 1970; Schwartz, 1974; Uilenbroek and Arendsen de Wolff-Exalto, 1977). Importantly, the level of FSH secretion necessary for maintenance and/or formation of FSH receptors during the infantile period is no greater than that observed during the late juvenile period (~200 ng/ml); suppression of serum FSH from the high, infantile levels down to the low juvenile values, using DHTP, failed to affect ovarian FSH receptor content even though serum LH was reduced to undetectable values (Smith and Ojeda, 1986). Further suppression of serum FSH resulted in disrupted follicular development (Uilenbroek and Arendsen de Wolff-Exalto, 1977). During the infantile period FSH becomes able to induce a

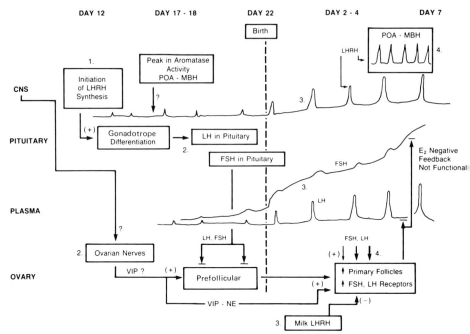

FIG. 6. Summary of some neuroendocrine events postulated to occur during the feto-neonatal period of the female rat. Although pituitary gonadotropins can be detected in the gland by gestational day 17, the ovary is insensitive to gonadotropins until at least 2 days after birth. Before birth it responds to VIP, a peptide contained in ovarian nerves, with increased aromatase activity; fetal regulation of ovarian function may, therefore, be exclusively neurogenic. Between postnatal day 2 and 4 primary follicles begin to develop. Gonadotropin control is initiated some time after postnatal day 2 when estrogen negative feedback on gonadotropin secretion is not yet functional. The increasing FSH levels observed after birth may result from a low frequency of "meaningful" LHRH discharges and a lack of estrogen negative feedback. However, by day 6 the isolated POA-MBH has already developed the capacity of discharging LHRH at regular 30 minute intervals (inset). ⊤, Not operative; (+), stimulatory; (−), inhibitory; NE, norepinephrine. Numbers indicate the sequence in which these events may occur.

substantial increase in ovarian aromatase activity (George and Ojeda, 1986) and the ovary unequivocally demonstrates the capacity to respond to endogenous increases in serum gonadotropins with steroid release (Andrews et al., 1981a) (Fig. 7).

Circulating FSH levels appear to be elevated tonically whereas serum LH increases as sporadic bursts of secretion (Döhler and Wuttke, 1975). Some authors have attributed this latter observation to nascent expression of E_2 positive feedback (Wuttke and Gelato, 1976) but others, includ-

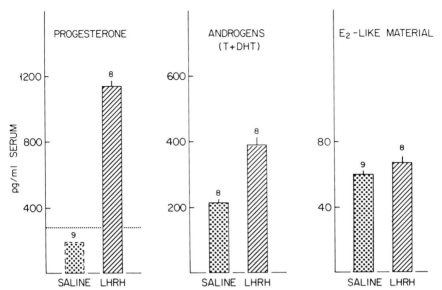

FIG. 7. Serum progesterone, androgens, and estradiol (E_2)-like material (mean ± SEM) in female rats adrenalectomized on day 10 (1600–1800 hours) and injected iv, 90 minutes before sacrifice on the morning of day 11 (0800–1000 hours) with LHRH (100 ng/100 g body weight). The horizontal dotted line in the first panel represents the sensitivity of the progesterone assay in this experiment. Serum T is expressed as androgens and serum E_2 is expressed as E_2-like material because they were measured in unchromatographed samples. Numbers above bars indicate the number of animals per group. From Andrews et al. (1981a), reprinted with permission.

ing ourselves (Andrews et al., 1981b; Frawley and Henricks, 1979; MacKinnon et al., 1976), consider this to be a moot point on the grounds that (1) E_2 is unable to induce an LH surge before postnatal day 15 (Andrews et al., 1981b), (2) passive immunization against E_2 fails to inhibit the LH surges (Frawley and Henricks, 1979), and (3) environmental disturbance inhibits their occurrence (MacKinnon et al., 1976). It seems clear, however, that such sporadic increases in LH release reflect activation of a central event as noradrenergic turnover in the POA of infantile rats increases at the time of the LH secretory episodes (Honma et al., 1979). Our recent observations on the transfer of maternal LHRH to the offspring suggest that a contributing factor may reside in milk LHRH which, upon absorption and attainment of a threshold level in plasma, may induce abrupt, short-lived increases in LH release.

The pattern of high FSH levels and sporadic increases in LH release that characterizes the infantile period of the female rat can be interpreted as a reflection of maturational events that occur at the levels of both the brain and anterior pituitary. One may also assume that the steroidal environment plays a definitive role in the modulation of these events.

Our attempts to gain insight into this issue have been based on the assumption that the function of the system can be elucidated by examining its components both individually and in relation to each other. In pursuing this strategy we first focused our attention on the anterior pituitary gland, then we defined the relative contribution of steroid feedback mechanisms, and, more recently, we have initiated the direct exploration of those mechanisms controlling the hypothalamic LHRH secreting machinery.

Pituitary responsiveness to LHRH is exceedingly high in infantile rats (Debeljuk *et al.*, 1972; Ojeda *et al.*, 1977) (Fig. 8) and decreases steadily after postnatal day 15. This increased responsiveness appears to be related, at least in part, to a direct facilitory effect exerted by nonaromatizable androgens and P on the pituitary gland. We observed that ovariectomy reduced the FSH response to LHRH whereas DHT or P treatment restored it (Ojeda *et al.*, 1977). Although plasma levels of DHT are low in female rats, the anterior pituitary of infantile rats has an enhanced 5α-reductase acitvity, the developmental pattern of which closely follows that of serum FSH (Denef *et al.*, 1974).

An important contributing factor to the elevated gonadotropin levels in infantile rats is the relative inefficiency of E_2 negative feedback. The earlier observation that administration of E_2 was less effective in depressing FSH levels in ovariectomized infantile rats than in juvenile animals (Ojeda and Ramirez, 1973/74) received the support of additional results from our own laboratory (Ojeda *et al.*, 1975; Andrews and Ojeda, 1977) and from other investigators (Meijs-Roelofs and Kramer, 1979; Frawley and Henricks, 1979). The findings of these latter authors were particularly supportive of our conclusions as they demonstrated that immunoneutralization of serum E_2 failed to induce a rise in circulating gonadotropin levels as would be expected if E_2 negative feedback was operative in infantile rats. Our own observations that administration of RU 2858, a synthetic estrogen that does not bind to AFP, resulted in a striking inhibition of the postovariectomy rise in serum gonadotropins (Andrews and Ojeda, 1977) (Fig. 9) conclusively established that E_2 negative feedback fails to operate in infantile rats because of the presence of AFP, and not because of lack of specific E_2 receptors. By day 16, AFP levels have decreased sufficiently, allowing free E_2 levels to be detected for the first time (Germain *et al.*, 1978; Meijs-Roelofs and Kramer, 1979; Andrews *et al.*, 1981b). In remarkable coincidence with such a shift, the peak level of serum FSH, seen at day 12, begins to decrease. Thus, it appears that both a facilitatory

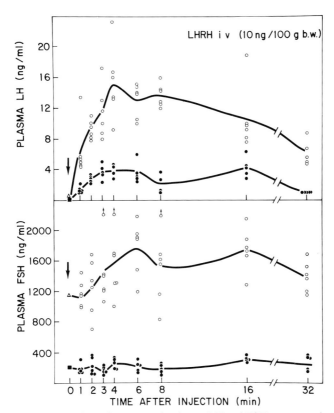

FIG. 8. Comparison of the time course in plasma LH and FSH concentration following an iv injection of LHRH (10 ng/100 g body weight in 10-day (○) and in 28-day-old female rats (●). Each point represents an individual value. Arrows indicate the time of injection. From Ojeda et al. (1977), reprinted with permission.

effect of 5α-reduced androgens, originated locally in the anterior pituitary, and the failure of E_2 negative feedback to operate at full capacity play a pivotal role in determining the infantile increase in serum FSH levels.

The repeated finding that ovariectomy of infantile rats results in a prompt increase in serum gonadotropin levels does not appear to support this hypothesis; if ovarian E_2 is the only gonadal steroid involved in controlling gonadotropin release then a postovariectomy rise should not have occurred. However, infantile ovaries produce measurable amounts of both testosterone and androstenedione (Döhler and Wuttke, 1975; Andrews et al., 1981a; Andrews and Ojeda, 1981a; Matthews et al., 1986). When 10-day-old rats were ovariectomized and given sc Silastic capsules

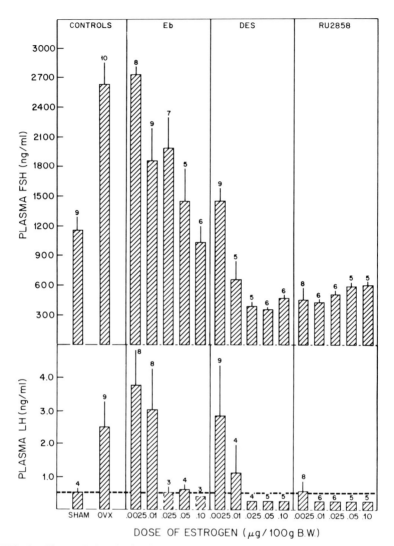

FIG. 9. Plasma LH and FSH in ovariectomized female rats (ovariectomy on day 10) treated with various doses of estradiol benzoate (Eb), diethylstilbestrol (DES), or the synthetic estrogen RU2858 once daily for 3 days and sacrificed on day 13 between 1000 and 1100 hours. Vertical lines represent the SEM and the numbers above the bars indicate the number of animals per group. Bars outlined with dashes represent groups in which the plasma of 2 or 3 animals was pooled to obtain the indicated number of determinations. The sensitivity of the LH assay is shown by the horizontal dashed line. Determinations which were below this level were assigned a value of 0.25 ng/ml. OVX, Ovariectomized oil-treated controls. From Andrews and Ojeda (1977), reprinted with permission.

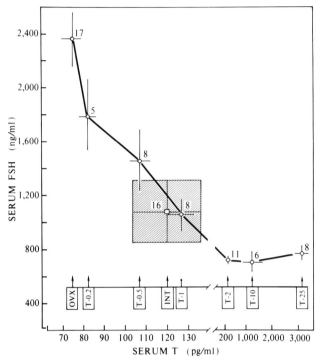

FIG. 10. Decrease in serum FSH titers of 12-day-old ovariectomized rats induced by different levels of serum T. The animals were ovariectomized on day 10 and immediately received sc Silastic capsules containing different concentrations of T dissolved in oil. The numbers inside the boxes indicate the different concentrations of T (mg/ml) used. INT, Intact sham-operated controls. The SEM for serum FSH and serum T are represented by vertical and horizontal lines, respectively. The hatched rectangle represents the SEM of serum FSH and T levels of intact sham-operated animals. The numbers above the means indicate the number of animals per group. From Andrews and Ojeda (1981a), reprinted with permission.

containing different amounts of T it was observed that as mean serum T levels increased, serum FSH (Fig. 10) and LH levels decreased. When the implants produced serum T levels[4] similar to those normally seen in intact rats, serum FSH titers were also normal. These observations permitted the conclusion (Andrews and Ojeda, 1981a) that during the infantile period of female development the bulk of steroid negative feedback on gonadotropin release is exerted by aromatizable androgens. Whether the effect of T is due to its androgenic capacity per se or to prior local aromatization to estrogens is unknown. Estradiol, itself, if given at high

[4] Produced by Silastic capsules containing 0.5 to 1 mg T/ml of corn oil.

enough doses, which presumably bypass binding to AFP, can effectively inhibit gonadotropin release in infantile rats (Ojeda *et al.*, 1975; Andrews and Ojeda, 1977; Meijs-Roelofs and Kramer, 1979).

Disclosure of such an inhibitory capacity of E_2 is in direct contrast with the absolute inability of the steroid to elicit a surge of gondotropin release before postnatal day 15 (Andrews *et al.*, 1981b; Puig-Duran and MacKinnon, 1978). Production, via Silastic capsules, of serum E_2 levels that exceeded the E_2 binding capacity of AFP by several times consistently failed to evoke LH release in 10- to 14-day-old rats. On the other hand, an increase in E_2 levels, amounting to only twice the level of E_2 necessary to induce an LH surge in juvenile rats, resulted in a blunted, but clearly demonstrable surge in LH release in 16- to 20-day-old rats.

These observations indicated to us that a complex array of steroid-dependent mechanisms is involved in determining the neonatal–infantile increase in gonadotropin release (Fig. 11), but gave us no clue as to the existence of steroid-independent components, which, if existed, we reasoned had to be originated within the CNS (Fig. 11). Indeed, evidence exists that the infantile hypothalamus of the female rat responds with more LHRH release to a depolarizing stimulus than does the hypothalamus of juvenile rats (Hompes *et al.*, 1982). On the other hand, the pattern of gonadotropin release in infantile females is highly reminiscent of that reported to occur under experimental conditions in the Rhesus monkey (Wildt *et al.*, 1981), when low-frequency pulses of LHRH are administered. It is conceivable, therefore, that while the reproductive hypothalamus becomes rapidly activated during infantile development, LHRH is released as infrequent discharges which maintain elevated FSH secretion, and generate short-lived bursts of LH release (Fig. 11). This hypothesis also implies that the decrease in FSH levels that follows the peak of secretion on day 12 is due to acceleration of LHRH discharges. Initial support for such an assumption is provided by the recent observation (Urbanski and Ojeda, unpublished) that sc pulsatile administration of LHRH[5] to infantile, ovariectomized rats via osmotic minipumps resulted in a decrease in FSH levels. More interestingly, FSH levels in untreated rats increased after ovariectomy (on day 6) to high levels and decreased thereafter to intermediate values at an age (about postnatal day 16) which corresponds to the age at which serum FSH levels normally begin to decline in intact rats. These results not only suggest that the frequency of

---

[5] Assuming that 1 LHRH pulse about every 3 hours maintains constantly high FSH levels (Wildt *et al.*, 1981) attempts to decrease FSH titers were made by providing ovariectomized rats with osmotic minipumps which had attached a polyethylene tubing containing segments of LHRH (separated by oil drops) at a distance which resulted in a 30 minute LHRH pulse delivered sc every hour.

MECHANISM OF FEMALE PUBERTY

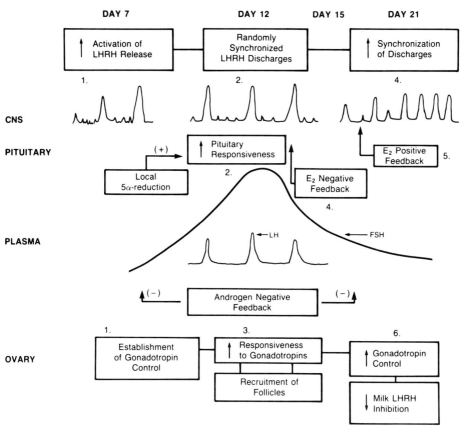

FIG. 11. Summary of main events occurring during infantile development of the female rat. The activity of LHRH neurons is increased, but LHRH neurons may discharge asynchronously. Random synchronization of these discharges may evoke infrequent "meaningful" LHRH pulses, which can maintain elevated plasma FSH levels and sporadically high LH values. Pituitary response to LHRH is high, in part because of a facilitatory effect of locally formed 5α-reduced androgens. The elevated plasma FSH levels are responsible for recruiting primordial follicles into a proliferative pool. After day 12, E_2 negative feedback begins to operate, LHRH neurons become more synchronized, and plasma FSH levels start declining. E_2 positive feedback becomes established after day 15. (−), Inhibitory; (+), stimulatory. Numbers refer to the sequence in which these events may occur.

LHRH discharge may play a substantial role in determining the pattern of changes in FSH release during infantile days but also provides initial evidence that such changes are, to a significant extent, the consequence of steroid-independent, centrally originated events (Fig. 11). These events, however, do not appear to originate within the POA-MBH area, as judged by the finding that infantile hypothalami *in vitro*, like those of neonatal rats, are capable of generating a rhythmic, pulsatile release pattern of LHRH which, by all criteria, appears indistinguishable from that of juvenile or peripubertal hypothalami (Urbanski and Ojeda, unpublished).

In any event, these purported centrally driven, steroid-modulated changes in gonadotropin secretion are not without transcendence as they appear to profoundly influence ovarian maturation, and hence the timing of puberty. Morphological evidence indicates that completion of ovarian follicular development takes no less than 15–19 days and that twice as many follicles start to grow in infantile as in juvenile rats (Hage et al., 1978). It then follows that at least some of the follicles recruited by the high serum FSH levels during infantile development (Schwartz, 1974) must be destined to ovulate at puberty, or at least to mature to a preovulatory, estrogen-secreting condition.

It may, therefore, be concluded from these observations that the first series of events with direct relevance to the onset of puberty does not occur close to the time of first proestrus, but rather during the infantile phase of development.

D. THE JUVENILE PERIOD

Once the turmoil of the infantile activational period subsides and the animal becomes juvenile it appears clear that the hypothalamic–pituitary unit has become capable, for the first time, of responding to estrogen stimulation with an increase in gonadotropin secretion. It is also apparent that at this time all basic components of the system (hypothalamus, pituitary, and ovaries) can respond with enhanced secretory activity to appropriate stimulatory inputs. Why then does a preovulatory surge of gonadotropin secretion fail to occur? The most obvious answer, i.e., "because the ovary is unable to generate estrogen levels of preovulatory magnitude," is undoubtedly an oversimplification of the situation. It is becoming clear that the acquisition of such ovarian competence is not a simple process but rather depends upon the integration of a complex series of mechanisms.

The beginning of the juvenile period (day 21) appears relatively quiescent as FSH levels decline, the bursts of LH release disappear, and AFP titers decrease further freeing more E_2 for biological activity. Promptly, however, a multiplicity of changes begins to occur and these establish definitive functional connections between the hypothalamic–pituitary unit and the ovary. On the one hand, the capacity of the animal to respond with LH release to E_2 levels of proestrous magnitude becomes fairly established by day 22 (Andrews et al., 1981b) (Fig. 12). On the other,

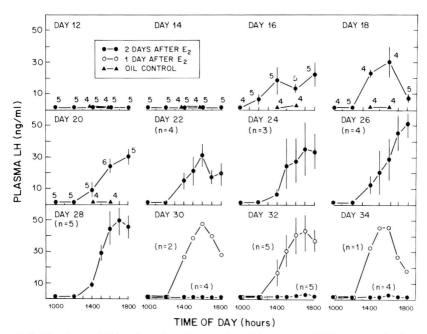

FIG. 12. Serum LH levels produced in prepubertal female rats of different ages by the sc implantation of Silastic capsules (20 mm/100 g body weight) containing E_2 dissolved in corn oil (400 μg/ml). Animals older than 20 days were bled every 2 hours via indwelling jugular cannulas. Between 12 and 20 days of age (weaning age, 21 days), animals were decapitated at the times indicated on the abscissa. Animals received the capsules 1 (○) or 2 (●) days before bleeding or decapitation. From Andrews et al. (1981b), reprinted with permission.

tonic gonadotropin secretion comes under firm estrogen inhibitory control which replaces the aromatizable androgens that played a predominant role during infantile days (Andrews and Ojeda, 1981a) (Fig. 13). Androgens, however, do contribute to the inhibitory control of gonadotropin secretion in juvenile rats but their role is relatively minor as replacement of physiological levels of T in ovariectomized, juvenile rats reduced gonadotropin titers only to levels seen in intact 12-day-old rats. In contrast, replacement of preovariectomy E_2 levels prevented the postovariectomy rise in LH and, to a significant extent, that of FSH (Fig. 13).

Throughout juvenile days waves of follicular development and atresia occur with overlapping frequency (Dawson and McCabe, 1951; Rennels, 1951; Richards, 1980), but in no instance does a crop of follicles reach the ovulatory stage. Acquisition of this capacity at puberty appears to be brought about by a multiplicity of hormonal and neurogenic factors operating under strict CNS direction.

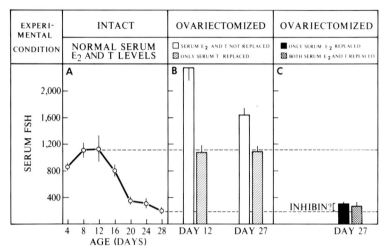

FIG. 13. Development of ovarian steroid negative feedback in the female rat. (A) Serum FSH levels in intact animals (days 4–28). (B) serum FSH levels in ovariectomized rats (open columns) and in ovariectomized rats in which precastration serum levels of T were quantitatively restored via Silastic capsules (hatched columns). (C) The effect of quantitatively replacing the precastration levels of serum E_2 and/or T on serum FSH of ovariectomized 27-day-old rats. In all cases, FSH was measured 2 days after ovariectomy, and the steroid replacement therapy was initiated immediately after castration. Each point or bar represents the mean of 7–12 animals and SEM are represented by vertical lines. From Andrews and Ojeda (1981a), reprinted with permission.

1. Hormonal Influences

The pivotal role played by LH and FSH in ovarian development is well established. Although serum LH levels are low during juvenile development, they are clearly pulsatile (Kimura and Kawakami, 1982; Andrews and Ojeda, 1981b). Careful examination of the pattern of LH using an automated bleeding system (Urbanski *et al.*, 1984) revealed that LH pulses occurred regularly every 30 minutes (Urbanski and Ojeda, 1985a) (Fig. 14). A similar pattern of release was observed in the morning and afternoon.

At the end of the juvenile period and heralding, in our view, the initiation of the peripubertal phase of development, basal LH levels begin to increase (Meijs-Roelofs *et al.*, 1983) and an afternoon increase in LH pulse amplitude becomes evident (Fig. 15) (Urbanski and Ojeda, 1985a). In the latter study some animals were found to exhibit an afternoon, more sustained episode of LH release (Fig. 16) which we have termed "LH minisurge" and that amounts to approximately 10% of the proper, proestrous surge of LH secretion.

The importance of these two modes of LH release for ovarian steroido-

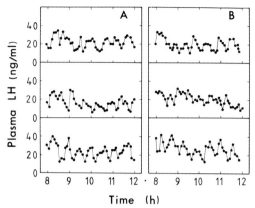

FIG. 14. Representative morning plasma LH profiles from juvenile (A; 27–29 days old) and peripubertal (B; 30–38 days old) female rats bled continuously for 4 hours. Six individual profiles from a total of 11 are depicted. From Urbanski and Ojeda (1985a), reprinted with permission.

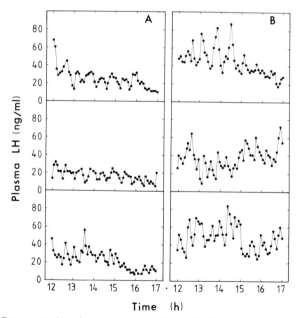

FIG. 15. Representative afternoon plasma LH profiles from juvenile (A; 27–29 days old) and peripubertal (B; 30–38 days old) female rats bled continuously for 5 hours. Six individual profiles from a total of 16 are depicted. Pulses of LH secretion are indicated by arrows. From Urbanski and Ojeda (1985a), reprinted with permission.

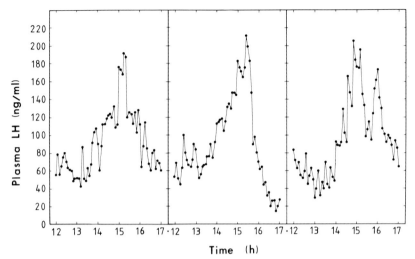

FIG. 16. Plasma LH profiles from three peripubertal (30–38 days old) female rats, showing a midafternoon minisurge of LH secretion. From Urbanski and Ojeda (1985a), reprinted with permission.

genesis was demonstrated by *in vitro* experiments (Urbanski and Ojeda, 1985b) in which peripubertal ovaries were exposed in a perifusion system to a pattern of LH in which the pulses resembled those previously observed *in vivo* either in the mornings or afternoons, or to a pattern comprising an LH minisurge.[6] Both large amplitude LH pulses (Fig. 17) and minisurges of LH stimulated E_2 and P release from the ovary. The effectiveness of LH given as pulses or minisurges was substantially greater than when LH was given continuously at the peak level, because less LH was needed to produce the same response. This demonstrates that interrupted delivery of LH to the gonad represents a much more efficient stimulatory signal for steroid release than exposure to continuous levels.

These observations permitted the conclusion that diurnal changes in the mode of LH release are of substantial importance for the ovary to become capable of producing a steroidal signal that can activate a surge of LH secretion. As we will discuss later in this review large amplitude LH pulses and LH minisurges are originated by activation of different mechanisms with independent anatomical localization.

Albeit of fundamental importance, gonadotropins are not the only ante-

[6] In addition to the LH pulses the perifusion medium contained FSH at a basal level (200 ng/ml) which is similar to that of late juvenile rats. Since the ovine FSH employed contains an equivalent of 20 ng LH/200 ng FSH, this amount closely mimicked basal LH levels observed in juvenile rats.

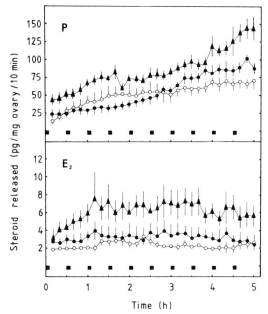

FIG. 17. Release of P and E_2 from ovaries exposed to various LH profiles *in vitro:* basal perfusion medium (○; $n = 5$), LH pulses of 20 ng/ml (●; $n = 4$), or LH pulses of 80 ng/ml (▲; $n = 5$). Each point represents the mean steroid level ± SEM (vertical lines). In some cases, the SEMs are so small that they fall within the symbol width. The horizontal bars above the abscissa indicate the periods of exposure to elevated LH concentrations in the perfusion medium. From Urbanski and Ojeda (1985b), reprinted with permission.

rior pituitary hormones involved in the regulation of ovarian maturation. Two additional hormones, prolactin (Prl) and growth hormone (GH) appear to play a supportive role, facilitating the effects of gonadotropins. The secretion of both somatomamotropins is low at the beginning of the juvenile period and increases gradually thereafter (Döhler and Wuttke, 1974, 1975; Ojeda and Jameson, 1977a; Eden, 1979). More importantly, the episodic mode of release of both hormones also becomes established during this time. While the adult quotidian pattern of GH release becomes firmly established around the time of puberty (Eden, 1979) a pattern of Prl release characterized by Prl discharges occurring approximately every 3 hours appears well defined by the beginning of the juvenile period (Kimura *et al.*, 1983). Nevertheless, the most prominent Prl secretory episodes occur at midafternoon and during the early morning hours. As the animal approaches the end of the juvenile period the nocturnal increases in Prl levels disappear whereas the afternoon surge becomes more prominent (Kimura and Kawakami, 1980). Our own results indicate that the afternoon increase in Prl levels can be already detected by the third week of postnatal life, i.e., during the infantile period (Ojeda *et al.*, 1976). We will present evidence later on suggesting that this afternoon change in Prl secretion is initiated by a central, gonadal-independent mechanism, a signal that is amplified by E_2.

It has been known for several years that Prl administration accelerates the onset of puberty in female rats (Clemens *et al.*, 1969). We were also struck by the common observation in pediatric endocrinology that children with isolated GH deficiency not only show attenuated growth, but also experience a delayed puberty, alterations which are all reverted by GH administration. To create a hyperprolactinemic condition we provided juvenile rats with sulpiride, a dopaminergic receptor blocker, in the drinking water (Advis and Ojeda, 1978) and the resulting changes in plasma Prl levels, ovarian steroidogenesis, and time of puberty were evaluated. Plasma Prl levels rose within a few hours after the sulpiride was made available to the rats and remained elevated throughout the study. Both vaginal opening and first ovulation were significantly advanced, an effect which could not be attributed to changes in mean plasma levels of LH or FSH. However, when the capacity of the ovaries to secrete steroids in response to gonadotropins was assessed *in vitro* we found that ovaries from hyperprolactinemic rats were much more responsive than controls. Both the E_2 and P responses to human chorionic gonadotropin (hCG) and to FSH were strikingly increased (Fig. 18). In a subsequent study we were able to demonstrate that hCG (LH)[7] receptor content of granulosa cells from hyperprolactinemic rats was significantly greater than that of controls, suggesting that Prl contributes to the maintenance and/or formation of LH receptors in the developing ovary (Advis *et al.*, 1981a).

Administration of ovine Prl to hypophysectomized rats yielded essentially identical results in that the ovarian P response to gonadotropins was markedly enhanced by the Prl treatment. Likewise, hyperprolactinemia induced by transplantation of anterior pituitaries from donor rats under the kidney capsule of juvenile animals rats resulted in precocious puberty which was preceded by an increase in ovarian steroidal response to gonadotropins (Advis and Ojeda, 1979). We next treated juvenile rats with CB-154, an ergoline derivative which, by binding to dopamine receptors, inhibits Prl release. As before, the drug was provided in the drinking water from postnatal day 22 onward (Advis *et al.*, 1981c). The results were the opposite of our previous findings with sulpiride as the onset of

---

[7] Human chorionic gonadotropin (hCG) is used as the ligand to measure LH receptors.

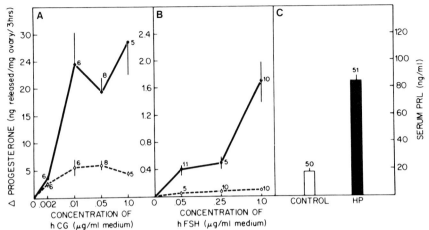

FIG. 18. Effect of different concentrations of highly purified human chorionic gonadotropin (hCG) (A) or human FSH (hFSH) (B) on *in vitro* P release (increment over basal release) from ovaries of intact or hyperprolactinemic (HP) 29-day-old immature rats. Hyperprolactinemia was induced by sulpiride treatment. PRL levels in HP and control animals are shown in (C). Vertical lines represent the SEM, and numbers next to means indicate the number of animals per group.(●) HP rats; (○) controls. From Advis *et al.* (1981a), reprinted with permission.

puberty was delayed, the ovarian E_2 and P response to gonadotropins was blunted, and the content of hCG (LH) receptors in the ovary was reduced.

We then concluded that one important function of the rising Prl levels during juvenile development is to facilitate the stimulatory actions of gonadotropins on ovarian steroidogenesis. A significant part of this effect appears to be exerted through a supportive action of Prl on the maintenance and/or formation of LH receptors. Nevertheless, a direct stimulatory effect of Prl on enzymes such as 20α-hydroxysteroid dehydrogenase (see Hsueh *et al.*, 1984, for a review) is likely to play a contributory role. A note of caution that should be introduced at this juncture concerns the fact that in all experiments in which Prl secretion was altered by dopaminergic agents we measured plasma LH using an heterologous RIA. The results showed that neither sulpiride nor CB-154 affected mean plasma LH levels; however, CB-154 slightly decreased the amplitude of the LH secretory episodes normally seen in immature rats. The possibility needs to be considered that more pronounced changes in plasma LH were not detected by the antiovine LH serum employed in the LH RIA. Very recently, using the NIADDK homologous rat RIA system we have confirmed (Urbanski and Ojeda, 1985a) the findings of Meijs-Roelofs *et al.*

(1983) who showed an increase in basal LH levels in animals approaching puberty. This elevation was unnoticed when the heterologous LH assay was employed (Ojeda et al., 1976; Andrews and Ojeda, 1981b).

Lacking a sufficient amount of pure GH preparation the approach of injecting GH into developing animals could not be used to assess the role of GH in the timing of puberty. We took advantage, however, of the fact that GH can control its own secretion through a short-loop negative feedback exerted on the hypothalamus (Katz et al., 1969; for review see Piva et al., 1979), and undertook studies to examine the effect of suppressing GH secretion on the onset of puberty. Implantation of GH into the MBH via a permanent stainless-steel cannula resulted in suppression of plasma GH levels, and, to our excitement, clearly delayed the age of vaginal opening and ovulation (Advis et al., 1981b). Examination of the ovaries from GH-deficient rats revealed a decreased content of LH receptors. Moreover, the ovarian P response to gonadotropins was significantly blunted suggesting that GH deficiency is associated with impaired ovarian function. This conclusion was further strengthened by the finding that in vivo administration of GH to hypophysectomized rats enhanced the subsequent in vitro P response of the ovaries to hCG and FSH (Fig. 19). In very recent studies Jia et al. (1985), using cultured granulosa cells from immature hypophysectomized rats, have shown that GH facilitates the capacity of FSH to induce LH receptors and to stimulate P secretion. Moreover, GH also facilitates the stimulatory effect of cyclic AMP and

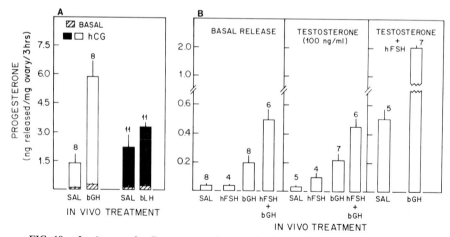

FIG. 19. In vitro ovarian P response to human chorionic gonadotropin (hCG) (A), T, or T together with human FSH (hFSH) (B) after in vivo administration of bovine LH (bLH), bovine GH (bGH), and/or hFSH. Numbers above bars indicate the number of animals per group. From Advis et al. (1981b), reprinted with permission.

forskolin on P secretion, indicating that the somatomamotrophic hormone acts at more than one biochemical step to positively regulate granulosa cell function.

The aforementioned studies have helped to clarify the role that GH and Prl play in the onset of female puberty. We believe that the juvenile ovary matures under the facilitatory influence of both Prl and GH, but primarily under the control of LH and FSH. As puberty approaches the ovaries gradually become more capable of responding to gonadotropin stimulation with estrogen release (Advis and Ojeda, 1978; Uilenbroek et al., 1983). Two events appear intimately associated with these changes in ovarian function. One of these is an increase in LH receptors, also occurring during infantile development but becoming more pronounced during the fourth week of life (Smith-White and Ojeda, 1981b). The other is a decrease in LHRH receptors which is initiated during the second half of the juvenile period (Smith-White and Ojeda, 1981a; Dalkin et al., 1981). This decrease becomes more pronounced during the days encompassing the first ovulation, as will be discussed later. Since LHRH has been shown to inhibit gonadal function under a variety of circumstances (for a review see Hsueh and Jones, 1981) one may argue that a decrease in LHRH receptor content reflects a decreased inhibitory influence of the peptide on ovarian development. The simultaneous increase in LH receptors would thus provide the necessary amplification for the stimulatory LH actions.

2. Neural Influences

It is now becoming increasingly clear that in addition to its hormonal control, the ovary is regulated by direct neural influences (for review see Burden, 1985; Ojeda and Aguado, 1985a). These neural inputs, we believe, provide the fine, minute to minute regulation of ovarian function.

The results of recent experiments in our laboratory have led us to conclude that the immature ovary is innervated by both adrenergic and peptidergic nerves. In regard to the adrenergic control, the developing ovary not only contains a well-defined population of β-adrenergic receptors of the β_2-subtype (Fig. 20) (Aguado et al., 1982) the content of which varies in relation to the phases of puberty, but it also exhibits a measurable amount of NE, which increases noticeably during juvenile development (Ben-Jonathan et al., 1984). These β_2-adrenergic receptors are coupled to progesterone and androgen release, an effect that can be demonstrated both by utilizing whole ovaries in short-term incubation and ovarian cells in culture (Ratner et al., 1980; Adashi and Hsueh, 1981; Aguado et al., 1982; Dyer and Erickson, 1985).

The sources of adrenergic inputs to the ovary appear to be (1) the

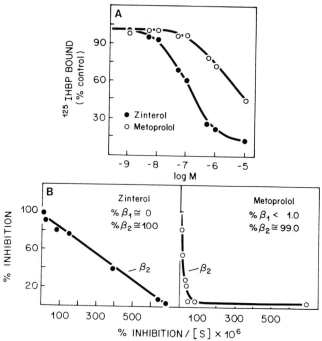

FIG. 20. (A) Competition for iodohydroxybenzylpindolol (IHBP) binding to ovarian membranes by zinterol, a β_2-agonist, or metoprolol, a β_1-antagonist. Membranes were incubated with IHBP (4 pM) in the presence or absence of adrenergic agents under the conditions indicated in the original report (Aguado et al., 1982). Each point represents the mean of duplicate determinations. The curves shown are representative of two experiments. The binding of IHBP in the absence of zinterol or metoprolol was 100%. (B) Hofstee plots for the inhibition of specific IHBP binding by zinterol and metoprolol to ovarian membranes from 30-day-old rats. The data were obtained using the assay conditions described in the original report. Each point represents the mean of duplicate determinations. In both cases, the presence of a single type of β-binding sites is indicated by the presence of a linear plot. From Aguado et al. (1982), reprinted with permission.

adrenergic nerves and (2) circulating epinephrine (EPI) of adrenal medullary origin. Based on the findings that adrenal medullectomy, which depressed plasma EPI levels, delayed the onset of puberty whereas EPI at nanomolar concentrations amplified the stimulatory effect of hCG and FSH on P secretion from granulosa cells in culture (Aguado and Ojeda, 1984a) we have hypothesized that under physiological conditions circulating EPI facilitates the effect of gonadotropins on P secretion and stimulates P secretion on its own.

The ovary is innervated by two main adrenergic nerves, the superior ovarian nerve (SON) which carries most of the noradrenergic fibers in-

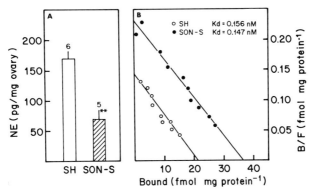

FIG. 21. Effect of superior ovarian nerve (SON) section on ovarian NE content (A) and the number and affinity of ovarian β-adrenergic binding sites (B). The nerve was transected on day 24, and the ovaries were collected 7 days later. SH, Sham-operated controls; SON-S, SON-sectioned animals. Vertical lines represent the SEM, and numbers above bars indicate the number of animals per group. **$p < 0.01$ vs SH control. From Aguado and Ojeda (1984b), reprinted with permission.

nervating the steroidogenic tissue of the gland, and the plexus nerve (PN) which primarily innervates the ovarian vasculature (Burden and Lawrence, 1980). Section of the SON, when performed in proestrous animals, resulted in an acute drop in P and E_2 secretion as measured in the ovarian vein effluent (Aguado and Ojeda, 1984c) suggesting that activation of neural inputs reaching the ovary via the SON is involved in maintaining and/or enhancing the increased steroid output that characterizes the day of proestrus. On a more chronic basis SON section resulted in a compensatory increase in β-adrenergic receptors (Fig. 21) (Aguado and Ojeda, 1984b). This increase was accompanied by hypersensitivity of steroid secretion to β-adrenergic stimulation as determined using an *in vitro* model for denervation,[8] which consists of granulosa cells cultured in the presence or absence of NE.

Thus, these results suggest that the adrenergic system is indeed involved in controlling specific functions of the immature ovary. The neural inputs arriving at the ovary, however, are not limited to noradrenergic fibers. In very recent experiments we have observed, utilizing immunohistofluorescence methods, the presence of delicate nerve fibers containing either substance P (SP), VIP, or neuropeptide Y (NPY). These fibers innervate the ovarian vasculature and interstitial tissue and are associated with the thecal layers of developing follicles (Dees *et al.*, 1985a; Ahmed *et al.*, 1985; McDonald *et al.*, 1986). We have also characterized biochemi-

---

[8] Details about this model are provided in the report by Aguado and Ojeda (1984b).

cally SP and VIP in the juvenile ovary (Ojeda et al., 1985a; Ahmed et al., 1985) and concluded that they are immunologically and chromatographically indistinguishable from the authentic peptides.

In searching for the function of these peptides we were unable to demonstrate any significant effect of SP on ovarian steroidogenesis in vitro, either using ovarian fragments or cultured granulosa cells (Ojeda et al., 1985a). Nevertheless, the histological localization of SP and NPY-containing fibers gives us a clue as to their possible function. Both peptides, in particular NPY in the ovary, innervate predominantly ovarian blood vessels, thus suggesting their participation in regulating ovarian blood flow. In contrast, VIP (which also innervates ovarian blood vessels) was found to be a potent stimulus for T, P, and E_2 secretion, as assessed by the in vitro incubation of immature ovaries with the peptide (Fig. 22) (Ahmed et al., 1985, 1986). This effect of VIP was not shared by any of the members of the VIP family (secretin, glucagon, and gastric inhibitory peptide) with the exception of peptide with N terminus histidine, C terminus isoleucine (PHI), which has the greatest degree of sequence homology with VIP and was 50% as effective as VIP at stimulating steroid

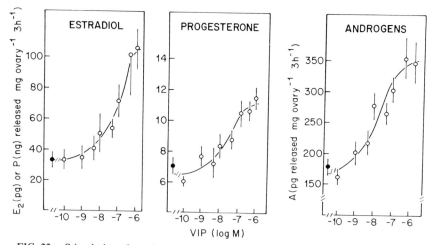

FIG. 22. Stimulation of ovarian steroid production in vitro by different concentrations of VIP. To examine the effect of VIP on E_2 and androgen (A) secretion 29-day-old rats were injected with 2 IU pregnant mare serum gonadotropin (PMSG) and killed 24 hours later. To examine the P response animals were killed 96 hours after PMSG, i.e., after ovulation, when corpora lutea had formed. The incubation was carried out as reported (see below). Each point represents the mean of at least 4 ovaries ± SEM. The closed circles represent basal levels of steroid in the absence of exogenous VIP. From Ahmed et al. (1986).

release. Motilin and GH-releasing factor, both of which have very little homology with VIP, were ineffective.

Of considerable interest was the finding that section of the SON completely eliminated immunofluorescent VIP-ergic nerve fibers in the ovary without altering SP-containing nerves (Dees et al., 1985b). Conversely, section of the plexus nerve was found to eliminate SP-immunoreactive fibers without affecting VIP immunoreactivity. These observations provide additional credence to the notion that SP and VIP play different roles in the control of ovarian development. In addition, the earlier surprising finding that sectioning of the SON resulted in a rapid decrease of E_2 secretion (Aguado and Ojeda, 1984c) can now be best attributed to the elimination of VIP-ergic fibers and not, as we originally believed, to transection of noradrenergic fibers. We should recall, at this point, that adrenergic stimulation of granulosa cells has consistently failed to induce E_2 release (Adashi and Hsueh, 1981; Aguado and Ojeda, 1984c), an outcome that has led to the conclusion that ovarian β-adrenergic receptors are coupled only to P and not to E_2 secretion.

It may be concluded from the foregoing observations that the nervous system directs the maturation of the ovary via two main routes, a hormonal and a neurogenic one (Fig. 23). While the former involves the secretion of hypothalamic factors that control the secretion of LH, FSH, Prl, and GH from the adenohypophysis, the latter directly links the CNS to the ovary via peptidergic and adrenergic nerves. Moreover, EPI of adrenal medullary origin also appears to facilitate ovarian function after reaching the gland via the blood stream.

It would appear that most of these regulatory mechanisms become firmly established during the juvenile period. While they develop the hypothalamic LHRH system is far from being quiescent, as one might wrongly assume based on the low juvenile levels of plasma gonadotropins. The turnover of hypothalamic catecholamines increases throughout juvenile development (Raum et al., 1980; Wuttke et al., 1980) and the capacity of the hypothalamus to release LHRH becomes more pronounced, as indirectly assessed in vivo by stimulating LH release with prostaglandin E_2 (PGE_2) (Andrews and Ojeda, 1978) or by directly measuring LHRH release in response to PGE_2 in vitro (Ojeda et al., 1986a) (Fig. 24). The LHRH response to PGE_2 may be considered a reliable index of LHRH neuron function as a sizable body of evidence indicates that PGE_2 is a physiological intracellular component in the process of neurotransmitter-induced LHRH release (for a review see Ojeda and Aguado, 1985b).

The age-related increase in the capacity of LHRH neurons to release

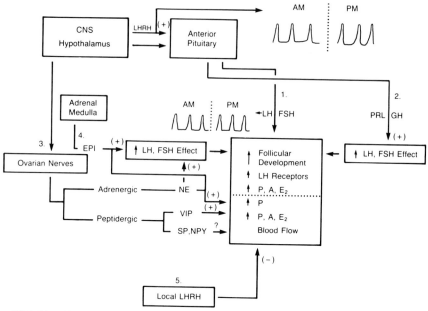

FIG. 23. Hormonal and neurogenic factors controlling ovarian development during the juvenile period of the female rat. A similar pattern of LH release in the mornings and afternoons is assumed to be the consequence of an unchanged pattern of LHRH release. Numbers indicate the different control mechanisms involved in regulating ovarian function. (+), Facilitatory; (−), inhibitory; ?, effect not known.

LHRH in response to PGE_2 appears to be, at least partially, maintained by circulating E_2 levels. Early ovariectomy (performed on day 22) blunted the LHRH response to PGE_2 on day 34. On the other hand, restoration of juvenile E_2 levels via E_2-containing Silastic capsules significantly reverted the effect of ovariectomy (Ojeda et al., 1986a).

These functional changes are accompanied by profound alterations in the morphology of LHRH neurons as demonstrated by Wray and Hoffman-Small (1984) who showed that the proportion of LHRH neurons having an irregular cell surface ("spiny" cells) increases markedly between postnatal day 26 to 32 while the "smooth" type of cell decreases. That these changes may be independent of estrogen action is suggested by the recent observation that they also occur in gonadectomized rats (Wray and Gainer, 1985).

In spite of these developmental alterations of the LHRH-secreting system the mode of LH release in juvenile rats is fairly constant, perhaps because as the capacity of the hypothalamus to release LHRH increases,

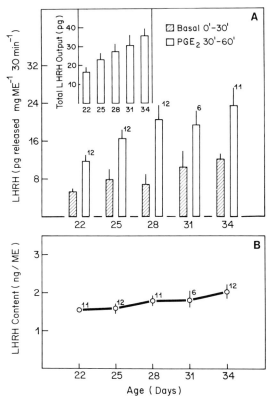

FIG. 24. (A) Changes in LHRH release from isolated median eminences *in vitro* in response to PGE$_2$ during juvenile and early peripubertal sexual development of the female rat. Following a 15-minute preincubation period the tissues were incubated for 30 minutes to determine basal LHRH release, and then for a further 30 minutes in the presence of PGE$_2$ (2.8 μ*M*). The inset depicts the sum of LHRH released under basal conditions and during PGE$_2$ stimulation at the different ages studied. (B) Changes in LHRH content of the median eminence during juvenile–early peripubertal development of the female rat. Vertical lines represent SEM and numbers next to means indicate the number of median eminences per group. From Ojeda *et al.* (1986a), reprinted with permission.

the pituitary LH response to LHRH paradoxically declines (Debeljuk *et al.*, 1972; Ojeda *et al.*, 1977; Andrews and Ojeda, 1978).

We have previously suggested that the end of the juvenile period, or rather the initiation of the peripubertal phase, is morphologically indicated by the appearance of intrauterine fluid (Ojeda *et al.*, 1980). This phenomenon, however, is only the result of earlier neuroendocrine events, one of which, we now believe, clearly establishes the end of juvenile days: the appearance of diurnal variations in the mode of LH release.

E. THE PERIPUBERTAL PERIOD

1. The Initial Events

Although strictly speaking the onset of puberty is determined by a multiplicity of interrelated events, some of which find their origin during the infantile period, direct neuroendocrine manifestations of the process itself only become evident after the fourth week of life. At this time plasma Prl and GH levels have increased significantly from juvenile values, ovarian responsiveness to gonadotropins is also increasing and the mode of LH release begins to change. As we have discussed earlier in this review the amplitude of the LH pulses increases in the afternoon, a time when an elevated basal release also becomes evident. Whether or not this afternoon increase in LH pulse amplitude is caused by a steroid-independent signal of neural origin or is the mere consequence of the changing steroid milieu is a matter of profound interest to us. We can confidently state that the phenomenon is not ovarian dependent and, more specifically, is not E_2 induced. Short-term ovariectomy of juvenile rats and concomitant restoration of precastration serum E_2 levels via sc Silastic capsules resulted in inhibition of pulsatile LH release, rather than in enhancement of LH pulse amplitude (Urbanski and Ojeda, 1986a). Stepwise increases in E_2 levels consistently failed to induce an increase in LH pulse amplitude, but rather resulted in the appearance of mini- and proper surges of LH secretion (vide infra).

The possibility that the afternoon increases in LH are, in fact, due to events occurring within the CNS and are not induced by ovarian influences received support from experiments in which the pattern of LH release was examined shortly after removal of the ovaries. Two days after ovariectomy plasma LH levels of prepubertal rats were found to be higher in the afternoon than in the morning. Close examination of the LH release pattern at this time, using an automated 5-minute bleeding paradigm (Urbanski *et al.*, 1984), revealed that these high LH values were not due to a sustained surge of LH release but were rather the consequence of an increased amplitude of LH secretory episodes (Fig. 25) (Urbanski and Ojeda, 1986b). Of considerable interest was the subsequent finding that occurrence of this diurnal change in mode of LH release could not be detected in 22- to 24-day-old juvenile rats, but it became demonstrable in 28- to 29-day-old rats and was clearly evident in 35- to 39-day-old animals. It thus seems that the strength of the signal responsible for the afternoon

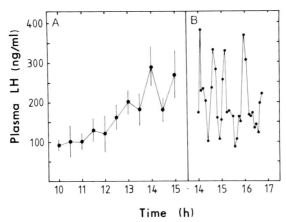

FIG. 25. Plasma LH levels in late juvenile (29-day-old) rats 48 hours after ovariectomy. The afternoon, gonadal-independent increase in mean LH levels (A, $n = 5$) was analyzed using a 5-minute automated bleeding paradigm which revealed that the elevated LH levels were due to an increased amplitude of episodic LH discharges (B) and not to a sustained surge of LH release. Vertical lines in A represent SEM. From Urbanski and Ojeda, unpublished data.

activation in LH release increases as the animal matures. This tentative conclusion has been supported by the results of other experiments which demonstrated that 4 days after ovariectomy the morning–afternoon difference in LH release is still unambiguous in peripubertal animals, in spite of a further increase in morning LH levels. In contrast, in late juvenile rats the morning increase in LH release, which presumably reflects removal of steroid negative feedback, has reached a point such that morning LH levels become similar to those in the afternoon. It would thus seem that removal of the ovaries relieves the LHRH–LH system from steroid negative feedback control and permits the disclosure of a gonadal-independent event which becomes expressed in the afternoon. As more time elapses after removal of the gonads the LH response to removal of steroid inhibitory control becomes stronger, overshadowing the diurnal changes in LH output. Since by 4 days after ovariectomy morning–afternoon differences are no longer seen in juvenile rats, but are clearly detected in peripubertal rats, the inevitable conclusion is that the strength of the central signal is greater in the older animals and can, therefore, be expressed in spite of the rising LH levels following removal of steroid inhibitory control.

Additional support for the concept that a change in "central drive" occurs as the female rat matures comes from the results of other recent experiments involving Prl release. It is known that Prl secretion occurs episodically (Kimura *et al.*, 1983) and that during development of the female rat Prl levels become more elevated in the afternoons than in the mornings (Ojeda *et al.*, 1977; Kimura and Kawakami, 1980). Kimura and Kawakami (1981) observed that ovariectomy of early juvenile rats did not abolish the afternoon surge of Prl indicating that the surges could occur in the absence of the ovaries. These findings however, did not completely rule out the possibility that the Prl surges are, in fact, induced by E_2, and that their expression persists for several days after ovariectomy (Gräf *et al.*, 1976) in a manner similar to that observed for LH (Legan *et al.*, 1975). In addressing this issue we have observed (Urbanski and Ojeda, 1985c) that if female rats are ovariectomized while neonates and their plasma pattern of Prl examined at different ages thereafter, at least 50% of the animals exhibit a mid-afternoon secretory episode of Prl release even as late as 40 days after ovariectomy. Restoration of juvenile serum E_2 levels via sc E_2-containing Silastic capsules resulted in amplification of the surge, which still occurred at the same time of the day. By using a more concentrated solution of E_2 in the capsules, proestrous serum levels of E_2 were produced and this resulted in a further amplification of the Prl surge, which closely resembled that normally seen at proestrus. Thus, our results are not only in harmony with those of Kimura and Kawakami (1981), but they also demonstrate that the neural mechanism responsible for the afternoon appearance of a Prl surge can develop in the absence of ovarian influences.

A quite different picture emerged when attempts were made to elucidate the mechanisms determining the peripubertal "minisurges" of LH secretion (vide supra). Such sustained episodes of release were never observed in either short- or long-term ovariectomized rats. Only when circulating E_2 levels were slightly increased over juvenile values via E_2-containing Silastic capsules did a minisurge of LH secretion occur (Fig. 26).

These observations lead us to believe that once the complex array of neuroendocrine regulatory mechanisms previously discussed becomes fully functional at the end of the juvenile period the hypothalamic LHRH-secreting system becomes activated[9] by hitherto unknown inputs, the operativity of which is accentuated in the afternoons (Fig. 27). The increased LHRH secretion then generates a diurnal pattern of LH release

---

[9] It is unclear whether this "activation" is caused by removal of inhibitory inputs impinging on LHRH neurons or by accentuation of stimulatory inputs. Other researchers espouse the view that puberty occurs as a result of disappearance (or attenuation) of a "central restraint" (see for instance Reiter and Grumbach, 1982; Plant, 1980). For reasons explained in detail elsewhere (Ojeda *et al.*, 1984a) we have preferred to use as a working hypothesis the alternative notion, i.e., that puberty is initiated by the activation of excitatory inputs (Ruf and Sharpe, 1979).

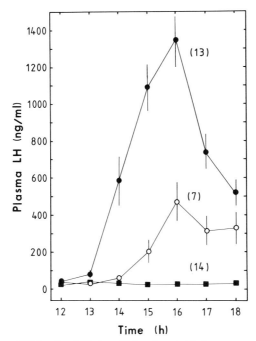

FIG. 26. Plasma LH levels of 28-day-old female rats bled once every hour from 1200–1800 hours. The animals were ovariectomized (at 26 days of age) and given sc Silastic capsules containing various concentrations of E_2 dissolved in corn oil. The plasma LH profiles have been grouped according to the type of response shown, irrespective of the actual concentration of E_2 used: basal release (■), minisurge (○), and the preovulatory-like surge (●). Each point represents the mean and the SEM are shown as vertical lines unless obscured by the symbol. Numbers in parentheses indicate number of animals per group. Drawn from data in Urbanski and Ojeda (1968a).

characterized by afternoon LH pulses of large amplitude. Such an event, we suspect, is the precipitating factor that determines the initiation of puberty, as under the influence of these LH secretory episodes the ovary is stimulated to produce more E_2 (Urbanski and Ojeda, 1985b). In turn, subtle increases in E_2 levels appear able to evoke minisurges of LH secretion (Urbanski and Ojeda, 1986a) which can induce further ovarian activation (Fig. 27).

2. *The Precipitation of Events*

From this point onward a new fast-moving cascade of events develops that culminates with the first preovulatory surge of gonadotropin and the first ovulation. In trying to elucidate the several components of this cascade we have considered it necessary to partition the various stages of

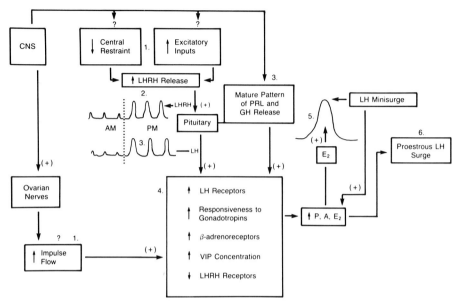

FIG. 27. Postulated cascade of initial events during the onset of puberty in the female rat. Activation of LHRH release in the afternoon may be due to elimination of a central restraint or to increase in excitatory inputs to LHRH neurons. Whether or not the activity of ovarian nerves also increases at this time is not known, but it is suggested by the changes in ovarian β-adrenergic receptors and ovarian VIP concentration. The numbers indicate sequence in which these events may occur.

puberty into well-circumscribed phases. According to this classification, which is mainly based on morphological criteria (Ojeda et al., 1976; Advis et al., 1979), puberty in the female rat can be divided into the following phases. *Anestrus* (A): this phase was meant to correspond to the late juvenile phase, but we now suspect it may well be the phase during which the changes in the mode of LH release begin to occur (Urbanski and Ojeda, 1985a). Animals in this phase are older than 30 days of age, their uteri are small (wet weight less than 100 mg), and, more importantly, no intrauterine fluid can be detected. The vagina is always closed. *Early proestrus* (EP): animals in this phase have larger uteri with intraluminal fluid; their vagina is closed. *Late proestrus* (LP): this phase corresponds to the day of first proestrus. Animals have large "ballooned" uteri, full of fluid with a wet weight greater than 200 mg. Their ovaries have large follicles. Most animals in this phase show closed vaginae. *Estrus* (E) is the day of first ovulation, uterine fluid has disappeared, fresh corpora lutea can be readily discerned, the vagina is open and vaginal cytology shows a predominance of cornified cells. *First diestrus* (D_1) is the phase of puberty characterized by a vaginal cytology showing a predominance of leukocytes, and by the presence of mature corpora lutea within the ovaries.

With the onset of puberty already determined by the activation of pulsatile LH release the single most important event that remains to be defined is the timing of the first preovulatory surge of gonadotropins, which in itself represents the culmination of female neuroendocrine reproductive maturation. There is little doubt, in our view, that both the occurrence and the timing of this final event depend primarily on the completion of ovarian maturation. Only when the ovary becomes capable of producing E_2 levels of sufficient magnitude and for a long enough period of time will the preovulatory LH surge occur.

A multitude of maturational changes appears to be involved in hastening the acquisition of preovulatory competence by the ovary. While FSH receptor content is already maximal by the end of the juvenile development, the number of LH receptors in granulosa cells increases dramatically between the A and LP phases of puberty (Smith-White and Ojeda, 1981b). Concomitant with this increase, an abrupt drop in LHRH receptor content takes place, the magnitude of the decrease being more pronounced between A and EP than at later times (Smith-White and Ojeda, 1983). We have already discussed the implications that these two divergent changes in hormone receptor may have for ovarian function. It is noteworthy that during the days preceding the preovulatory gonadotropin surge the steroidal responsiveness of the ovary to gonadotropins increases dramatically (Fig. 28) (Advis *et al.*, 1979; Uilenbroek *et al.*, 1983), most likely reflecting the progressive development of the follicles destined to ovulate at the first estrus.

The neurogenic component of the ovary also undergoes profound changes. Thus the content of β-adrenergic receptors increases between A and LP, and then declines abruptly at the time of the proestrous surge (Aguado *et al.*, 1982). In harmony with the increase in receptor content, P responsiveness to β_2-adrenergic stimulation becomes augmented between A and LP. However, the greatest change in response occurs after ovulation, more specifically during the first estrus, at which time the receptor content is surprisingly low.

The concentration of VIP in the ovary, which remains almost unchanged between the second postnatal day and the end of juvenile development, increases significantly during the early part of the peripubertal period (day 30–35) (Ahmed and Ojeda, unpublished). Moreover, the steroidogenic response to VIP undergoes profound changes at the time of puberty (Ahmed *et al.*, 1986). Particularly relevant to the acquisition of ovarian preovulatory capacity is the fact that the E_2 response to VIP, albeit evident in juvenile rats, increases strikingly during the EP and LP

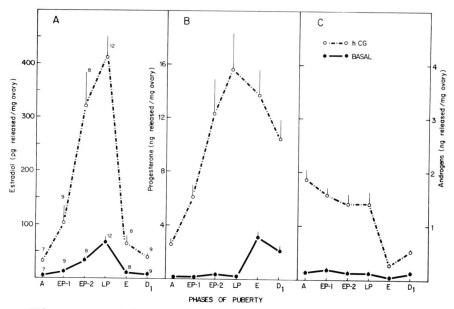

FIG. 28. Changes in the *in vitro* ovarian responsiveness to human chorionic gonadotropin (hCG) during the time of puberty in female rats. The numbers next to the means indicate the number of ovaries per group. From Advis *et al.* (1979), reprinted with permission.

phases of puberty (Fig. 29). The P response to the peptide increases only moderately at this time, and then abruptly after ovulation. Radioimmunoassayable SP content in the ovary also increases between A and LP (Ojeda *et al.*, 1985b). Although we do not know what function these changes in SP content may have, we suspect that they may be implicated in the edematization of the ovary that occurs at puberty (Osman, 1975) and/or in changes in blood flow which occur during the preovulatory period (Niswender *et al.*, 1976; Zeleznick, 1982).

Viewed in the light of these observations (Fig. 27), it is not surprising that the pattern of steroid production by the ovary changes so dramatically. Serum E_2 levels increase markedly between A and LP (Meijs-Roelofs *et al.*, 1975; Andrews *et al.*, 1980) reaching values of about 80 pg/ml during the morning of the latter phase (Andrews *et al.*, 1980). Serum P increases moderately before the LH surge but serum T levels do so more prominently (Andrews *et al.*, 1980). This increase in T (or in more general terms aromatizable androgen) appears to have relevance in the mechanism of vaginal opening. The production of early proestrous levels of T via T-containing Silastic capsules in late juvenile rats resulted in precocious vaginal opening, but not in advancement of the first ovulation

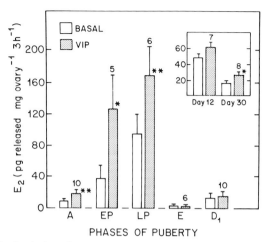

FIG. 29. VIP stimulation of E_2 production *in vitro* from ovaries at different phases of development. Ovary halves were preincubated in Krebs-Ringer bicarbonate buffer containing 1 mg glucose/ml for 30 minutes, then with or without 1 μM VIP for three hours in Krebs-Ringer bicarbonate buffer with 0.1 mg BSA/ml and 5×10^{-5} M bacitracin. The number of animals per group is indicated above each bar. Values shown are means ± SEM. $*p < 0.05$; $**p < 0.01$ vs basal levels. A, Anestrus, early peripubertal animals; EP, early proestrus; LP, late (first) proestrus; E, first estrus; D_1, first diestrus. From Ahmed *et al.* (1986).

(Mathews *et al.*, 1986). Examination of serum E_2 levels in animals treated with T revealed that E_2 was not increased by the exposure to elevated T levels. Nevertheless, the presence of a small, but measurable level of aromatase activity was detected in the vaginal epithelium suggesting that the hastening effect of physiological levels of T on vaginal opening is, at least in part, due to local estrogen production by aromatization.

While the secretion of E_2, P, and T increases, the secretion of 3α-androstanediol diminishes, a change that becomes much more pronounced during the hours encompassing the first preovulatory LH surge and that seems to be induced, at least in part, by the rising Prl levels (Advis *et al.*, 1981d; Ojeda *et al.*, 1984b). This decrease, however, does not appear to relieve gonadotropin release from a 3α-androstanediol inhibitory control (Ojeda *et al.*, 1984b) as was originally proposed by Eckstein *et al.* (1967).

The changing levels of steroids in the blood stream profoundly affect hypothalamic reproductive functions. Noradrenergic and serotoninergic activity increases before the proestrous surge of gonadotropins with the former increasing more noticeably during early proestrus and the latter on the day of late proestrus (Advis *et al.*, 1978). Evidence exists that at least the enhancement in NE activity results from an E_2 action, as E_2 has been shown to increase NE turnover in the hypothalamus (for review see Barclough et al., 1984; Ramirez et al., 1984) and to promote NE release from hypothalamic slices in vitro (Paul et al., 1979).

The capacity of the hypothalamus to synthesize PGE_2 from [^{14}C]arachidonic acid increases during the days preceding the preovulatory LH surge (Fig. 30; Ojeda and Campbell, 1982). Such an increase may be of considerable importance for the proestrous surge of LHRH to occur since PGE_2

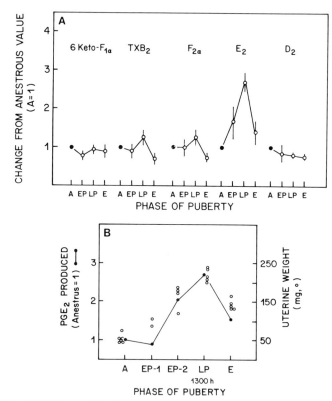

FIG. 30. (A) Changes in hypothalamic production of prostaglandins (PGs) from [^{14}C]arachidonic acid during the time of puberty. For each PG the area under the radioactive peak in the HPLC profile was measured with a planimeter. The areas of the PG peaks of anestrous (A) rats were assigned a value of 1 and the corresponding areas at the other phases of puberty were expressed in relation to the A values. Each point represents the mean of three determinations ± SEM. At early proestrus (EP) the mean shown derives from values observed at EP-1 (1 determination) and EP-2 (two determinations). LP, Late proestrus; E, estrus. (B) Correlation between changes in uterine weight (an index of estrogenic activity) and changes in hypothalamic production of PGE_2 during the time of puberty. From Ojeda and Campbell (1982), reprinted with permission.

has been implicated as an obligatory intermediate in the mechanism by which NE induces LHRH release (for reviews see Ojeda and Aguado, 1985b; Ramirez et al., 1985). The augmented capacity of the medial basal hypothalamus to produce PGE_2 may be a consequence of E_2 action as administration of E_2 to juvenile rats reproduced the increase in PGE_2 synthesis from [^{14}C]arachidonic acid observed in normal rats at puberty (Ojeda and Campbell, 1982).

To gain further insight into this phenomenon juvenile rats were provided with Silastic capsules containing E_2 dissolved in corn oil at a concentration (400 μg/ml) that produces serum levels of E_2 similar to those seen on the day of proestrus (Andrews et al., 1980). Subsequent incubation of their median eminences with various concentrations of NE or PGE_2 revealed that E_2 facilitates LHRH release by acting at two different biochemical steps (Ojeda et al., 1986a). On the one hand, it enhances the sensitivity of the PGE_2 synthesizing machinery to NE stimulation without increasing its responsiveness. On the other, it increases both the sensitivity and the responsiveness of the LHRH terminals to PGE_2. As expected from these latter results, maximal LHRH response to NE was also greater in E_2-treated rats than in controls, and NE elicited significant LHRH release at doses that were ineffective in untreated animals (Fig. 31).

Thus it appears that E_2 exerts its stimulatory action on LHRH release

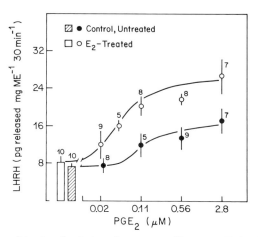

FIG. 31. Effect of in vivo simulation of proestrous-like serum E_2 levels on the in vitro release of LHRH from the median eminence of juvenile 28-day-old rats in response to PGE_2. Estradiol (E_2) was provided in sc Silastic capsules at a concentration (400 μg/ml corn oil) that reproduces serum E_2 levels found on the day of first proestrus. The E_2-containing capsules were implanted 48 hours before the experiment. From Ojeda et al. (1986a), reprinted with permission.

by activating a NE-PGE$_2$ dependent pathway. Ramirez et al. (1985) and we (Ojeda et al., 1974, 1985a) have provided evidence that cyclic AMP is involved in the process of LHRH secretion, and have concluded that formation of cyclic AMP occurs at a step subsequent to NE-induced synthesis of PGE$_2$. This view stems from the findings that PGE$_2$ increases cAMP production from hypothalamic fragments *in vitro* and that methoxamine, an α_1-adrenergic agonist, enhances both cyclic AMP formation and LHRH release, an effect that is blocked by indomethacin, an inhibitor of prostaglandin synthesis (Ramirez et al., 1985). Indomethacin also blocks NE-induced LHRH release without altering the stimulatory effect of PGE$_2$ on LHRH release (Ojeda et al., 1979). Furthermore, when adenylate cyclase activity was stimulated by forskolin, cholera toxin, or pertussis toxin, LHRH release was enhanced but PGE$_2$ formation, if anything, was depressed (Ojeda et al., 1985a). This finding and the observation that simultaneous exposure of median eminence nerve terminals to PGE$_2$ and forskolin did not result in a greater LHRH response than that elicited by either agent individually further suggested that PGE$_2$ and cyclic AMP act along a common pathway. Moreover, these results lend additional credence to the view that formation of cyclic AMP follows changes in PGE$_2$ synthesis, which in turn are induced by binding of NE to α-receptors (Ojeda et al., 1982) of the α_1-subtype (Heaulme and Dray, 1984).

Since E$_2$ can also increase cyclic AMP formation in the hypothalamus (Gunaga and Menon, 1973; Weissman and Skolnick, 1975) the suggestion can be made that the first preovulatory surge of LHRH at puberty (Sarkar and Fink, 1979) involves the activation of a PGE$_2$–cyclic AMP pathway. This mechanism, however, may not be the only one that operates at the first proestrus. In very recent experiments (Ojeda et al., 1986b) we have found that activation of protein kinase C, a Ca^{2+} activated, phospholipid-dependent kinase (Nishizuka, 1983, for a review) by either a synthetic diacylglycerol or a phorbol ester resulted in LHRH release (Fig. 32). Moreover, phospholipase C which in intact cells catalyzes the hydrolysis of membrane polyphosphoinositides to generate diacylglycerol also induced LHRH release. Activation of this protein kinase C-dependent pathway was found to induce release of LHRH independently from the PGE$_2$-cyclic AMP pathway. This conclusion was mainly derived from three observations. One of them showed that blockade of prostaglandin synthesis failed to suppress diacylglycerol-induced LHRH release. The second finding was that phorbol ester induced LHRH release without affecting PGE$_2$ formation. The last one demonstrated that exposure of median eminences to maximally effective doses of diacylglycerol or phorbol ester

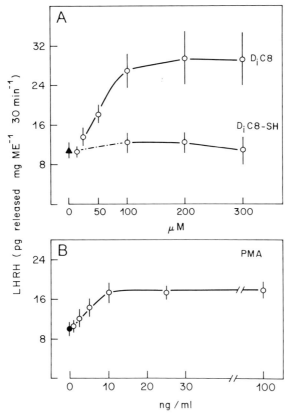

FIG. 32. Stimulatory effect of two activators of protein kinase C on *in vitro* LHRH release from isolated median eminences (MEs) of 28-day-old juvenile female rats. Both dioctanoylglycerol (DiC_8, A) and the phorbol ester 4, β-phorbol 12β-myristate 13α-acetate (PMA, B) stimulated LHRH release. A DiC_8 analog in which the 3′ hydroxyl group was replaced by a sulfhydryl moiety (DiC_8-SH, A) was ineffective. Each point represents the mean of 5–10 individual MEs and SEMs are shown as vertical lines. From Ojeda *et al.* (1986b).

together with PGE_2 or forskolin resulted in an additive effect on LHRH release, suggesting coexistence of both pathways.

Based on these observations we have suggested that simultaneous activation of both pathways may be required for the proestrous surge of LHRH to occur. Such a mechanism implies that either NE itself acts on two subtypes of adrenergic receptor, or that NE and another neurotransmitter act on different receptors to provide the extracellular signal for activation of both pathways and the corresponding increase in LHRH release. In any event, while the intracellular mechanisms leading to the proestrous surge of LHRH are just being scrutinized, it is clear that the surge is an E_2-dependent phenomenon and that its inevitable consequence is the preovulatory discharge of gonadotropins.

We have deliberately left the anterior pituitary out of these considerations based on the belief that the most important determinants of the initiation and timing of puberty are the hypothalamus and the ovary, respectively. Nevertheless, changes in the responsiveness of pituitary gonadotrophs to LHRH are of considerable importance for the occurrence of the proestrous gonadotropin surge. While pituitary responsiveness to LHRH is markedly elevated during the juvenile period and declines throughout juvenile development (vide supra), it increases again abruptly on the day of proestrus (Castro-Vazquez and Ojeda, 1977; Sarkar and Fink, 1979). This increase appears to be brought about by both the rising plasma estrogen titers and by the initial LHRH discharge which induces the LH surge. As a consequence of the release of LHRH and/or LHRH-related peptides (Seeburg and Adelman, 1984) available pituitary LHRH receptors decline in the afternoon of proestrus, a phenomenon that can be reversed by preventing the expression of neural events leading to LHRH release (Adams and Spies, 1981; Barkan *et al.*, 1983). The decrease in LHRH receptors, rather than representing a true loss of receptors, appears to reflect ligand-induced unavailability of receptor for binding (Smith-White and Ojeda, 1982, 1985).

IV. Conclusions

The foregoing considerations have made us keenly aware of the fact that the developmental process that leads to puberty in the female rat is composed of an extraordinary, complex series of interrelated events (Fig. 27 and 33). The CNS plays the pivotal role in the process by controlling both anterior pituitary function through the secretion of hypothalamic factors and the ovary via pituitary hormones and direct neural inputs. Pituitary gonadotropins play a decisive role at almost all stages of sexual development, with the possible exception of the fetal period during which ovarian development appears to be primarily under direct neural control. Based on our results and those of other investigators we propose the existence of three major periods of activation of gonadotropin secretion during postnatal development, all of them having profound repercussions on the initiation and completion of puberty (Fig. 33). *The first activational period* occurs during infantile development and is expressed as an enhancement in FSH secretion, with sporadic elevation in LH levels. Although this dramatic activation of gonadotropin release results, to a sig-

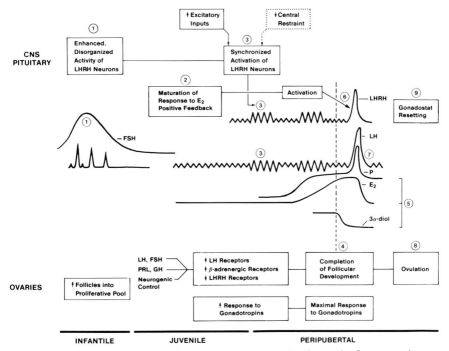

FIG. 33. Proposed sequence of developmental events leading to the first preovulatory LH surge in the female rat. The numbers indicate the sequence in which the events may occur. The dotted line represents 1200 hours on the day of first proestrus. The box outlined by interrupted lines indicates that a loss in central restraint may not be a predominant factor for the synchronized activation LHRH-secreting neurons. Modified from Ojeda *et al.* (1984a), reprinted with permission.

nificant extent, from the interplay of steroidal influences, its primary determinant appears to be of central origin.[10] At this time ovarian development comes under tight gonadotropic control. As a consequence of increased FSH secretion, a large crop of primordial follicles, some of which are destined to ovulate at puberty, is incorporated into a proliferative pool and begin to grow. *The second activational period* signals the end of juvenile development and represents the first neuroendocrine man-

---

[10] We postulate that this period results from a "disorganized" activity of LHRH neurons which under the influence of variable extrahypothalamic inputs fail to discharge synchronously. Infrequent, randomly occurring synchronization may result in low-frequency, high-amplitude LHRH discharges which result in maintenance of high plasma FSH levels.

[11] We postulate that this activation represents the synchronization of LHRH discharges, perhaps as a consequence of completion of both the synaptic circuitry impinging upon LHRH neurons, and the morphological maturation of the LHRH-secreting cells.

ifestation of the onset of puberty. This activational period is also determined by a centrally driven, gonadal-independent mechanism and is expressed as a diurnal change in pulsatile LH release.[11] Both basal release of LH and the magnitude of LH secretory episodes increase markedly in the afternoon. These changes stimulate the ovary to produce more E_2 which then evokes minisurges of LH secretion that in turn further stimulate ovarian development and steroidogenesis. *The third and final activational period* occurs more abruptly, and is predominantly determined by an increased output of ovarian steroids, especially E_2. This period corresponds to the preovulatory discharge of LHRH which directly promotes the first surge of gonadotropins. Figure 34 depicts a composite of plasma LH profiles obtained from conscious, free-moving animals in different developmental stages, and which have been arranged in an order that faithfully represent the above postulated sequence of events.

While there is little doubt that the last activational phase represents the overt manifestation of E_2 positive feedback, there is no evidence that the preceding activational phase is due to a decrease in hypothalamic sensi-

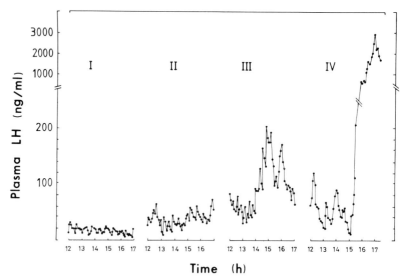

FIG. 34. Postulated sequence of changes in the mode of LH release during the onset of puberty in the female rat. Roman numerals indicate the phases in which different afternoon patterns of LH release were observed in conscious free-moving peripubertal animals, bled every 5 minutes using an automated bleeding technique. Each profile is derived from a different animal. I, Low amplitude pulses similar to those seen in the morning; II, increased basal LH release and LH pulse amplitude; III, minisurge of LH secretion; IV, proper, proestrous surge of LH. From Ojeda and Urbanski (1986), reprinted with permission.

tivity to E_2 negative feedback. If such were the case, one would have to postulate that the sensitivity to steroid inhibitory control has a diurnal rhythm. Moreover, experiments in which the sensitivity to E_2 negative feedback was assessed in peripubertal rats distinctly indicated that a decrease in sensitivity did not occur until after ovulation (Andrews *et al.,* 1981c). Estradiol positive feedback, on the other hand, develops by the end of the infantile period. At this time the ovary is unable to produce sufficient levels of E_2 to produce a surge of LH. Completion of ovarian maturation is, therefore, essential for the induction of the preovulatory gonadotropin discharge.

The ovaries grow under the influence of the gonadotropins, which is modulated by milk LHRH during neonatal-infantile development, and facilitated afterward by GH and Prl. An additional facilitatory control mechanism is provided by adrenergic and peptidergic nerves, and by EPI of adrenal medullary origin. The adrenergic inputs to the ovary facilitate P secretion either directly or by amplifying the effect of gonadotropins. VIP-ergic nerves appear to be involved in stimulating the secretion of all the main steroids (E_2, P, and aromatizable androgens). SP and NPY-ergic nerves, on the other hand, may be involved in regulation of blood flow.

Under the dynamic interaction and influence of all these factors, which appear to be capable of functioning long before the first proestrus, the steroid output of the ovary increases more rapidly as the proestrous day approaches. The pace of follicular development is accelerated by the increase in LH receptors, and by a possible decrease in a local LHRH inhibitory influence. Finally, serum E_2 levels become elevated for a sufficiently long period of time to activate the central component of E_2 positive feedback. An intracellular PGE_2–cyclic AMP pathway, which appears to mediate the stimulatory effect of NE on LHRH release, becomes activated, possibly in conjunction with an independent, diacylglycerol–protein kinase C-mediated mechanism. Simultaneous activation of both pathways may then trigger the proestrous surge of LHRH, which in turn, by acting on a pituitary sensitized by E_2, elicits the first preovulatory surge of gonadotropins. The gonadotropin discharge induces the first ovulation on the early morning of estrus. On this day, the marked estrogenic stimulation of the vaginal epithelium has resulted in vaginal opening, which constitutes the first somatic manifestation that reproductive competence has been finally attained.

Perhaps the most formidable challenge that we now face is the elucidation of the factors responsible for the two major gonadal-independent changes in central "drive." Needless to say, disclosure of the mechanisms underlying these activational periods is intimately linked to the solution of the mystery of puberty.

ACKNOWLEDGMENTS

We would like to express our gratitude to Drs. V. D. Ramirez and S. M. McCann for their help and encouragement during the formative years of the scientific career of S.R.O. We are grateful to Drs. J. P. Advis, W. W. Andrews, S. White, and L. Aguado who contributed extensively to our quest for the understanding of the onset of female puberty. We also appreciate the fruitful collaboration of Drs. W. B. Campbell, P. M. Conn, A. Castro-Vazquez, F. W. George, L. Hersh, M. D. Lumpkin, and E. Viyayan. We are indebted to Ms. K. H. Katz and M. E. Costa for their excellent technical assistance and invaluable dedication. We thank Mrs. Judy Scott for typing the manuscript. These studies have been supported by grants from NIH (HD-09988, project IV) and NSF (BNS-8318017).

REFERENCES

Adams, T. E., and Spies, H. G. (1981). *Biol. Reprod.* **25**, 298.
Adashi, E. Y., and Hsueh, A. J. W. (1981). *Endocrinology* **108**, 2170.
Advis, J. P., and Ojeda, S. R. (1978). *Endocrinology* **103**, 924.
Advis, J. P., and Ojeda, S. R. (1979). *Biol. Reprod.* **20**, 879.
Advis, J. P., Andrews, W. W., and Ojeda, S. R. (1979). *Endocrinology* **104**, 653.
Advis, J. P., Simpkins, J. W., Chen, H. T., and Meites, J. (1978). *Endocrinology* **103**, 11.
Advis, J. P., Richards, J. S., and Ojeda, S. R. (1981a). *Endocrinology* **108**, 1333.
Advis, J. P., Smith-White, S. S., and Ojeda, S. R. (1981b). *Endocrinology* **108**, 1343.
Advis, J. P., Smith-White, S. S., and Ojeda, S. R. (1981c). *Endocrinology* **109**, 1321.
Advis, J. P., Wiener, S. L., and Ojeda, S. R. (1981d). *Endocrinology* **109**, 223.
Aguado, L. I., and Ojeda, S. R. (1984a). *Biol. Reprod.* **31**, 605.
Aguado, L. I., and Ojeda, S. R. (1984b). *Endocrinology* **114**, 1845.
Aguado, L. I., and Ojeda, S. R. (1984c). *Endocrinology* **114**, 1944.
Aguado, L. I., Petrovic, S. L., and Ojeda, S. R. (1982). *Endocrinology* **110**, 1124.
Ahmed, C. E., Dees, W. L., and Ojeda, S. R. (1985). *Prog. 65th Annu. Meet. Endocrinol. Soc.* p. 206.
Ahmed, C. E., Dees, W. L., and Ojeda, S. R. (1986). *Endocrinology*, in press.
Amarant, T., Fridkin, M., and Koch, Y. (1982). *Eur. J. Biochem.* **127**, 647.
Andrews, W. W., and Ojeda, S. R. (1977). *Endocrinology* **101**, 1517.
Andrews, W. W., and Ojeda, S. R. (1978). *J. Endocrinol.* **78**, 281.
Andrews, W. W., and Ojeda, S. R. (1981a). *Endocrinology* **108**, 1313.
Andrews, W. W., and Ojeda, S. R. (1981b). *Endocrinology* **109**, 2032.
Andrews, W. W., Advis, J. P., and Ojeda, S. R. (1980). *Proc. Soc. Exp. Biol. Med.* **163**, 305.
Andrews, W. W., Heiman, M., Porter, J. R., and Ojeda, S. R. (1981a). *Biol. Reprod.* **24**, 597.
Andrews, W. W., Mizejewski, G. J., and Ojeda, S. R. (1981b). *Endocrinology* **109**, 1404.
Andrews, W. W., Advis, J. P., and Ojeda, S. R. (1981c). *Endocrinology* **109**, 2022.
Araki, S., Torand-Allerand, C. D., Ferin, M., Vande-Wiele, R. L. (1975). *Endocrinology* **97**, 693.
Aubert, M. L., Begeot, M., Winiger, B. P., Morel, G., Sizonenko, P. C., and Dubois, P. M. (1985). *Endocrinology* **116**, 1565.
Barkan, A. L., Regiani, S. R., Duncan, J. A., and Marshall, J. C. (1983). *Endocrinology* **112**, 1042.
Barraclough, C. A., Wise, P. M., and Selmanoff, M. K. (1984). *Recent Prog. Horm. Res.* **40**, 487.

Begeot, M., Morel, G., Rivest, R. W., Aubert, M. L., Dubois, M. P., and Dubois, P. M. (1984). *Neuroendocrinology* **38**, 217.
Ben-Jonathan, N., Arbogast, L. A., Rhoades, T. A., and Bahr, J. M. (1984). *Endocrinology* **115**, 1426.
Benno, R. H., and Williams, T. H. (1978). *Brain Res.* **142**, 182.
Ben-Or, S. (1963). *J. Embryol. Exp. Morphol.* **11**, 1.
Bourguignon, J-P., and Franchimont, P. (1984). *Endocrinology* **114**, 1491.
Burden, H. W. (1985). In "Catecholamines as Hormone Regulators" (N. Ben-Jonathan, J. M. Bahr, and R. I. Weiner, eds.), p. 261. Raven, New York.
Castro-Vazquez, A., and Ojeda, S. R. (1977). *Neuroendocrinology* **23**, 88.
Chiappa, S. A., and Fink, G. (1977). *J. Endocrinol.* **72**, 211.
Chowdhury, M., and Steinberger, E. (1976). *J. Endocrinol.* **69**, 381.
Clemens, J. A., Minagushi, H., Storey, P., Voogt, J. L., and Meites, J. (1969). *Neuroendocrinology* **4**, 150.
Critchlow, V., and Bar-Sela, M. E. (1967). In "Neuroendocrinology" (L. Martini, and W. F. Ganong, eds.), Vol II, p. 101. Academic Press, New York.
Dalkin, A. C., Bourne, G. A., Pieper, D. R., and Marshall, J. C. (1981). *Endocrinology* **108**, 1658.
Dawson, A. B., and McCabe, M. (1951). *J. Morphol.* **88**, 543.
Debeljuk, L., Arimura, A., and Schally, A. V. (1972). *Endocrinology* **90**, 1499.
Dees, W. L., Kozlowski, G. P., Dey, R., and Ojeda, S. R. (1985a). *Biol. Reprod.* **32**, 471.
Dees, W. L., Ahmed, C. E., and Ojeda, S. R. (1985b). *Neurosci. Abstr.* p. 802.
Denef, C., Magnus, C., and McEwen, B. S. (1974). *Endocrinology* **94**, 1265.
Döhler, K. D., and Wuttke, W. (1974). *Endocrinology* **94**, 1003.
Döhler, K. D., and Wuttke, W. (1975). *Endocrinology* **97**, 898.
Dyer, C. A., and Erickson, G. F. (1985). *Endocrinology* **116**, 1645.
Eckstein, B., Yehud, S., Shani, J., and Goldhaber, G. (1976). *J. Endocrinol.* **70**, 25.
Eden, S. (1979). *Endocrinology* **105**, 555.
Eshkol, A., Lunenfeld, B., and Peters, H. (1970). In "Gonadotropins and Ovarian Development" (W. R. Butt, A. C. Crooke, and M. Ryle, eds.), p. 249. Livingstone, Edinburgh.
Foster, D. L., Olster, D. H., and Yellon, S. (1985). In "Adolescence in Females" (S. Venturoli, C. Flamigni, and J. R. Givens, eds.), p. 1. Year Book Medical Publishers, Chicago.
Frawley, L. S., and Henricks, D. M. (1979). *Endocrinology* **105**, 1064.
Funkenstein, B., and Nimrod, A. (1982). In "Development and Function of Reproductive Organs" (A. G. Byskov and H. Peters, eds.), p. 307. Excerpta Medica, Amsterdam.
Funkenstein, B., and Nimrod, A., and Lindner, H. R. (1980). *Endocrinology* **106**, 98.
George, F. W., and Ojeda, S. R. (1982). *Endocrinology* **111**, 522.
George, F. W., and Ojeda, S. R. (1984). *Proc. Int. Congr. Endocrinol., 7th* p. 719.
George, F. W., and Ojeda, S. R. (1986). In preparation.
Germain, B. J., Campbell, P. S., and Anderson, J. N. (1978). *Endocrinology* **103**, 1401.
Goldman, B. D., Grazia, Y. R., Kamberi, I. A., and Porter, J. C. (1971). *Endocrinology* **88**, 771.
Gräf, K.-J., Esch, B., and Horowski, R. (1976). In "Cellular and Molecular Basis of Neuroendocrine Processes" (E. Endröczi, ed.), p. 197. Akadémiai Kiadó, Budapest.
Gunaga, K. P., and Menon, K. M. J. (1973). *Biochem. Biophys. Res. Commun.* **54**, 440.
Hage, A. J., Groen-Klevant, A. C., and Welschen, R. (1978). *Acta Endocrinol. (Copenhagen)* **88**, 375.

Heaulme, M., and Dray, F. (1984). *Neuroendocrinology* **39**, 403.
Hisaw, F. L. (1947). *Physiol. Rev.* **27**, 95.
Hompes, P. G. A., Vermes, I., and Tilders, F. J. H. (1982). *Neuroendocrinology* **35**, 8.
Honma, K., Hohn, K. G., and Wuttke, W. (1979). *Brain Res.* **77**, 277.
Hsueh, A. J. W., and Jones, P. B. C. (1981). *Endocr. Rev.* **2**, 437.
Hsueh, A. J. W., Adashi, E. Y., Jones, P. B. C., and Welsh, T. H., Jr. (1984). *Endocr. Rev.* **5**, 76.
Hunzicker-Dunn, M., and Birnbaumer, L. (1976). *Endocrinology* **99**, 198.
Jia, X. C., Kalmijn, J., and Hsueh, A. J. W. (1985). *Prog. 65th Meet. Endocr. Soc.* p. 205.
Katz, S. H., Molitch, M., and McCann, S. M. (1969). *Endocrinology* **85**, 725.
Kimura, F., and Kawakami, M. (1980). *Endocrinology* **107**, 172.
Kimura, F., and Kawakami, M. (1981). *Endocrinol. Jpn.* **28**, 647.
Kimura, F., and Kawakami, M. (1982). *Neuroendocrinology* **35**, 128.
Kimura, F., Tsai, C.-W., and Kawakami, M. (1983). *Endocrinol. Jpn.* **30**, 297.
Kragt, C. L., and Dahgren, J. (1972). *Neuroendocrinology* **9**, 30.
Kraiem, Z., Eshkol, A., Lunenfeld, B., and Ahren, K. (1976). *Acta Endocrinol.* **82**, 388.
Lamprecht, S. A., Zor, U., Tsafriri, A., and Lindner, H. R. (1973). *J. Endocrinol.* **57**, 217.
Lamprecht, S. A., Kohen, F., Ausher, J., Zor, U., and Lindner, H. R. (1976). *J. Endocrinol.* **68**, 343.
Larsson, L.-I., Fahrenkrug, J., and Schaffalitzky de Muckadell, O. B. (1977). *Science* **197**, 1374.
Lawrence, I. E., and Burden, H. W. (1980). *Anat. Rec.* **196**, 51.
Legan, S. J., Coon, G. A., and Karsch, F. J. (1975). *Endocrinology* **96**, 50.
McDonald, J., Dees, W. L., Ahmed, C. E., and Ojeda, S. R. (1986). In preparation.
MacKinnon, P. C. B., Mattock, J. M., and ter Haar, M. B. (1976). *J. Endocrinol.* **70**, 361.
Matsumoto, A., and Arai, Y. (1976). *Cell Tissue Res.* **169**, 143.
Matthews, Andrews, W. W., Parker, C. R., Jr., and Ojeda, S. R. (1986). In preparation.
Meijs-Roelofs, H. M. A., and Kramer, P. (1979). *J. Endocrinol.* **81**, 199.
Meijs-Roelofs, H. M. A., Uilenbroek, J. Th. J., de Jong, F. H., and Welschen, R. (1973). *J. Endocrinol.* **59**, 295.
Meijs-Roelofs, H. M. A., Uilenbroek, J. Th. J., de Greef, W. J., de Jong, F. H., and Kramer, P. (1975). *J. Endocrinol.* **67**, 275.
Meijs-Roelofs, H. M. A., Kramer, P., and Sander, H. J. (1983). *J. Endocrinol.* **98**, 241.
Melrose, P. A. (1985). *Biol. Reprod. Suppl. 1* **32**, 107.
Naftolin, F., and Brawer, J. R. (1977). *J. Steroid Biochem.* **8**, 339.
Nishizuka, Y. (1983). *Trends Biochem. Sci.* **8**, 13.
Niswender, G. D., Reimers, T. J., Diekman, M. A., and Nett, T. M. (1976). *Biol. Reprod.* **14**, 64.
Nunez, E., Engelman, F., Benassayag, C., and Jayle, M. F. (1971). *C.R. Acad. Sci. (Paris)* **273**, 831.
Ojeda, S. R., and Aguado, L. I. (1985a). In "Catecholamines as Hormone Regulators" (N. Ben-Jonathan, J. M. Bahr, and R. I. Weiner, eds.), p. 293. Raven, New York.
Ojeda, S. R., and Aguado, L. I. (1985b). In "Handbook of Pharmacologic Methodologies for the Study of the Neuroendocrine System" (R. W. Steger and A. Johns, eds.), p. 205. CRC Press, Boca Raton, Florida.
Ojeda, S. R., and Campbell, W. B. (1982). *Endocrinology* **111**, 1031.
Ojeda, S. R., and Jameson, H. E. (1977a). *Endocrinology* **100**, 881.
Ojeda, S. R., and Jameson, H. E. (1977b). *Endocrinology* **101**, 475.
Ojeda, S. R., and Ramirez, V. D. (1972). *Endocrinology* **90**, 466.

Ojeda, S. R., and Ramirez, V. D. (1973/74). *Neuroendocrinology* **13**, 100.
Ojeda, S. R., and Urbanski, H. F. (1986). *In* "The Physiology of Reproduction" (E. Knobil and J. D. Neill, eds.). Raven, New York, in press.
Ojeda, S. R., Krulich, L., and McCann, S. M. (1974). *Neuroendocrinology* **16**, 342.
Ojeda, S. R., Kalra, P. S., and McCann, S. M. (1975). *Neuroendocrinology* **18**, 242.
Ojeda, S. R., Krulich, L., and Jameson, H. E. (1976a). *Endocr. Res. Commun.* **3**, 387.
Ojeda, S. R., Wheaton, J. E., Jameson, H. E., and McCann, S. M. (1976b). *Endocrinology* **98**, 630.
Ojeda, S. R., Jameson, H. E., and McCann, S. M. (1977). *Endocrinology* **100**, 440.
Ojeda, S. R., Negro-Vilar, A., and McCann, S. M. (1979). *Endocrinology* **104**, 617.
Ojeda, S. R., Andrews, W. W., Advis, J. P., and Smith-White, S. (1980). *Endocr. Rev.* **1**, 228.
Ojeda, S. R., Smith-White, S. S., Urbanski, H. F., and Aguado, L. I. (1984a). *In* "Neuroendocrine Perspectives" (E. E. Müller and R. M. MacLeod, eds.), p. 225. Elsevier, Amsterdam.
Ojeda, S. R., Katz, K. H., Costa, M. E., and Advis, J. P. (1984b). *Neuroendocrinology* **39**, 19.
Ojeda, S. R., Urbanski, H. F., Katz, K. H., and Costa, M. E. (1985a). *Endocrinology* **117**, 1175.
Ojeda, S. R., Costa, M. E., Katz, K. H., and Hersh, L. B. (1985b). *Biol. Reprod.* **32**, 286.
Ojeda, S. R., Urbanski, H. F., Katz, K. H., and Costa, M. E. (1986a). *Neuroendocrinology,* in press.
Ojeda, S. R., Urbanski, H. F., Katz, K. H., Costa, M. E., and Conn, P. M. (1986b). *Proc. Natl. Acad. Sci. U.S.A.,* in press.
Osman, P. (1975). *J. Endocrinol.* **67**, 259.
Paul, S. M., Axelrod, J., Saavedwa, J. M., and Skolnick, P. (1979). *Brain Res.* **178**, 499.
Pedersen, T. (1970). *Acta Endocrinol. (Copenhagen)* **64**, 304.
Peluso, J. J., Steger, R. W., and Hafez, E. S. E. (1976). *J. Reprod. Fertil.* **47**, 55.
Peters, H. (1979). *In* "Ovarian Follicular Development and Function" (A. R. Midgley and W. A. Sadler, eds.), p. 1. Raven, New York.
Peters, H., Byskov, A. G., Lintern-Moore, S., Faber, M., and Andersen, M. (1973). *J. Reprod. Fertil.* **35**, 139.
Piva, F., Motta, M., and Martini, L. (1979). *In* "Endocrinology" (L. J. de Groot, ed.), p. 41. Grune & Stratton, New York.
Plant, T. M. (1980). *Endocrinology* **106**, 1451.
Plapinger, L., and McEwen, B. S. (1975). *Steroids* **26**, 255.
Puig-Duran, E., and MacKinnon, P. C. B. (1978). *J. Endocrinol.* **76**, 321.
Ramirez, V. D. (1973). *In* "Handbook of Physiology" (R. O. Greep and E. B. Astwood, eds.), Vol II, Section 7, p. 1. Amer. Physiol. Soc., Washington, D.C.
Ramirez, V. D., Feder, H. H., and Sawyer, C. H. (1984). *Front. Neuroendocrinol.* **8**, 27.
Ramirez, V. D., Kim, K., and Dluzen, D. (1985). *Recent Prog. Horm. Res.* **41**, 421.
Ratner, A., Weiss, G. K., and Sanborn, G. R. (1980). *J. Endocrinol.* **87**, 123.
Raum, W. J., Glass, A. R., and Swerdloff, R. S. (1980). *Endocrinology* **106**, 1253.
Raynaud, J. P., Mercier-Bodard, C., and Baulieu, E. E. (1971). *Steroids* **18**, 767.
Reiter, E. O., and Grumbach, M. M. (1982). *Annu. Rev. Physiol.* **44**, 595.
Rennels, E. G. (1951). *Am. J. Anat.* **88**, 63.
Richards, J. S. (1980). *Physiol. Rev.* **60**, 51.
Ruf, K. B., and Sharpe, M. S. (1979). *In* "Clinical Endocrinology" (G. Tolis, ed.), p. 239. Raven, New York.

Salisbury, R. L., Dudley, S. D., and Weisz, J. (1982). *Neuroendocrinology* **35**, 265.
Sarkar, D. K., and Fink, G. (1979). *J. Endocrinol.* **83**, 339.
Schwartz, N. B. (1974). *Biol. Reprod.* **10**, 236.
Seeburg, P. H., and Adelman, J. P. (1984). *Nature (London)* **311**, 666.
Siebers, J. W., Peters, F., Zenzes, M. T., Schmidtke, J., and Engel, S. (1977). *J. Endocrinol.* **73**, 491.
Smith-White, S. S., and Ojeda, S. R. (1981a). *Endocrinology* **108**, 347.
Smith-White, S. S., and Ojeda, S. R. (1981b). *Endocrinology* **109**, 151.
Smith-White, S. S., and Ojeda, S. R. (1982). *Endocrinology* **111**, 353.
Smith-White, S. S., and Ojeda, S. R. (1983). *Neuroendocrinology* **36**, 449.
Smith-White, S. S., and Ojeda, S. R. (1984). *Endocrinology* **115**, 1973.
Smith, S. S., and Ojeda, S. R. (1985). *Neuroendocrinology* **41**, 246.
Smith, S. S., and Ojeda, S. R. (1986). *Biol. Reprod.* **34**, 219.
Soto, A. M., and Sonnenschein, C. (1980). *Proc. Natl. Acad. Sci. U.S.A.* **77**, 2084.
Szabo, K., and Csanyi, K. (1982). *Cell Tissue Res.* **224**, 563.
Tanner, J. M. (1974. *In* "The Control of the Onset of Puberty" (M. M. Grumbach, G. D. Grave, and F. E. Mayer, eds.), p. 448. Wiley, New York.
Toran-Allerand, C. D. (1976). *Brain Res.* **106**, 407.
Uilenbroek, J. Th. J., and Arendsen de Wolff-Exalto, E. (1977). *J. Endocrinol.* **72**, 56.
Uilenbroek, J. Th. J., Woutersen, P. J. A., and van der Linden, R. (1983). *J. Endocrinol.* **99**, 469.
Urbanski, H. F., and Ojeda, S. R. (1985a). *Endocrinology* **117**, 644.
Urbanski, H. F., and Ojeda, S. R. (1985b). *Endocrinology* **117**, 638.
Urbanski, H. F., and Ojeda, S. R. (1985c). *Neurosci. Abstr.* **11**, 160.
Urbanski, H. F., and Ojeda, S. R. (1986a). *Endocrinology* **118**, 1187.
Urbanski, H. F., and Ojeda, S. R. (1986b). In preparation.
Urbanski, H. F., Urbanski, D., and Ojeda, S. R. (1984). *Neuroendocrinology* **38**, 403.
Vannier, B., and Raynaud, J. P. (1975). *Mol. Cell. Endocrinol.* **3**, 323.
Weissman, B. A., and Skolnick, P. (1975). *Neuroendocrinology* **18**, 27.
Wildt, L., Hausler, A., Marshall, G., Hutchison, J. S., Plant, T. M., Belchetz, P. E., and Knobil, E. (1981). *Endocrinology* **109**, 376.
Wray, S., and Gainer, H. (1985). *Neurosci. Abstr.* **11**, 900.
Wray, S., and Hoffman-Small, G. (1984). *J. Steroid Biochem.* **20**, 1420.
Wuttke, W., and Gelato, M. (1976). *Ann. Biol. Anim. Biochem. Biophys.* **16**, 349.
Wuttke, W., Honma, K., Lamberts, R., and Hohn, K. G. (1980). *Fed. Proc., Fed. Am. Soc. Exp. Biol.* **39**, 2378.
Zeleznick, A. J. (1982). *In* "Follicular Maturation and Ovulation" (R. Rolland, E. V. van Hall, S. G. Hillier, K. P. McNatty, and J. Schoemaker, eds.), p. 37. Excerpta Medica, Amsterdam.

DISCUSSION

B. Eckstein. First of all I would like to congratulate you for a very nice presentation. I wonder why you haven't mentioned a very important change that takes place at puberty: steroid production. Before the onset of puberty there is a qualitatively different type of steroid production in the ovary, which changes with the first ovulation. You haven't mentioned this. This is a profound change, and I would like your comment on this.

S. R. Ojeda. I imagine that you are referring to the decrease in 3α-androstanediol secretion that occurs around the time of puberty. We have confirmed your finding of an abrupt decrease in the capacity of the ovary to produce 3α-androstanediol at puberty (Advis *et al., Endocrinology* **109**, 223, 1981; Ojeda *et al., Neuroendocrinology* **39**, 19, 1984). However, in attempting to define the importance of this change for the timing of puberty we have failed either to advance the first ovulation by blocking 5α-reductase activity or to delay it by administering physiological doses of 3α-androstanediol. Therefore we don't know at this point whether or not 3α-diol plays any role in determining the time of puberty in the female rat.

B. Eckstein. It's not only the 3α-androstanediol that is produced and is present in the blood stream in high quantities and drops at the first ovulation, but the whole steroidogenesis that goes on in the immature ovary which is different qualitatively from what appears right after the ovulation. That's a very dramatic change and it changes with the first ovulation. Even if you don't know what the physiological significance of it is I think that there has to be some significance in it, and it is worthwhile to speculate on such a dramatic change that takes place within hours of the first ovulation.

S. R. Ojeda. Yes, I agree, but I think the subject is sufficiently complex not to introduce still another group of events that I am sure are very important, i.e., the events that occur between the first gonadotropin surge and ovulation.

H. Kulin. I wonder if you could elaborate a bit on the development of positive feedback during the juvenile period? What occurs to allow LH release with sufficient estogen exposure? Is that prostaglandin mediated or are there any anatomic correlates to the phenomenon? Is positive feedback a separate neuroendocrine mechanism which requires a separate hypothalamic nucleus in the rat?

S. R. Ojeda. We do not know because during the past few years we concentrated most of our efforts on trying to understand how the ovaries become activated at puberty. Only recently are we returning to the hypothalamus in trying to understand how estradiol induces the first LHRH surge. Thus, we do not know whether the lack of LH response to estradiol stimulation before day 15 is due to factors such as a lack of prostaglandin response or the absence of specific neurotransmitter receptors on the LHRH neurons.

O. Pescovitz. The first question relates to your suggestion that growth hormone has a role in the induction of sex steroid secretion. In view of some structural homology between VIP and growth hormone releasing hormone have you looked at any possible effect of GHRH on sex steroid secretion? Second, I found it most intriguing that you found messenger RNA for LHRH in mammary gland tissue and postulate its secretion in milk. I wonder if you've manipulated that system by changing feeding times in the neonate to see if perhaps you could give the neonate a different schedule of LHRH administration and if that might in some way affect receptors in the ovary?

S. R. Ojeda. In regard to your first question we have tested GRF in our system and found it to be ineffective. It behaves like motilin, i.e., it does not alter steroid secretion at all. In regard to your second question, we have not played with the feeding times. As a matter of fact, we suspect that milk LHRH plays a more general role than a reproductive one. The peptide may be involved in other processes such as regulation of specific functions within the mammary gland. This possibility derives from a number of observations but one of them appears as particularly relevant in this context. Miller *et al.* (*Nature* **313**, 231, 1985) have recently demonstrated that an LHRH agonist inhibits cell growth of a human breast tumor cell line, thus suggesting an involvement of LHRH in regulating the growth of mammary gland cells.

E. Terasawa. My first question is whether or not increases in LHRH output from hypothalamic slices (*in vitro* preparation) during the peripubertal period correlate with peripubertal LH changes, especially the afternoon increase in LH amplitude, that you observed in *in vivo* preparations.

S. R. Ojeda. Yes, we are now examining that issue. We are studying both the development of the afternoon changes in LH release and the changes in LHRH output from the medial basal hypothalmus preoptic area *in vitro* but we don't have this information as yet.

E. Terasawa. What mechanism do you think turns on the increased release of LHRH during the onset of puberty in the female rat?

S. R. Ojeda. If I could answer that question we probably wouldn't have another meeting on puberty. I think we are all asking the same question.

K. Yoshinaga. By day 15 of age the pituitary responsiveness to LHRH is established and you can see fluctuations of LH in the plasma is response to LHRH pulses. Doesn't milk LHRH influence the LH secretion?

S. R. Ojeda. That is an interesting question. If we add milk LHRH to hemipituitaries *in vitro* it is extremely effective in inducing LH and FSH release. However, all our attempts to induce LH release by giving milk LHRH intravenously have failed to induce a consistent increase in LH levels. Thus I don't know whether milk LHRH is quickly degraded in the plasma or when it binds to the receptors is unable to produce microaggregation necessary for LH release.

K. Yoshinaga. However, milk LHRH is effective in inhibiting ovarian activity *in vivo*.

S. R. Ojeda. Yes, that is true.

Pubertal Growth: Physiology and Pathophysiology

GORDON B. CUTLER, JR., FERNANDO G. CASSORLA,
JUDITH LEVINE ROSS,[1] ORA H. PESCOVITZ,[2] KEVIN M. BARNES,
FLORENCE COMITE,[3] PENELOPE P. FEUILLAN, LOUISA LAUE,
CAROL M. FOSTER,[4] DANIEL KENIGSBERG,[5]
MANUELA CARUSO-NICOLETTI, HERNAN B. GARCIA,
MERCEDES URIARTE, KAREN D. HENCH, MARILYN C. SKERDA,
LAUREN M. LONG, AND D. LYNN LORIAUX

Developmental Endocrinology Branch, National Institute of Child Health and Human Development, National Institutes of Health, Bethesda, Maryland

I. Introduction

The pubertal growth spurt is a two-edged sword. On the one hand, it causes an approximate doubling of the prepubertal growth rate (Fig. 1), and contributes 15–18% to adult height (Frisch and Revelle, 1971; Tanner *et al.*, 1976a). On the other hand, it sets in motion the events that cause epiphyseal fusion and cessation of growth. When it occurs prematurely, it threatens premature cessation of growth and diminished adult height.

Our interest in pubertal growth arose from three groups of patients. The first group have Turner syndrome. Untreated, these girls are short during childhood, lack a pubertal growth spurt, and are severely short as adults (mean untreated adult height, 146.8 cm) (Brook *et al.*, 1974; Ranke *et al.*, 1983). We wished to induce a pubertal growth spurt in these girls, and if possible to do so in a way that would increase adult height. The second group of patients have precocious puberty. We wished to prevent their early epiphyseal fusion and diminished adult height (Jolly, 1955; Thamdrup, 1961; Sigurjonsdottir and Hayles, 1968; Lee, 1981). The third group

---

[1] Present address: Department of Pediatrics, Hahnemann University, Philadelphia, Pennsylvania 19102.
[2] Present address: Department of Pediatrics, University of Minnesota, Minneapolis, Minnesota 55455.
[3] Present address: Department of Pediatrics, Yale University, New Haven, Connecticut 06510.
[4] Present address: Department of Pediatrics, University of Michigan, Ann Arbor, Michigan 48109.
[5] Present address: Department of Obstetrics and Gynecology, State University of New York at Stony Brook, Stony Brook, New York 11794.

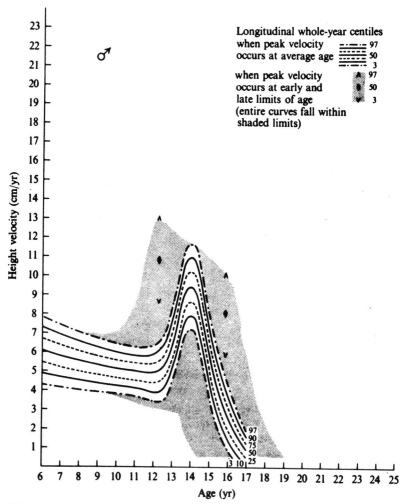

FIG. 1. Tanner–Whitehouse velocity standards for British boys. A child having his pubertal spurt at the average time will follow one of the centile lines shown. If he is earlier or later his curve will be of similar shape and located within the stippled areas which define 97th and 3rd centile limits of age for the pubertal growth spurt. From Eveleth and Tanner (1976).

have short stature for a number of reasons, such as Russell–Silver dwarfism, and a normal pubertal onset (Tanner et al., 1975). Physicians have asked us to delay the normal puberty of such children for the purpose of enhancing adult height.

Thus, our work has addressed three questions. First, how can we induce optimal pubertal growth in girls with Turner syndrome? Second, how can we prevent premature pubertal growth in children with precocious puberty? And, third, what is the potential for enhancing height by delay of normal puberty?

II. Induction of Pubertal Growth in Turner Syndrome

To stimulate a pubertal growth spurt in girls with Turner syndrome, we needed to know which hormones cause pubertal growth. When we began this work, the prevailing hypotheses in the literature were that pubertal growth in boys is the result of testicular androgen, and that pubertal growth in girls is the result of adrenal androgen (Blizzard et al., 1974; Tanner, 1975). It seemed unlikely, however, that adrenal androgen was driving pubertal growth in girls because pubertal growth rates can occur despite low adrenal androgen levels in young girls with precocious puberty (Comite et al., 1981) and in girls with Addison disease (New et al., 1981). Moreover, isolated adrenarche fails to cause pubertal growth rates in premature adrenarche (Cutler et al., 1986), in Kallmann syndrome (Copeland et al., 1977; Sizonenko, 1981), and in Turner syndrome itself, in which adrenarche occurs normally (Lucky et al., 1979).

We postulated that estrogen causes the female pubertal growth spurt. We predicted that the dose of estrogen would be critical, for two reasons. First, the pubertal growth spurt is the earliest sign of puberty in girls (Frisch and Revelle, 1971). In fact, Tanner et al. (1976b) estimated that approximately 20% of the female pubertal growth spurt is completed before the first appearance of breast budding or pubic hair. Thus, if estrogen is driving pubertal growth, it is probably doing so at very low concentrations.

The second reason that we predicted that estrogen dose would be critical is that even moderate doses of estrogen inhibit growth, as shown in the study of van den Bosch et al. (1981). Fifty micrograms per day orally of ethinyl estradiol (a dose conventionally used in birth control pills) for 6 weeks decreased ulnar growth rate by more than 75% ($p < 0.001$) in pubertal girls receiving treatment for tall stature (Fig. 2). Withdrawal of estrogen in a subset of the girls led to a resumption of growth. Thus, we postulated that the dose–response relationship for estrogen effects on growth would be biphasic, and that too high a dose would actually inhibit growth.

To evaluate this hypothesis, we studied the effect of estrogen dose on short-term growth rate in 19 patients with Turner syndrome, ages 5–15 (Ross et al., 1983). Patients underwent the following protocol. Baseline measurements of ulnar length were made monthly for 2 months. The mean of the two 1-month velocities defined the ulnar growth rate during the pretreatment period. Patients then received oral ethinyl estradiol daily

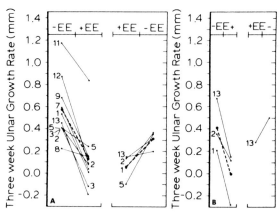

FIG. 2. The decrease in the mean and the individual ulnar growth rates during ethinyl estradiol treatment (+EE, 50 μg orally per day) of pubertal girls with tall stature, and the increase after stopping ethinyl estradiol treatment. (A) The ulnar growth rates during the 6 weeks before and after starting (−EE, +EE) and before and after stopping (+EE, −EE) ethinyl estradiol. (B) Similar data for the 3 weeks before and after starting or stopping ethinyl estradiol. The numbers indicate individual girls. The dotted lines refer to the change in mean ulnar growth rates for the whole group. From van den Bosch et al. (1981).

for 1 month. Ulnar measurements were repeated at 1 and 2 months after starting estrogen treatment. The mean of these two ulnar velocities defined the treatment period ulnar growth rate. This rate included the first month following estrogen treatment because the growth response to estrogen persisted during this month.

Children with bone age <10 years received 0, 50, 100, 200, or 400 ng of ethinyl estradiol per kg per day for 28 days. Children with bone ages greater than 10 years received one of these same doses or 800 ng/kg/day. For a 40 kg patient, this corresponded to 0, 2, 4, 8, 16, or 32 μg of ethinyl estradiol per day. The doses in each patient were assigned in random sequence and in a double-blind fashion.

Ulnar length was measured with a device that immobilizes the forearm and measures the distance between the elbow and the styloid process of the ulna (Valk, 1971). Each determination of ulnar length was the mean of 6–9 measurements. The SD of the individual measurements was 0.47 mm. The SEM of the 6–9 measurements averaged 0.17 mm. For comparison, the control growth rate in Turner syndrome was about 0.6 mm/month.

Increasing doses of ethinyl estradiol had a biphasic effect on ulnar growth (Fig. 3). The relatively low dose of 100 ng/kg/day (approximately 4 μg/day) of ethinyl estradiol produced the maximal growth response, an 83% increase above the baseline ulnar growth rate. The doses corre-

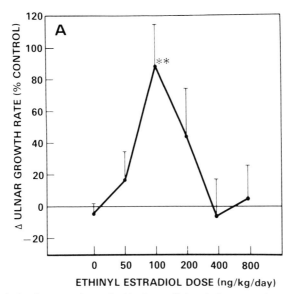

FIG. 3. Relation between dose of ethinyl estradiol and ulnar growth rate (mean ± SE). The double asterisk indicates $p < 0.025$, as compared with the 0 dose group. The number of subjects studied was 6 at the dose of 0, 7 to 10 at 50 to 400 ng/kg/day, and 3 at 800 ng/kg/day. From Ross et al. (1983). Reprinted by permission of the *New England Journal of Medicine*.

sponding to approximately 16 and 32 μg ethinyl estradiol per day actually produced no growth response.

These observations indicated that estrogen has a biphasic effect on growth, that the optimal ethinyl estradiol dose for growth is about 4 μg/day, and that standard ethinyl estradiol doses in Turner syndrome (10–50 μg/day) are not optimal for ulnar growth and may conceivably impair growth potential.

At this point we asked: Would the short term growth seen at 100 ng/kg/day be sustained over a longer period? Would this low dose of estrogen stimulate breast development? Would it increase the rate of bone maturation and would it alter predicted adult height?

These questions were addressed using the same subjects who had participated in the previous study (Ross et al., 1985a). The study design involved administration of ethinyl estradiol, at a dose of 100 ng/kg/day, or placebo for 6 months. The sequence of estradiol and placebo was selected randomly and the study was double-blind. After the initial 6 months there was a period of 2 months without treatment, and then the subjects were crossed over to the other treatment group for an additional 6 months.

The ethinyl estradiol-treated group had a significantly increased height velocity compared to the placebo group ($p < 0.001$). The 61% increase in height velocity observed over the 6-month period of this study was similar to the 83% increase in 1-month ulnar velocity observed in the original study.

To evaluate whether the effects of estrogen on growth might be mediated in part by alterations in serum somatomedin C, we measured serum somatomedin C levels at the end of each treatment period. Serum somatomedin C concentration at the end of the estrogen treatment period, however, did not differ from the concentration at the end of the placebo period [1.01 ± 0.11 (SE) vs 1.03 ± 0.10 U/ml, $p = $ NS].

Breast budding occurred in only 6 of the 16 girls during the 6 months of estrogen treatment, whereas increased growth rate occurred in 15 of the 16. Thus, growth was a more sensitive index of estrogen effect than breast development. This is consistent with Tanner's observation that the growth spurt is the earliest sign of puberty in girls.

Skeletal maturation, expressed as Δ bone age/Δ chronological age, did not differ between the 6-month estrogen treatment period and the placebo period (0.99 ± 0.25 vs 0.99 ± 0.31, $p = $ NS).

We used the Roche–Wainer–Thissen (or RWT) method (Roche et al., 1975) to predict adult height because the bone ages of some of our patients were too low for the Bayley–Pinneau method (Bayley and Pinneau, 1952). Over the 6-month treatment periods of our study, predicted height decreased significantly (-0.93 ± 0.40 cm, $p < 0.05$) during the placebo period but increased slightly during the estrogen treatment period ($+0.32 \pm 0.42$ cm). This yielded a net improvement of 1.25 cm in the estrogen compared to the placebo period, which was significant at $p < 0.03$. The fall in predicted height during the placebo period was expected because the RWT method overestimates adult height in Turner syndrome (Zachmann et al., 1978), and thus predicted height by this method decreases with time in untreated patients.

We conclude that ethinyl estradiol at 100 ng/kg/day for 6 months increases growth rate without bone age acceleration and improves predicted height in girls with Turner syndrome. The effect on growth did not appear to be mediated by changes in circulating somatomedin C.

III. Effect of Estrogen on Ulnar Growth Rate in Boys

The sensitivity of growth to estrogen in girls with Turner syndrome, and the biphasic dose–response curve, led us to ask whether estrogen influences growth in boys with a similar dose–response curve. To examine this we tested the effects of three different doses of estradiol (4, 20,

TABLE I

Three-Week Ulnar Growth Rate (TUG) and Serum Estradiol (E_2) Concentrations (Mean ± SEM) during Estradiol Infusions in Five Normal Boys[a,b]

	Dose E_2					
	Basal	4 µg/day	Basal	20 µg/day	Basal	90 µg/day
TUG (mm/3 weeks)	0.45 ± 0.11	1.38 ± 0.51[c]	0.49 ± 0.11	1.0 ± 0.4	0.46 ± 0.1	0.84 ± 0.12
E_2 (pg/ml)	<8	10 ± 2.3	<8	16 ± 2.3	<8	96 ± 12

[a] From Caruso-Nicoletti et al. (1985).
[b] The basal estradiol values were measured once in each patient immediately before starting the infusion. The estradiol levels during the infusions were the mean of two measurements at 48 and 96 hours for each patient. Normal range of estradiol for adult men: <10–58 pg/ml.
[c] $p < 0.05$ (two-tailed paired t test).

and 90 µg/day) on the ulnar growth rate of five normal boys (Caruso-Nicoletti et al., 1985).

The study design involved ulnar length measurements every 3 weeks to determine ulnar growth rate. Four-day intravenous infusions of estradiol at doses of 4, 20, and 90 µg/day were then given in a randomized, double-blind sequence, with a period of 6–12 weeks between infusions.

The plasma estradiol levels corresponding to the three doses infused were 10 pg/ml at the 4 µg dose, 16 pg/ml at the 20 µg dose, and 96 pg/ml at the 90 µg dose. Levels of 10–16 pg/ml are well within the range observed during the normal pubertal growth spurt in boys (Lee and Migeon, 1975).

The highest rate of growth occurred after the 4 µg/day infusion, and was more than triple the baseline rate (Table I). The ulnar growth rates after the 20 and 90 µg/day infusions appeared smaller and were not significantly greater than the baseline rate. Thus, low dose estradiol can stimulate ulnar growth in boys and may play a role in the male pubertal growth spurt.

IV. The Contribution of Growth Hormone and Thyroid Hormone to the Pubertal Growth Spurt

Current concepts of the role of growth hormone in pubertal growth come principally from clinical studies by Tanner et al. (1976b) and by Aynsley-Green et al. (1976). Tanner examined the pubertal growth spurt of boys with isolated growth hormone deficiency who entered puberty during growth hormone treatment. As soon as they entered puberty, the growth hormone replacement regimen was changed from continuous treatment to treatment periods of 3 months alternating with no treatment for 3 months. During the periods of no GH treatment, pubertal growth rate was only about 2/3 of the rate observed when GH was given (Fig. 4). The growth rate during the periods when GH was given was similar to that observed during the normal pubertal growth spurt. Other settings in which a blunted pubertal spurt have been associated with deficient growth hormone action include familial growth hormone deficiency (Rimoin et al., 1968) and Laron dwarfism (Merimee et al., 1968). Thus, sex steroids can produce a blunted pubertal spurt despite deficient growth hormone or growth hormone action. Growth hormone is required, however, to achieve normal pubertal growth.

Thyroid hormone is essential for normal growth. Its role in pubertal growth is illustrated by the syndrome of precocious puberty associated with untreated primary hypothyroidism. Children with this syndrome have marked retardation of height and bone age, and absence of a pubertal growth spurt, despite other signs of puberty (Van Gelderen, 1962; Bergstrand, 1955; Laron et al., 1970).

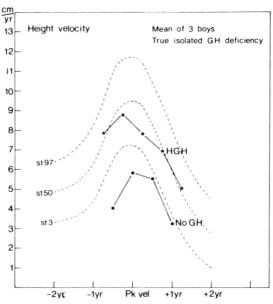

FIG. 4. Mean height velocities of three boys with isolated growth hormone deficiency plotted against normal longitudinal height velocity centiles for British boys. The time scale for normals and growth hormone deficiency has been made equivalent by centering each on their own peak velocity age. From Tanner et al. (1976b).

V. Suppression of Premature Pubertal Growth in Precocious Puberty

Precocious puberty can be classified, according to its mechanism, into two basic categories: central precocious puberty, which is pituitary gonadotropin dependent, and peripheral precocious puberty, which is pituitary gonadotropin independent (Pescovitz *et al.*, 1985a). I will first discuss central precocious puberty.

The mechanism of early puberty in these children is an early onset of the pubertal program of hypothalamic LHRH secretion. Thus, a logical approach to treatment would be to block the LHRH signal. The key insight into how to achieve this came from the work of Belchetz *et al.* (1978), who showed the critical requirement for pulsatility of the LHRH signal. When pulsatile LHRH was administered to rhesus monkeys whose endogenous LHRH-secreting neurons had been ablated, normal gonadotropin levels were maintained. When pulsatile LHRH at a frequency of 1 pulse per hour was changed to continuous LHRH administration, gonadotropin secretion was suppressed. This gave rise to the concept that intermittent or pulsed LHRH stimulates gonadotropin secretion, whereas continuous LHRH inhibits gonadotropin secretion.

At about the same time that Belchetz *et al.* (1978) were making these observations, Tharandt *et al.* (1977) and Crowley *et al.* (1979) were trying to use long-acting agonist analogs of LHRH to induce puberty in patients who lacked endogenous LHRH. The result was a very transient activation of the pituitary and gonad followed by complete refractoriness to further stimulation, of the sort that Belchetz *et al.* had observed. This suggested that long-acting LHRH analogs could be used to treat precocious puberty. At the encouragement of Dr. William F. Crowley, Jr., we undertook such a project (Cutler *et al.*, 1985).

The analog that we used had a D-tryptophan substitution at position 6 and an N-ethylamide substitution at position 10, and was provided by Drs. Wylie Vale and Jean Rivier of the Salk Institute.

The first precocious puberty patient to be treated with LHRH analog began menstruating at age 6 months, and by the time of her second birthday had a height age of 3.5 years. Prior to treatment, she had large LH pulses at night, typical of the pubertal pattern (Boyar *et al.*, 1974), and an LH response to LHRH that exceeded the FSH response (Fig. 5). After 8 weeks of treatment, at a dose of 4 µg/kg/day by sc injection, serum LH and FSH were suppressed and the response to exogenous LHRH was blocked. Eight weeks after discontinuing the drug, the gonadotropins returned to pretreatment levels.

These favorable hormonal effects, and those of other children (Comite

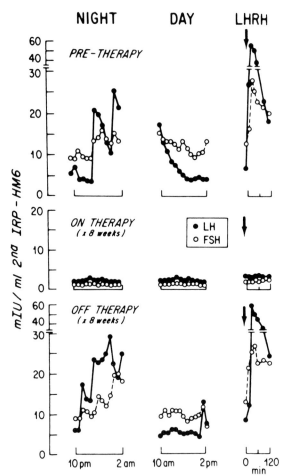

FIG. 5. LH (●) and FSH (○) levels in a 2-year-old girl with idiopathic precocious puberty, expressed as mIU/ml of the 2nd International Reference Preparation of Human Menopausal Gonadotropin (2nd IRP-HMG). Levels were determined on samples collected from 10 PM to 2 AM (Night), 10 AM to 2 PM (DAY) and following administration of 100 μg of synthetic LHRH (LHRH). Following a baseline assessment (Pre-therapy), the patient was treated with daily injections of 4 μg/kg of D-Trp^6-Pro^9-NEt-LHRH. Repeat evaluation occurred following 8 weeks of treatment (On therapy). Injections were then discontinued for another 2 months following which repeat evaluation was performed (Off therapy). From Crowley et al. (1981).

et al., 1981), led us to start a large, long-term study of LHRH analog treatment of precocious puberty. What follows is an update on the current results of this effort to normalize the growth and development of these children.

The subjects of this update are the first 27 children (21 girls, 6 boys) admitted to the study, who have now been treated for at least 3 years. The mean age at treatment was 5.1 years, the mean bone age was 9.0 years, the mean breast stage in the girls was 3.3, and the mean testis volume of the boys was 16.0 ml.

LHRH analog treatment suppressed both basal and LHRH-stimulated LH and FSH levels (Fig. 6). Estradiol levels fell to essentially undetectable levels in the girls, and testosterone fell to approximately 40 ng/dl in

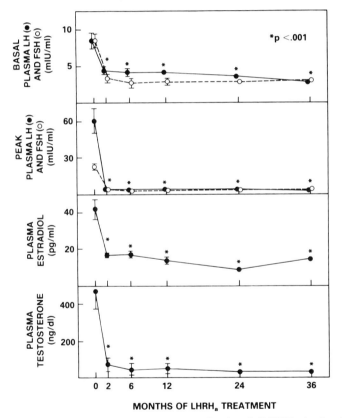

FIG. 6. Effect of LHRH analog (LHRH$_a$) on basal and peak (LHRH-stimulated) plasma LH (●) and FSH (○), plasma estradiol (in the girls), and plasma testosterone (in the boys) in children with central precocious puberty. The asterisks indicate $p < 0.001$ by paired two-tailed Student's t test compared to the pretreatment levels.

the boys, a level which was slightly above normal prepubertal levels. There was no evidence of escape from the suppressive effect of LHRH analog over the 3 years of treatment. Additionally, none of the patients has developed antibodies against LHRH analog.

Growth rate fell nearly 50% during the first 6 months, to within the normal prepubertal range, then declined slightly over the next 2.5 years (Fig. 7). The rate of bone age advancement, which was initially 2 years for each chronologic year, fell over the first 6 months to just above the normal

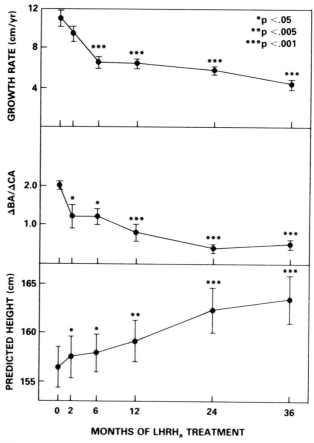

FIG. 7. Effect of LHRH analog (LHRH$_a$) on growth rate, rate of bone maturation (ΔBA/ΔCA: Δ bone age/Δ chronological age), and predicted adult height in 27 children with central precocious puberty. The asterisks indicate significant differences by paired two-tailed Student's t test relative to the pretreatment values. Data from 3 children were not available at 3 years of treatment, and 7 children were not included in the predicted height results because their initial bone ages were too low for use of the Bayley–Pinneau method.

value of 1. Thereafter it continued to fall, and by 2 years of treatment was less than 0.5 years per chronologic year. Thus, the chronologic age was actually catching up to the advanced bone age.

Predicted adult height was calculated by the Bayley–Pinneau method in the children who had bone ages of 7 or more at the start of treatment. Predicted height increased progressively over the 3 years by 7 cm, or approximately 1 inch per year. Thus, the mean predicted height rose from about 5'2" to about 5'5".

The gain in predicted height over the 3 years of treatment tended to be greatest in children with the lowest bone ages at the start of treatment (Fig. 8). The 7 youngest children could not be included in this analysis because the Bayley–Pinneau method cannot be used for bone ages less than 7. On the other hand, the girl and boy with the highest bone ages, which were 13 years at the start of treatment, had increases in predicted height over the 3 years of treatment of 5 and 8 cm. Thus, an advanced bone age at the start of treatment did not preclude a favorable response of predicted height.

We conclude that LHRH analog treatment reduces the accelerated growth rate and bone maturation, and improves the predicted height, of children with central precocious puberty. Whether treatment will normalize final height, however, will not be known for a number of years.

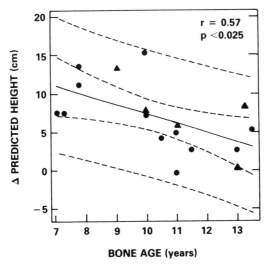

FIG. 8. The effect of bone age at the start of treatment on the change in predicted height during 3 years of LHRH analog treatment. The solid line is the linear regression line by least-squares analysis. The inner dashed lines denote the 95% confidence interval for the regression line; outer dashed lines indicate the 95% confidence interval for a single observation. ●, Girls; ▲, boys.

Whether the treatment is safe will require surveillance until normal reproduction has been demonstrated.

VI. The Mechanism of Increased Growth Rate in Precocious Puberty

Our studies on the mechanism of increased growth velocity in precocious puberty have addressed three questions: First, is growth hormone secretion increased in children with precocious puberty? Second, is serum somatomedin C increased in children with precocious puberty? And, third, does growth rate in children with precocious puberty correlate better with the levels of sex steroids or of somatomedin C?

To examine the first question we measured GH every 20 minutes for 24 hours in 42 girls with idiopathic precocious puberty and in 9 age-matched prepubertal girls (Ross et al., 1985b). Mean 24-hour growth hormone levels were significantly higher in the girls with idiopathic precocious puberty than in normal girls [5.0 ± 0.5 (SE) ng/ml versus 3.2 ± 0.2 ng/ml, $p < 0.005$]. In a subset of 30 of the girls with precocious puberty in whom somatomedin C levels were measured, serum somatomedin C correlated significantly with the 24-hour mean GH level ($r = 0.41$, $p < 0.025$).

A more detailed evaluation of somatomedin C in precocious puberty required extensive normative data because of the marked age dependence of somatomedin C levels. In collaboration with Ron Rosenfeld and Ray Hintz, we examined somatomedin C levels in 40 patients with true precocious puberty, 87 prepubertal controls, and 110 pubertal controls (Pescovitz et al., 1985b).

The children with precocious puberty had markedly elevated somatomedin C levels compared to age-matched prepubertal children (Fig. 9). Although elevated for age, the somatomedin C levels of children with precocious puberty were generally appropriate for pubertal stage (Fig. 10).

How did LHRH analog treatment affect serum somatomedin C, and did decreased somatomedin C appear to explain the beneficial effect of LHRH analog on growth rate? Over the first 6 months of treatment, serum somatomedin C fell only 10% (1.75 ± 0.20 to 1.57 ± 0.18 U/ml, $p < 0.025$) and remained elevated for age ($p < 0.002$), whereas growth rate fell by 40% (11.5 ± 0.6 to 6.9 ± 0.4 cm/year, $p < 0.0025$) and became normal for age. Growth rates before and during $LHRH_a$ failed to correlate with serum somatomedin C but did correlate significantly with estradiol levels in the girls and with testosterone levels in the boys (Table II).

From these studies it is apparent that children with precocious puberty, like children with normal puberty, have increased levels of growth hormone (Finkelstein et al., 1972; Plotnick et al., 1975; Howse et al., 1977;

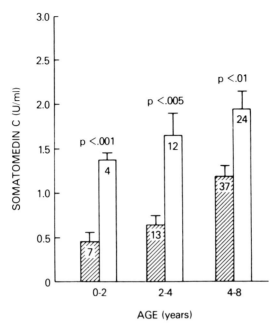

FIG. 9. Somatomedin C concentrations (mean ± SE) in patients with precocious puberty and in age-matched prepubertal children. Hatched bars indicate normal children; open bars, patients with precocious puberty. Number of subjects is shown in each bar. p values represent comparisons within each age group between normal children and children with precocious puberty (two-tailed Student's t test). From Pescovitz et al. (1985b).

TABLE II

Correlation between Growth Rate and Plasma Somatomedin C, Estradiol (Girls), and Testosterone (Boys) in Children with Central Precocious Puberty[a]

Hormone	n^b	Correlation coefficient	p^c
Estradiol	61	0.39	<0.0005
Testosterone	16	0.52	<0.025
Somatomedin C	77	0.04	0.37

[a] From Pescovitz et al. (1985b).
[b] Analysis combines data before and during LHRH analog treatment.
[c] One-tailed t test.

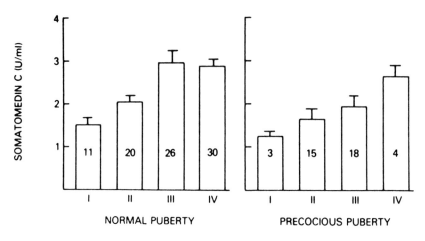

FIG. 10. Somatomedin C levels (mean ± SE) in normal pubertal subjects and in patients with precocious puberty, grouped according to stage of pubic hair development in order to combine the data of boys and girls. Number of subjects in each stage is shown in each bar. Normal subjects with stage I (absent) pubic hair are prepubertal children ≥10 years. Subjects with precocious puberty with stage I pubic hair had other clinical and hormonal evidence of central precocious puberty, such as growth and bone age acceleration and a pubertal response to LHRH. From Pescovitz et al. (1985b).

Miller et al., 1982; Zadik et al., 1985) and somatomedin C (Zapf et al., 1981; Luna et al., 1983; Rosenfield et al., 1983). In assessing the physiological importance of these changes in causing the pubertal growth spurt, however, it must be remembered that normal pubertal growth rates can be achieved by sex steroid replacement in growth hormone-treated hypopituitary children (Aynsley-Green et al., 1976). Sex steroids presumably do not change growth hormone levels in such children, and recent studies have shown that they also do not increase somatomedin C levels (Craft and Underwood, 1984; Rosenfield and Furlanetto, 1985). Since normal pubertal growth rates can occur without a rise of growth hormone or somatomedin C in growth hormone-replaced, sex steroid-treated hypopituitary children, the importance of increased growth hormone and somatomedin C for the normal pubertal growth spurt appears questionable.

VII. Suppression of Premature Pubertal Growth in McCune–Albright Syndrome and in Familial Male Precocious Puberty

With the introduction of LHRH analog therapy, it was hoped that the problem of controlling the premature growth spurt in precocious puberty would be solved. However, two rare forms of precocious puberty—the McCune–Albright syndrome (Foster *et al.*, 1984b; Comite *et al.*, 1984) and familial male precocious puberty (Egli *et al.*, 1985; Wierman *et al.*, 1985)—proved resistant to LHRH analog treatment. Thus, these two disorders required different approaches to treatment.

The McCune–Albright syndrome refers to the unexplained association of polyostotic fibrous dysplasia, hyperpigmented skin spots, and precocious puberty (McCune, 1936; Albright *et al.*, 1937). Our previous studies had indicated that McCune–Albright syndrome is a form of pseudopuberty, with sex steroid secretion by ovarian cysts in the absence of pubertal hypothalamic–pituitary activation (Foster *et al.*, 1984a; Shawker *et al.*, 1984).

Given this mechanism, how could one control the accelerated growth in this disorder? We postulated that ovarian estrogen secretion was the principal mechanism driving pubertal growth, and thus predicted that the aromatase inhibitor testolactone would reduce the increased growth rate by inhibiting estrogen production (Foster *et al.*, 1985; Feuillan *et al.*, 1985).

The study involved four girls with McCune–Albright syndrome who underwent a pretreatment period, a treatment period of 2 months followed by 2 months without treatment, and then a treatment period of 6 months followed by 6 months without treatment. Evaluations were performed every 2 months initially and every 3 months during the 6-month treatment periods, and the data from all of the evaluations before, all evaluations during, and all evaluations after testolactone were each combined for purposes of analysis.

Testolactone caused a significant fall in mean estradiol from about 130 to about 30 pg/ml ($p < 0.05$). Since patients with McCune–Albright syndrome often have large ovarian cysts, ovarian size was measured by ultrasound. Testolactone caused a 50% decrease in mean ovarian volume ($p < 0.05$), which suggests that inhibition of estradiol synthesis may have hindered the formation of ovarian cysts. Ovarian size increased after testolactone was discontinued.

Growth rate fell from about 14 cm/year to a normal prepubertal rate of about 6 cm/year ($p < 0.025$). Bone maturation fell significantly from more than 2 years per year to about 1.3 years per year ($p < 0.01$). These bone age and growth results were similar to what was observed with LHRH analog after a similar period of treatment of central precocious puberty. We conclude, therefore, that reduction of estrogen levels by testolactone decreases the accelerated growth rate and bone maturation in the McCune–Albright syndrome.

The second form of precocious puberty that continued to pose a therapeutic challenge was familial male precocious puberty, an autosomal dominant disorder that causes testicular activation in the absence of pubertal gonadotropin secretion. This disorder, like McCune–Albright syndrome, is refractory to LHRH analog.

How could one control the growth in this disorder? We first decided to try to block the effect of androgen with the antiandrogen spironolactone (Kenigsberg et al., 1984) (Fig. 11). Androgenic signs, such as acne and the frequency of spontaneous erections, were diminished, but growth rate and bone maturation remained increased. We then postulated that the continuing secretion of estrogen might be driving pubertal growth, and added testolactone to inhibit estrogen synthesis (Fig. 12).

Baseline growth rate while receiving either LHRH analog or no treatment for 6 months was 17 cm/year in the first 3 boys studied. Growth rate fell slightly during 6 months of spironolactone to 12 cm/year. Addition of testolactone for 6 months caused a significant fall relative to the pretreatment growth rate to a normal prepubertal rate of 6 cm/year ($p < 0.025$). Bone maturation fell from a baseline value of 2.5 to 1.2 years per year, although this change did not achieve statistical significance.

Preliminary data in one patient suggested that 6 months treatment with testolactone alone only partially corrected the accelerated growth rate, similar to the effect of spironolactone alone. If these data are confirmed in additional patients, it would suggest that inhibition of estrogen alone is not sufficient to control pubertal growth in this disorder. From these observations it would appear that control of pubertal growth in familial

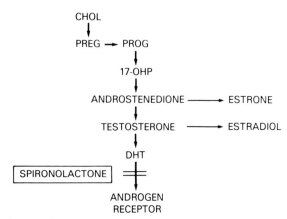

FIG. 11. Schematic diagram of the postulated effect of the antiandrogen spironolactone on sex steroid action in familial male precocious puberty. The access of dihydrotestosterone (DHT) to the androgen receptor is blocked, but formation of estrone and estradiol is not altered.

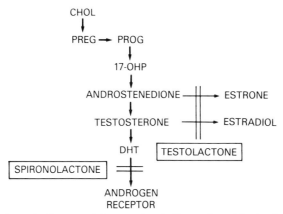

FIG. 12. Schematic diagram of the postulated effects of combined spironolactone and testolactone treatment of familial male precocious puberty. The combined treatment inhibits the action of androgens and the formation of estrogens.

male precocious puberty requires removal of the effects of both androgen and estrogen.

VIII. Enhancement of Height by Delay of Normal Puberty

Lastly, I would like to address the problem of the short boy in whom delay of puberty is being contemplated as a means to increase adult height. What is the realistic potential of this approach?

It occurred to us that patients with isolated hypogonadotropic hypogonadism often experience pubertal delay as a result of delay in diagnosis and treatment, and thus provide a natural experiment on the effect of pubertal delay on adult height. We therefore reviewed the final adult height of 46 men who met rigorous criteria for the diagnosis of isolated hypogonadotropic hypogonadism and who had been studied at the NIH over the past 24 years.

Final adult height was significantly higher in isolated hypogonadotropic hypogonadism than in normal men, by 4.2 cm, or approximately 1.66 inches [$p < 0.001$ compared to the National Center for Health Statistics (NCHS) standards (Hamill et al., 1979)]. Adult height also correlated significantly with age at onset of treatment (a measure of the duration of pubertal delay) ($r = 0.42$, $p < 0.01$). However, all of the increased height of this population was attributable to the patients whose therapy was delayed until age 18 or greater. The adult height of patients whose treatment began between ages 12 and 17 was essentially identical to that of the normal population.

Midparental heights were available for 25 of the 46 men studied and did not differ significantly from the normal midparental height according to the NCHS standards. The height of these 25 patients significantly exceeded the height predicted from their midparental heights. Thus, the enhanced adult height in isolated hypogonadotropic hypogonadism (IHH) could not to be attributed to a family background of tall stature, as this would have been reflected by increased midparental height. It also did not appear to be attributable to an effect of the IHH gene that was distinct from the delay of puberty, since this should have caused increased height in subjects who were treated between ages 12 and 17. From these observations we conclude that the potential to enhance adult height by delay of normal puberty appears modest, and that clinically significant gains in height are likely to require long periods of delay.

IX. Summary

Our current hypothesis is that pubertal growth in girls is driven by low levels of estrogen, on the order of about 4 μg/day, whereas pubertal growth in boys is driven both by low levels of estrogen and by high (i.e., late pubertal) levels of androgen.

Earlier investigators have established that growth hormone and thyroxine play a permissive role in pubertal growth. The contribution of increased growth hormone and somatomedin C secretion at puberty to the pubertal growth spurt, however, is uncertain. Normal pubertal growth rates can be achieved without increased circulating growth hormone or somatomedin C in growth hormone-replaced, sex steroid-treated hypopituitary children.

Clinical evidence suggests that the adrenal androgens contribute little if anything to pubertal growth.

In girls with Turner syndrome, who normally lack a pubertal growth spurt, low dose estrogen (100 ng/kg/day orally of ethinyl estradiol) increases growth rate and improves predicted height. Growth hormone also increases growth rate (Ross *et al.*, 1985c,d) and studies are needed to assess the effects on growth of low dose estrogen combined with growth hormone.

In children with central precocious puberty, long-term LHRH analog treatment decreases growth rate and bone maturation and improves predicted height. In boys with familial male precocious puberty, control of pubertal growth appears to require removal of the effects of both androgens and estrogens.

The potential to enhance height by delay of normal puberty is likely to be modest. In our view, this approach to the treatment of short stature should not be used widely until its effectiveness has been established through controlled trials such as are in progress at the NIH and other centers.

REFERENCES

Albright, F., Butler, A. M., Hampton, A. O., and Smith, P. (1937). *N. Eng. J. Med.* **216**, 727–746.
Aynsley-Green, A., Zachmann, M., and Prader, A. (1976). *J. Pediatr.* **89**, 992–999.
Bayley, N., and Pinneau, S. R. (1952). *J. Pediatr.* **40**, 432–441.
Belchetz, P. E., Plant, T. M., Nakai, Y., Keogh, E. J., and Knobil, E. (1978). *Science* **202**, 631–633.
Bergstrand, C. G. (1955). *Acta Endocrinol.* **20**, 338–342.
Blizzard, R. M., Thompson, R. G., Baghdassarian, A., Kowarski, A., Migeon, C. J., and Rodriguez, A. (1974). In "Control of the Onset of Puberty" (M. M. Grumbach, G. D. Grave, and F. E. Mayer, eds.), pp. 342–359. Wiley, New York.
Boyar, R. M., Rosenfeld, R. S., Kapen, S., Finkelstein, J. W., Roffwarg, H. P., Weitzman, E. D., and Hellman, L. (1974). *J. Clin. Invest.* **54**, 609–618.
Brook, C. G. D., Murset, G., Zachmann, M., and Prader, A. (1974). *Arch. Dis. Child.* **49**, 789–795.
Caruso-Nicoletti, M., Cassorla, F. G., Skerda, M. C., Ross, J. L., Loriaux, D. L., and Cutler, G. B., Jr. (1985). *J. Clin. Endocrinol. Metab.* **61**, 896–898.
Comite, F., Cutler, G. B., Jr., Rivier, J., Vale, W. W., Loriaux, D. L., and Crowley, W. F., Jr. (1981). *N. Engl. J. Med.* **305**, 1546–1550.
Comite, F., Shawker, T. H., Pescovitz, O. H., Loriaux, D. L., and Cutler, G. B., Jr. (1984). *N. Engl. J. Med.* **311**, 1032–1036.
Copeland, K. C., Paunier, L., and Sizonenko, P. C. (1977). *J. Pediatr.* **91**, 985–991.
Craft, W. H., and Underwood, L. E. (1984). *Clin. Endocrinol.* **20**, 549–554.
Crowley, W. F., Vale, W. W., Beitins, I. Z., Rivier, J., Rivier, C., McArthur, J. (1979). Presented at the 61st annual meeting of the Endocrine Society, Anaheim, California, June 13–15.
Crowley, W. F., Jr., Comite, F., Vale, W., Rivier, J., Loriaux, D. L., and Cutler, G. B., Jr., (1981). *J. Clin. Endocrinol. Metab.* **52**, 370–372.
Cutler, G. B., Jr., Hoffmann, A. R., Swerdloff, R. S., Santen, R. J., Meldrum, D. R., and Comite, F. (1985). *Ann. Int. Med.* **102**, 643–657.
Cutler, G. B., Jr., Schiebinger, R. J., Albertson, B. D., Cassorla, F. G., Chrousos, G. P., Comite, F., Booth, J. D., Levine, J., Hobson, W. C., and Loriaux, D. L. (1986). In "The Control of the Onset of Puberty II" (M. M. Grumbach, P. C. Sizonenko, and M. Aubert, eds.). Academic Press, Orlando, in press.
Egli, C. A., Rosenthal, S. M., Grumbach, M. M., Montalvo, J. M., and Gondos, B. (1985). *J. Pediatr.* **106**, 33–39.
Eveleth, P. B., and Tanner, J. M. (1976). "Worldwide Variation in Human Growth," p. 13. The Cambridge University Press, Cambridge, England.
Feuillan, P., Foster, C., Pescovitz, O., Hench, K., Shawker, T., and Cutler, G. B., Jr. (1985). Presented at the 67th annual meeting of the Endocrine Society, Baltimore, Maryland, June 19–21, Abstract 1012, p. 253.
Finkelstein, J. W., Roffwarg, H. P., Boyar, R. M., Kream, J., and Hellman, L. (1972). *J. Clin. Endocrinol. Metab.* **35**, 665–670.

Foster, C. M., Ross, J. L., Shawker, T., Pescovitz, O. H., Loriaux, D. L., Cutler, G. B., Jr., and Comite, F. (1984a). *J. Clin. Endocrinol. Metab.* **58,** 1161–1165.
Foster, C. M., Comite, F., Pescovitz, O. H., Ross, J. L., Loriaux, D. L., and Cutler, G. B., Jr. (1984b). *J. Clin. Endocrinol. Metab.* **59,** 801–805.
Foster, C. M., Pescovitz, O. H., Comite, F., Feuillan, P., Shawker, T. H., Loriaux, D. L., and Cutler, G. B., Jr. (1985). *Acta Endocrinol.* **109,** 254–257.
Frisch, R. E., and Revelle, R. (1971). *Human Biol.* **43,** 140–159.
Hamill, P. V. V., Drizd, T. A., Johnson, C. L., Reed, R. B., Roche, A. F., and Moore, W. M. (1979). *Am. J. Clin. Nutr.* **32,** 607–629.
Howse, P. M., Rayner, P. H. W., Williams, J. W., Rudd, B. T., Bertrande, P. V., Thompson, C. R. S., and Jones, L. A. (1977). *Clin. Endocrinol.* **6,** 347–359.
Jolly, H. (1955). "Sexual Precocity," pp. 10–11. Charles C Thomas, Springfield, Illinois.
Kenigsberg, D., Pescovitz, O., Comite, F., Hench, K., Loriaux, D. L., and Cutler, G. B., Jr. (1984). Presented at the 7th International Congress of Endocrinology, July 1–7, Quebec City, Excerpta Medica (International Congress Series 652), Amsterdam, Abstract 1593.
Laron, Z., Karp, M., and Dolberg, L. (1970). *Acta Paediatr. Scand.* **59,** 317–322.
Lee, P. A. (1981). *Am. J. Dis. Child.* **135,** 443–445.
Lee, P. A., and Migeon, C. J. (1975). *J. Clin. Endocrinol. Metab.* **41,** 556–562.
Lucky, A. W., Marynick, S. P., Rebar, R. W., Cutler, G. B., Jr., Glenn, M., Johnsonbaugh, R. E., and Loriaux, D. L. (1979). *Acta Endocrinol.* **91,** 519–528.
Luna, A. M., Wilson, D. M., Wibbelsman, C. J., Brown, R. C., Nagashima, R. J., Hintz, R. L., and Rosenfeld, R. G. (1983). *J. Clin. Endocrinol. Metab.* **57,** 268–271.
McCune, D. J. (1936). *Am. J. Dis. Child.* **52,** 743–744.
Merimee, T. J., Hall, J., Rabinowitz, D., McKusick, V. A., and Rimoin, D. L. (1968). *Lancet* **2,** 191–193.
Miller, J. D., Tannenbaum, G. S., Colle, E., and Guyda, H. J. (1982). *J. Clin. Endocrinol. Metab.* **55,** 989–994.
New, M. I., Levine, L. S., and Pang, S. (1981). In "The Biology of Normal Human Growth" (M. Ritzen *et al.,* eds.), pp. 285–295. Raven, New York.
Pescovitz, O., Cutler, G. B., Jr., and Loriaux, D. L. (1985a). In "Neuroendocrine Perspectives" (E. E. Muller, R. M. MacLeod, and L. A. Frohman, eds.), Vol. 4, pp. 73–93. Elsevier Science Publishing, New York.
Pescovitz, O. H., Rosenfeld, R. G., Hintz, R. L., Barnes, K., Hench, K., Comite, F., Loriaux, D. L., and Cutler, G. B., Jr. (1985b). *J. Pediatr.* **107,** 20–25.
Plotnick, L. P., Thompson, R. G., Kowarski, A., DeLacerda, L., Migeon, C. J., and Blizzard, R. M. (1975). *J. Clin. Endocrinol. Metab.* **40,** 240–247.
Ranke, M. B., Pfluger, H., Rosendahl, W., Stubbe, P., Enders, H., Bierich, J. R., and Majewski, F. (1983). *Eur. J. Pediatr.* **141,** 81–88.
Rimoin, D. L., Merimee, T. J., Rabinowitz, D., and McKusick, V. A. (1968). *Recent Prog. Horm. Res.* **24,** 365–429.
Roche, A. F., Wainer, H., and Thissen, D. (1975). *Pediatrics* **56,** 1026–1033.
Rosenfield, R. L., and Furlanetto, R. (1985). *J. Pediatr.* **107,** 415–417.
Rosenfield, R. L., Furlanetto, R., and Bock, D. (1983). *J. Pediatr.* **103,** 723–728.
Ross, J. L., Cassorla, F. G., Skerda, M. C., Valk, I. M., Loriaux, D. L., and Cutler, G. B., Jr. (1983). *N. Engl. J. Med.* **309,** 1104–1106.
Ross, J. L., Long, L. M., Skerda, M. C., Cassorla, F. G., Kurtz, D., and Cutler, G. B., Jr. (1985a). *Pediatr. Res.* **19,** 192A.
Ross, J. L., Pescovitz, O. H., Hench, K., Barnes, K. M., Loriaux, D. L., and Cutler, G. B., Jr. (1985b). Presented at the International Symposium on Recent Developments in

the Study of Growth Factors, GHRH, and Somatomedins, Paris, France, May 31–June 1.
Ross, J. L., Long, L. M., Loriaux, D. L., and Cutler, G. B., Jr. (1985c). *J. Pediatr.* **106,** 202–206.
Ross, J. L., Long, L., Skerda, M., Cassorla, F., Pence, P., and Cutler, G. B., Jr. (1985d). Presented at the 67th annual meeting of the Endocrine Society, Baltimore, Maryland, June 19–21, Abstract 490, p. 123.
Shawker, T. H., Comite, F., Rieth, K. G., Dwyer, A. J., Cutler, G. B., Jr., and Loriaux, D. L. (1984). *J. Ultrasound Med.* **3,** 309–316.
Sigurjonsdottir, T. J., and Hayles, A. B. (1968). *Am. J. Dis. Child.* **115,** 309–321.
Sizonenko, P. C. (1981). In "The Biology of Normal Human Growth" (M. Ritzen *et al.,* eds.), pp. 297–308. Raven, New York.
Tanner, J. M. (1975). In "Endocrine and Genetic Diseases of Childhood and Adolescence" (L. I. Gardner, ed.), pp. 14–64. Saunders, Philadelphia.
Tanner, J. M., Lejarraga, H., and Cameron, N. (1975). *Pediatr. Res.* **9,** 611–623.
Tanner, J. M., Whitehouse, R. H., Marubini, E., and Resele, L. F. (1976a). *Ann. Human Biol.* **3,** 109–126.
Tanner, J. M., Whitehouse, R. H., Hughes, P. C. R., and Carter, B. S. (1976b). *J. Pediatr.* **89,** 1000–1008.
Thamdrup, E. (1961). "Precocious Sexual Development: A Clinical Study of 100 Children," pp. 44–63. Charles C. Thomas, Springfield, Illinois.
Tharandt, L., Schulte, H., Benker, G., Hackenberg, K., and Reinwein, D. (1977). *Neuroendocrinology* **24,** 195–207.
Valk, I. M. (1971). *Growth* **35,** 297–310.
Van den Bosch, J. S. G., Smals, A. G. H., Kloppenborg, P. W. C., and Valk, I. M. (1981). *Acta. Endocrinol.* **98,** 156–160.
Van Gelderen, H. H. (1962). *Arch. Dis. Child.* **37,** 337–339.
Wierman, M. E., Beardsworth, D. E., Mansfield, M. J., Badger, T. M., Crawford, J. D., Crigler, J. F., Jr., Bode, H. H., Loughlin, J. S., Kushner, D. C., Scully, R. E., Hoffman, W. H., and Crowley, W. F., Jr. (1985). *N. Engl. J. Med.* **312,** 65–72.
Zachmann, M., Sobradillo, B., Frank, M., Frisch, H., and Prader, A. (1978). *J. Pediatr.* **93,** 749–755.
Zadik, Z., Chalew, S. A., McCarter, R. J., Jr., Meistas, M., and Kowarski, A. A. (1985). *J. Clin. Endocrinol. Metab.* **60,** 513–516.
Zapf, J., Walter, H., and Froesch, E. R. (1981). *J. Clin. Invest.* **68,** 1321–1330.

DISCUSSION

S. L. Kaplan. We have evaluated the long-term effect of 5 to 10 μg of ethinyl estradiol on growth rate and final height of girls with gonadal dysgenesis (Turner syndrome). An acceleration in growth rate was observed only for the initial 12 to 18 months of therapy. The final height of these girls was not significantly different from that of patients who received no estradiol therapy. Therefore we must be cautious in the extrapolation of short-term growth rate measurements for prediction of final height.

G. B. Cutler. I agree with that entirely, and it's why I'm careful to use the term predicted height. We agree fully that what is needed are studies of actual height and long-term controlled trials. Our hope is to use the short-term studies to define an optimal short-term regimen, which we currently anticipate will involve both growth hormone and low-dose estrogens, and then to use that regimen in a long-term controlled trial to determine its effect on ultimate height.

B. D. Albertson. I would like to ask if you might speculate on the role of androgens and estrogens and their synergistic roles with growth hormone, somatomedin C, or other growth factors in the role of normal growth in children before puberty?

G. B. Cutler. That's an interesting question. I don't really have any evidence for this. An interesting observation that's been made in Turner syndrome is that the bone age tends to be delayed from early childhood. The question has been raised whether there might be a contribution of sex steroid deficiency to the growth reduction and delayed bone maturation. But I'm not aware of any compelling evidence about the role of sex steroids in normal prepubertal growth, although there clearly is some sex steroid secretion by the prepubertal gonads.

B. D. Albertson. Is the growth profile in children with testicular feminization normal?

G. B. Cutler. I don't have any personal experience with this.

K. Ho. Can you reassure us that your observation of a biphasic effect of low doses of estrogens on growth rate is not due to a chance assignment of a higher dose of estrogen to an older group of patients with Turner syndrome. Second, I think you stopped short of postulating that the way estrogens enhance growth may be due to stimulation of growth hormone release from the pituitary gland. With that in mind, have you had a chance to look at the pulsatile release of GH and correlate it with the doses of estrogens given or to look at whether the inhibition of estrogen secretion by GnRH analogs causes a parallel reduction in pulsatile growth hormone secretion?

G. B. Cutler. Those are important questions. First, there were approximately 7–10 subjects randomized to each estrogen dose. Thus, I think it would be unlikely that an assignment bias contributed to biphasic dose–response relationship. Replication of the study by others, however, will be the ultimate test of whether it's valid. The second question is an important one: does the low dose estrogen replacement increase growth hormone secretion? We are in the process of assessing this. We did measure somatomedin, and there was no change in serum somatomedin in response to low dose estrogen. My speculation, since you've asked, would be that the increase in growth hormone that occurs during puberty is probably not critical to the growth acceleration that occurs. I suppose one experiment that could be done is to administer to somebody who is prepubertal an amount of growth hormone equivalent to pubertal secretion rates, or drive the somatomedin by exogenous growth hormone to the levels that ordinarily occur during puberty. My prediction would be that this would not result in a pubertal growth rate, and that there will be an independent effect of estrogen beyond its effect on growth hormone and somatomedin C. Obviously further studies need to be done.

R. P. Kelch. I want to applaud your cautious approach to the use of agents such as LHRH analogs in normal children. Your review of the factors controlling pubertal growth should make us all adopt a cautious approach in that regard. I wanted to, if I may, refer back to earlier comments made on LH pulse frequency during human puberty. We've studied approximately 2 dozen boys 15 years of age with a bone age of 12 years at the time of the study to determine when and how fast the brain begins to stimulate the pituitary to release LH. Almost without exception, around 10:00–11:00 PM, we saw the beginning of low amplitude LH pulses, and then a rapid burst of high-frequency, high-amplitude LH pulses which decrease after about 4:00 AM. We also found that serum testosterone in the evening is only about 0.3 ng/ml and that following the rapid discharge of LH, it rises to about 0.9 ng/ml. We also studied 21 pubertal boys. These indices of gonadotropin secretion, the mean LH pulse frequency, the mean LH amplitude, and the mean LH, were out of phase by approximately 1 to 2 hours. The mean LH peak frequency increased from approximately one pulse every 3 hours, during the evening and daytime hours, to nearly a circhoral pattern for approximately 3–4 hours, and then decreased as morning approached. Serum testosterone increases the latest in this series of indices, but our short-term studies thus far tell us that an acute infusion of testosterone will not prevent the striking increase of LH pulse frequency which occurs at the onset of sleep. Also, opiate antagonism does not mimic this during the daytime. Although we do not know what the responsible mechanisms are, we have demonstrated a sleep-entrained pattern of increased LH pulse frequency during normal human puberty.

G. B. Cutler. I think these are the best data that we have of what happens in the human with respect to LH pulse frequency. Having looked at a lot of these patterns myself, I would caution that when gonadotropin levels are changing, it is difficult to determine whether the apparent lower pulse frequency that one sees at low gonadotropin levels is a true decrease in frequency or whether it's due to a failure to detect pulses as readily at lower levels. I think that this question is going to take some additional work by those who are experts at figuring out what is and isn't a pulse before we can interpret these data in a way that will achieve widespread acceptance.

R. Osathanondh. Besides modifying the physical appearance with the use of sex steroids in the dose–response trial, can you say something about the behavioral aspect, i.e., the development of their behavior for example, personality profile and mental development, I.Q. scores, and so on? Have you or your colleagues at NIH followed them mentally?

G. B. Cutler. In the Turner syndrome children during the 6-month study?

R. Osathanondh. In the groups of normal boys receiving the doses of ethinyl estradiol, or is it not ethical to perform behavioral studies on the subjects?

G. B. Cutler. We only did a 4-day study because we wanted to be sure that we didn't feminize any of these boys. Thus, we chose the minimum period of estrogen administration that we believed from previous studies would be able to stimulate growth. We haven't studied the effects of estrogen administration on behavioral development in either the boys or girls. We have collaborated with others to carry out behavioral studies with precocious puberty. There are behavioral changes, but I am not enough of a behavioral expert to want to try to define these for you. The changes are somewhat different in the boys and the girls, and some of these findings have been published this year by Bill Sonis in the *Journal of Pediatrics*. I think it's difficult in this area to determine whether the changes observed are the result of sex steroids per se acting on the brain or whether the phenotypic or other accompaniments of precocious puberty, such as the parental concern and the doctor visits, are also playing an important role.

M. Raj. I am interested in knowing whether you measure the androgen level in the girls when you give the aromatase inhibitor. Did you measure androgens free or total androgens? Did it have any such untoward side effects in your studies?

G. B. Cutler. We measured total testosterone and androstenedione levels, and they were between 15 and 20 ng/dl. These were relatively low levels, below adult female levels, and they did not change during testolactone therapy. We weren't too surprised by this because in this setting the LHRH oscillator is turned off. The production of estradiol and estrone is only a small fraction of total production of testosterone and androstenedione, and we had predicted before the study that testolactone would not alter testosterone or androstenedione levels.

C. S. Nicoll. The accelerated growth that you observed in response to the gonadal steroids appears to be associated with no increase in either growth hormone or somatomedins. If these hormones do not mediate the response, how do you think it is mediated?

G. B. Cutler. We really don't have any studies or data on this, nor have I seen any compelling data. There have been efforts to look for sex steroid receptors in growing cartilage. There are, in abstract form, some observations from a French group that such receptors exist, but there has not yet been convincing published work. My current postulate would be, in the absence of such published data, that sex steroids are able to influence growing cartilage directly.

C. S. Nicoll. My next question is how do you think the pubertal growth spurt is terminated?

G. B. Cutler. This is very important. I think of the pubertal growth spurt as involving two very different processes: one is an acceleration of growth and the other is the degeneration and fusion of the growth plate and the cessation of growth. Our hope in developing a very low dose sex steroid treatment regimen is that it might be possible to dissociate these two processes. Most therapeutic regimens in the past have not done so, that is, regimens that promoted growth invariably advanced bone age as well. At least over a 6-month period it appears that a daily dose of approximately 4 μg accelerates growth rate without advancing bone maturation.

W. D. Drucker. You have said that the adrenal androgens probably play no role in the acceleration of growth. I would be curious to know, however, since the adrenal androgens can give rise, by peripheral conversion, to increased estrogen levels, whether in premature adrenarche with estrogen levels as high as you have noted are they capable of accelerating bone growth?

G. B. Cutler. I expected to be challenged regarding the adrenal androgens. I may have somewhat overstated the evidence against their role in pubertal growth as a reaction to previous conclusions that they were the principal hormones driving pubertal growth. I would like to see a series of children with premature adrenarche, confirmed by modern diagnostic methods, whose growth and bone age maturation were carefully studied. To date there has not been such a study. Several years ago I reviewed all cases of alleged premature adrenarche that I could find in the literature. In these older studies, late-onset 21-hydroxylase deficiency, 11-hydroxylase deficiency, 3β-hydroxysteroid dehydrogenase deficiency, or even early cases of true precocious puberty were not adequately excluded. There was a rather small height age advance of the order of about 0.6 years, so that these children were about 4 cm taller than average children for that age, in a group of about 50 cases. On the other hand children with precocious puberty at the same age will often be 15 cm taller than normal age-matched children. Thus increased height of children with premature adrenarche was much smaller than what the pubertal growth spurt produces. Also this small increase may not be entirely attributable to adrenarche because of the potential contamination of other diagnoses in the reported cases. Some of these early cases of alleged premature adrenarche had bone age advancement of as much as 4 years, which we don't see now in newly diagnosed cases.

Your other point about estrogen is very important. If premature adrenarche or normal adrenarche contributes to growth, I think the mechanism is likely to involve the estrogen that is derived from peripheral conversion rather than the androgen. Quantitatively, the normal adult adrenal produces about 100 μg/day of testosterone. With a peripheral conversion ratio from testosterone to estradiol of about 0.5%, this contributes about half a microgram to the blood production rate of estradiol. The adult adrenal also produces about 35 μg of estrone from either direct secretion or from aromatization of adrenal androstenedione. The peripheral conversion ratio of estrone to estradiol is about 7% which contributes about 2 μg to estradiol blood production rate. Thus in the adult one would predict that about 2.5 μg/day of estradiol is being derived from adrenal sources. At age 10, when one is about one-third to one-half of the way through adrenarche and when the pubertal growth spurt is initiated in girls, the adrenal contribution to estradiol blood production rate would be of the order of 1 μg/day. That's close enough to the doses of 4 μg/day that caused increased growth in Turner syndrome that I would be willing to believe that one could be getting some growth acceleration through the adrenal contribution to estradiol.

M. O. Thorner. I have really enjoyed your presentation. I would like to come back to the question of estrogen. The first question is why you used testolactone rather than using an estrogen receptor blocking agent because I would have anticipated that you probably could have more antiestrogen effect by that mechanism? The second part of the question is at what level of estrogen do you think that estrogen facilitates growth and is there a critical level at which normal growth occurs? You have told us about the higher end of the spectrum not the lower end.

G. B. Cutler. We gave a lot of thought to an antiestrogen and I carefully studied the data for the available ones. To my knowledge there is not yet an antiestrogen that is devoid of estrogenic activity. Tamoxifen, for example, suppresses gonadotropin levels in postmenopausal women and stimulates uterine growth in immature rats. Thus we were concerned, because of the marked sensitivity of growth to estrogen, that the residual amount of estrogen activity in all of the available antiestrogens might confound us. Nonetheless I think it is a very worthwhile approach, and it might work just as well as aromatase inhibitor. When there are antiestrogens that only block estrogen and have no intrinsic activity, I think they will normalize the increased growth in McCune–Albright syndrome.

In your second question, if I understand it correctly, you would like to know the plasma levels of estradiol that correlate with growth. From the study of the boys, the lowest infusion rate that we gave of 4 μg/day resulted in estradiol levels that were mostly undetectable. If we were to extrapolate back, since 90 μg/day of iv estradiol gave a concentration of 96 pg/ml, I would estimate that if we are able to measure it accurately 4 μg/day should give a level of about 4 pg/ml. Thus, the estrogen level that produces optimal growth is probably about 4 pg/ml, but our assay doesn't measure that low.

D. Keefe. You showed the diurnal rhythm of LH secretion in a child with precocious puberty. Is that an invariable finding in children with central precocious puberty? Do you see the development of the circadian rhythm of LH even in this pathologic condition?

G. B. Cutler. Yes, the children with central precocious puberty exhibit the same diurnal pattern, with the striking nocturnal augmentation, that is seen during normal puberty. I would not say that it's an invariable finding, because I have not looked at all the data for that particular feature, but it's a very common finding that nocturnal levels are higher.

M. Grumbach. I wonder if you would comment on the use of ketoconazole for treating familial testotoxicosis or gonadotropin-independent puberty.

G. B. Cutler. That's a good question. We don't have any personal experience using the drug. I favor our own approach, and the reasons are the following. First, ketoconazole also blocks adrenal function. Second, and more importantly, from what I've heard, some patients treated with ketoconazole get elevations of liver function tests, of the order of about 2% of patients, that would require discontinuation of therapy. I worry about whether we have enough data on ketoconazole to begin treating a child at age 3 who is going to need treatment for almost 10 years. With testolactone and spironolactone one is dealing with drugs that have been marketed in the United States for 20 years and with which we have a lot of experience concerning safety. Thus, I would rather start with these drugs than with a relatively new drug, the safety of which over a long time period has not been well established.

E. Rall. I'm a little confused about your results with precocious puberty, apparently due to ovarian cysts which cause elevated production of estradiol. You didn't, I think, say what their estradiol production rates were but I have a vague recollection that you said that estradiol levels were 80 or 90. Now this would imply that they were secreting estradiol at such a high level that it would have, if anything, inhibited growth.

G. B. Cutler. That's a very astute question. I think it's one of the fascinating things about these girls that they have cyclical ovarian function. I didn't go into the details of this, but our published data from one such girl showed cycles of estrogen secretion in which estradiol levels went from near the detection limit of about 20 up to 1000 pg/ml, levels higher than we ever see in adult normal women. She continued to have several more of these cycles over a period of 40 weeks, despite suppression of gonadotropins by LHRH analog. Thus, these children have estradiol levels that range over the whole dose–response curve, which makes it hard to interpret the average levels that we present. I suspect that if we could follow such patients with our lower leg measuring device, which can now measure growth over a 1-week period, it would show that they may cycle through periods of growth inhibition and periods of growth stimulation.

Seasonal Breeding in a Marsupial: Opportunities of a New Species for an Old Problem

C. H. TYNDALE-BISCOE, L. A. HINDS, AND S. J. MCCONNELL

Division of Wildlife and Rangelands Research, Commonwealth Scientific and Industrial Research Organisation, Lyneham, Australia

I. Introduction

The comparative approach in biology has a long and distinguished history but has often suffered the disapprobation of those who consider it to be merely derivative—confirming in one species principles already established in another. It is evident, however, that the members of the Organizing Committee for this Conference did not hold such a view of comparative endocrinology or they would not have sought an article about a little known species from the other side of the world. We study unusual species primarily for what they can tell us about the fundamental processes of biology; they provide a necessary dialectic for the many studies on laboratory and domesticated species on which our knowledge of endocrinology is so largely based. In Australia we have the special opportunity to study a group of mammals that have had a separate evolutionary history for the past 100 million years—plenty of time in which to have evolved different endocrine mechanisms. In this article we present the results of our attempts to understand how one of these species, the tammar wallaby, perceives its environment and regulates its reproduction to the annual changes of the seasons.

In recent years the neuroendocrine control of seasonal reproduction in the sheep has been reviewed at these Conferences, beginning with Gerald Lincoln and Roger Short (1980) and followed by Fred Karsh and colleagues (1984) and Doug Foster (1986) at this Conference. These have demonstrated that the central feature of the control system in the sheep is the generation of pulsatile releases of LH, modulated by the circadian profile of melatonin, secreted by the pineal gland. The central importance of the LH pulse generator in the control of reproduction in mammals has received much support from the results of these elegant studies in the sheep (see Lincoln *et al.* 1985). Many other mammals are also seasonal breeders and in all that have been investigated the pineal gland has been shown to play an important role in transducing photoperiodic information into reproductive response. However, in none of these species, including the sheep, has the connection between reading the melatonin message and responses in the pituitary been established. This is because the interval between the photic signal and the first detectable response is very long, as much as several weeks. The tammar wallaby is, like the sheep, a short day breeder, but there the resemblance ends. The LH pulse generator appears to be unimportant in the tammar, whereas prolactin, which is unimportant in the sheep (McNeilly and Land, 1979), is profoundly important in the tammar. Of greater potential interest, however, is the fact that the melatonin message is read by the tammar in about 3 days and translated into a detectable pituitary response within a further 2 days.

Because the reproductive patterns of marsupials differ in several respects from more conventional mammals, it is necessary to begin by reviewing these features, before reviewing the evidence for the control of reproduction in the tammar.

II. Marsupial Reproduction

For the first two-thirds of their long history mammals were represented by small insectivorous creatures, no larger than a mouse (Crompton, 1980). From such ancestors the two main groups of living mammals (eutherian and marsupial) diverged in the early Cretaceous, and it is inferred that their distinctive patterns of reproduction evolved then. However, the marked differences seen today may be associated with the concurrent adaptive radiations of the Tertiary, when many species of both groups evolved to large size (Tyndale-Biscoe and Renfree, 1986). To accommodate the longer growing phase implicit in large size the eutherian mammals extended the intrauterine phase, while the large marsupials extended the postnatal phase of lactation.

Most species of marsupial are polyestrous and ovulate spontaneously to form active corpora lutea, which induce a luteal phase in the uteri. Pregnancy is generally short and, with a very few exceptions, is accommodated within a single estrous cycle, and subsequent cycles are inhibited during the relatively long period of lactation. Four broad patterns of reproduction can be distinguished according to the relationship between the length of gestation, estrous cycle, and the pattern of lactation inhibition of reproductive activity (Tyndale-Biscoe, 1984; Tyndale-Biscoe and Renfree, 1986). However, all but a few species exhibit one of two basic patterns. In the first pattern, which is found in the majority of species, the females are polyovular or monovular and the gestation period occupies less than 60% of the estrous cycle, so that the subsequent estrus and ovulation are suppressed during lactation (Tyndale-Biscoe and Renfree, 1986). The best known examples of this pattern are the polyovular Virginian opossum, *Didelphis virginiana* (Hartman, 1923; Reynolds, 1952; Harder and Fleming, 1981), and the monovular brush possum, *Trichosurus vulpecula* (Pilton and Sharman, 1962; Shorey and Hughes, 1973). The second pattern is found in most species of kangaroo and is restricted to this family (Macropodidae). The females are exclusively monovular and the gestation period is almost the same length as the estrous cycle (91–109%), so that pre- or postpartum estrus and ovulation normally occur. However, growth of the newly formed corpus luteum is arrested during the ensuing lactation. If conception occurs then development of the embryo is arrested at the unilaminar blastocyst stage and this state of diapause persists until the corpus luteum is reactivated by loss of the pouch young or at the end of lactation.

Because of their long separate evolution and independent radiation since the Cretaceous, modern marsupials afford special opportunities for understanding the common principles that underlie the endocrine control of seasonal reproduction, pregnancy, lactation, and embryo development of all mammals. However, to achieve more than a superficial comparison of these processes between marsupials and eutherians it has been necessary to select suitable species, which are amenable to captivity and domestication, to describe their reproductive biology, growth, and development, and to validate assays specific for the important hormones of reproduction. Three species have been established as self-sustaining laboratory colonies, namely *Monodelphis domestica* (VandeBerg, 1983), *Sminthopsis crassicaudata* (Smith *et al.*, 1978), and *Macropus eugenii* (Renfree and Tyndale-Biscoe, 1978), and five other species are nearing this state, namely *Didelphis virginiana* (Jurgelski and Porter, 1974), *Isoodon macrourus* (Gemmell, 1982), *Trichosurus vulpecula* (Renfree and Tyndale-Biscoe, 1978), *Macropus rufogriseus* (Merchant and Calaby, 1981), and *Setonix brachyurus* (Shield, 1968). The study of the hormonal control of reproduction in several of these species began in the last decade (Tyndale-Biscoe and Renfree, 1986), but the development of hormone assays has only been reached to any substantial degree for the tammar wallaby, *M. eugenii*, in which validated assays have been published for both the gonadotropins (Sutherland *et al.*, 1980; Evans *et al.*, 1980; Tyndale-Biscoe and Hearn, 1981), prolactin (Hinds and Tyndale-Biscoe, 1982b), melatonin (McConnell and Tyndale-Biscoe, 1985), prostaglandin FM (Tyndale-Biscoe *et al.*, 1983; Shaw, 1983), and four steroids, progesterone (Sernia *et al.*, 1980; Ward and Renfree, 1984), estradiol (Shaw and Renfree, 1984; Harder *et al.*, 1984), testosterone (Catling and Sutherland, 1980; Inns, 1982), and cortisol (Janssens and Hinds, 1981).

Not only is the tammar now the best studied marsupial but it exhibits a pattern of seasonal breeding that is among the most precise of any mammal. Furthermore, the response to altered photoperiod is extraordinarily rapid (3 days) when compared to the best-studied eutherian seasonal breeders, such as the golden hamster (Reiter, 1975) and the domestic sheep (Karsch *et al.*, 1984). The tammar is thus an excellent species in which to investigate the photoperiod control of reproduction, more especially the way in which information generated by the pineal influences the hypothalamic centers regulating the pituitary.

To appreciate the control of seasonal breeding in the tammar we begin by reviewing the salient features of its reproduction and pituitary–ovarian interactions, and the unusual relationship between this and the phenomenon of embryonic diapause, which is central to the species' reproductive strategy. Then we will present the evidence for the control of lactational quiescence and conclude with the evidence for the control of seasonal quiescence and the way that the tammar perceives photoperiodic change and transduces it into a reproductive response.

III. The Annual Cycle of Reproduction of the Tammar

In the southern hemisphere the great majority of females give birth in late January or early February, about 6 weeks after the austral summer solstice, and the single young is then suckled in the pouch for the next 8 months (Fig. 1). The young emerges from the pouch and is weaned in September or October, that is to say after the winter rains and in the spring, which is the most favorable time of the year. As with other season-

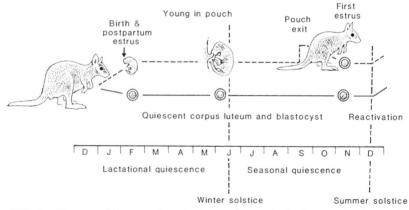

FIG. 1. Diagram of the annual cycle of reproduction in the tammar, in the southern hemisphere.

ally breeding marsupials, it is not parturition that is the ultimate factor controlling seasonal breeding, but the time of emergence of the young from its mother's pouch or the den; to the contrary, parturition may occur at an unfavorable time, such as high summer or mid-winter, depending on the particular length of lactation for the species.

The female tammar comes into estrus soon after parturition and the majority conceive then, the embryo developing to a blastocyst in the next 9 days. As in other kangaroos, if the pouch is occupied and the newborn young becomes attached to one of the teats, neither the blastocyst nor its associated corpus luteum develops any further; but if the young is lost or experimentally removed before June, the corpus luteum and blastocyst resume their development and birth occurs 26–27 days later. This condition, distinctive of kangaroos, is termed lactational quiescence. At the same time of the year unmated females undergo a succession of estrous cycles of 30 days. However, after the winter solstice in June, removal of the pouch young from lactating females does not reactivate the corpus luteum or its associated blastocyst, and unmated females cease to cycle as the corpus luteum of their last cycle also becomes quiescent. Thus from mid-winter to mid-summer all adult females have a quiescent corpus luteum in one ovary and only small antral follicles in the other, and the majority will also have in one uterus a blastocyst in diapause. This condition is termed seasonal quiescence because it is controlled by seasonal factors rather than lactation. Both states are also different from anestrus because all females are potentially capable of ovulating at all times of the year but are prevented by the presence of the quiescent corpus luteum (see below). This is nicely illustrated by the young females, which emerge from the pouch in October and come into first estrus then and conceive (Fig. 1). Although they have never lactated, their corpora lutea also become quiescent and their embryos enter diapause (Andrewartha and Barker, 1969; Tyndale-Biscoe and Hawkins, 1977).

Shortly after the summer solstice the corpora lutea in all these females, both young and adult, spontaneously resume their growth and the embryos reactivate, so that births occur at the end of January. Thus, although the active period of gestation lasts for less than 1 month most of the young born so synchronously in January–February were conceived 12 months before but, for 11 months, were awaiting the appropriate signal to complete development (Berger, 1966). As with other seasonally breeding species transferred across the equator, such as red deer and sheep (see Lincoln and Short, 1980), tammars that were transferred to the northern hemisphere immediately readjusted their annual cycle and gave birth in July or August, after the boreal summer solstice (Berger, 1970). It was therefore reasonable to conclude that photoperiod cues might be used by tammars to maintain quiescence after the winter solstice and to initiate reactivation after the summer solstice.

Whereas lactational quiescence is common to all but one species of kangaroo (Tyndale-Biscoe et al., 1974; Tyndale-Biscoe and Renfree, 1986), only one subspecies of one other kangaroo, *M. rufogriseus*, is known to display seasonal quiescence (Catt, 1977; Merchant and Calaby, 1981; Loudon et al., 1985), so it may be a relatively recent adaptation in these two species, which has been superimposed on the phylogenetically older pattern of lactational quiescence. This thought dictated our approach to investigating the phenomenon in the tammar. We began by investigating the role of the corpus luteum in regulating embryo development and folliculogenesis, then the way in which the corpus luteum is maintained in a quiescent state during lactation, and finally how photoperiod influences have been superimposed on lactational quiescence.

IV. Hormones in the Estrous Cycle and Pregnancy

The tammar is polyestrous and exclusively monovular, with ovulation occurring from alternate ovaries in successive cycles. Because of this it is possible to determine the separate contributions of the corpus luteum and Graafian follicle to peripheral plasma levels of progesterone, estradiol, and other steroids by sampling from the ovarian vein on each side (Harder et al., 1984). The short-term changes in these and other hormones have also been determined in females undergoing parturition and postpartum estrus and ovulation (Tyndale-Biscoe et al., 1983; Shaw and Renfree, 1984), and the relationships examined by selective ablation in late pregnancy of the corpus luteum, the Graafian follicle or both ovaries (Harder et al., 1985). The estrous cycle is 30 days and gestation 29 days (Merchant, 1979) and the profiles of progesterone are indistinguishable until the last day. The account that follows refers to the pregnant cycle and the events after parturition.

A. PREGNANCY AND POSTPARTUM ESTRUS

The end of the previous cycle is heralded by a decline in progesterone and estrus occurs about 8 hours after it reaches basal level (Fig. 2). The decline in progesterone occurs a day or so earlier and is more precipitate than that at the end of the nonpregnant cycle (Tyndale-Biscoe et al., 1983); it is preceded by a transient pulse of prolactin and is followed by birth and postpartum estrus. The time of the drop in progesterone and birth is a useful point of reference for the subsequent events to be described.

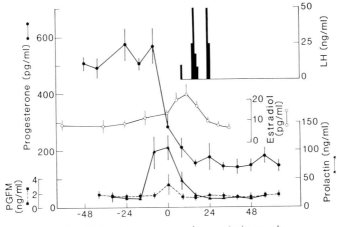

FIG. 2. Peripheral plasma concentrations (mean + SEM) of progesterone, prostaglandin FM, and prolactin for 7 females sampled through the period of parturition and postpartum estrus and ovulation. The occurrences of the preovulatory pulses of LH in the individual females are shown as bars, from Tyndale-Biscoe et al. (1983). Estradiol sampled similarly in 5 females from Harder et al. (1985).

The rapid decline in progesterone at the end of pregnancy is associated with a marked decline in the total progesterone content of the corpus luteum 1 day before parturition (Fig. 3) and a corresponding decline in the concentration of progesterone in the ovarian vein draining the corpus luteum (Fig. 4). The decline has been mimicked by lutectomy 3 days before parturition (Findlay et al., 1983; Harder et al., 1985), so we conclude that the decline in progesterone is due to luteolysis immediately prior to parturition; the transient pulse of prolactin that occurs immediately before parturition appears at present to be the most likely agent, as it has been shown to be in the rat (Ensor, 1978). Furthermore, when nonpregnant tammars were injected with a bolus of prolactin on the day that the decline occurs in pregnant animals, they showed the same rapid decline in progesterone (authors' unpublished results).

The metabolite of prostaglandin F2α, 13,14-dihydro-15-keto-PGF2α (PGFM), is either undetectable or at very low levels in peripheral plasma through late pregnancy, but was transiently elevated immediately after parturition (Tyndale-Biscoe et al., 1983; Shaw, 1983). In a subsequent study P. Lewis, T. P. Fletcher, and M. B. Renfree (personal communication) found peak concentrations of >3 ng/ml in plasma collected within 20 minutes before or after parturition but none was detected 2 hours before or after parturition.

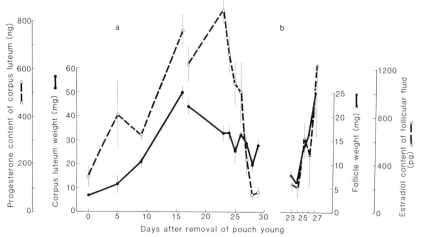

FIG. 3. Weight and hormone content (mean + SEM) of (a) the corpora lutea and (b) the Graafian follicles of groups of 5 tammars sampled through pregnancy and parturition. Data from Hinds et al. (1983), Harder et al. (1984), and unpublished results of authors.

The Graafian follicle, destined to ovulate postpartum, can be recognized in the ovary contralateral to the corpus luteum from about 3 days before parturition (Fig. 3) and it reaches its maximum diameter (4 mm) 40 hours after parturition, at ovulation. The concentration of estradiol-17β in the peripheral circulation is less than 10 pg/ml until 1 day before parturition when it begins to rise (Fig. 2); it reaches a peak concentration of 15–30 pg/ml 8–12 hours after parturition and the fall in progesterone, and coincides with postpartum estrus (Shaw and Renfree, 1984; Harder et al., 1985). This peak in peripheral circulation reflects the very high concentrations of estradiol in follicular fluid (Fig. 3) and in the ovarian vein draining the follicle-ovary at this time (Fig. 4) and indicates that the Graafian follicle is the main source of this steroid (Harder et al., 1984). This conclusion was confirmed by removal of the ovary bearing the Graafian follicle 3 days before parturition, when the rise was abolished and estrus did not occur (Harder et al., 1985).

Luteinizing hormone is low or undetectable throughout pregnancy except for a brief but large surge 7–8 hours after estrus and the estradiol peak, or 16 hours after parturition (Fig. 2). This surge of LH was also prevented by removal of the follicle 3 days before parturition (Harder et al., 1985). Other evidence also supports the inference that estradiol provides a positive feedback to the hypothalamus, as in other mammals, to induce the preovulatory pulse of LH (Horn et al., 1985). Ovulation follows the LH pulse by about 24 hours, as in the ewe, and it failed to occur

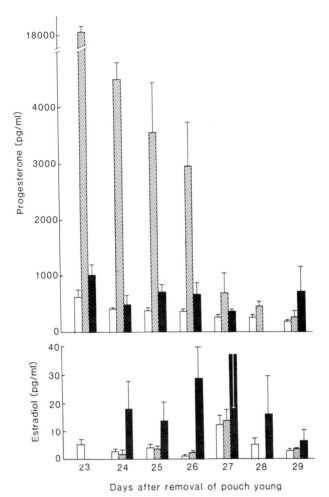

FIG. 4. Concentrations (mean + SEM) of progesterone and estradiol-17β in the peripheral circulation (open bars), the ovarian vein draining the ovary with the corpus luteum (hatched bars), and the ovarian vein draining the ovary bearing the Graafian follicle (closed bars) through late pregnancy, parturition, and postpartum estrus for some of the same group of tammars as in Fig. 3. Data for estradiol from Harder *et al.* (1984) and for progesterone from unpublished results of authors.

in females that did not show a LH pulse. It is of interest that estradiol concentrations in the ovarian vein and peripheral circulation declined soon after estrus and about 30 hours before ovulation, suggesting that the follicle ceases to secrete estradiol at or immediately after the LH pulse, as Goodman et al. (1981) demonstrated in the sheep. This may be due to a direct effect of the LH redirecting steroid synthesis from estradiol to progesterone by inhibiting aromatase in granulosa cells. This has not yet been investigated, but it is supported by the evidence of a rise in plasma progesterone in the ovarian vein draining the ovary with the preovulatory follicle or newly formed corpus luteum (Fig. 4).

Estrus and mating occur 30 hours before ovulation and the spermatozoa travel through the genital tract rapidly, reaching the fimbria of the oviduct within 12 hours (Tyndale-Biscoe and Rodger, 1978), so that they can be assumed to be awaiting the egg immediately after it is shed from the follicle. The zona pellucida is readily penetrated by sperm but the mucoid layer, which is laid down around it, is an effective barrier to sperm, so that only those sperm that are at the fimbria when the egg is shed are capable of fertilizing it. As in all marsupials passage through the oviduct is rapid and the egg enters the uterus within 24 hours, or day 2 postestrus, while it is still uncleaved. The first six to seven cleavage divisions are accomplished by day 8 when the embryo has become a unilaminar blastocyst of 80–100 identical cells. This degree of development will occur in females ovariectomized on day 2 (Sharman and Berger, 1969), so that the corpus luteum formed at ovulation is not required up to this stage. However, further development, which involves a rapid expansion of the vesicle by absorption of uterine secretions, depends on the presence of an active corpus luteum (Renfree and Tyndale-Biscoe, 1973).

B. POSTOVULATORY CHANGES IN THE CORPUS LUTEUM

After ovulation the peripheral concentration of progesterone remains at about 200 pg/ml until day 7 postestrus when there is a transient but marked increase to >500 pg/ml for 1 day, which subsequently reverts to the previous level until about day 10 (Hinds and Tyndale-Biscoe, 1982a). From day 10 the concentration steadily increases to a maximum plateau of 500–600 pg/ml, which is maintained to the end of pregnancy or the estrous cycle (Fig. 2) on day 29 or 30. The early pulse is an invariable component of the progesterone profile of the cycle and the interval from it to parturition is remarkably constant at 21–22 days (Table I). The occurrence of the pulse is therefore a first indication that a female is undergoing a reproductive cycle and, in a female known to be pregnant, it is an excellent predictor of the day of parturition.

TABLE I

Interval to Early Progesterone Pulse and Birth after Different Treatments in the Tammar

| | Interval to | | | |
Treatment	Progesterone pulse	Birth	Pulse to birth	Reference[a]
Estrus	7.0 ± 0.5	29.3 ± 0.3	22.3	2, 3
Removal of pouch young	5.6 ± 0.3	27.4 ± 0.4	21.8	2
Bromocryptine, 5 mg/kg	5.2 ± 0.2	26.3 ± 0.3	21.1	4
Hypophysectomy	7.3 ± 0.2			4
Melatonin, 12 hour total	8.3 ± 0.3	29.9 ± 0.3	21.6	4
Photoperiod change, 15L:9D to 12L:12D	10.0 ± 1.3	33.5 ± 1.1	23.5	1
After summer solstice, 22 December	20.6 ± 0.7	42.6 ± 0.8	22.0	4

[a] (1) Hinds and den Ottolander (1983); (2) Hinds and Tyndale-Biscoe (1982a); (3) Merchant (1979); (4) L. A. Hinds and C. H. Tyndale-Biscoe (unpublished results).

In the lactating female the early pulse of progesterone does not occur at the expected time on day 7 and the embryo ceases to develop past the stage reached on day 8–9. Since gestation uninterrupted by lactation is 29 days, the progesterone pulse could be expected to occur immediately after removal of the pouch young (day 0) and birth to occur 21–22 days later. This is never seen; the early pulse of progesterone does not occur until day 5–6 and birth not until day 26 or 27 after removal of the pouch young (Fig. 5). Since the interval from the pulse to birth is the same as in pregnancy uninterrupted by lactation (Table I), the additional 5–6 days before the pulse occurs is apparently required for the corpus luteum to recover from inhibition and for the blastocyst to reawaken from diapause. Shaw and Renfree (1984) have observed that a transient pulse of estradiol occurs concurrently with the early pulse of progesterone on day 5–6 and presumably this originates from the corpus luteum also.

C. THE ROLE OF THE CORPUS LUTEUM IN REACTIVATION

After removal of the pouch young the corpus luteum increases in size from 10 mg to a maximum of 60 mg by day 20 (Renfree *et al.*, 1979) but for the first 13 days the concentration of progesterone in luteal tissue remains unchanged at 10–11 ng/mg, rising thereafter to 32 ng/mg by day 22. However, Hinds *et al.* (1983) found that the secretion rate of the corpus luteum *in vitro* was significantly higher on day 5 than on day 0, 9, or 16 and concluded that this could account for the transient pulse seen in the

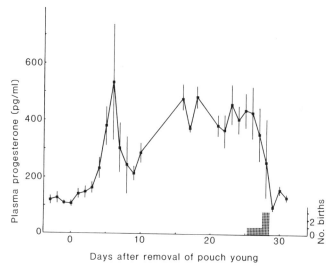

FIG. 5. Concentrations (mean + SEM) of progesterone in peripheral plasma of 5 tammars through delayed pregnancy, initiated by removing the pouch young. Note the pulse of progesterone on days 5–6, and the rapid decline at parturition. After Hinds and Tyndale-Biscoe (1982a).

peripheral circulation on day 5 or 6. On the other hand the subsequent rise to a plateau after day 16 could be accounted for by the increasing size and progesterone content of the corpus luteum at this time. The first indication of blastocyst reactivation occurs on day 5 when RNA metabolism increases (Moore, 1978; Thornber et al., 1981; Shaw and Renfree, 1986). This is followed by resumption of cell division and growth of the blastocyst by day 8 (Renfree and Tyndale-Biscoe, 1973). From the effects of lutectomy or ovariectomy after reactivation it is clear that the corpus luteum provides a first signal by day 3, which commits the blastocyst to further development; ovariectomy before this time, on day 0 or day 2, blocked reactivation and the blastocyst remained in diapause, while ovariectomy on day 4 was followed by blastocyst expansion and subsequent collapse, and ovariectomy on day 6 or 8 resulted in reactivation and fetal development to full term (Berger, 1970; Sharman and Berger, 1969). These results, performed before the early pulse had been described, are now interpreted as indicating that the reactivated blastocyst is unable to maintain its growth unless the pulse of progesterone occurs on day 5–6, presumably because this induces secretory activity in the endometrium, which is necessary for the embryo's nourishment.

We conclude from the foregoing that the corpus luteum is essential to the completion of pregnancy and hence is the key to the control of breeding in this species, and that the progesterone pulse, being an essential component of this, is a useful criterion of reactivation and a precise indicator of corpus luteum function. We will now consider the role of the pituitary in the events just considered.

V. Pituitary–Ovarian Interactions

Apart from the preovulatory pulse of LH referred to, the levels of LH in intact female tammars remain consistently low or undetectable throughout the year (Fig. 6). Nevertheless, as in other mammals, the development of the Graafian follicle and ovulation are dependent on gonadotropic stimulation; after hypophysectomy follicular development is blocked at the stage of small antral follicles (Panyaniti *et al.*, 1985) and, if performed at the end of the estrous cycle or pregnancy when a Graafian follicle has already developed, neither estrus nor ovulation occurs, nor parturition in pregnant animals (Hearn, 1974). Likewise in tammars immunized against GnRH in early pregnancy follicular development, estrus, and ovulation were prevented, although their corpora lutea grew, pregnancy was sustained, and parturition occurred (Short *et al.*, 1985). Both these studies indicated that, while the preovulatory events in the tammar are gonadotropin dependent, the corpus luteum appears to be independent of pituitary support. This latter aspect will be discussed in a later section.

From the response to bilateral ovariectomy it is clear that the very low levels of LH and FSH in peripheral plasma of intact female tammars are maintained by a negative feedback from the ovaries (Fig. 7). In three studies the basal concentration of both gonadotropins, measured daily, rose steadily to reach a plateau at about 2 weeks after bilateral ovariec-

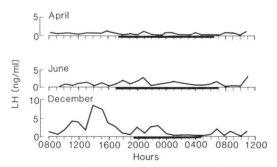

FIG. 6. Concentrations of LH, sampled once an hour for 28 hours, in a female tammar during lactational quiescence (April, June) and at the end of seasonal quiescence (December). Unpublished data of C. A. Horn and C. H. Tyndale-Biscoe.

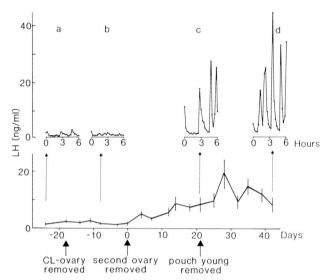

FIG. 7. Concentrations of LH in 8 lactating females sampled twice per week to show the successive responses to lutectomy, bilateral ovariectomy, and removal of the pouch young. Upper profiles (a,b,c,d), LH concentrations sampled at 15 minute intervals for 6 hours shown for one of these tammars before each treatment, to demonstrate pulsatile release of LH after bilateral ovariectomy. Authors' unpublished data.

tomy (Evans et al., 1980; Tyndale-Biscoe and Hearn, 1981; Horn et al., 1985). Because the sample intervals were long in these studies, short-term fluctuations might have been missed, so in another study (authors' unpublished results) samples were taken at intervals of 15 minutes for 6 hours from intact lactating females in the breeding season and after lutectomy and bilateral ovariectomy. No pattern of pulsatile release could be discerned in the intact females (Fig. 7) or after removal of the ovary bearing the corpus luteum. However, after removal of the second ovary and subsequently the pouch young the basal level of LH rose and a marked pulsatile release was observed, as has been described in the sheep and several other species. In order to determine which component of the nonluteal tissue of the ovary is responsible for the negative feedback, grafts of cortex or interstitial tissue were placed under the skin of ovariectomized females and only in those in which the grafts of cortex became established was the rise in gonadotropins reversed (Evans et al., 1980; Tyndale-Biscoe and Hearn, 1981).

After bilateral ovariectomy neither LH nor FSH was reduced with progesterone treatment but estradiol-17β was effective in reducing both gonadotropins (Evans et al., 1980). Horn et al. (1985) investigated this further. At a dose which did not alter the level in peripheral circulation, estradiol reduced both LH and FSH to levels seen in intact females, while a dose that raised estradiol in circulation to estrous levels of 50–70 pg/ml first depressed FSH and LH and then induced a large transient pulse of LH, in a classic example of positive feedback (Fig. 8). This result also confirmed that the assays used were differentiating between the two gonadotropins and that, as in other mammals, the two hormones respond differently to estradiol.

In the tammar, unlike in the sheep, no seasonal differences in the peripheral concentrations of FSH or LH were detected (Fig. 6) nor in the response of females to a standard challenge of GnRH (C. A. Horn and C. H. Tyndale-Biscoe, unpublished results). Similarly, no seasonal differences were detected in the gonadotropic response to ovariectomy or estradiol injections; the responses were the same when the experiments were performed in the first or the second half of the year (Fig. 8), that is to say during the period when unmated females undergo regular estrous cycles, and in seasonal quiescence when they do not (Horn *et al.*, 1985). These results in the tammar are strikingly different from the response of the ovariectomized ewe, in which the negative feedback effects of estra-

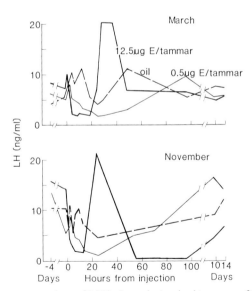

FIG. 8. Mean concentrations of LH in 8 ovariectomised tammars after a single injection of saline (dashed lines), 0.5 µg estradiol, and 2.5 µg estradiol, during the period of declining day length (March) and the period of increasing day length (November). No difference is apparent in either the negative or positive feedback effects of estradiol between the two seasons. After Horn *et al.* (1985).

diol on concentrations of LH in peripheral plasma are marked during anestrus but negligible during the breeding season (Legan and Karsch, 1980; Haresign and Friman, 1983; Karsch et al., 1984). This has led us to conclude that the inhibition of estrus and ovulation during seasonal quiescence is controlled by a different mechanism than that in the anestrous sheep and does not involve changes in hypothalamic sensitivity to ovarian feedback or changes in the LH pulse generator. This is not to say that negative feedback is not a component of the whole mechanism, only that it is not subject to seasonal change. What prevents estrus and ovulation in the tammar after the winter solstice is the quiescent corpus luteum.

During the first half of the reproductive cycle, follicular growth is inhibited by the corpus luteum, but after day 12 lutectomy does not affect the time of next estrus and ovulation (Tyndale-Biscoe and Hawkins, 1977), even though this is the period of elevated progesterone secretion from the corpus luteum (Fig. 5). Similarly, removal of the quiescent corpus luteum from lactating females is followed by follicular growth and ovulation from the contralateral ovary within 12–18 days and the same response is observed after lutectomy in seasonal quiescence (Tyndale-Biscoe and Hawkins, 1977). Thus it appears that the quiescent and immature corpus luteum, which secretes low levels of progesterone, can inhibit the follicle but that the mature corpus luteum cannot. Evans et al. (1980) and Renfree et al. (1982) demonstrated that estradiol but not progesterone could inhibit follicular growth and ovulation after lutectomy. From this it might be concluded that the immature corpus luteum inhibits follicle growth by tonic secretions of estradiol, which inhibit gonadotropin secretion. However, two aspects of this hypothesis are unresolved. First, as mentioned earlier, no increase in plasma LH or its pulsatile release occurred after lutectomy (Fig. 7), as might be expected, and second the luteal cells have not been shown to be capable of aromatizing steroid precursors to estradiol, nor has estradiol been shown to be secreted by the corpus luteum either *in vitro* or *in vivo* (Renfree et al., 1984). Notwithstanding these unresolved points, the evidence so far indicates that follicular growth and ovulation can potentially occur throughout the year in the tammar but are prevented by the presence of the quiescent corpus luteum.

We mentioned earlier that the corpus luteum continued to function after hypophysectomy performed during the reproductive cycle, even though follicular development was suppressed (Hearn, 1974). Similarly, in females immunized against GnRH during pregnancy the corpus luteum continued to function, although follicular growth did not continue (Short et al., 1985). The conclusion that the corpus luteum is not dependent on luteotropic stimulation from the pituitary for normal growth and function is one of the more remarkable and certainly unexpected aspects of the reproductive endocrinology of this species. In order to determine whether this is peculiar to the macropodids or is more general among marsupials, another species, the brushtail possum, which displays the first pattern of reproduction referred to in the introduction, was investigated (authors' unpublished results). After hypophysectomy as early as day 1 after estrus, the corpus luteum of this species continued to grow, to secrete progesterone in a normal manner, and to sustain pregnancy to full term in the same way as in the tammar. We conclude that the autonomy of the corpus luteum may be a common feature of marsupials.

Not only is the developed and active corpus luteum of the tammar autonomous but so is the quiescent corpus luteum once it is released from inhibition; hypophysectomy performed during lactational or seasonal quiescence is followed promptly by reactivation of the corpus luteum and the resumption of pregnancy (Hearn, 1974; Tyndale-Biscoe and Hawkins, 1977). Furthermore, in the tammars that reactivated in response to hypophysectomy the progesterone pulse occurred on day 7–8 (Fig. 9, Table I). This presumably reflects an altered secretion rate by the reactivated corpus luteum, even though it can have received no luteotropic stimulation since the pituitary was removed 7–8 days before. This is consistent with the demonstration that the tammar corpus luteum lacks specific bind-

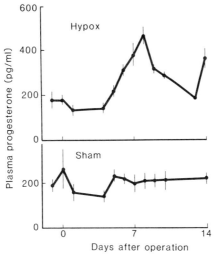

FIG. 9. Responses in the concentrations (mean + SEM) of progesterone in lactating tammars subjected to hypophysectomy ($n = 4$) or sham hypophysectomy ($n = 5$). Removal of the pituitary released the corpora lutea from inhibition and they underwent normal development, including a change in secretion rate at day 7–8, whereas the sham-operated females remained in lactational quiescence. Authors' unpublished results.

ing sites for LH, while having a higher concentration of specific binding sites for prolactin than any other tissue, including lactating mammary gland (Sernia and Tyndale-Biscoe, 1979; Stewart and Tyndale-Biscoe, 1982).

Thus the main role of the pituitary, in this respect, is not to stimulate the corpus luteum but to tonically inhibit it. For its part the corpus luteum, despite its undeveloped state, inhibits further ovulation and so a pattern of seasonal reproduction is imposed on an essentially nonseasonal system. Since lactational quiescence is probably the primary reproductive mechanism in this, as in other species of kangaroo, we will now examine the way in which it is controlled.

VI. Endocrine Control of Lactational Quiescence

The proximate stimulus in lactational quiescence is the sucking by the pouch young (Fig. 19). Using the red kangaroo, which has the same pattern as the tammar, Sharman (1965) showed that it is the frequency of sucking rather than the amount of milk or the size of the young which is the effective stimulus, and Renfree (1979) confirmed this in tammars suckling small young by denervating the sucked mammary gland and its teat. Lactation was not impaired as the young continued to grow during the next 3 weeks, but the inhibition on the corpus luteum was removed immediately; it grew and induced reactivation in the associated blastocyst and supported the ensuing pregnancy. This response is the same as that after hypophysectomy (Hearn, 1974) or adenohypophysectomy (Tyndale-Biscoe and Hawkins, 1977), and suggests that the sucking stimulus acts via the anterior pituitary. Sharman (1965) showed in the red kangaroo that oxytocin injections prolonged quiescence after the pouch young was removed but Tyndale-Biscoe and Hawkins (1977) found that it did not do so in hypophysectomized tammars. Conversely they showed that prolactin, injected 3 times a day, delayed reactivation of the corpus luteum for the duration of the treatment in both intact and hypophysectomized tammars, and they concluded that sucking increases the release of prolactin from the anterior pituitary which, in turn, inhibits the growth of the corpus luteum or its secretion of progesterone. The observation that the luteal cell membranes contain high concentrations of prolactin-specific receptors led to the idea that prolactin might be acting directly on the corpus luteum to suppress progesterone secretion. However, *in vitro* studies by Sernia *et al.* (1980) and Hinds *et al.* (1983) failed to demonstrate any inhibitory or stimulatory effect of prolactin or LH on steroidogenesis by luteal tissue.

Experiments with the dopamine agonist, bromocryptine (Sandoz, CB

154), have added further support to the hypothesis that prolactin is the inhibitor of luteal function (Tyndale-Biscoe and Hinds, 1984). During lactational quiescence a single injection of bromocryptine of 5 mg/kg body weight invariably causes immediate reactivation (Table II), with the progesterone pulse on day 5–6 and birth and/or estrus on day 26–27 (Table I), the same interval as after removing the pouch young. These results appear to be consistent with bromocryptine suppressing prolactin secretion and thereby removing the inhibition on the corpus luteum. However, the peripheral plasma concentrations of prolactin, measured once a day, are consistently low throughout lactational quiescence and not different from concentrations in nonlactating tammars at this time of year (Hinds and Tyndale-Biscoe, 1982b, 1985). Furthermore, the injections of bromocryptine, which induced such rapid responses in the corpora lutea, caused no detectable difference in the short-term levels of prolactin (Tyndale-Biscoe and Hinds, 1984). New evidence to be presented later (Fig. 15) in reference to seasonal quiescence may provide an explanation for this paradox, namely that it is not the basal levels of prolactin that are important but one

TABLE II

Response of Tammars in Lactational and Seasonal Quiescence to a Single Injection of Bromocryptine (5 mg/kg Body Weight), at Different Months of the Year

Month	Number treated	Number reactivated	Percentage reactivated
Lactational quiescence			
February	5	5	100
March	5	5	100
May	9	9	100
June	5	5	100
Seasonal quiescence			
June	20	20	100
July	47	38[b]	81
August 12	6	4[b]	67
August 26	6	6	100
September 9	6	2	33
September 22	16	0	0
October	5	0	0
November	5	1	20
December	5	0	0

[a] Data from Tyndale-Biscoe and Hinds (1984) and additional unpublished results for May through September. Reactivation determined by occurrence of progesterone pulse, birth, or estrus.

[b] Reactivation determined only by occurrence of birth; others in these groups may have undergone a nonpregnant cycle that was not recorded.

or more transient pulses each day that provide the tonic inhibition of the corpus luteum. If the bromocryptine treatment abolished such a pulse it would not have been detected by daily sampling. In the experiment reported by Tyndale-Biscoe and Hawkins (1977) the injection of prolactin 3 times a day may have been effective, not because it maintained a high level of prolactin, but because it mimicked a transient pulse that inhibits the corpus luteum. Nevertheless, this does not alter the fact that the corpus luteum contains an unusually high concentration of prolactin receptors and the effect may be a direct inhibition of the luteal cells.

To conclude this section, the major control of ovulation is exercised by the quiescent corpus luteum. During the period of the year when daylength is declining the sucking stimulus of the pouch young is the proximate stimulus, acting via a neural arc to induce the secretion of prolactin, possibly in a pulsatile fashion, from the anterior pituitary. Prolactin is the most likely agent of the tonic inhibition of the corpus luteum, which in turn prevents the completion of embryo development and inhibits further ovulation until after it has completed its own development. The rate of recovery of the corpus luteum, as determined by the occurrence of the progesterone pulse (Table I), suggests that bromocryptine and hypophysectomy are affecting the same end pathway of inhibition as the removal of the pouch young and that therefore most of the interval before the progesterone pulse must be required for the corpus luteum to recover from inhibition and resume its interrupted life. Such a long lag time has the hallmark of a developmental step, rather than a direct endocrine effect on steroidogenesis, which might be expected to be more rapid. Having concluded that the mechanism just reviewed represents the basic control of quiescence we can now address the question of how seasonality is perceived and superimposed on it in the tammar.

VII. Endocrine Control of Seasonal Quiescence

In the red kangaroo and most other species of kangaroo, removal of the young at any stage of pouch life is sufficient to induce reactivation of the corpus luteum, but in the tammar it is only effective during the period of declining daylength. After the winter solstice it is ineffective, even in females carrying very small young, which are still continuously attached to the teat. The response to bromocryptine also has a remarkable seasonal component; while a single dose of 5 mg/kg induced reactivation in all tammars treated from February through June the same dose was ineffective in all but one animal treated during September through December (Tyndale-Biscoe and Hinds, 1984). In a subsequent series we have observed that it was effective in all tammars treated in July and August, while a reduced number responded in early September and it was ineffective in all animals treated at the vernal equinox on 22 September. This confirmed the observations made at this time in 3 previous years (Table II). On the other hand, hypophysectomy in October and November resulted in immediate reactivation, as it did in lactational quiescence before the winter solstice (Hearn, 1974; Tyndale-Biscoe and Hawkins, 1977). This suggested that another seasonal influence is operating on the pituitary after the winter solstice and a photoperiodic cue seemed to be the most likely.

A. RESPONSE TO EXPERIMENTAL CHANGE IN PHOTOPERIOD

Because the normal time of reactivation is shortly after the summer solstice, studies on photoperiod responses were designed to test the influence of mid-summer day length against equinoctial day length (Sadleir and Tyndale-Biscoe, 1977). These showed that exposure to a constant photoperiod of 15L:9D from the vernal equinox did not stimulate reactivation and the tammars gave birth at the same time as the control tammars shortly after the summer solstice (Fig. 10c), whereas, after exposure to an equinoctial photoperiod (12L:12D), the tammars reactivated in a haphazard manner (Fig. 10d), which suggested that this photoperiod was neither inhibitory nor stimulatory but permissive (Sadleir and Tyndale-Biscoe, 1977; Hinds and den Ottolander, 1983). While a long-day photo-

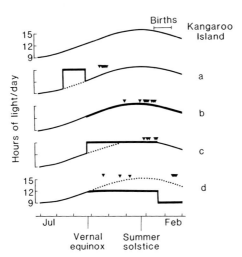

FIG. 10. Photoregimens and results compared to the normal pattern of births of tammars on Kangaroo Island. Closed lines represent the natural photoperiod for 36°S, heavy lines the experimental photoperiods in light-proof rooms, and arrows the time of birth of individual tammars. From Sadleir and Tyndale-Biscoe (1977), with permission.

period was thus shown to be inhibitory, a sudden reduction from 15L:9D to 12L:12D stimulated premature reactivation in all tammars (Fig. 10a), and this has now become our standard photoregimen for examining the tammar's response to photoperiod. A similar response followed a change from 12L:12D to 9L:15D although in this case only 4/9 animals responded (Fig. 10d). The response time after photoperiod change is slower than that seen after removal of pouch young, the progesterone pulse occurring on day 10 and birth and/or estrus on day 33 (Fig. 11). The difference is due to a longer interval to the progesterone pulse (Table I), not to a change in the subsequent period from the pulse to birth and/or estrus, which was 21–24 days. When tammars on a permissive photoregimen, initially 12L:12D and then 9L:15D, had given birth, Hinds and den Ottolander (1983) removed the new young and on this occasion the progesterone pulse occurred on day 6 and birth on day 28, the same intervals as after removing the pouch young during declining day length before the winter solstice (Table I). Conversely, when tammars were exposed to

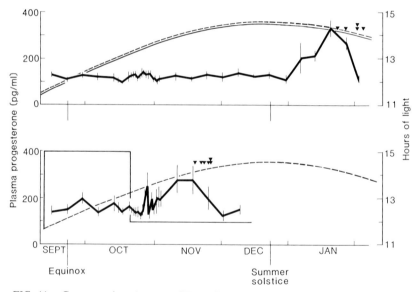

FIG. 11. Concentrations (mean + SEM) of plasma progesterone in two groups of tammars which experienced artificial photoregimens of either increasing day length (upper) or 15L:9D to 12L:12D. The former group reactivated after the solsticial photoperiod and the other group after the sudden decrease in photoperiod. Note the early pulse of progesterone on day 10 in the latter group. Dashed lines represent the natural photoperiod, closed lines the experimental photoregimen, and arrows the time of birth for individual tammars. After Hinds and den Ottolander (1983).

natural increasing day length after they had been exposed to the photoregimen of 15L:9D to 12L:12D, further reactivation did not occur after their new pouch young were removed (McConnell and Tyndale-Biscoe, 1985). Similarly, when tammars were transferred to a photoperiod of 18L:6D at the autumnal equinox (March 22) and had their young removed they did not reactivate, as do all tammars on natural photoperiod at this time of year. Together these results indicate that there is an additional photoperiod-controlled component in the inhibition of the corpus luteum when tammars experience increasing day length or long days, which is not present when they experience decreasing day length or short days (Fig. 12). Furthermore, the factor is abolished when tammars experience artificially short day length after the vernal equinox, and is induced when tammars experience artificially long days after the autumnal equinox.

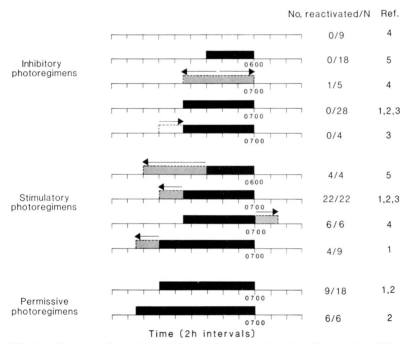

FIG. 12. Summary of experimental photoregimens that have been found to be inhibitory, stimulatory, and permissive in tammars. Closed block, initial dark phase; hatched block, extension of dark phase after photoperiod change. Numbers on right represent the totals of all tammars treated and their responses. References are (1) Sadleir and Tyndale-Biscoe (1977), (2) Hinds and den Ottolander (1983), (3) McConnell and Tyndale-Biscoe (1985), (4) McConnell et al. (1986), and (5) authors' unpublished results.

B. ROLE OF THE PINEAL GLAND

There is now much evidence that the pineal gland of mammals is involved in transducing photoperiod information, and thereby the control of seasonal breeding, through its secretion of melatonin (Lincoln and Short, 1980; Karsch et al., 1984; Tamarkin et al., 1985). In the tammar, as in other mammals, the concentration of melatonin in peripheral plasma shows marked circadian fluctuations, with the rise coinciding with the onset of the dark phase and the fall with the onset of the light phase each day (McConnell and Tyndale-Biscoe, 1985; McConnell, 1986). These fluctuations are abolished in the tammar (Fig. 13), as in other species such as the sheep (Kennaway et al., 1977), by pinealectomy (McConnell and Hinds, 1985) and denervation of the pineal by bilateral cervical sympathetic ganglionectomy (Renfree et al., 1981; McConnell, 1984).

The longer interval to the progesterone pulse after a light change compared to the interval after removal of the pouch young (Table I) might be due either to the time required for the pineal to adjust its secretion of melatonin to the new photoperiod, or to the time required for the melatonin message to be read. These two possibilities have been investigated by using the design described above, in which the tammars were first subjected to a photoregimen of 15L:9D for 40 days and then changed abruptly to 12L:12D by extinguishing the lights 3 hours earlier in the evening. Under this regimen the nocturnal rise of melatonin advanced on the first night to almost the same extent as it reached on the fifth night (Fig. 14c), as observed in the sheep (Bittman et al., 1983), and the births that followed occurred on day 32, similar to the earlier experiments (McConnell and Tyndale-Biscoe, 1985). This suggested that the lag is not due

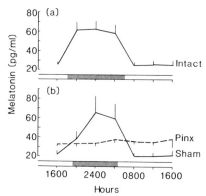

FIG. 13. Concentrations (mean + SEM) of plasma melatonin in relation to the scotophase (shaded) in intact tammars and in the same tammars after pinealectomy (hatched line) or sham pinealectomy. From McConnell and Hinds (1985).

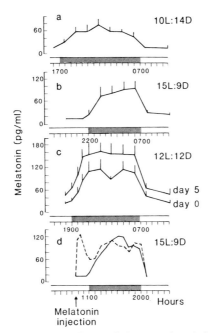

FIG. 14. Concentrations (mean + SEM) of plasma melatonin in tammars exposed successively to the photoperiods indicated. The response in melatonin to the reduced light phase was substantially apparent on day 1 (c). The profile obtained after a single injection of melatonin 2.5 hours before lights off on 15L:9D mimicked the effect of a change to 12L:12D (d). From McConnell and Tyndale-Biscoe (1985).

to a slow response by the pineal to the photoperiod change, and to test this further another group of tammars was subjected to 15L:9D for 15 days but 2.5 hours before lights off they were injected with sufficient melatonin to raise the plasma concentration to nocturnal levels (Fig. 14d). All the melatonin-treated tammars gave birth 32 days after the start of the injections, while the controls did not reactivate until the photoregimen was altered to 12L:12D. These results supported the idea that the photoperiodic cue is read as a change in melatonin secreted by the pineal during the dark phase, but they did not resolve the nature of the message. In birds (Follett, 1984) and some mammals, such as *Peromyscus leucopus* (Margolis and Lynch, 1981), the response is thought to be due to a coincidence between one phase of an endogenous circadian rhythm and elevated concentrations of melatonin in circulation. In these species the photoperiodic response can be mimicked in pinealectomized animals by exposure to melatonin at a particular time of the day. For other species, such as the sheep (Karsch *et al.*, 1984) and the Djungarian hamster (Carter and Goldman, 1983a,b), the total duration of elevated melatonin experienced each day is the critical factor. For the tammar these results did not differentiate between change in duration of the nocturnal melatonin level, increased concentration of melatonin, or exposure to an elevated concentration of melatonin at a particular time of the 24 hours, in this case 1930–2200 hours.

To test the duration and time of day hypotheses two experiments were set up (Fig. 15). In the first, the normal endogenous melatonin elevation was abolished by exposure to 24L:0D for 15 days and the tammars were injected at 1930 hours each day for 23 days with either melatonin or oil vehicle (McConnell et al., 1986). After 15 days the photoperiod was changed to 15L:9D, with the dark phase beginning 2.5 hours after the injection, so that the tammars experienced 9 h or 12 hours of elevated melatonin. None of the oil-treated animals reactivated under either regimen, whereas all the melatonin-treated tammars reactivated after the change to 15L:9D. Thus, exposure to melatonin from 1930 hours by itself did not abolish the inhibition but, in combination with a nonstimulatory dark period of 9 hours, it did. The result in the control group also showed that a change in duration from no dark to 9 hours dark does not abolish the inhibition (Fig. 12), so we conclude that change of duration is only an effective signal if the dark phase or period of elevated melatonin is greater than 9 hours. In the second experiment the dark phase was increased from 9 to 12 hours in the morning. This was equally as effective as extending the dark phase in the evening (Fig. 12), and is thus a further refutation of the time of day hypothesis for the tammar. Another attempt to induce reactivation by injecting melatonin at the time of lights on instead of

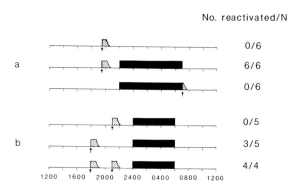

FIG. 15. Diagram to indicate the patterns of melatonin achieved by injections (hatched blocks) at various times before or after the scotophase (closed blocks), and the responses in groups of tammars so treated. From McConnell et al. (1986) and authors' unpublished results.

extending the dark phase in the morning failed to induce reactivation. On analysis of the melatonin profiles, however, it was evident that in these tammars the endogenous elevated melatonin profile of 9 hours had been prolonged for only about 1 hour (Fig. 15), which was presumably an insufficient signal for reactivation.

Since the tammar is shown by these results to respond to duration of elevated melatonin, the next question was whether melatonin must be elevated for the full period or only at the two ends. This question has been addressed in recent experiments by exposing animals to 18L:6D and then injecting melatonin or oil for 10 days at 6, 3, or 6 and 3 hours before the lights went off at midnight (Fig. 15). None of the control group or those given melatonin 3 hours before lights off, but all those injected 6 and 3 hours before lights off, reactivated and gave birth on day 30, thereby confirming the duration hypothesis (Fig. 15). In the intermediate group 3/5 reactivated, even though the melatonin declined to basal levels between the injection and the endogenous rise 6 hours later. These 3 gave birth or showed estrus on day 34, 4 days later than the other group. Presumably, under this condition the signal was not clear, so that only some animals could read it as a long duration of melatonin, and they took 4 days longer to do so.

The last question was how many days does the tammar need to read the melatonin signal before it can reactivate. In previous experiments the treatment was continued for 15 days but since the progesterone pulse occurred on day 8 or 9 (Table I) it could be assumed that the signal had been read by then. Conversely, in preliminary trials to determine the dose of melatonin required to elevate the plasma concentration (see Fig. 14d), one injection of melatonin did not cause reactivation. Therefore tammars were subject to the inhibitory photoregimen of 15L:9D and groups of two were injected for 3 to 8 successive days with melatonin 3 hours before lights off (Fig. 16). All these animals reactivated, had a progesterone pulse on day 6–9, and gave birth on day 29–31 (Table I), from which we conclude that for some tammars the signal need only be read for 3 days. This is an extraordinarily rapid response to a change in the melatonin profile and leads to the next question. How is the signal translated into pituitary response?

C. THE PITUITARY RESPONSE AFTER PHOTOPERIOD CHANGE

Since the progesterone pulse occurs 10.5 days after the light change but only 7.5 days after hypophysectomy (Table I), it may be presumed that the pituitary inhibition of the corpus luteum is abolished about 3 days after

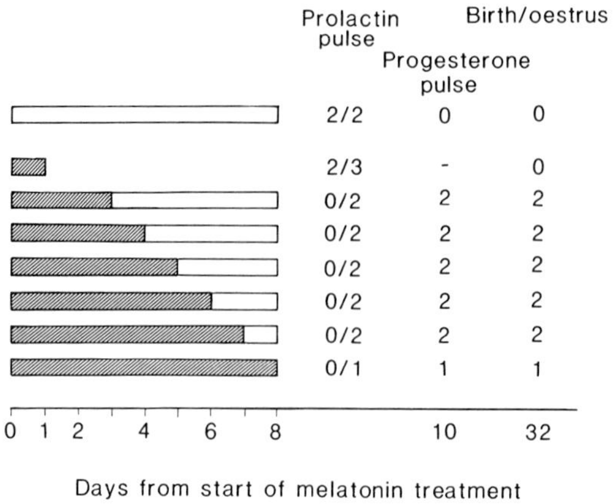

	Prolactin pulse	Progesterone pulse	Birth/oestrus
☐	2/2	0	0
▨	2/3	-	0
▨▨	0/2	2	2
▨▨▨	0/2	2	2
▨▨▨▨	0/2	2	2
▨▨▨▨▨	0/2	2	2
▨▨▨▨▨▨	0/2	2	2
▨▨▨▨▨▨▨	0/1	1	1

0 1 2   4   6   8   10   32

Days from start of melatonin treatment

FIG. 16. Experimental design to determine the number of days required for a tammar to rspond to a stimulatory treatment of melatonin 3 hours before lights off on 15L:9D. Responses observed in the occurrence of the prolactin peak, the progesterone pulse, and birth are indicated. Authors' unpublished results.

the light change. However, more direct evidence is now available to support the conclusion that the pituitary has responded by day 5 after the light change. In tammars exposed to an inhibitory photoperiod of 15L:9D we observed a marked pulse of prolactin of less than 2 hours duration coinciding with the time of lights on (McConnell *et al.*, 1986). However, 5 days after the change to 12L:12D the pulse was absent from all 6 tammars, which reactivated and gave birth 30–39 days later (Fig. 17). Similarly, in the experiments in which tammars were injected with melatonin 6 and 3 hours before lights off there was a prolactin pulse on day 8 only in those tammars that did not reactivate (Fig. 15). We conclude that the prolactin pulse is the most likely agent of the inhibition of the corpus luteum during inhibitory photoperiod and its abolition by day 5 is followed by reactivation, the subsequent pulse of progesterone, and birth. An interesting aspect of this is that the prolactin pulse was still absent on day 8 in those tammars that reactivated after only receiving melatonin for 3–7 days (Fig. 16), which suggests that the abolition is permanent once the melatonin signal has been read. This supports the idea that melatonin does not affect the pituitary directly but operates through a neural mechanism in the hypothalamus.

To conclude, the tammar responds to the change in duration of the dark phase by closely synchronized secretion of melatonin from the pineal and this is essentially achieved on the first night of change. Nevertheless, it takes about 3 successive days for the message to be acted upon (Fig. 18). Then the response before day 5 is to abolish permanently the early morn-

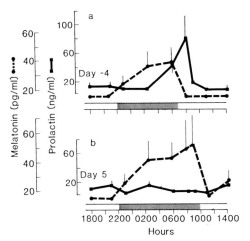

FIG. 17. Concentrations (mean + SEM) of plasma melatonin and prolactin in a group of 6 tammars sampled through the scotophase on an inhibitory photoperiod of 15L:9D (a) and after the change to 12L:12D, (b), when they all reactivated and gave birth. From McConnell et al. (1986).

ing peak of prolactin, presumably by increasing the secretion of dopamine. The absence of the prolactin peak in turn releases the corpus luteum from its long inhibition and enables it to secrete the early pulse of progesterone, complete its development autonomously, and induce embryo reactivation and birth (Fig. 19). We can now relate these results from tammars under artificial photoregimens to the events that occur each year after the summer solstice.

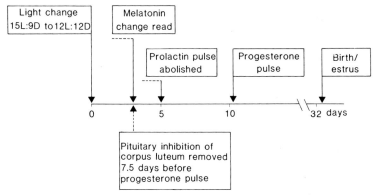

FIG. 18. Schematic diagram to indicate the sequence of events that may follow the stimulatory change of photoperiod based on the data reviewed in the text.

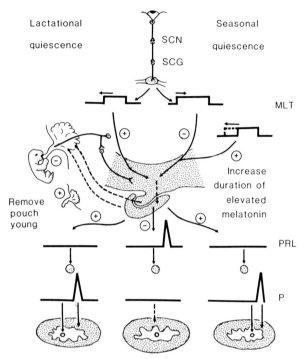

FIG. 19. Summary of the pathways involved in the control of lactational and seasonal quiescence. MLT, Melatonin; P, progesterone; PRL, prolactin; SCG, superior cervical ganglion; SCN, suprachiasmatic nucleus; +, stimulatory; −, inhibitory.

D. NATURAL RESPONSE AFTER THE SUMMER SOLSTICE

The period during which births occur after the summer solstice is remarkably constant from year to year in the same locality and for populations of tammars in different localities (see Tyndale-Biscoe and Renfree, 1986). For the Kangaroo Island population, and the Canberra colony derived therefrom, the mean date of birth in successive years has ranged from 28 January to 3 February (Fig. 21). The actual date of reactivation of the corpus luteum has been determined more precisely by measuring the early pulse of progesterone in a group of 19 tammars from early December until birth and estrus in January or February. The progestrone pulse occurred between 9 and 17 January and births between 31 January and 8 February. As in other conditions already described the interval from the pulse to birth in each tammar was very similar (22 days) and the variation between individuals was largely due to variation in the interval from the solstice to the progesterone pulse, the mean of which was 20.6 days

(Table I). This is much longer than the interval after an experimental light change of 3 hours or an increase in the duration of the period of elevated melatonin, previously discussed. Since the minimum time required for a corpus luteum to recover from inhibition is 5–6 days, as after removal of the pouch young or treatment with bromocryptine during lactational quiescence, we may conclude that it took about 15 days for these tammars to respond to the change in day length after the solstice, compared to 4–5 days after the experimental light change of 3 hours (Table I). The change in day length during this period is about 6 minutes, most of which occurs after 1 January.

If the hypothesis that photoperiodic change is read as a change in the duration of elevated melatonin is correct, then the clearest signal, and the one that can be read most rapidly, will be exogenous melatonin added to the endogenous melatonin rise to provide a continuous total of 12 hours; the next clearest signal will be that provided by a very abrupt and large change in photoperiod because of the time required for the shift in elevated melatonin to be complete; and the least clear signal, and the one that will take longest to read, will be a very gradual change in photoperiod, such as occurs after the summer solstice. If female tammars do indeed read the change in duration of day length after the summer solstice in this way, it implies an extraordinarily sensitive mechanism for recording and storing information about time. Another possibility is that there is a circannual rhythm, which determines the approximate time of reactivation and the changes in day length after the solstice provide the fine adjustment. The results from pinealectomy and pineal denervation provide some support for this idea.

E. EFFECTS OF PINEALECTOMY

Having demonstrated the importance of melatonin in the transduction of photoperiod change and the control of seasonal quiescence one might suppose that pinealectomy would abolish photoperiodic control and that such tammars would revert to the nonseasonal pattern of lactational quiescence of the red kangaroo. This is only partially so. Renfree *et al.* (1981) reported that denervation of the pineal by superior cervical sympathetic ganglionectomy, performed during lactational quiescence in March–May, abolished the subsequent seasonal quiescence, since the ganglionectomized tammars produced a succession of 3 young each after the winter solstice, from August through November, whereas the sham-operated tammars only gave birth at the normal time after the summer solstice in February. However, when McConnell and Hinds (1985) pinealectomized tammars during seasonal quiescence in October, none reactivated until

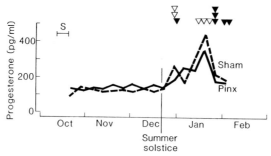

FIG. 20. Concentrations (mean + SEM) of plasma progesterone and times of birth in two groups of tammars that were subjected to pinealectomy (closed line, open arrows) or sham pinealectomy (dashed line, closed arrows) in seasonal quiescence in October. S, Period of surgery for both groups. From McConnell and Hinds (1985).

the normal time after the summer solstice and they gave birth at the same time as the sham-operated controls (Fig. 20). Subsequent experiments, designed to determine whether these differences were due to the type of surgical intervention or to the time of year when the operations were done (McConnell, 1984), have not resolved the discrepancy but have added a further complexity to the topic. In each series, whether the intervention employed was pinealectomy or ganglionectomy, about half of the group behaved like the tammars reported by Renfree *et al.* (1981) and about half retained seasonality and gave birth at the end of January, with the controls (Table III). Some of these tammars have now lived for 3 years since they were operated on and a diminishing but substantial number of them have continued to retain seasonal quiescence and give birth in January–March (Table III).

TABLE III

Effect of Pinealectomy (PINX) or Superior Cervical Ganglionectomy (SCGX) on Retention of Seasonality in Tammars[a]

Year	Treatment	N	Reactivation in seasonal quiescence	Reactivation after summer solstice	Percentage retained seasonal quiescence
1	PINX/SCGX	29	12	17	59
	Sham	17	2	15	88
2	PINX/SCGX	26	12	12	48
	Sham	15	0	15	100
3	PINX/SCGX	14	6	4	29
	Sham	9	0	9	100

[a] From McConnell (1984) and authors' unpublished results.

One possibility is that there is yet another control system beyond the pineal, a circannual rhythm, which continues to operate after pinealectomy in some animals. The only evidence in support of this is the result reported by Sadleir and Tyndale-Biscoe (1977) that tammars held on long day length for 3 months from the equinox to the summer solstice gave birth at the same time as the control group, despite having experienced no change in photoperiod during that time (Fig. 10c). Clearly this is an area for future investigation. However the most important conclusion to emerge from these results is the opportunity that the tammar now offers for investigating the central control site where the melatonin message is read so expeditiously and translated into a command to the pituitary.

F. SEASONALITY IN THE MALE TAMMAR

It is clear that the female tammar is a seasonal breeder and is sensitive to photoperiod, but what of the male tammar? It might be expected that the formidable demands of the very brief sexual season from mid-January to mid-February, when all the adult females experience estrus, would require synchronous recrudescence of the testes and accessory organs. In the wild population on Kangaroo Island, Inns (1982) found no significant increase in testicular size throughout the year, but he found a significant elevation of plasma testosterone and increase in prostate weight of the adult males during December–February. He also observed a smaller but significant rise in plasma testosterone and prostate weight in October at the time that the young females experience first estrus. In order to determine whether the male is responding to changes in the females, or to environmental factors such as photoperiod, Catling and Sutherland (1980) measured LH, FSH, and testosterone in males from the vernal equinox until after the summer solstice and the subsequent sexual season. One group was isolated from females throughout the experiment and the other group was run with females that reactivated after the solstice, as judged by the increase in plasma progesterone, and gave birth and underwent estrus in January–February (Fig. 21). In the group isolated from females all three hormones remained unchanged at basal levels, whereas in the second group LH and testosterone, but not FSH, became significantly elevated in December and remained so until after the females had given birth in February (Fig. 21). This study indicates that the male tammar is probably unresponsive to photoperiodic changes but responds to changes occurring in the female at the time of reactivation. It is of interest that the hormonal changes in the males began about 3 weeks before any female gave birth or came into estrus, which suggests that the cue that the male responds to is pheromonal, possibly controlled by progesterone or estra-

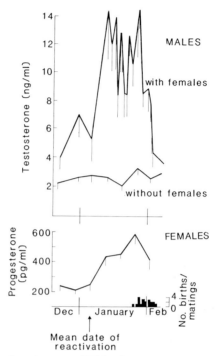

FIG. 21. Concentrations (mean + SEM) of plasma testosterone in 8 male tammars associated with females during the period of reactivation, birth, and postpartum estrus and in 5 male tammars that were isolated from females throughout this period. Concentrations (mean + SEM) of plasma progesterone in the females and the times of birth are shown also. After Tyndale-Biscoe (1980).

diol from the reawakened corpus luteum. Flint and Renfree (1982) showed that there is a transient elevation of estradiol in females shortly after the summer solstice, and at about the same time as the early progesterone pulse.

The final question raised by this exceptional species is when does the mechanism of photoperiodic control of reproduction develop in the life of the female? The mechanism is fully operational at puberty in October but it may differentiate much earlier, while the young is in the pouch or even before the eyes open at 120 days. The suprachiasmatic nucleus, which has been shown to drive the pineal rhythm in the rat (Moore and Klein, 1974), is innervated from the retina by day 65 in the tammar (Wye-Dvorak, 1984), 2 months before the eyes open. Histological differences have been reported in the suprachiasmatic nuclei of male and female rats at birth (LeBlond et al., 1982), which are thought to be related to the differentiation of the hypothalamus at an early age in this species, so it will be of interest to learn whether similar differences can be identified in young tammars, which are related to the differentiation of photosensitivity.

ACKNOWLEDGMENTS

Many people have been involved in the work reviewed here. We wish especially to acknowledge the help of our colleagues, Dr. Terry P. Fletcher, Dr. Brian F. Green, Dr. Mervyn E. Griffiths, Dr. John D. Harder, Mrs. Carol A. Horn, Dr. Peter A. Janssens, Mr. Ray Leckie, Professor Richard F. Mark, Mr. James C. Merchant, Dr. Kevin R. Nicholas, Dr. Francesca Stewart, and Mr. John Wright. We also thank Mr. Graeme Chapman and Mr. Frank Knight for the preparation of the illustrations for the presentation and this paper. The Ingram Trust of Victoria provided financial support in 1984 for S.J.M. Finally we thank the Organizing Committee of the Laurentian Hormone Conference for inviting us to participate in the Conference, and for the financial assistance that made it possible.

REFERENCES

Andrewartha, H., and Barker, S. (1969). *Trans. R. Soc. S. Aust.* **93**, 127–132.
Berger, P. J. (1966). *Nature (London)* **211**, 435–436.
Berger, P. J. (1970). PhD thesis, Tulane University.
Bittman, E. L., and Karsch, F. J. (1984). *Biol. Reprod.* **30**, 585–593.
Bittman, E. L., Dempsey, R. J., and Karsch, F. J. (1983). *Endocrinology* **113**, 2276–2283.
Carter, D. S., and Goldman, B. D. (1983a). *Endocrinology* **113**, 1261–1267.
Carter, D. S., and Goldman, B. D. (1983b). *Endocrinology* **113**, 1268–1273.
Catling, P. C., and Sutherland, R. L. (1980). *J. Endocrinol.* **86**, 25–33.
Catt, D. C. (1977). *New Zealand J. Zool.* **4**, 401–411.
Crompton, A. W. (1980). *In* "Comparative Physiology: Primitive Mammals" (K. Schmidt-Nielsen, L. Bollis, and C. R. Taylor, eds.), pp. 1–12. Cambridge Univ. Press, London and New York.
Ensor, D. M. (1978). "Comparative Endocrinology of Prolactin." Chapman & Hall, London.
Evans, S. M., Tyndale-Biscoe, C. H., and Sutherland, R. L. (1980). *J. Endocrinol.* **86**, 13–23.
Findlay, L., Ward, K. L., and Renfree, M. B. (1983). *J. Endocrinol.* **97**, 425–436.
Flint, A. P. F., and Renfree, M. B. (1982). *J. Endocrinol.* **95**, 293–300.
Follett, B. K. (1984). *In* "Marshall's Physiology of Reproduction" (G. E. Lamming, ed.), 4th Ed., Vol. 1, pp. 283–350. Churchill Livingstone, Edinburgh.
Foster, D. L. (1986). *Recent Prog. Horm. Res.* **42**.
Gemmell, R. T. (1982). *Aust. Mammalogy* **5**, 187–193.
Goodman, R. L., Reichert, L. E., Legan, S. J., Ryan, K. D., Foster, D. L., and Karsch, F. J. (1981). *Biol. Reprod.* **25**, 134–142.
Harder, J. D., and Fleming, M. W. (1981). *Science* **212**, 1400–1402.
Harder, J. D., Hinds, L. A., Horn, C. A., and Tyndale-Biscoe, C. H. (1984). *J. Reprod. Fertil.* **72**, 551–558.
Harder, J. D., Hinds, L. A., Horn, C. A., and Tyndale-Biscoe, C. H. (1985). *J. Reprod. Fertil.* **75**, 449–459.
Haresign, W. and Friman, B. R. (1983). *J. Reprod. Fertil.* **69**, 469–472.

Hartman, C. G. (1923). *Am. J. Anat.* **32**, 353-421.
Hearn, J. P. (1974). *J. Reprod. Fertil.* **39**, 235-241.
Hinds, L. A., and den Ottolander, R. C. (1983). *J. Reprod. Fertil.* **69**, 631-639.
Hinds, L. A., and Tyndale-Biscoe, C. H. (1982a). *J. Endocrinol.* **93**, 99-107.
Hinds, L. A., and Tyndale-Biscoe, C. H. (1982b). *Biol. Reprod.* **26**, 391-398.
Hinds, L. A., and Tyndale-Biscoe, C. H. (1985). *J. Reprod. Fertil.* **74**, 173-183.
Hinds, L. A., Evans, S. M., and Tyndale-Biscoe, C. H. (1983). *J. Reprod. Fertil.* **67**, 57-63.
Horn, C. A., Fletcher, T. P., and Carpenter, S. (1985). *J. Reprod. Fertil.* **73**, 585-592.
Inns, R. W. (1982). *J. Reprod. Fertil.* **66**, 675-680.
Janssens, P. A., and Hinds, L. A. (1981). *Gen. Comp. Endocrinol.* **45**, 56-60.
Jurgelski, W., and Porter, M. E. (1974). *Lab. Anim. Sci.* **24**, 412-425.
Karsch, F. J., Bittman, E. L., Foster, D. L., Goodman, R. L., Legan, S. J., and Robinson, J. E. (1984). *Recent Prog. Horm. Res.* **40**, 185-225.
Kennaway, D. J., Frith, R. G., Phillipou, G., Mathews, C. D., and Seamark, R. F. (1977). *Endocrinology* **101**, 119-127.
Le Blond, C. B., Morris, S., Karakiulakis, G., Powell, R., and Thomas, P. J. (1982). *J. Endocrinol.* **95**, 137-145.
Legan, S. J., and Karsch, F. J. (1980). *Biol. Reprod.* **23**, 1061-1068.
Lincoln, G. A., and Short, R. V. (1980). *Recent Prog. Horm. Res.* **36**, 1-52.
Lincoln, D. W., Fraser, H. M., Lincoln, G. A., Martin, G. B., and McNeilly, A. S. (1985). *Recent Prog. Horm. Res.* **41**, 369-419.
Loudon, A. S. I., Curlewis, J. D., and English, J. (1985). *J. Zool.* **206**, 35-39.
McConnell, S. J. (1984). Ph.D. thesis, Australian National University.
McConnell, S. J. (1986). *J. Pineal Res.* **3**(2).
McConnell, S. J., and Hinds, L. A. (1985). *J. Reprod. Fertil.* **75**, 433-440.
McConnell, S. J., and Tyndale-Biscoe, C. H. (1985). *J. Reprod. Fertil.* **73**, 529-538.
McConnell, S. J., Tyndale-Biscoe, C. H., and Hinds, L. A. (1986). *J. Reprod. Fertil.* **77**.
McNeilly, A. S., and Land, R. B. (1979). *J. Reprod. Fertil.* **56**, 601-609.
Margolis, D. J., and Lynch, G. R. (1981). *Gen. Comp. Endocrinol.* **44**, 530-537.
Merchant, J. C. (1979). *J. Reprod. Fertil.* **56**, 459-463.
Merchant, J. C., and Calaby, J. H. (1981). *J. Zool. (London)* **194**, 203-217.
Moore, G. P. M. (1978). *J. Cell. Physiol.* **94**, 31-36.
Moore, R. Y., and Klein, D. C. (1974). *Brain Res.* **71**, 17-33.
Panyaniti, W., Carpenter, S. M., and Tyndale-Biscoe, C. H. (1985). *Aust. J. Zool.* **33**, 303-311.
Pilton, P. E., and Sharman, G. B. (1962). *J. Endocrinol.* **25**, 119-136.
Reiter, R. J. (1975). *J. Exp. Zool.* **191**, 111-120.
Renfree, M. B. (1979). *Nature (London)* **278**, 549-551.
Renfree, M. B., and Tyndale-Biscoe, C. H. (1973). *Dev. Biol.* **32**, 28-40.
Renfree, M. B., and Tyndale-Biscoe, C. H. (1978). In "Methods in Mammalian Reproduction" (J. C. Daniel, ed.), pp. 307-331. Academic Press, New York.
Renfree, M. B., Green, S. W., and Young, I. R. (1979). *J. Reprod. Fertil.* **57**, 131-136.
Renfree, M. B., Lincoln, D. W., Almeida, O. F. X., and Short, R. V. (1981). *Nature (London)* **293**, 138-139.
Renfree, M. B., Wallace, G. I., and Young, I. R. (1982). *J. Endocrinol.* **92**, 397-403.
Renfree, M. B., Flint, A. P. F., Green, S. W., and Heap, R. B. (1984). *J. Endocrinol.* **101**, 231-240.
Reynolds, H. C. (1952). *Univ. Calif. Publ. Zool.* **52**, 223-284.
Sadleir, R. M. F. S., and Tyndale-Biscoe, C. H. (1977). *Biol. Reprod.* **16**, 605-608.
Sernia, C., and Tyndale-Biscoe, C. H. (1979). *J. Endocrinol.* **83**, 79-89.

Sernia, C., Hinds, L. A., and Tyndale-Biscoe, C. H. (1980). *J. Reprod. Fertil.* **60**, 139–147.
Sharman, G. B. (1965). *Excerpta Med.* **83**, 669–674.
Sharman, G. B., and Berger, P. J. (1969). *Adv. Reprod. Physiol.* **4**, 211–240.
Shaw, G. (1983). *J. Reprod. Fertil.* **69**, 429–436.
Shaw, G., and Renfree, M. B. (1984). *J. Reprod. Fertil.* **72**, 29–37.
Shaw, G., and Renfree, M. B. (1986). *J. Reprod. Fertil.* **76**, 339–347.
Shield, J. (1968). *J. Zool. (London)* **155**, 427–444.
Shorey, C. D. and Hughes, R. L. (1973). *Aust. J. Zool.* **21**, 477–489.
Short, R. V., Flint, A. P. F., and Renfree, M. B. (1985). *J. Reprod. Fertil.* **75**.
Smith, M. J., Bennett, J. H., and Chesson, C. M. (1978). *Aust. J. Zool.* **26**, 449–463.
Stewart, F., and Tyndale-Biscoe, C. H. (1982). *J. Endocrinol.* **92**, 63–72.
Sutherland, R. L., Evans, S. M., and Tyndale-Biscoe, C. H. (1980). *J. Endocrinol.* **86**, 1–12.
Tamarkin, L., Baird, C. J., and Almeida, O. F. X. (1985). *Science* **227**, 714–720.
Thornber, E. J., Renfree, M. B., and Wallace, G. I. (1981). *J. Embryol. Exp. Morphol.* **62**, 325–338.
Tyndale-Biscoe, C. H. (1980). In "Endocrinology, 1980" (I. A. Cumming, J. L. Funder, and F. A. O. Mendelsohn, eds.), pp. 277–282. Australian Academy of Science, Canberra.
Tyndale-Biscoe, C. H. (1984). In "Marshall's Physiology of Reproduction" (G. E. Lamming, ed.), 4th Ed., Vol. 1, pp. 386–454. Churchill Livingstone, Edinburgh.
Tyndale-Biscoe, C. H., and Hawkins, J. (1977). In "Reproduction and Evolution" (J. H. Calaby and C. H. Tyndale-Biscoe, eds.), pp. 245–252. Australian Academy of Science, Canberra.
Tyndale-Biscoe, C. H., and Hearn, J. P. (1981). *J. Reprod. Fertil.* **63**, 225–230.
Tyndale-Biscoe, C. H., and Hinds, L. A. (1984). *Gen. Comp. Endocrinol.* **53**, 58–68.
Tyndale-Biscoe, C. H., and Renfree, M. B. (1986). "Reproduction in Marsupials." Cambridge Univ. Press, London and New York.
Tyndale-Biscoe, C. H., and Rodger, J. C. (1978). *J. Reprod. Fertil.* **52**, 37–43.
Tyndale-Biscoe, C. H., Hearn, J. P., and Renfree, M. B. (1974). *J. Endocrinol.* **63**, 589–614.
Tyndale-Biscoe, C. H., Hinds, L. A., Horn, C. A., and Jenkin, G. (1983). *J. Endocrinol.* **96**, 155–161.
VandeBerg, J. L. (1983). *ILAR News* **26**, 9–12.
Ward, K. L., and Renfree, M. B. (1984). *J. Reprod. Fertil.* **72**, 21–28.
Wye-Dvorak, J. (1984). *J. Comp. Neurol.* **288**, 491–508.

DISCUSSION

I. Callard. You mentioned that the wallaby corpus luteum was autonomous. Have you tried to determine whether the corpus luteum is receiving any positive luteotropic stimuli of any kind from the uterus? For example, does it still reactivate after hypophysectomy if you hysterectomize the animal? Are there any uterine luteotropic stimuli of a neural or chemical nature?

C. H. Tyndale-Biscoe. We have not hysterectomized tammars to see whether the corpus luteum would reactivate, because all the evidence we have suggests that the first changes appear in the corpus luteum and that the uterine changes follow. However, we can say that the corpus luteum does not require a specific stimulus from the embryo, because the same changes occur in a nonpregnant animal as occur in the pregnant one after reactivation.

I. Callard. My second question relates to your statement that early pregnancy goes on in the absence of progesterone, yet you showed a small postovulatory progesterone peak. Is that derived from the ruptured follicle and is that small progesterone surge necessary for the first 7 days of blastocyst maintenance?

C. H. Tyndale-Biscoe. I would not like to put too much emphasis on that small peak of progesterone, since we have not done any experiments to see whether it has an effect on embryo survival or not. All I would say is that it appears to come from the ovary bearing the Graafian follicle, rather than from the corpus luteum, of the preceding cycle (see Fig. 4).

M. O. Thorner. That was a beautiful presentation—really fascinating. I have two questions. When you did the hypophysectomy I gather it took 7.3 days before you got your progesterone pulse while when you gave bromocryptine it was only 5 days. Was that a significant difference? If so what explanation do you have for that?

C. H. Tyndale-Biscoe. We are not sure why the progesterone pulse is delayed after hypophysectomy. According to the hypothesis that we are now entertaining there must be a pulse of prolactin at sometime each day, which we have not yet detected, and the bromocryptine may abolish that pulse. That could account for the paradox that we do not see a change in the basal level of prolactin after injection of bromocryptine (Tyndale-Biscoe and Hinds, 1984). If that is so, then after hypophysectomy the prolactin pulse should disappear and reactivation occur immediately. The only explanation I can offer for the longer interval to the progesterone pulse is that hypophysectomy is a more traumatic experience for the animal than a single injection of bromocryptine and it may take longer for the corpus luteum to recover, but that is conjectural.

M. O. Thorner. You pointed out that the bromocryptimic response was only effective at one season and not at the other season. I am not sure whether you gave an explanation. I wonder if you would go into that a little further.

C. H. Tyndale-Biscoe. I do not understand that either. When we began this study we could not assay prolactin and so we used bromocryptine as a way of testing the role of prolactin in the inhibition of the corpus luteum indirectly. As shown in Table II, bromocryptine was effective in inducing reactivation until the vernal equinox but not thereafter. Initially we thought this result might be because the prolactin concentration might be so high after the winter solstice as it is in the sheep during anestrus, that the bromocryptine was unable to depress it sufficiently to remove the inhibition. However, when prolactin was measured (Hinds and Tyndale-Biscoe, 1982b), it was evident that the change in peripheral concentration was not great and, at present, we can only conclude that at this stage of the year the photoperiod influence is beginning to overrule the residual effects of lactational quiescence.

M. O. Thorner. Does this not mean that there must be another inhibitory response?

C. H. Tyndale-Biscoe. Yes it does. I can say one other thing, that some, but not all, pinealectomized tammars responded to bromocryptine after the vernal equinox, which again suggests that in the intact female there is a photoperiod influence, which is moderating the response to bromocryptine at this time of the year.

M. Raj. We know of situations in which progesterone dramatically rises without any apparent increase in either the tonic LH levels; what we really see is an increase to the response to LH or the LH binding ability by the corpus luteum. In your prolactin-inhibited animals instead of giving bromocryptine, if you were to give LH for example, can you override the inhibition?

C. H. Tyndale-Biscoe. We cannot detect any receptor binding of LH in luteal tissue (Stewart and Tyndale-Biscoe, 1982). Neither incubation with LH nor prolactin had any effect on the secretion rate *in vitro* of corpora lutea, either on day 0 (lactational quiescence) or on day 5 after removal of the pouch young and reactivation of the corpus luteum. Many people to whom we have reported this work have been very skeptical of the conclusion that the corpus luteum is apparently independent of luteotropic or LH support. Because of this we felt that we may be dealing with a very peculiar species and should look at another species of marsupial. So, as mentioned, we did the same experiments on the brushtail possum, which has a different reproductive pattern from the tammar, and in this species also the corpus luteum continued to grow, to secrete progesterone, and to sustain pregnancy, even after hypophysectomy on the day of ovulation. Some years ago B. Cook and A. V. Nalbandov (*J. Reprod. Fert.* **15**,267, 1968) showed that the corpora lutea of the Virginian opossum also only required an initial LH stimulus and then could secrete progesterone without further stimulation. We now think that the condition described in the tammar may be common to all marsupials, which thus display the primitive pattern of corpus luteum regulation, and that what is seen in many eutherian species, such as rodents and primates, but not possibly in the Carnivora, is an overlay on that simpler pattern of luteotropic and luteolytic regulatory mechanisms, which are necessary adaptations for prolonged gestation.

F. Karch. I would like to thank you very much for a most interesting talk which emphasizes the different strategies animals use to accomplish the process of seasonal reproduction. I have two questions related to the photoperiod part of your talk. The first is related to the basis for the reactivation of the corpus luteum shortly after the summer solstice when day length just begins to decline. You suggest that this small decrease in day length provides a signal for reactivation of the corpus luteum. Further, you pointed out that photoperiod does not really drop very much, only a couple of minutes. My first question is whether you know that reactivation of the corpus luteum is actually triggered by this very small decrease in day length? Alternatively could the reactivation result from the fact that day length stops increasing around the time of the summer solstice and the animals become insensitive, or refractory, to the long days of summer? Have you every moved, for instance, your animals indoors on the summer solstice and held them on that photoperiod to see if the corpus luteum reactivates at the usual time, or whether it is delayed?

C. H. Tyndale-Biscoe. Yes in the first experiments that were done by Richard Sadleir 10 years ago (Sadleir and Tyndale-Biscoe, 1977), that question was addressed (see Fig. 10). We thought that it is the long mid-summer day that they respond to and so we held a group of animals on 15L:9D from the vernal equinox through January and they gave birth at the same time as the control group on changing photoperiod. Hinds and den Ottolander (1983) repeated the experiment with the same result, which indicates that long day length by itself is inhibitory but if animals are held on it for up to 3 months, they then do give birth, so there is another factor, which at the moment we can only say is either a circannual rhythm that is imposed on top of the photoperiod or, as you suggest, a refractory influence, which is declining to zero. The reason that we concluded that the change in photoperiod may be important is that it seems to fit so nicely in the series: they respond quickly to an abrupt melatonin change, they respond less quickly to an abrupt photoperiod change, and they respond even slower to the photoperiod change after the summer solstice. It is possible, I suppose, that they are using both mechanisms and that a response to declining photoperiod after the solstice is a fail-safe mechanism in those tammars whose circannual pattern or refractory decline is insufficiently precise. I agree that we must keep an open mind at the moment on whether they really read that very small change in photoperiod after the solstice.

F. Karch. My second question pertains to melatonin, specifically which component of the melatonin secretory profile provides the signal for day length. You addressed two possibilities, how long each day melatonin is elevated and when melatonin is elevated during the course of the 24 hour light–dark cycle. All the studies you showed pertaining to this involved injecting melatonin into animals with intact pineal glands. Your rationale is based on the fact that the animal's own melatonin will summate with the melatonin you inject. It is clear that the animal's own melatonin rise will occur during the night. Thus, if there is a responsive phase, it could be that at least a portion of the melatonin rise would have overlapped with it in all of your studies. To really rule out the phase or rule in duration, it may well be necessary to administer melatonin to animals that do not have their own melatonin and to get the melatonin rise completely out of synchrony with the light–dark cycle. My question is whether you have ever tried injections into pinealectomized animals to see if there is a time of day when melatonin fails to induce a response?

C. H. Tyndale-Biscoe. No, we have not treated pinealectomized tammars with a full melatonin profile, but in all the experiments in which we have injected melatonin in conjunction with a dark phase, we have measured the melatonin profile on at least one night through the experimental period, so that we have been able to confirm, as we showed in Fig. 14d, that the nocturnal profile of melatonin was elevated for the whole period. There was one experiment in which Steve McConnell shifted the dark phase to the daytime and the responses were the same. Would you like to respond to that question Steve?

S. J. McConnell. That experiment as you pointed out was so that I could sleep at a more normal time of the day, and the idea was to just look at the melatonin profile resulting from a 15L:9D photoperiod with an injection of melatonin 2.25 hours before dark. This was only done for 1 day, and from the evidence that we now have we would not expect that 1 day of treatment with melatonin would be effective. In answer to your original question, no we haven't addressed whether or not by shifting the melatonin profile 180° out of phase, for example, you get the same response.

F. Karch. I agree that, with all of the different injections you have done, an important role for the duration is certainly implicated. Yet, I think there is a chance the time of day is important and we really cannot exclude this unless you can repeat your findings in the absence of the pineal gland.

C. H. Tyndale-Biscoe. Thank you, we will attempt that experiment.

B. F. Rice. This is a fascinating talk, and for those of us who have worked with primate corpus luteum over the years it is really a challenge to try to accept some of these new concepts. The one thing that I was not totally certain about, in some of your experiments, was whether blastocyst itself has been excluded as a source of regulation since Chris Ikness a number of years ago, showed that the rabbit blastocyst could produce steroid hormones and maybe even polypeptide hormones?

C. H. Tyndale-Biscoe. We have tended to exclude the blastocyst as having a role in the control of the cycle and the reactivation because the progesterone profiles are so similar between the pregnant and the nonpregnant animal. However, M. H. Cake, F. J. Owen, and S. D. Bradshaw (*J. Endocrinol.* **84**, 153, 1980) reported results from a different species of wallaby, *Setonix brachyurus,* in which they only observed the progesterone pulse in the pregnant animals, and they concluded that the pulse was induced by the presence of the blastocyst. Because this was so unusual we decided to look at this critically in the tammar (Hinds and Tyndale-Biscoe, 1982b). Female tammars were mated to vasectomized males for one cycle and then to intact males for the next cycle, so that we were matching everything in the same group of animals. We found that there was no significant difference in the time or occurrence of the pulse in either situation and so we concluded that the blastocyst is not involved in the stimulation of the corpus luteum. At the end of pregnancy, however, there was evidence that the placenta may have an endocrine function. Since this is a monovular species with two separate uteri, only one uterus is gravid and it can be compared to the condition of the adjacent nonpregnant uterus. The endometrium of the pregnant uterus is heavier and its secretion has different protein components from the nonpregnant uterus (Renfree and Tyndale-Biscoe, 1973). This suggests that the placenta or the fetus is having some influence on the endometrium, and the placenta has been shown to be able to synthesize progesterone from steroid precursors (R. B. Heap, M. B. Renfree, and R. D. Burton. *J. Endocrinol.* **87**, 339, 1980), so there is evidence at the late stage of pregnancy, but no evidence for the early blastocyst, being involved in an endocrine response.

G. J. MacDonald. I have two questions. First, I would like to know the manner in which you assayed the serum gonadotropins? Second, I would like to know your observations about lactation by these animals following hypophysectomy?

C. H. Tyndale-Biscoe. On the second question, after hypophysectomy lactation ceases within 2 or 3 days. Dr John Hearn (1974), who did the first of these experiments, kept the young in the pouch, and observed that they ceased to grow, and died within about 5–10 days after the operation. The alveoli of the mammary glands also shrunk to the nonlactational state, and he considered that the pituitary is essential for the maintenance of lactation at both early and late stages. To answer your first question, we use heterologous assays, using Niswender's antiserum for LH (Sutherland *et al.*, 1980) and rabbit anti-ovine antiserum for FSH (Evans *et al.*, 1980), both validated for the tammar using tammar pituitary and purified tammar gonadotropins, which were prepared in Harold Papkoff's laboratory (S. W. Farmer *et al. Gen. Comp. Endocrinol.* **43**, 336, 1981).

T. Horton. Has anyone done any lesions of the suprachiasmatic nuclei? What is known about the involvement of the SCN in the circadian system of these animals?

C. H. Tyndale-Biscoe. No work has been done on surgical intervention at the level of the SCN, but as mentioned in the text, the development of the visual system of the tammar has been investigated by Wye-Dvorak (1984).

T. Horton. In the information processing theory there is a complex relationship between the signal-to-noise ratio, such that when you have a weak signal you require a longer exposure to the information in order for it to be read as a signal. In physics you have the principle of heterodyning, or the beating phenomenon, such that when you have two oscillators imposed upon each other, which link in the appropriate phase, you get amplification of the beat amplitude. Is there any possibility that with the different strengths of your signal there may be some heterodyning phenomenon in the circadian system or the reading of the signal for these animals?

C. H. Tyndale-Biscoe. That is what we are thinking about for the next stage. Much of what I have reported here is work that we have done this year, and particularly the evidence that they only need to read the message for about 3 days was quite surprising to us and exciting because it means that we now have a means to investigate that part of the chain between the pineal message and the hypothalamus.

S. Cohen. My question relates to the role of estrogen during pregnancy of your animals. Gallagher and his group at the Sloan-Kettering Institute postulated that estrone could be hydroxylated in one of two positions, either the 16 position which leads to the formation of estriol or in the 2 position which led to the formation of a catechol estrogen. Now we know that estriol is not formed by these animals but the catechol estrogens are much more difficult to assay because of their instability. I wonder if you have made any attempt to get at this problem?

C. H. Tyndale-Biscoe. Dr John Harder from Ohio State University investigated the role of estrogens in the reproduction of the tammar in our laboratory in 1982–3. He attempted to assay for estrone, estradiol, and estriol but could find negligible quantities of all except estradiol and so he concentrated on developing an assay for this steroid (Harder *et al.*, 1984). The ovarian steroids have also been investigated in the brushtail possum by J. D. Curlewis, M. Axelson, and G. M. Stone (*J. Endocrinol.* **105**, 53, 1985) and they found that estradiol is by far the most abundant estrogen in the circulation of this species. To answer the first part of your comment, I do not think we can conclude that estrogen is not involved in reproduction in the tammar. We think it may be very important in stimulating progesterone receptors in the endometrium prior to the early pulse of progesterone, which is so important in stimulating the secretory processes of the uterus. Also, we mentioned in the text, Shaw and Renfree (1984) have shown that a transient pulse of estradiol occurs at the same time as the progesterone pulse. Since then Dr. Renfree and her colleagues at Monash University have attempted to test the importance of estradiol in the initiation of development of the blastocyst, but they have so far not found unequivocal evidence for a role for estradiol in this.

I. Callard. You stated that the corpus luteum of the wallaby might represent a primitive condition for mammals with regard to its autonomy, and you suggested that eutherian mammals have added controls to extend the life span of the corpus luteum. Would you accept instead that this may be a secondary rather than primitive condition in the Metatheria, which have lost some controls in the development of their particular adaptations in which the gestational luteal phase is shortened with the corollary of the long lactation period? I say this simply because we have evidence that both reptiles and elasmobranchs have corpora lutea which are pituitary regulated, that is these species have pituitary luteotrophic mechanisms which stimulate production of progesterone. Since reptiles are ancestral to mammals it would seem that the presence of controls would seem to be more primitive.

C. H. Tyndale-Biscoe. Yes, I suppose the reason I have tended to go along with the idea that the autonomous corpus luteum may be the primitive one is that if we consider the early mammals, all of them were very small animals and, generally speaking, small mammals have a short gestation period, a short luteal phase, and a short period of lactation. In the adaptive radiation that occurred in the Tertiary, when both marsupial groups and eutherian groups expanded into the large size range, they needed to accommodate a longer period of maternal nurture. The eutherians opted for extended gestation and the marsupials opted for extended lactation, so for marsupials there was no strong adaptive reason to extend the life of the corpus luteum, because most of the adaptations were the extension and refining of lactation. Conversely, in the process of extending gestation the eutherians adopted all sorts of different ways of regulating the corpus luteum, from the Carnivora, which extended the life of the corpus luteum of the estrous cycle, to others which used the uterus or the embryo to modulate the life of the corpus luteum.

I. Callard. Isn't there a species of marsupial which has a gestation period longer than the estrous cycle? What is the situation there? Have you any ideas whether the pituitary is involved in extending the life of the corpus luteum?

C. H. Tyndale-Biscoe. There is one species, the swamp wallaby (*Wallabia bicolor*), in which the length of gestation is 3–5 days longer than the estrous cycle and it has a prepartum estrus. However, apart from this, its pattern is like that of other kangaroos.

K. Yoshinaga. You referred to the work of Dr. Renfree on denervation of the mammary gland that resulted in reactivation of the corpus luteum. Would you speculate on possible mechanisms involved in that reactivation because prolactin secretion was not inhibited by the denervation?

C. H. Tyndale-Biscoe. At the moment we only have direct evidence in seasonal quiescence (Fig. 17), but our working hypothesis is that a similar early morning pulse of prolactin may occur similarly in early lactation, which is when Dr. Renfree (1979) did the denervation experiment. If the hypothesis is correct, the basal level of prolactin was sufficient to maintain lactogenesis during early lactation, when the suckled mammary gland has an increased concentration of prolactin receptors (Sernia and Tyndale-Biscoe, 1979) so that it can bind more of the available prolactin in circulation. However, if the corpus luteum is inhibited by exposure to a pulse of prolactin once a day, then denervation of the mammary gland may have removed that pulse and so enabled the corpus luteum to be released from inhibition while lactation was maintained on the basal level of prolactin. I must emphasize that this is just a working hypothesis because we have not yet shown that there is a pulse of prolactin during lactational quiescence.

Interactions of Catecholamines and GnRH in Regulation of Gonadotropin Secretion in Teleost Fish

RICHARD E. PETER, JOHN P. CHANG, CAROL S. NAHORNIAK,
ROBERT J. OMELJANIUK, M. SOKOLOWSKA, STEPHEN H. SHIH,
AND ROLAND BILLARD

Department of Zoology, University of Alberta, Edmonton, Alberta, Canada

I. Introduction

In mammals a key element in the neuroendocrine regulation of luteinizing hormone (LH) and follicle-stimulating hormone (FSH) secretion is the decapeptide neurohormone luteinizing hormone-releasing hormone (LHRH). The actions of LHRH, and various synthetic analogs on pituitary LH and FSH cells, have been extensively studied, using the laboratory rat as the primary experimental model (e.g., Bex and Corbin, 1984; Kalra and Kalra, 1985). In view of the central role of LHRH in regulation of secretion of LH and FSH in mammals, the factors that regulate activity of LHRH cells are clearly of fundamental importance in understanding the neuroendocrine regulation of reproduction. Studies in mammals, again primarily on the rat, indicate that a number of amine neurotransmitters are involved in regulation of LH secretion, presumably by influencing the activity of LHRH neurons (e.g., Barraclough *et al.*, 1984; Ramirez *et al.*, 1984); the current interpretation is that amines do not act directly at the level of the pituitary to influence LH or FSH secretion. In contrast to this, our recent studies on teleost fishes, primarily the goldfish, indicate that direct actions of the catecholamines dopamine (DA) and norepinephrine (NE) on the pituitary are of major importance in the regulation of gonadotropin (GtH) secretion. Most importantly we have found that DA acts as a gonadotropin release inhibitory factor (GRIF) to modulate the action of gonadotropin-releasing hormone (GnRH), as well as spontaneous release of GtH. In this review we shall discuss supporting evidence for this, and draw parallels with situations in mammals, primarily the human, in which a similar interaction of DA and LHRH may occur.

II. Nature of Teleost Gonadotropins

The purpose of this section is to provide some information on the nature of teleost GtH, to enable workers in other fields to better relate the discussion to the neuroendocrine regulation of LH and FSH secretion.

It was thought for some time that in teleost fishes only one GtH was responsible for the full range of gonadal actions ascribed to LH and FSH (for review, Peter and Crim, 1979). However, it is now apparent that there are two pituitary hormones that act on the gonads in teleost fishes (for review, Idler and Ng, 1983). One of these, isolated from pituitary glands from carp, salmonids, plaice, flounder, and pike eel, is a classical GtH in that it has a high carbohydrate content, and is made up of α- and β-subunits (for review, Burzawa-Gerard, 1982; Idler and Ng, 1983). The β-subunit from carp GtH will associate with the α-subunit of bovine LH, restoring biological activity to a level similar to that of intact carp GtH in a cyclic AMP bioassay on eel ovary (Burzawa-Gerard, 1982). These and other data suggest a high level of homology between this form of teleost GtH, and LH and FSH from other vertebrates. The teleost GtHs with high carbohydrate content have a broad spectrum of biological activity, including stimulation of cyclic AMP production in gonadal tissue, stimulation of steroidogenesis, spermatogenesis, spermiation, some aspects of oocyte growth, and oocyte maturation and ovulation (for review, Idler and Ng, 1983; Ng and Idler, 1983). Specific radioimmunoassays have been developed for "high carbohydrate" GtH in cyprinid teleosts such as goldfish and carp, various salmonid species (for review, Peter, 1981), and the African catfish (de Leeuw et al., 1985b); the β-subunit of carp GtH has been used to develop a radioimmunoassay for GtH in the European eel (Dufour et al., 1983). A second GtH, isolated from pituitaries from common carp, chum salmon, plaice, and flounder, has a low carbohydrate content, and acts to stimulate incorporation of yolk protein into developing oocytes and to stimulate steroidogenesis, to a greater or lesser degree depending on the species (for review, Idler and Ng, 1983; Ng and Idler, 1983). A subunit structure for this "low carbohydrate" GtH has not yet been demonstrated. Immunohistochemical studies have demonstrated that both the "low" and "high" carbohydrate GtHs originate from basophils, in some cases the same basophils, in the pars distalis of the pituitary (for review, Van Oordt and Peute, 1983), which suggests some relatedness of the two hormones. Unfortunately, radioimmunoassays for the "low carbohydrate" GtHs have not yet been developed. Thus, for the remainder of this review where measurements of GtH are reported, we refer to only the "high carbohydrate" GtH with the broad spectrum of actions on the gonads.

III. Gonadotropin-Releasing Hormone (GnRH) in Teleosts

Several early studies on a number of teleost species demonstrated the presence of a factor in crude hypothalamic extracts that stimulated GtH release (for review, Peter, 1983a,b). Radioimmunoassay studies on crude hypothalamic extracts using antisera with characterized specificity against LHRH suggested similarity in the forms of GnRH present in the chicken, a tortoise, and a teleost (tilapia, *Sarotherodon mossambicus*) (King and Millar, 1979, 1980). The behavior of immunoreactive "LHRH" fractions on cation-exchange and affinity chromatography columns, and in high-pressure liquid chromatography (HPLC) further indicated that the GnRH in a teleost (*S. mossambicus*), an elasmobranch, reptiles, and birds is similar if not identical; the GnRH present in amphibians appeared to be identical to LHRH (King and Millar, 1979, 1980). Chromatography and immunological studies on hypothalamic extracts from cod, *Gadus morhua* (Barnett *et al.*, 1982), winter flounder, *Pseudopleuronectus americanus* (Idler and Crim, 1985), and chum salmon *Onchorhynchus keta* (Sherwood *et al.*, 1983) confirmed that GnRH in teleosts is similar but not identical to LHRH; furthermore, two forms of GnRH similar in size to LHRH were detected by Barnett *et al.* (1982) and Sherwood *et al.* (1983). Some results in these studies also suggested the presence of an immunoreactive high-molecular-weight form of GnRH, which may represent a precursor of the authentic GnRH. On the basis of immunoreactive properties, King and Millar (1980) and Barnett *et al.* (1982) suggested that the GnRH present in tilapia and cod, respectively, differed from LHRH by substitutions at positions seven and/or eight of the decapeptide. Sherwood *et al.* (1983) carried out amino acid analysis of the major chum salmon GnRH immunoreactive fraction isolated by HPLC, and by comparison of the chromatographic behavior of various synthetic peptides with native GnRH determined that the structure of the major peptide was [Trp^7,Leu^8]-LHRH (tGnRH). More recently Sherwood *et al.* (1984) and Breton *et al.* (1984) have demonstrated that this form of GnRH is present in the brain of a wide range of teleost species.

A number of studies have shown that injection of LHRH or superactive analogs of LHRH into teleosts stimulates an increase in the circulating levels of GtH (for review, Peter, 1983a,b; Peter and Crim, 1979). In our *in vivo* studies on goldfish the response to a single intraperitoneal injection of LHRH or [D-Ala^6,Pro^9-NEt]-LHRH (LHRH-A) was poor; however, a relatively large GtH release–response was found following a pair of injections of LHRH (compare Fig. 1A to Fig. 1B) or LHRH-A, indicating that self-potentiation of the action can occur (Peter, 1980; Lin *et al.*, 1985b; Peter *et al.*, 1985; Sokolowska *et al.*, 1984, 1985b). Notably, a clear dose-

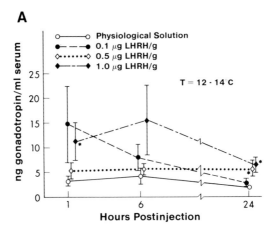

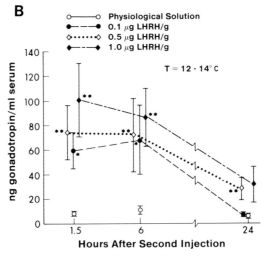

FIG. 1. Serum gonadotropin levels (mean ± SE, $n = 6$ or 7) of mature male goldfish, held at 12–14°C, following a single intraperitoneal injection (A) or a pair of injections given 12 hours apart (B) of LHRH. Significant differences from controls at same sample time were determined by the U test (*$p < 0.05$; **$p < 0.01$). Reproduced from Peter (1980).

dependent response to LHRH or LHRH-A was not evident; furthermore, the magnitude of the responses, in terms of the peak concentrations of serum GtH reached, were similar between LHRH and the superactive analog, although the responses to the analog had a longer duration. A clear dose-dependent response to LHRH or a superactive analog was not found in *in vivo* experiments with common carp (Weil *et al.*, 1975) and coho salmon (Van Der Kraak *et al.*, 1983); however, a dose-dependent response has been reported in other investigations on common carp (Breton and Weil, 1973), brown trout (Crim et al., 1981), and coho salmon (Van Der Kraak et al., 1985).

Similar to our findings in the goldfish, Van Der Kraak et al. (1983) found in coho salmon that similar dosages of LHRH and LHRH-A caused increases in plasma GtH of similar magnitude, but that the duration of the response to the analog was longer. Results from *in vitro* studies reveal a different aspect of the responsiveness of teleost pituitaries to GnRH. Rainbow trout pituitaries in static culture display a significant dose-related response to LHRH (Crim and Evans, 1980; Crim et al., 1981; Fahraeus-van Ree et al., 1983); however, there was no indication of superactivity of analogs compared to LHRH (Crim et al., 1981). Likewise, goldfish pituitary glands in perifusion culture show a dose-dependent response of similar magnitude and duration to LHRH, tGnRH, and LHRH-A (MacKenzie et al., 1984). Several aspects of these results require further explanation. For example, why are dose–response relationships not readily apparent *in vivo* in teleosts? Why is the magnitude of the GtH release–response, in terms of the peak GtH levels accomplished, to LHRH and superactive analogs similar *in vivo* and *in vitro*? In latter sections evidence will be presented that DA acts as a gonadotropin release-inhibitory factor (GRIF) directly at the level of the gonadotrope in the pituitary. Furthermore, when the GRIF activity of endogenous DA is blocked, dose–response relationships and superactivity of various analogs become evident *in vivo*.

IV. Evidence for a Gonadotropin Release-Inhibitory Factor

Peter et al. (1978) reported that preovulatory female goldfish receiving large radiofrequency current lesions in the ventrobasal hypothalamus that destroyed the nucleus lateralis tuberis (NLT; homologous to arcuate nucleus) in the pituitary stalk region had a marked increase in serum GtH levels and ovulation; serum GtH levels were significantly elevated for at least 12 days postlesioning. These results were interpreted to indicate the presence of a gonadotropin release inhibitory factor (GRIF), and that blocking the action of GRIF on the pituitary allowed a prolonged bout of spontaneous secretion of GtH. In additional radiofrequency current brain lesioning studies on male and female goldfish, Peter and Paulencu (1980) mapped the location of brain lesions which were effective in causing a significant increase in serum GtH levels. On the basis of these studies, we suggested that GRIF originates from the anterior-ventral nucleus preopticus periventricularis, in the region of the preoptic recess, and that a pathway courses from this center through the lateral preoptic region, and

518   RICHARD E. PETER ET AL.

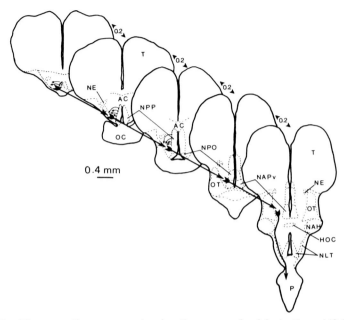

FIG. 2. Diagrammatic summary showing the proposed origin in the goldfish of the gonadotropin release-inhibitory factor in the anterior preoptic region (shaded area) and the proposed pathway of the factor (arrows) in the lateral preoptic region, lateral anterior hypothalamic region, and pituitary stalk. Distances between the cross-sections given in millimeters above the drawings. AC, Anterior commissure; HOC, horizontal commissure; NAH, nucleus anterioris hypothalami; NAPv, nucleus anterioris periventricularis; NE, nucleus entopeduncularis; NLT, nucleus lateralis tuberis; NPO, nucleus preopticus; NPP, nucleus preopticus periventricularis; OC, optic chiasma; OT, optic tract; P, pituitary; T, telencephalon. Reproduced, by permission of S. Karger AG, from Peter and Paulencu, *Neuroendocrinology* **31**, 133–141 (1980).

lateral-anterior hypothalamic region to the pituitary (Fig. 2). In teleost fishes there is no median eminence per se, as neurosecretory fibers directly innervate the anterior pituitary (for review, Peter and Fryer, 1983). Lesions in the ventrobasal hypothalamus which destroy the pituitary stalk caused a marked increase in serum GtH levels (Peter and Paulencu, 1980). Immunohistochemical studies on goldfish brain using antisera against LHRH (Kah *et al.*, 1982) or native tGnRH (O. Kah, J. Dulka and R. E. Peter, unpublished results) demonstrate a major concentration of GnRH perikarya in the lateral preoptic region adjacent to the most lateral aspects of the nucleus preopticus periventricularis, nucleus preopticus, and the anterior part of the nucleus anterioris periventricularis. This distribution of GnRH perikarya does not overlap the location of the proposed GRIF center. However, large lesions in the anterior preoptic region which destroy the GRIF center would also destroy a major number of GnRH cell bodies. In any case, it is highly unlikely that the prolonged increases in GtH secretion found following lesions of either the pituitary stalk, the proposed GRIF center, or the proposed GRIF pathway are due to release of GnRH from terminals of lesioned cells remaining in the pituitary, as such fibers are phagocytozed within a few days (R. E. Peter, unpublished observations). Furthermore, small lesions in the ventrobasal hypothalamus of goldfish which destroy the nucleus lateralis tuberis in the posterior pituitary stalk region cause a significant decrease in the gonadosomatic index, due to onset of ovarian atresia, and abolition of significant daily cycles in GtH secretion (Peter and Paulencu, 1980; Peter, 1982). These results are consistent with early evidence interpreted to indicate that this particular region of the ventrobasal hypothalamus is involved in stimulation of gonadal activity via GnRH (Peter, 1970).

Additional evidence for the presence of GRIF comes from studies in which the pars distalis of the pituitary was transplanted to various locations (Peter et al., 1984). In these studies the pars distalis was transplanted from one goldfish to another goldfish, of matched size, sex, and stage of gonadal development, to either beside the brain, or into the third ventricle in the preoptic region, or into the brain ventricle underlying the optic tectum. Such a transplanted pars distalis releases GtH spontaneously, irrespective of location of the transplant, indicating that GtH secretion is normally under tonic inhibition. Pars distalis transplants beside the brain released more GtH than those transplanted into the brain ventricles, indicating the presence of GRIF in the brain. Pars distalis transplant studies done with male and female goldfish at different stages of seasonal gonadal development indicate that GtH secretion is under tonic inhibition at all stages of gonadal development, although pituitaries from fish that are in the latter stages of ovarian or testicular recrudescence prior to spawning have the greatest potential for spontaneous secretion of GtH.

Brain lesioning studies have been carried out on male and female goldfish at different stages of seasonal gonadal development to further explore the changes in ability of the pituitary to spontaneously secrete GtH (R. E. Peter and C. S. Nahorniak, unpublished results). Radiofrequency current lesions were stereotaxically placed in the preoptic region of goldfish to destroy the GRIF center, following which blood samples were taken for measurement of serum GtH levels by radioimmunoassay; the pituitary concentration of GtH was also measured in a number of experiments. Preoptic lesions resulted in increased serum GtH levels in male and female goldfish at all stages of seasonal gonadal development, indicating that GRIF is present continuously. Female goldfish had the greatest difference in potential to spontaneously secrete GtH, with sexually regressed females having the least potential and females in latter stages of ovarian recrudescence having the greatest potential (Fig. 3). In fish held at 12°C, but not at 20°C, there was a decrease in the concentration of GtH in the pituitary following preoptic lesions that abolished GRIF (R. E. Peter and C. S. Nahorniak, unpublished results). In other studies, the ultrastructural appearance of the pituitary following brain lesions that abolish GRIF suggested that the GtH cells in the pituitary were very actively secreting and synthesizing protein-like material (Nagahama and Peter, 1982). The depletion of pituitary GtH following a preoptic lesion in fish held at 12°C suggests that at this temperature the rate of synthesis cannot keep pace with the secretion rate. Overall these data provide evidence that GRIF is of importance in regulating GtH secretion in goldfish at all stages of the seasonal sexual cycle.

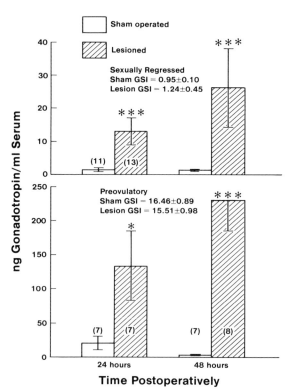

FIG. 3. Effects on serum gonadotropin levels of radiofrequency current lesions of the preoptic region, to destroy the gonadotropin release-inhibitory factor, in sexually regressed and sexually mature (preovulatory) female goldfish held at 12°C. Significant differences from sham operated controls were determined by the U test (*$p < 0.05$; ***$p < 0.001$). R. E. Peter and C. S. Nahorniak, unpublished results.

V. Involvement of Catecholamines in Neuroendocrine Regulation of GtH Release in Goldfish

A. DOPAMINE

1. Dopamine (DA) as an Inhibitor of GtH Release

The involvement of catecholamines in the neuroendocrine regulation of GtH release in the goldfish was first demonstrated by Chang *et al.* (1983). Long-term treatment with 6-hydroxydopamine as well as a single intraperitoneal injection of reserpine increased circulating GtH levels. Intraperitoneal injections of α-methyl-*p*-tyrosine or carbidopa, drugs known to block L-dopa and DA synthesis, respectively, elevated serum GtH levels; on the other hand, injections of diethyl-dithiocarbamate, an agent capable of inhibiting the conversion of DA to NE, did not alter circulating GtH concentrations. These observations suggest that DA may have an inhibitory influence on GtH secretion in the goldfish.

The ability of DA to inhibit GtH release in the goldfish has since been confirmed. Intraperitoneal injections of DA and its agonist apomorphine decreased, whereas intraperitoneal injections of the DA antagonists pimozide and metoclopramide elevated serum GtH concentrations (Chang and Peter, 1983a,b; Chang *et al.*, 1984b; Sokolowska *et al.*, 1984). Although intraperitoneal injection of DA decreased serum GtH levels, injection of DA into the third cranial ventricle was without effect (Chang and Peter, 1983a). DA does not cross the blood–brain barrier in mammals and a blood–brain barrier for DA also seems to exist in the goldfish; systemically administered radiolabeled DA does not accumulate in the brain of goldfish (J. P. Chang and R. E. Peter, unpublished results, cited in Chang, 1983). The ability of peripherally, but not centrally administered DA to alter serum GtH levels suggests that it acts outside the blood–brain barrier to decrease GtH release in the goldfish. Since the pituitary lies outside of the blood–brain barrier, and the gonadotropes of goldfish are directly innervated by neurons that are morphologically similar to aminergic neurons (for review see Ball, 1981; Peter and Fryer, 1983), these results suggest that DA may act at the level of the pituitary to inhibit GtH secretion.

Evidence for the direct action of DA on the goldfish pituitary to inhibit spontaneous GtH release was provided by *in vitro* and *in vivo* studies. Goldfish pituitary fragments and dispersed cells, in a column perifusion system, have a high rate of spontaneous GtH release (Chang *et al.*, 1984c). Addition of DA to the perifusate reduced the level of GtH released spontaneously from dispersed goldfish pituitary cells as well as pituitary

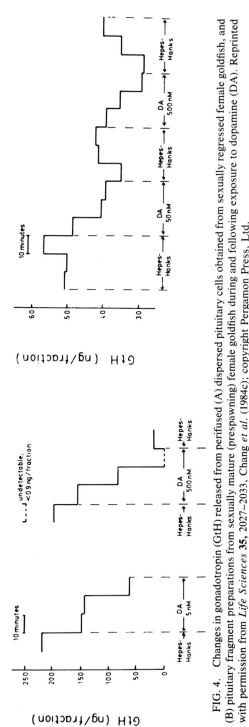

FIG. 4. Changes in gonadotropin (GtH) released from perifused (A) dispersed pituitary cells obtained from sexually regressed female goldfish, and (B) pituitary fragment preparations from sexually mature (prespawning) female goldfish during and following exposure to dopamine (DA). Reprinted with permission from *Life Sciences* **35**, 2027–2033, Chang et al. (1984c); copyright Pergamon Press, Ltd.

fragments (Fig. 4). Goldfish in which the anterior preoptic region has been lesioned have increased serum GtH levels, due to abolition of endogenous GRIF allowing a high rate of spontaneous GtH secretion (see Section IV); intraperitonal injection of either DA or apomorphine significantly depressed the lesion-induced increase in serum GtH levels (Chang and Peter, 1983a). Pars distalis transplants have a high spontaneous GtH release rate, causing marked increases in circulating GtH levels (see Section IV); intraperitoneal injection of DA or apomorphine significantly reduced the highly elevated serum GtH values in fish bearing pars distalis transplants (Chang et al., 1984a).

Notably, DA also inhibits the GnRH-stimulated GtH release in goldfish. Intraperitoneal injection of DA, or its agonist apomorphine, reduced the magnitude of the LHRH-A stimulated increase in serum GtH levels on a dose-dependent basis (Chang and Peter, 1983a; Chang et al., 1984b; Fig. 5). Similarly, another DA agonist, bromocryptine also decreased the LHRH-A stimulated increase in serum GtH concentrations on a dose-dependent basis (Chang et al., 1984a). Application of DA completely blocked the LHRH-A stimulation of GtH release in vitro (Chang et al., 1984c; Fig. 6). Injections of the DA antagonists, pimozide (Fig. 7) or metoclopramide (Fig. 8), potentiated the LHRH-A-induced increase in serum GtH levels (Chang et al., 1984b; Sokolowska et al., 1984, 1985a,b; Peter et al., 1985; R. E. Peter, R. Billard, and C. S. Nahorniak, unpub-

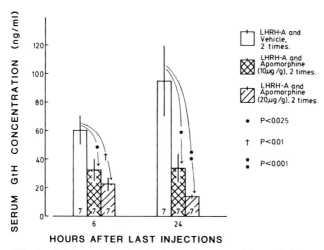

FIG. 5. Effects of intraperitoneal injections of apomorphine, administered simultaneously with both the first and second injections of LHRH-A given 12 hours apart, on serum gonadotropin (GtH) concentrations in female goldfish held at 12°C (mean ± SE are plotted). Significant differences determined by the U test. Reproduced, by permission of S. Karger AG, from Chang and Peter, Neuroendocrinology 36, 351–357 (1983a).

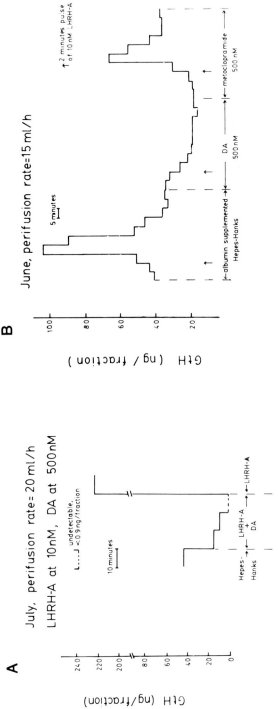

FIG. 6. Changes in gonadotropin (GtH) released from perifused (A) dispersed pituitary cells obtained from sexually regressed female goldfish in response to LHRH-A and dopamine (DA), or LHRH-A alone, and (B) pituitary fragments prepared from sexually mature (prespawning) female goldfish in response to a 2 minute pulse (arrow) of LHRH-A during perifusion with either albumin-supplemented Hepes-Hanks' solution alone, 500 nM DA, or 500 nM metoclopramide. Reprinted with permission from *Life Sciences* **35**, 2027–2033, Chang *et al.* (1984c); copyright Pergamon Press, Ltd.

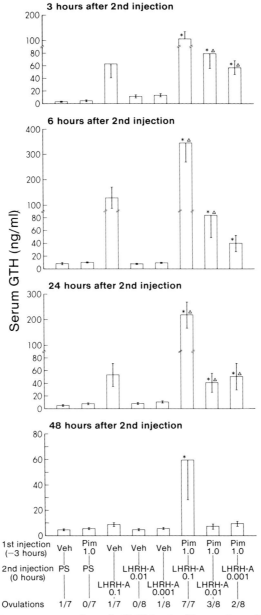

FIG. 7. Effects on serum gonadotropin (GtH) levels of intraperitoneal injection of pimozide (Pim: 1.0 µg/g body weight) and different dosages of LHRH-A in sexually mature (prespawning) female goldfish held at 20°C. Dosages of LHRH-A given in µg/g body weight on the figure. Values are mean ± SE. Number of animals ovulating in each group are given at the base of each column. Significant differences determined by the U test. Adapted from Sokolowska et al. (1985b).

526   RICHARD E. PETER ET AL.

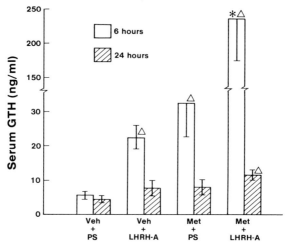

FIG. 8. Effects of intraperitoneal injection of metoclopramide (Met: 100 μg/g body weight) on the LHRH-A- (0.1 μg/g) induced increase in serum gonadotropin (GtH) levels in female goldfish undergoing ovarian recrudescence, held at 20°C. Values are mean ± SE (n = 8 or 9). Significant differences determined by the U test ($p < 0.05$; Δ, versus Veh + PS group; *, versus Veh + LHRH-A group). R. E. Peter, R. Billard, and C. S. Nahorniak, unpublished results.

lished results). Pimozide injection also potentiated the LHRH- and tGnRH-induced increase in serum GtH concentrations (Peter, 1983a; Peter et al., 1985). Notably, the effects of pimozide (Chang et al., 1984b; Peter et al., 1985; Sokolowska et al., 1985b) and metoclopramide (Chang et al., 1984b; R. E. Peter, R. Billard, and C. S. Nahorniak, unpublished results) in potentiating the action of LHRH-A are dose dependent.

If DA acts as a GRIF, then drugs that block the synthesis of DA or cause depletion of DA from presynaptic terminals should also influence the responsiveness of goldfish to injected LHRH-A. Intraperitoneal injection of reserpine, 6-hydroxydopamine, α-methyl-p-tyrosine, and carbidopa all potentiated the GtH releasing activity of LHRH-A in goldfish (Fig. 9); however, injection of diethyl-dithiocarbamate did not influence the responsiveness to LHRH-A. This lends further support to the idea that DA serves as GRIF in goldfish.

By use of immunohistochemistry, Kah et al. (1984) demonstrated a dopaminergic group of perikarya in the anterior-ventral nucleus preopticus periventricularis, in the region of the preoptic recess, of the goldfish; these DA neurons were found to project to the pituitary via bilateral tracts in the lateral preoptic and anterior hypothalamus. The location of this dopaminergic nucleus and pathway coincides with the proposed origin of

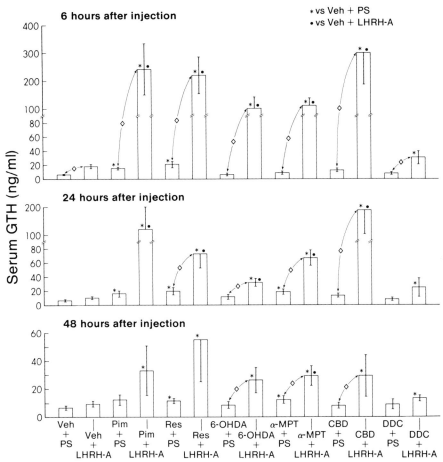

FIG. 9. Effects of intraperitoneal injections of pimozide (Pim; 10 µg/g body weight), reserpine (Res; 100 µg/g), 6-hydroxydopamine (6-OHDA; 50 µg/g), α-methyl-p-tyrosine (α-MPT; 100 µg/g), carbidopa (CBD; 50 µg/g), and diethyl-dithiocarbamate (DDC; 2.5 µg/g) on the LHRH-A- (0.1 µg/g) induced increase in serum gonadotropin levels in sexually mature (prespawning) female goldfish held at 20°C. Values are mean ± SE (n = 8 or 9). Significant differences by the U test. C. S. Nahorniak and R. E. Peter, unpublished results.

GRIF and the GRIF pathway to the pituitary (Peter and Paulencu, 1980). Radiofrequency lesions that interrupted the neural pathway carrying GRIF also caused degeneration of DA terminals adjacent to gonadotropes (O. Kah, J. Dulka, and R. E. Peter, unpublished results). Together these data lend additional strong support for the idea that DA serves as GRIF in goldfish.

2. Importance of DA as GRIF

Several lines of evidence indicate that the GRIF actions of DA play an integral role in the neuroendocrine regulation of GtH release in the goldfish. In teleosts, germinal vesicle migration is a preliminary step to ovulation (Yamazaki, 1965; Billard et al., 1978) and final oocyte maturation and ovulation are induced by a preovulatory GtH surge (for review see Peter, 1981, 1983a,b). In sexually matured (prespawning) female goldfish, injection of the DA antagonist pimozide not only potentiates the actions of LHRH, tGnRH, and LHRH superactive analogs in stimulating GtH secretion, but treatment with the combination of pimozide and LHRH-A can stimulate increases in blood levels of GtH similar to or higher than those observed during spontaneous ovulation, and induce a high rate of ovulation (Chang and Peter, 1983b; Chang et al., 1984b; Sokolowska et al., 1984, 1985a,b). The ovulatory response in the goldfish is probably dependent on both the magnitude of the serum GtH concentration as well as the rate of increase in circulating GtH levels, since a large but slow increase in serum GtH levels, as by injection of a high dosage of LHRH-A alone (Chang and Peter, 1983b; Sokolowska et al., 1984, 1985a,b; Fig. 7) or by implants of pellets containing a large dosage of LHRH-agonist (Sokolowska et al., 1984), were not effective in inducing ovulation in the goldfish. Thus, in the normal *in vivo* situation, removal of the DA inhibition and an increase in GnRH stimulation of GtH secretion are probably both required to allow for the rapid increase in circulating GtH levels that trigger ovulation.

By following the effects of intraperitoneal injection of pimozide on basal, as well as LHRH-A stimulated, serum GtH levels throughout the reproductive cycle of the goldfish, it is evident that the inhibitory DA tone on GtH release is present continuously. The magnitude of the increase in basal and LHRH-A-stimulated serum GtH levels following pimozide injection can be correlated with the degree of ovarian development in the goldfish (Sokolowska et al., 1985a), suggesting that the inhibitory DA tone increases in intensity with advancement in the degree of reproductive readiness. Additional support for an increase in the degree of DA inhibition with ovarian maturation in the female goldfish comes from the observations that the preoptic lesion-induced elevations in serum GtH become greater as the fish progress through the various stages of ovarian recrudescence (see Section IV). This increase in the intensity of DA inhibition is probably responsible for the maintenance of relatively low circulating GtH levels in the face of increases in the readily releasable GtH pool as the goldfish enters the reproductive season. The seasonal changes in DA inhibitory tone may be due to influences of sex steroids, as treatment of goldfish with estradiol and testosterone results in a greater GtH release–response to pimozide and LHRH-A (R. E. Peter and C. S. Nahorniak, unpublished results).

As discussed in Section III, LHRH and its superactive agonists stimulate GtH release in goldfish, but dose–response relationships to injections of various concentrations of these peptides, as well as superactivity, cannot be readily demonstrated *in vivo*. However, when the DA inhibition of GtH release was blocked by injection of pimozide, *in vivo* applications of various concentrations of LHRH-A in sexually mature (prespawning) females (Sokolowska *et al.*, 1985b; Fig. 7), and LHRH-A and [D-Arg^6,D-Trp^7,D-Leu^8,Pro^9-NEt]-LHRH in mature (prespawning) males (Peter *et al.*, 1985) increased serum GtH levels in a more clearly defined dose-dependent manner. When combined with pimozide, several LHRH analogs that were superactive in mammals were also shown to be superactive in increasing serum GtH in goldfish; the magnitudes of the increase in GtH levels were up to 8-fold greater than that resulting from treatment with a similar dose of LHRH plus pimozide (Peter *et al.*, 1985). These observations imply that not only is the dopaminergic GRIF system used to modulate stimulatory influences on GtH release, but the DA inhibitory tone is a powerful component in the neuroendocrine regulatory mechanisms controlling GtH secretion; only when this DA inhibition on GtH release is blocked, such as by the addition of pimozide, can dose relationships between GtH release and concentrations of releasing peptide, and superactivity of analogs, be demonstrated. It is interesting to note that although superactivity of LHRH-A in stimulating GtH release, relative to LHRH, can be demonstrated with the use of pimozide *in vivo*, LHRH-A, LHRH, and tGnRH are equipotent in inducing GtH secretion *in vitro* (MacKenzie *et al.*, 1984). It may be that the structure–activity requirements of the GnRH receptor of the goldfish are different from that in mammals, and that the *in vivo* superactivity of various LHRH analogs is due primarily to a difference in the degradation or metabolism of these peptides, rather than differences in binding affinity or receptor–ligand interactions.

3. Specificity of the DA Inhibition on GtH Release

Results from the available literature indicate that the DA inhibitory influence on GtH secretion is likely mediated by dopaminergic, but not α- or β-adrenergic receptors. As discussed previously, DA and its agonists, apomorphine and bromocryptine, inhibit, whereas the DA antagonists, pimozide and metoclopramide, increase spontaneous as well as GnRH-stimulated GtH release *in vivo*. Recent studies demonstrate that other DA

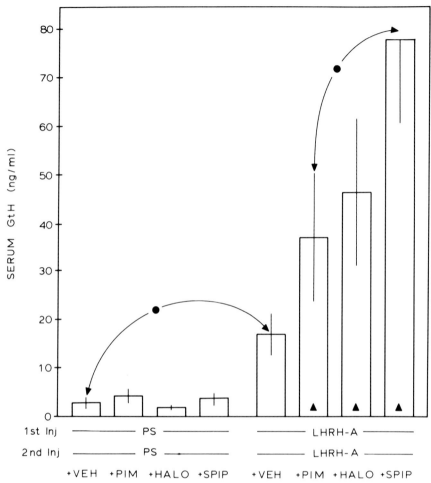

FIG. 10. Effects of intraperitoneal injection of pimozide (10 µg/g; PIM), haloperidol (10 µg/g; HALO), and spiperone (10 µg/g; SPIP) on serum gonadotropin (GtH) concentrations in LHRH-A-injected goldfish undergoing gonadal recrudescence and held at 12°C. LHRH-A (0.1 µg/g) was injected twice at an interval of 12 hours. PIM, HALO, and SPIP were injected at the time of the second LHRH-A injection. Control injections were acidified physiological saline (PS). Serum GtH levels are mean ± SE at 6 hours after the last set of injections. Significant differences were determined by the U test ($p < 0.05$; ∆, versus LHRH-A + Veh group). R. J. Omeljaniuk, S. H. Shih, M. Sokolowska, and R. E. Peter, unpublished results.

antagonists including haloperidol, spiperone (Fig. 10), and domperidone (R. J. Omeljaniuk, S. H. Shih, M. Sokolowska, and R. E. Peter, unpublished results) are also effective. However, injection of the α-receptor antagonist, phentolamine, and the β-receptor antagonist, propranolol, did not alter basal or LHRH-A-stimulated serum GtH concentrations (Chang et al., 1984a, Fig. 11). The ability of the specific D-2 dopaminergic agonist bromocryptine (Kebabian and Calne, 1979) to inhibit, and the ability of DA antagonists with D-2 receptor specificity, pimozide, metoclopramide, spiperone, and domperidone (Creese et al., 1981, 1983; Kebbabian and Calne, 1979), to enhance the LHRH-A-induced GtH response suggest that DA acts at a receptor resembling the mammalian D-2 type to exert its GRIF activity. (For definitions and properties of D-2 dopaminergic receptor, see Cote et al., 1982; Crease, 1982; Creese et al., 1981, 1983; Kebabian et al., 1983; Seeman, 1984; Grigoriadis and Seeman, 1984.)

Results from preliminary experiments on the properties of goldfish pituitary DA receptors demonstrate that these receptors have characteristics

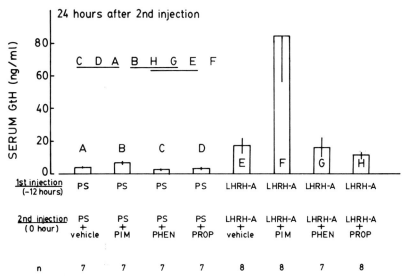

FIG. 11. Effects of intraperitoneal injection of pimozide (10 μg/g; PIM), phentolamine (10 μg/g; PHEN), or propranolol (10 μg/g; PROP) on serum gonadotropin (GtH) concentrations in LHRH-A-injected goldfish undergoing gonadal rescrudescence. LHRH-A (0.1 μg/g) was injected twice at an interval of 12 hours. PIM, PROP, or PHEN was injected at the time of the second LHRH-A injection. Control injections were acidified physiological saline (PS). Serum GtH levels are mean ± SE. Significant differences were determined by one-way analysis of variance and Duncan's multiple range test. Serum GtH concentrations of groups underlined by the same line are not significantly different ($p > 0.05$). Reproduced from Chang et al. (1984b).

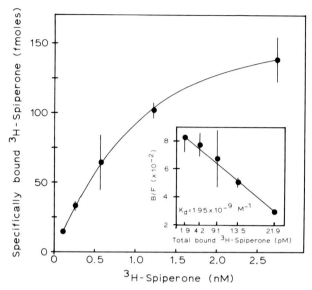

FIG. 12. Binding of [³H]spiperone to goldfish whole pituitary homogenate. Specific binding of [³H]spiperone was calculated as the difference in binding of [³H]spiperone in the absence (total binding) or presence (nonspecific binding) of 10 μM domperidone, a specific dopamine D-2 receptor blocker. Data were expressed as specifically bound [³H]spiperone (fmol) as a function of [³H]spiperone concentration. Inset: Scatchard analysis of these same data ($r = -0.996$). R. J. Omeljaniuk and R. E. Peter, unpublished results.

consistent with mammalian D-2 dopaminergic receptors (R. J. Omeljaniuk and R. E. Peter, unpublished results). In an *in vitro* radioreceptor assay, [³H]spiperone binds to a goldfish pituitary tissue preparation in a saturable fashion and with high affinity (Fig. 12). The specific binding of [³H]spiperone to goldfish pituitary tissue preparation is heat labile and is differentially displaceable by dopamine agonists and antagonists in a manner similar to that reported for mammalian D-2 dopaminergic receptors. In addition, the order of potency of DA antagonists, injected intraperitoneally, in elevating serum GtH levels is consistent with that of DA antagonists binding to mammalian D-2 receptor sites (e.g., domperidone > pimozide).

B. OTHER CATECHOLAMINES

There is relatively little information available on the possible involvement of NE and epinephrine (E) in the neuroendocrine regulation of GtH release in fish. Deery (1975) reported that the adenylate cyclase activity of the pars distalis of goldfish, measured *in vitro*, was increased by NE and E. Chang *et al.* (1983) reported that intraperitoneal injection of the α-agonist clonidine increased serum GtH levels in goldfish, and proposed that NE may stimulate GtH release.

Confirmation of the stimulatory influence of NE on GtH secretion was subsequently provided by Chang and Peter (1984). Intraperitoneal injection of NE was found to increase serum GtH levels in female goldfish with regressed ovaries or ovaries undergoing early stages of recrudescence, but not in female fish at other stages of the reproductive cycle. Since systemically applied radiolabeled NE was not taken up by brain tissues (Chang, 1983), a blood-brain barrier for NE probably exists in the goldfish, similar to mammals. Thus, the results from the *in vivo* studies suggest that NE probably acts directly on the pituitary to stimulate GtH release. Chang *et al.* (1984c) found that NE stimulated GtH release from perifused pituitary fragments, as well as dispersed pituitary cells, obtained from goldfish with regressed gonads (Fig. 13), confirming that NE acts directly on GtH cells. NE has been detected in the goldfish pituitary (approximately 110 pg/pituitary; Chang, 1983). Although the pituitary NE is presumed to be of neural origin, the source of these NE neurons is still unknown.

Variations in the ability of intraperitoneal injections of clonidine to increase serum GtH concentrations in female goldfish at different stages of the reproductive cycle paralleled those of intraperitoneal injections of NE. Clonidine increased circulating GtH concentrations in fish with ovaries at early stages of recrudescence, but not in those having gonads at late stages of ovarian recrudescence (Chang *et al.*, 1983; Chang and Peter, 1984). The similarity in the seasonal variation in the ability of NE and clonidine to stimulate GtH release, as well as the observation that the NE-stimulated elevation in serum GtH levels was reversed by concurrent applications of phentolamine (Chang *et al.*, 1984c), indicates that NE acts through α-adrenergic receptors to directly stimulate GtH secretion.

Although NE can directly stimulate pituitary GtH secretion, a central site of action for NE in stimulating GtH release is also possible. Chang and Peter (1984) reported that injection of a low dose of NE into the third cranial ventricle increased goldfish serum GtH concentrations. Interestingly, this stimulatory effect of centrally administered NE on serum GtH levels was also demonstrable in female goldfish at early, but not at later stages of ovarian recrudescence. The significance of the seasonality of NE to stimulate GtH release is not known.

In a recent experiment using female goldfish at late stages of ovarian recrudescence, it was demonstrated that intraperitoneal injection of E

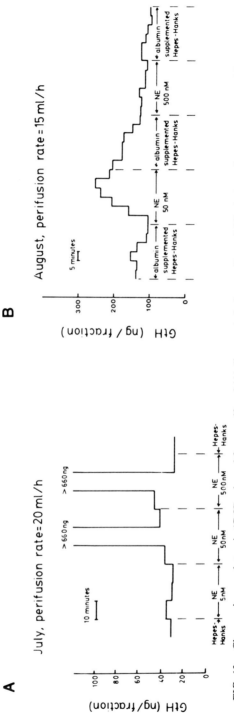

FIG. 13. Changes in gonadotropin (GtH) released from perifused (A) dispersed pituitary cells, and (B) pituitary fragments prepared from sexually regressed female goldfish during and following exposure to norepinephrine (NE). Reprinted with permission from *Life Sciences* **35**, 2027–2033, Chang et al. (1984c); copyright Pergamon Press, Ltd.

increased serum GtH levels (J. P. Chang and R. E. Peter, unpublished results). Therefore, it remains possible that E may also participate in the neuroendocrine regulation of GtH release in goldfish.

C. DA AS GRIF IN OTHER TELEOSTS

Few studies have been performed to systematically investigate the role of neurotransmitters in the neuroendocrine regulation of GtH release in other teleost species. However, information available in the literature suggests that an inhibitory DA influence on spontaneous and/or GnRH-stimulated GtH release exists in teleost species other than the goldfish.

Lin *et al.* (1985a,c) reported that in the Chinese loach (*Paramisgurnus dabryanus*), injection of pimozide or reserpine potentiated the action of LHRH-A in stimulating increases in serum GtH as well as the frequency of induced ovulation. Injection of α-methyl-*p*-tyrosine and carbidopa, but not diethyl-dithiocarbamate, also potentiated the action of LHRH-A in stimulating GtH release and in inducing ovulation (Lin *et al.*, 1985c). Similarly, in common carp (*Cyprinus carpio*) (Billard *et al.*, 1983; Lin *et al.*, 1985c) and bream (*Parabramis pekinensis*) (Lin *et al.*, 1985c) injection of pimozide potentiated the ability of superactive analogs of LHRH in stimulating GtH release, and in inducing oocyte maturation and ovulation. Lin *et al.* (1985c) also reported that reserpine treatment enhanced the ability of LHRH-A to cause increased circulating levels of GtH and induce ovulation in the common carp. These results demonstrate that in the Chinese loach and common carp, depletion of DA by blockage of synthesis or blockage of DA receptors removes the inhibitory influence of DA on GtH release, resulting in enhancement of the GtH release-response to LHRH-A.

De Leeuw *et al.* (1984) reported that apomorphine inhibits spontaneous GtH release from cultured pituitaries of the African catfish (*Clarias larzera*). In the catfish injection of pimozide was found to increase plasma GtH levels in juvenile fish, and to potentiate the ability of LHRH-A to stimulate GtH release in both juvenile and mature fish (de Leeuw *et al.*, 1985a,b). In sexually immature European eel (*Anguilla anguilla*) treatment with estradiol stimulates the synthesis of GtH and increases pituitary GtH content; injection of LHRH-A together with pimozide, but not LHRH-A alone, stimulates release of GtH in the estradiol-treated eels (Dufour *et al.*, 1984). In the white sucker (*Catostomus commersoni*), injection of pimozide increased, whereas apomorphine decreased, basal serum GtH levels (Chang, 1983).

Although GtH measurements were not available, results from studies on the effects of pimozide on the response to injection of LHRH agonists in the walleye (*Stizostedion vitreum*) are consistent with the existence of an inhibitory DA influence on GtH release. Injection of pimozide alone was found to be effective in advancing germinal vesicle migration and oocyte maturation in walleye (Pankhurst *et al.*, 1985).

In salmonids, it is well established that injection of LHRH superactive analogs alone stimulates a marked increase in blood levels of GtH and ovulation (Donaldson *et al.*, 1981; Sower *et al.*, 1982; Crim *et al.*, 1983a,b; Van Der Kraak, *et al.*, 1983, 1985). However, in spite of this greater sensitivity to LHRH and superactive analogs than is apparent in other teleosts, an inhibitory DA influence on spontaneous, as well as GnRH-stimulated GtH release is also present. Billard *et al.* (1985) reported that pimozide increased the GtH release–response to injections of an LHRH agonist in rainbow trout (*Salmo gairdneri*) and brown trout (*S. trutta*). In rainbow trout, injection of apomorphine was reported to inhibit the postovulatory GtH release (Gielen, unpublished results, cited by de Leeuw *et al.*, 1985b), and DA was found to inhibit the release of GtH from pituitaries incubated *in vitro* (Crim, 1981). In the coho salmon (*Oncorhynchus kisutch*), pimozide injection also increased the basal as well as the LHRH analog-stimulated GtH secretion *in vivo* (Van Der Kraak, 1984).

It is clear from these studies that the GRIF action of DA is found in a wide phylogenetic range of teleosts, perhaps in all teleosts. The apparent difference between salmonids and other teleosts in the degree of GRIF activity exerted by DA suggests that the relative importance of both DA and GnRH in regulating GtH secretion in teleosts may be altered through evolution.

VI. DA as GRIF in Mammals?

Several excellent and extensive reviews on the involvement of catecholamines and other neurotransmitters in the regulation of LH secretion in mammals are available (Krulich, 1979; Sawyer, 1979; Barraclough and Wise, 1982; Kalra and Kalra, 1983, 1985; Barraclough *et al.*, 1984) and it would be inappropriate for us to attempt another review. In brief, although the influence of DA on LH secretion is still controversial, catecholamines have been shown to be involved in the modulation of pulsatile LH release and the preovulatory LH surge in mammals. Based on studies in rats, it is generally believed that, in mammals, neurotransmitters influence LH release by altering LHRH release from the median eminence. This is in contrast with the situation in fish where DA and NE exert direct inhibitory and stimulatory influences, respectively, on GtH release.

Although it has been demonstrated, *in vitro*, that DA does not directly influence pituitary LH release in rats (e.g., Kamberi *et al.*, 1969, 1970, 1971; Denef and Andries, 1983), similar information on other mammalian species is not available. Results from a number of reports are not at variance with a possible pituitary level inhibitory action of DA on LH release in mammals. Metoclopramide has been reported to increase circulating LH levels in patients with hypothalamic amenorrhea and prolactin-secreting microadenomas (Quigley *et al.*, 1979, 1980). Metoclopramide treatment has also been reported to be effective in increasing plasma LH concentrations in patients with normoprolactinemic amenorrhea (Hagen *et al.*, 1983, 1984). These stimulatory effects of metoclopramide on LH release in women with various pathological conditions have always been attributed to the blockage, by metoclopramide, of a central inhibitory DA tone on LHRH secretion since metoclopramide readily crosses the blood–brain barrier and the infusion of DA, which does not cross the blood–brain barrier, has been reported not to alter LH secretion in normal, as well as hyperprolactinemic female subjects (Connell *et al.*, 1984; Ho *et al.*, 1984). However, Esquifino *et al.* (1984) reported that in rats in which hyperprolactinemia was experimentally induced by ectopic pituitary transplants, acute as well as chronic treatment with metoclopramide enhanced the LH response to LHRH administration. These data clearly suggest that metoclopramide acts at an extrahypothalamic site to potentiate LHRH effects.

Recently, DA infusion has been shown to significantly decrease circulating LH levels in normal men, castrated men, and normal men treated with clomiphene citrate (Fig. 14; Foresta *et al.*, 1984; Delitala *et al.*, 1983). Similarly, DA infusion causes a decrease in circulating LH levels in women with polycystic ovarian disease (Paradisi *et al.*, 1985), and in normal as well as hyperprolactinemic women pretreated with α-methyl-*p*-tyrosine (Nicholetti *et al.*, 1984). Although Pehrson *et al.* (1983) reported that infusion of DA did not alter the LH response to simultaneous infusion of LHRH in postmenopausal women, Judd *et al.* (1978) reported that DA infusion attenuated the LH release response to a bolus of LHRH in women at the early follicular phase. DA infusion also attenuates the LH release–response to LHRH in normal men (Leebaw *et al.*, 1978; Huseman *et al.*, 1980; Fig. 15). These observations suggest that under certain pathological conditions, and possibly also in the normal human, a direct inhibitory effect of DA on LH release may be present. Steroid condition of the test subjects seems to be of importance; the depression of circulating LH levels by DA infusion was more pronounced in castrated than in clomiphene citrate-treated men, which in turn showed a greater response than normal men (Foresta *et al.*, 1984). Also, the metoclopramide-stimu-

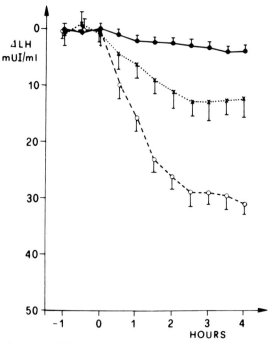

FIG. 14. Net decrease (Δ) in plasma luteinizing hormone (LH) levels during dopamine infusion (0.4 μg/kg/minute) in normal men (●), normal men after clomiphene citrate treatment (×), and castrated subjects (○). Values are mean ± SD ($n = 4$). From C. Foresta, G. Scanelli, S. Marra, and C. Scandellari: The influence of gonadal steroids on the dopamine inhibitory effect on gonadotropin release in man. *Fertil. Steril.* **42**, 942 (1984). Reproduced with permission of the publisher, The American Fertility Society.

lated increase in serum LH levels was only observed in normal menstruating women during the mid-luteal phase, but not in the early or late follicular phases (Ropert *et al.*, 1984).

The possible existence of a direct DA inhibitory influence on LH release in mammals is not restricted to humans. Dailey *et al.* (1981) reported that DA directly inhibits the actions of LHRH on LH release in normal and stalk-sectioned rabbits (Fig. 16). DA infusion has also been shown to reduce plasma LH concentrations in a dose-dependent manner in ovariec-

---

FIG. 15. Mean serum luteinizing hormone (LH) concentrations in five normal men during a 4 hour constant intravenous infusions of luteinizing hormone-releasing hormone (GnRH), dopamine (DA) plus GnRH, oral haloperidol (HA), and DA and after saline infusion. Values are mean ± SD ($n = 5$). Reproduced from Huseman *et al.* (1984). Copyright © 1984 The Endocrine Society.

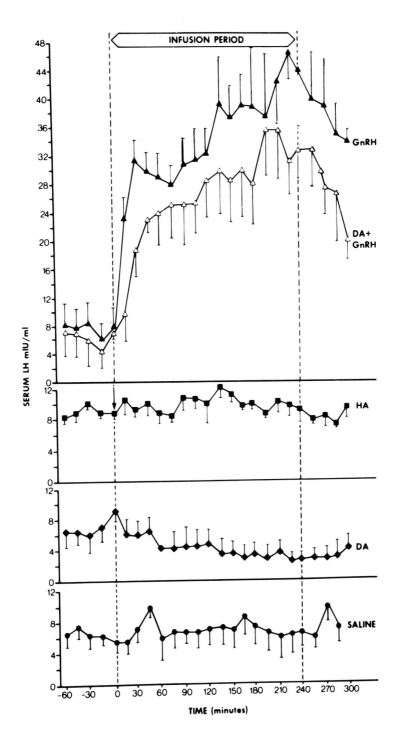

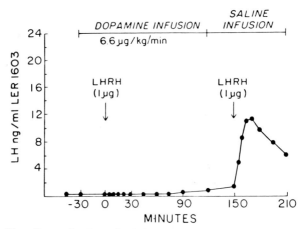

FIG. 16. The effects of a dopamine infusion followed by a saline infusion on LHRH-induced LH release in an acutely pituitary stalk-sectioned rabbit. Reprinted with permission from *Life Sciences* 22, 1491–1498, Dailey *et al.* (1981); copyright 1981, Pergamon Press, Ltd.

tomized (Deaver and Dailey, 1982) and cycling ewes during luteal regression (Deaver and Dailey, 1983), but not in anestrous ewes (Domanski *et al.*, 1975; Deaver and Dailey, 1982).

Further studies are required to systemically prove or disprove the existence of direct influences of DA and/or other neurotransmitters on gonadotropin release in mammals. It is possible that direct DA inhibition on gonadotropin secretion is an ancient evolutionary component of the neuroendocrine regulation of release of reproductive (gonadotropic) hormones; however, the intensity of this inhibition may differ between classes of vertebrates, as well as within a smaller phylogenetic unit as the Subclass Teleostei. During the evolution of higher vertebrates such as mammals, the importance of a direct inhibitory influence of DA on gonadotropin release may have been diminished so that it is now of minor significance or nonfunctional under normal conditions. On the other hand, this action of DA on gonadotropin release may be manifested only under special hormonal or pathological conditions.

VII. Conclusions

Available information on the neuroendocrine regulation of gonadotropin in teleost fish reviewed in this article indicates that GtH release is under the stimulatory influence of GnRH and the inhibitory control of DA. DA acts as a GRIF to directly inhibit spontaneous GtH release and modulate GnRH action (Fig. 17). The intensity of the DA inhibitory tone

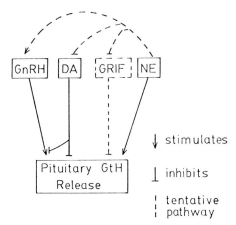

FIG. 17. Summary diagram of the influences of dopamine (DA) and norepinephrine (NE) on gonadotropin (GtH) release in goldfish.

differs between species and the intensity may also change along with sexual recrudescence or maturation within the same species. In species such as the goldfish, the inhibitory DA influence is intense and the removal of this inhibitory tone is part of the mechanism regulating the ovulatory GtH increase. Although DA acts as a GRIF in a number of teleost species, other GRIFs may also exist. In the goldfish, although NE can directly stimulate GtH release, it may also act centrally to increase GtH secretion (Fig. 17). The stimulatory influence of NE on GtH secretion is seasonal, dependent on the stage of gonadal recrudescence. Information on the possible participation of other neurotransmitters in the neuroendocrine regulation of GtH release in teleosts is not available. Although it is generally believed that neurotransmitters do not exert a direct influence on the pituitary to modify gonadotropin release in mammals, results from some studies on humans, rabbits, and ewes are strongly suggestive of a direct inhibitory DA influence on LH release.

ACKNOWLEDGMENTS

We thank Dr. Glen Van Der Kraak for reviewing the manuscript. Unpublished work cited herein was supported by grants from the National Sciences and Engineering Research Council of Canada, and the International Development Research Centre of Canada.

REFERENCES

Ball, J. N. (1981). *Gen. Comp. Endocrinol.* **44,** 135–170.
Barnett, F. H., Sohn, J., Reichlin, S., and Jackson, I. M. D. (1982). *Biochem. Biophys. Res.* **105,** 209–216.

Barraclough, C. A., and Wise, P. S. (1982). *Endocrinol. Rev.* **3**, 91–118.
Barraclough, C. A., Wise, P. M., and Selmanoff, M. K. (1984). *Recent Prog. Horm. Res.* **40**, 487–529.
Bex, F. J., and Corbin, A. (1984). *Front. Neuroendocrinol.* **8**, 85–151.
Billard, R., Breton, B., Fostier, A., Jalabert, B., and Weil, C. (1978). *In* "Comparative Endocrinology" (P. J. Jaillard and H. H. Boer, eds.), pp. 37–48. Elsevier, Amsterdam.
Billard, R., Alagarswami, K., Peter, R. E., and Breton, B. (1983). *C.R. Acad. Sci. Paris Ser. III* **296**, 181–184.
Billard, R., Reinaud, P., Hollenbecq, M. G., and Breton, B. (1985). *Aquaculture* **43**, 57–66.
Breton, B., and Weil, C. (1973). *C.R. Hebd. Seances Acad. Sci., Ser.* D **277**, 2061–2064.
Breton, B., Motin, A., Kah, O., Lemenn, F., Geoffre, S., Precigoux, G., and Chambolle, P. (1984). *C.R. Hebd. Seances Acad. Sci. Paris, Ser. III* **299**, 383–388.
Burzawa-Gerard, E. (1982). *Can. J. Fish. Aquat. Sci.* **39**, 80–91.
Chang, J. P. (1983). Ph.D. thesis. University of Alberta.
Chang, J. P., and Peter, R. E. (1983a). *Neuroendocrinology* **36**, 351–357.
Chang, J. P., and Peter, R. E. (1983b). *Gen. Comp. Endocrinol.* **52**, 30–37.
Chang, J. P., and Peter, R. E. (1984). *Gen. Comp. Endocrinol.* **55**, 89–95.
Chang, J. P., Cook, A. F., and Peter, R. E. (1983). *Gen. Comp. Endocrinol.* **49**, 22–31.
Chang, J. P., Peter, R. E., and Crim, L. W. (1984a). *Gen. Comp. Endocrinol.* **55**, 347–350.
Chang, J. P., Peter, R. E., Nahorniak, C. S., and Sokolowska, M. (1984b). *Gen. Comp. Endocrinol.* **55**, 351–360.
Chang, J. P., MacKenzie, D. S., Gould, D. R., and Peter, R. E. (1984c). *Life Sci.* **359**, 2027–2033.
Connell, J. M. C., Ball, S. G., Inglis, G. C., Beastall, G. H., and Davies, D. L. (1984). *Clin. Sci.* **67**, 219–223.
Cote, T. E., Eskay, R. L., Frey, E. A., Grewe, C. W., Munemura, M., Stoof, J. C., Tsuruta, K., and Kebabian, J. W. (1982). *Neuroendocrinology* **35**, 217–224.
Creese, I. (1982). *Trends Neurosci.* **5**, 40–43.
Creese, I., Sibley, D. R., Leff, S., and Hamblin, M. (1981). *Fed. Proc.* pp. 147–152.
Creese, I., Morrow, A. L., Hamblin, M. N., Leff, S. E., and Sibley, D. R. (1983). *Adv. Biosci.* **44**, 1–50.
Crim, L. W. (1981). *In* "Neurosecretion" (D. S. Farner and K. Lederis, eds.), p. 446. Plenum, New York.
Crim, L. W., and Evans, D. M. (1980). *Gen. Comp. Endocrinol.* **40**, 283–290.
Crim, L. W., Evans, D. M., Coy, D. H., and Schally, A. V. (1981). *Life Sci.* **28**, 129–135.
Crim, L. W., Evans, D. M., and Vickery, B. H. (1983a). *Can. J. Fish, Aquat. Sci.* **40**, 61–67.
Crim, L. W., Sutterlin, A. M., Evans, D. M., and Weil, C. (1983b). *Aquaculture* **35**, 299–307.
Dailey, R. A., Tsou, R. C., Tindall, G. T., and Neill, J. D. (1981). *Life Sci.* **22**, 1491–1498.
Deaver, D. R., and Dailey, R. A. (1982). *Biol. Reprod.* **27**, 624–632.
Deaver, D. R., and Dailey, R. A. (1983). *Biol. Reprod.* **28**, 870–877.
Deery, D. J. (1975). *Gen. Comp. Endocrinol.* **25**, 395–798.
De Leeuw, R., Kamphuis, W., Goos, H. J. Th., and Van Oordt, P. G. W. J. (1984). *Gen. Comp. Endocrinol.* **53**, 438 (Abstr. 18).
De Leeuw, R., Goos, H. J. Th., Richter, C. J. J., and Eding, F. H. (1985a). *Aquaculture* **44**, 295–302.
De Leeuw, R., Resink, J. W., Rooyakkers, E. J. M., and Goos, H. J. Th. (1985b). *Gen. Comp. Endocrinol.* **58**, 120–127.
Delitala, G., Devilla, L., and Musso, N. R. (1983). *J. Clin. Endocrinol. Metab.* **56**, 181–184.

Denef, C., and Andries, M. (1983). *Endocrinology* **112**, 813-822.
Domanski, E., Prezekop, F., Skubiszewki, B., and Wolinska, E. (1975). *Neuroendocrinology* **17**, 265-273.
Donaldson, E. M., Hunter, G. A., and Dye, H. (1981). *Aquaculture* **26**, 129-141.
Dufour, S., Delerue-Le Belle, N., and Fontaine, Y.-A. (1983). *Gen. Comp. Endocrinol.* **52**, 190-197.
Dufour, S., Delerue-Le Belle, N., and Fontaine, Y.-A. (1984). *C.R. Acad. Sci. Paris Ser. III* **299**, 231-234.
Esquifino, A. I., Ramos, J. A., and Tresguerres, J. A. F. (1984). *J. Endocrinol.* **100**, 141-148.
Fahraeus-van Ree, G. E., van Vlaardingen, M., and Gielen, J. T. (1983). *Cell Tissue Res.* **232**, 157-176.
Foresta, C., Scanelli, G., Marra, S., Scandellari, C. (1984). *Fertil. Steril.* **42**, 942-945.
Grigoriadis, D., and Seeman, P. (1984). *Can. J. Neurol. Sci.* **11**, 108-113.
Hagen, C., Djursing, H., Petersen, K., and Carstensen, L. (1983). *Lancet* **1**, 422-423.
Hagen, C., Petersen, K., Kjursing, H., and Andersen, A. N. (1984). *Acta Endocrinol.* **106**, 8-14.
Ho, K. Y., Smythe, G. A., and Lazarus, L. (1984). *Clin. Endocrinol.* **20**, 53-63.
Huseman, C. A., Kugler, J. A., and Schneider, I. G. (1980). *J. Clin. Endocrinol. Metab.* **51**, 209-214.
Idler, D. R., and Crim, L. W. (1985). In "Current Trends in Comparative Endocrinology" (B. Lofts and W. N. Holmes, eds.), Vol. 1, pp. 81-84. Univ. of Hong Kong Press, Hong Kong.
Idler, D. R., and Ng, T. B. (1983). In "Fish Physiology" (W. S. Hoar, D. J. Randall, and E. M. Donaldson, eds.), Vol. 9, pp. 187-221. Academic Press, New York.
Judd, S., Rakoff, J., and Yen, S. S. C. (1978). *J. Clin. Endocrinol. Metab.* **47**, 494-498.
Kah, O., Chambolle, P., Dubourg, P., and Dubois, M. P. (1982). *Proc. Int. Symp. Reprod. Physiol. Fish* p. 56.
Kah, O., Chambolle, P., Thibault, J., Geffard, M. (1984). *Neurosci. Lett.* **48**, 293-298.
Kalra, S. P., and Kalra, P. S. (1983). *Endocr. Rev.* **4**, 311-351.
Kalra, P. S., and Kalra, S. P. (1985). In "The Pituitary Gland" (H. Imura, ed.), pp. 189-220. Raven, New York.
Kamberi, I. A., Mical, R. S., and Porter, J. C. (1969). *Science* **166**, 388-390.
Kamberi, I. A., Mical, R. S., and Porter, J. C. (1970). *Endocrinology* **87**, 1-12.
Kamberi, I. A., Mical, R. S., and Porter, J. C. (1971). *Endocrinology* **88**, 1003-1011.
Kebabian, J. W., and Calne, D. B. (1979). *Nature (London)* **277**, 93-96.
Kebabian, J. W., Beaulieu, M., Cote, T. E., Eskay, R. L., Frey, E. A., Goldham, M. E., Grewe, C. W., Munemura, M., Stoof, J. C., and Tsuruta, K. (1983). In "Dopamine Receptors" (C. Kaiser and J. W. Kebabian, eds.), pp. 33-52. Amer. Chem. Soc., Washington, D.C.
King, J. A., and Millar, R. P. (1979). *Science* **206**, 67-69.
King, J. A., and Millar, R. P. (1980). *Endocrinology* **106**, 707-717.
Krulich, L. (1979). *Annu. Rev. Physiol.* **41**, 603-615.
Leebaw, W. F., Lee, L. A., and Woolf, P. D. (1978). *J. Clin. Endocrinol. Metab.***47**, 480-487.
Lin, H. R., Peng, C., Lu, L. Z., Zhou, X. J., Van Der Kraak, G., and Peter, R. E. (1985a). *Aquaculture* **46**, 333-340.
Lin, H.-R., Peter, R. E., Nahorniak, C. S., and Bres, O. (1985b). In "Current Trends in Comparative Endocrinology" (B. Lofts and W. N. Holmes, eds.), Vol. 1, pp. 77-79. Univ. of Hong Kong Press, Hong Kong.

Lin, H. R., Van Der Kraak, G., Liang, J. Y., Peng, C., Li, G. Y., Lu, L. Z., Zhou, X. J., and Peter, R. E. (1985c). *Proc. Int. Symp. Aquacult. Carp Related Species* (in press).
MacKenzie, D. S., Gould, D. R., Peter, R. E., River, J., and Vale, W. W. (1984). *Life Sci.* **35**, 2019-2026.
Nagahama, Y., and Peter, R. E. (1982). *Cell Tissue Res.* **225**, 259-265.
Ng, T. B., and Idler, D. R. (1983). *In* "Fish Physiology" (W. S. Hoar, D. J. Randall, and E. M. Donaldson, eds.), Vol. 9, pp. 373-404. Academic Press, New York.
Nicholetti, I., Filipponi, P., Sfrappini, M., Fedeli, L., Petrelli, S., Gregorini, G., Santensanio, F., and Brunetti, P. (1984). *Horm. Res.* **19**, 158-170.
Pankhurst, N. W., Van Der Kraak, G., and Peter, R. E. (1985). *Fish Physiol. Biochem.* (in press).
Paradisi, R., Margrini, O., Venturoli, S., Fabbri, R., Porcu, E., Stanzani, L., and Flamigni, C. (1985). *Horm. Metab. Res.* **17**, 29-31.
Pehrson, J. J., Jaffee, W. L., and Vaitukaitus, J. L. (1983). *J. Clin. Endocrinol. Metab.* **56**, 889-892.
Peter, R. E. (1970). *Gen. Comp. Endocrinol.* **14**, 334-356.
Peter, R. E. (1980). *Can. J. Zool.* **58**, 1100-1104.
Peter, R. E. (1981). *Gen. Comp. Endocrinol.* **45**, 294-305.
Peter, R. E. (1982). *Can. J. Fish. Aquat. Sci.* **39**, 48-55.
Peter, R. E. (1983a). *Am. Zool.* **23**, 685-695.
Peter, R. E. (1983b). *In* "Fish Physiology" (W. S. Hoar, D. J. Randall, and E. M. Donaldson, eds.), Vol. 9, pp. 97-135. Academic Press, New York.
Peter, R. E., and Crim, L. W. (1979). *Rev. Physiol. A* **41**, 323-335.
Peter, R. E., and Fryer, J. N. (1983). *In* "Fish Neurobiology. Higher Brain Areas and Functions" (R. E. Davis and R. G. Northcutt, eds.), Vol. 2, pp. 165-201. Univ. of Michigan Press, Ann Arbor.
Peter, R. E., and Paulencu, C. R. (1980). *Neuroendocrinology* **31**, 133-141.
Peter, R. E., Crim, L. W., Goos, H. J. Th., and Crim, J. W. (1978). *Gen. Comp. Endocrinol.* **35**, 391-401.
Peter, R. E., Nahorniak, C. S., Chang, J. P., and Crim, L. W. (1984). *Gen. Comp. Endocrinol.* **55**, 337-346.
Peter, R. E. Nahorniak, C. S., Sokolowska, M., Chang, J. P., Rivier, J. E., Vale, W. W., King, J. A., and Millar, R. P. (1985). *Gen. Comp. Endocrinol.* **58**, 231-242.
Quigley, M. E., Judd, S. J., Gilliland, G. G., and Yen, S. S. C. (1979). *J. Clin. Endocrinol. Metab.* **48**, 718-720.
Quigley, M. E., Sheehan, K. L., Casper, R. F., and Yen, S. S. C. (1980). *J. Clin. Endocrinol. Metab.* **50**, 949-954.
Ramirez, V. D., Feder, H. H., and Sawyer, C. H. (1984). *Front. Neuroendocrinol.* **8**, 27-84.
Ropert, J. F., Quigley, M. E., and Yen, S. S. C. (1984). *Life Sci.* **34**, 2067-2073.
Sawyer, C. H. (1979). *Can. J. Physiol. Pharmacol.* **57**, 667-680.
Seeman, P. (1984). *In* "Brain Receptor. Methodologies. Part A" (P. J. Marangos, I. C. Campbell, and R. M. Cohen, eds.), pp. 285-307. Academic Press.
Sherwood, N., Eiden, L., Brownstein, M., Spiess, J., Rivier, J., and Vale, W. (1983). *Proc. Natl. Acad. Sci. U.S.A.* **80**, 2794-2798.
Sherwood, N. M., Harvey, B., Brownstein, M. J., and Eiden, L. E. (1984). *Gen. Comp. Endocrinol.* **55**, 174-181.
Sokolowska, M., Peter, R. E., Nahorniak, C. S., Pan, C. H., Chang, J. P., Crim, L. W., and Weil, C. (1984). *Aquaculture* **36**, 71-83.
Sokolowska, M., Peter, R. E., Nahorniak, C. S., and Chang, J. P. (1985a). *Gen. Comp. Endocrinol.* **57**, 472-479.

Sokolowska, M., Peter, R. E., Nahorniak, C. S. (1985b). *Can. J. Zool.* **63**, 1252–1256.
Sower, S. A., Schreck, C. B., and Donaldson, E. M. (1982). *Can. J. Fish. Aquat. Sci.* **39**, 627–632.
Van Der Kraak, G. (1984). Ph.D. thesis, University of British Columbia.
Van Der Kraak, G., Lin, H.-R., Donaldson, E. M., Dye, H. M., and Hunter, G. A. (1983). *Gen. Comp. Endocrinol.* **49**, 470–476.
Van Der Kraak, G., Dye, H. M., Donaldson, E. M., and Hunter, G. A. (1985). *Can. J. Zool.* **63**, 824–833.
Van Oordt, P. G. W. J., and Peute, J. (1983). In "Fish Physiology" (W. S. Hoar, D. J. Randall, and E. M. Donaldson, eds.), Vol. 9, pp. 137–186. Academic Press, New York.
Weil, C., Breton, B., and Reinaud, P. (1975). *C.R. Acad. Sci. Paris* **280D**, 2469–2472.
Yamazaki, F. (1965). *Mem. Fac. Fish. Hokkaido Univ.* **13**, 1–65.

DISCUSSION

G. B. Cutler. I was interested in the increases in gonadotropin secretion you found with pituitary transplants since in mammals the loss of LHRH support usually results in low levels. Do you think that the unusual anatomy in the goldfish with the projection of the median eminence into the pituitary results in your transplanting LHRH activity along with the pituitary?

R. E. Peter. I think that in mammals it depends on when you look at the LH levels following pituitary transplantation, because if LH levels are measured very shortly after transplantation the LH levels are in fact quite high. In other words a situation may occur in mammals similar to what we find in goldfish, that, over the short term following pituitary transplantation, there is quite a lot of gonadotropin release.

G. Aurbach. I enjoyed your presentation very much. Your results with dopamine inhibitors suggests a D-2 receptor-mediated phenomenon. I presume that cyclic AMP mediates release of gonadotropin and that a cyclic AMP rise in response to some signal would be inhibited by the D-2 mechanism. Do you think this might be a calcium-mediated event? Have you looked into this?

R. E. Peter. No, we have not yet looked at postreceptor mechanisms. We hope to do some studies on GnRH receptors, but we haven't yet gotten to the stage of looking at the kinds of things that you suggest. It would be very interesting and I think most appropriate.

M. V. Nekola. Do the gonads have any effect on modulating gonadotropin release from the pituitary?

R. E. Peter. Of course, there are steroid actions on the brain and the pituitary in telecost fish. There is evidence for positive as well as negative feedback actions of sex steroids. We found in recent experiments that when we treat goldfish with steroids and test their responsiveness to the combination of LHRH analog plus pimozide at 72 and 48 hours after steroid treatment, in fact the response to the combination is potentiated by steroid treatment.

M. V. Nekola. Have you tried reducing the level of LHRH that you give to see if you could get a dose–response? The gonadotropin response may have been maximal with the doses you gave.

R. E. Peter. Yes, the dosages we administered in those experiments were relatively high. However, where we had a dose dependency of LHRH analog in the presence of pimozide. With LHRH alone at a low dose without pimozide we get no response.

M. V. Nekola. Then have you tried sampling at an earlier time frame? You were looking at 1.5 hours and in the rat by that time the effect of LHRH has already disappeared.

R. E. Peter. Yes, we have looked at earlier times as well. If the experiments are done with intravenous infusion of low dosages of LHRH or LHRH analog a release response occurs within a few minutes. However, the experiments I showed you all involved intraperitoneal injections which gives a longer time frame of response. Also, the LHRH analog apparently stays in circulation in the goldfish for a very long period of time.

G. Anogianakis. I would like to go back over the pituitary transplants that you described. I noticed that when you transplanted the pituitary in the preoptic area you did not have as great an effect as when you transplanted it to other areas. If I recall correctly the existence of the blood–brain barrier in the preoptic area is disputed and it may not exist. Does this mean that whatever inhibitory factor you have coming from the anterior hypothalamus and from the preoptic area, gets to your transplant more readily?

R. E. Peter. I think the reason why the pituitary transplants in the brain ventricles release less gonadotropin is because of the high levels of dopamine in the cerebrospinal fluid. If the cerebrospinal fluid is perfusing the transplanted pituitary, the dopamine in the CSF will suppress gonadotropin release.

G. Anogianakis. Then this means that when you refer to inhibition you are not really referring to direct inhibitory innervation, but to a release factor. I was not clear on that.

S. J. McConnell. You said that there was a seasonality in the response to GnRH. I may have missed this, but what photoperiod conditions did you keep the fish under?

R. E. Peter. We have mimicked the seasonal photoperiod changes throughout the year during our experiments.

S. J. McConnell. You demonstrated that a photoperiod shift alters the period of ovulation during the day. Do you have any idea which factor provides this photic information?

R. E. Peter. Via the eyes, not the pineal.

N. B. Schwartz. I was curious about what the gonadotropin you are measuring is doing in the ovary. You did tantalize us with the observation that there is a second factor which you are not measuring. What kinds of bioassays have you used for the immunoassayable gonadotropin and what kind of effects do these gonadotropins have on the ovaries of these animals?

R. E. Peter. The gonadotropin that we are measuring by radioimmunoassay, we have bioassayed by *in vitro* steroidogenesis. It has been bioassayed in a number of laboratories for both cyclic AMP and steroidogenic actions on the gonads.

N. B. Schwartz. What action does it have on follicular maturation? I assume there has to be precedent follicular maturation in the ovary before the rather broad LH surge occurs.

R. E. Peter. Yes, the migration of the germinal vesicle to the periphery of the oocyte and break down of the germinal vesicle are the first steps in the process of oocyte maturation. At 5 hours following a preoptic lesion the germinal vesicle has already migrated to the periphery and is starting to break down.

N. B. Schwartz. Are all of those events occurring from the time in which the goldfish are put back into an aquarium which is stimulatory?

R. E. Peter. The time frame is similar to the sequence of events that we find during spontaneous ovulation. In other words, the timing of the maturation events during spontaneous ovulation seems to be on the same time course as we find following injections of pimozide and LHRH analog, or following brain lesions in the preoptic region.

R. M. Carey. Have you had an opportunity to study the effects of dopamine on GnRH binding in the pituitary gonadotrope?

R. E. Peter. No, we are just starting that work now.

S. Y. Ying. Did you look at other types of hormones in your lesioned animals? Such as GH.

R. E. Peter. There is a major somatostatin area in the preoptic region of the brain, and if we do large preoptic lesions we will also cause increases in growth hormone levels. The time course, however, of the increases in growth hormone and gonadotropin is quite different.

G. T. Campbell. I have a comment as it relates to the first question by Dr. Cutler. We have been studying the cellular composition and secretory activity of long-term pituitary allografts beneath the renal capsule in the hamster. Depending upon the age of the donor tissue, cells which contain LH actually end up being the numerically dominant cell in the allograft. They don't release very much LH but that may be a consequence of other intraallograft factors. I think the question of what happens to the pituitary cell types in malregulated adenohypophyseal tissue in mammals may need to be reinvestigated. Second, although I cannot quote the year in which this was published in abstract form, the late Dr. E. M. Bogdanove once published some results having to do with seminal vesicle hypertrophy following lesioning of various parts of the hypothalamus of the rat. One of many interpretations of that experiment is that he was lesioning out neurons which contain an LH inhibiting factor.

R. E. Peter. Thank you very much. I can add that a colleague in France, Dr. Oliviere Kah, has found a dopaminergic nucleus in the anterior preoptic region of goldfish, in the area that I interpreted as being the origin of GRIF. A pathway comes from that nucleus through to the pituitary stalk. Also, in recent work that Dr. Kah and I did in my laboratory we found that if we lesioned the preoptic brain area, tyrosine hydroxylase activity disappeared from the area of the pituitary containing the gonadotropin cells.

I. M. Spitz. Thank you for your elegant presentation. I wondered if prolactin and TSH are also under dopamine control in the goldfish?

R. E. Peter. Dopamine does inhibit prolactin release in teleost fish, but at this point we have no evidence that dopamine inhibits TSH release in goldfish.

I. M. Spitz. Can you down-regulate the gonadotropin receptor *in vivo* with repeated or continuous administration of LHRH or dopamine antagonists? If you can down-regulate with one of these stimuli is the response to the second maintained?

R. E. Peter. In short, I can't tell you about down-regulation of dopamine receptors. However, regarding the response to LHRH analog, we see that this response goes on for literally days in the pimozide-blocked animal. We know that there is a protein in plasma that binds the GnRH and seems to result in a reservoir of circulating GnRH in the plasma. GnRH does seem to be very stable in the fish. Likewise the GnRH receptors do not down-regulate very readily, which is quite different from the situation occurring in mammals.

T. Horton. You made a couple of interesting allusions to some evolutionary biology and a statement about the median eminence moving into the pituitary. I wondered if you could comment on the embryology of the median eminence over a phylogenetic scale. Does the median eminence in fact move into the pituitary in teleosts or out of the pituitary in higher vertebrates?

R. E. Peter. The situation in teleosts is a derived condition. In more primitive bony fishes a true median eminence is present, but the teleosts are the oddballs in this situation by having direct innervation of the pituitary. The standard condition in vertebrates is the median eminence with a hypothalamo–hypophyseal portal system.

J. D. Veldhuis. Do you have any information in this relatively small animal as to whether the short-term release of dopamine is periodic, i.e., episodic or nonuniform over time?

R. E. Peter. No, although I wish we did, we don't have such information.

J. D. Veldhuis. What central mechanisms do you envision in this species as being linked to the inhibitory dopaminergic system? For example, in a number of other species, the opiate system is inhibitory of gonadotropin release. Have you had occasion to investigate that in the goldfish?

R. E. Peter. The opiate system seems, if anything, to be stimulatory to gonadotropin release. As to what may regulate the activity of the dopaminergic input to the gonadotropin cells in the pituitary, I really don't know. We are hoping to try and address that question subsequently, but it is going to be a difficult problem to investigate.

H. Kulin. One experiment of nature that may bear on the possible dual control, i.e., inhibitory and facilitory, of gonadotropins in man is the recent observation in a small group of children who received head and neck irradiation that hypopituitarism developed in the setting of precocious puberty. This paradox could be explained by the radiation-induced destruction of facilitory centers for hGH, TSH, and ACTH but an inhibitory center for gonadotropin control. This is admittedly speculative but of interest with respect to your own data in the fish.

R. E. Peter. Thank you very much.

J. M. Hershman. I noted in the data showing that pimozide potentiated the action of GnRH that many of the species listed are grown commercially in culture. I wonder if there has been application of either pimozide or other dopamine blockers in commercial culture of fish. Could you elaborate on that?

R. E. Peter. At this point it is under development. I have collaborative work going on in China and in Poland on cultured fish to test the system to be sure that the combination is appropriate for inducing ovulation, and also that the treatment of the animals with pimozide doesn't cause some undesirable secondary side effects.

G. Anogianakis. I will go back to what you mentioned about the dopaminergic nucleus in the anterior hypothalamic area. Now in higher mammals this nucleus is under certain serotoninergic control from the raphe nuclei and also projects down the CNS axis through serotoninergic neurons. Did you try any experiments with serotonin?

R. E. Peter. We haven't yet investigated serotonin actions. I know the work you refer to, but I should also add that recently a dopaminergic pathway has been described from the anterior preoptic region in mammals to the median eminence.

B. M. Dobyns. Quite some years ago we were particularly interested in exophthalmos in fish using goldfish and Fundulus. The exophthalmos response was produced by pituitary extract. In the course of experiments we found that the response was much better using pituitary extracts that were rich in the LH. I wonder if by any chance you've observed exophthalmos in any of your studies in these goldfish. Variations in exophthalmos responses were particularly true with respect to seasonality and to salinity of the environment.

R. E. Peter. No, we haven't found any cases of exophthalmos. We measured thyroid hormone levels in some of the preoptic lesioned animals and the thyroid hormone levels don't change.

M. O. Thorner. I would like to come back to the whole thesis that you are proposing and ask you whether there are dopamine receptors on the gonadotrophs. I wonder if you would speculate about that. I know that in some fish the pituitary is divided up into neat lobes. In the fish that you are working with are there only gonadotropes in that area or are there lactotropes too? Have you considered the possibility of an interaction between the lactotrope and the gonadotrope to produce this effect?

R. E. Peter. In the model we presented which shows dopamine acting directly on the gonadotropin cells, we infer that there are dopamine receptors on the gonadotropin cells. We believe the action is direct and not indirect. The answer to your second question is that this particular region of the pituitary contains gonadotropin cells and growth hormone cells. Dopamine does not have a similar action on growth hormone and gonadotropin release. The prolactin cells are located in a different region of the pituitary, so there isn't the opportunity, so far as we can envisage, of interaction of prolactin and gonadotropin cells.

Evolutionary Aspects of the Endocrine and Nervous Systems

DEREK LEROITH, GEORGE DELAHUNTY, GAYE LYNN WILSON,
CHARLES T. ROBERTS, JR., JOSHUA SHEMER, CELESTE HART,
MAXINE A. LESNIAK, JOSEPH SHILOACH, AND JESSE ROTH

Diabetes Branch, National Institute of Arthritis, Diabetes, and Digestive and Kidney Diseases, Bethesda, Maryland

I. Introduction

In mammals and other vertebrates, the major systems of intercellular communication are the endocrine and nervous systems. Traditionally, these two systems have been considered separate entities both anatomically and functionally, and the messenger molecules used by each system are considered unique to that system. In the classic concept of the endocrine system the chemical messenger molecule, the hormone, is produced in a localized region, released into the general circulation, and acts on a target cell at a distance (Fig. 1). In the nervous system, on the other hand, the secretory cell is a neuron and the messenger molecule a neurotransmitter. Because the typical endocrine system was assumed to arise at the level of the vertebrates and since glandular structures such as the pancreas, pituitary, and thyroid were found only in vertebrate species, it was thought that their hormonal products, such as insulin, ACTH, and thyroxine, also were limited to vertebrate species (Fig. 2). In contrast, neurons and nervous systems began considerably earlier at the level of simple multicellular organisms. Therefore it was concluded that the neurotransmitters and other neuroactive molecules probably began at this level. Indeed, when searched for, classic neurotransmitter molecules, namely, the biogenic amines, were found in nonvertebrate animals.

Over the past few decades, a gradual erosion of the boundaries between the endocrine and nervous systems has occurred, in that hormones have been found in nonendocrine tissues (Table I). Initially, hormone-like messenger molecules were found in cancer cells and nervous tissue, but, more recently, synthesis of these hormones has been demonstrated in noncancer and nonneural tissues as well (Table II). On the other hand, classic neuropeptides are also produced by nonneural tissues. Many of the mes-

Shared Aqueous Compartment
CELL 1
SIGNAL
CELL 2

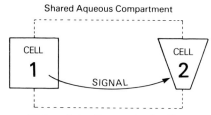

FIG. 1. Basic Features of Intercellular Communication Systems. The secretory cell (cell 1) synthesizes and releases a soluble messenger into the shared fluid (aqueous) compartment that it shares with the target cell (cell 2).

senger molecules assigned to the endocrine or nervous systems reach their target tissues by other systems of intercellular communication such as exocrine and paracrine systems. Furthermore, a number of nonhormonal peptides which act on a target tissue have structural similarities to classic hormones and act on the target tissues in a manner almost identical to hormones (Roth *et al.*, 1982, 1983; LeRoith *et al.*, 1982b).

More recently, typical vertebrate-type peptide hormones have been found in nonvertebrate tissues, suggesting that, despite the restriction of vertebrate-type endocrine glands to vertebrate species, biochemical elements such as messenger molecules may indeed have an earlier phyloge-

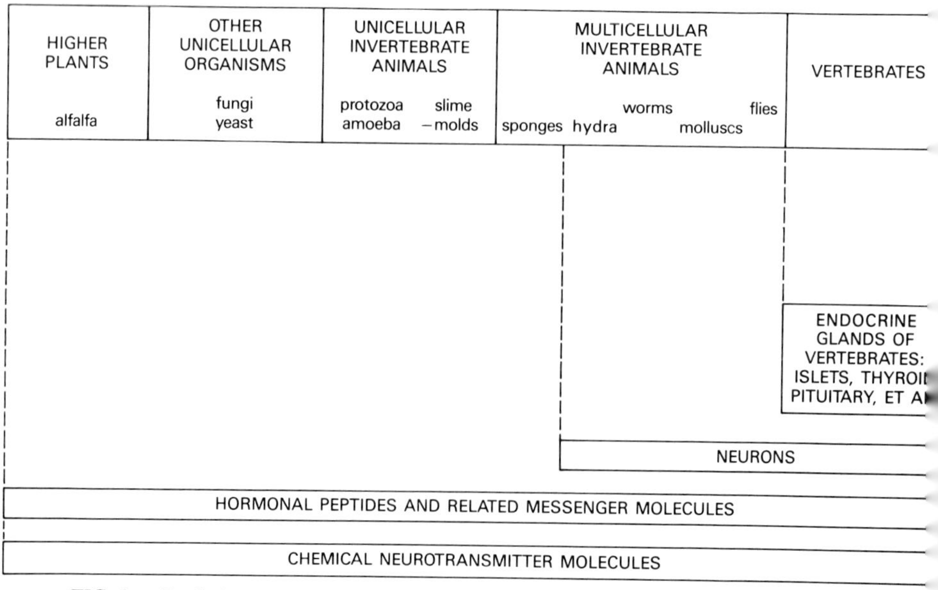

HIGHER PLANTS
alfalfa
OTHER UNICELLULAR ORGANISMS
fungi
yeast
UNICELLULAR INVERTEBRATE ANIMALS
protozoa
amoeba
slime
–molds
MULTICELLULAR INVERTEBRATE ANIMALS
sponges
hydra
worms
molluscs
flies
VERTEBRATES
ENDOCRINE GLANDS OF VERTEBRATES: ISLETS, THYROI
PITUITARY, ET A
NEURONS
HORMONAL PEPTIDES AND RELATED MESSENGER MOLECULES
CHEMICAL NEUROTRANSMITTER MOLECULES

FIG. 2. Evolutionary origins of the biochemical elements of the endocrine and nervous systems. (Adapted from LeRoith *et al.*, 1982b.)

TABLE I

Example of Overlaps between Systems of Intercellular Communication

1. Classic gastroentero-pancreatic hormones are synthesized in nervous tissue, e.g., cholecystokinin
2. Neuropeptides are found in nonneural tissues, i.e., somatostatin in gut-related tissues and bombesin in frog skin
3. "Nonhormonal messengers" have structural and functional similarities to hormones, i.e., IGF, NGF, and EGF
4. Hormones may reach their target site via exocrine secretions, e.g., somatostatin in gastric juice

netic origin than has been supposed previously, and may be used as hormones in nonvertebrate species (Fritsch *et al.*, 1976; Thorpe and Duve, 1984; Van Noorden, 1984; Kramer, 1985). In this review we also will present evidence that the biochemical elements of intercellular communication (signal messenger molecules and receptors) that we have classically associated with vertebrate-type endocrine systems and nervous systems are present in unicellular organisms. These studies support the idea that true intercellular communication began at the level of unicellular organisms.

II. Peptide Hormones in Multicellular Nonvertebrates

Following the demonstration that nonendocrine vertebrate tissues produce hormonal peptides, a search for these same compounds was undertaken in nonvertebrate species. A large number of vertebrate-type peptide

TABLE II

Multiple Tissue Sites of Possible Production of Canonical Hormones in Vertebrates

ACTH/ β-endorphin	hCG	Insulin	Vasopressin
1. Pituitary	1. Placenta	1. Pancreas	1. Hypothalamus
2. Brain	2. Pituitary	2. Central and peripheral nerves	2. Brain
3. Tumors	3. Kidney	3. Salivary gland	3. Adrenal
4. Placenta	4. Liver	4. Embryo	4. Gonads
5. Lung	5. Tumors	5. Placenta	
6. Spleen	6. Fetus		
7. Adrenal	7. Testes		
8. Gonads			
9. Pancreas			

TABLE III

Vertebrate-Type Hormone-Related Peptides Found in Multicellular Invertebrate Tissues[a]

Hormone	Ch	E	I	C	M	A	F	H
Insulin	+	+	+	+	+			
Glucagon	+	+	+	+	+	+		
Somatostatin	+		+	+	+		+	
Cholecystokinin	+		+		+	+		+
Pancreatic polypeptide	+		+		+	+		
Substance P	+		+	+	+			+
Vasotocin			+		+			+
ACTH	+		+	+	+			
Endorphin	+		+		+	+		
Calcitonin	+				+			+
TRH			+					
LHRH	+							
Bombesin					+			+
Neurotensin								+
Secretin	+		+		+			
VIP	+		+		+	+		

[a] Chordates (Ch), Echinoderm (E), Insects (I), Crustaceans (C), Molluscs (M), Annelids (A), Flatworm (F), *Hydra* (H). Studies utilized immunocytochemical techniques as well as immunoassay and bioassay of partially purified tissue extracts.

hormones are present in almost every class of multicellular nonvertebrates studied (Table III) (Fritsch *et al.*, 1976; Van Noorden, 1984; Kramer, 1985). Many of the studies demonstrated the presence of these hormonal peptides using immunocytochemical techniques with antisera raised against vertebrate peptide hormones. In other studies, the hormonal peptide was extracted from the tissues, partially purified, and its presence demonstrated by radioimmunoassay and bioassay techniques. In this section we describe some studies in which the presence of the hormonal peptide is related to a physiological function in multicellular nonvertebrates.

A. INSULIN-RELATED MATERIALS IN INSECTS

Insulin immunoreactivity has been found in a wide variety of insect species including *Hymenoptera* and *Orthoptera* (Tager *et al.*, 1976; LeRoith *et al.*, 1981a; Thorpe and Duve, 1984; Kramer, 1985). Localization of insulin immunoreactivity to the brain, corpora allata, and corpora cardiaca has been noted in the tobacco hornworm *Manduca sexta*.

Duve and co-workers demonstrated insulin-related material in the head of the blowfly, *Calliphora vomitoria* (Duve, 1976; Duve and Thorpe, 1979;

Duve et al., 1979). Using anti-bovine insulin antiserum, the immunoactivity was localized to the median neurosecretory cells of the brain (Fig. 3). Characterization of this material was performed by purification of blowfly brain homogenates using acid-alcohol extraction, gel filtration, and ion-exchange chromatography. The purified material cross-reacted with bovine insulin antiserum, was capable of displacing ^{125}I-labeled insulin from rat liver plasma membranes, and stimulated [^{3}H]glucose uptake by rat adipose tissue in a dose-dependent fashion. In addition, purification of this material to homogeneity using HPLC, and analysis of its amino acid composition, showed it to be very similar to, but not identical with, mammalian insulins.

Interestingly, the possible physiological role of this brain-related insulin was shown by extirpation of the median neurosecretory cells; this treatment caused elevation of trehalose and glucose levels in the hemolymph of the blowflies (Fig. 3). Replacement therapy with purified mammalian insulin or with purified blowfly brain homogenates reversed these changes, suggesting that brain insulin may be important in the carbohydrate homeostasis of the blowfly.

B. INSULIN RECEPTORS IN *DROSOPHILA*

For peptide hormones like insulin to function biologically, specific cell surface receptors must be present. The mammalian insulin receptor (MW 350K–400K) is a glycoprotein tetramer composed of 2 α-subunits and 2 β-subunits linked together by disulfide bonds. The α-subunit (MW 135,000) binds insulin with high affinity and specificity while the β-subunit (MW 90,000) consists of the transmembrane portion and a cytoplasmic tail. Insulin binding to the α-subunit results in autophosphorylation of the β-subunit via a tyrosine-specific kinase. Petruzzelli et al. (1985) recently described the presence of insulin receptor molecules in *Drosophila melanogaster*. ^{125}I-labeled insulin bound specifically to membrane preparations from both head and body regions with the highest concentration of receptors in the head region. Specificity studies demonstrated that insulin analogs (IGF-II, proinsulin, and guinea pig insulin) were capable of displacing ^{125}I-labeled insulin from the *Drosophila* receptor preparation in a manner similar to mammalian insulin receptors. Also, peptides known not to bind mammalian insulin receptors (desoctapeptide insulin and glucagon) did not bind to the *Drosophila* receptors. These results suggest that there is indeed a membrane component that specifically binds insulin with high affinity.

Further characterization of the *Drosophila* insulin receptor finds other similarities to the mammalian receptor. The molecular weight of the insu-

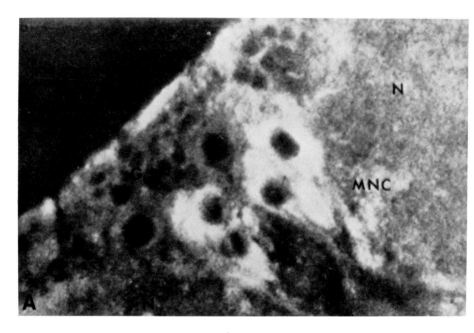

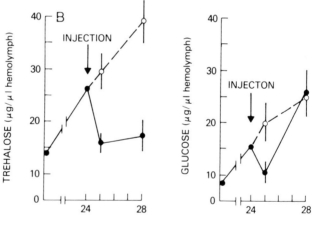

FIG. 3. Insulin-related material in the insect. (A) Median neurosecretory cells (MNC) in the blowfly brain demonstrate specific reactivity with anti-bovine insulin antiserum. N, Neuropil; G, glial cells. (B) After extirpation of the MNC cells, glucose and trehalose concentrations in hemolymph rise. Injection (arrow) of partially purified extracts from MNC normalizes the concentration of these carbohydrate intermediates (solid circles). In saline-treated controls (open circles) the levels continue to rise. (Adapted from Duve and Thorpe, 1979; Duve et al., 1979.)

lin binding moiety (α-subunit), as determined by cross-linking the receptor preparation to ^{125}I-labeled insulin and electrophoresis on polyacrylamide gels, was determined to be of MW 130,000 slightly smaller than the mammalian insulin subunit. The functional aspects of the *Drosophila* insulin receptor also appear similar to those in mammals. When stimulated by insulin, a tyrosine-specific kinase activity can be demonstrated by insect insulin receptors purified on wheat germ agglutinin. Interestingly, the insulin receptor of *Drosophila* is not recognized by anti-receptor antisera directed toward insulin receptors of human placenta or rat liver, nor by spontaneously occurring anti-insulin receptor antibodies obtained from patients exhibiting insulin resistance. These immunological differences may be due to differences in glycosylation patterns between the *Drosophila* and mammalian molecules or differences in the protein parts.

These studies suggest that the insulin receptors in *Drosophila* are similar to the mammalian insulin receptor with regard to affinity, specificity, size, and enzymatic activity. The lack of immunoreactivity to mammalian antireceptor antisera, however, suggests that structural differences exist between the mammalian and insect insulin receptors.

C. PROTHORACICOTROPIC HORMONE OF THE SILKWORM

Prothoracicotropic hormone (PTTH) is a regulatory peptide from insect brain which controls metamorphosis by promoting the release of ecdysone from the prothoracic glands (Fig. 4). Recent analysis of PTTH isolated from the brains of adult silkworms (*Bombyx mori*) finds both large (22K PTTH) and small (4K PTTH) forms to be present. Three types of the 4K PTTH molecule have now been identified and their sequences are illustrated in Fig. 4 (Nagasawa *et al.*, 1984). PTTH-II has 50% homology with the A chain of human insulin and IGF-I as well as 58% homology with the A chain of cod and toadfish insulin (Nagasawa *et al.*, 1984). These findings suggest that PTTH belongs to the family of insulin-related peptides. PTTH does not, however, cross-react with anti-porcine insulin antisera, nor does vertebrate insulin stimulate molting in pupae (the typical biological action of PTTH).

D. INSULIN-RELATED MATERIAL IN MOLLUSCS

Using immunocytochemical techniques, insulin-related material has been localized to gut-related cells of mussels and snails and to the hepatopancreas of oysters, as well as to the central nervous system of the pond snail. Plisetskaya *et al.* (1978), demonstrated the presence of insulin in the alimentary tract of two fresh water snails (*Anodonta cygnea* and *Unio*

A

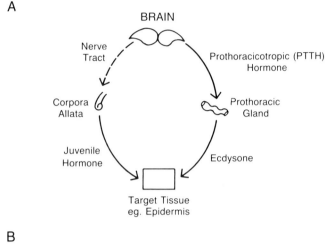

B

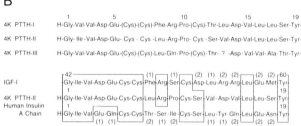

FIG. 4. (A) The endocrine control of moulting and metamorphosis in insects. Brain neurosecretory cells release prothoracicotropic hormone (PTTH) which stimulates the production of ecdysone from the prothoracic gland. Ecdysone acts on target tissues (e.g., epidermis) to cause moulting. The specific character of the moult is determined by the presence of juvenile hormone. The corpora allata secrete juvenile hormone under control of nerve tracts from the brain. The presence of juvenile hormone with ecdysone results in the expression of juvenile characteristics. (B) Upper half illustrates the amino acid sequence of the known 4K prothoracicotropic hormones of insects. A multistep purification procedure was used to isolate material from 648,000 heads of male adult *Bombyx mori*. Lower portion illustrates homology between 4K PTTH-II and the A chain of human insulin and insulin-like growth factor I (IGF-I). Identical amino acids are indicated by boxed sections and the numbers in parentheses note the minimum number of nucleotide base changes required for the amino acid substitution from that in 4K PTTH-II. (Adapted from Nagasawa *et al.*, 1984.)

pictorum) and examined the role of insulin in carbohydrate metabolism in these species. Using antibodies to mammalian insulin, these investigators localized insulin-related material to cells of the epithelial layer of the intestine. To study the possible physiological role of insulin, *Anodonta* was treated with mammalian insulin and with anti-insulin sera. Insulin treatment resulted in increased muscle glycogen synthase activity and increased muscle glycogen stores when followed by a glucose load. Injection of anti-insulin sera resulted in hyperglycemia, a decrease in glycogen synthase, and a decrease in muscle glycogen.

E. THE PRESENCE OF PEPTIDE HORMONES IN COELENTERATES

Hydra, a small freshwater polyp, consists of a cylindrical body with a mouth at one end (hypostome and tentacles) and a sticky foot at the other. The cell types in this animal are few and simple, consisting of layers of ectoderm and endoderm separated by mesoglia. A simple nerve net, associated mostly with ectoderm, extends throughout the body with higher nerve density in the head and foot. There are no ganglia or glial tissue associated with the nerve net, and all nerve cells appear to be multifunctional as they are involved in sensory, interneuronal, motor, and neurosensory activity. Chemical synapses have been observed in all classes of Coelenterates. The nervous system is apparently involved in the control of locomotion and growth and differentiation of cells. The nervous system of *Hydra* has been examined for the presence of vertebrate-type peptides. Immunocytochemical studies and immunoassays demonstrate the presence of at least 6 peptides in *Hydra* which are similar to their vertebrate counterparts (Grimmelikhuijzen *et al.*, 1982). Gastrin/CCK, substance P, neurotensin, bombesin/gastrin-releasing factor, FMRF amide, and oxytocin/vasopressin have been identified by immunocytochemistry in various body regions (Fig. 5). Generally (with the exception of FMRF amide), only certain select antisera for the peptides being studied give positive results in *Hydra*, suggesting that the peptides in Hydra are similar to those in mammals, but not identical. Another interesting aspect of the nervous system in *Hydra* is that these neuropeptides may coexist in the same neurons (Grimmilikhuijzen, 1983). Using a double labeling technique, bombesin and oxytocin were found to coexist in neurons of the basal disk, lateral tentacle junctions, and gastric regions of Hydra.

Besides the typical vertebrate-type peptides, other materials have been isolated from *Hydra* which act as head and foot activators, as well as head and foot inhibitors (Schaller *et al.*, 1979). The two activators are peptides with molecular weights of about 1000, while the two inhibitors are nonpeptides of molecular weight about 500. The head activator has recently been sequenced and found to contain eleven amino acids (Schaller and Bodenmiller, 1981). These four substances act specifically and at physiological concentrations (below $10^{-8}M$). It is also of interest to note that these activators and inhibitors can be produced outside the nervous system (Schaller *et al.*, 1980). In populations of *Hydra* devoid of a nervous system and containing almost exclusively epithelial cells, the four mor-

558   DEREK LEROITH ET AL.

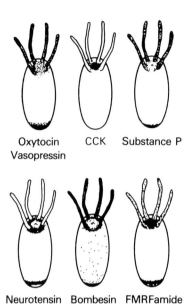

FIG. 5. Distribution of five vertebrate-type peptide hormones and a nonvertebrate hormone (FMRF-amide) in hydra as localized using specific antisera. (Adapted from Grimmelikhuijzen *et al.*, 1982.)

phological substances are produced at concentrations much greater than normal. The nonnervous system epithelial cells have the potential to produce these morphological factors, but this ability seems to be repressed in the presence of nerve cells. Interestingly, the head activator substance of *Hydra* has been found in the hypothalamus (cow, rat, pig), intestine (rat), and plasma (rat, human) of mammals (Schaller, 1979; Bodenmuller and Roberge, 1985). This activator may play a role in digestive function; initial studies found that plasma levels of this substance increased following feeding in man. It has also been shown to cause an increase in rat pancreatic amylase.

F. SUMMARY

Classic vertebrate-type peptide hormones are widespread throughout multicellular nonvertebrate species. In some instances, a physiological role for these hormones has been identified. In addition, hormones originally identified and characterized in nonvertebrates also are found in higher animals, although their physiological roles in the higher animals await elucidation.

III. Vertebrate-Type Peptide Hormones in Unicellular Organisms

We as well as others have found a large number of hormone-related materials in unicellular eukaryotes and bacteria (Table IV). Macchia *et al.* (1967) detected material in *Clostridium perfringens* with a molecular weight of 30,000 that had both *in vitro* and *in vivo* biological properties of thyroid-stimulating hormone (TSH). Following a 300-fold purification using column chromatography, this material, like TSH, stimulated [^{14}C]glucose oxidation, phospholipid synthesis, and the formation of intracellular colloid droplets in dog and bovine thyroid slices. Furthermore, when injected into newborn chicks, it caused the release of ^{131}I from the thyroid in a manner analogous to TSH. In addition, these workers showed that polyvalent *Clostridia* antitoxins contained antibodies directed against this TSH-like material and that these antibodies were capable of removing the thyroid-stimulating activity from the bacterial extract. Thus, these investigators suggested that cultures of *Clostridium perfringens* contain a material that resembles mammalian TSH.

Several laboratories have isolated bacteria from cancer patients and were able to detect CG-related material in these microbes (Acevedo *et al.*, 1978; Maruo *et al.*, 1979). The latter group also showed that the CG-like material extracted from one of these bacteria was very similar to hCG in

TABLE IV

Microbial Materials Closely Resembling Messenger Peptides of Vertebrates

Hormone-related materials	Microbe
Insulin	*Tetrahymena, E. coli, Halobacterium solinarium, Acinetobacter calcoaceticus, Bordetella pertussis Neurospora crassa, Aspergillus fumigatus*
Somatostatin	*Tetrahymena, E. coli, B. subtilis*
ACTH, β-endorphin	*Tetrahymena*
Relaxin	*Tetrahymena*
Calcitonin (human-type)	*Candida, E. coli*
Calcitonic (salmon-type)	*Tetrahymena*
Neurotensin	*E. coli, Caulaobacter, Rhodopseudomonas,*
TSH	*Clostridium perfringens*
hCG	Many bacteria

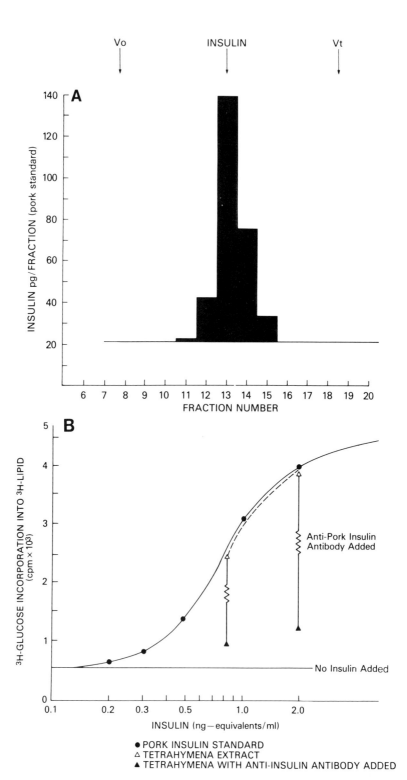

its interaction with ion-exchange, Sephadex, and concanavalin A columns. The material also cross-reacts in a radioreceptor and bioassay specific for hCG. Acevedo and co-workers immunized rabbits with some of these bacterial strains that contain hCG, and caused the generation of antibodies that cross-reacted with authentic hCG.

In extending these studies, we found insulin-related material in unicellular organisms grown in synthetic, defined medium in the absence of serum or macromolecules (LeRoith *et al.*, 1980, LeRoith *et al.*, 1981b, 1985c). When the cells were processed, insulin-related material was found in *Tetrahymena pyriformis*, a ciliated protozoan, *Neurospora crassa*, and *Aspergillus fumigatus* (unicellular fungi), as well as a number of bacteria, including *Escherichia coli* (Table IV). The insulin-related material from these organisms was extracted and purified using a number of chromatographic systems (Figs. 6, 7, and 8). These included Sep-Pak C_{18}, octadecasilylsilica columns (Waters Associates), and ion-exchange systems (DEAE-cellulose and DEAE-Sephadex G-50) as well as Sephadex G-50 gel chromatography and reverse-phase high-performance liquid chromatography (HPLC) using a phenyl Bondapak column. The material behaved similarly to, but not identical with, vertebrate insulins on these chromatographic systems, and cross-reacted with a specific anti-porcine antiserum used in the radioimmunoassay. Furthermore, the partially purified materials stimulated glucose metabolism in adipocytes isolated from young male rats, a classic bioassay for insulin that measures the incorporation of [^{3}H]glucose into lipids. The amount of biological activity was predicted by the amount of immunoreactivity. In addition, the bioactivity was neutralized in the presence of anti-insulin antibody, strongly suggesting that immunoactivity and bioactivity reside on the same molecule. Further specificity was demonstrated by using anti-insulin receptor block-

---

FIG. 6. Insulin-like material found in *Tetrahymena pyriformis*. (A) Cells were grown in defined medium, harvested by continuous centrifugation, and homogenized, using a Brinkman Polytron, in ice-cold acid–ethanol (0.2 N HCl/75% ethanol). After centrifugation, the supernatant was diluted with five volumes of 1M acetic acid and applied to Sep-Pak C_{18} cartridges. The bound materials were eluted with 75% ethanol, 0.01 M HCl, and the ethanol was removed from the eluate by gel filtration on a Sephadex G-25 column in 0.1 M acetic acid. The eluate was then neutralized and filtered on a Sephadex G-50 (fine) column in 0.05 M $(NH_4)_2CO_3$. The effluent fractions were tested for insulin immunoactivity using a porcine insulin radioimmunoassay. The left arrow represents the void volume and the right arrow represents salt elution. (B) Biological activity of the pooled peak material was measured as stimulation of glucose incorporation into lipids by isolated epididymal fat cells from the rat. (●) Porcine insulin standard curve; (△) increasing amounts of purified *Tetrahymena* extract represented as immunoassayable equivalents. Neutralization of the bioactivity in the presence of anti-insulin antibody is also shown (▲). (Reproduced from LeRoith *et al.*, 1980.)

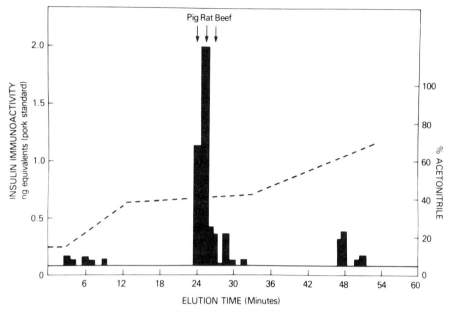

FIG. 7. Further purification of *T. pyriformis* insulin-like material. The pooled immunoactive insulin material from Sephadex G-50 fractions (Fig. 6) was applied to a DEAE-Sephadex ion-exchange column in 0.5 M Tris–HCl, 7 M urea, and eluted using a NaCl gradient from 0.01 to 0.02 M. Fractions from this column were tested in the insulin immunoassay and those containing insulin were pooled, lyophilized, and resuspended in 1 ml 0.1% TFA. The material was applied to a Waters Associates phenyl μBondapak column in 15% acetonitrile, 0.1% TFA. An elution gradient of 37 to 40% acetonitrile in 0.1% TFA was used to elute the material from the column. When fractions were tested for immunoactive insulin, the peak of activity eluted in the region typical for authentic mammalian insulins. (Reproduced from LeRoith *et al.*, 1981a, 1985c.)

ing antibody, which decreased the bioactivity of the purified material from the extracts, confirming that these materials exert their insulin-like bioactivity through the insulin receptor.

Extracts of *Tetrahymena pyriformis*, as well as *Escherichia coli* and *Bacillus subtilus*, contain materials resembling somatostatin (Table IV) (Berelowitz *et al.*, 1982; LeRoith *et al.*, 1985a, 1985b). Following initial purification on Sep-Pak C_{18} cartridges, the immunoactive somatostatin from the extracts eluted on HPLC with a profile similar to synthetic somatostatin (Figs. 9 and 10). The purified extracts from *Tetrahymena* demonstrated biological activity, i.e., inhibited release of growth hormone from dispersed pituitary cells from rat in a manner similar to synthetic somatostatin. The biological activity was neutralized in the presence of antisomatostatin antibody. In extracts of *Escherichia coli* and

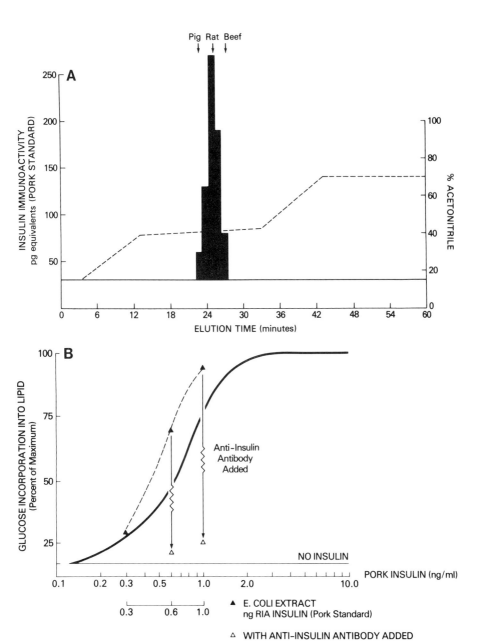

FIG. 8. Insulin-related material from *E. coli* was extracted and purified as described for *Tetrahymena pyriformis*. (A) Further purification on HPLC revealed a major peak of insulin immunoactivity again eluting in a position similar to the standard mammalian insulins. (B) A pool of the peak fractions from a Sephadex G-50 column was tested for biological activity by measuring the incorporation of glucose into lipids. The solid line represents the porcine insulin standard; ▲ represents increasing amounts of purified *E. coli* extract, and △ shows the neutralization of bioactivity by anti-porcine insulin antiserum. (Reproduced from LeRoith *et al.*, 1981, 1985c.)

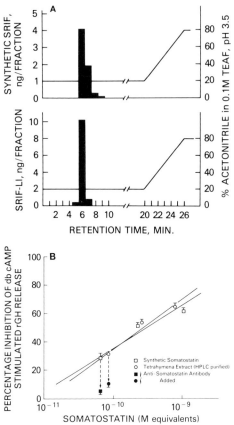

FIG. 9. Somatostatin-like material in *Tetrahymena pyriformis*. Acid-ethanol (0.2 N HCl, 750 μl/ml ethanol) extract of cells was applied to Sep-Pak C_{18} cartridges and eluted with acetonitrile/TFA. (A) The eluate was lyophilized and applied to reverse-phase hydrophobic HPLC in 20% acetonitrile, 0.1 M triethylammonium formate, pH 3.5. Fractions eluting during the isocratic 20% gradient were tested for SRIF-like immunoactive material, and showed a retention time similar to that of synthetic somatostatin. (B) The peak of SRIF-like immunoactivity was also tested for SRIF-like bioactivity by measuring its ability to inhibit the release of growth hormone from dispersed rat pituitary cells. The addition of anti-SRIF antibody neutralized the SRIF-like bioactivity. (Reproduced from Berelowitz *et al.*, 1982.)

Bacillus subtilis, the HPLC chromatograms demonstrated two major peaks of somatostatin immunoactivity (Fig. 10). One peak eluted with a retention time close to that of SRIF-28 and a second peak corresponded to the region of SRIF-14. In *Bacillus subtilis* extracts, the earlier peak reacted in both C-terminal and N-terminal specific immunoassays, suggesting the possible presence of a somatostatin-like molecule that resembles

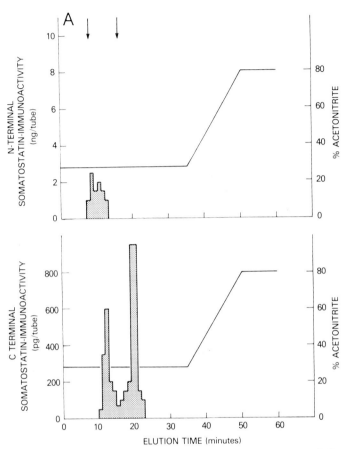

FIG. 10. Somatostatin-like activity in extracts of *Bacillus subtilis*. *B. subtilis* grown in defined medium was extracted in acid-ethanol (0.2 N HCl, 75% ethanol) and purified by Sep-Pak C_{18} chromatography using acetonitrile in 0.1% TFA, and BioGel P-6 chromatography in 0.01% formic acid. BioGel fractions containing somatostatin-like material were then applied to reverse-phase HPLC and eluted with 28% acetonitrile in 0.1% TFA. (A) SRIF-like material in the fractions was measured using both a C-terminal specific immunoassay and an N-terminal specific assay. Two major forms were detected, the early peak which reacted in both assays and eluted with a retention time like somatostatin-28, and a later peak that reacted only in the C-terminal specific assay, similar to somatostatin-14. (B) A bioassay for somatostatin, measuring the effect on guinea pig oxyntic cell hydroxyl ion production by quantitative cytochemistry, was used to test the HPLC-purified material. [Synthetic somatostatin (●), SRIF-28-like early HPLC peak (□), SRIF-14-like late HPLC peak (△).] The biological activity was inhibited in the presence of anti-somatostatin antiserum. (Reproduced from LeRoith *et al.*, 1985b.)

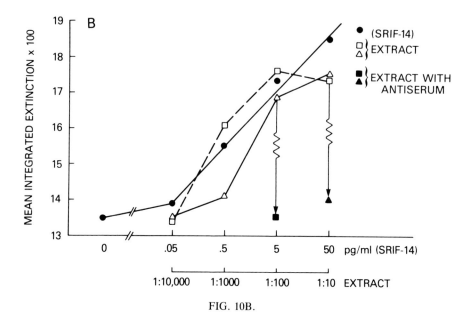

FIG. 10B.

SRIF-28. The second peak cross-reacted only in the C-terminal specific immunoassay, and thus resembled SRIF-14. Furthermore, the somatostatin-like material purified from *Bacillus subtilus* demonstrated biological activity as measured by hydroxyl ion production in gunea pig oxyntic cells, and this biological activity also was inhibited by anti-somatostatin antiserum (Fig. 10).

Extracts of *Tetrahymena* also contain substances that resemble ACTH (1–39) (LeRoith et al., 1982a). This material, following chromatography on Sephadex G-50 and electrophoresis on SDS–polyacrylamide gels, cross-reacted with two different specific anti-ACTH antisera. The partially purified material was biologically active in an assay measuring corticosterone release from dispersed rat adrenal cells (Fig. 11). Biological activity (which was predicted from the demonstration of ACTH immunoactivity) was removed by exposure to antibodies produced against synthetic ACTH (1–39). In addition to the ACTH-like material, β-endorphin-like material was detected in these extracts by a specific β-endorphin immunoassay as well as a radioreceptor assay for opiate peptides. Higher molecular weight material with approximate size equal to proopiomelanocortin was found which had both ACTH and β-endorphin-related moieties on the same molecule (Fig. 12). Thus, this material resembles the biosynthetic precursor of ACTH and β-endorphin in vertebrates.

Material closely resembling relaxin (Schwabe et al., 1983) was identi-

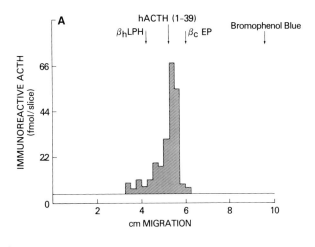
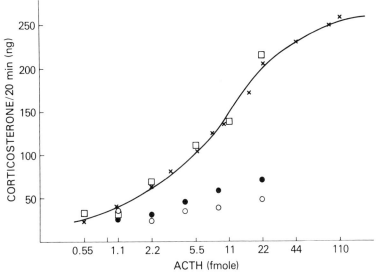

FIG. 11. ACTH-like material in *Tetrahymena pyriformis*. Cells were grown in defined medium, centrifuged, and extracted in 10 volumes of 0.1 M HCl, 0.22 M formic acid by homogenizing and mixing overnight. The extract was concentrated on Sep-Pak C_{18} cartridges, eluted with 1% formic acid, 60% acetonitrile, evaporated to dryness and applied to Sephadex G-50 in 0.1 M HCl, 0.22 M formic acid. (A) Fractions containing a peak of ACTH-like immunoactivity were pooled and concentrated on a Sep-Pak C_{18} cartridge. The concentrated pool of Sephadex G-50 peak fractions was then submitted to SDS-polyacrylamide gel electrophoresis. (B) Slices (0.2 cm) of the gel were eluted and assayed for immunoactive ACTH, human β-lipoprotein, and camel β-endorphin. This material was tested for bioactivity in an assay measuring release of corticosterone from dispersed rat adrenal cells. Extract (□), synthetic human ACTH (×), extract affinity-purified in the presence of anti-ACTH immunoglobulin (●), extract after Sepharose-Protein A immunodepletion in the presence of anti-ACTH immunoglobulin (○). (From LeRoith *et al.*, 1982a.)

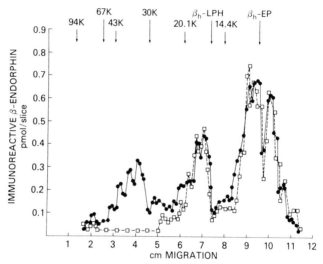

FIG. 12. Presence of substances in *Tetrahymena* extract that contain the antigenic determinants of both ACTH and β-endorphin. Equal aliquot of *Tetrahymena* extract was passed through microcolumns of affinity-purified ACTH antiserum covalently bound to Sepharose 4B, in the absence (column A, □) and presence (column B, ●) of 25μg of synthetic hACTH. The effluent fractions (unretarded material) from both columns were subjected to SDS–polyacrylamide gel electrophoresis. After 1-mm slices of the gels were eluted, individual fractions were quantified in the β-endorphin radioimmunoassay. High-molecular-weight material is present in the column A profile and is absent from the column B profile. The high-molecular-weight β-endorphin presumably consists of material containing ACTH and β-endorphin antigenic determinants. Most of this material could be specifically eluted from column A by the addition of an excess of synthetic hACTH (25 μg), further strengthening this hypothesis. The two other major peaks probably represent β-lipotropin (β-LPH)- and β-endorphin (β-EP)-like material. (Reproduced From LeRoith *et al.*, 1982a.)

fied in extracts of *Tetrahymena* and was purified on ion-exchange and Sephadex G-50 columns as well as HPLC; in these systems the material behaved in a manner similar to vertebrate relaxins (Table IV). Other hormone-like peptides found in unicellular organisms include human and salmon-type calcitonin (Deftos *et al.*, 1985), neurotensin (Bhatnagar and Carraway, 1981) as well as arginine vasotocin-like material (Table IV).

A. SPECIFICITY OF THESE FINDINGS

We have demonstrated the presence of materials closely resembling vertebrate-type peptide hormones in microbes. These hormonal peptides behaved both immunologically, biologically, and chromatographically in a manner similar to their vertebrate counterparts. To exclude the possibility that our findings were due to assay artifacts, a number of control experiments were performed. We were able to show for each hormone that the partially purified extract reacted in the radioimmunoassay in a specific manner (LeRoith et al., 1980, 1981a). That the biological activity was predicted by the content of immunoactive material and that specific antisera neutralized the biological activity strongly suggests that our findings were not due to artifacts in the various assay systems.

Having excluded the possibility of artifacts and demonstrated that the reactivity in the assays was specific, we performed experiments to decide whether these materials were native to the organisms, or due to exogenous contamination of our extracts by vertebrate materials. For each hormone studied, fresh unconditioned medium was carried through the fermentation, extraction, and purification procedures in a manner identical to the cells and was found to be devoid of any immunoactivity. In the case of insulin-related material, the content of both cells and media was clearly affected by the duration of fermentation (Fig. 13). The strongest evidence against contamination comes from the studies performed by outside laboratories in collaboration with us (LeRoith et al., 1985c). Cells were grown in an independent fermentation unit, partially purified by four collaborating laboratories that do not ordinarily study insulin, and the insulin immunoactivity was assayed by a contract laboratory using our immunoassay. In each study, results similar to ours were obtained, strongly suggesting that our results were not due to inadvertent contami-

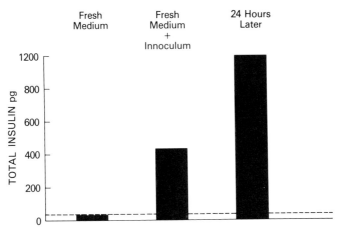

FIG. 13. To exclude the possibility of exogenous contamination, a time course study followed the net production of insulin. Aliquots were removed from the fermenter at various times, before inoculation of medium with cells, immediately after inoculation, and after 24 hours of cell growth. Each aliquot was processed separately by Sep-Pak C_{18}, Sepadex G-25, as well as Sephadex G-50 chromatography and immunoactive insulin measured. The duration of fermentation clearly affected the insulin content. (Reproduced From LeRoith et al., 1985c.)

nation of the extracts by purified vertebrate insulin. Thus we are convinced that the peptide hormones normally considered unique to vertebrates are, in fact, also native to unicellular organisms.

B. SUMMARY

We have presented evidence for the presence of vertebrate-type hormonal peptides in unicellular organisms. Similar findings have been described for other hormones such as steroids and classic biogenic amines (Sandor and Mehdi, 1979; Janakidevi, *et al.*, 1966a,b). We have however, used techniques that demonstrate "activity"; final confirmation of our findings requires the demonstration of either the amino acid sequence of the peptide or the nucleotide sequence of the DNA or messenger RNA corresponding to that peptide. Though we cannot exclude the possibility that these genes were transferred from vertebrates to microbes by a relatively recent recombinant DNA event, we do suggest that they probably arose evolutionarily in unicellular organisms and progressed phylogenetically through nonvertebrate metazoa to vertebrates and mammals.

Though we have demonstrated the presence of vertebrate-type peptide hormones in unicellular organisms, we have not as yet demonstrated either specific receptors for these particular hormones or a specific function. We speculate, however, that these hormonal peptides (including insulin, ACTH, somatostatin, and relaxin) do function as messenger molecules at the unicellular level. The strong similarity between the hormones found in unicellular organisms and their vertebrate counterparts suggests conservation of the molecules through evolution, probably as a consequence of some important function. Furthermore, unicellular organisms exchange information by means of soluble messenger molecules. Survival and reproduction of these unicellular organisms are major reasons for communication between microbes (Table V) (Sarkar *et al.*, 1979; Dunny *et al.*, 1979; Stotzler and Duntze, 1976; Stephens *et al.*, 1982).

TABLE V

Intercellular Communication Systems in Unicellular Organisms

	Organism	Function
Eukaryotes	*Saccharomyces cerevisae*	Sex
	Dictyostelium	Food
Prokaryotes	*Streptococcus fecalis*	Sex
	Myxobacterium	Food
	Bacillus brevis	Food

IV. Intercellular Communication in Unicellular Organisms

In *Bacillus brevus*, a gram-positive bacterium, conversion from vegetative form to spore form is stimulated by gramicidin, a small peptide produced by these cells. Mutant strains that do not produce gramicidin produce defective spores that are suseptible to elevated temperatures. These defects are correctable by adding gramicidin (Sarkar *et al.*, 1979).

For reproduction as well as other functions, unicellular organisms use soluble sex pheromones. Transfer of plasmid occurs during conjugation of cells of the species *Streptococcus faecalis* and this transfer is facilitated by sex pheromones (Fig. 14). *Streptococcus faecalis* cells which lack the plasmids ("recipient") secrete soluble peptides which act on the plasmid-containing ("donor") cells, causing them to adhere to plasmid lacking cells. The amino acid sequence of one of these pheromones has recently been determined (Fig. 14). This pheromone has been shown to induce the presence of a new antigen on the surface of the donor cells, and this aggregation substance binds to specific binding substances on the surface of both recipient and donor cells (Mori *et al.*, 1985).

Saccharomyces cerevisiae, a unicellular budding yeast, exists either as

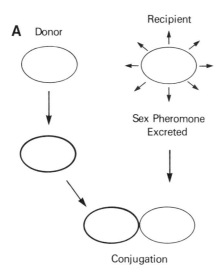

B H-Leu-Phe-Ser-Leu-Val-Leu-Ala-Gly-OH

FIG. 14. (A) A variety of plasmids contained in *S. faecalis* are transferred by a conjugation process. Plasmid-lacking "recipient" strains produce sex pheromones which cause plasmid-containing "donor" strains to adhere, thus facilitating mating and transfer of plasmids. (B) Amino acid sequence of the sex pheromone of *Streptococcus faecalis* that facilitates transfer of a drug resistance plasmid. (Adapted from Mori *et al.*, 1985.)

haploid or diploid cells (Fig. 15). The haploids have two mating types, α and a, and each mating type produces a specific oligopeptide mating pheromone. Type α produces α mating factor and a type produces a mating factor. Each mating factor affects the opposite mating type by interacting with specific cell surface receptors. Mating is brought about by a change in the surface properties allowing α and a cells to agglutinate, followed by fusion of the cytoplasm and formation of diploid zygotes. The amino acid sequence of α-mating factor has significant homology with a sex factor from the mammalian hypothalamus (gonadotropin-releasing

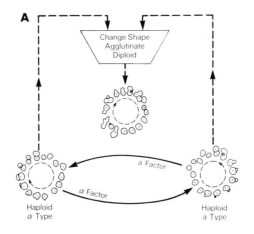

FIG. 15. (A) Sex pheromones in *S. cerevisiae*. The life cycle of *S. cerevisiae* involves vegetatively growing haploids of two mating types (α and a) as well as sporulating diploids. Each haploid produces mating pheromones that induce specific changes in the other haploid cell. This allows sexual conjugation to occur with the formation of diploid zygotes. (B) Similarity of vertebrate and microbial sex factor. Amino acid sequences of mammalian luteinizing hormone-releasing hormone (LHRH) and the N-terminus of α mating factor of common yeast *S. cerevisiae*. (Adapted from Stotzler and Duntze, 1976; Hunt and Dayhoff, 1979.)

hormone) (Fig. 15). α-Mating factor also binds to GnRH receptors on the rat pituitary and causes a dose-dependent release of lutenizing hormone from cultured rat pituitary cells, although at a much lower potency than natural GnRH. These findings thus demonstrate a strong structural and functional homology between a sex factor of a unicellular eukaryote and a sex factor from mammals (Stotzler and Duntze, 1976; Hunt and Dayhoff, 1979; Loumaye et al., 1982).

A. SUMMARY

Unicellular organisms communicate with each other. The presence of typical vertebrate-type messenger molecules suggests to us that these hormones also are probably involved in this process of intercellular communication at the level of unicellular organisms.

B. PREDICTIONS ARISING FROM THESE FINDINGS

The presence of these peptides in unicellular organisms has allowed us to predict that (1) receptor-like molecules may be present in unicellular organisms, (2) vertebrate-type peptide hormones may be present in higher plants, since plants apparently arose evolutionary from unicellular organisms, and (3) similar hormonal peptides may be present and function in early embryos even before development of glands (Table VI).

TABLE VI

Evidence That Endogenous Insulin Functions in Embryos
Insulin Production before Emergence of Pancreatic

1. Insulin is found in unfertilized eggs and in chick embryos at days 2–3, although pancreatic islets, and pancreatic production of insulin do not develop until later (de Pablo et al., 1983)
2. Receptors that prefer insulin to IGFs, as well as receptors that prefer IGFs to insulin, are present in chick embryos; at early times (1–2 days of embryogenesis) the latter dominate (Bassas et al., 1985)
3. Insulin at low doses (ng/embryo) accelerates growth as well as morphological and biochemical development by receptor-mediated pathways, whereas insulin at high doses (μg/embryo) causes a high ratio of death and developmental abnormalities which are not prevented by coadministration of glucose (de Pablo et al., 1985)
4. Anti-insulin antibodies administered to embryos at day 2 cause a high incidence of death; in survivors by days 3 and 4, growth as well as morphological and biochemical development are retarded. Since those antibodies are not reactive with insulin-like growth factors, it is most probably the neutralization of insulin or some closely related molecules which interfere with normal embryogenesis (de Pablo et al., 1985)

C. RECEPTOR-LIKE MOLECULES IN UNICELLULAR ORGANISMS

We have previously postulated that the presence of vertebrate-type peptide hormones in unicellular organisms and their similarity to their vertebrate counterparts suggest that they probably function in these unicellular organisms as messenger molecules for intercellular or intracellular communication (Roth et al., 1982, 1983; LeRoith et al., 1982b). Since activation of a target cell by a messenger molecule requires the presence of a receptor, one would predict the presence of receptor-like molecules in these unicellular organisms (Csaba, 1984). Indeed, a number of materials are present in microbes that are capable of binding hormones, and the structures of these binding materials resemble vertebrate-type receptors (Table VII). Richert and Ryan (1977) showed that ^{125}I-labeled hCG bound to components of *Pseudomonas maltophilia* but not to other microorganisms. This binding was of high affinity and with a degree of specificity which resembled the hCG receptor of vertebrates, i.e., unlabeled hCG competed better for the labeled hCG than did TSH, LH, and FSH. TSH-binding proteins are found in membranes prepared from *Yersinia enterocolitica* and *Escherichia coli* (Weiss et al., 1983). The binding of TSH was inhibited by sera from patients with Graves' disease but not with sera from controls, strongly suggesting a homology between the TSH binding protein of bacteria and that found in human thyroid membranes.

Feeding behavior, or endocytosis, of food particles by *Amoeba proteus* was affected by opioid alkaloids and peptides at very low (physiological) concentrations (Josefsson and Johansson, 1979). This effect was inhibited by several specific bioactive isomers of naloxone, suggesting that the amoebae probably contain a μ-type opiate receptor similar to that found

TABLE VII

Hormone-Binding Materials in Microbes That Resemble Vertebrate-Type Receptors

Vertebrate hormone bound	Microbe
hCG	*Pseudomonas maltophilia*
TSH	*E. coli, Yersinia enterocolitica*
Opioid	*Amoeba proteus*
Corticosterone	*Candida albicans*
Estrogen	*Saccharomyces cerevisiae*
	Paracoccidioides brasiliensis
"Antheridiol"	*Achlya ambisexualis*

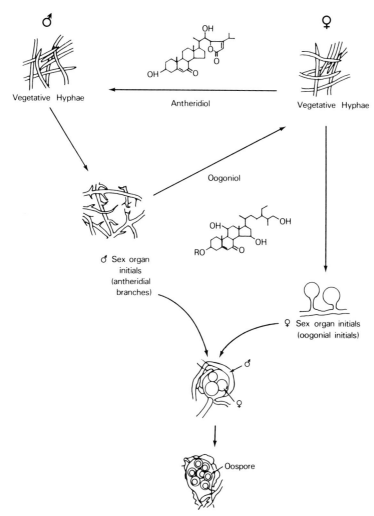

FIG. 16. The female secretes the sterol pheromone antheridiol inducing the formation of antheridial branches in the male, which grow toward the female and a second pheromone, oogoniol, causing the female to form oogonial initials. Nuclei within each sex organ undergo meiosis, forming oospheres and sperm. Sperm pass through a fertilization tube formed between the sex organs and fuse with oospheres to produce zygotes called oospores, which germinate and produce new diploid individuals. (Adapted from Horgen, 1981.)

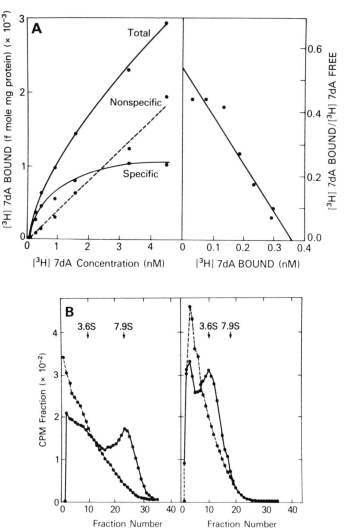

FIG. 17. (A) Binding of [^{3}H]7dA to *Achlya*. Mycelia of *Achlya ambisexualis* were minced with scissors and homogenized using a Polytron. Homogenates were filtered through glass wool and centrifuged at 250,000 g for 2 hours. The supernatant was clarified by filtration through 0.22-μm membranes. Aliquots (286 μg) of cytosol protein were incubated at 0°C for 1 hour in the presence of [^{3}H]7dA (a radiolabeled analog of antheridiol) in the range of 0.1–4.5 nM. Similar incubations were performed in the presence of 50-fold molar excess of antheridiol. After the incubation period bound and free steroid were separated by dextran-coated charcoal. Nonspecific binding is the amount of [^{3}H]7dA bound in the presence of 50-fold excess of unlabeled antheridiol. Specific binding is calculated by subtracting nonspecific from total at each concentration (left panel). Scatchard analysis (right panel) revealed an equilibrium dissociation constant of 0.65 nM and maximum binding capacity of 1245 fmol/mg protein. (Adapted from Reihl and Toft, 1984; Reihl *et al.*, 1984, with permission.) (B) Sucrose gradient analysis of [^{3}H]7dA binding in *Achlya* cytosol. *Achlya* cytosol

in higher organisms and, furthermore, that interaction with this receptor elicits a biological response.

Feldman and co-workers (Loose et al., 1981; Loose and Feldman, 1982) have described an intracellular protein in *Candida albicans* which binds corticosterone with high affinity. This high-affinity binding protein is similar to both the corticosterone-binding globulin and the glucocorticoid receptor of mammals. This group also found a cytosolic protein in *Saccharomyces cerevisiae* which selectively binds with high affinity the estrogenic hormone, 17β-estradiol (Burshell et al., 1984). They also showed that *Saccharomyces cerevisiae* contains a lipid-soluble material that binds to the endogenous steroid binding site of *Saccharomyces*, as well as to the estrogen receptors of mammals (Feldman et al., 1984). Another yeast, *Paracoccidioides brasiliensis*, contains specific binding sites for estradiol (Loose et al., 1983). Interestingly, estradiol inhibits the organism's transformation from the mycelial form to the yeast form, which is the initial step in the infection of humans. These authors further speculate that the disease caused by this organism (South American blastomycosis) is much more virulent in men than in women (despite equal exposure) because the endogenous estrogen of women inhibits a key step in the infective process by its interaction with the estrogen binding protein of *Paracoccidiodes*.

Achlya, a unicellular water mold, has two distinct sexual types designated male and female (Horgen, 1981) (Fig. 16). The female secretes antheridiol, a steroid-type hormone, which affects the male. The male produces another pheromone, oogoniol, which has the female as its target cell. These pheromones are both similar in structure to steroid hormones found in vertebrates. The soluble intracellular protein that binds antheridiol specifically is found in high concentrations in the male but not detected in the female. Characterization of this receptor-type molecule shows that it has many of the exceptional physical and chemical characteristics that are typical of the steroid receptors of vertebrates (Fig. 17). In this system, then, both the hormones themselves, as well as their specific

---

containing [³H]7dA, either with or without 200 nM antheridiol, were layered on 5–20% w/v sucrose gradients and centrifuged for 16 hours at 150,000 g. Sedimentation profile in low ionic strength gradients containing molybdate (left panel) and high ionic strength gradients without molybdate (right panel). (●——●) Profile in presence of [³H]7d alone; (●---●) profile of [³H]7dA in the presence of a 50-fold excess of unlabeled antheridiol. These results were interpreted as suggesting that the *Achlya* cytosolic binding protein can be observed as apparent aggregated (8 S) and dissociated (4 S) states and highly sensitive to the stabilizing action of low ionic strength and sodium molybdate. 3.65 arrow represents peak radioactivity from [¹⁴C]ovalbumin and 7.95 from [¹⁴C]aldolase. (From Riehl and Toft, 1984; Riehl et al., 1984.)

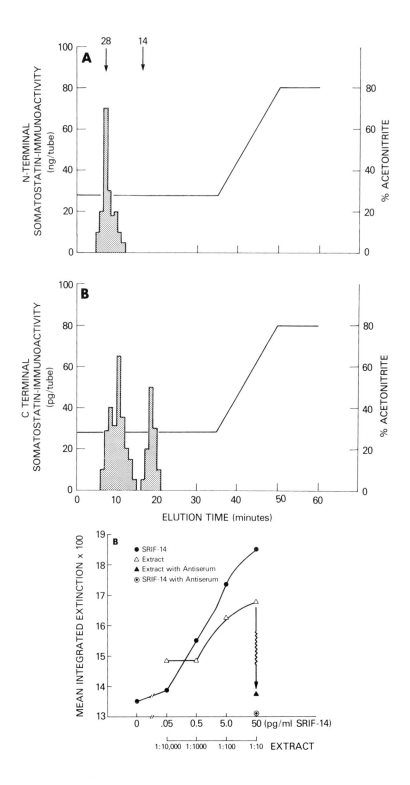

ENDOCRINE AND NERVOUS SYSTEMS 579

receptors, are structurally analogous to similar vertebrate molecules (Riehl and Toft, 1984; Riehl et al., 1984).

D. SUMMARY

Unicellular eukaryotes and prokaryotes contain hormone-binding proteins that resemble binding proteins or receptors found in vertebrates. What physiological role these binding proteins may play in the organism remains to be established.

E. VERTEBRATE-TYPE PEPTIDE HORMONES IN PLANTS

The presence of peptide hormones, neuropeptides, biogenic amines, and steroids, as well as hormone binding proteins in unicellular organisms, suggests to us that the biochemical elements of the endocrine, nervous, and other systems of intercellular communication had evolutionary origins at the level of unicellular organisms, and advanced phylogenetically in a Darwinian fashion. Thus, we have speculated that the nervous system and endocrine system had common origins and that one system did not give rise to the other as has been postulated by others.

Since unicellular organisms are supposedly the evolutionary precursors of both the animal and plant kingdoms, we should anticipate (in light of our proposal) the presence of these vertebrate-type hormonal peptides in multicellular plants. Indeed, a number of investigators (Zioudrov et al.,

---

FIG. 18. The plant material was homogenized in 5 volumes of cold methanol, chloroform, water, formic acid buffer, and stored overnight at 4°C. The extract was then centrifuged for 20 minutes at 7000 rpm at 4°C, and the supernatant was filtered through glass fiber filters. The filtrate was extracted once by addition of chloroform and water. The aqueous phase was collected, and again filtered through glass fiber filters. After boiling for 20 minutes and clarifying, the supernatant was made 0.2% TFA and applied to Sep-Pak C_{18} cartridges. The material that adsorbed to the column was eluted with a 20, 40, and 80% step gradient of acetonitrile in 0.1% TFA. Aliquots of each fraction were tested for SRIF-LI in the C-terminal immunoassay. The material eluting with 20–49% acetonitrile/0.1% TFA was applied to a BioGel P-6 column and eluted in 0.1% formic acid. Fractions containing SRIF-like material were further purified using C_{18} μBondapak, reverse-phase HPLC, with an isocratic gradient of 28% acetonitrile in 0.1% TFA. (A) Fractions were tested for SFIF-LI using both N-terminal and C-terminal specific radioimmunoassays. SRIF-28 and SRIF-14 standards were also applied to the column under identical conditions. (B) The HPLC purified SRIF-LI material was combined and tested for biological activity by measuring hydroxyl ion production by guinea pig gastric oxyntic cells. Results are shown as the mean integrated extinction × 100. (△) Dilutions of HPLC purified extract; (●) parallel dilutions of SRIF-14 standard; (▲) extract with bioactivity neutralized using anti-somatostatin antibody; (○) SRIF-14 plus anti-somatostatin antibody. (From LeRoith et al., 1985d.)

TABLE VIII

Materials in Higher Plants That Resemble
Messenger Peptides of Vertebrates

Vertebrate analogs	Plant
LHRH	Avena
TRH	Alfalfa
Opioid	Wheat
Interferon	Tobacco
Insulin	Spinach, Lemna
Somatostatin	Spinach, Lemna, tobacco

1979; Morley et al., 1980; Jackson, 1981; Sela, 1981), including ourselves, have found materials very similar to vertebrate-type peptide hormones in different plants (Table VIII). Material strongly resembling somatostatin-28 and somatostatin-14 was found in spinach (LeRoith et al., 1985d) (Fig. 18). To exclude the possibility that bacteria which contaminate spinach could be the reason for these findings, we similarly studied Lemna (grown under sterile conditions) and similar results were obtained. Multiple molecular weight forms of somatostatin also have been found in tobacco (Werner et al., 1986).

V. Conclusions

Intercellular communication is not unique to organisms which possess endocrine and nervous systems, rather it is essential to all forms of life including microbes. Interestingly, the biochemical elements necessary for intercellular communication, i.e., messenger molecules, their receptors, as well as postreceptor intracellular components, are present in unicellular organisms and show distinct structural and functional similarities to their counterparts in vertebrate tissues.

ACKNOWLEDGMENTS

This review is dedicated to Dorothy T. Krieger, a close friend and colleague of the members of the Diabetes Branch, and whose generous help and support in many aspects of the work will be missed. Thanks to Violet Katz for excellent secretarial assistance.

REFERENCES

Acevedo, H. F., Slifkin, M., Pouchet, G. R., and Pardo, M. (1978). Cancer **41**, 1217–1219.
Bassas, L., de Pablo, F., Lesniak, M. A., and Roth, J. (1985). Endocrinology **117**, 2321–2329.

Berelowitz, M., LeRoith, D., Von Schenk, H., Newgard, C., Szabo, M., Frohman, L. A., Shiloach, J., and Roth, J. (1982). *Endocrinology* **110**, 1939–1944.
Bhatnagar, Y. M., and Carraway, R. (1981). *Peptides* **2**, 51–59.
Bodenmuller, H., and Roberge, M. (1985). *Biochim. Biophys. Acta* **825**, 261–267.
Burshell, A., Stathis, P. A., Do, Y., Miller, S. C., and Feldman, D. (1984). *J. Biol. Chem.* **259**, 3450–3456.
Csaba, G. (1984). *Horm. Metab. Res.* **16**, 329–335.
Deftos, L., LeRoith, D., Shiloach, J., and Roth, J. (1985). *Horm. Metab. Res.* **17**, 82–85.
de Pablo, F., Roth, J., Hernandez, E., and Pruss, R. M. (1982). *Endocrinology* **111**, 1909–1915.
de Pablo, F., Girbau, M., Gomez, J. A., Hernandez, E., and Roth, J. (1985). *Diabetes* **34**, 1063–1067.
Dunny, G. M., Craig, R. A., Carron, R. L., and Clewell, D. B. (1979). *Plasmid* **2**, 454–465.
Duve, H. (1978). *Gen. Comp. Endocrinol.* **36**, 102–110.
Duve, H., and Thorpe, A. (1979). *Cell Tissue Res.* **200**, 187–191.
Duve, H., Thorpe, A., and Lazarus, N. R. (1979). *Biochem. J.* **184**, 221–227.
Feldman, D., Stathis, P. A., Hirst, M. A., Stover, E. P., and Do, Y. S. (1984). *Science* **224**, 1109–1111.
Fritsch, H. A. R., Van Noorden, S., and Pearse, A. G. E. (1976). *Cell Tissue Res.* **165**, 365–369.
Grimmelikhuijzen, C. J. P. (1983). *Neuroscience* **9**, 837–845.
Grimmelikhuijzen, C. J. P., Dierickx, K., and Baer, G. J. (1982). *Neuroscience* **7**, 3191–3199.
Horgen, P. A. (1981). *In* "Sexual Interactions in Eukaryotic Microbes" (D. H. O'Day and P. A. Horgen, eds.), pp. 155–178. Academic Press, New York.
Hunt, L. T., and Dayhoff, M. D. (1979). *In* "Peptides: Structure and Biological Functions" (E. Gross and J. Meienhofer, eds.), pp. 757–760. Pierce Chemical Co., Rockford, IL.
Jackson, I. M. D. (1981). *Endocrinology* **108**, 344.
Janakidevi, K., Dewey, V. C., and Kidder, G. W. (1966a). *J. Biol. Chem.* **241**, 2576–2578.
Janakidevi, K., Dewey, V. C., and Kidder, G. W. (1966b). *Arch. Biochem.* **113**, 758–759.
Josefsson, J. O., and Johansson, P. (1979). *Nature (London)* **282**, 78–80.
Kramer, K. J. (1985). *In* "Comprehensive Insect Physiology, Biochemistry and Pharmacology" (G. A. Kerkat and L. I. Gilbert, eds.), Vol 7, Ch. 10. Pergamon, Oxford.
LeRoith, D., Shiloach, J., Roth, J., and Lesniak, M. A. (1980). *Proc. Natl. Acad. Sci. U.S.A.* **77**, 6184–6188.
LeRoith, D., Lesniak, M. A., and Roth, J. (1981a). *Diabetes* **30**, 70–76.
LeRoith, D., Shiloach, J., Roth, J., and Lesniak, M. A. (1981b). *J. Biol. Chem.* **256**, 6533–6536.
LeRoith, D., Liotta, A. S., Roth, J., Shiloach, J., Lewis, M. E., Pert, C. B., and Krieger, D. T. (1982a). *Proc. Natl. Acad. Sci. U.S.A.* **79**, 2086–2090.
LeRoith, D., Shiloach, J., and Roth, J. (1982b). *Peptides* **3**, 211.
LeRoith, D., Pickens, W., Crosby, L. K., Berelowitz, M., and Shiloach, J. (1985a). *Biochim. Biophys. Acta* **838**, 335–342.
LeRoith, D., Pickens, W., Vinik, A. I., and Shiloach, J. (1985b). *BBRC* **127**, 713–719.
LeRoith, D., Shiloach, J., Heffron, R., Rubinovitz, C., Tanenbaum, R., and Roth, J. (1985c). *Can. J. Biochem. Cell Biol.* (in press).
LeRoith, D., Pickens, W., Wilson, G. L., Miller, B., Berelowitz, M., Vinik, A. I., Collier, E., and Cleland, C. F. (1985d). *Endocrinology* (in press).
Loose, D. S., and Feldman, D. (1982). *J. Biol. Chem.* **257**, 4925–4930.

Loose, D. S., Schurman, D. J., and Feldman, D. (1981). *Nature (London)* **293**, 477–479.
Loose, D. S., Stover, E. P., Restrepo, A., Stevens, D. A., and Feldman, D. (1983). *Proc. Natl. Acad. Sci. U.S.A.* **80**, 7659–7663.
Loumaye, E., Thorner, J., and Catt, K. J. (1982). *Science* **218**, 1324–1325.
Macchia, V., Bates, R. W., and Pastan, I. (1967). *J. Biol. Chem.* **242**, 3726–3730.
Maruo, T., Cohen, H., Segal, S. J., and Koide, S. S. (1979). *Proc. Natl. Acad. Sci. U.S.A.* **76**, 6622–6626.
Mori, M., Sakagami, Y., Narita, M., Isogai, A., Fujino, M., Kitada, C., Craig, R. A., Clewell, D. B., and Suzuki, A. (1985). *FEBS Lett.* **178**, 97–100.
Morley, J. E., Meyer, N., Pekary, A. E., Melmed, S., Carlson, H. E., Brigg, J. E., and Hershman, J. M. (1980). *BBRC* **96**, 47–50.
Nagasawa, H. H., Kataoka, A., Isogai, S., Tamura, A., Suzuki, H., Ishizake, A., Mizoguchi, A., Fujiwara, Y., and Suzuki, A. (1984). *Science* **226**, 1344–1345.
Petruzzelli, L., Herrera, R., Garcia, R., and Rosen, O. M. (1985). *Cancer Cells* **3**, 115–121.
Plisetskaya, E., Kazakov, V. K., Solititskaya, L., and Leibson, L. G. (1978). *Gen. Comp. Endocrinol.* **35**, 133–145.
Richert, N. D., and Ryan, R. J. (1977). *Proc. Natl. Acad. Sci. U.S.A.* **73**, 878–882.
Riehl, R. M., and Toft, D. O. (1984). *J. Biol. Chem.* **259**, 15324–15330.
Riehl, R. M., Toft, D. O., Meyer, M. D., Carlson, G. L., and McMorris, T. C. (1984). *Exp. Cell. Res.* **153**, 544–549.
Roth, J., LeRoith, D., Shiloach, J., Rosenzweig, J. L., Lesniak, M. A., and Havrankova, J. (1982). *N. Engl. J. Med.* **306**, 523–525.
Roth, J., LeRoith, D., Shiloach, J., and Rubinovitz, C. (1983). *Clin. Res.* **31**, 354–359.
Sandor, T., and Mehdi, A. Z. (1979). *In* "Hormones and Evolution" (E. J. W. Barrington, ed.), pp. 1–72. Academic Press, New York.
Sarkar, N., Langley, D., and Paulus, H. (1979). *Biochemistry* **18**, 4536–4541.
Schaller, H. C. (1979). *Trends Neurosci.* **2**, 120–122.
Schaller, H. C., and Bodenmuller, H. (1981). *Proc. Natl. Acad. Sci. U.S.A.* **78**, 7000-7004.
Schaller, H. C., Schmidt, T., and Grimmelikhuijzen, C. J. P. (1979). *Wilhelm Roux's Arch. Dev. Biol.* **186**, 139–149.
Schaller, H. C., Rau, T., and Bode, H. (1980). *Nature (London)* **283**, 589–591.
Schwabe, C., LeRoith, D., Thompson, R. P., Shiloach, J., and Roth, J. (1983). *J. Biol. Chem.* **258**, 2778–2782.
Sela, I. (1981). *Adv. Virus Res.* **26**, 201.
Stephens, K., Hegeman, G. D., and White, D. (1982). *J. Bacteriol.* **149**, 739–747.
Stotzler, D., and Duntze, W. (1976). *Eur. J. Biochem.* **65**, 257–262.
Tager, H. S., Markese, J., Kramer, K. J., Spiers, R. D., and Childs, C. N. (1976). *Biochem. J.* **156**, 515–520.
Thorpe, A., and Duve, H. (1984). *Mol. Physiol.* **5**, 235–260.
Van Noorden, S. (1984). *In* "Evolution and Tumor Pathology of the Neuroendocrine System" (S. Falkmer, R. Hakanson, and F. Sundler, eds.), pp. 7–28. Elsevier, Amsterdam.
Weiss, M., Ingbar, S. H., Winblad, S., and Kasper, D. L. (1983). *Science* **219**, 1331–1333.
Werner, H., Fridkin, M., Aviv, D., and Koch, Y. (1986). *Peptides* (in press).
Zioudrov, C., Streaty, R. A., and Klee, W. A. (1979). *J. Biol. Chem.* **254**, 2446.

DISCUSSION

C. S. Nicoll. Since Dr. LeRoith didn't mention my favorite hormone (i.e., prolactin) I should point out that a prolactin-like material has been detected by immunohistochemistry in the neural gland of a protochordate by M. Pestarino (*Gen. Comp. Endocrinol.* **54**, 444–449, 1984). Furthermore, the sparganum stage of a tape worm (*Spirometra mansonoides*) secretes a protein that is biologically similar to human growth hormone because it has both growth hormone and prolactin-like activities (see C. K. Phares and W. R. Ruegamer, *Proc. Soc. Exp. Biol. Med.* **143**, 147–151, 1973; W. R. Ruegamer and J. F. Mueller, *Proc. Soc. Expt. Biol. Med.* **143**, 133–137, 1973).

H. Kronenberg. That's certainly a beautiful summary of the way cells communicate with each other. I had a question about the last issue that you mentioned, evidence that microorganisms are synthesizing the peptides you've detected. In addition to defining the genes using expression libraries probed with your antibodies, a perhaps less ambitious study would be to use those antibodies either by radioactively labeling the *Tetrahymena* or *E. coli* and immunoprecipitating the appropriate proteins or doing the same thing with the messenger RNA translation products with the same antibodies. Have those studies been done with insulin, somatostatin, or ACTH?

D. LeRoith. In fact, we considered three ways of proving that unicellular organisms do produce these peptide hormones: (1) incorporation of labeled amino acids into the protein or (2) alternatively by isolating the messenger RNA or the gene using cDNA probes, and (3) sequencing the protein itself. We decided to use the second and third technique because they would allow us to then follow up on further biological studies once we have the protein sequence or the nucleotide sequence. As I mentioned we are looking for the nucleotide sequence at the moment, using the cDNAs that are available from vertebrate systems. As you know these are difficult studies because there may not be sufficient homology at such phylogenetic distances. In terms of labeled amino acid studies, we decided not to undertake such studies since the amount of protein in the organisms is very small. We would need enormous cultures, tremendous amount of purification, and a lot of luck, and so we decided to look for the nucleotide sequence and at the same time to purify the protein.

H. Kronenberg. I can imagine good reasons why using cDNA probes might not work. The degeneracy of the genetic code might make it hard to detect things that are really there. But you have antibodies that definitely detect specific proteins, so that I would think that using lambda gt 11 or another expression system would be, in fact, the more straightforward way to do the cloning. You've got an antibody that detects something. If you can't detect that something with an expression library, then it really calls into question what the antibody's signal means in your experiments.

D. LeRoith. Those studies have been tried by some good groups, though not extensively, and were unsuccessful, so we've decided to take a different route, which is to go for the genomic DNA, not the message, since the message can be very unstable in *Tetrahymena* and other organisms. We are presently looking at the Southern blots with the vertebrate cDNAs available to us.

M. A. Greer. I would like to explore the possibility that the reverse phenomenon of what you have beautifully presented might be true. In addition to very specialized forms of intercellular communication having developed at an early phase of evolution, there may be a remnant of a single form of communication that's still going on even in the higher vertebrates. When we initially started out as cells it was probably as just some kind of muck and some water surrounded by a membrane. Opportunities for elaborate intercellular communication were limited.

Serendipitously we have found that there seems to be a very simple intercellular stimulus for secretion which works even in mammalian cells, namely, water. If one adds as little as 2% water to certain endocrine cell systems there is an immediate stimulation of secretion which is indistinguishable from that induced by specific secretagogues. There is the same dose-response relationships, desensitization, and latency periods with water as with hypothalamic hormones when one examines pituitary cells, for instance.

Since as far as I can determine we're the only people in the world interested in this phenomenon, I'd like to know whether you have looked or anyone has examined the effect of osmotic changes in the simpler organisms and whether osmotic stimulation of secretion may be a universal phenomenon which has persisted through phylogeny and possibly has some physiologic significance in mammals and other higher vertebrates at the present time?

D. LeRoith. I'm afraid I'm going to be of very little help with that question; the answer is, we haven't done any studies. I'm not aware of the other studies that may have been done, but I would like to take this opportunity of mentioning that a lot of other studies have been performed, and I probably haven't given credit to all of them, and I do apologize for that. Obviously we're very biased in our studies and look only at peptide hormones. There may well be studies that are buried somewhere in the literature that nobody is aware of.

R. M. Blizzard. My questions relate to growth hormone and also to juvenile hormone, which you mentioned. I did not see growth hormone listed anywhere, and I'm certain there must be some data available. I would appreciate a comment, and also any information you have concerning juvenile hormone.

D. LeRoith. Regarding the growth hormone we just heard about some studies on growth hormone in the more primitive animals. We've actually looked for growth hormone and prolactin in very preliminary studies and failed to demonstrate their presence in microorganisms. We've looked mostly at small peptides. Regarding growth hormone, I don't think anything is known other than the studies Dr. Nicoll mentioned. Regarding the juvenile hormone, all I can tell you is that the juvenile hormone in fact is an acyclic terpenoid derivative. It does work to inhibit maturation of the insect. I don't think it's been found in higher animals than the insect, and it's an unusual terpenoid derivative.

W. D. Odell. I enjoyed your paper very much. Over the years, our findings during the study of cancer endocrine syndromes have forced us to reexamine Starling's classic concept of endocrinology. We have identified several protein hormones in normal nonendocrine tissues of rat and human: liver, kidney, skeletal muscle, gut, and lungs. These hormones have included proACTH, growth hormone, prolactin and hCG. In trying to study the control, if any exists, for these substances we have as yet been unable to modify these concentrations with what would be considered the classical control systems, for example, glucocorticoid suppression or elevation does not modify proACTH concentrations. You didn't give us any control data for the single cellular organism. Do you have any data or evidence to suggest that you can stimulate or suppress secretion or synthesis of these peptides?

D. LeRoith. That's an excellent question. First, in the unicellular organisms we have tried a number of different conditions, such as changing the glucose to galactose and other different substrates and starvation. We can't actively change the secretion but we have noticed that secretion alters somewhat with a change in nutritional state of the organism. So in answer to your question, no we have been unable to actually change secretion rate or the release of the hormone into the media.

W. D. Odell. When you add glucocorticoid to medium, for example, do the concentrations change inside the organisms?

D. LeRoith. We haven't tried that study. I should mention also that the culture conditions that we have used are generally between 15 to 100 liter fermentors, and we even had to use 1000 liter fermentors, so we haven't tried putting in anything but the essential amino acids and glucose or other substrates.

M. I. New. Thank you for a most stimulating presentation. I just want to add to your repertoire of the sexual relationship between plants and man. I was called as a consultant to Puerto Rico because there was an epidemic of precocious breast development in young girls and though the source of this epidemic has not yet been proven, one of the things that was discovered is that the wheat which is fed to the cattle is contaminated with fusarium, a yeast that produces estrogens, and this estrogen enters the beef which the children eat and it was suspected that the estrogen in the beef produced by the yeast caused the breast development.

D. LeRoith. Thank you for an interesting study. In the study of David Feldman, South American blastomycosis, which I showed as paracoccidiomycosis, seems to have a predominance in the male and not in the female. It was suggested at one stage that this is because the men worked outside in the fields and the women were in the house making food. The paracoccidioides have an estrogen binding receptor, and it has been suggested that perhaps the estrogen from the female binds to the receptor and prevents the paracoccidioides from undergoing a change to a more ineffective type and that is why males are predominantly affected by blastomycosis and not females.

G. Aurbach. Thank you for a really very interesting and thought provoking presentation. Microorganisms present the opportunity to examine mutants and, of course, an array of mutants is known in yeast. I wonder if there are any studies suggesting that a mutant form is incapable of producing some substance which then interferes with the life cycle or the reproduction of the yeast. Also, I presume we are now forced to consider still other possibilities in terms of transfer of the information from primitive organisms to sophisticated organisms. One does not necessarily have to think of a progression through evolution of development of hormones in unicellular organisms and then on to multicellular organisms but, in fact, the possibility of viral transfection of genes from one species to another. Perhaps genes can be transferred from multicellular complex organisms to unicellular ones.

D. LeRoith. To comment on your first question, initially what we decided to do was to look at *E. coli* for examples of mutants that didn't make insulin or other peptides and see what was different in their behavior. Unfortunately there are many mutants, and it was a rather expensive way to go about it. What we are planning to do instead is to find the cDNA or the gene that is producing the particular hormone and then we hope to make mutants. With respect to your second remark, it's a very good idea and, in fact, one of our first questions we had was whether unicellular organisms were getting the gene for ACTH or insulin from some other animal, say mammalian species. We have no proof against that, though we have looked for plasmids and we found that even when there is no plasmid in the bacteria we've been able to show that it's producing insulin, suggesting that the gene for the insulin or the somatostatin is not on a plasmid but is probably in the genome. It will be necessary to find the nucleotide sequence and protein sequence throughout evolution and show that there is an orderly progression, which would suggest that in fact the gene did originate in the unicellular organisms.

R. E. Peter. In your presentation of vertebrate peptides found in the various nonvertebrate animal, the list for mollusks seemed a bit longer than the list for some other groups, such as the insects. Have you considered whether or not there is divergence of the kind of peptides and hormones that are found in protostome versus the deuterostome animals?

D. LeRoith. I believe that in a few years many more investigators will complete that list. I think the reason the list for mollusks is longer than that for the insects is a bias of the particular study, i.e., an interest in the organism. I believe that all the hormones and many of the receptors may begin at the unicellular level and that if we look at every species we are going to find all those hormones.

R. E. Peter. On the basis of evolution, however, you should expect divergence between the deuterostome and the protostome lines. The advanced insects are not liable to be as similar to the line of vertebrate hormones and the mollusks.

D. LeRoith. I think you are talking in terms of macroevolution. In terms of peptides and steroid hormones, I believe they began early and are to be found in all species. Notice our findings (as well as others) of these hormones in plants.

D. K. Pomerantz. I am also struck by the presence of the receptor, or the binding sites at least, for estrogens in the unicellular organisms. For obvious reasons I am wondering whether that estrogen binding agent is in the cytoplasm or the nucleus of the eukaryotes.

M. R. Sherman. In response to the question of Dr. Pomerantz, I can comment on a few of the characteristics of the estrogen-binding protein of *Saccharomyces* that have been published by David Feldman. The binding protein is, indeed, cytosolic, but its physicochemical parameters are different from those of the mammalian estrogen receptors. For example, it sediments at approximately 4 S in buffers of both low and high ionic strength and has not been shown to form the 9–10 S complexes that are characteristic of the untransformed mammalian receptors. In addition, the steroid-binding specificities of the protein from *Saccharomyces* and the mammalian estrogen receptors are somewhat different.

T. Horton. I have another example of an interaction between plants, insects, and mammals. In 1961 Pat Berger, Norman Nesus, and their collaborators at the University of Utah, published a report of a compound from grasses that is a food resource of small mammal populations in the Great Basin, which stimulates reproduction in small mammal populations. It was subsequently found that this compound, 6-methoxybenzoxazolinone stimulates reproduction in mink and bobwhite quail. After further searching the literature they found that the same compound was discovered in the mid-1960s by people working with the European cornborer. The reason I mention this compound is that its structure is similar to melatonin. It is hypothesized to interact with the melatonin signal in regulating seasonal reproduction.

D. LeRoith. Thank you.

H. Kronenberg. One of the big differences between eukaryotic unicellular organisms and prokaryotic ones has to do with the ability of even very primitive eukaryotic organisms to, for example, cleave precursor proteins at dibasic residues and glycosylate proteins. When you look at the types of proteins or the peptides that you are finding in *Tetrahymena* as opposed to *E. coli*, is there any correlation with the need or the lack of the need for maturation–posttranslational maturation or cotranslational maturation—of proteins found in one versus the other.

D. LeRoith. We have not studied whether there are any differences related to the hormones we are finding in prokaryote vs eukaryote regarding, say, glycosylation. Thus we have no evidence for a difference in either the processing or the posttranslational modifications.

C. S. Nicoll. Dr. LeRoith's presentation has clearly substantiated an old axiom from the annals of comparative endocrinology which states that it is not the hormones that have evolved so much, rather, the uses to which they have been put have changed. I would like to consider a question that is related to Dr. LeRoith's topic, namely the origin and coevolution of hormones and their receptors. Two recent reports are particularly pertinent to that question. K. L. Bost, E. M. Smith, and J. E. Blalock (*Proc. Natl. Acad. Sci. U.S.A.*, **82**, 1372, 1985) found that a peptide encoded by an RNA that is complementary to the mRNA of ACTH can bind to ACTH with high affinity. They suggest that such complementary peptides might arise if the complementary untranslated DNA sequence was translocated to a new site in the genome in such a way that it could be expressed independently. The two complementary DNA sequences could then be expressed by different cell types or in the same cell type and would thus form the basis for the development of a hormone–receptor regulatory mechanism. One of the peptides might be inserted into the cell membrane, possibly in association with a functional component (e.g., a transporter or cyclase regulator). The complementary peptide could then regulate the cell functions in an autocrine, paracrine, or endocrine manner if it were secreted by the cells in which it was expressed.

David B. Kaback (*Nature* **316,** 490, 1985) has suggested another plausible mechanism to explain the emergence of hormone–receptor regulatory mechanisms. His suggestion is based on an earlier one made by C. N. Hales (*Nature* **314,** 20, 1985) that peptide effectors are produced by proteolytic cleavage of membrane proteins that are evolutionarily related. Kaback proposed that the hormone and its receptor were originally part of a single membrane protein that carried out some regulatory function. By mutation part of the gene that encoded for that protein was translocated but the two parts could still be expressed. Hence, the separate peptide chain could combine noncovalently to form a functional complex. By differential suppression of the genes that now encode for the membrane protein complex in different cell types the possibility of endocrine regulation would emerge if one of the components is secreted.

S. Cohen. My question concerns the time Dr. Marrian was scheduled to visit the University of Toronto. It was shortly after he had discovered, isolated, and determined the structure of one of the three classical estrogens, estriol. Tomy Jukeo, who was at the University of Toronto working for a degree at that time, thought he should give a seminar on estrogens the year before Dr. Marrian arrived so we would thus be well prepared. In this seminar he talked about the estrogen content of plants, and pointed out that the pussywillow has a great deal of estrogen. My question is I seem to recall reading of some bacteria that could use cholesterol for mobilizing sodium. Do you have anything on that?

D. LeRoith. There are studies showing that bacteria can both metabolize steroid hormones and convert them, and some other studies that even suggest that bacteria can synthesize many of the steroid hormones.

Physiological and Clinical Studies of GRF and GH

MICHAEL O. THORNER,* MARY LEE VANCE,* WILLIAM S. EVANS,*
ROBERT M. BLIZZARD,* ALAN D. ROGOL,* KEN HO,* DENIS A.
LEONG,* JOAO L. C. BORGES,* MICHAEL J. CRONIN,* ROBERT M.
MACLEOD,* KALMAN KOVACS,† SYLVIA ASA,† EVA HORVATH,†
LAWRENCE FROHMAN,‡ RICHARD FURLANETTO,§ GEORGEANNA JONES
KLINGENSMITH,** CHARLES BROOK,†† PATRICIA SMITH,†† SEYMOUR
REICHLIN,‡‡ JEAN RIVIER,§§ AND WYLIE VALE§§

*University of Virginia Medical School, Charlottesville, Virginia, † Department of Pathology, St. Michael's Hospital, Toronto, Ontario, Canada, ‡ University of Cincinnati Medical Center, Cincinnati, Ohio, § Children's Hospital of Philadelphia, Philadelphia, Pennsylvania, ** The Children's Hospital and The University of Colorado School of Medicine, Denver, Colorado, †† The Middlesex Hospital, London, England, ‡‡ Tufts New England Medical Center, Boston, Massachusetts, and §§ The Clayton Foundation Laboratories for Peptide Biology, The Salk Institute, San Diego, California

I. Historical Perspective of GRF

Almost 40 years after the pioneering work of Green and Harris (1949), who proposed the role of hypothalamic factors secreted into the hypothalamo-hypophyseal portal circulation in the regulation of anterior pituitary function, major progress has been made in the isolation and chemical characterization of these hypothalamic hormones. By 1980, all of the originally proposed hypothalamic hormones except corticotropin-releasing factor (CRF) and growth hormone-releasing factor (GRF),[1] had been identified through the remarkable efforts of Guillemin and Schally and their colleagues. Subsequently, hypothalamic CRF was isolated by Vale et al. (1981). GRF was the last of the classical factors to be structurally identified. Reichlin (1960, 1961) demonstrated that lesions of the ventromedial nucleus of the rat hypothalamus resulted in cessation of growth as a result of growth hormone (GH) deficiency; this was the first evidence of the existence of a GRF. Although many groups attempted to isolate GRF activity from hypothalamic extracts, there was always insufficient material (with one exception) to achieve a sequence. Moreover, when the

---
[1] The abbreviation, GRF, for growth hormone releasing factor has been used throughout this text. However, because of editorial policies of different journals, some figures are labeled as hpGRF-40, hGRF-40, and GHRH. All clinical studies carried out by our groups have been performed using hGRF-(1-40)-OH.

sequence data were available, the peptide did not meet the necessary biologic criteria for a GRF (reviewed by Guillemin et al., 1984).

A major problem in isolating GRF was the minute amount of peptide present in the hypothalamus, which, in addition, contained high concentrations of the GH release-inhibiting hormone, somatostatin. In 1968, Kruhlich et al. advanced the hypothesis that the hypothalamus contained somatostatin. This was confirmed when the 14 amino acid sequence of somatostatin was reported by Brazeau and colleagues (1973). Thus, to elucidate the structure of hypothalamic GRF it was essential either to prepare hypothalamic extracts without somatostatin or to find a rich source of GRF which was devoid of somatostatin. Burger and his colleagues (Beck et al., 1973) observed that extracts of various human tumors were able to stimulate the release of GH from rat pituitary fragments. In addition, cases of ectopic GRF secretion associated with acromegaly were described (Southern, 1960; Weiss and Ingram, 1961; Buse et al., 1961; Dabek, 1974; Sonksen et al., 1976; Ballard et al., 1964; Caplan et al., 1978; Shalet et al., 1979; Uz-Zafar et al., 1979; Leveston et al., 1981). Frohman and his colleagues reported partial purification of GRF from an extrapituitary tumor causing acromegaly (Frohman et al., 1980). These rare tumors were usually noted only in retrospect to contain GRF.

In October 1980 we were referred a 21-year-old woman with Turner's Syndrome and acromegaly (Thorner et al., 1982). Because of the unusual association of these two conditions, this patient was studied in detail. Our investigations suggested that she had a pituitary tumor causing her acromegaly, and she therefore underwent transsphenoidal surgery. However, this therapy did not result in cure. The pituitary tissue was interpreted by Dr. Kalman Kovacs in Toronto as characteristic of somatotroph hyperplasia rather than a pituitary adenoma. This conclusion was based upon (1) the interspersion of other pituitary cell types among somatotrophs; (2) the preservation or exaggeration of the reticulin fibers maintaining the normal acinar pattern of the anterior pituitary architecture (Fig. 1); and (3) electron microscopic evidence which suggested that the somatotrophs were extremely active (i.e., stimulated), evidenced by the prominent rough endoplasmic reticulum and Golgi apparatus, and the dense accumulation of cytoplasmic secretory granules. Thus, based on persistent acromegaly and the pituitary histology of hyperplasia, we suspected that this patient could have an ectopic tumor secreting GRF. Accordingly, an abdominal CT scan was performed and a 5-cm tumor in the tail of the pancreas was identified. This mass was successfully resected in August 1981. The serum GH levels fell from 70 to 2 ng/ml within 2 hours after the removal of the tumor (Fig. 2) and the patient has remained cured of her

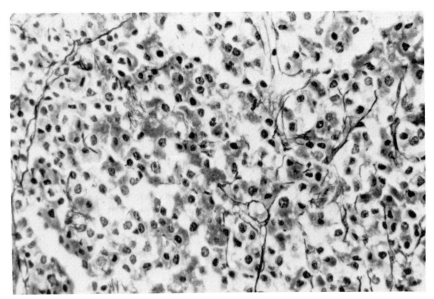

FIG. 1. Pituitary somatotroph hyperplasia in a patient with GRF-secreting pancreatic tumor. Large acini containing more cells than in the normal gland are surrounded by reticulin fibers. Gordon-Sweet silver technique ×250.

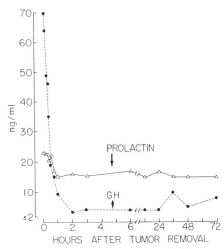

FIG. 2. Serum GH and prolactin levels during and after removal of GRF-secreting pancreatic tumor. Note the rapid fall of GH levels with no change of prolactin levels following removal of the tumor. Reprinted with kind permission from Thorner et al. (1982).

acromegaly. We carefully planned how to optimally use this tumor since it had the potential of being a concentrated source of GRF which might be free of somatostatin. After discussions with W. Vale and L. A. Frohman, we decided to divide the tumor at surgery. Half was frozen in liquid nitrogen and the other half was subdivided and sent to leading laboratories in the United States for tissue culture. The latter strategy was oriented toward achieving a cell line to produce large quantities of GRF. We did not use all of the tissue for extraction as it was judged, at that time, that the tumor probably would not contain sufficient material for successful isolation, characterization, and sequencing of GRF, assuming that it was a peptide.

By September 1981, Dr. Vale had extracted a portion of the tissue and demonstrated that it was rich in a biological peptide which released GH from primary cultured rat pituitary cells. This peptide had chromatographic characteristics similar to those of rat hypothalamic extract rich in GRF (Thorner *et al.*, 1982). Furthermore, both Dr. Cronin at the University of Virginia and Dr. F. Zeytin were able to culture the tumor tissue. At that time (September 1981) Dr. Zeytin moved from the laboratory of Dr. Armen Tashjian, Jr. in Boston to that of Dr. Roger Guillemin in San Diego. She asked permission to take the tissue (which we had sent to Dr. Tashjian) with her and to make it available to Dr. Guillemin. We agreed and, in addition, gave Dr. Guillemin over half of our stock of the frozen tissue. Furthermore, Dr. Vale, at the request of Dr. Guillemin, gave to Dr. Guillemin half (approximately 5 g) of his stored tissue. Dr. Guillemin was to proceed to isolate, characterize, and sequence the GRF. The first steps of determining the amino acid composition, but not the sequence, were achieved within a few weeks (October 24, 1981). However, before attempting to sequence the peptide, Dr. Guillemin wished to install and optimize his new microsequencer in view of the great scarcity and rarity of the material. In November 1981, Dr. Guillemin became aware of a probable GRF-secreting pancreatic tumor in a patient in France. In the spring of 1982, he obtained from Dr. Geneviève Sassolas of Lyon a large amount of tissue from another acromegalic patient with a pancreatic tumor and multiple metastases (Sassolas *et al.*, 1983). Dr. Guillemin proceeded to isolate GRF from the Lyon tissue. Simultaneously, Drs. Vale, Rivier, and Spiess worked with the remaining 5 g of the Charlottesville tumor.

Subsequently, four papers were published within the space of 3 weeks, detailing the isolation of GRF from two pancreatic tumors (Esch *et al.*, 1982; Guillemin *et al.*, 1982; Rivier *et al.*, 1982; Spiess *et al.*, 1982). From the Charlottesville tumor both Vale's group and Guilleman's group isolated a single 40 amino acid peptide GRF(1–40)-OH. From the Lyon tumor three GRFs, GRF(1–37)-OH, GRF(1–40)-OH, and GRF(1–44)-NH_2, were isolated by Guillemin's group. The 1–37 and 1–40 peptides were homologous with the first 37 and 40 amino acids of the 1–44 peptide. It was therefore proposed by Guillemin's group that the smaller forms represented degradation products of the GRF(1–44). However, it is to be noted that in the Charlottesville tumor, all of the GRF activity was in the form of 1–40 while in the Lyon tumor 64% was in the form of 1–40, 12% in the form of 1–37, and about 22% in the form of 1–44. Messenger RNA from both the Charlottesville and Lyon tumors was isolated by the groups at the Salk Institute (Charlottesville tumor) and Roche Institute (Lyon tumor) and cDNA probes were made (Gubler *et al.*, 1983; Mayo *et al.*, 1983). More recently, Mayo and colleagues sequenced the hGRF gene which spans 10 kilobases and consists of 5 exons (Mayo *et al.*, 1985) (Fig. 3). It is located on human chromosome 20. The third exon codes for hGRF 1–31. It is therefore of particular interest that the biological activity of hGRF resides in the first 27 amino acids from the amino terminus of the molecule. It is now clear that the human hypothalamus contains both GRF(1–40) and GRF(1–44) (Rivier *et al.*, 1983; Spiess *et al.*, 1983; Bohlen *et al.*, 1983a; Ling *et al.*, 1984a,b).

The structures of several hypothalamic GRFs from other species have now been defined using probes developed during the human GRF isolation. The first hypothalamic GRF sequence identified was that of rat GRF, which is a 43 amino acid peptide with a free carboxy-terminus (Spiess *et al.*, 1983). This peptide differed significantly from hGRF(1–44), with 15

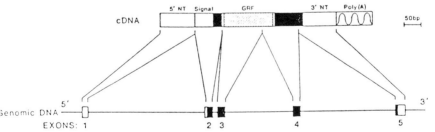

FIG. 3. Structure of the human GRF cDNA and gene. Top line shows a schematic of the GRF cDNA. The 5′ nontranslated region (5′ NT), the signal-peptide-encoding region, the GRF-encoding region, the 3′ nontranslated region (3′ NT), and the poly(A) tract are indicated. Shaded regions encode the NH_2-terminal and COOH-terminal flanking peptides. Bottom line shows a schematic of the GRF gene. Open boxes indicate noncoding exon regions; dark boxes indicate coding exon regions. Connecting lines indicate the relationship between structural regions of the cDNA and exons of the gene. Notice that the two scales are different. Reprinted with kind permission from Mayo *et al.* (1985).

amino acid substitutions or deletions. The sequence of porcine (Bohlen *et al.*, 1983b) and bovine (Esch *et al.*, 1983) hypothalamic GRF has also been described as a 44 amino acid structure, amidated at the carboxy-terminus.

II. Distribution of GRF in Man

A. GRF IN NORMAL TISSUE

GRF cell bodies have been identified in the infundibular and ventromedial hypothalamic nuclei of the human and other primates (Bloch *et al.*, 1983a,b, 1984; Bresson *et al.*, 1984) (Fig. 4). As expected, GRF staining was found in the processes and terminals of the external layer of the median eminence. These GRF axon boutons appeared to be in close apposition to the capillary loops that coalesce to form the portal vessels. In other species, cell bodies staining positively for GRF have been identified in the hippocampus, amygdala, and putamen, near the fornix, and within the medial forebrain bundle and zona incerta as well as the hypothalamic dorsomedial and periventricular nuclei (Jacobowitz *et al.*, 1983; Bugnon *et al.*, 1983; Merchenthaler *et al.*, 1984).

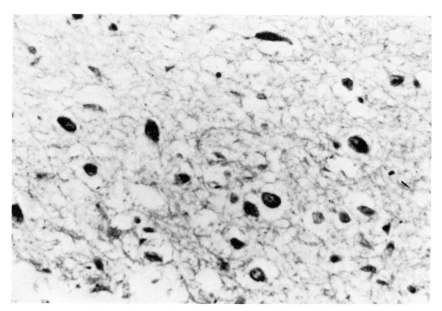

FIG. 4. Infundibular nucleus of the human hypothalamus is composed of neurons containing immunoreactive GRF (avidin–biotin–peroxidase complex technique with hematoxylin counterstain; original magnification ×80).

Using the technique of extraction followed by radioimmunoassay, Christofides and colleagues (1984) measured immunoreactive GRF (IR-GRF) concentrations in various areas of the human brain and gut. They found IR-GRF in the greatest concentration (pmol/g wet weight; ± SEM) in the hypothalamus (5.9 ± 4.1), but detectable levels were also observed in the septum (0.58 ± 0.45) and substantia innominata (0.58 ± 0.52). Further, IR-GRF was present in the upper intestine, particularly in the jejunum (2.2 ± 0.64), with lesser quantities in the duodenum (0.47 ± 0.09) and the stomach (0.25 ± 0.04). None was found in the ileum or colon. The IR-GRF was confined to the mucosa. IR-rat GRF has been detected in rat duodenum (Blum et al., 1985). In addition IR-GRF has been reported to be present in the human pancreas and placenta, when measured by radioimmunoassay, although not by immunocytochemistry (Shibasaki et al., 1984a; Tanaka et al., 1984). This IR-GRF has not, to date, been demonstrated to be authentic GRF(1–40) or (1–44). However, Sopwith and colleagues (1984, 1985) have demonstrated that plasma IR-GRF levels increased after food ingestion in normal subjects and hypopituitary (presumably GRF-deficient) subjects. The function of GRF in the periphery is unknown and the patients with ectopic GRF secretion appear to have no endocrine disturbance other than acromegaly. A tantalizing preliminary report demonstrates that in the isolated perifused canine pancreas, hGRF(1–40)-OH stimulated, in a dose-dependent fashion (5–30 nM), insulin, glucagon, and somatostatin secretion. The greater the concentration of glucose in the infusion medium, the greater the insulin and somatostatin and the smaller the glucagon responses. Of even greater interest was that hGRF(1–44)-NH$_2$ was ineffective at the highest (30 nM) concentration (Hermansen et al., 1985). Pandol and colleagues (1984) have reported that low concentrations of rat GRF stimulate pancreatic enzyme secretion and suggested that this effect may be mediated through interaction of GRF with vasoactive intestinal peptide (VIP) receptors rather than GRF receptors (Pandol et al., 1984; Waelbroeck et al., 1985; Bruhn et al., 1985). Using a clonal cell line derived from a human medullary carcinoma of the thyroid, Zeytin and Brazeau (1985) found that high concentrations of GRF stimulated calcitonin and neurotensin release. However, in normal men, we found no effect of hGRF-40 on serum calcitonin levels (M. O. Thorner and H. Heath III, unpublished observations).

B. GRF IN HUMAN TUMORS

As mentioned above, GRF was initially isolated from two pancreatic islet cell tumors and GRF immunoreactivity has been detected in extracts of human pancreas although it is not detectable by immunocytochemistry.

We have previously described the morphologic appearances of GRF-secreting tumors in patients with acromegaly secondary to ectopic GRF secretion (Thorner et al., 1982; Kovacs et al., 1984) (Fig. 5). These tumors contain argyrophilic cytoplasmic granules, as demonstrated by the Grimelius technique, and exhibit neuron-specific enolase immunoreactivity. With the immunoperoxidase technique the tumors were positive for GRF, and were negative for GH. On electron microscopy, epithelial cells contained well-developed rough endoplasmic reticulum and Golgi complexes and membrane bound secretory granules (Fig. 6). These tumor cells did not correspond to those of islet cells in nontumorous pancreas.

Occasionally hypothalamic gangliocytomas may be associated with acromegaly. We have described GRF immunocytochemical staining of such tumors and suggested that these tumors should be considered as an unusual cause of acromegaly (Asa et al., 1984) (Fig. 7). Rarely these tumors are intrasellar and GH adenoma cells are in close anatomic association with neurons which suggests that GRF not only stimulates GH secretion but may also cause adenoma formation (Fig. 8).

Although ectopic GRF secretion appears to be a rare cause of acromegaly, the precise prevalence of this entity is not known. For this reason, and in order to effect proper therapy, it is important to determine how often patients with acromegaly have detectable levels of plasma immuno-

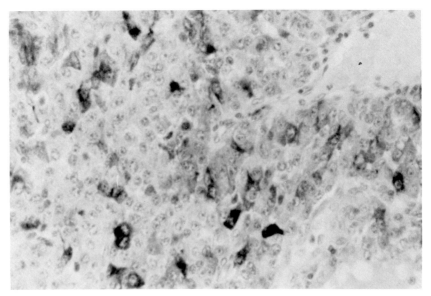

FIG. 5. Pancreatic endocrine tumor associated with GH excess and acromegaly. The cytoplasm of many tumor cells show positive immunostaining for GRF (avidin–biotin–peroxidase complex technique; hematoxylin counterstain ×250).

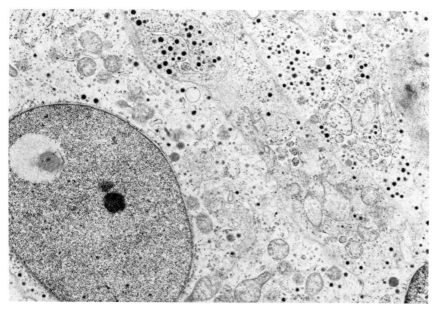

FIG. 6. Electron micrograph of a pancreatic endocrine tumor associated with GH excess and acromegaly. The tumor cells contain well developed cytoplasm, prominent rough endoplasmic reticulum, an active Golgi apparatus, and several secretory granules. Original magnification ×9600.

reactive GRF. In an attempt to determine the prevalence of ectopic GRF secretion in acromegaly, 177 specimens from unselected acromegalic patients were measured for GRF immunoreactivity (Thorner et al., 1984). GRF levels in all 177 samples were within the normal range (less than 100 pg/ml in normal men and women). However, in three patients with known ectopic GRF secretion (2 from Charlottesville and 1 patient under the care of Professor G. Michael Besser), all had nanogram per milliliter levels of plasma GRF. In a similar study Penny and co-workers (1984) found 4 out of 80 patients had elevated plasma GRF levels. However, in 3 of these patients no ectopic tumor could be identified.

Since less than 20 cases of ectopic GRF secretion and acromegaly have been described in the world literature, it is surprising how frequently immunopositive staining for GRF is found in neuroendocrine tumors. Asa and colleagues (1985) found positive immunostaining for GRF in 4 of 24 pancreatic endocrine tumors, 1 of 5 bronchial carcinoid tumors, 2 of 15 gut carcinoid tumors, 1 of 2 thymic carcinoid tumors, 2 of 20 medullary carcinomas of the thyroid, 1 of 12 pheochromocytomas, and 5 of 20 small cell carcinomas. Only 2 of 24 pancreatic endocrine tumors and the bronchial carcinoid tumor were associated with clinical acromegaly. Christo-

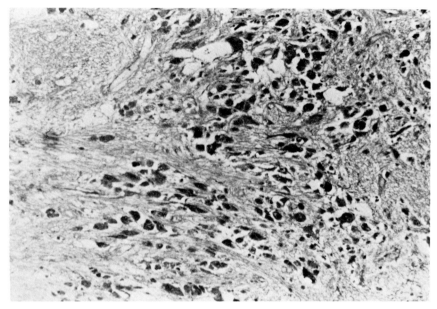

FIG. 7. Immunoreactive GRF is localized within the cell bodies and axonal processes of neurons in a hypothalamic gangliocytoma associated with acromegaly (avidin–biotin–peroxidase complex technique with hematoxylin counterstain; original magnification ×32).

fides et al. (1984) measured GRF by radioimmunoassay in extracts of tumors and found immunoreactive GRF in 11 of 20 pheochromocytomas, 1 of 10 ganglioneuroblastomas, 4 of 8 medullary carcinomas of the thyroid, 4 of 12 insulinomas, 3 of 8 glucagonomas, 4 of 10 gastrinomas, 3 of 15 vasoactive intestinal peptide secreting tumors, 5 of 8 small cell carcinomas, and 0 of 6 adenocarcinomas of the pancreas. In addition, we recently studied a patient with an ectopic ACTH-secreting thymic carcinoid tumor which was rich in GRF immunoreactivity but which also contained messenger RNA for both proopiomelanocortin (POMC) and hGRF (Liberman et al., 1985). This patient had no stigmata of acromegaly but had ectopic ACTH syndrome. The serum growth hormone level in a single random sample was undetectable.

It appears likely that GRF or a closely related peptide is frequently produced by neuroendocrine tumors, but that resultant clinical acromegaly is rare. This suggests that the peptide is either biologically inactive or is not secreted into the plasma in sufficeint quantity to stimulate excessive GH secretion. However, since the discovery of GRF is so recent and the incidence of ectopic GRF secretion has not yet been determined, it is imperative that a prospective study of patients with potential endocrine

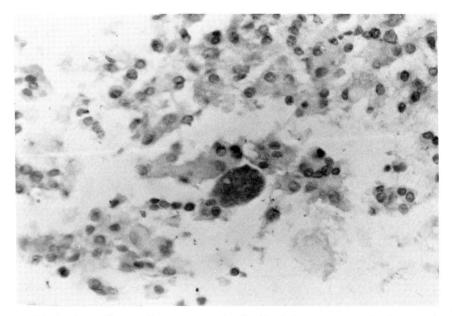

FIG. 8. Intrasellar gangliocytoma associated with pituitary somatotroph adenoma. A neuron contains immunoreactive GRF; none is detected in small round cells of the adenoma (avidin–biotin–peroxidase complex technique with hematoxylin counterstain; original magnification ×128).

tumors be carried out to determine the clinical significance of these pathologic findings.

III. GRF-Stimulated GH Secretion

A. STUDIES IN NORMAL MAN

Following isolation, characterization, and synthesis of hGRF-40 by Dr. Vale's group, we undertook clinical studies to determine the biologic effects and specificity of the peptide in normal human volunteers. In December 1982 a group of 6 normal young men were given hGRF-40, 1 µg/kg, intravenously and multiple hormone responses were compared to those after administration of vehicle for GRF. On the vehicle (control) day, two subjects had small spontaneous pulses of GH secretion. Following GRF administration, all subjects had an increase in serum GH levels. There was, however, considerable variability in the degree of responsiveness (Thorner et al., 1983) (Fig. 9). GRF specificity was demonstrated by measurement of serum prolactin, cortisol (a reflection of ACTH), LH,

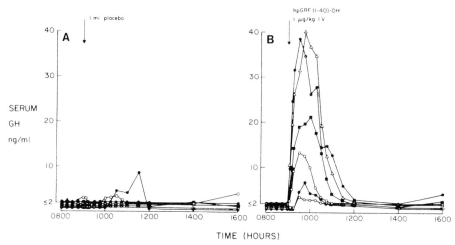

FIG. 9. Serum GH before and after placebo (A) and GRF-40 (B) in 6 normal men. Note the variability of response among the subjects. Reprinted with kind permission from Thorner et al. (1983).

and TSH; no change in these hormone levels occurred following GRF. Additionally, there was no effect on levels of blood glucose, insulin, pancreatic glucagon, pancreatic polypeptide, cholecystokinin, gastric inhibitory peptide, motilin, or somatostatin. Thus, GRF specifically stimulated GH secretion in normal men. A dose–response study to determine the minimum effective stimulatory intravenous dose of GRF was then carried out (Vance et al., 1984; Evans et al., 1985b). Normal men were given either vehicle or the following doses of GRF: 0.003, 0.01, 0.03, 0.1, 0.33, 1.0, 3.3, and 10 μg/kg as a single intravenous bolus injection. Doses of 0.1 to 10 μg/kg resulted in significant stimulation of GH secretion; again, variable responsiveness among individuals occurred. The lower doses of 0.1, 0.33, and 1.0 μg/kg produced a monophasic response while the higher doses of 3.3 and 10 μg/kg resulted in more prolonged elevation of GH which was characterized by a second increase 2 hours after the bolus was given (biphasic response). Additionally, serum somatomedin C levels increased 24 hours after GRF administration in 11 of 13 studies and indicated that the GH stimulated by GRF had a biologic effect.

B. STUDIES IN NORMAL WOMEN

That there is no major difference in sensitivity to GRF between men and women has been suggested by Gelato and colleagues (1984) who reported that half-maximal GH release was obtained with 0.4 and 0.2 µg/kg hGRF-40 in men and women, respectively. In addition, neither this study nor our studies detected differences in GH release in response to a supramaximal GRF dose during several phases of the menstrual cycles of normal women (Evans *et al.*, 1984). The apparent lack of differences in GRF-stimulated GH release between the sexes and during the menstrual cycle is of interest when consideration is given to studies which have reported that GH secretion in the human is clearly dependent upon gender and gonadal hormones. For example, Frantz and Rabkin (1965) found that basal GH levels in men and women were not different; however, serum GH levels during ambulation were significantly higher in premenopausal women than in men. That circulating estrogen might exert a major influence was suggested by the findings that ambulatory levels of GH were lower in postmenopausal than in premenopausal women and that diethylstilbestrol administration to men resulted in a marked increase in ambulatory GH levels. These studies were confirmed by Thompson and colleagues (1972) who found that GH secretory rates were higher in premenopausal women than in men and lower in postmenopausal women. Moreover, premenopausal women taking an estrogen-containing oral contraceptive exhibited strikingly higher rates of GH secretion. Studies attempting to examine the effect of the gonadal hormones on GH secretion during the menstrual cycle have led to similar conclusions. Investigations in which serum levels of GH have been measured in ambulatory women during the menstrual cycle have suggested that secretion of the hormone is maximal during the late follicular and periovulatory phases of the cycle (Frantz and Rabkin, 1965).

It must be recognized that our data and those of others which suggest no difference in GRF-stimulated GH secretion between the sexes and during the menstrual cycle must be interpreted with caution. A confounding factor in all studies performed in the human is the inability to eliminate endogenous factors which may affect the response to exogenously administered GRF, e.g., somatostatin. Thus, concerns have been voiced that apparently "negative" studies may, in fact, reflect inadequacies in experimental design, rather than physiology. *In vitro* techniques in which pituitary tissue or individual somatotrophs are studied in the absence of both hypothalamic and blood-borne factors are useful within this context.

C. EFFECTS OF GENDER AND GONADAL HORMONES ON GRF-STIMULATED GH RELEASE: *IN VITRO* STUDIES

We have used the techniques of both perifusion of dispersed anterior pituitary cells and the reverse hemolytic plaque assay to assess the effects of gender and the gonadal hormone environment on GRF-stimulated GH release by rat pituitary tissue. It is known that anterior pituitary glands of male rats contain more GH than do those from female rats (Birge et al., 1967), and that GH synthesis is greater in pituitaries from male compared with female rats (MacLeod et al., 1969; Burek and Frohman, 1970; Yamamoto et al., 1970; Tannenbaum and Martin, 1976; Steiner et al., 1982; McCormick et al., 1984). Moreover, in vivo studies in which circulating levels of GH have been measured and subjected to pulse analysis have demonstrated a male pattern consisting of high amplitude pulses with relatively low basal levels (Tannenbaum and Martin, 1976) and a female pattern showing irregular, low amplitude pulses superimposed on a relatively high baseline (Eden, 1979). In order to define the effect of gender and the gonadal hormone environment of GRF-stimulated GH release in vitro, we defined concentration–response relationships between hGRF-40 and GH release in dispersed perifused cells from intact male and female animals, castrate male animals, and castrate male animals treated with either testosterone or 17β-estradiol (Evans et al., 1985a). The results are depicted in Fig. 10. A concentration–response curve was documented in cells from each experimental group. Maximal GRF-stimulated GH release

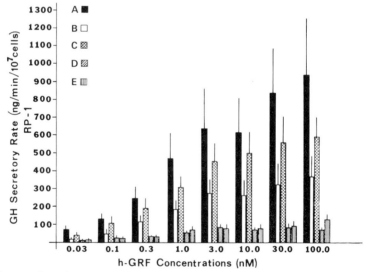

FIG. 10. Growth hormone (GH) secretion (ng/minute per 10^7 cells; above basal) by perifused rat anterior pituitary cells in response to 2.5-minute pulses of 8 concentrations of human GH-releasing factor (hGRF-40). Experimental groups include intact males (A), castrate males (B), castrate males treated with testosterone (C), castrate males treated with 17β-estradiol (D), and diestrus day 2 females (E). Reprinted with kind permission from Evans et al. (1985a).

occurred in cells from intact male rats. Castration resulted in a significant reduction in the GH response to GRF and this reduction was partially obviated by testosterone replacement. Compared with intact male rats, cells from intact female rats had a markedly diminished response to GRF. Of note was the observation that cells from castrate male rats treated with 17β-estradiol had GRF-stimulated GH release which was indistinguishable from that of intact female rats. From these data we concluded that in the rat, testosterone and 17β-estradiol are associated with enhanced and diminished hGRF-40-stimulated GH release, respectively.

The data from the pituitary cell perifusion studies do not permit determination of the basis of the remarkable sex-related difference in GRF-stimulated GH release since secretory responses in this system reflect secretion from a mixed cell population. These gender-associated differences could arise from differences in somatotroph cell population or differences in the secretory capacity of somatotrophs between the sexes. To address this question we used the technique of reverse hemolytic plaque assay which was introduced into endocrinology by Neill (Neill and Frawley, 1983). The basis of the assay is shown in Fig. 11. Briefly a mixture of dispersed pituitary cells and ovine red blood cells precoated with Protein A is plated into a shallow incubation chamber to which GH antiserum is added. Protein A allows the binding of GH antibody to the surface of the erythrocytes. This results in the formation of antigen (GH)–antibody complexes on the surface of erythrocytes surrounding GH-secreting pituitary cells. Addition of complement triggers lysis of erythrocytes which results in the formation of a lucent zone or plaque (see Fig. 12). Thus the somatotroph is identified by the appearance of a plaque and the secretory activity by the area of the plaque.

This technique allows the identification and quantitation of the proportion of somatotrophs among dispersed pituitary cells as well as the assessment of hormone secretion from individual somatotrophs. The latter was accomplished by measuring plaque area with the use of a computer-assisted optical system (Videoplan, Carl Zeiss, Kontron, Munich). The male pituitary contained a significantly greater percentage of somatotrophs than the female pituitary (53 ± 3 vs 30 ± 0.5%; $p < 0.005$; $N = 6$) (Leong et al., 1985). GRF concentration response studies using 0.01, 0.1, 1, 10, 100, and 1000 nM hGRF were carried out. Somatotrophs from the male had greater GH secretion than the female rat in the unstimulated state and at all concentrations of GRF tested. At maximal levels of GRF stimulation, GH secretion in the male was approximately twice that of the female (mean plaque area ± SEM μm^2: 29,983 ± 1944 vs 14,950 ± 2102; $p < 0.001$). The EC$_{50}$ was significantly lower in the male than in the female (0.25 ± 0.09 vs 1.78 ± 0.64 nM; $p < 0.001$), i.e., somatotrophs from the

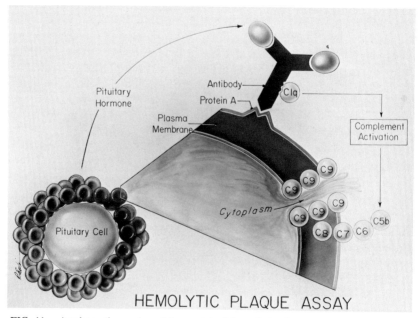

FIG. 11. A schematic version of the protein A hemolytic plaque assay. Dispersed pituitary cells are mixed with protein A-coated sheep erythrocytes and plated as a monolayer on the surface of a glass slide. The slide contains a "lawn" of immobilized sheep erythrocytes with occasional pituitary cells in it. A small chamber is constructed above the glass slide, allowing changes in the medium bathing the cells. The monolayer of cells is incubated with a specific GH antiserum; if GH is released from a pituitary cell, this hormone combines with the specific antibody that is, in turn, bound to the erythrocyte plasma membrane. Hormone-specific antibody is bound to all of the erythrocytes since protein A binds antibodies of the IgG class. Under these conditions, the pituitary cell is immobilized and GH secreted by the cell diffuses outward and coats surrounding sheep erythrocytes. After the sheep erythrocytes are coated with hormone–antibody complexes, they lyse once complement is added. A GH-secreting pituitary cell is indicated by a clear spot, or plaque, in an opaque layer of sheep erythrocytes. While GH antiserum is bound to all of the sheep erythrocytes, hormones that do not react with GH antibody do not form antigen–antibody complexes and cannot activate complement. Reprinted with kind permission from Leong et al. (1985).

male were approximately five times more sensitive to GRF than were those from the female rat. These observations suggest that at least three mechanisms are responsible for greater GH secretion in the male than in the female rat: greater proportion of somatotrophs, greater secretory capacity, and greater sensitivity to GRF in the male. However, we have not excluded the possibility that gender-related differences in responses to somatostatin may also be important and studies to investigate this possibility are currently in progress.

STUDIES OF GRF AND GH 605

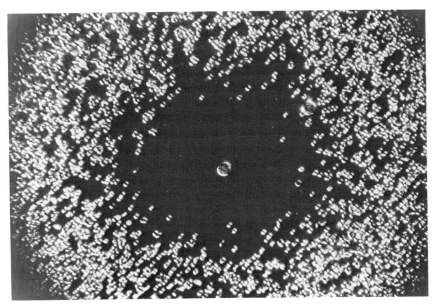

FIG. 12. High power magnification of an individual cultured GH cell as identified by the hemolytic plaque assay. This GH cell is surrounded by smaller hemolyzed sheep erythrocytes that have hemolysed; the lysed erythrocytes form a hemolytic plaque. Magnification, ×600. Reprinted with kind permission from Leong et al. (1985).

D. EFFECTS OF FASTING ON PULSATILE GH SECRETION

Since in both rat and man sex differences in GH secretion occur, it is curious that an ultradian rhythm is present in the rat but is reportedly absent in the human. In Fig. 13 a schematic representation of some factors regulating GH secretion is shown. In this scheme IGF-1 inhibits GH secretion, a process well documented in the rat. In man it is clear that, in conditions of protein–calorie malnutrition (e.g., anorexia nervosa, starvation), IGF-1 levels are low and may be associated with high GH levels. In addition it has recently been demonstrated that 8 days of fasting by normal men results in a profound reduction in IGF-1 levels (Clemmons et al., 1981). These results stimulated us to question whether we could uncover a more basic pattern of GH secretion during fasting. We present preliminary data on the effects of a prolonged fast on GH secretion in a healthy 37-year-old male volunteer of ideal body weight. In this study blood samples were taken every 20 minutes for 24 hours on a control day with meals, and on days 1 and 5 of a 5-day fast. GH samples were assayed in triplicate using the Hybritech immunoradiometric assay which was modified to give an enhanced sensitivity of 0.1 ng/ml. A modification of the

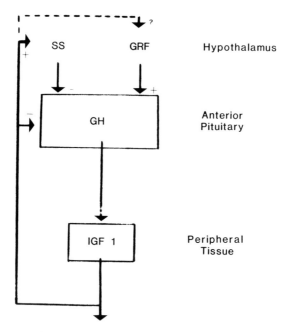

Growth Metabolic effect

FIG. 13. Schematic representation of the control of GH secretion. Growth hormone secretion is regulated by two hypothalamic hormones, somatostatin and GRF; these inhibit and stimulate growth hormone secretion, respectively. The release of GH results in the generation of IGF-1 (somatomedin C) from peripheral tissues which produces some of the metabolic effects of growth hormone. IGF-1 exerts a negative feedback directly at the pituitary and also hypothalamus by stimulating somatostatin secretion and possibly by inhibiting GRF secretion.

Santen and Bardin method for GH pulse identification was used (Veldhuis et al., 1985). As shown in Fig. 14, fasting resulted in an increase in GH pulse frequency and in integrated GH secretion.

The number of pulses increased from 2 during the control day to 5 during the first day of the fast, and 10 during the fifth day of the fast. These 10 pulses occurred throughout the 24-hour period and were not confined to the night time. Integrated GH increased from 0.33 on the control day to 3.75 and 5.59 µg/minute/ml on the first and fifth days of the fast, respectively. We recognize that many factors, such as low serum somatomedins, mild hypoglycemia, alterations in the concentrations of metabolic fuels, and stress, may contribute to enhanced GH secretion during fasting. However, it appears that fasting unmasks an inherent or intrinsic pattern of GH secretion. If this is correct, then a nutrient-de-

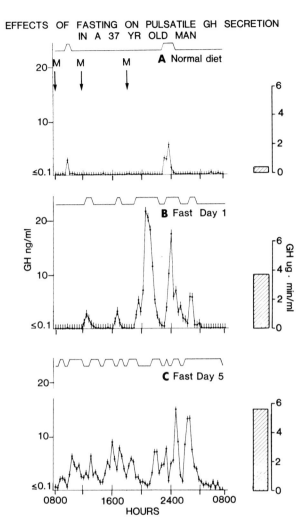

FIG. 14. Serum GH levels measured every 20 minutes for 24 hours in a normal 37-year-old male prior to (A), during day 1 (B) and day 5 (C) of a 5 day fast. Computer identification of GH pulses are shown above each profile. The shaded bars indicate integrated GH (µg minute/ml) for each of the 3 days.

prived state may be an important and necessary condition to conduct studies directed at understanding factors which control pulsatile GH release.

E. PHARMACOKINETICS OF hGRF-40

Most hypothalamic hormones are small peptides with a short half-life. It was therefore important to define the pharmacokinetics of hGRF(1–40)-OH in man. With L. A. Frohman and colleagues, we determined the metabolic clearance rate (MCR) and plasma disappearance rate ($t_{1/2}$) of GRF in normal men after both a single injection and a constant infusion of GRF. Immunoreactive GRF levels were measured by radioimmunoassay using an extracted plasma specimen with hGRF(1–40)-OH as radioiodinated tracer, an antibody raised against hGRF(1–20)-OH, and a second antibody separation method (Frohman et al., 1984).

Single iv injections of 1, 3.3, and 10 µg/kg of hGRF-40 were administered; plasma immunoreactive GRF (IR-GRF) levels were measured during the subsequent 180 minutes and biexponential curve analysis was performed. A constant infusion of hGRF-40 at increasing doses of 1, 3.3, 10, and 33 ng/kg/minute for 90 minutes each was administered and the MCR was calculated from measurement of steady-state plasma IR-GRF levels at each infusion dose. The postinfusion GRF disappearance rate was determined by linear regression analysis of plasma IR-GRF levels during the 120-minute period after cessation of the infusion. The $t_{1/2}$ is shown in Fig. 15.

The calculated MCR (1/m^2/day) during the single injection study was 194 ± 17.5 and was not significantly different from the calculated value during the continuous infusion study (202 ± 16). The plasma disappearance rate after the single GRF injection was subdivided into two linear phases: an initial equilibration phase (7.6 ± 1.2 minutes) and a subsequent elimination phase (51.8 ± 5.4 minutes). The latter was similar to the linear disappearance rate observed (41.3 ± 3.0 minutes) after cessation of the constant infusion. The chromatographic and biologic characteristics of plasma IR-GRF, 30 minutes after injection, were similar to those of synthetic hGRF-40.

F. EFFECTS OF CONTINUOUS GRF ADMINISTRATION IN MAN

It is known that continuous exposure of the gonadotroph to gonadotropin-releasing hormone (GnRH) results in diminished pituitary responsiveness to this releasing factor over time; the gonadotroph becomes refrac-

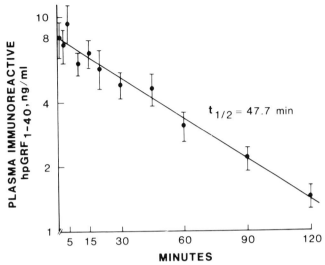

FIG. 15. Plasma IR-GRF levels after discontinuation of the constant infusion of hGRF(1–40). Shown are the mean ± SEM of results in 8 normal subjects. Reprinted with kind permission from Frohman et al. (1984).

tory to continued stimulation by GnRH. In order to determine if the somatotroph remains responsive to GRF stimulation or also becomes refractory to continued stimulation, continuous infusions of GRF were administered to normal men. In an initial study, performed in collaboration with L. A. Frohman, subjects received a 90 minute infusion of GRF in increasing doses of 1, 3, 10, and 33 ng/kg/minute. Additionally, a placebo infusion was given on a separate day; GH levels rose toward the end of that day (perhaps a result of prolonged fasting). A GRF infusion of 1 ng/kg/minute resulted in increased GH levels and higher doses had a greater stimulatory effort. However, the increase in GH was not sustained at the highest dose of 33 ng/kg/minute and suggested that partial desensitization of the somatotroph had occurred (Webb et al., 1984).

In order to further investigate this possibility, 6 hour infusions of GRF were given to normal men; doses of 1, 3, and 10 ng/kg/minute and placebo were administered on 4 separate occasions (Vance et al., 1985a) (Fig. 16). After 5.5 hours of infusion, a supramaximal dose of GRF (3.3 µg/kg) was given as an intravenous bolus. GH levels were measured every 15 minutes over a total of 10 hours. All three doses of continuously infused GRF stimulated GH secretion, however, the GH levels did not remain consistently elevated through the 6 hours; some subjects had intermittent pulsatile GH secretion during the infusions. The GH response to the

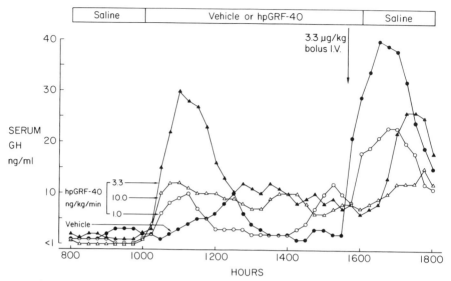

FIG. 16. Mean serum GH (ng/ml) before, during, and after infusion of vehicle or hGRF(1–40) (1.0, 3.3, and 10 ng/kg/minute) and after iv bolus injection of 3.3 μg/kg hGRF(1–40) in six normal men. Reprinted with kind permission from Vance et al. (1985a).

supramaximal bolus dose of GRF was inversely related to the continuous infusion dose, i.e., the greatest GH response to the bolus dose occurred on the day of placebo infusion (Fig. 17). However, the total amount of GH secreted during the infusion and after bolus injection was similar, approximately 1 mg, on all 4 study days. Since the pituitary gland contains between 10 and 15 mg of immunoreactive GH, only a small portion of the total GH was released. These findings were suggestive of two possibilities: either that there is only a limited pool of GH which is available for release, or that response attenuation, i.e., partial desensitization, likely occurs in the somatotroph. Because it appeared that there was a limited releasable amount of GH available, the 6 hour infusion study was repeated using placebo and one dose of GRF (10 ng/kg/minute) and either a supramaximal dose of GRF (3.3 μg/kg) or regular insulin (0.15 μg/kg) was administered after 5.5 hours of infusion. Insulin-induced hypoglycemia was used as an alternate stimulus of GH secretion because it acts indirectly to effect GH release. The GRF bolus following GRF infusion resulted in an attenuated response compared with that administered following saline infusion. However, the GH response to insulin-induced hypoglycemia during GRF infusion was significantly greater than on the other 3 study days, i.e., the growth hormone response to insulin-induced hypoglycemia was augmented during GRF infusion (Vance et al., 1986;

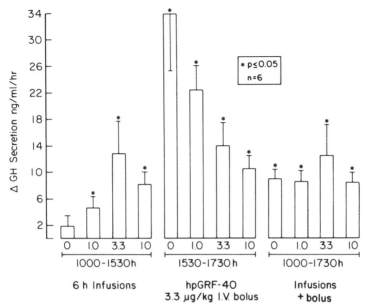

FIG. 17. Change (Δ) over baseline in GH secretion (ng/ml/hour) during continuous infusion of vehicle (○) or hGRF(1–40) (1.0, 3.3, and 10 ng/kg/minute), after an iv bolus of 3.3 μg/kg, and during the two study periods. The change in GH secretion is expressed as the mean ± SEM and compared by paired t test. Reprinted with kind permission from Vance *et al.* (1985a).

Fig. 18). These studies suggested that pituitary reserve was not decreased by prior exposure to GRF but that the somatotroph becomes partially refractory to GRF over time and that the hypoglycemia-induced GH response during GRF infusion likely occurred via other mechanisms, including possibly a reduction in hypothalamic somatostatin secretion.

A more prolonged infusion of GRF, 2 ng/kg/minute, for 24 hours in six normal men supported our previous findings of partial desensitization. Maximal GH stimulation by GRF occurred not at the beginning of the GRF infusion but during sleep when the naturally occurring GH pulses were augmented by GRF. Again, the response to a supramaximal dose of GRF at the end of the infusion was diminished by prior GRF administration and the total amount of GH released during placebo or GRF infusions followed by a GRF bolus was the same. Thus, it appears that continuous stimulation of the somatotroph by GRF results in continued secretion of GH in normal man despite the occurrence of partial desensitization.

That the somatotroph retains the ability to continuously secrete GH during protracted GRF stimulation is perhaps best illustrated by an "experiment of Nature." Patients with ectopic GRF secretion and acromeg-

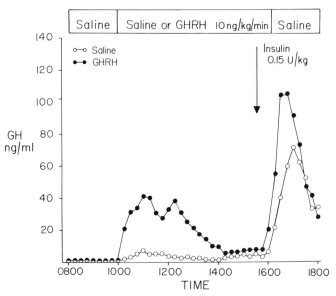

FIG. 18. Mean serum GH in 6 normal men during infusion of saline or hGRF-40 and after administration of regular insulin. Note the augmented GH response to insulin during hGRF-40 infusion.

aly have persistent pulsatile GH secretion despite sustained elevation of plasma immunoreactive GRF levels. We have evaluated an acromegalic patient who has acromegaly secondary to ectopic GRF secretion from a metastatic carcinoid tumor. This patient had pulsatile secretion of both GRF and GH throughout a 24-hour period (Fig. 19). GRF levels were in the 5–30 ng/ml range which is an order of magnitude above minimum

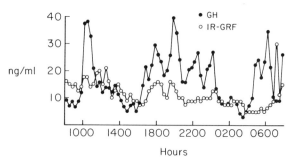

FIG. 19. Serum GH (ng/ml) and plasma immunoreactive GRF (ng/ml) levels in a patient with acromegaly and ectopic GRF secretion. ●, GH; ○, IR-GRF. Reprinted with kind permission from Vance et al. (1985b).

levels of GRF necessary to stimulate GH secretion in normal subjects. Serum GH levels ranged from 10 to 40 ng/ml and at no time did either GH or GRF levels become undetectable (Vance et al., 1985b).

From these studies, it is evident that the regulation of GH secretion is a complex process which has yet to be completely defined. Since the inhibitory peptide, somatostatin, plays an equally important role in regulating GH secretion, it is reasonable to conclude that GH secretion is likely effected by concomitant secretion of GRF and withdrawal of somatostatin secretion which results is pulsatile GH release and the ability of the somatotroph to continuously secrete GH.

G. STUDIES USING THE CONTINUOUS PERIFUSION SYSTEM OF DISPERSED RAT ANTERIOR PITUITARY CELLS

The results of *in vitro* studies which address the question of somatotroph desensitization are conflicting. While Bilezekian and Vale (1984) and Ceda and Hoffman (1985) demonstrated that preincubation of rat pituitary cells in primary culture with GRF for 24 hours resulted in significant depletion of GH and partial desensitization to subsequent GRF stimulation, Dieguez *et al.* (1984) using a similar *in vitro* system, did not. These latter investigators, however, found that LH responses to GnRH in the same pituitary cells were markedly desensitized after GnRH pretreatment. Because of these discrepant observations and the problems of using a static incubation system to study this question, we have used the pituitary cell perifusion system. It is known from our clinical studies, from our earlier work using the perifusion system (Borges *et al.*, 1983), and from the *in vivo* studies in animals after passive immunization with somatostatin antibodies that during continuous GRF administration the initial burst of GH release is not sustained and GH secretion becomes markedly attenuated over time (Wehrenberg *et al.*, 1982). We therefore set out to determine whether or not the GH response to a known maximal stimulatory concentration of GRF was reduced compared to that in control cell column. Additionally, we measured the GH response to a GH secretagogue, GRP, which induces GH release without interacting with the GRF receptor. In these experiments, pituitaries from young adult male Sprague–Dawley rats were enzymatically dispersed and mixed thoroughly with Bio-Gel matrix before placement in the perifusion column (Borges *et al.*, 1983). Typically, the dispersion was performed in the afternoon before the experiment and the cells were perifused overnight with medium. Four or five male pituitaries were used per column and culture medium (M199, GIBCO, NY) containing 0.1% BSA, 1 nM triiodotyronine, and 5 nM dexamethasone was used as the perifusate, and fractions were col-

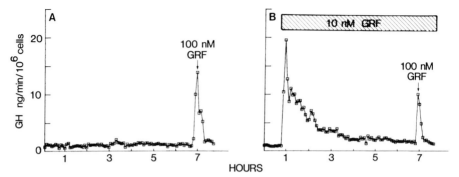

FIG. 20. Results from a typical experiment comparing the effects of a 5 minute pulse of 100 nM hGRF-40 in rat pituitary cells perifused with medium alone (A) or with medium containing 10 nM hGRF-40 for 6 hours (B). GH secretion is expressed as ng/minute/10^6 cells.

lected at 5 minute intervals. The columns were perifused with media alone or media containing 10 nM GRF for 6 hours prior to the administration of a 5 minute pulse of 100 nM GRF or 100 nM GRP. GRP was used as an alternate stimulus of GH release. It is a synthetic hexapeptide structurally unrelated to GRF but which induces GH release via an unknown receptor (Sartor et al., 1985). The results from a typical experiment demonstrating the effects of this dose of GRF in a control and GRF-treated column are shown in Fig. 20.

The GH response, induced by the supramaximal pulse of GRF during GRF infusion, was not significantly different from that of the control (18.2 ± 4.1 vs 18.7 ± 6 ng/10^6 cells, $n = 5$, Fig. 21). Thus, 6 hours of pretreat-

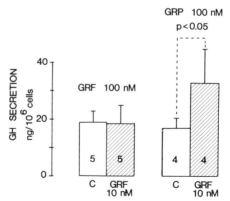

FIG. 21. Comparison of the mean ± SEM GH response following a 5 minute pulse of 100 nM GRF (left) or 100 nM GRP (right) in cells perifused with medium alone or medium containing 10 nM GRF for 6 hours (ng/10^6 cells). $p < 0.05$ by Wilcoxon rank testing.

ment with 10 n*M* GRF did not result in GH response attenuation to a subsequent supramaximal GRF stimulus. In parallel studies with GnRH, the LH response was markedly attenuated in the GnRH pretreated column (data not shown).

In contrast, GRP administration induced significantly greater GH release in the GRF perifused cells than in the control cells (Fig. 22). These data indicate that the failure of pituitary cells to constantly secrete GH during continuous GRF administration is not a result of depletion of a releasable pool of GH. Additionally, the releasable pool of GH is augmented by prior GRF treatment as uncovered by GRP administration which is similar to our *in vivo* findings using insulin-induced hypoglycemia as an alternative stimulus. Although prior treatment with GRF did not alter the GH response to a subsequent maximal stimulatory dose of GRF, some degree of desensitization likely occurred as this dose of GRF caused release of significantly less GH than did GRP under similar conditions. These data suggest that in cells continuously treated with GRF, an increase in the releasable pool of GH may compensate for a reduction in the secretory response due to partial desensitization such that the overall changes in GH secretion are not significant. We recognize that the situation *in vivo* is more complex since the feedback effects of somatomedins, or alterations in endogenous somatostatin secretion, may significantly alter somatotroph function. Nevertheless, these observations provide an explanation for the occurrence of acromegaly or maintenance of GH release in the face of chronic GRF secretion.

To further investigate this phenomenon *in vivo*, we have administered chronic intravenous 14 day infusions of GRF-40 (10 ng/kg/minute) to 4 normal men. The preliminary results in one subject are shown in Fig. 22. Each of the four subjects studied demonstrated enhanced GH secretion and augmentation of the nocturnal GH pulses. This was associated with a doubling of serum IGF-1 levels which then fell to pretreatment values 14 days after discontinuing the infusion. Thus, in normal subjects chronic continuous GRF administration can augment GH secretion. These data are supportive of the theory that a delayed release preparation of GRF may be efficacious in the treatment of GH-deficient subjects.

IV. GRF as a Diagnostic Agent

A. COMBINED ANTERIOR PITUITARY HORMONE TEST

The possibility that GRF may be administered concurrently with other hypothalamic hormones for the purpose of anterior pituitary testing has been examined by Sheldon and colleagues (1985). In the study hGRF-40

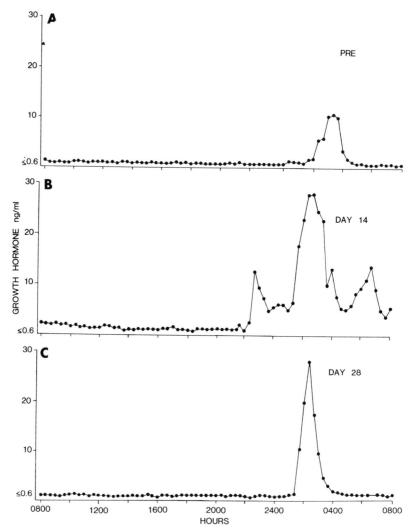

FIG. 22. Serum GH levels measured every 20 minutes in a normal young adult male prior to (A), on day 14 of a 14 day intravenous infusion of hGRF-40 (10 ng/kg/minute) (B), and 14 days after stopping the GRF infusion (C). Note basal growth hormone levels are higher during the 14th day of the infusion and the greater number and amplitude of pulses during the nighttime hours. Following withdrawal of the GRF the pattern of growth hormone secretion returned to that seen prior to the infusion.

(1 μg/kg) was combined with corticotropin-releasing factor (CRF; 1 μg/kg), gonadotropin-releasing hormone (GnRH; 100 μg), and thyrotropin-releasing hormone (TRH; 200 μg) and given as an intravenous bolus to normal young men. The administration of this combination of hormones resulted neither in synergism nor inhibition of responses when compared to the results of each hormone given individually. The results of these initial studies and those performed in normal young women during the early follicular phase of the menstrual cycle are detailed in Table I. No significant differences between men and women were noted with the exception of a slightly higher luteinizing hormone response in men and a slightly higher prolactin response in women. Although these results are encouraging, the clinical role of this particular test in assessing patients with hypothalamic and/or pituitary disease remains to be elucidated.

B. ACROMEGALY

In Table II are listed results of several studies in which patients with acromegaly have been given GRF. Gelato and colleagues (1985) found that of 29 patients with acromegaly given hGRF-40 (1 μg/kg iv), all had an increase in GH except 4 patients who had previously received external pituitary radiation. However, there was significant variability of responsiveness among patients, an observation also made by Shibasaki and colleagues (1984b). However, Gelato and colleagues did not find, as did Wood and colleagues (1983), a relationship between suppression of GH

TABLE I

Maximum Increments in Plasma or Serum IR Hormone Levels in 14 Men and 12 Women after Receiving the Combination of All Four Hypothalamic-Releasing Hormones[a,b]

	Men	Women
IR-ACTH (pg/ml)	40 ± 7.6	40 ± 6.6
IR-cortisol (μg/dl)	9.5 ± 1.0	9.8 ± 1.2
IR-FSH (mIU/ml)	2.7 ± 0.6	3.0 ± 0.2
IR-LH (mIU/ml)	21 ± 2.6	13.9 ± 1.9[c]
IR-TSH (uU/ml)	12 ± 1.2	15 ± 2.1
IR-PRL (ng/ml)	46 ± 5.0	92 ± 15[d]
IR-GH (ng/ml)	23 ± 7.1	55 ± 20

[a] Reprinted with kind permission from Sheldon *et al.* (1985).
[b] Values are the mean ± SEM.
[c] $p < 0.025$.
[d] $p < 0.005$.

TABLE II

GH Responses in Acromegalic Patients to Bolus iv Injections of hGRF(s)

Reference	hGRF	Dose	Number		Response
Wood et al. (1983)	hGRF-44	100 μg	6	2 untreated 4 treated (99 Y or external) irradiation	3 normal increment 3 "exaggerated" responses
Shibasaki et al. (1984b)	hGRF-44	100 μg	10	5 untreated 5 treated by surgery but not cured	3 exaggerated (4-fold increase) 5 "normal" (2-fold increase) 2 with < 2-fold increase
Gelato et al. (1985)	hGRF-40	1.3 μg/kg	29	8 untreated 21 previously treated	25/29 increase in GH 4 no response (all previously irradiated)
Chiodini et al. (1985)	hGRF-44	100 mg	35		22/35 > 100% increase from basal

levels (by greater than 20%) with oral glucose and the GH response to hGRF-44. Also of interest were the findings of Chiodini and colleagues (1985) who noted that acromegalic patients who had a decrease in GH after bromocriptine had a poor GH response to GRF.

It remains to be determined whether or not GRF will be useful in defining the pathophysiologic mechanism responsible for acromegaly. Three preliminary reports have suggested that GRF does not stimulate GH release in patients with the ectopic GRF syndrome (Losa et al., 1984; Lytras et al., 1984; Schulte et al., 1985). However, the fact that GRF does stimulate GH secretion by pituitary tumors has been documented both *in vivo* (Thorner et al., 1982; Gelato et al., 1985; Vance et al., 1985b) and *in vitro* (Webb et al., 1983; Daniels et al., 1984; Lamberts et al., 1984; Spada et al., 1984; Adams et al., 1983).

C. CHILDREN WITH SHORT STATURE

Although short stature, height-for-age below the third percentile, is by definition common, GH deficiency is rarely the cause. The common non-GH-deficient etiologies of short stature include genetic predisposition (i.e., short parents) and constitutional delay of growth. Many chronic systemic illnesses, chromosomal abnormalities, chondrodystrophies, and intrauterine growth retardation may also result in short stature. Additionally, endocrine diseases which result in poor growth include hypothyroidism, Cushing's Syndrome, pseudohypoparathyroidism, and GH deficiency.

The basis of GH deficiency may be organic hypopituitarism in which a tumor or infiltrative process of the hypothalamus or pituitary is present or follows destructive (surgical or radiologic) therapy in this area. It may more frequently be idiopathic (sometimes familial) in which no such lesion is identified. We evaluated a series of 23 short children for GH deficiency by standard pharmacological tests and by the infusion of a single intravenous dose of GRF. Subjects were tested on 2 successive days. On the first day a combined test was performed with administration of L-arginine (0.5 g/kg for 30 minute) intravenously and L-DOPA (9 mg/kg) orally. On the following day GRF, 3.3 μg/kg, was injected as a single intravenous bolus. The responses to the pharmacological tests were by definition normal (peak GH level > 7 ng/ml) in those classified as having intrauterine growth retardation or constitutional delay of growth and adolescence and/or familial short stature. As shown in Fig. 23A and B all GH-sufficient children responded to the single dose of GRF with increased GH levels (A). Nineteen of 23 (83%) achieved levels of 10 ng/ml or more. These percentages have been confirmed in our now larger series of 71 children (unpublished data). These results were similar to those of Schriok and colleagues (1984) and Takano and co-workers (1984). As in normal adults, there was wide variability in the GH responses to GRF administration. There was no particular pattern of response that distinguishes children with intrauterine growth retardation from those with constitutionally delayed growth or familial short stature.

Ten children with idiopathic GH deficiency and 7 with organic hypopituitarism were also given a single intravenous dose of GRF. The mean peak GH response to GRF is shown in Fig. 24A. In only the children with organic hypopituitarism (B) were the responses to GRF significantly greater than after arginine/L-DOPA. The peak GH values after GRF were significantly less for the GH-deficient children than for the control group of GH-sufficient short children (see above). These studies indicate that a majority of hypopituitary subjects have somatotrophs which are responsive to GRF; thus, a hypothalamic site of dysfunction is strongly implicated. The GH responses of hypopituitary children were less than those of normal volunteers or for GH-sufficient short children and the GH responses of the children with organic hypopituitarism were less than those with idiopathic GH deficiency. Whether these poorer responses are related to their primary disease process or to the destructive therapy aimed at their tumors is not clear.

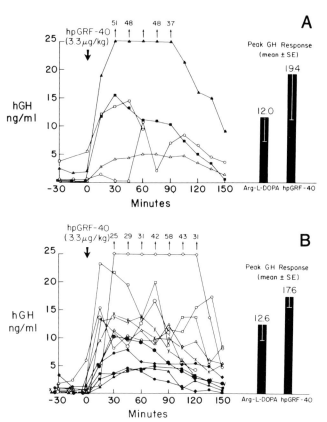

FIG. 23. GH release in response to hGRF-40 in children with short stature. Each symbol represents an individual patient. (A) IUGR; (B) CD and/or familial short stature. Bars at the right of each panel are the mean ± SEM of the peak GH responses to the arginine/L-dopa and hGRF-40. Reprinted with kind permission from Rogol *et al.* (1984).

V. GRF as a Therapeutic Agent

When synthetic GRF became available for clinical research, we had the opportunity to determine if it could stimulate GH secretion in GH-deficient subjects. Initially, 12 adults with GH deficiency in childhood (Borges *et al.*, 1983) were studied. Seven had isolated growth hormone deficiency, 4 multiple anterior pituitary deficiencies, and 1 had Hand-Schuller-Christian disease. Eight subjects had no growth hormone response to a single intravenous dose of hGRF-40 (10 µg/kg) while 4 had small responses (Fig. 25). Despite this, 7 of 10 had an increase in serum IGF-1. To explore this further we administered to six similar subjects

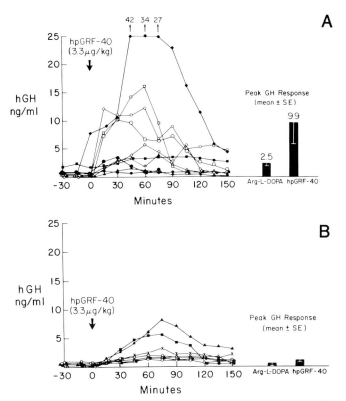

FIG. 24. GH release in response to hGRF-40 in children with short stature. Each symbol represents an individual patient. (A) IGHD; (B) organic hypopituitarism. Bars at the right of each panel are the mean ± SEM of the peak GH responses to the arginine/L-dopa and hGRF-40. Reprinted with kind permission from Rogol *et al.* (1984).

hGRF-40, 0.3 μg/kg, intravenously every 3 hours for 5 days (Borges *et al.*, 1984). There was a progressive rise in IGF-1 levels from a mean of 0.24 ± 0.07 to 0.78 ± 0.32 U/ml (Fig. 26). Prior to GRF treatment only 2 of the 6 subjects had an increase in serum GH levels, while GH levels rose in 5 of 6 subjects after the 5 days of GRF administration. These data indicate that the somatotroph can be primed by GRF over 5 days. Thus, these preliminary results suggested that GRF had potential as a therapeutic agent for treatment of GH deficiency.

A. ROUTES OF ADMINISTRATION

Since GRF may be useful in the treatment of some children with GH deficiency, questions necessarily arise with regard to the potential routes

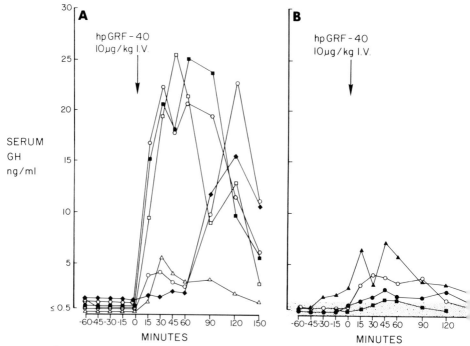

FIG. 25. Serum GH concentrations in 6 normal men (A) and 12 adult patients with GH deficiency (B) in response to intravenous hhGRF-40. Shaded area in B represents GH levels in 8 patients who exhibited no response to hGRF-40. Reprinted with kind permission from Borges et al. (1983).

of peptide administration. In order to address this issue, we studied GH responses in normal men after intravenous, subcutaneous, and intranasal hGRF-40 administration (Fig. 27) (Evans et al., 1983, 1985b). As chronic therapy, intravenous administration of GRF is not practical. Moreover, if GRF is given several times per day, then the intranasal route is more acceptable than is subcutaneous injection. Growth hormone is stimulated when GRF is given by the intranasal route. However, a 300-fold higher dose is required to effect stimulation of GH release comparable to that after intravenous administration. Moreover, 30-fold more GRF is required when given subcutaneously to stimulate a comparable amount of GH as compared with intravenous administration. The most plausible explanation for these divergent effects on GH release is depicted in Fig. 28. Following intravenous administration of GRF, plasma immunoreactive GRF levels are 60- and 500-fold higher than those after administration of the same subcutaneous or intranasal dose. Although these data are consistent with the hypothesis that the peptide is incompletely absorbed

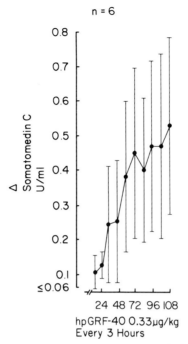

FIG. 26. Mean (± SEM) increments in serum somatomedin C in 6 adults with idiopathic GH deficiency during 5 days of intravenous hGRF-40 administration every 3 hours. Reprinted with kind permission from Borges et al. (1984).

when administered subcutaneously and intranasally, it is impossible to exclude the possibility that the bioactivity of the peptide is reduced as the result of degradation in subcutaneous tissues and/or nasal mucosa. Therefore, although it is encouraging that intranasal administration of GRF is a viable option, it appears that the quantity of GRF required is, at this time, economically unfeasible. However, if less expensive forms of GRF are developed or development of a formulation which allows for improved absorption occurs, then intranasal administration will likely be the preferred mode of delivery.

B. GRF-ASSOCIATED LINEAR GROWTH

Many GH-deficient children have a hypothalamic neurosecretory disturbance rather than somatotroph defect (Spiliotis et al., 1984). As demonstrated above, and by many others, GRF stimulates GH secretion in the majority. We reasoned that GRF would most likely be effective in stimulating GH secretion and possibly linear growth if it were adminis-

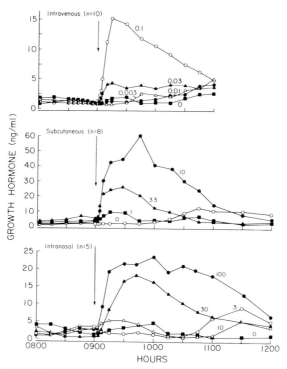

FIG. 27. Effect of intravenous, subcutaneous, or intranasal administration of hGRF-40 on mean serum GH levels in normal men. Note that the y axes have different scales. Reprinted with kind permission from Evans et al. (1985b).

tered to mimic the physiologic pattern of GH secretion. In the first study GRF was administered for 6 months using a Pulsamat (Ferring) pump which delivered 1 or 3 μg/kg every 3 hours. The children were initially treated for 2 months with 1 μg/kg hGRF-40 and then for an additional 4 months with 3 μg/kg. Each of the two children grew at an increased rate (cm/year) during therapy although the second child had a more dramatic response, i.e., from 2.2 to 13 cm/year during the 6 months of GRF treatment. The first child had an increase in growth rate to 7.4 cm/year from a pretreatment rate of 4.6 cm/year. This child grew maximally during the first 3 months of treatment and then had a deceleration (Fig. 29). The reasons for this decline in growth rate are unclear, but may be related to the development of antibodies to GRF (low titer which disappeared after cessation of GRF therapy). An alternative explanation is that there were a large number of previously unrecognized psychosocial problems during therapy which may have played a role in his growth response.

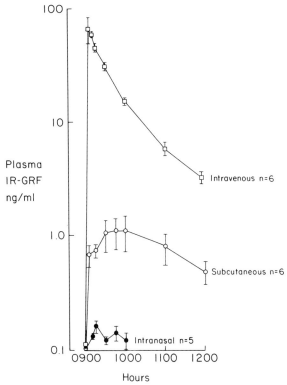

FIG. 28. Mean (± SEM) plasma IR-GRF levels following 10 μg hGRF-40 administered intravenously, subcutaneously, or intranasally. Note the logarithmic scale. Reprinted with kind permission from Evans et al. (1985b).

In collaboration with Dr. Georgeanna Klingensmith at Denver Children's Hospital and Drs. Seymour Reichlin and Boris Senior at Tufts-New England Medical Center, we have treated an additional 5 children; three received twice daily subcutaneous injections of GRF (4 μg/kg). The results (including the first two children) are shown in Fig. 30. Although all children grew more rapidly with GRF treatment than previously, there was great variability in the growth rates.

In collaboration with Drs. Charles Brook and Patricia Smith at the Middlesex Hospital in London, 5 children with idiopathic growth hormone deficiency were treated with GRF administered subcutaneously at night for 4 pulses at 3-hour intervals using a Zyklomat pump (Fig. 31). They were initially treated with 1 μg/kg/pulse for 3 months and then with 2 μg/kg/pulse for the second 3 months. Two of the children had a growth rate of greater than 4.5 cm/year before starting therapy and some sponta-

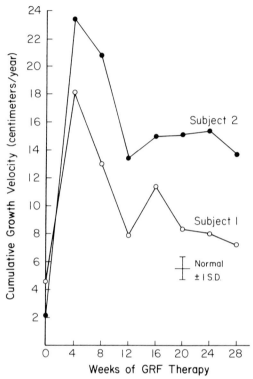

FIG. 29. Cumulative linear growth rate before and during 6 months of GRF therapy in 2 GH-deficient children. Note acceleration of the growth rate in both children, which was sustained in Patient 2 but waned after 16 weeks in Patient 1. The normal (± SD) growth rate is shown for an 8-year-old boy. Reprinted with kind permission from Thorner et al. (1985).

neous GH secretion before the start of therapy; neither had a response to GRF treatment. Three children had accelerated growth rates during therapy and the growth rate increased on the higher dose regimen. We have yet to define the optimal dose of GRF for the treatment of children with growth hormone deficiency.

VI. Conclusions

GRF is the last of the classical hypothalamic hormones isolated. Its application to animal and human studies has opened a new horizon in GH physiology and pathophysiology. The pulsatile pattern of GH secretion in several mammalian species is accounted for by a dynamic interaction of hypothalamic somatostatin and GRF at the somatotroph. The increased GH secretion of the male rat is accounted for at least in part by increased

STUDIES OF GRF AND GH 627

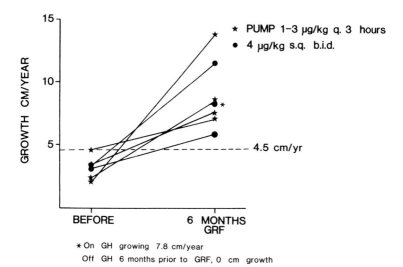

*On GH growing 7.8 cm/year
Off GH 6 months prior to GRF, 0 cm growth

FIG. 30. Growth velocity in 7 GH-deficient children treated with GRF in the United States by the University of Virginia group, by Dr. Georgeanna Klingensmith (Denver Children's Hospital, Denver, Colorado) and Drs. Seymour Reichlin and Boris Senior (Tufts New England Medical Center). All children demonstrated accelerated growth rate. One child (asterisk) grew at the same rate during GRF therapy as on growth hormone. During the 6 months prior to GRF therapy, this child did not take growth hormone and did not grow (data not shown). Note that 4 children were treated with subcutaneous injections every 3 hours administered by Pulsamat pump (Ferring) and 3 received GRF, 4 µg/kg, subcutaneously twice daily.

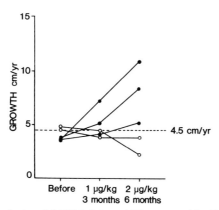

FIG. 31. Growth velocity of 5 GH-deficient children treated by Drs. Charles Brook and Patricia Smith at the Middlesex Hospital in London. These children were treated with GRF administered by Zyklomat pump (Ferring) administered every 3 hours, 4 pulses at night only. Note that three of the children demonstrated accelerated growth. The 2 children who did not have acceleration of growth had the highest growth rate prior to treatment.

percentage of somatotrophs, increased GH secretion per cell, and increased sensitivity to GRF.

The isolation of GRF was only possible by the identification of peripheral GRF-secreting tumors. While acromegaly is usually a result of a pituitary tumor, it is occasionally caused by excessive eutopic or ectopic GRF secretion. Studies with GRF in GH-deficient subjects have demonstrated that the majority of these patients do not suffer from a pituitary defect, but instead from a hypothalamic neurosecretory defect giving rise to GRF deficiency. Preliminary data demonstrate that GRF accelerates growth in GH-deficient children. GRF offers the prospect of restoring endogenous GH secretion to children and adults with GH deficiency. In addition, it appears that to restore pulsatile GH secretion GRF may be administered either in a pulsatile or continuous fashion. The latter observation raises the prospect of development of an effective delayed release preparation which would revolutionize therapy for GH-deficient children. In the adult the physiologic function of GH is poorly understood. The tight regulation of GH by nutritional factors and the known metabolic effects of GH suggest that it is an important regulator of metabolism. We suggest that the metabolic state dictates the GH requirements of the body in the same manner as the ovary dictates its gonadotropin requirements. During the coming years we believe the role of both GRF and GH in children and adults will be more clearly delineated with consequent therapeutic advances.

ACKNOWLEDGMENTS

We thank Ms. Sandra W. Jackson and the staff of the Clinical Research Center at the University of Virginia and the staff of the Clinical Research Center at the University of Cincinnati for their excellent nursing care. We also thank Ms. Jean Chitwood, Ms. Pattie Hellmann, Ms. Ellen Limber, Ms. Catherine Wolfe, Mr. S. K. Lau, and Mr. Tom Brannigan for organizational and technical assistance at the University of Virginia, and Ms. Carole Wieshart at the Children's Hospital of Philadelphia for technical assistance. We appreciate the help of Ms. Donna Harris and Ms. Ina Hofland in the preparation of this manuscript.

These studies were supported in part by a grant from the March of Dimes (MOT) and by the following U.S. Public Health Service or other grants: AM 32632 and RR 00847 (MOT), 1 R23 HD 17120 (MLV), HD 19170 (WSE), AG 04303 (RMB), AM 35937 (DAL), RCDA 1K04NS00601 (MJC), CA 07535 (RMM), AM 30667 (LAF), American Cancer Society BC 386B, CA 38981 (RF), GCRC RR 69 (GJK), GCRC RR 00054 and 2R01 AMI-16684 (SR), and AM 26741 (JR, WV).

REFERENCES

Adams, E. F., Winslow, C. L., and Mashiter, K. (1983). *Lancet* **1,** 1100.
Asa, S. L., Scheithauer, B. W., Bilbao, J. M., Horvath, E., Ryan, N., Kovacs, K., Randall, R. V., Laws, E. R., Singer, W., Linfoot, J. A., Thorner, M. O., and Vale, W. (1984). *J. Clin. Endocrinol. Metab.* **58,** 796.

Asa, S. L., Kovacs, K., Thorner, M. O., Leong, D. A., Rivier, J., and Vale, W. (1985). *J. Clin. Endocrinol. Metab.* **60**, 423.
Ballard, H. S., Frame, B., and Hartsock, R. J. (1964). *Medicine (Baltimore)* **43**, 481.
Beck, C., Larkins, R. G., Martin, T. J., and Burger, H. G. (1973). *J. Endocrinol. (London)* **59**, 325.
Bilezikjian, L. M., and Vale, W. W. (1984). *Endocrinology* **115**, 2032.
Birge, C. A., Peake, G. T., Mariz, I. K., and Daughaday, W. H. (1967). *Endocrinology* **81**, 195.
Bloch, B., Brazeau, P., Bloom, F., and Ling, N. (1983a). *Neurosci. Lett.* **37**, 23.
Bloch, B., Brazeau, P., Ling, N., Bohlen, P., Esch, F., Wehrenberg, W. B., Benoit, R., Bloom, F., and Guillemin, R. (1983b). *Nature (London)* **301**, 607.
Bloch, B., Gaillard, R. C., Brazeau, P., Lin, H. D., and Ling, N. (1984). *Regul. Pept.* **8**, 21.
Bohlen, P., Brazeau, P., Bloch, B., Ling, N., Gaillard, R., and Guillemin, R. (1983a). *Biochem. Biophys. Res. Commun.* **114**, 930.
Bohlen, P., Esch, F., Brazeau, P., Ling, N., and Guillemin, R. (1983b). *Biochem. Biophys. Res. Commun.* **116**, 726.
Borges, J. L. C., Blizzard, R. M., Gelato, M. C., Furlanetto, R., Rogol, A. D., Evans, W. S., Vance, M. L., Kaiser, D. L., MacLeod, R. M., Merriam, G. R., Loriaux, D. L., Spiess, J., Rivier, J., Vale, W., and Thorner, M. O. (1983). *Lancet* **2**, 119–124.
Borges, J. L. C., Blizzard, R. M., Evans, W. S., Furlanetto, R., Rogol, A. D., Kaiser, D. L., Rivier, J., Vale, W., and Thorner, M. O. (1984). *J. Clin. Endocrinol. Metab.* **59**, 1.
Brazeau, P., Vale, W., Burgus, R., Ling, N., Butcher, M., Rivier, J., and Guillemin, R. (1973). *Science* **179**, 77.
Bresson, J. L., Clavequin, M. C., Fellmann, D., and Bugnon, C. (1984). *Neuroendocrinology* **39**, 68.
Bruhn, T., Mason, R., and Vale, W. (1985). *Endocrinology* **117**, 1710.
Bugnon, C., Gouget, A., Fellmann, D., and Clavequin, M. C. (1983). *Neurosci. Lett.* **38**, 131.
Burek, C. L., and Frohman, L. A. (1970). *Endocrinology* **86**, 1361.
Buse, J., Buse, M. G., and Roberts, W. J. (1961). *J. Clin. Endocrinol. Metab.* **21**, 735.
Caplan, R. H., Koob, L., Avellera, R. M., Pagliara, A. S., Kovacs, K., and Randall, R. V. (1978). *Am. J. Med.* **64**, 874.
Ceda, G. P., and Hoffman, A. R. (1985). *Endocrinology* **116**, 1330.
Chiodini, P. G., Liuzzi, A., Dallabonzana, D., Oppizzi, G., and Verde, G. G. (1985). *J. Clin. Endocrinol. Metab.* **60**, 48.
Christofides, N. D., Stephanou, A., Suzuki, H., Yiangou, Y., and Bloom, S. R. (1984). *J. Clin. Endocrinol. Metab.* **59**, 747.
Clemmons, D. R., Klibanski, A., Underwood, L. E., McArthur, J. W., Ridgway, E. C., Beitins, I. Z., and Van Wyk, J. J. (1981). *J. Clin. Endocrinol. Metab.* **53**, 1247.
Dabek, J. T. (1974). *J. Clin. Endocrinol. Metab.* **38**, 329.
Daniels, M., Turner, S. J., Cook, D. B., Mathias, D., Kendall-Taylor, P., Teasdale, G., and White, M. C. (1984). *Int. Congr. Endocrinol., 7th* Abstr. No. 678.
Dieguez, C., Foord, S. M., Shewring, G., Edwards, C. A., Heyburn, P. J., Peters, J. R., Hall, R., and Scanlon, M. F. (1984). *Biochem. Biophys. Res. Commun.* **121**, 111.
Eden, S. (1979). *Endocrinology* **105**, 555.
Esch, F. S., Bohlen, P., Ling, N. C., Brazeau, P. E., Wehrenberg, W. B., Thorner, M. O., Cronin, M. J., and Guillemin, R. (1982). *Biochem. Biophys. Res. Commun.* **109**, 152.
Esch, F. S., Bohlen, P., Ling, N. C., Brazeau, P. E., and Guillemin, R. (1983). *Biochem. Biophys. Res. Commun.* **117**, 772.
Evans, W. S., Borges, J. L. C., Kaiser, D. L., Vance, M. L., Sellers, R. P., MacLeod, R.

M., Vale, W., Rivier, R., and Thorner, M. O. (1983). *J. Clin. Endocrinol. Metab.* **57,** 1081.
Evans, W. S., Borges, J. L. C., Vance, M. L., Kaiser, D. L., Rogol, A. D., Furlanetto, R., Rivier, J., Vale, W., and Thorner, M. O. (1984). *J. Clin. Endocrinol. Metab.* **59,** 1006.
Evans, W. S., Krieg, R. J., Limber, E. R., and Kaiser, D. L. (1985a). *Am. J. Physiol.* **249,** (*Endocrinol. Metab.* **12**), E276.
Evans, W. S., Vance, M. L., Kaiser, D. L., Sellers, R. P., Borges, J. L. C., Downs, T. R., Frohman, L. A., Rivier, J., Vale, W., and Thorner, M. O. (1985b). *J. Clin. Endocrinol. Metab.* **61,** 846.
Frantz, A. G., and Rabkin, M. T. (1965). *J. Clin. Endocrinol. Metab.* **25,** 1470.
Frohman, L. A., Szabo, M., Berelowitz, M., and Stachura, M. E. (1980). *J. Clin. Invest.* **65,** 43.
Frohman, L. A., Thominet, J. L., Webb, C. B., Vance, M. L., Uderman, H., Rivier, J., Vale, W., and Thorner, M. O. (1984). *J. Clin. Invest.* **73,** 1304.
Gelato, M. C., Pescovitz, O. H., Cassorla, F., Loriaux, D. L., and Merriam, G. R. (1984). *J. Clin. Endocrinol. Metab.* **59,** 197.
Gelato, M. C., Merriam, G. R., Vance, M. L., Goldman, J. A., Webb, C., Evans, W. S., Rock, J., Oldfield, E. H., Molitch, M. E., Rivier, J., Vale, W., Reichlin, S., Frohman, L. A., Loriaux, D. L., and Thorner, M. O. (1985). *J. Clin. Endocrinol. Metab.* **60,** 251.
Green, J. D., and Harris, G. W. (1947). *J. Endocrinol.* **5,** 136.
Gubler, U., Monahan, J. J., Lomedico, P. T., Bhatt, R. S., Collier, K. J., Hoffman, B. J., Bohlen, P., Esch, F., Ling, N., Zeytin, F., Brazeau, P., Poonian, M. S., and Gage, L. P. (1983). *Proc. Natl. Acad. Sci. U.S.A.* **80,** 4311.
Guillemin, R., Brazeau, P., Bohlen, P., Esch, F., Ling, N., and Wehrenberg, W. B. (1982). *Science* **218,** 585.
Guillemin, R., Brazeau, P., Bohlen, O., Esch, F., Ling, N., Wehrenberg, W. B., Bloch, B., Mougin, C., Zeytin, F., and Braid, A. (1984). *Recent Prog. Horm. Res.* **40,** 233.
Hermansen, K., Kappelgaard, A. M., and Orskov, H. (1985). *Acta Endocrinol. Congr. 15th,* Abstr. No. 109.
Jacobowitz, D. M., Schulte, H., Chrousos, G. P., and Loriaux, D. L. (1983). *Peptides* **4,** 521.
Kovacs, K., Ryan, N., Horvath, E., Asa, S. L., Thorner, M. O., Leong, D. A., Vale, W., Rivier, J., Scheithauer, B. W., Randall, R. V., Carpenter, P. C., and Caplan, R. H. (1984). *Arch. Pathol. Lab. Med.* **108,** 355.
Krulich, L., Dhariwal, A. P. S., and McCann, S. M. (1968). *Endocrinology* **83,** 783.
Lamberts, S. W., Verleun, T., and Oosterom, R. (1984). *J. Clin. Endocrinol. Metab.* **58,** 250.
Leong, D. A., Lau, S. K., Sinha, Y. N., Kaiser, D. L., and Thorner, M. O. (1986). *Endocrinology* **116,** 1371.
Leveston, S. A., McKeel, D. W., Buckley, P. J., Deschryver, K., Greider, M. H., Jaffe, B. M., and Daughaday, W. H. (1981). *J. Clin. Endocrinol. Metab.* **53,** 682.
Liberman, B., Mayo, Y., Borges, J., Kovacs, K., Vale, W., and Thorner, M. O. (1985). In preparation.
Ling, N., Baird, A., Wehrenberg, W. B., Ueno, N., Minegumi, T., Chiang, T. C., Regno, M., and Brazeau, P. (1984a). *Biochem. Biophys. Res. Commun.* **122,** 304.
Ling, N., Esch, F., Bohlen, P., Brazeau, P., Wehrenberg, W. B., and Guillemin, R. (1984b). *Proc. Natl. Acad. Sci. U.S.A.* **81,** 4302.
Losa, M., Stalla, G. K., Muller, O. A., Hartl, R., and von Werder, K. (1984). *Int. Congr. Endocrinol., 7th* Abstr. No. 1441.

Lytras, N., Grossman, A., Wass, J. A. H., Coy, D. H., Rees, L. H., and Besser, G. M. (1984). *Int. Congr. Endocrinol., 7th* Abstr. No. 1444.
McCormick, G. F., Millard, W. J., Badger, T. M., and Martin, J. B. (1984). *Annu. Meet. Soc, Neurosci., 14th, Anaheim, CA* p. 1214.
MacLeod, R. M., Abad, A., and Eidson, L. L. (1969). *Endocrinology* **84,** 1475.
Mayo, K. E., Vale, W., Rivier, J., Rosenfeld, M. G., and Evans, R. M. (1983). *Nature (London)* **306,** 86.
Mayo, K. E., Cerelli, G. M., Lebo, R. V., Bruce, B. N., Rosenfeld, M. G., and Evans, R. M. (1985). *Proc. Natl. Acad. Sci. U.S.A.* **82,** 63.
Merchenthaler, I., Vigh, S., Schally, A. V., and Petrusz, P. (1984). *Endocrinology* **114,** 1082.
Neill, J. D., and Frawley, L. S. (1983). *Endocrinology* **112,** 1135.
Pandol, S. J., Seifert, H., Thomas, M. W., Rivier, J., and Vale, W. (1984). *Science* **225,** 326.
Penny, E., Penman, E., Price, J., Rees, L. H., Sopwith, A. M., Wass, J. A. H., Lytras, N., and Besser, G. M. (1984). *Br. Med. J.* **289,** 453.
Reichlin, S. (1960). *Endocrinology* **67,** 760.
Reichlin, S. (1961). *Endocrinology* **69,** 225.
Rivier, J., Spiess, J., Thorner, M., and Vale, W. (1982). *Nature (London)* **300,** 276.
Rivier, J., Spiess, J., and Vale, W. (1983). In "Peptides: Structure and Function" (V. J. Hubry and D. H. Rich, eds.), p. 853. Pierce Chemical Company, Rockford, IL.
Rogol, A. D., Blizzard, R. M., Johanson, A. J., Furlanetto, R. W., Evans, W. S., Rivier, J., Vale, W. W., and Thorner, M. O. (1984). *J. Clin. Endocrinol. Metab.* **59,** 580.
Sartor, O., Bowers, C. Y., and Chang, D. (1985). *Endocrinology* **116,** 952.
Sassolas, G., Chayviatte, J. A., Partensky, C., Berger, G., Troushas, J., Berger, F., Claustrat, B., Cohen, R., Girod, C., and Guillemin, R. (1983). *Ann. Endocrinol (Paris)* **44,** 347.
Schriock, E. A., Lustig, R. H., Rosenthal, S. M., Kaplan, S. L., and Grumbach, M. M. (1984). *J. Clin. Endocrinol. Metab.* **58,** 1043.
Schulte, H. M., Benker, G., Windeck, R., Olbricht, T., and Reinwein, D. (1985). *J. Clin. Endocrinol. Metab.* **61,** 585.
Shalet, S. M., Beardwell, C. G., MacFarlane, I. A., Ellison, M. L., Norman, C. M., Rees, L. H., and Hughes, M. (1979). *Clin. Endocrinol.* **10,** 61.
Sheldon, W. R., Jr., DeBold, C. R., Evans, W. S., DeCherney, G. S., Jackson, R. V., Island, D. P., Thorner, M. O., and Orth, D. N. (1985). *J. Clin. Endocrinol. Metab.* **60,** 623.
Shibasaki, T., Kiyosawa, Y., Masuda, A., Nakahara, M., Imaki, T., Wakabayashi, I., Demura, H., Shizume, K., and Ling, N. (1984a). *J. Clin. Endocrinol. Metab.* **59,** 263.
Shibasaki, T., Shizume, K., Masuda, A., Nakahara, M., Hizuka, N., Miyakawa, M., Takano, K., Demura, H., Wakabayashi, I., and Ling, N. (1984b). *J. Clin. Endocrinol. Metab.* **58,** 215.
Sonksen, P. H., Ayrrs, A. B., Braimbridge, M., Corrin, B., Davies, D. R., Jeremiah, G. M., Oaten, S. W., Lowy, C., and West, T. E. T. (1976). *Clin. Endocrinol.* **5,** 503.
Sopwith, A. M., Penny, E. S., Besser, G. M., and Rees, L. H. (1984). *J. Endocrinol.* **104,** (Suppl.), 84.
Sopwith, A. M., Penny, E. S., Besser, G. M., and Rees, L. H. (1985). *Clin. Endocrinol.* **22,** 337.
Southern, A. L. (1960). *J. Clin. Endocrinol. Metab.* **20,** 298.
Spada, A., Vallar, L., and Giannattasio, G. (1984). *Endocrinology* **115,** 1203.
Spiess, J., Rivier, J., Thorner, M., and Vale, W. (1982). *Biochemistry* **21,** 6037.
Spiess, J., Rivier, J., and Vale, W. (1983). *Nature (London)* **303,** 532.

Spiliotis, B., August, G., Hung, W., Sonis, W., Mendelsohn, W., and Bercu, B. (1984). *J. Am. Med. Assoc.* **152**, 2223.
Steiner, R. A., Bremner, W. J., and Clifton, D. K. (1982). *Endocrinology* **111**, 2055.
Takano, K., Hizuka, N., Shizume, K., Asakawa, K., Miyakawa, M., Hirose, N., Shibasaki, T., and Ling, N. C. (1984). *J. Clin. Endocrinol. Metab.* **58**, 236.
Tanaka, Y., Saito, S., Namba, O., Ohashi, S., Sawano, S., and Irie, M. (1984). *Int. Congr. Endocrinol., 7th* Abstr. No. 2335.
Tannenbaum, G. S., and Martin, J. B. (1976). *Endocrinology* **98**, 562.
Thompson, R. G., Rodriquez, A., Kowarski, A., and Blizzard, R. M. (1972). *J. Clin. Invest.* **51**, 3193.
Thorner, M. O., Perryman, R. L., Cronin, M. J., Rogol, A. D., Draznin, M., Johanson, A., Vale, W., Horvath, E., and Kovacs, K. (1982). *J. Clin. Invest.* **70**, 965.
Thorner, M. O., Rivier, J., Spiess, J., Borges, J. L. C., Vance, M. L., Bloom, S. R., Rogol, A. D., Cronin, M. J., Kaiser, D. L., Evans, W. S., Webster, J. D., MacLeod, R. M., and Vale, W. (1983). *Lancet* **1**, 24.
Thorner, M. O., Frohman, L. A., Leong, D. A., Thominet, J., Downs, T., Hellmann, P., Chitwood, J., Vaughan, J. M., and Vale, W. (1984). *J. Clin. Endocrinol. Metab.* **59**, 846.
Thorner, M. O., Reschke, J., Chitwood, J., Rogol, A. O., Furlanetto, R., Rivier, J., Vale, W., and Blizzard, R. M. (1985). *N. Engl. J. Med.* **312**, 4.
Uz Zafar, M. S., Mellinger, R. C., Fine, G., Szabo, M., and Frohman, L. A. (1979). *J. Clin. Endocrinol. Metab.* **48**, 66.
Vale, W., Spiess, J., Rivier, C., and Rivier, J. (1981). *Science* **213**, 1394.
Vance, M. L., Borges, J. L. C., Kaiser, D. L., Evans, W. S., Furlanetto, R., Thominet, J. L., Frohman, L. A., Rogol, A. D., MacLeod, R. M., Bloom, S., Rivier, J., Vale, W., and Thorner, M. O. (1984). *J. Clin. Endocrinol. Metab.* **58**, 838.
Vance, M. L., Kaiser, D. L., Evans, W. S., Thorner, M. O., Furlanetto, R., Rivier, J., Vale, W., Perisutti, G., and Frohman, L. A. (1985a). *J. Clin. Endocrinol. Metab.* **60**, 370.
Vance, M. L., Kaiser, D. L., Evans, W. S., Furlanetto, R., Vale, W., Rivier, J., and Thorner, M. O. (1985b). *J. Clin. Invest.* **75**, 1584.
Vance, M. L., Kaiser, D. L., Rivier, J., Vale, W., and Thorner, M. O. (1986). *J. Clin. Endocrinol. Metab.*, in press.
Veldhuis, J. D., Rogol, A. D., and Johnson, M. L. (1985). *Am. J. Physiol.* **248**, E475.
Waelbroeck, M., Robberecht, P., Coy, D., Camus, J. C., DeNeef, P., and Christophe, J. (1985). *Endocrinology* **116**, 2643.
Webb, C. B., Thominet, J. L., and Frohman, L. A. (1983). *J. Clin. Endocrinol. Metab.* **56**, 417.
Webb, C. B., Vance, M. L., Thorner, M. O., Perisutti, G., Thominet, J., Rivier, J., Vale, W., and Frohman, L. A. (1984). *J. Clin. Invest.* **74**, 96.
Wehrenberg, W. B., Ling, N., Brazeau, P., Esch, F., Bohlen, P., Baird, A., Ying, S., and Guillemin, R. (1982). *Biochem. Biophys. Res. Commun.* **109**, 382.
Weiss, L., and Ingram, M. (1961). *Cancer* **14**, 161.
Wood, S. M., Chng, J. L., Adams, E. F., Webster, J. D., Joplin, G. F., Mashiter, K., and Bloom, S. R. (1983). *Br. Med. J.* **286**, 1687.
Yamamoto, K., Taylor, L. M., and Cole, F. E. (1970). *Endocrinology* **87**, 21.
Zeytin, F., and Brazeau, P. (1985). *Biochem. Biophys. Res. Commun.* **123**, 496.

DISCUSSION

R. M. Blizzard. I would like to ask you to comment on the possible incidence of antibodies against GRF.

M. O. Thorner. Those of you who are following the clinical literature will be aware that one of the first two children that we treated in Virginia did not grow as well in the second 3 months as he did in the first, and in any case as part of the requirement of the FDA and also because of our own interest we obviously screened these children for the possible development of antibodies to GRF. One of them did develop low titer antibody to GRF and, in fact, 3 months after withdrawal of the GRF therapy the antibody titer fell and became undetectable. When one is giving a peptide one should be concerned about the possibility of antibody development. On the other hand I think we have to say that we don't know what the significance of that antibody was, whether it had anything to do with the apparent slowing of the rate of growth in that child. I think there are many other factors that may have been involved, and we really have no idea of the significance of the GRF antibodies. I should also remind those of you who are not clinicians that the incidence of antibody development to insulin is rather high, probably 30 or 40% in diabetic patients, even in those receiving human insulin. I think we have to follow this further, but not overreact to the development of the antibodies.

R. M. Blizzard. It should be added that this particular patient was subsequently given and did respond very nicely to native growth hormone. Therefore the development of antibodies, if responsible for the growth inhibition, did not negate the response to growth hormone itself.

M. A. Greer. I have a couple of questions regarding the interpretation of some of the studies on the continuous infusion or the *in vitro* perifusion that Ken Ho did.

In studies that we've been doing with *in vitro* pituitary perifusions, we've found that if one gives a continuous perifusion of either LHRH or TRH (we don't have any experience with GRF), there is an immediate high-amplitude pulse with a subsequent lower amplitude sustained secretion of hormone, as most other people have found. However, we've additionally found that if the continuous perifusion of secretagogue is maintained, but the secretagogue concentration is increased, one again gets a high-amplitude secretory burst followed by a sustained lower amplitude release. This is similar to your data going from 10 to 100 nM GRF, except you gave a bolus injection of the higher dose rather than a higher concentration of a continuous infusion.

It also may relate to your studies in which insulin given at the end of a continuous GRF infusion produced a greater GH response than you would obtain without the preceding GRF infusion. Possibly this is because the insulin is stimulating endogenous GRF secretion to add to the exogenous infused GRF rather than some independent mechanism to stimulate GH secretion.

Does all this seem plausible?

In relation to the therapy of children with growth hormone deficiency, if this sort of phenomenon with sustained low-level release of GH with nonpulsatile administration of GRF occurs *in vivo*, one might get an overall increase in GH secretion without having to set up an elaborate scheme for pulsatile GRF administration. I think that would fit in with your data indicating that somatomedin increases whether you see the actual GH secretory pulses or not.

The last question I would like to ask relates to terminology. As a nonparticipant in the studies, I am confused by the abbreviations used. Why was GRF chosen, since G can refer to gonadotropins even more easily than to growth hormone. I noticed apparently even you are unconvinced and have used GHRH. Since "somatotropin" seems to be a generally accepted synonym for growth hormone (I noticed you were using SS for somatostatin), why not call it SRF rather than GRF if you want to economize by using only three letters?

M. O. Thorner. You know I wouldn't do a thing like that! Maybe I can discuss the last question first. There is a lot of confusion as you say with various differences in the nomenclature.

The reason that you see GRF or GHRH is because *Endocrinology* wants one and the *Journal of Clinical Endocrinology and Metabolism* wants another. As you know the expense of art work is such that it is not very easy to change these back and forth and we ask for your forgiveness. I agree with you that I think it is very confusing and it would be much better if everybody could agree on a nomenclature. I think whether you call it factor or hormone to me is immaterial, and I would be happy with any one that you suggest. Perhaps we should have a meeting to decide what everybody is going to call this substance. Now if I can just go back to the column experiment. At the same time we also gave GnRH in exactly the same paradigm. The same cells at the same time were exposed to both GRF and GnRH. The LH response was blunted by the previous infusion of GnRH so that is why we believe the somatotrophs and the gonadotrophs are different. The interpretation is extremely complicated. The story of the gonadotropin-releasing hormone has been so impressive and we have all become "supersensitized," if I can use that term, to the idea that desensitization is going to occur that everybody has been saying the same is going to happen to GRF. I think the message that I want to give you is to be skeptical. From experiments we know that probably is not true because you would never get acromegaly due to ectopic GRF secretion if it were, if the desensitization were as profound as after the gonadotropin-releasing hormone. I think our data, whether it be on the column or whether it be infusions in man, would also suggest that my conclusion is correct. The other point that I really want to make is that I believe that pulsatile growth hormone secretion is probably very important to the biologic effect of growth hormone but again we must not be locked into the idea that because we want to get pulses of growth we have to give GRF in a pulsative fashion. That may not be true. Our data would suggest that it is not true. I am afraid I missed the third question.

R. M. Blizzard. Whether there is any evidence that there is suppression of endogenous growth hormone.

M. O. Thorner. We have obviously been concerned about that. The reason that I think that stimulation of endogenous GRF per se as an explanation for the effect of hypoglycemia may not be right is that the levels of GRF after giving 3 μg/kg is something like 20 to 30 ng/ml. Although we have not obtained measurements in human portal blood or for that matter in animal portal blood taken on a second by second basis, which is what one would need in order to be able to decide that. I think it is rather unlikely that you would find these types of concentrations or concentrations greater than that. Therefore I think that hypoglycemia probably acts through different mechanisms. But that is an opinion.

S. Kaplan. In our studies in adult men, we carried out continuous daytime infusions with GRF using 0.15 μg/kg/hour for 5 hours followed by 0.75 μg/kg/hour for 5 hours. We noted a decline in the GH levels before the infusion terminated. Did you observe a similar pattern in your studies?

In addition, we administered pulses of GRF 1 μg/kg every 2 hours during a 12-hour daytime and a 12-hour nighttime period. We did not observe a significant difference in the GH response to GRF between the daytime or nighttime schedule. This contrasts with your study using prolonged infusions. Do you have any comments concerning the differences in GH responsiveness?

M. O. Thorner. I think that your doses are μg/kg/hour?

S. Kaplan. No, per minute.

M. O. Thorner. So yours were what doses?

S. Kaplan. Well the total dose they received was 5.5 μg over a 24-hour period.

M. O. Thorner. Did you think they were higher concentrations than we used?

S. Kaplan. I wondered if you noticed any difference in the day–night time responses at the lower compared to the higher doses?

M. O. Thorner. We found augmentation with both 2 ng/kg/minute dose which was the one that we used for the 24-hour studies and then the 10 ng/kg/minute, which we used for the 14-day studies in the nighttime secretion. I think that we have many problems with GRF. One that I tried to emphasize was that you get great variability in the response; you not only get variability between subjects but within subjects at different times. Therefore, unless you are using a large number of subjects, I think the data are very difficult to interpret.

S. Kaplan. I was just trying to figure out whether there was some difference related to the dose.

M. O. Thorner. I would have to work out the doses.

S. Kaplan. You didn't comment on the declining responsiveness at the end of the infusion; I wonder if you'd do so?

M. O. Thorner. I think that is probably a time-related phenomenon and as time goes on it diminishes. I did not show the results of giving a supramaximal dose at the end of 14 days. In fact the response is the same or greater than it is before beginning the infusion. Again, whether we do 6-hour or 24-hour infusions we find the diminished response. In contrast, when we gave a 14-day infusion we did not find a diminished response. It is a very complicated system which is going to keep us busy for many many years.

J. M. Nolin. Is it really true then that we need not worry anymore about the recent Creutzfeldt–Jakob scare?

M. O. Thorner. I don't want to comment on this. If your question was whether it is likely that GRF is going to have therapeutic potential, the answer would be yes. In addition it would be a synthetic hormone.

N. Samaan. In the patients in which you find GRF in the tumor, did you find other peptide hormones in these tumors? The reason I ask is because we have 2 patients with pancreatic tumors secreting CRF, but we find excess CRF and TRF as well in the tumorous extract. Clinically they presented with Cushing's syndrome but did not present any symptoms of excess TRF. Did you find any other hormones in the tumors you described?

M. O. Thorner. In this particular patient there was no other hormone in the tumor of the patient from which the GRF was isolated. In another patient who had ACTH secretion (which is when we questioned ectopic CRF production) we sent the tumor to Wylie Vale and he assayed it for GRF and found that it contained a large amount of GRF. In fact that particular tumor also was provided to Kelly Mayo and Ron Evans and they found messenger RNA for POMC and to GRF in that tumor. I think most of us who have occasionally had these patients with ectopic hormone production recognize that they can often secrete multiple hypothalamic hormones, in fact they can be veritable hypothalami.

N. Samaan. Did I hear correctly that the growth hormone level during fasting goes above normal, especially at night?

M. O. Thorner. I should say these are very preliminary data, but they were so marked that I can't help believe that they're probably going to be representative. In this particular individual subject, growth hormone secretion increased. The integrated, 24-hour growth hormone went up 10-fold from 0.3 to between 3 and 4 μg/minute/ml. After 5 days it went up to 6 μg/minute/ml so it went up 20-fold. The number of pulses that were present before fasting was 2; at 12–36 hours of fasting the number went up to 5 pulses, and at the end of 5 days it was 10 pulses. Those 10 pulses were spread throughout the 24-hour period. There are some data in the literature that indicate that growth hormone goes up as you go to sleep at night. As you all fell asleep during my lecture, many of you may have gotten a growth hormone pulse. However, I think that we have to really question all those data, because they are based on very few subjects.

N. Samaan. Did you measure the GRF levels in the peripheral circulation?

M. O. Thorner. We did not measure GRF levels in that subject, believing the peripheral immunoreactive GRF levels really represent secretion of GRF from the hypothalamus. I think it would be interesting to do so.

N. Samaan. We have shown in the past that fasting produces a significant increase in the nonsuppressible insulin-like activity, and lowers the blood sugar level in obese diabetic subjects, so this is a complex subject (Samaan *et al., Br. Med J.* **5455**, 195–198, 1958).

I. M. Spitz. Have you studied the response to repeated pulses of GRF in normal subjects over 14 days? It would be of interest to compare this with your constant infusion studies. When LHRH is infused constantly, down-regulation of LH is seen. This does not however occur with repeated pulses of LHRH.

M. O. Thorner. These are studies that have only just been completed in the last few days, and we have not yet done the pulsatile, but I think that they will need to be done to compare pulsatile versus continuous GRF.

R. A. Hoffman. I was particularly impressed by the studies in which you showed pulsatile growth hormone secretion occurring despite constant infusion of GRF, and also the one study you had on fasting where pulsations of GH appeared, presumably because of the decrease in serum insulin and insulin-like growth factor. My question is whether you believe that growth hormone is primarily under inhibitory control and that GRF merely acts in a permissive fashion to allow growth hormone secretion.

M. O. Thorner. Well, again it would be entirely speculation. Recalling Dr. Knobil's now classic statement that the ovary dictates its LH requirements to the pituitary, I think the same may be true for growth hormone. In fact we've been thinking that growth hormone is driven from the hypothalamus working down. However, we know that growth hormone has metabolic effects. However, the metabolic effects are really very poorly worked out, because they usually involve administration of pharmacologic doses of growth hormone (plus contaminants). I wonder now whether we really have to open up the whole question again. To be very speculative I would say that I think we need to question whether growth hormone secretion is a function of our nutritional state. In protein–calorie malnutrition you want to have a minimal anabolic effect, maximal lipolysis, and minimal carbohydrate utilization. The best way to achieve this is to have very low levels of IGF_1 to obtain minimal anabolism and high growth hormone levels for its lipolytic effects and antiinsulin effects. So I think it would make a lot of sense that nutrition regulates GH secretion.

E. A. Schriock. While at the University of California, San Francisco with Drs. Kaplan and Grumbach, my colleagues and I administered GRF every 2 hours for a period of 24 hours to a group of growth hormone-deficient children and adolescents. I have two comments I'd like to make. First is to emphasize that in evaluating the growth hormone response to a single dose GRF test, caution needs to be exercised in interpreting an absent response as indicating a primary pituitary problem. A couple of our patients who initially did not respond to GRF subsequently developed significant growth hormone levels with the administration of additional pulses of GRF. Second, I would like to confound further the discussion of the day/night differences in growth hormone secretion with GRF by a significant difference between day and night in the magnitude of the growth hormone responses. Looking at 12-hour integrated areas of growth hormone secretion we found that the GRF pulses administered during the night resulted in significantly greater growth hormone secretion than those GRF pulses given during the day.

M. O. Thorner. You will note that I was careful not to say what the use of the GRF test was. I really do not believe that the administration of hypothalamic hormones for testing really tells you a great deal except whether there is or is not a response. I think that we have to define much more carefully how we administer GRF, what dose we give, what condition the patient is in, the time of day, and the nutritional state. I think that if you do find a response all you can say is that presumably there are some functional somatotrophs. I was not trying to claim anymore than that.

E. A. Schriock. I agree.

J. M. Hershman. Perhaps your list of causes of acromegaly was not meant to be all inclusive. Melmed *et al.* (*N. Engl. J. Med.* **312,** 9–17, 1985) recently reported on a patient with acromegaly who had an ectopic pancreatic islet cell tumor which was removed and shown to secrete growth hormone in culture. It did not secrete GRF. Dr. Calvin Ezrin, who has taken care of this patient, told me recently that though the acromegaly was apparently cured by the removal of the tumor, this patient has had a recurrence of the acromegaly associated with high growth hormone levels and recurrence of the metastatic islet cell tumor which secretes growth hormone. My question relates to your subject who fasted for 5 days. I believe that you and others have reported that the responses to GRF are attenuated in obesity. I wonder whether this may have influenced the results?

M. O. Thorner. I agree entirely with your first comment. It was not all inclusive. Second, this patient was 99% of the ideal body weight. In fact the study will be on subjects within 10% of ideal body weight.

H. Kronenberg. I wanted to know a little bit more on the elliptic comments you just made about the possible usefulness of the GRH test, diagnostically or clinically. As I understand it, it looks as if currently the test isn't of much help. I wondered if you could speculate a bit as to what types of strategies you were alluding to that might conceivably make it a useful test, or do you think that that's just going to be impossible?

M. O. Thorner. In the human it is not possible to give antibodies to somatostatin, and I think that we're going to be confounded by somatostatin until we find a mechanism either to switch off endogenous hypothalamic somatostatin, whether it be by fasting or whatever; alternatively, a somatostatin antagonist could be used if we had one. At present there is none. Until then I am very skeptical as to whether the GRF test will be useful, other than to tell you that somatotrophs are present. The other question that is of major importance in pediatric endocrinology at the present time, even though I'm not a pediatrician, is the definition of what a growth hormone deficiency is. We have become very wary of the tests that we've used for which, when all is said and done, there is very little normative data. We often do not take into account the body habitus of the child. I believe we have to start by judging the normal pattern of growth hormone secretion in children at all different stages of development, from infancy on. This is obviously very difficult. It presents ethical questions. I know that at NIH they are doing such a study, and I think that's going to be extremely important. We will then compare short children with average height children. And again in the past, for other reasons, the "normal" children used as the controls were children who were short but had so-called normal GH responses. They are not normal because if they were they would never have undergone the testing. And so I really think that we have to start again with a clean notebook.

J. Geller. I was intrigued by your data, particularly concerning the growth hormone responses in patients who responded to insulin with a big release of growth hormone but did not respond to GRF. Since growth hormone is one of the so-called stress hormones and we know that ACTH and prolactin also increase with hypoglycemic stress, I wondered if you would care to speculate on what other mechanisms could be involved in the response to insulin since you have indicated it is probably not endogenous GRF.

M. O. Thorner. Again I have to say it is speculation and you did ask me to speculate. I think that there is a real possibility that hypoglycemia could lead to suppression of endogenous hypothalamic somatostatin. That is the mechanism I think the most likely. There is also a possibility that there may be more than one growth hormone releasing factor. If we recognize that the hypothalamus contains more and more different peptides which have more and more different functions it may be that there may be other GRFs. Those would be the two speculations.

R. Rittmaster. Dr. George Merrian and I at the NIH examined the growth hormone response to GRF in perifused rat anterior pituitary cells. In relation to these studies I would like to make the following comments. Several groups have looked at the *in vitro* response of pituitary cells to GRF, and nearly all of these investigators have found that there is a decrease in the growth hormone response to GRF after preincubation with GRF. Bileziekjian and Vale have shown that continuous GRF exposure causes somatotroph desensitization. We examined this desensitization further by giving continuous 6-hour infusions of 10 nM GHRH with or without 10 nM somatostatin in the media during the first 3 hours. The amount of GH released during the fourth hour of the infusions was the same, whether or not GH release was blocked by somatostatin during the first 3 hours. In other words this apparent GRF-induced somatotroph desensitization occurred independent of GH secretion. These findings are clearly different from what have been seen during prolonged GRF infusions in normal volunteers. Can you reconcile these findings?

M. O. Thorner. I agree with you that there is conflict in the literature and that you can cite examples for whichever case you wish. I think that one has to be very careful as to the experimental protocols being used: the donor animal, the culture conditions whether the culture medium contains dexamethasone and T3 etc. I don't want to go into great detail about this but it is one possible reason for the differences in responses. The second is that in your and our experiments only very short-term effects were studied. In the GRFoma patient, we are looking at a long-term effect. Indeed, there may be short-term effects and long-term effects. I think that even if there were some degree of desensitization, it would be relatively small and, in fact, the receptor studies have shown only a small shift.

M. Saffran. You did try nasal administration. Hiquchi and his co-workers in Lawrence, Kansas have been working on rectal absorption of peptides. They have found that peptides as large as insulin are quite well absorbed from the rectum. Moreover, the rate of absorption is greatly enhanced by mixing the peptide with something as innocuous as an acid derivative. I may be able to offer you an oral salicylic form of administration in the near future. With a colleague at Bowling Green State University, Doug Neckers, we have developed a polymer that is impervious to water and protects peptides against destruction in the stomach and the small intestine. The same polymer is opened by bacterial action and releases the peptide in the colon, from which it's absorbed.

M. O. Thorner. I think that would be very exciting.

W. VanderLaan. I would like to ask you to comment a little further on the protein–calorie malnutrition and anorexia nervosa aspects of this topic. Dr. Granner referred to the fact that insulin suppresses the gene expression for growth hormone. Protein–calorie malnutrition and anorexia nervosa are attended by low rates of insulin secretion and therefore might be attended by enhanced gene expression for growth hormone. It has long been of interest to me that it is in these situations that one obtains so-called paradoxical growth hormone response to glucose. Finally, in terms of the gene expression, Parker, Rossman, and E. F. VanderLaan in our group performed sleep studies on identical twins. Their growth hormone releases are nearly identical. I presume that not only somatostatin but also other modulating factors may influence the response to GRF. Perhaps you would comment on these points.

M. O. Thorner. I am not sure that there is a great deal that I can add to what you have said. I think that the observations about the effects of insulin *in vitro* on growth hormone messenger RNA in the pituitary in the rat system would fit very nicely with what we are proposing. What we are talking about are correlations, not necessarily cause and effect; one has to be a little cautious. There are other problems. In the rat, the response to hypoglycemia and hyperglycemia is the reverse of what is found in man. Also the sex differences are really curious. In the young human female, growth hormone secretion is much higher than in the male. In the male rat, growth hormone secretion is greater than in the female. Thus, there clearly are species differences, and I don't think we should forget about these. In terms of the so-called paradoxical responses, what Dr. VanderLaan is referring to here is that when a dose of glucose is administered to normal adults growth hormone secretion will be suppressed. If you administer TRH to a normal adult, there is no rise in growth hormone, but when you give L-dopa there is a rise in growth hormone in normal men. In the acromegalic patients the reverse occurs in each case: L-dopa will suppress growth hormone secretion, glucose does not. TRH will stimulate growth hormone. We had the opportunity of looking at this in our initial patient with GRFoma. She had paradoxical responses to dopamine, to glucose, and to TRH. On removal of the tumor all these paradoxical responses disappeared. These paradoxical responses occur in a number of pathological situations as well. They occur in diabetes mellitus, in renal failure, in hepatic failure, and in depression. One of the explanations in the past was that maybe there are abnormal receptors in the GH secreting tumors in the acromegalics. I have never found that a satisfactory answer because they shouldn't occur in all these other conditions. The way that I see it is that the somatotroph probably has TRH receptors. The TRH response is not normally expressed because the somatotroph is dominantly suppressed by somatostatin. Perhaps in these pathological situations somatostatin drive is not as great as it is in the normal situation.

M. I. New. I wondered if you would comment on the possibility that upper cerebral centers may be controlling the release of GRF in a parallel way in other behavioral disorders, such as anorexia nervosa or psychosocial dwarfism. There is evidence that emotional disturbances might affect the secretion of pituitary hormones through hypothalamic-releasing factors. My question is, how do you know that the man after 5 days of fasting didn't have a high level of anxiety?

M. O. Thorner. The subject claims that he was not stressed. However, he had lost about 7 kg over that 5-day period and, interestingly, he had what appeared to be a massive diuresis which apparently is well described with fasting. The reason that we did the study was that I was at a meeting a few months ago and Lou Underwood presented data on IGF-1 which are absolutely astonishing. IGF-1 levels fall into the hypopituitary range within 5 days of fasting. I asked him what happened to growth hormone. They had not measured growth hormone, so that was why we did the study. We thought that at least we had a mechanism by which IGF-1 levels would fall. I don't know whether it is the suppression of IGF-1 that caused this rise of GH but at least we now have a model system. We can study this by administering recombinant DNA IGF-1. We will then be able to infuse IGF-1 in man and demonstrate once and for all whether IGH-1 really is important in feedback. That was one of the purposes of doing these studies but I agree with you, we don't know whether it's fasting or whether it's a whole series of other factors. These obviously will have to be looked into. In addition, I should add that the glucose levels fell by about 10 to 15 mg/dl so there was a substantial fall in blood glucose as well during that period of the fast.

G. Segre. If I understood your data correctly, you demonstrated that in response to GRF you observed an increase in somatomedin C levels without, in some instances, an increase in growth hormone levels, and that you took this to indicate that the elevation in somatomedin C had been due to growth hormone. If my understanding is correct, is it clear that GRF did not act directly or indirectly through another hormone to influence somatomedin C levels?

M. O. Thorner. First, your understanding of what I said is correct. We don't know for sure that GRF does not have a direct effect on IGF-1 production and after everything that we have heard one has to consider that as a possibility. I think, however, that as the work of Dr. VanderLaan and of many others has indicated, growth hormone is secreted in multiple different forms. When we are using immunoradiometric assay we are presumably only detecting one growth hormone, and I think that it is quite conceivable that other growth hormones are being missed. Perhaps a whole family of growth hormones is being secreted in response to GRF. Again that is speculation and we have no data to show that at the present time but it is something that obviously needs to be investigated further. Also we would really like to know of a method with which to study the direct production of IGF-1 in response to GRF. We have discussed this with Dr. Underwood and also with Dr. Furlanetto. I gather there are some liver cell lines that do produce IGF-1, but we have not done that study yet.

S. Cohen. Can you give any explanation of why the tail of the pancreas should be chosen for the site of production of GRH or GH?

M. O. Thorner. No, is the simple answer. It is interesting that the pancreas seems to be the most common site for this very unusual tumor, and, in fact, both our patient and the French patient had tumors that were primarily in the pancreas and Dr. Hoffman has another such patient. In fact I think there are now about 15 or 18 patients reported in the literature, about 5 having primarily pancreatic tumors. So, I don't have an explanation. There is also a suggestion that perhaps GRF may be found outside the hypothalamus, but the data are conflicting and I don't really want to discuss it. I suppose that there is always a possibility that it could be eutopic production and GRF may be normally produced in the pancreas.

Structure and Expression of the Human Parathyroid Hormone Gene

HENRY M. KRONENBERG, TETSUYA IGARASHI, MASON W. FREEMAN, TOMOKI OKAZAKI, STEPHEN J. BRAND, KRISTINE M. WIREN, AND JOHN T. POTTS, JR.

Endocrine Unit, Massachusetts General Hospital, and Department of Medicine, Harvard Medical School, Boston, Massachusetts

I. Introduction

Parathyroid hormone (PTH) is an 84-rsidue protein which, along with vitamin D, regulates the level of ionized calcium in blood. PTH stimulates the release of calcium from bone, the reabsorption of calcium from the kidney's glomerular filtrate, and, through the 1-hydroxylation of 25-hydroxyvitamin D, the absorption of calcium across the intestinal mucosa. Each of these actions serves to increase the level of ionized calcium in blood. Calcium, in turn, is the major modulator of PTH release from the parathyroid gland. A fall in blood calcium stimulates PTH secretion; a rise in blood calcium suppresses PTH secretion. Thus, a negative feedback loop between calcium and PTH tends to stabilize the blood calcium within a narrow range (reviewed in Habener *et al.*, 1984).

While the regulation of secretion of PTH has been extensively studied, biosynthetic events in the parathyroid cell have been more difficult to examine. The pathway leading to PTH synthesis, like that leading to synthesis of other secreted peptides, involves a series of proteolytic cleavages of a large precursor (Cohn and Elting, 1983; Habener *et al.*, 1977b). The primary translation product, preproparathyroid hormone (prepropPTH), is cleaved in the rough endoplasmic reticulum to remove the amino-terminus "pre" or signal sequence, leaving behind the intermediate precursor, proparathyroid hormone (proPTH). ProPTH is cleaved to PTH in the Golgi apparatus and secretory granule (Figs. 1 and 2).

The messenger RNA that encodes prepropPTH is itself the product of cleavages of a precursor (Fig. 1). The gene encoding prepropPTH is first

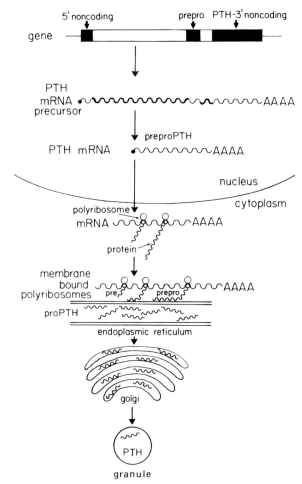

FIG. 1. Synthesis and secretion of parathyroid hormone in the parathyroid cell.

transcribed into a large precursor RNA molecule. Two stretches of sequence are then removed from the middle of the precursor, yielding the mature messenger RNA. Thus, the biosynthesis of PTH involves two complex processes—one nuclear and one cytoplasmic. This disucssion will concern itself with these two aspects of the biosynthetic machinery of the individual parathyroid cell. A summary of our current understanding of the structure and expression of the PTH gene will be followed by a description of the systems being used to analyze the functions of the peptide precursors of PTH.

HUMAN PARATHYROID HORMONE GENE 643

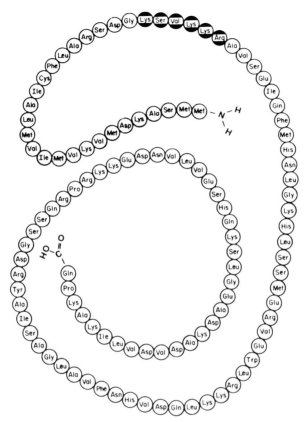

FIG. 2. Amino acid sequence of bovine preproparathyroid hormone. The "pro" sequence, in black, separates the amino-terminal "pre" or signal sequence, from the 84-residue mature hormone.

II. Parathyroid Hormone Gene: Structure and Function

A. ISOLATION OF CLONED DNA ENCODING PTH

The gene encoding parathyroid hormone (see Fig. 3), like most other genes, is flanked by DNA sequences controlling the gene's expression. The transcribed region contains two types of sequences—exons (black in Fig. 3) and introns, or intervening sequences (white). RNA transcribed from the exons is found in mature messenger RNA; RNA corresponding to the introns is transcribed along with the exons to form the primary RNA transcript, but the intronic sequences are subsequently spliced out

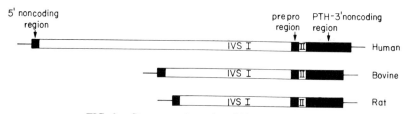

FIG. 3. Structure of prarthyroid hormone genes.

from the middle of the transcript to yield mature messenger RNA. The isolation of the PTH gene and determination of its detailed structure required the screening of a gene library using recombinant DNA techniques. Finding the PTH gene in such a library presented a formidable task that could not, at the time, be addressed directly. Since the parathyroid hormone gene represents only one part in a million of the mammalian genome, the parathyroid hormone gene necessarily represented a very small fraction of any gene library. In contrast to a library of genomic DNA, the parathyroid cell, uniquely specialized for the transcription of the PTH gene, contains relatively large amounts of mRNA encoding PTH. This mRNA, though missing the introns and flanking sequences, contains all of the sequences of the PTH gene's exons. By seeking to isolate PTH mRNA (several percent of the mRNA in the parathyroid cell) instead of PTH genomic DNA (one part in a million in the human genome), the problem of nucleic acid isolation could be simplified by several orders of magnitude. Characterizing the mRNA sequence was thus both a worthwhile goal in its own right, as well as a step toward obtaining the sequence of the entire PTH gene. Using standard techniques (Kronenberg *et al.*, 1979), mRNA from bovine parathyroid glands was partially purified and then transcribed into double-stranded DNA (cDNA) using a series of enzymatic steps. The DNA was then introduced into the plasmid pBR322 and amplified in *E. coli*. A plasmid encoding bovine PTH was identified by the DNA's ability to bind mRNA which could be translated into prepropTH.

With this first bovine cDNA clone in hand, the tremendous specificity and affinity of single-stranded DNA sequences for their complementary strands allowed us to use radioactive bovine PTH cDNA to identify and isolate the closely related human PTH cDNA (Hendy *et al.*, 1981). In turn, the human PTH cDNA was then used to screen hundreds of thousands of recombinant phage for those containing human genomic DNA encoding PTH (Vasicek *et al.*, 1983). A similar strategy allowed Heinrich *et al.* (1984) to isolate the rat PTH gene, and Kemper and co-workers

(Weaver et al., 1982, 1984) to isolate bovine PTH cDNA and genomic DNA.

B. SEQUENCES OF THE PTH GENES

By comparing the DNA sequences of cDNA with sequences of genomic DNA, it was possible to deduce which genomic sequences are found in the mature mRNA (exons) and which sequences are removed (introns). A comparison of the human, bovine, and rat PTH genes shows striking conservation of the precise locations of introns (Fig. 3). Each gene contains two introns. The first intron in each species comes precisely 5 base pairs before the sequence encoding the prepropPTH sequence. The second intron in each species interrupts the codon encoding the fourth amino acid of the "pro" sequence. The first intron is large in each PTH gene, though the human gene's first intron (3400 base pairs) is considerably larger than the first intron of the bovine (1714 bp) or rat (1600 bp) genes. The second introns are short and of similar size across the three species: human, 103 bp; bovine 119 bp; and rat 111 bp.

The pattern of intron insertion in genes often suggests that introns roughly separate discrete functional domains (Gilbert, 1978). Certainly the PTH gene's introns follow this rule. The first intron separates the sequences corresponding to the 5'-noncoding region of the mRNA from the rest of the gene. The second intron separates the sequence encoding the precursor-specific "prepro" region from that encoding mature PTH. A similar pattern of intervening sequence has been found in a number of genes encoding hormones such as gastrin (Wiborg et al., 1984) and insulin (Bell et al., 1980).

Not surprisingly, the three mammalian PTH genes resemble each other at the nucleotide level. The bovine and human proteins are identical in 85% of their amino acid residues. The rat protein has diverged somewhat from the other two proteins, but still maintains 75% identity with each of the other two proteins. A quantitatively similar relationship exists among the nucleotide sequences encoding the three proteins: the bovine and human sequences are identical over 85% of their bases; the rat/bovine identity is 78% and the rat/human identity is 77%. A similar pattern holds in the second intervening sequence: the bovine/human identity is 80%; rat/bovine, 74%; rat/human 70%. The 5'-noncoding regions of the mRNAs are somewhat less well conserved than the coding regions and the second intervening sequences. Because only the bovine first intervening sequence has been determined in its entirety, comparisons in this region cannot be made. The 3'-noncoding regions of the mRNAs are not well conserved. Even the more closely related human and bovine genes share only 48% homology.

C. GENE NUMBER AND CHROMOSOMAL LOCATION

The parathyroid hormone gene is represented only once in the haploid genomes of humans, rats, and cows (Vasicek *et al.*, 1983; Heinrich *et al.*, 1984; Weaver *et al.*, 1984). This conclusion follows from two kinds of data: first, the pattern of restriction enzyme generated bands on Southern blots of genomic DNA is no more complicated than predicted from the restriction enzyme cleavage sites in and near the cloned PTH genes; second, the intensities of restriction enzyme generated bands on Southern blots are most consistent with the presence of one gene/haploid genome. A systematic search for genes related to the PTH gene but substantially different in sequence has not been reported.

By analyzing a series of mouse–human hybrid cell lines, each containing a different group of human chromosomes, Naylor *et al.* (1983) determined that the human PTH gene is located on the short arm of chromosome 11. More recent studies using radioactive PTH genes to bind to chromosomal squashes (*in situ* hybridization) have confirmed the gene's assignment to the short arm of chromosome 11 (Zabel *et al.*, 1985) and further suggest that the gene is located near the end of the chromosome in region 11p15.

Genetic analysis of human families has allowed the mapping of the PTH gene with respect to other genes located on the short arm of chromosome 11. Polymorphic markers closely linked to the PTH gene, detected using restriction enzyme digestion of genomic DNA, make such an analysis possible. Antonarakis *et al.* (1983) noted that cleavage of the human PTH gene with the restriction endonuclease *Pst*I results in a pattern of cleavage sites that varies across the human population. *Pst*I always cleaves the human PTH gene at a particular point in the third exon. In 70% of human chromosomes, the next *Pst*I site is in the DNA flanking the 3'-end of the gene and is 2800 bp away (see Fig. 4). In 30% of human chromosomes, the nearest *Pst*I site is only 2200 bp away. Thus, about 40% of people are heterozygotes for the *Pst*I-generated polymorphism adjacent to the PTH gene. Schmidtke *et al.* (1984) discovered a second useful polymorphism in the human PTH gene generated by the restriction enzyme, *Taq*I. The variable *Taq*I site is due to a point mutation in the gene's second intervening sequence (Igarashi *et al.*, unpublished). The inheritance of these polymorphic markers and of similar markers near other genes on chromosome 11 was traced by a number of groups (Antonarakis *et al.*, 1983; Kittur *et al.*, 1985). A genetic map could be constructed by noting how often ge-

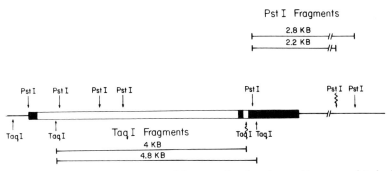

FIG. 4. Restriction-enzyme generated fragment length polymorphisms associated with the human PTH gene.

netic recombination occurred between the various markers over several generations in human families. Figure 5 shows the results of these studies. The PTH gene is most closely linked to the β-glogin gene and is also relatively near to the calcitonin, c-Harvey-*ras* I oncogene, and insulin genes. These mapping studies provide a framework for the mapping of other genetic markers on chromosome 11 (inherited diseases, in particular). The PTH gene polymorphisms allow us to determine whether inherited diseases of calcium metabolism are linked to the PTH gene (and are, therefore, possibly caused by abnormal PTH genes) in particular families (Levine *et al.*, 1984).

D. PROMOTER SEQUENCES

The functional analysis of the PTH gene is just beginning; consequently, little is yet understood about sequences important in regulating transcription of the PTH gene. Two types of regulation need greater definition and then explanation: first, the mechanism whereby the PTH gene is transcribed in the parathyroid cell and only in the parathyroid cell; second, the means whereby calcium, $1,25(OH)_2$ vitamin D, and possibly

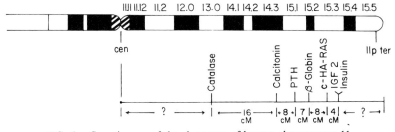

FIG. 5. Genetic map of the short arm of human chromosome 11.

other modulators influence the amount of PTH mRNA in the parathyroid cell. Early studies suggested that protein synthesis in the parathyroid gland was regulated by the level of extracellular calcium (Roth and Raisz, 1966; Roth *et al.*, 1968). More recent studies (Habener *et al.*, 1975; Chu *et al.*, 1973), measuring rates of PTH and proPTH synthesis, showed little influence of calcium on these rates. Heinrich *et al.* (1983) further showed that bovine PTH mRNA levels were unaffected by changes in extracellular calcium over 7 hours of *in vitro* culture. More recently, however, Russell *et al.* (1983) and Silver *et al.* (1985) have studied the effects of changes in extracellular calcium and $1,25(OH)_2$ vitamin D on cultured bovine parathyroid cells over several days. They found that elevations of either calcium or $1,25(OH)_2$ vitamin D led to a reversible fall in PTH mRNA levels to about 50% of baseline over several days.

The sequences that determine the tissue-specific transcription of particular genes as well as those that regulate the rate of gene transcription in response to environmental signals are often located upstream from the transcribed sequences. When these upstream sequences of the human, bovine, and rat PTH genes are compared, substantial conservation of sequence is found. In the 108 bases adjacent to the transcription start site, the human and bovine genes share 80% homology, the rat/human homology is 73%, and the rat/bovine homology is 64%. This sequence conservation suggests that the 5'-flanking region of the PTH genes contains important regulatory sequences.

The precise sites of initiation of transcription have been determined for each gene. Weaver *et al.* (1982) first showed that the bovine gene contains two functional start sites of transcription. Each start site is located about 30 bp downstream from a different TATA sequence, suggesting that these TATA sequences, like those near transcription initiation sites of many genes, determine the PTH gene's transcription initiation sites. The TATA sequences are separated by 30 bp. Igarashi *et al.* (1986) have recently demonstrated that the two homologous TATA sequences flanking the human PTH gene direct the synthesis of two human PTH gene transcripts both in normal parathyroid glands and in parathyroid adenomas. In contrast, Heinrich *et al.* (1984) showed that the rat PTH gene contains only one transcription start site, corresponding to the bovine and human upstream start sites. The downstream TATA sequence in the bovine and human genes is changed in the rat sequence from TATATATA to TGCA-TATG. The alteration of this concensus sequence may be responsible for the absence of a second initiation site in the rat gene. The functional importance of two different start sites for transcription in unknown. In other systems, the TATA sequences have been shown to bind a protein required for accurate and efficient *in vitro* gene transcription (Davison *et al.*, 1983). The use of two different start sites might lead to a greater rate of transcription of the PTH gene. Alternatively, each start site might be regulated in a different fashion. A few other genes have been shown to initiate transcription at two closely clustered locations. These genes include the chicken lysozyme gene (Grez *et al.*, 1981) and the avian very low-density apolipoprotein II gene (Hache *et al.*, 1983).

To examine the possible functions of the two transcription start sites in the PTH gene, the regulation of transcription of the gene in response to changes in extracellular calcium and $1,25(OH)_2$ vitamin D, and the determinants of cell-specific transcription, Igarashi *et al.* (1986) recently introduced the PTH gene into cultured cells. Normally behaving, cultured parathyroid cells would have been the preferred hosts for PTH gene expression, but unfortunately a cell line with these properties has not yet been established. Consequently, we inserted the PTH gene into a number of other established cell lines, looking for one with the following properties: first, that the cell line actively transcribe the inserted PTH gene, and, second, that the cell line contain intracellular second messengers similar to those of postulated importance in parathyroid cells. In order to rapidly screen cell lines for their ability to transcribe the PTH gene, a selectable marker, the *E. coli* gene encoding resistance to neomycin, was joined to a short portion of the PTH gene, just downstream from the 5'-flanking region of the PTH gene (see Fig. 6). The neomycin analog, G418, kills normal mammalian cells, but cells making sufficient aminoglycoside 3"-phosphotransferase II (the product of the *neo* gene) survive exposure to G418. By making *neo* expression dependent on activity of the PTH promoter (transcription control region), we could rapidly screen for cells which direct transcription from this promoter.

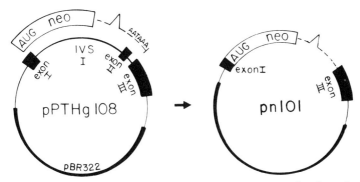

FIG. 6. Construction of plasmid pn101. Plasmid pPTHg108 contains the entire human PTH gene. The *E. coli neo* gene was substituted for a portion of the PTH gene, so that *neo* expression depended on the activity of the PTH gene's regulatory region.

As expected, cell lines varied substantially in their ability to survive G418 selection, after the plasmid pn101 (containing the fused PTH/neo gene) was introduced into the cells. HeLa cells failed completely to recognize the PTH promoter. In contrast, rat pituitary GH4 cells, which produce prolactin and growth hormone, became G418 resistant with high efficiency. When the start points of PTH gene transcription in these cells were directly assessed by SI nuclease analysis, the start sites precisely corresponded to the two clusters used by normal human parathyroid cells. The ability of GH4 cells to accurately recognize the PTH gene promoter was particularly fortunate, because the GH4 cells contain an elaborate, extensively characterized apparatus for regulating prolactin gene transcription. Specifically, swings of free cytoplasmic calcium and $1,25(OH)_2$ vitamin D, both potential modulators of transcription in the parathyroid cell, regulate the level of prolactin mRNA in GH4 cells. Further, both parathyroid and GH4 cells contain adenylate cyclases and modulate phosphotidylinositol synthesis and protein kinase C activity in response to extracellular signals. Thus, the GH4 cells expressing the inserted fusion PTH/*neo* gene may prove a useful model for studying regulation of transcription of the PTH gene. In preliminary studies, Igarashi *et al*. (1985) have shown that elevations of extracellular calcium and $1,25(OH)_2$ vitamin D lead to a fall in fusion PTH/neo mRNA in these cells. Future studies of these cell lines, as well as of other cells expressing the cloned PTH gene, should allow a definition of the signals regulating PTH gene transcription.

III. Intracellular Processing of Preproparathyroid Hormone

The pioneering studies of Habener *et al*. (1977b) and Cohn's group (Cohn and Elting, 1983) led to the definition of the precursors of PTH—prepropTH and propTH (see Fig. 2). PrepropTH is the primary translation product of the PTH mRNA. During protein synthesis the "pre" or signal sequence is cleaved from the precursor protein. A signal peptidase removes the signal as the protein traverses the membrane of the rough endoplasmic reticulum, leaving the intermediate precursor, propTH, in the cisternae of the endoplasmic reticulum. The cleavage is so rapid that very little intact prepropTH can be found in parathyroid cells. Habener and Potts (1979) showed that the small amount of prepropTH found in parathyroid cells is located on the cytoplasmic side of the endoplasmic reticulum and might well be degraded without tranversing the secretory pathway.

The relatively more stable propTH is the major precursor protein seen in parathyroid cells after brief pulses of radioactive amino acids. After a lag of several minutes, the proPTH moves to the Golgi apparatus where the dibasic residues at the end of the "pro" region direct another protease to cleave off the "pro" sequence. PTH, the mature hormone, is then packaged in secretory granules, and leaves the cell by exocytic fusion of the granule with the plasma membrane.

The "pre" or signal sequence of prepropTH resembles signal sequences found at the amino-termini of virtually all precursors of secreted protein. Signals differ widely in their primary amino acid sequences, but all contain certain general features. They all contain a core sequence of 10–15 hydrophobic, uncharged, nonpolar amino acid residues in a row. Preceding this hydrophobic core is a hydrophilic stretch of amino acids of variable length which usually contains at least one positively charged amino acid (if one counts the amino-terminal amino acid as positively charged, then a positive charge is invariable). The signal cleavage site is located five to six residues after the end of the hydrophobic core. Small amino acids are usually found just before the signal cleavage site (residue −1) and also at residue −3 (von Heijne, 1983). Studies from many laboratories have firmly established that the signal sequence serves to direct the polyribosome synthesizing secreted proteins to the rough endoplasmic reticulum (reviewed in Blobel, 1983). As the signal sequence emerges from the ribosome, it binds an 11 S signal recognition particle (SRP) (see Fig. 7). The SRP contains 6 peptides and a 7 S RNA molecule. In cell-free extracts of wheat germ, the binding of the signal sequence to an SRP results in a temporary halt in the elongation of the peptide chain. Whether this arrest in protein synthesis occurs in the intact cell is unknown. The

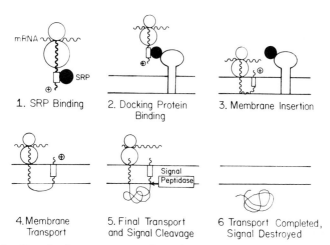

FIG. 7. Steps leading to transport of proteins across the endoplasmic reticulum.

SRP–polyribosome complex then binds to the endoplasmic reticulum. An integral membrane protein of the endoplasmic reticulum, called docking protein or SRP receptor, binds directly to the SRP (Meyer *et al.*, 1982). This binding is associated with the release of the polyribosome from the SRP and the subsequent insertion of the nascent polypeptide chain into the membrane of the endoplasmic reticulum.

While these early steps in the secretory pathway have a firm biochemical underpinning, the more distal steps associated with protein transport across the membrane of the endoplasmic reticulum remain largely uncharacterized. Two groups of hypotheses have been proposed to explain the mechanism of the transport of largely hydrophilic proteins through the hydrophobic membrane. One group suggests that a partially hydrophilic protein-lined pore guides the transport of protein across the membrane. The other group of hypotheses argue that proteins can cross the lipid bilayer directly, driven by some combination of the energy of protein synthesis, hydrophobic interactions with membrane constituents, and the energy of protein folding inside the cisternae of the endoplasmic reticulum.

As the precursor protein crosses the membrane of the endoplasmic reticulum, the signal sequence is cleaved by signal peptidase, an endopeptidase located on the inner surface of the reticular membrane. The number of signal peptidases is not clearly defined, but one signal peptidase, that found in canine pancreatic microsomes, can cleave virtually any signal sequence presented to it. This observation suggests that cells may need only one signal peptidase.

Figure 7 emphasizes that the transport of proteins across the membrane of the endoplasmic reticulum is a multistep process. The signal sequence is likely to have several discrete functions: binding SRPs (reversibly), inserting into the membrane, and presenting an appropriate cleavage site to signal peptidase. The distinct domains of the signal sequence may well play differing roles in these serial functions.

The "pro" sequence of preproPTH has a function considerably more obscure than the function of the "pre" sequence. The "pro" sequence contains only six residues, the last two of which allow the cleavage of the "pro" sequence from the mature hormone. No proPTH is secreted from the parathyroid cell. The "pro" peptide is rapidly degraded in the parathyroid cell after cleavage from PTH, making unlikely the functional importance of the free hexapeptide or further cleavage products of the "pro" peptide (Habener *et al.*, 1977a). The short, exclusively intracellular life span of proPTH suggests that the "pro" sequence may play a role in guiding PTH through the secretory pathway. The "pro" sequence may pariticpate with the "pre" sequence in directing the precursor across the membrane of the endoplasmic reticulum. The "pro" sequence could be important for assuring accurate and efficient cleavage of the "pre" sequence by signal peptides. Conceivably, the "pro" sequence is required for appropriate folding of the PTH molecule. The "pro" sequence might also be required for efficient transport of PTH from the endoplasmic reticulum to the Golgi apparatus and beyond.

To determine the functional importance of the "pre" and "pro" sequences of prepropTH, we have used a genetic approach. We have systematically altered the "pre" and "pro" sequences of prepropTH and then tested the functional implications of these alterations. Two types of expression systems have been used to determine the functional consequences of mutations in the prepropTH molecule: a cell-free, linked transcription–translation system in which plasmids direct the synthesis of proteins, and intact cultured cells, into which cDNA encoding prepropTH has been inserted.

Figure 8 outlines the strategy of cell-free protein synthesis. cDNA encoding prepropTH is placed distal to the *lac* promoter, an *E. coli* sequence that directs *E. coli* RNA polymerase to initiate RNA synthesis. The *lac* promoter–cDNA combination is introduced into a bacterial plasmid so that large amounts of the sequence can be amplified in *E. coli* and then isolated. The plasmid codes for one additional protein, pre-β-lactamase, the precursor of the protein that makes the bacterium resistant to ampicillin. In *E. coli*, pre-β-lactamase is cleaved to β-lactamase and the β-lactamase is transported across the plasma membrane into the periplasmic space. This process bears some resemblance to the mechanism used to transport mammalian proteins across endoplasmic reticulum. In fact, we (Kronenberg *et al.*, 1983) and others (Muller *et al.*, 1982) have shown that pre-β-lactamase is transported across endoplasmic reticulum from dog pancreas *in vitro*, in association with accurate cleavage of the bacterial signal sequence. Thus, the plasmid in Fig. 8 encodes two proteins which can be transported across the membrane of the endoplasmic reticulum. The pre-β-lactamase serves as a useful internal positive control for evaluating the dysfunction associated with mutant prepropTH molecules.

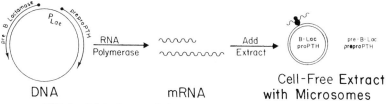

FIG. 8. Linked transcription–translation of prepropTH cDNA.

Following slight modifications of a protocol first developed by Brian Roberts and colleagues (Roberts et al., 1975), the plasmid is transcribed into RNA by simply adding salts, nucleotide triphosphates, and RNA polymerase. After adjusting buffer conditions, a cell-free protein-synthesizing extract is then added to the mixture. The cell-free extract, made either from wheat germ or from a lysate of rabbit reticulocytes, can then be supplemented with endoplasmic reticulum from dog pancreas. The proteins are made using radioactive amino acids, and their structures can be subsequently analyzed biochemically and immunologically. The location of the proteins—inside or outside of the membrane vesicles—can be ascertained by a proteolytic protection assay. Proteins susceptible to digestion with protease are outside the vesicles, while proteins that resist proteolysis are inside the vesicles. The inherent susceptibility of protected protein to proteolysis is demonstrated in control experiments in which the reticular membrane is solubilized with detergent.

The cell-free system offers several important advantages. Because pre-β-lactamase and prepropTH represent the major proteins synthesized, biochemical analysis of the proteins' cleavage sites and the sequestration of proteins inside vesicles are straightforward. Further, because amounts of SRPs, endoplasmic reticulum, mRNA and ribosomes can all be independently varied and added at different times, each step in the transport across the reticular membrane can be analyzed separately and quantitatively. The *in vitro* system has serious limitations, however. Most importantly, the system does not allow transport of proteins out of the endoplasmic reticulum. Consequently, the behavior of mutant proteins traversing the more distal portions of the secretory pathway cannot be examined. Further, though the cell-free systems do mimic events in the intact cell rather closely, one must always worry that the behavior of mutant proteins might partly result from some unanticipated artifactual limitation of the cell-free system (which, after all, uses artificial mRNA added to mixtures of components from different species and, in the case of wheat germ, even different kingdoms).

Consequently, we complement the use of the cell-free system with the use of intact cells. We use retroviral vectors to carry prepropTH cDNA into cultured cells and direct the expression of the cDNA. Figure 9 illustrates the relevant points of the retroviral life cycle that make retroviral vectors particularly advantageous. The RNA genome is transcribed in the cell into a double-stranded DNA, which is then inserted into the host cell's genome. The conversion of single-stranded RNA into double-stranded DNA uses reverse transcriptase in a multistep process that leads to the presence of duplicated sequences, called long terminal repeats (LTRs), at each end of the proviral DNA. The proviral DNA is always

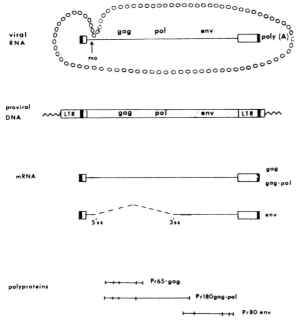

FIG. 9. Overview of retroviral biology.

inserted into the host genome in such a way that the ends of the LTRs form the junctions with the host cell genome. The LTRs contain enhancer and promoter elements that direct RNA polymerae II to transcribe the proviral DNA into RNA; the LTRs also determine the end points of viral RNA. The full-length RNA transcript serves as the viral genome and also serves as the messenger RNA encoding *gag* and *pol* functions—the genes responsible for encapsidating the viral RNA, for reverse-transcribing the RNA into DNA, and for subsequently integrating the proviral DNA into the host cell's genome. A shorter transcript, formed by splicing out the *gag* and *pol* genes, contains the *env* gene, which encodes proteins of the virus' membrane coat. A short sequence near the start of the *gag* gene (and therefore removed during the formation of *env* mRNA) directs the viral capsid proteins to package the full-length genomic RNA. Without this ψ (psi) or packaging sequence, the viral RNA fails to enter the virus budding from the surface of the cell.

To use cloned retroviral provirus as a cloning vector, Richard Mulligan and co-workers (Mann *et al.*, 1983) retained the regulatory sequences (the LTRs, the psi sequences, and the sequences bordering the intervening sequence) but removed the regions encoding the *gag, pol,* and *env* proteins. In place of the *gag–pol* sequences, we inserted the human prepro-

PTH cDNA. In place of the *env* sequence, we inserted the selectable marker gene encoding *E. coli* xanthine-guanine phosphoribosyltransferase (*gpt*). To convert this DNA sequence into an RNA viral genome and to provide the viral genome with a full complement of viral proteins, the DNA was inserted into an established cell line that was already producing all the protein encoded by Moloney murine leukemia virus. This particular cell line, called psi-2, contains a Moloney viral genome that is missing the psi sequence (see Fig. 10). Thus, it cannot package its own viral RNA into infectious viral particles. RNA transcribed from the recombinant PTH retrovirus, however, can be packaged by the proteins of the psi-2 cell. The resultant virus can then be used to infect almost any rodent cell line. After infecting a host cell line, the viral RNA is reverse transcribed into proviral DNA, and the DNA is integrated into the host cell's genome via the viral LTRs. At that point. free virus can no longer be

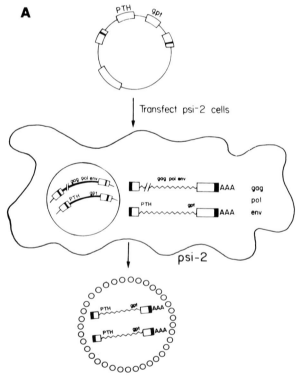

FIG. 10. (A) Passage of recombinant PTH retroviral DNA through psi-2 cells, resulting in production of helper-free, infectious PTH retrovirus. (B) Infection of rodent cells with PTH retrovirus.

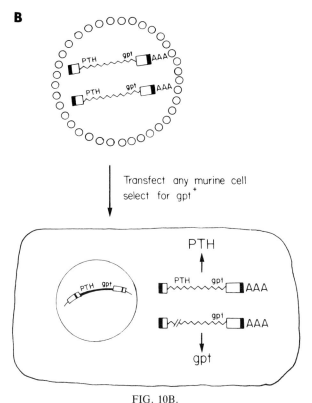

FIG. 10B.

produced. The viral control sequences can only direct the synthesis of prepropTH and the selectable protein, xanthine-guanine phosphoribosyltransferase. The viral proteins that made the initial infection possible cannot be replicated, since the infecting virus contained none of the *gag, pol,* or *env* sequences.

This use of retroviral vectors has proven advantageous for several reasons. The use of infectious virus allows the infection of a wide variety of cell lines, including those that are poor hosts for standard transfection methods. The use of the psi-2 line allows the use of viral proteins without generating cells that chronically bud virus from their plasma membranes. Since we are particularly interested in studying the effects of mutations on the ability of proteins to transverse the secretory pathway, we felt it worthwhile to avoid potential complications caused by the constant budding of virus into the medium bathing the cells. Finally, the retrovirus-directed insertion of one or a few proviral sequences into the host cell's genome, predictably via the LTR sequences, allows greater control of insertion of genetic material into cell lines than that afforded by standard transfection methods.

The host cell line that we have used most extensively is the rat pituitary GH4 line, a line that secretes both prolactin and growth hormone (Dannies and Tashjian, 1973). The PTH retrovirus was used to infect GH4 cells and clones of cells expressing the xanthine-guanine phosphoribosyltransferase were selected (Hellerman et al., 1984). These resultant lines were shown, as expected, to contain retroviral DNA inserted into the cell's genome via the LTRs and to produce the predicted unspliced and spliced RNA molecules. No evidence of infectious virus was found in the cells or the medium. The cells were pulse-labeled for 15 minutes with [^{35}S]methionine, then "chased" with cold methionine. Cell extracts and medium were examined for PTH-related peptides at various times, using immunoprecipitation followed by gel electrophoresis. Figure 11 shows the results of such an experiment. Almost no prepropTH was found in the cells. At the earliest time point proPTH and PTH were found in the cells. With time, the proPTH was converted to PTH, and after a lag of about 30 minutes, the PTH was secreted from the cell. The accuracy of the cleavages of prepropTH to proPTH and proPTH to PTH were verified by amino-terminal sequence determination using an automated Edman degradation procedure.

Since thyrotropin-releasing hormone (TRH) stimulates the secretion of prolactin from GH4 cells, we could ask whether TRH stimulated the secretion of PTH as well. The affirmative result allowed us to conclude that PTH left the cell using the same secretory apparatus used by prolactin.

The retroviral expression vector thus provides a useful, physiologic complement to the cell-free translation system for the analysis of the function of mutant prepropTH molecules. We have recently constructed a number of such mutant genes and have begun to analyze their pheno-

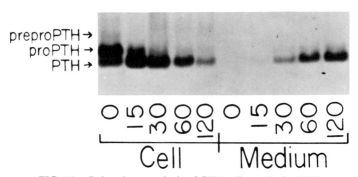

FIG. 11. Pulse-chase analysis of GH4 cells producing PTH.

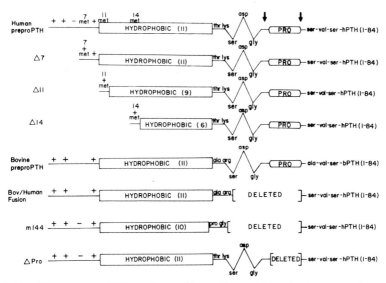

FIG. 12. Mutant prepropPTH gene inserted into cells to analyze signal sequence function.

types. Figure 12 schematically illustrates the mutants that we have constructed so far. The Δ7 mutant, missing the first 6 amino acids of the signal sequence, functions indistinguishably from normal prepropPTH. In contrast, all of the other mutants present abnormal phenotypes, ranging from rapid precursor degradation without any peptide secretion, to inefficient or altered signal cleavage, associated with diminished peptide secretion. The precise definition of the biochemical underpinnings of these mutant phenotypes should result in a functional characterization of the discrete domains in the "prepro" region of prepropPTH.

REFERENCES

Antonarakis, S. E., Phillips, J. A., Mallonee, R. L., Kazazian, H. H., Jr., Fearon, E. R., Waber, P. G., Kronenberg, H. M., Ullrich, A., and Meyers, D. A. (1983). *Proc. Natl. Acad. Sci. U.S.A.* **80,** 6615–6619.
Bell, G. I., Pictet, R. L., Rutter, W. J., Cordell, B., Tischer, E., and Goodman, H. M. (1980), *Nature (London)* **284,** 26–32.
Blobel, G. (1983). *In* "Methods in Enzymology" (S. Fleischer and B. Fleischer, eds.), Vol. 96, pp. 663–682. Academic Press, New York.
Chu, L., MacGregor, R., Anast, C., Hamilton, J., and Cohn, D. (1973). *Endocrinology* **93,** 915–924.
Cohn, D. V., and Elting, J. (1983). *Recent Prog. Horm. Res.* **39,** 181–209.
Dannies, P. S., and Tashjian, A. H. (1973). *J. Biol. Chem.* **248,** 6174–6179.
Davison, B. L., Egly, J.-M., Mulvihill, E. R., and Chambon, P. (1983). *Nature (London)* **301,** 680–686.
Gilbert, W. (1978). *Nature (London)* **271,** 501.

Grez, M., Land, H., Giesecke, K., and Schutz, G. (1981). *Cell* **25**, 743–752.
Habener, J. F., and Potts, J. T., Jr. (1979) *Endocrinology* **104**, 265–275.
Habener, J. F., Kemper, B., and Potts, J. T., Jr. (1975). *Endocrinology* **97**, 431–441.
Habener, J. F., Chang, H. T., and Potts, J. T., Jr. (1977a). *Biochemistry* **16**, 3910–3917.
Habener, J. F.,Kemper, B. W., Rich, A., and Potts, J. T., Jr. (1977b). *Recent Prog. Horm. Res.* **33**, 249–308.
Habener, J. F., Rosenblatt, M., and Potts, J. T., Jr. (1984). *Physiol. Rev.* **64**, 985–1053.
Hache, R. J. G., Wiskocil, R., Vasa, M., Roy, R. N. Lau, P. C. K., and Deeley, R. G. (1983). *J. Biol. Chem.* **258**, 4556–4564.
Heinrich, G., Kronenberg, H. M., Potts, J. T., Jr., and Habener, J. F. (1984). *J. Biol. Chem.* **259**, 3320–3329.
Hellerman, J. G., Cone, R. C., Segre, G. V., Potts, J. T., Jr., Rich, A., Mulligan, R. C., and Kronenberg, H. M. (1984). *Proc. Natl. Acad. Sci. U.S.A.* 5340–5344.
Hendy, G. N., Kronenberg, H. M., Potts, J. T., Jr., and Rich. A. (1981). *Proc. Natl. Acad. Sci. U.S.A.* **78**, 7365–7369.
Igarashi, T., Okazaki, T., Gaz, R. D., Potts, J. T., Jr., and Kronenberg, H. M. (1985). *Program Endocr. Soc.* Abstr No. 647.
Igarashi, T., Okazaki, T., Potter, H., Gaz, R. D., and Kronenberg, H. M. (1986). *Mol. Cell. Biol.* **6**, in press.
Kittur, S. D., Hoppener, J. W. M., Antonarakis, S. E., Daniels, J. D. J., Meyers, D. A., Maestri, N. E., Jansen, M., Korneluk, R. G., Gelkin, B. D., and Kazazian, H. H., Jr. (1985). *Proc. Natl. Acad. Sci. U.S.A.* **82**, 5064–5067.
Kronenberg, H. M., McDevitt, B. E., Majzoub, J. A., Nathans, J., Sharp, P. A., Potts, J. T., Jr., and Rich, A. (1979). *Proc. Natl. Acad. Sci. U.S.A.* **76**, 4981–4985.
Kronenberg, H. M., Fennick, B. J., and Vasicek, T. J. (1983). *J. Cell. Biol.* **96**, 1117–1119.
Levine, M. A., Antonarakis, S., and Kronenberg, H. M. (1984). *Annu. Meet. Am. Soc. Bone Min. Res.*, 6th Abstr. No. A56.
Mann, R., Mulligan, R. C., and Baltimore, D. (1983). *Cell* **33**, 153–159.
Meyer, D. I., Krause, E., and Dobberstein, B. (1982). *Nature (London)* **297**, 647–650.
Muller, M., Ibrahimi, I., Chang, C. N., Walter, P., and Blobel, G. (1982). *J. Biol. Chem.* **257**, 11860–11863.
Naylor, S. L., Sakaguchi, A. Y., Szoka, P., Hendy, G. H., Kronenberg, H., Rich, A., and Shows, T. B. (1983). *Somat. Cell Genet.* **9**, 609–616.
Roberts, B. E., Gorecki, M., Mulligan, R. C., Danna, K. J., Rozenblatt, S., and Rich, A. (1975). *Proc. Natl. Acad. Sci. U.S.A.* **72**, 1922–1926.
Roth, S. I., and Raisz, L. G. (1966). *Lab. Invest* **15**, 1187–1211.
Roth, S. I., Au, W., Kunin, A., Krane, S., and Raisz, L. G. (1968). *Am. J. Pathol.* **53**, 631–638.
Russell, J., Lettieri, D., and Sherwood, L. M. (1983). *J. Clin. Invest.* **72**, 1851–1855.
Schmidtke, J., Pape, B., Krengel, U., Langenback, U., Cooper, D. N., Breyel, E., and Meyer, H. (1984). *Hum. Genet.* **67**, 428–431.
Silver, J., Russell, J., and Sherwood, L. M. (1985). *Proc. Natl. Acad. Sci. U.S.A.* **82**, 4270–4273.
Vasicek, T., McDevitt, B., Hendy, G., Freeman, M., Potts, J. T., Jr., Rich, A., and Kronenberg, H. M. (1983). *Proc. Natl. Acad. Sci. U.S.A.* **80**, 2127–2131.
von Heijne, G. (1983). *Eur. J. Biochem.* **133**, 17–21.
Weaver, C. A., Gordon, D. F., and Kemper, B. (1982). *Mol. Cell. Endocrinol.* **28**, 411–424.
Weaver, C. A., Gordon, D. F., Kissel, M. S., Mead, D. A., and Kemper, B. (1984). *Gene* **28**, 319–329.
Weiss, R., Teich, N., Varmus, H., and Coffin, J., eds. (1982). "RNA Tumor Viruses." Cold Spring Harbor Laboratory , Cold Spring Harbor, New York.

Wiborg, O., Berglund, L., Boel, E., Norris, F., Norris, K., Rehfeld, J. F., Marcker, K. A., and Vuust, J. (1984). *Proc. Natl. Acad. Sci. U.S.A.* **81,** 1067–1069.
Zabel, B. U., Kronenberg, H. M., Bell, G. I., and Shows, T. B. (1985). *Cytogenet. Cell Genet.* **30,** 200–205.

DISCUSSION

L. A. Fitzpatrick. I am interested in your GH4 cells that produce both prolactin and PTH. You indicated that TRH increases both prolactin secretion and parathyroid hormone secretion. Have you tried raising intracellular calcium to see if this will produce secretion of both prolactin and parathyroid hormone in a manner opposite to that of a normal parathyroid cell?

H. M. Kronenberg. We haven't done that. It might well be. Describing our experiments is always a bit confusing, because we use GH4 cells for two dramatically different purposes. In these cells which you are asking about, all the sequences controlling protein synthesis are provided by retroviral vectors, so that we wouldn't expect the synthesis of PTH to be affected at all by calcium. We would expect, in fact, that calcium signals that cause the secretion of prolactin would similarly cause the secretion of PTH, but we haven't tested that prediction directly.

L. A. Fitzpatrick. My second question has to do with something that is of particular interest, namely that you had two parathyroid hormones start sites for transcription. Is this a usual situation? And would you care to speculate on what this redundancy may mean?

H. M. Kronenberg. It is not usual, but it's not without precedent. There are five or six other examples of two rather tightly clustered TATA boxes, and there are several examples of genes in which there are two rather widely dispersed start sites that end up getting together after splicing in the intact message. There are, broadly speaking, two possible explanations of the two start sties. One is that the phenomenon doesn't mean much of anything at all, that it doesn't take too much to be a functional TATA sequence, and that there's just nothing wrong with it. That the precise start site of the 5' noncoding region doesn't matter an awful lot, and that it's sort of evolutionary garbage. Another possibility is that 2 start sites may allow you to make more message from the same number of regulatory molecules: that's just speculation. The most interesting possibility, for which we have no data whatsoever as yet, is that the two start sites allow potential for differential regulation. What we know is that in adenomas, hyperplasia, and normal parathyroid glands, there's no dramatic change in the ratio of the two start sites. But we are just beginning to look in the GH4 cells, and we hope soon in parathyroid cells to see whether calcium and vitamin D signals might regulate one start site more than the other. We have no information whatsoever yet to report.

D. Goltzman. What physiologic or pathophysiologic situations would you envisage in which there would be inappropriate or erroneous processing so that either larger forms of PTH would be released or perhaps no PTH would be released at all?

H. M. Kronenberg. It's a very intriguing question, of course. We always hope that our work will lead to functional correlations, not just in cells but perhaps in patients with one problem or another. Bovine/human protein, which is cleaved very inefficiently and when uncleaved is not secreted, just stays, we think, in the endoplasmic reticulum. (There are preliminary data that suggest that conclusion.) Data such as this make us think that failure of signal cleavage leads to failure of secretion. The implication is that there may be somebody somewhere with hypoparathyroidism due to a similar defect in the structural prepropTH gene. We're looking for such patients, but the problem is that we don't have access to those hundreds and thousands of people who have sugar and insulin levels checked for no apparent reason. Inherited isolated hypoparathyroidism obviously is an unusual disease, and there are lots of potential causes for it. We haven't found any cases caused by secretory defects yet, but we're looking, and I think it's a very intriguing possibility.

M. L. Brandi. Could you speculate about the possible importance of the fact that insulin, PTH, and calcitonin genes are located near one another. My question is related to the possible implications in the multiple endocrine neoplastic syndrome.

H. M. Kronenberg. That's obviously a very intriguing observation, and we don't know what it means. The PTH gene and the calcitonin gene have nothing apparently in common with each other, and, of course, the PTH gene is closer to the β-globin gene than it is to the calcitonin gene. There are only so many chromosomes, and while I say that these genes are all together in the short arm of 11, they still are millions and millions of base pairs away from each other. There are only 23 pairs of chromosomes, so that a certain number of genes are going to be near each other coincidentally. Whether in fact there's something wrong with this chromosome in some of the M.E.N. 2 patients is possible, but you know this chromosome 11 is not the apparently abnormal chromosome that has been reported in some of the families with M.E.N. 2. It's certainly possible, but there's no data to suggest anything like that yet.

M. L. Brandi. Only a suggestion. There is another cell line that probably would be useful for you, and it's the Fisher rat thyroid line labeled as FRTL-S. In fact in these cells growth and differentiation are controlled by TSH and iodine uptake is modulated by calcium ions. They have receptors for the 1-25(OH)$_2$D$_3$, which regulates both the growth and the cyclic nucleotide accumulation in these cells. Finally, phosphatidylinositol controls the growth and differentiation of these cells.

H. M. Kronenberg. You obviously specialize in studying cells that are of great interest to us. I appreciate the comment.

M. Melner. I'm a bit concerned about the cell model system that you described in terms of the regulation experiments with neomycin resistance. My concern is that the few regulatory elements in DNA which have been described show very many properties in common with enhancers, and enhancers have the property of being highly cell specific, so that although you can find a cell model that has regulation, its ultimate physiological meaning may be questionable in terms of what you're interested in, regulation.

H. M. Kronenberg. You're certainly right that we only go into nonparathyroid cells with jeopardy, because, no matter what we find it will eventually be important to show that whatever happens in GH4 cells, also happens in the authentic parathyroid cell. We're not as worried as you are, however, about the model. The reason is that there are many genes that have been studied successfully using an analogous strategy. Much of the progress in analyzing the function of corticoid receptors, for instance, has resulted from taking genes that respond to glucocorticoids and putting them in virtually any cell that responds to glucocorticoids, and seeing what seems to be appropriate glucocorticoid regulation. It's always possible that the results will be unrelated to events in the parathyroid cell or that there will be important quantitative differences among cell lines. It's possible that we'll see a regulatory phenomenon that won't occur in the parathyroid cell, or will only superficially resemble what happens in the parathyroid cell. We'll have to be careful. It is equally likely, however, and testable, that at least some of the regulatory elements won't be absolutely confined to one particular cell, and that we, like many other investigators, will be able to exploit not just the cell of a gene's origin but also cells that are similar in relevant ways.

M. Melner. While many cells respond to glucocorticoids, very few cells in the body produce parathyroid hormone, so that is a source of concern.

H. M. Kronenberg. A more precise statement of that analogy illustrates my point. The growth hormone gene is normally expressed only in the somatotroph. Yet, when the growth hormone gene is introduced into many cells containing glucocorticoid receptors, the regulation of the growth hormone gene by glucocorticoids can be demonstrated. PTH responses to 1,25(OH)$_2$ vitamin D may be closely analogous. You're quite right, however, that these cells were chosen because there were no better ones around. We're cautiously optimistic, but have to wait and see whether any regulatory phenomenon that we find actually corresponds to the precise regulation in the parathyroid cell.

G. D. Aurbach. Your observations on the transformed GH4 cells were very interesting as are the effects of TRH thereon. In the usual GH cells, TRH action is mediated through a calcium effect that's positive on secretion. This is the reverse of the normal effect on parathyroid secretion, where calcium inhibits. I wonder if in your TRH experiments with transfected GH cells the calcium-dependent phenomena were positive or negative on secretion.

H. M. Kronenberg. We didn't look to see whether the response to TRH was dependent, though we expect it was. We think that the parathyroid hormone was in the secretory granule with prolactin and that it was the host cell, and not anything magic about the parathyroid hormone molecule itself, that determined the physiology of secretion. So, while we haven't tested it, we'd be flabbergasted if there were something unique about the parathyroid hormone molecule that resulted in a different pattern of regulated secretion from that determined by the host cell's secretory machinery.

G. D. Aurbach. I had just one question about the system that perhaps is somewhat far afield and may actually impinge on Dr. Goltzman's presentation. Ed Krebs once pointed out to me that the terminal sequence of parathyroid hormone actually represented a substrate for cyclic AMP stimulated protein kinase and that positions 1 and 3 could serve as sites for phosphorylation. Your experiments manipulating the amino terminal portion of the molecule and thereby changing the efficiency of conversion to parathyroid hormone raise the question of whether biochemical mechanisms control conversion of prepro- or pro-hormone to the final product.

H. M. Kronenberg. That's certainly a very inmportant observation. We're particularly intrigued by the question of what the "pro" sequence is for. The "pro" sequence is a funny little 6 amino acid sequence that is rapidly degraded in the cell. ProPTH is not like the ACTH precursor, which contains several bioactive molecules. The "pro" piece also seems unlikely to serve a function like that of the connecting peptide of insulin, which helps bring together two different chains to allow proper folding. One of the possibilities that we would like to test is whether the "pro" sequence functions in the secretory pathway. If we wiped out the "pro" sequence, would prePTH be processed normally? We know, not just from the kind of work you're describing, but from all the relationships of structure and function that have been established in Dr. Potts' lab and elsewhere that the amino-terminus of PTH is very important for its biological activity. Those amino acids are invariably small at position number 1. We wondered whether the "pro" sequence was present to allow the amino-terminus of PTH to interact with its receptor, go through its own evolutionary constraints, and yet allow the signal to follow its own drummer, with constraints determined by the specificity of signal peptidase. This hypothesis would imply that signals cannot function completely without regard to adjacent sequences. When Kris Wiren made prePTH, we wondered whether the cleavage would be abnormal. Preliminary results suggest that, at least *in vitro,* the signal is cleaved off normally. We need data from intact cells and measures of the efficiency of original cleavage to more fully examine the function of the "pro" sequence. These experiments are in progress.

Studies of the Multiple Molecular Forms of Bioactive Parathyroid Hormone and Parathyroid Hormone-Like Substances

DAVID GOLTZMAN, HUGH P. J. BENNETT, MICHAEL KOUTSILIERIS, JANE MITCHELL, SHAFAAT A. RABBANI, AND MARIE F. ROULEAU

Departments of Medicine, McGill University and the Royal Victoria Hospital, Montreal, Canada

I. Introduction

Considerable advances have been made in recent years in our knowledge of the chemistry of parathyroid hormone (PTH), and in the correlation of the chemistry with hormonal function. Within the parathyroid cell the biosynthetic pathway of PTH is known to conform to that of other secretory peptides (Cohn and Elting, 1983; Habener *et al.*, 1977). Thus, the translation product of messenger RNA, preproparathyroid hormone, has been shown to contain an NH_2-terminal hydrophobic leader sequence which is removed as the nascent hormone enters the cisternae of the endoplasmic reticulum (Habener *et al.*, 1978; Kemper *et al.*, 1974); the remaining peptide, proparathyroid hormone (ProPTH) (Cohn *et al.*, 1972; Kemper *et al.*, 1972), is extended at the NH_2-terminus by a basic hexapeptide which is removed by a tryptic-like cleavage (Goltzman *et al.*, 1976a), presumably in the Golgi complex. The major glandular molecular entity, thus formed, an 84 amino acid straight chain peptide, is eventually released from the gland, mainly via a hypocalcemic stimulus.

The use of sequence-specific radioimmunoassays and gel filtration analysis led to the observation that PTH within the circulation also exists in heterogeneous forms. Studies by a number of workers have shown that the major circulating immunoreactive forms of the hormone consist of mid-region and COOH fragments and that relatively little of the circulating hormonal material is composed of the major glandular form of PTH, PTH-(1–84) (Arnaud *et al.*, 1974; Berson and Yalow, 1968; Canterbury and Reiss, 1972; Habener *et al.*, 1971; Segre *et al.*, 1972; Silverman and Yalow, 1973). The source of these fragments is believed to be peripheral degradation in sites such as liver (Canterbury *et al.*, 1975; D'Amour *et al.*, 1979; Neuman *et al.*, 1975; Singer *et al.*, 1975), as well as the parathyroid gland itself (Flueck et al., 1977; Hanley et al., 1978; Mayer et al, 1979), where intraglandular degradation may be a function of the ambient calcium level. To date these techniques have failed to convincingly demonstrate the presence of NH_2-terminal fragments within the circulation. In addition to being degraded by tissues such as liver, PTH-(1–84) is capable as well of being cleared from the circulation by the kidney (Freitag et al., 1978; Hruska et al., 1975). The clearance of mid-region and COOH fragments seems even more dependent on renal mechanisms; consequently, exceptionally high levels of immunoreactive PTH have been detected in the presence of renal failure.

In view of the complexity of the metabolism of PTH and the existence of multiple circulating molecular forms, studies correlating the chemical structure of PTH with its function assumed increasing importance in an effort to interpret the significance of hormonal metabolism. Such studies demonstrated that despite the presence of 84 residues within the molecule, a synthetic peptide composed of the NH_2-terminal 34 residues appeared to contain all of the bioactivity of the molecule (Potts et al., 1971; Tregear et al., 1973) in a variety of bioassay systems. Extension of the molecule at the NH_2-terminus, as in ProPTH, resulted in a sharp drop in bioactivity (Goltzman et al., 1975; Peytremann et al., 1975). Deletion of the COOH end of PTH to position 26 could apparently be tolerated before an essentially inert peptide resulted (Goltzman et al., 1978; Tregear et al., 1973). In contrast, at the NH_2-terminus, deletion of the first two residues not only resulted in a peptide which was essentially inactive, but which in vitro, at least, appeared capable of acting as a hormonal antagonist through its continued ability to bind to PTH receptors. (Goltzman et al., 1976b). Subsequently, modification of this deleted peptide has resulted in more potent peptides (Rosenblatt et al., 1977), such as the analog [norleucine-8,18,tyrosine-34]bovine PTH-(3–34)-amide ([$Nle^{8,18}$,Tyr^{34}]bPTH-(3–34)-NH_2), which has been widely used as an in vitro peptide antagonist.

In view of these findings from structure–function studies, which localized the bioactivity of the molecule to the NH_2-terminal region, the possibility that an NH_2-terminal fragment could enter the circulation as a consequence of the metabolism of PTH-(1–84) took on increasing importance. Although in previous studies we demonstrated that cleavage of PTH-(1–84) to an NH_2-terminal fragment was not an essential requirement for action (adenylate cyclase stimulation) of the hormone in either kidney (Goltzman et al., 1976c) or bone (Goltzman, 1978), the possibility still remained that such an NH_2-terminal fragment might represent either an additional bioactive form or an entity with a different spectrum of bioactivity from PTH-(1–84).

In our recent studies, we have therefore examined molecular forms of PTH and PTH-like substances at three levels. We examined the hormone synthesized within the parathyroid gland to determine if potential chemical modifications of the peptide chain might occur; we reassessed the heterogeneity of circulating PTH or PTH-like substances in four clinical states, employing bioassay techniques to evaluate the biologically active forms; and we examined the early interaction of PTH with its target tissues, both to examine the precise cellular locus of this interaction *in vivo* as well as to determine if there might be a role for the mid-region or COOH end of the PTH-(1–84) molecule in such interactions.

II. Intraglandular Modification of Parathyroid Hormone

We first examined the possibility that PTH might be phosphorylated within the parathyroid gland (Rabbani *et al.*, 1984). Human parathyroid glands obtained at surgery or bovine glands from a local abattoir were incubated in short-term culture with inorganic ^{32}P and radiolabeled amino acids. PTH-(1–84) was subsequently purified from these glands by gel filtration, ion-exchange chromatography, and reverse-phase high-performance liquid chromatography (HPLC), monitoring purification with NH_2-terminal and COOH-terminal radioimmunoassays. The purified peak of immunoreactive PTH contained ^{32}P, suggesting that phosphorylation of the hormone was indeed occurring under these *in vitro* conditions. To determine the extent of phosphorylation, similar incubations were performed but purification was carried out entirely by reverse-phase HPLC (Bennett *et al.*, 1981). This permitted separation of the phosphorylated and unphosphorylated peaks (Fig. 1), both of which were found to have the amino acid composition of PTH-(1–84). Phosphorylated material accounted for approximately 10–20% of the total hormone extracted.

In view of the fact that both the bovine and the human hormones were phosphorylated, and that the bovine hormone lacks a threonine and the human hormone lacks a tyrosine, it appeared likely that phosphorylation was occurring on serine residues within the two molecules. Consequently, parathyroid glands were incubated with [^{3}H]serine, and hormone was extracted and purified. After subjecting the hormone, internally labeled with [^{3}H]serine, to mild acid hydrolysis (to prevent chemical dephosphorylation), the amino acid mixture was derivatized by dansylation. The derivatized residues were then resolved by reverse-phase HPLC. Two peaks of ^{3}H radioactivity were detected in this way, one coeluting with the derivatized serine standard and the other coeluting with the derivatized phosphoserine standard; these studies therefore indicated that phosphorylation was occurring on serine residues. To determine the re-

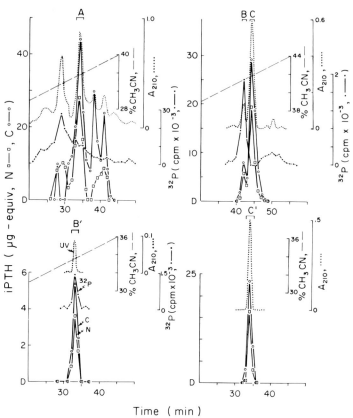

FIG. 1. Analysis by reverse-phase HPLC of ^{32}P-labeled peptides extracted from bovine parathyroid tissue after incubation of the tissue with inorganic ^{32}P for 4 hours. PTH immunoreactivity was determined in eluted fractions with both NH$_2$-directed (N, □) and COOH-directed (C, ○) assays and expressed as microgram equivalents (μg equiv) of bovine (b) PTH-(1–84). The detection limit of the assays is represented by D. UV absorbance was determined at 210 nm (A_{210}, ····). After initial separation (upper left), the major peak (A) containing immunoreactivity, ^{32}P (●), and UV absorbance, was then rerun on the same C$_{18}$ μBondapak column (upper right). The peak (B) of coeluting ^{32}P, immunoreactivity, and UV absorbance was then rerun (lower left) resulting is an apparently homogeneous peak (B′) which retained ^{32}P, and had the amino acid composition of bPTH-(1–84). Similarly, the peak (C) of coeluting immunoreactivity and UV absorbance which was devoid of ^{32}P (upper right) was rerun (lower right) resulting in an apparently homogeneous peak (C′) which also had the amino acid composition of bPTH-(1–84). (From Rabbani et al., 1984.)

gion of the molecule phosphorylated, glands were incubated with inorganic ^{32}P and [^{35}S]methionine, and ^{32}P-labeled hormone with incorporated [^{35}S]methionine was purified. The purified hormone was then subjected to dilute acid hydrolysis which is known to cleave the molecule at aspartic acid residues (Keutmann et al., 1971). Several fragments are thereby produced, the most NH_2-terminal of which is a 1–29 fragment. Such a fragment contains both methionines, which are present at positions 8 and 18 of the human and bovine molecules, and would be the only fragment labeled with ^{35}S. After subjecting the hormone to dilute acid hydrolysis, the mixture of fragments was resolved by reverse-phase HPLC. Only a single peak of ^{32}P was seen, which coeluted with the single peak of ^{35}S. Consequently, the molecule appeared to be phosphorylated on serine residues within the NH_2-terminal region of the molecule.

Two serine residues are conserved, within the NH_2-terminal 29 residues of the molecule, in all species of PTH whose sequence has been determined to date (Heinrich et al., 1984). One serine, at position three, would be present within the sequence lysine-X-X-serine, if the molecule were still in its precursor form (Fig. 2). Such a sequence has been demonstrated to be a favorable substrate for cyclic AMP-dependent kinases. Phosphorylation of this serine, close to the prohormone-to-hormone cleavage site, could help regulate the rate of formation of the 1–84 peptide. The second serine, at position 17, is located within the sequence serine-X-glutamic acid which is a recognition site for the non-cyclic AMP-dependent kinases ("physiological casein kinases"). Further studies are required to determine whether phosphorylation is, in fact, a posttranslational or cotranslational event (Hatfield et al., 1982) and to define the

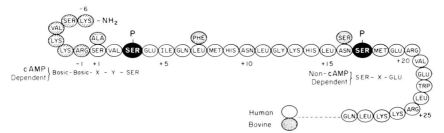

FIG. 2. Amino acid sequence of the 1–29 fragment of human (h) PTH (open circles) extended at the NH_2-terminus by the prohormone hexapeptide (hatched circles). Amino acid substitutions in the bovine molecule are depicted by shaded circles. Serine residues at positions 3 and 17 are conserved in both species, and one or both are the likely sites of phosphorylation of the molecule. The serine at position 3 when located within the prohormone sequence, Lys-Arg-X-Y-Ser, could be a substrate for a cyclic AMP-dependent kinase, whereas the serine at position 7, located within the sequence Ser-X-Glu, could be a substrate for a non-cyclic AMP-dependent protein kinase ("physiological casein kinase").

precise site and mechanism of this modification. The presence of phosphorylation within the NH_2-terminal portion of the molecule needs also to be explored in terms of the potential effect of this modification on the bioactivity of the hormone.

III. Examination of the Heterogeneity of Circulating Bioactive PTH and PTH-Like Substances in Disease States

Sequence-specific radioimmunoassays have demonstrated that mid-region and COOH fragments circulate in hyperparathyroid states. Additionally, microsequencing studies have determined that the earliest cleavage site of the PTH-(1–84) molecule which generates such fragments occurs between residues 33 and 34, and at more COOH-terminal sites (Segre *et al.*, 1974; Segre *et al.*, 1976), permitting the production of a bioactive NH_2-terminal fragment. Nevertheless, radioimmunoassay studies have generally failed to detect circulating NH_2-terminal PTH fragments. To more directly examine the nature of the circulating bioactive forms of PTH, we employed a cytochemical bioassay for PTH, similar to that previously described by Chayen and co-workers (Chayen *et al.*, 1976). This assay is based upon dose-dependent stimulation by PTH of glucose-6-phosphate dehydrogenase activity in guinea pig kidney tubules maintained in short-term organ culture (Goltzman *et al.*, 1980; Goltzman, 1983). Hormonal stimulation of enzyme activity is then assessed by quantitative cytochemistry. The validity of the assay for PTH measurement has now been confirmed by several groups (Chambers *et al.*, 1978; Fenton *et al.*, 1978). Although limited in its applicability because of its low through-put, the major advantage of the assay is its exquisite sensitivity. We utilized this assay, in conjunction with other biochemical techniques, to examine the nature of bioactive PTH, or PTH-like substances in four clinical conditions: hyperparathyroidism with normal renal function, hyperparathyroidism with diminished renal function, pseudohypoparathyroidism, and the syndrome of hypercalcemia associated with malignancy.

A. HYPERPARATHYROIDISM WITH NORMAL RENAL FUNCTION

In initial studies, the nature of circulating bioactive forms of PTH in hyperparathyroid states associated with normal renal function was examined (Goltzman *et al.*, 1980). Plasma was obtained from patients with either primary hyperparathyroidism, or X-linked hypophosphatemic rickets and secondary hyperparathyroidism. These samples were then subjected to gel chromatography on columns of BioGel P-100, and effluent fractions were bioassayed in the renal cytochemical bioassay. The majority of the bioactivity was found to coelute with the column marker for PTH-(1–84), and no appreciable bioactivity was found eluting in the position of the column marker for the NH_2-terminal fragment PTH-(1–34) (Fig. 3). Consequently, most of the circulating bioactivity appeared similar or identical to PTH-(1–84), the major glandular form of the hormone. These studies were therefore consistent with the thesis that the major cause of increased circulating bioactive PTH in hyperparathyroid states associated with normal renal function was accelerated secretion of the major glandular form of the hormone; there appeared to be little or no contribution, in these disease states, of intra- or extraglandular metabolism of PTH-(1–84) to the circulating complement of hormonal bioactivity.

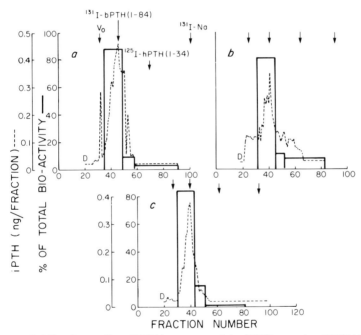

FIG. 3. Gel filtration profiles of PTH immunoreactivity (iPTH, - - -) and PTH bioactivity in a renal cytochemical assay (———) after chromatography on BioGel P-100 of plasma from two patients (a) and (c) with primary hyperparathyroidism and one patient (b) with X-linked hypophosphatemic rickets and secondary hyperparathyroidism. All patients had normal renal function. For bioactivity the height of the solid bars indicates the percentage of total bioactivity represented by pooled eluted fractions and the width of the bars, the size of the pools assayed. Vertical arrows from left to right denote, respectively, the elution position of the void volume (V_0), the intact hormone [^{131}I-bPTH-(1–84)], an active fragment [^{125}I-hPTH-(1–34)], and the salt volume (^{131}I-Na). (From Goltzman et al., 1980).

B. HYPERPARATHYROIDISM WITH DIMINISHED RENAL FUNCTION

We then utilized the same approach to assess the nature of the circulating bioactivity in hyperparathyroidism with chronic renal failure as had been employed in our studies with hyperparathyroidism and normal renal function; that is, plasma obtained from patients on hemodialysis with end-stage renal disease and secondary hyperparathyroidism was subjected to gel filtration on BioGel P-100, and eluted fractions were bioassayed in the renal cytochemical assay (Goltzman et al., 1980). In these studies, again, a substantial amount of bioactivity coeluted with the marker for the major glandular form of the hormone. However, in contrast to the previous studies, a considerable proportion of the total bioactivity recovered eluted later than PTH-(1–84), in the position of PTH-(1–34) (Fig. 4). The bioactivity of this material could be inhibited by the PTH antagonist,

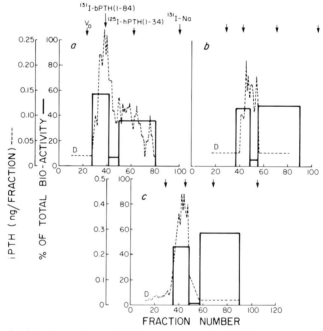

FIG. 4. Gel filtration profiles of PTH immunoreactivity (iPTH,- - -) and PTH bioactivity in a renal cytochemical assay (———) after chromatography on BioGel P-100 of plasma from three patients, (a) to (c), with uremic secondary hyperparathyroidism while on chronic hemodialysis. Bioactivity is depicted as described in the legend to Fig. 3, and vertical arrows represent the regions defined in the legend to Fig. 3. (From Goltzman et al., 1980.)

[Nle^{8,18}, Tyr^{34}]bPTH-(3–34)-NH$_2$, confirming its relationship to PTH. Consequently, in patients with chronic uremia, but not in those with normal renal function, small-molecular-weight bioactivity could be detected, in addition to bioactivity eluting in the position of PTH-(1–84).

Further studies were then initiated to determine the origin of this small-molecular-weight bioactivity. To assess whether reduced clearance could account for increased circulating concentrations, the levels of total bioactivity in plasma were determined sequentially after parathyroidectomy in patients with chronic renal failure, in patients with normal renal function, and in patients who had received a renal transplant to normalize kidney function (Goltzman *et al.,* 1984). In patients with chronic renal failure, when compared to patients with normal renal function or to those who had received renal transplants, clearance of mid-region and COOH-terminal immunoreactive PTH forms was markedly delayed. In contrast, the rates of disappearance of total levels of bioactivity were not significantly different among the three groups of patients (Fig. 5). These studies therefore confirmed the predominant role of the kidney in the metabolism of mid-region and COOH-terminal immunoreactive fragments. More importantly, they also indicated that the kidney may play a rather minor role in the disposal of bioactive hormone once renal failure is long-standing.

Computer analysis of the disappearance curves indicated that a single component model provided the best fit of the bioactivity data whether from patients with chronic renal failure or normal renal function; therefore, two components, one related to each of the two major forms of bioactivity detected by gel filtration analysis, could not be detected in the disappearance curves of the uremic patients. These findings suggest that the disappearance rates of the two bioactive entities noted in chronic renal failure are sufficiently similar so that they could not be resolved with the data available and with the mathematical approach used. The overall conclusions from these studies, therefore, were that delayed clearance could not account for the substantial levels of small-molecular-weight bioactivity observed in the plasma of patients with hyperparathyroidism and chronic renal failure.

To determine the site of origin of the small-molecular-weight bioactive material, plasma from the parathyroid venous effluent of a patient with chronic renal failure was subjected to gel filtration analysis (Goltzman *et al.,* 1980). In contrast to the marked levels of small-molecular-weight bioactivity found in peripheral uremic plasma, the major bioactive moiety in the parathyroid venous drainage of the uremic patient was found, by gel filtration, to be similar or identical to PTH-(1–84); only minor amounts of small-molecular-weight bioactivity were observed (Fig. 6). Consequently,

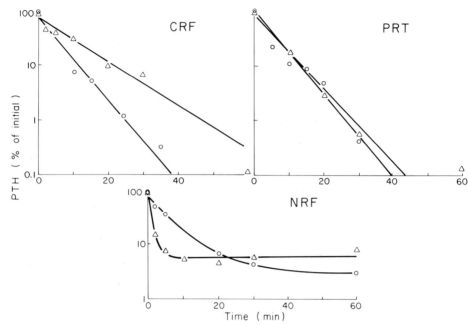

FIG. 5. Comparison of the disappearance of circulating bioactive PTH with time: after total parathyroidectomy in two patients each with secondary hyperparathyroidism and chronic renal failure (CRF); after subtotal parathyroidectomy in two patients each with persistent hyperparathyroidism after renal transplantation (PRT); after removal of a parathyroid adenoma from each of two patients with primary hyperparathyroidism and normal renal function (NRF). PTH was determined in plasma samples by renal cytochemical bioassay, and continuous lines were derived by EXPFIT computer program to produce best fit to data points. Bioactivity disappeared according to a single component model in patients with CRF and PRT, with $t_{\frac{1}{2}}$ values of 3.9 ± 0.13 to 7.6 ± 0.7 minutes. In patients with NRF, best fit of the data was obtained with a single component model plus a baseline with $t_{\frac{1}{2}}$s of 0.66 ± 0.08 and 4.9 ± 0.3 minutes. No significant differences in the half-times of disappearance of bioactivity was observed in any of the patients. (From Goltzman et al., 1984.)

the parathyroid gland seems not to be the source of this material in uremia; instead, the major site of origin would appear to be increased peripheral production from PTH-(1–84).

To assess the nature of the bioactive PTH-related entities occurring at different stages of renal insufficiency, a model of progressive renal failure in dogs (Slatopolsky et al., 1971) was employed. By sequential two-thirds infarction, first of one kidney, then of the other, at 7–10 day intervals, followed by complete removal of the first kidney, a model of renal failure of progressive severity is produced. This model can be shown to be accompanied by hyperparathyroidism of increasing severity with circulating

ANALYSIS OF BIOACTIVE PARATHYROID HORMONE 675

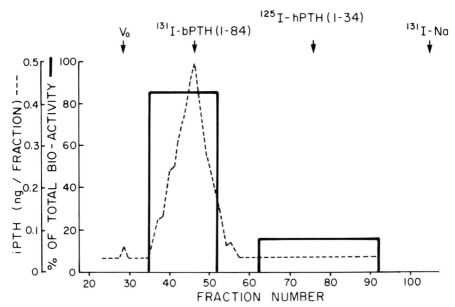

FIG. 6. Gel filtration profile of PTH immunoreactivity (iPTH, – – –) and bioactivity in a renal cytochemical assay (———) after chromatography on BioGel P-100 of plasma from the venous effluent of a parathyroid autograft in a uremic patient on chronic hemodialysis. Bioactivity is depicted as described in the legend to Fig. 3; vertical arrows represent the regions defined in the legend to Fig. 3. (From Goltzman *et al.*, 1980).

levels of both immunoreactive and bioactive PTH rising dramatically at each stage of renal failure (Grunbaum *et al.*, 1984). Plasma was obtained at successive stages for gel filtration analysis of bioactive forms. In the first and second stages (associated with exogenous creatinine clearances of 58 ± 12 and 20 ± 3 ml/min, respectively), the majority of the bioactivity eluted in the position of PTH-(1–84). Only in the third and final stage (associated with an exogenous creatinine clearance of 8 ± 1 ml/min) was a modest amount of small-molecular-weight bioactivity first seen (Fig. 7). Consequently, these results indicate that the pattern of PTH metabolism in the presence of normal renal function is altered to provide circulating small-molecular-weight entities only when renal insufficiency is severe and of sufficient duration. In view of the fact that the renal and presumed skeletal effects of PTH were most marked in the final stage of renal failure in this canine model, when small-molecular-weight material was first noted, further studies are required to test the possibility that the potency of small-molecular-weight bioactive material appearing in more advanced uremia exceeds that of the larger compound.

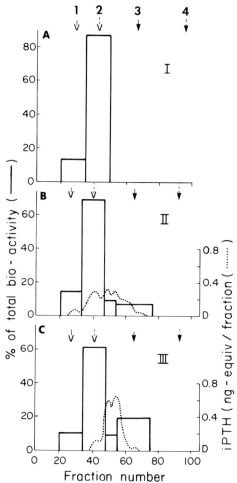

FIG. 7. Gel filtration profiles of PTH COOH-terminal immunoreactivity (iPTH, broken line) and of PTH bioactivity (solid line) after chromatography on BioGel P-100 of plasma from dogs with stages I (A), II (B), and III (C) of progressive renal failure. Exogenous creatinine clearances in normal dogs and in dogs with Stages I, II, and III of renal failure were 78 ± 12, 58 ± 7, 20 ± 3, and 8 ± 1 ml/minute. For bioactivity, which was determined in a renal cytochemical assay, the height of the solid bars indicates the percentage of total bioactivity in pooled eluted fractions, and the width of the bars indicates the size of pools assayed. Vertical arrows from left to right denote, respectively, the elution positions of the void volume (1), of the intact hormone, ^{131}I-labeled bPTH-(1–84) (2), of an active fragment, ^{125}I-labeled hPTH-(1–34) (3), and of the salt volume, ^{131}I-Na (4). (From Grunbaum et al., 1984.)

C. PSEUDOHYPOPARATHYROIDISM

Although pseudohypoparathyroidism (Albright et al., 1942) is an entity characterized by biochemical hypoparathyroidism, we have operationally defined the disorder as a hyperparathyroid state, in view of the high circulating levels of PTH which have been reported. Pseudohypoparathyroidism is a heterogeneous disorder in which resistance to the renal effect of injected exogenous bioactive PTH is recognized as a hallmark (Chase et al., 1969). This resistance is characterized by failure to increase urinary cAMP and phosphate output in response to administered hormone, resulting in the condition known as pseudohypoparathyroidism type I, or may be characterized by a normal increase in urinary cAMP excretion in response to administered PTH, but a low, or absent, phosphaturic response, a condition referred to as pseudohypoparathyroidism type II (Drezner et al., 1973). The type I disorder may (in the type IA form), or may not (in the type IB form), be associated with a variety of somatic disorders known collectively as Albright's hereditary osteodystrophy.

Recent attempts to understand the syndrome have been advanced by the discovery that most, but not all, patients with the type IA syndrome have a deficiency in the guanine nucleotide regulatory protein (G_S or N_S) of the adenylate cyclase system (Farfel et al., 1980; Levine et al., 1980). This observation could explain the resistance to several hormones, including PTH, which may occur in this syndrome but does not appear to explain the preponderance of hypoparathyroidism as a manifestation of hormonal ineffectiveness. Other studies have suggested that bioineffective PTH might be produced by these patients (Nagant de Deuxchaisnes et al., 1981) accompanied by inhibition of the effects of exogenously administered hormone (Loveridge et al., 1982).

We examined characteristics of circulating PTH in three patients with type IA pseudohypoparathyroidism, all of whom were normocalcemic on vitamin D therapy (Mitchell and Goltzman, 1985). All three had persistently elevated levels of mid-region and COOH-terminal immunoreactivity, but reduced levels of bioactivity, as assessed in the renal cytochemical bioassay (Fig. 8). Although the reduced levels of bioactivity could be considered appropriate suppression in the face of normocalcemia induced by vitamin D therapy, the elevated immunoreactive PTH levels seemed clearly inappropriate; in view of the presence of normal renal function, the parathyroid gland appeared to be the source of the persistent immunoreactivity. Analysis of the plasma immunoreactivity of these patients by reverse-phase HPLC, and comparison with patterns in patients with primary hyperparathyroidism, or with secondary hyperparathyroidism and vitamin D deficiency, indicated that the profiles were different from those

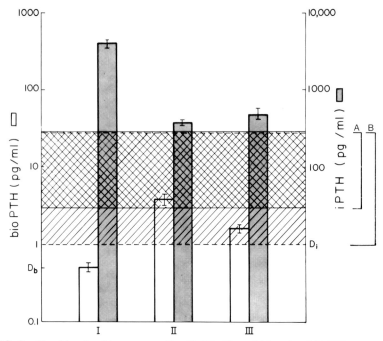

FIG. 8. Basal levels of immunoreactive (iPTH, ■) and bioactive (bioPTH, □) PTH in three patients with pseudohyparathyroidism type IA. Plasma bioactivity was determined in a renal cytochemical assay. The normal range in the bioassay is depicted by A and the normal range in the immunoassay is depicted by B. Each bar represents the mean ± SE of triplicate determinations. D_b and D_i represent the detection limits of the bioassay and immunoassay, respectively. (From Mitchell and Goltzman, 1985.)

seen in the latter two disorders. We then determined that plasma from these patients could inhibit the effect of exogenous bovine PTH in two *in vitro* bioassays for PTH, a renal cytochemical assay and a renal adenylate cyclase assay. The locus of this inhibition appeared to be the target tissue per se, since circulating inhibitors to bioactive PTH were not detected.

Although synthetic antagonists to PTH action have been produced (Horiuchi *et al.*, 1983), to date there is no evidence for the existence *in vivo* of fragments of endogenous PTH which might serve as hormonal inhibitors. Our *in vitro* studies suggested that target tissue antagonists to PTH action might occur in pseudohypoparathyroidism; however, when immunoreactive PTH levels were reduced by over 50%, no change in inhibitory effect of plasma was seen (Fig. 9). Consequently, either residual PTH fragments were responsible for the inhibition, or the inhibition was unrelated to circulating PTH forms.

Although the nature of the inhibitory material and its *in vivo* role remain

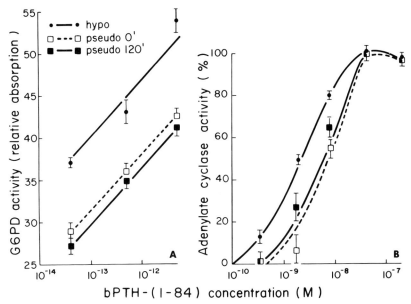

FIG. 9. Effect on bPTH-(1–84) bioactivity of hypoparathyroid plasma or serum (●) and of plasma or serum from a patient with pseudohypoparathyroidism obtained before calcium infusion (□) and 120 minutes after beginning the calcium infusion (■). After 120 minutes of infusion of calcium chloride, circulating levels of immunoreactive PTH in this patient were reduced by over 50% from baseline levels but were still above normal. The bioactivity of bPTH-(1–84) was assessed in the *in vitro* renal cytochemical assay (A) and in the *in vitro* renal adenylate cyclase assay (B). Adenylate cyclase activity is expressed as the enzyme activity above basal stimulated by each concentration of hormone, divided by the maximally stimulated activity above basal (%). Basal and maximal activities were 220.5 ± 0.6 and 610.7 ± 3.4 pmol cyclic AMP/mg protein/10 minutes, respectively. (From Mitchell and Goltzman, 1985.)

unclear, these studies do provide evidence that abnormal parathyroid function and abnormal PTH metabolism occur in pseudohypoparathyroidism. If the defect in the regulatory subunit of the adenylate cyclase system described in a variety of tissues (Bourne *et al.*, 1981; Downs *et al.*, 1983; Farfel and Bourne, 1980) in the type IA syndrome extends as well to the parathyroids (as seems likely), and if the parathyroid adenylate cyclase system is involved in the regulation of PTH metabolism and/or secretion [for which there is *in vitro* evidence (Abe and Sherwood, 1972; Brown *et al.*, 1978)], then the alteration in parathyroid gland function could be a result of the same defect observed in peripheral tissues. Increased production of PTH moieties, which are poorly active and serve as target tissue inhibitors, might then contribute to the predominance of hypoparathyroid manifestations in this disorder.

D. HYPERCALCEMIA OF MALIGNANCY

Malignancies can alter skeletal function in several ways. Two major consequences of the effects of malignancies on the skeleton are excess bone resorption and excess bone formation. The most common malignancy causing new bone formation, apparently as a result of osteoblastic stimulation by bone metastases, is prostatic carcinoma. We have recently extracted peptides from both prostatic carcinoma and benign prostatic hyperplasia tissue which are mitogenic in osteoblast-like cell systems (Koutsilieris et al., 1985). These basic peptides appear to be different from a variety of known growth factors and can be purified by reverse-phase HPLC. Peptides purified in this way retain their metogenic activity in osteoblast-related systems but not in fibroblasts, and can also stimulate alkaline phosphatase activity in osteoblasts. These studies may, therefore, provide a humoral basis for the osteogenic response to prostatic metastases.

A more common skeletal response to malignant neoplasms is bone resorption (Stewart et al., 1982) a response which may frequently lead to hypercalcemia. Such bone loss may result from focal metastatic lesions or from more widespread bone resorption. Although both lesions presumably occur through the action of humoral mediators, the more generalized resorption may frequently be produced by factors active at a distance, that is, by humoral factors released from malignancies outside of the skeleton and capable of withstanding peripheral degradation to carry out their resorptive actions upon reaching skeletal target cells. The latter syndrome has, in the past, been referred to as ectopic hyperparathyroidism or pseudohyperparathyroidism (Lafferty, 1966), although it is now clear, through several lines of evidence (Powell et al., 1973; Simpson et al., 1983), that PTH, per se, is not the mediator of this disorder in the majority of cases.

We began a series of studies in a subset of patients with malignancies and hypercalcemia and in animal models which simulate the human condition. This subset of patients (Stewart et al., 1980) had certain biochemical features in common with hyperparathyroidism, namely increased nephrogenous cyclic AMP production (Stewart et al., 1980; Kukreja et al., 1980; Rude et al., 1981) and reduced renal phosphate threshold; other biochemical features, however, differed from those found in hyperparathyroidism, including low or undetectable immunoreactive PTH levels, increased fractional calcium excretion, and low circulating levels of the active metabolite of vitamin D, 1,25-dihydroxycholecalciferol. We determined that the majority of these patients also had elevated plasma levels of activity in the renal cytochemical bioassay for PTH (Goltzman et al.,

1981). We then found, upon analysis of plasma bioactivity by gel filtration, that, in contrast to hyperparathyroidism, the majority of the bioactivity eluted earlier than PTH-(1–84), and that the bioactivity pattern was more heterogeneous (Fig. 10). These studies therefore illustrated that an *in vitro* bioassay for PTH could detect activity in the plasma of these patients; however, the cross-reacting factor was both immunochemically and chromatographically distinct from PTH. In subsequent studies, both

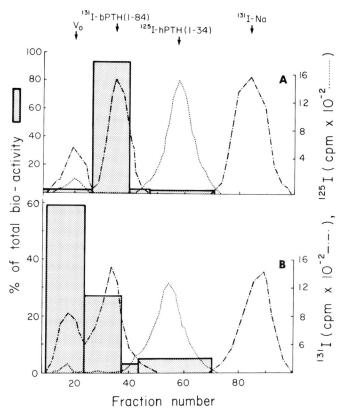

FIG. 10. Gel filtration profiles of bioactivity in plasma from a patient with primary hyperparathyroidism (A) and from a representative patient with hypercalcemia associated with malignancy (B). Chromatography was performed on columns of BioGel P-100 and bioactivity was determined in a renal cytochemical assay. ^{131}I and ^{125}I profiles were obtained by chromatographing the 3-ml plasma samples with ^{131}I-bPTH-(1–84), ^{125}I-hPTH-(1–34), and ^{131}I-Na. Vertical arrows from left to right denote, respectively, the elution postion of the void volume (V_0), labeled intact PTH, labeled active fragment, and salt. The height of the shaded bars indicates the percentage of total bioactivity represented by pooled eluted fractions and the width of the bars indicates the size of the pools assayed.

renal (Stewart et al., 1983; Strewler et al., 1983) and skeletal (Rodan et al., 1983) in vitro adenylate cyclase bioassays were shown to detect PTH-like activity in extracts of tumors associated with the disorder and in the conditioned medium of tumors grown in tissue culture. Additionally, the activity in these in vitro assays could be inhibited by the synthetic PTH antagonist, [Nle^{8,18}, Tyr^{34}]bPTH-(3–34)-NH$_2$, suggesting that the tumor-derived material was acting via PTH receptors. Cross-reactivity at target tissue receptor sites in the absence of immunological cross-reactivity is known to occur with other biologically active peptides (Goltzman and Mitchell, 1985).

In more recent studies we examined the in vivo effects of partially purified material extracted from human or rodent tumors or from conditioned medium of rodent tumors, maintained in tissue culture (Rabbani et al., 1985). All tumors were associated with the hypercalcemia syndrome. Partially purified extracts, when infused into parathyroidectomized rats, induced phosphaturia, increased urinary cyclic AMP excretion, and prevented the fall in serum calcium which occurred after parathyroidectomy (Fig. 11). Thus, extracts of tumor and conditioned medium both simulated effects of PTH in assay systems in vivo as well as in vitro. Furthermore, the activity was not only present within the tumors but could be released into the medium.

To ascertain whether this material was indeed a secretory peptide, polyadenylated messenger RNA was extracted from several human and rodent malignancies and microinjected into Xenopus oocytes which served as a surrogate secretory system (Broadus et al., 1985). Employing the most sensitive of the bioassays for PTH-like bioactivity, the renal cytochemical assay, it was possible to detect activity in the conditioned medium of cultured oocytes which had received messenger RNA from tumors associated with hypercalcemia, but not in the medium of oocytes injected with messenger RNA from tumors unassociated with hypercalcemia. Furthermore, the bioactivity could be inhibited by the synthetic PTH antagonist. Consequently, these studies provided evidence that the material obtained from the tumors associated with the syndrome and detected by PTH bioassays was indeed a secretory peptide synthesized by the tumors.

In more recent studies, using reverse-phase chromatography techniques, we have purified several peptides from a rat Leydig cell tumor (Rice et al., 1971) and a human squamous cell carcinoma, both associated with hypercalcemia, which react in bioassays for PTH-like substances (Rabbani et al., 1985). The in vitro assay employed for monitoring purification was an adenylate cyclase assay performed in a cell line of osteoblast-derived osteosarcoma cells. The major active form in the rat, and

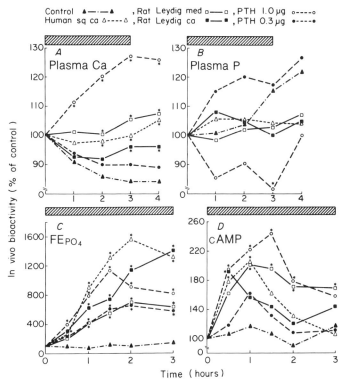

FIG. 11. Effect of infusion into thyroparathyroidectomized rats of partially purified extracts of a human squamous cell carcinoma associated with hypercalcemia (△), of a rat Leydig cell tumor associated with hypercalcemia (■), of rat Leydig cell conditioned medium (□), of vehicle alone (▲), and of bPTH-(1–84), 1.0 μg/hour (○) or 0.3 μg/hour (●). Parameters measured in the rats were plasma calcium (A), plasma phosphorus (B), fractional excretion (FE) of phosphate (C), and cyclic AMP (cAMP) excretion (D). Each parameter is expressed as a percentage of the control value, determined immediately prior to beginning the 3-hour infusion of test substance (0 time), which was 3 hours after thyroparathyroidectomy. When the partially purified extract of a colon adenocarcinoma not associated with hypercalcemia was infused, values were not significantly different from those obtained with infusion of vehicle alone. (From Rabbani *et al.*, 1985.)

sole form purified from the human tumor, had approximate molecular weights of 9500 and 9000, respectively. A large form of 28,000 MW was also obtained from the rat tumor (Fig. 12). Amino acid analyses revealed compositions of these peptides which were different from any form of PTH whose structure has been determined to date.

Beginning, therefore, with observations from clinical investigation, and proceeding from there to more basic investigation involving studies of tumors, analysis of conditioned medium, and use of molecular biology

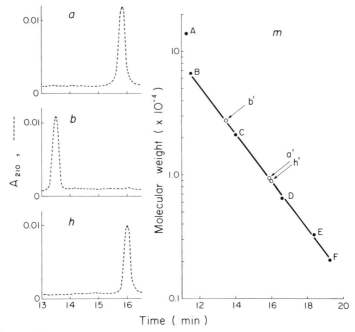

FIG. 12. Gel permeation HPLC of material from a rat Leydig cell tumor and a human squamous cell carcinoma, both associated with hypercalcemia. The material (peaks a and b from the rat tumor and peak h from the human tumor) was previously purified by reverse-phase HPLC, employing a skeletal adenylate cyclase assay, performed in rat osteosarcoma cells, to monitor bioactivity. The columns (Waters I-125 and Waters Protein Pak 300-SW connected in series) were monitored for UV absorbance at 210 nm (- - -). Calibration (m) was with (A) bovine γ-globulin, (B) bovine serum albumin, (C) human growth hormone, (D) aprotinin, (E) insulin B-chain, and (F) bovine γ-melanocyte stimulating hormone. The molecular weights, a', b', and h', of peaks a, b, and h, respectively, were estimated using the retention times of standards B, C, D, E, and F.

techniques, it has been possible to mount a large body of evidence which indicates that peptides with PTH-like bioactivity contribute to the pathogenesis of the hypercalcemia of malignancy.

IV. Examination of the Interaction of Bioactive PTH with Target Tissues

The presence of methionine residues within the NH_2-terminal region of the PTH molecule renders it particularly susceptible to inactivation by oxidation (Tashjian *et al.*, 1964; Goltzman *et al.*, 1976b). This has resulted in considerable difficulty in obtaining a high specific-activity ^{125}I-labeled radioligand which retains bioactivity and which may be useful for studies of the interaction of the hormone with its receptor. In recent years, several approaches have been employed to circumvent this problem. One approach has been the use of synthetic sulfur-free analogs of PTH which are oxidation resistant and which can be labeled with ^{125}I by conventional means (Nusbaum et al., 1980; Rosenblatt et al., 1976; Segre et al., 1979). A second approach has been to minimize oxidation by employing electrolytic iodination; this method can be used to obtain a bioactive ^{125}I-labeled preparation of the unsubstituted NH$_2$-terminal fragment (Nissenson and Arnaud, 1979; Teitelbaum et al., 1982). A third approach, and one which we have employed (Kremer et al., 1982), involves gentle labeling of either the intact hormone, or its unsubstituted fragment with ^{125}I by the lactoperoxidase technique. To optimize the specific activity of the radiolabeled preparation, iodinated hormone is separated from uniodinated hormone by reverse-phase HPLC (Fig. 13). We have employed such radioactive

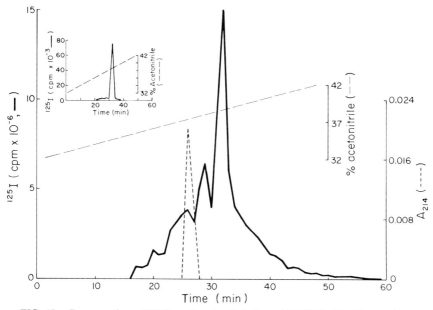

FIG. 13. Reverse-phase HPLC profile of radioiodinated bPTH-(1–84). The peptide was labeled with ^{125}I by the lactoperoxidase technique and extracted with cartridges of octadecylsilyl silica prior to chromatography. The C$_{18}$ μBondapak column eluates were monitored for UV absorbance at 210 nm (– – –) and for radioactivity (———). Fractions containing the main peak of radioactivity were pooled and rechromatographed on the same column, eluting with the same gradient, as shown in the inset. The interrupted line indicates the elution position of unlabeled bPTH-(1–84). No absorbance was detected during the second chromatographic step shown in the inset. (From Kremer et al., 1982).

probes of PTH for examination of the interaction of PTH with its target tissues, both *in vivo* and *in vitro*.

A. *IN VIVO* BINDING STUDIES

Although there have been a number of reports examining the characteristics of PTH-receptor binding *in vitro*, there have been few studies (Silve *et al.*, 1982) attempting to provide information about the nature of the cell whose receptors are being probed. We have employed an *in vivo* method (Warshawsky *et al.*, 1980; Rouleau *et al.*, 1984) to examine the binding of PTH to its morphologically identifiable target cells. This method also permits simultaneous analysis of several tissues which serve as targets for the action of the hormone.

As for *in vitro* binding studies, high specific activity bioactive radiolabeled hormone is required. The hormone is injected alone into experimental animals or with excess unlabeled hormone (related or unrelated) into control animals, generally rodents. After permitting the hormone to circulate for only a short period, so as to observe the earliest events in hormone action, "bound" is separated from "free" hormone by perfusion of the circulation with Ringer's lactate. Hormone and tissues are then fixed by perfusion with gluturaldehyde, and corresponding tissues are then removed from experimental and control animals and processed for light and electron microscope autoradiography. Autoradiographs are then analyzed both qualitatively and quantitatively to localize sites of specific and saturable binding.

Employing this approach, we assessed the interaction of both intact PTH and its synthetic NH_2-terminal fragment with three tissues of the rat: liver, bone, and kidney (Rouleau *et al.*, 1986). In liver, competitive binding of both forms of PTH to hepatocytes was observed which correlated with the capacity of the hormone to stimulate adenylate cyclase in this tissue (Bergeron *et al.*, 1981). The precise significance of the action of PTH in hepatocytes is uncertain. High capacity binding of both forms of PTH, presumably related to hormonal degradation, was also seen on sinusoidal cells. Binding of intact PTH, but not of the synthetic NH_2-terminal fragment, was seen over Kupffer cells. This sequence-specific interaction appears to be related to the cleavage of PTH-(1–84) into NH_2 and COOH fragments (D'Amour *et al.*, 1979). Therefore, depending on the specific cell type which binds circulating hormone, the interaction between PTH and liver cells may result in a biological response, nonspecific degradation or sequence-specific cleavage.

In bone, specific, competitive PTH binding was seen over osteoblasts (Fig. 14). In contrast to calcitonin, which (in similar studies) binds in-

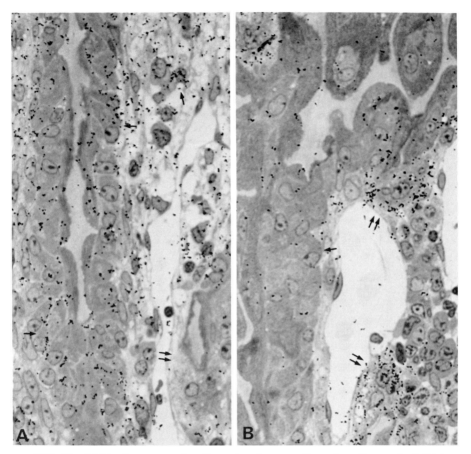

FIG. 14. Light microscope autoradiographs of osseous tissue from the proximal tibial metaphysis of rats injected with ^{125}I-labeled bPTH-(1–84) alone (A) or with ^{125}I-labeled salmon calcitonin alone (B). In the autoradiograph from the experimental animal which received only labeled PTH (A), silver grains can be seen related to osteoblasts (single arrows on left of A) aligned on the mixed spicule and to a mononuclear connective tissue cell (single arrow on right of A) between the sinusoid and the bone spicule. No labeling is seen over an osteoclast (double arrow). The density of PTH binding was markedly diminished in corresponding autoradiographs from the control animal. In animals which received ^{125}I-labeled salmon calcitonin (B), the autoradiographic reaction is exclusively localized to osteoclast profiles (double arrows). Exposure time for the radioautograph shown in the experiments depicted in A and B were 21 days and 7 days, respectively. ×600.

tensely to osteoclasts, no binding of PTH to osteoclasts was observed. Whether the binding of PTH to osteoblasts results in anabolic effects, antianabolic effects, or catabolic effects, associated with stimulation of osteoclasts, is unclear. Our *in vivo* studies do, nevertheless, correlate with multiple *in vitro* studies which have identified osteoblasts as target cells for PTH action (Dietrich *et al.*, 1976; Howard *et al.*, 1981; Majeska *et al.*, 1980; Partridge *et al.*, 1981; Rizzoli *et al.*, 1983). Although PTH binding to osteoclasts was not observed, competitive binding was seen on a connective tissue mononuclear cell lying between the skeletal sinusoids and layers of osteoblasts (Fig. 14). Whether this cell type functions as an osteoclast precursor and serves to initiate PTH-induced osteoclastic osteolysis must await further study. Lesser degrees of binding over sinusoidal lining cells were also seen which may be sites of hormonal degradation.

In kidney, binding of PTH to multiple cell types along the nephron was observed (Fig. 15). Specific competitive binding was seen on the primary foot processes of glomerular podocytes, and on the antiluminal surface of cells in the proximal tubule. Less intense competitive binding was also seen on the antiluminal surface of the thick ascending limb of the loop of Henle and of the distal tubule. In contrast, specific competitive calcitonin binding was localized only to the antiluminal portion of the ascending limb of Henle's loop and the distal tubule. Very intense, noncompetitive, binding of both PTH and calcitonin was observed on the periluminal portion of the first part of the proximal tubule which, by electron microscopy, was seen to be related to the apical vacuolar system. This high capacity binding is most likely associated with hormonal degradation.

These studies have therefore demonstrated the multiplicity of loci for competitive PTH binding along the nephron; the distribution of binding sites appears to correlate well with studies of PTH-stimulated adenylate cyclase activity in microdissected kidney tubules (Morel, 1983) and with micropuncture and microperfusion techniques (Agus *et al.*, 1981) which have localized PTH-mediated inhibition of phosphate and bicarbonate reabsorption, and PTH-stimulated calcium reabsorption to various portions of the nephron. These studies have also shown that individual cells, such as the polar proximal tubule cell, may communicate in more than one way with the same PTH molecule, illustrating the high degree of specialization of the cell membrane of such cells in interacting with bioactive peptides in the microenvironment.

In vivo binding studies can, therefore, provide a morphological correlate for the biochemical and physiologic studies of PTH action *in vivo* and *in vitro,* and illustrate the diversity which characterizes the relationships between PTH and its target cells.

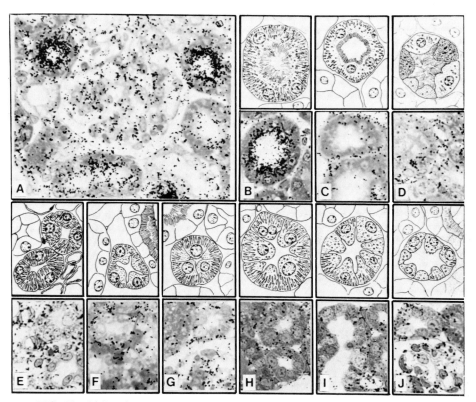

FIG. 15. Light microscope autoradiographs from the kidney of rats injected with ^{125}I-labeled bPTH-(1–84). Shown above the corresponding radiographs are diagrammatic representations of tubules from segment 1, segment 2, and segment 3 of the proximal tubule (B, C, and D), from the medullary (E and F) and cortical (G) portions of the thick ascending limb of Henle's loop, from the convoluted and connecting segments of the distal tubule (H and I), and from the pars arcuata of the collecting duct (J). Silver grains can be seen over glomerular tissue (A), on the antiluminal aspect of all three segments (B–D) of the proximal tubule, on the antiluminal aspect of type 1 cells of the thick ascending limb of Henle's loop (E–G), and on the antiluminal aspect of the distal part of the nephron (H–J). The autoradiographic reaction was lowest in the distal region of the nephron (H–J). In autoradiographs from corresponding control animals, silver grains were reduced over all of these areas. Intense labeling can be seen on the luminal aspect of segment 1 of the proximal tubule (B), which was not reduced in control animals. Similar autoradiographs were obtained in experiments with the ^{125}I-labeled bPTH-(1–34) analog. Exposure time was 7 days for the autoradiograph shown in B and was 21 days for the autoradiographs shown in A and C–J. ×600. (From Rouleau *et al.*, 1986.)

B. IN VITRO BINDING STUDIES

We initiated a series of *in vitro* binding studies to compare the characteristics of binding of the major glandular and circulating bioactive form of the hormone with the characteristics of binding of synthetic NH_2-terminal fragments. Canine kidney membranes and rat osteoblast-derived osteosarcoma cells (Partridge et al., 1980) served as sources of renal and skeletal receptors, respectively. Intact bovine (b) PTH-(1–84) and the synethetic NH_2-terminal fragments bPTH-(1–34) and rat (r) PTH-(1–34), labeled with the lactoperoxidase technique and purified by reverse-phase HPLC, served as hormonal probes of receptor binding. The highest concentration of specific binding was found with ^{125}I-labeled rPTH-(1–34) (Fig. 16), and this peptide appeared, by Scatchard analysis of equilibrium binding data, to have the highest affinity of the three peptides. Additionally, the rat NH_2-terminal fragment appeared to be the most potent in stimulating adenylate cyclase activity in renal and skeletal systems.

When the characteristics of binding of intact PTH-(1–84) were compared to those of the two synthetic NH_2-terminal fragments, distinct differences were seen. Thus, unlabeled rPTH-(1–34) and bPTH-(1–34) were both more potent than bPTH-(1–84) in inhibiting the binding of both ^{125}I-

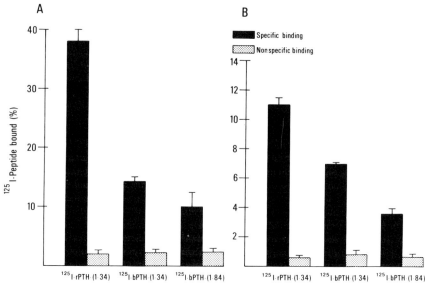

FIG. 16. Comparative specific and nonspecific receptor binding in renal membranes (A) and in UMR 108 osteosarcoma cells (B) of ^{125}I-rPTH-(1–34), ^{125}I-bPTH-(1–34), and ^{125}I-bPTH-(1–84). Each bar is the mean ± SE of 12 determinations from four experiments. (From Demay et al., 1985.)

labeled NH_2-terminal fragments to renal membranes and osteosarcoma cells. This correlated with the greater potency of the NH_2-terminal fragments than of bPTH-(1–84) in stimulating adenylate cyclase activity in the two systems. However, neither unlabeled fragment was as potent as unlabeled bPTH-(1–84) in inhibiting the binding of ^{125}I-labeled bPTH-(1–84) to the renal and skeletal systems, suggesting that sequences beyond position 34 might contribute to binding of the intact hormone.

To pursue this finding, bPTH-(1–34) and bPTH-(1–84) were oxidized; in our studies, this procedure reduced the capacity of the two peptides to

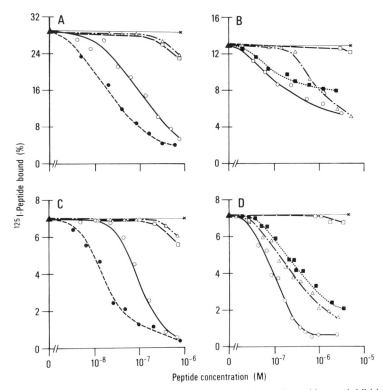

FIG. 17. Effect of increasing concentrations of PTH-related peptides on inhibition of receptor binding of ^{125}I-rPTH-(1–34) (A and C) and on inhibition of receptor binding of ^{125}I-bPTH-(1–84) (B and D) in canine renal membranes (A and B) and in a cell line (UMR 108) of osteoblast-derived rat osteosarcoma cells (C and D). Unlabeled peptides added were rPTH-(1–34) (●), bPTH-(1–84) (○), oxidized bPTH-(1–34) (□), oxidized bPTH-(1–84) (△), and hPTH-(28–48), hPTH-(44–68), and hPTH-(53–84), all represented by "×." (▲) Initial radioligand binding in the absence of unlabeled peptide. Oxidized bPTH-(1–34) and oxidized bPTH-(1–84) which inhibited ^{125}I-rPTH-(1–34) binding poorly (A and C) stimulated adenylate cyclase activity only weakly when assayed in the same renal membranes and osteosarcoma cells. (From Demay et al., 1985.)

stimulate adenylate cyclase and to inhibit the binding of ^{125}I-labeled NH$_2$-terminal fragments to renal and skeletal systems. These findings, thus reemphasized the importance of the NH$_2$-terminal region of the molecule in stimulating adenylate cyclase activity. The oxidized bPTH-(1–84), now no longer capable of binding to NH$_2$-terminal sites, could nevertheless be used as a probe of potential receptor binding sequences in the middle and COOH ends of the molecule; indeed, oxidized bPTH-(1–84) was capable of inhibiting ^{125}I-labeled bPTH-(1–84) binding to both renal membranes and osteosarcoma cells, demonstrating directly the participation of these regions in the interaction of this form of the molecule with target tissues (Fig. 17). In view of our previous studies showing that the major, if not the only, circulating bioactive form of PTH in most hyperparathyroid states is similar or identical to PTH-(1–84) (Goltzman *et al.*, 1980), examination of the interaction of this form of the hormone with target tissues is of major importance in understanding the physiology of PTH, and the functional consequences of the binding of the mid-region and COOH ends of the molecule to receptor sites warrant further investigation.

V. Summary

Our studies have emphasized the complexity of the biosynthesis and processing of parathyroid hormone within the parathyroid gland, which includes not only multiple cleavage steps but also, at least to some extent, modification of the peptide chain by phosphorylation. Once released by the parathyroid cell, the extraglandular metabolism of the gland is equally complex and may be altered in disease states (Table I). Thus, in hyperparathyroidism with normal renal function, the major circulating bioactive form of the hormone is similar or identical to the major glandular form, consistent with glandular hypersecretion. However, in end-stage renal disease, in addition to hypersecretion of the major glandular form of PTH, altered metabolism of the hormone appears to occur and contribute smaller material to the total complement of circulating bioactive hormone. In examining circulating bioactive forms of PTH, a discordance between bioactive and immunoreactive levels of hormone or hormone-like entities has been found in selected cases of both pseudohypoparathyroidism and hypercalcemia associated with malignancy. In the former condition, elevated immunoreactive PTH levels may be associated with low bioactive levels and altered forms of PTH appear to circulate. In the latter condition, immunoreactive PTH is undetectable and non-PTH peptides appear to exhibit PTH-like bioactivity.

At the target tissue level, *in vivo* studies have suggested that the interaction of PTH with target cells is as complex as intra- and extraglandular

TABLE I
Summary of Studies of PTH-Like Bioactivity in Disease States

Clinical disorder	RIA[a]	CBA[b]	Chromatographic pattern[c]	Postulated abnormality
Hyperparathyroidism with NRF[d]	Elevated	Elevated	Intact bioactive hormone	Increased secretion of PTH
Hyperparathyroidism with CRF[e]	Elevated	Elevated	Intact bioactive hormone and bioactive fragment(s)	Increased secretion of PTH and abnormal metabolism
Pseudohypoparathyroidism	Elevated	Decreased	Abnormal immunoreactive profile	Excess secretion of altered PTH
Hypercalcemia of malignancy	Decreased	Elevated	Large-molecular-weight PTH-like bioactivity	Secretion of material cross-reacting with PTH receptors

[a] RIA refers to immunoreactive levels in plasma.
[b] CBA refers to plasma bioactivity determined in a renal cytochemical bioassay for PTH.
[c] Chromatographic pattern refers to the pattern of bioactivity or immunoreactivity determined by gel filtration or by reverse-phase HPLC.
[d] NRF is normal renal function.
[e] CRF is chronic renal failure.

metabolism. Thus binding reactions to high and low capacity sites on discrete membrane regions of the same or different cells, and to sites demonstrating specificity for different sequential regions of the molecule, can be demonstrated. Finally, *in vitro* studies have suggested that the midregion or COOH end of the molecule may contribute to the binding of PTH-(1–84) to target cells, although the functional significance of this interaction is as yet unclear.

When examined at several levels therefore, analysis of PTH structure and function illustrates the intricacy of the control of this molecule by cellular processes. Alterations of such processes may substantially modify the role which the hormone plays in disease states.

ACKNOWLEDGMENTS

We thank Marie Demay, David Grunbaum, and Richard Kremer for essential contributions to parts of this work, and Dr. Hershey Warshawsky for important collaboration. We also extend our appreciation to Diane Allen for excellent secretarial assistance.

The research described in this article was funded by grants from the Medical Research Council (MRC) of Canada, the National Cancer Institute of Canada, and the National Institutes of Health of the U.S. Dr. Goltzman holds a Scientist Award from the MRC of Canada.

REFERENCES

Abe, M., and Sherwood, L. M. (1972). *Biochem, Biophys. Res. Commun.* **48**, 396.
Agus, Z. S., Wasserstein, A., and Goldfarb, S. (1981). *Annu. Rev. Physiol.* **43**, 583.
Albright, F., Burnett, C. H., Smith, P. H., and Parson, W. (1942). *Endocrinology* **30**, 922.
Arnaud, C. D., Goldsmith, R. S., Bordier, P. J., Sizemore, G. W., Larsen, J. A., and Gilkinson, J. A. (1974). *Am. J. Med.* **56**, 785.
Bennett, H. P. J., Solomon, S., and Goltzman, D. (1981). *Biochem. J.* **197**, 391.
Bergeron, J. J. M., Tchervenkov, S., Rouleau, M. F., Rosenblatt, M., and Goltzman, D. (1981). *Endocrinology* **109**, 1552.
Berson, S. A., and Yalow, R. S. (1968). *J. Clin. Endocrinol. Metab.* **28**, 1037.
Bingham, E. W., and Groves, M. L. (1979). *J. Biol. Chem.* **254**, 4510.
Bourne, H. R., Kaslow, H. R., Brickman, A. S., and Farfel, Z. (1981). *J. Clin. Endocrinol. Metab.* **63**, 636.
Broadus, A. E., Goltzman, D., Webb, A. C., and Kronenberg, H. M. (1985). *Endocrinology* **117**, 1661.
Brown, E. M., Gardner, D. G., Windeck, R. A., and Aurbach, G. D. (1978). *Endocrinology* **103**, 2323.
Canterbury, J. M., and Reiss, E. (1972). *Proc. Soc. Exp. Biol. Med.* **140**, 1393.
Canterbury, J. M., Bricker, L. A., Levey, J. S., Kozlovskis, R. L., Ruiz, E., Zull, J. E., and Reiss, E. (1975). *J. Clin. Invest.* **55**, 1245.
Chambers, D. J., Dunham, J., Zanelli, J. M., Parsons, J. A., Bitensky, L., and Chayen, J. (1978). *Clin. Endocrinol.* **9**, 375.
Chase, L. R., Melson, G. L., and Aurbach, G. D. (1969). *J. Clin. Invest.* **48**, 1832.
Chayen, J., Daly, J. R., Leveridge, N., and Bitensky, L. (1976). *Recent Prog. Horm. Res.* **32**, 33.

Cohn, D. V., and Elting, J. (1983). *Recent Prog. Horm. Res.* **39**, 181.
Cohn, D. V., MacGregor, R. R., Chu, L. L. H., Kimmel, J. R., and Hamilton, J. W. (1972). *Proc. Natl. Acad. Sci. U.S.A.* **69**, 1521.
D'Amour, P., Segre, G. V., Roth, S. I., and Potts, J. T., Jr. (1979). *J. Clin. Invest.* **63**, 89.
Demay, M., Mitchell, J., and Goltzman, D. (1985). *Am. J. Physiol.* **249**, E437.
Dietrich, J. W., Canalis, E., Maina, D. M., and Raisz, L. G. (1976). *Endocrinology* **98**, 943.
Downs, R. W., Jr., Levine, M. A., Drezner, M. K., Burch, W. M., Jr., and Spiegel, M. A. (1983). *J. Clin. Invest.* **71**, 231.
Drezner, M. K., and Burch, W. M., Jr. (1978). *J. Clin. Invest.* **62**, 1222.
Drezner, M., Neelon, F. A., and Lebovitz, H. E. (1973). *N. Engl. J. Med.* **289**, 1056.
Farfel, Z., and Bourne, H. R. (1980). *J. Clin. Endocrinol. Metab.* **51**, 1202.
Farfel, Z., Brickman, A. S., Kaslow, H. R., Brothers, V. M., and Bourne, H. R. (1980). *N. Engl. J. Med.* **303**, 237.
Fenton, S., Somers, S., and Heath, D. A. (1978). *Clin. Endocrinol.* **9**, 381.
Flueck, J. C., Di Bella, F. P., Edis, A. J., Kehrwald, J. M., and Arnaud, C. D. (1977). *J. Clin. Invest.* **60**, 1367.
Freitag, J., Martin, K. J., Hruska, K. A., Klahr, S., and Slatopolsky, E., (1978). *N. Engl. J. Med.* **298**, 29.
Goltzman, D. (1978). *Endocrinology* **102**, 1555.
Goltzman, D. (1983). *In* "Assay of Calcium-Regulating Hormones" (D. D. Bikle, ed.), p. 261. Springer-Verlag, New York.
Goltzman, D., and Mitchell, J. (1985). *Science* **227**, 1343.
Goltzman, D., Peytremann, E., Callahan, E. N., Tregear, G. W., and Potts, J. T., Jr. (1975). *Proc. Parathyroid Conf., 5th, Oxford* p. 172.
Goltzman, D., Callahan, E. N., Tregear, C. W., and Potts, J. T., Jr. (1976a). *Biochemistry* **15**, 5076.
Goltzman, D., Peytremann, E., Callahan, E. N., Tregear, G. W., and Potts, J. T., Jr. (1976b). *J. Biol. Chem.* **250**, 3199.
Goltzman, D., Peytremann, A., Callahan, E. N., Segre, G. W., and Potts, J. T., Jr. (1976c). *J. Clin. Invest.* **57**, 8.
Goltzman, D., Callahan, E. N., Tregear, G. W., and Potts, J. T., Jr. (1978). *Endocrinology* **103**, 1352.
Goltzman, D., Henderson, B., and Loveridge, N. (1980). *J. Clin. Invest.* **65**, 1309.
Goltzman, D., Stewart, A. F., and Broadus, A. E. (1981). *J. Clin. Endocrinol. Metab.* **53**, 899.
Goltzman, D., Gomolin, A., DeLean, A., Wexler, M., and Meakins, J. L. (1984). *J. Clin. Endocrinol. Metab.* **58**, 70.
Grunbaum, D., Wexler, M., Antos, M., Gascon-Barré, and Goltzman, D. (1984). *Am. J. Physiol.* **247**, E442.
Habener, J. F., Powell, D., Murray, T. M., Mayer, G. P., and Potts, J. T., Jr. (1971). *Proc. Natl. Acad. Sci. U.S.A.* **68**, 2986.
Habener, J. R., Kemper, B., Rich, A., and Potts, J. T., Jr. (1977). *Recent Prog. Horm. Res.* **33**, 249.
Habener, J. F., Rosenblatt, M., Kemper, B., Kronenberg, H. M., Rich, A., and Potts, J. T., Jr. (1978). *Proc. Natl. Acad. Sci. U.S.A.* **75**, 2616.
Hanley, D. A., Takatsuki, K., Sultan, J. M., Schneider, A. B., and Sherwood, L. M. (1978). *J. Clin. Invest.* **62**, 1247.
Hatfield, D., Diamond, A., and Dudock, B. (1982). *Proc. Natl. Acad. Sci. U.S.A.* **79**, 6215.
Heinrich, G., Kronenberg, H. M., Potts, J. T., Jr., and Habener, J. F. (1984). *J. Biol. Chem.* **259**, 3320.

Horiuchi, N., Holick, M. F., Potts, J. T., Jr., and Rosenblatt, M. (1983). *Science* **220**, 1053.
Howard, G. A., Bottemiller, B. L., Turner, R. T., Rader, J. T., and Baylink, D. L. (1981) *Proc. Natl. Acad. Sci. U.S.A.* **78**, 3208.
Hruska, K. A., Kopelman, R., Rutherford, W. E., Klahr, S., and Slatopolsky, E. (1975). *J. Clin. Invest.* **56**, 39.
Kemper, B., Habener, J. F., Potts, J. T., Jr., and Rich, A. (1972). *Proc. Natl. Acad. Sci. U.S.A.* **69**, 643.
Kemper, B., Habener, J. F., Mulligan, R. C., Potts, J. T., Jr., and Rich, A. (1974). *Proc. Natl. Acad. Sci. U.S.A.* **71**, 3731.
Keutmann, H. T., Aurbach, G. D., Dawson, B. F., Niall, H. D., Deftos, L. J., and Potts, J. T., Jr. (1971). *Biochemistry* **10**, 2779.
Koutsilieris, M., Rabbani, S. A., and Goltzman, D. (1986). *Prostate*, in press.
Kremer, R., Bennett, H. P. J., Mitchell, J., and Goltzman, D. (1982). *J. Biol. Chem.* **257**, 14048.
Kukreja, S. C., Shemerdiak, W. P., Lad, T. E., and Johnson, P. A. (1980). *J. Clin. Endocrinol. Metab.* **51**, 167.
Lafferty, F. W. (1966). *Medicine (Baltimore)* **45**, 247.
Levine, M. A., Downs, R. W., Jr., Singer, M., Marx, S. J., Aurbach, G. D., and Spiegel, A. M. (1980). *Biochem. Biophys. Res. Commun.* **94**, 1319.
Loveridge, N., Fischer, J. A., Nagant de Deuxchaisnes, C., Dambacher, M. A., Tschopp, F., Werder, E., Devogelaer, J.-P., De Meyer, R., Bitensky, L., and Chayen, J. (1982). *J. Clin. Endocrinol. Metab.* **54**, 1274.
Majeska, R. J., Rodan, S. B., and Rodan, G. A. (1980). *Endocrinology* **107**, 1494.
Mayer, G. P., Keaton, J. A., Hurst, J. G., and Habener, J. F. (1979). *Endocrinology* **104**, 1778.
Mitchell, J., and Goltzman, D. (1985). *J. Clin. Endocrinol. Metab.* **61**, 328.
Morel, F. (1983). *Recent Prog. Horm. Res.* **39**, 271.
Nagant de Deuxchaisnes, C., Fischer, J. A., Dambacher, M. A., Devogelaer, J.-P., Arber, C. E., Zanelli, J. M., Parsons, J. A., Loveridge, N., Bitensky, L., and Chayen, J. (1981). *J. Clin. Endocrinol. Metab.* **53**, 1105.
Neuman, W. F., Neuman, M. W., Lane, K., Miller, L., and Sammon, P. J. (1975). *Calcif. Tissue Res.* **18**, 271.
Nissenson, R. A., and Arnaud, C. D. (1979). *J. Biol. Chem.* **254**, 1469.
Nusbaum, S. R., Rosenblatt, M., and Potts, J. T., Jr. (1980). *J. Biol. Chem.* **255**, 10183.
Partridge, N. C., Frampton, R. J., Eisman, J. A., Michelangeli, V. P., Elms, E., Bradley, T. R., and Martin, T. J. (1980). *FEBS Lett.* **115**, 139.
Partridge, N. C., Kemp, B. E., Veroni, M. C., and Martin, T. J. (1981). *Endocrinology* **108**, 220.
Peytremann, A., Goltzman, D., Callahan, E. N., Tregear, G. W., and Potts, J. T., Jr. (1975). *Endocrinology* **97**, 1270.
Potts, J. T., Jr., Tregear, G. W., Keutmann, H. T., Niall, H. D., Sauer, R., Deftos, J., Dawson, B. F., Hogan, M. L., and Aurbach, G. D. (1971). *Proc. Natl. Acad. Sci. U.S.A.* **68**, 63.
Powell, D., Singer, F. R., Murray, T. M., Minkin, C., and Potts, J. T., Jr. (1973). *N. Engl. J. Med.* **289**, 176.
Rabbani, S. A., Kremer, R., Bennett, H. P. J., and Goltzman, D. (1984). *J. Biol. Chem.* **259**, 2949.
Rabbani, S. A., Mitchell, J., Roy, D. R., Kremer, R., Bennett, H. P. J., and Goltzman, D. (1985). *Endocrinology* **118**, 1200.
Rice, B. F., Ponthier, R. L., and Miller, M. C. (1971). *Endocrinology* **88**, 1210.

Rizzoli, R. E., Somerman, M., Murray, T. M., and Aurbach, G. D. (1983). *Endocrinology* **113**, 1832.
Rodan, S. B., Insogna, K. L., Vignery, M.-C., Stewart, A. F., Broadus, A. E., D'Souza, S. M., Bertolini, D. R., Mundy, G. R., and Rodan, G. A. (1983). *J. Clin. Invest.* **72**, 1511.
Rouleau, M. F., Warshawsky, H., and Goltzman, D. (1984). *Brain* **107**, 107.
Rouleau, M. F., Warshawsky, H., and Goltzman, D. (1986). *Endocrinology* **118**, 919.
Rosenblatt, M., Goltzman, D., Keutmann, H. T., Tregear, G. W., and Potts, J. T., Jr. (1976). *J. Biol. Chem.* **251**, 159.
Rosenblatt, M., Callahan, E. N., Mahaffey, J. E., Pont, A., and Potts, J. T., Jr. (1977). *J. Biol. Chem.* **251**, 159.
Rude, R. K., Sharp, C. F., Fredericks, R. S., Oldham, S. B., Elbaum, N., Link, J., Irwin, L., and Singer, F. R. (1981). *J. Clin. Endocrinol. Metab.* **52**, 765.
Segre, G. V., Habener, J. F., Powell, D., Tregear, G. W., and Potts, J. T., Jr. (1972). *J. Clin. Invest.* **51**, 3163.
Segre, G. V., Niall, H. D., Habener, J. F., and Potts, J. T., Jr. (1974). *Am. J. Med.* **56**, 774.
Segre, G. V., D'Amour, P., and Potts, J. T., Jr. (1976). *Endocrinology* **99**, 1645.
Segre, G. V., Rosenblatt, M., Reiner, B. L., Mahaffey, J. E., and Potts, J. T., Jr. (1979). *J. Biol. Chem.* **254**, 6980.
Silve, C. M., Hradek, G. T., Jones, A. L., and Arnaud, C. D. (1982). *J. Cell Biol.* **94**, 379.
Silverman, R., and Yalow, R. S. (1973). *J. Clin. Invest.* **52**, 1958.
Simpson, E. L., Mundy, G. R., D'Souza, S. M., Ibbotson, K. J., Bockman, R., and Jacobs, J. W. (1983). *N. Engl. J. Med.* **309**, 325.
Singer, F. R., Segre, G. V., Habener, J. F., and Potts, J. T., Jr. (1975). *Metabolism* **24**, 139.
Slatopolsky, E., Caglar, S., Pennell, J. P., Taggart, D. D., Canterbury, J. M., Reiss, E., and Bricker, N. S. (1971). *J. Clin. Invest.* **50**, 492.
Stewart, A. F., Horst, R., Deftos, L. J., Cadman, E. C., Lang, R., and Broadus, A. E. (1980). *N. Engl. J. Med.* **303**, 1377.
Stewart, A. F., Vignery, A., Silverglate, A., Ravin, N. D., LiVolsi, V., Broadus, A. E., and Baron, R. (1982). *J. Clin. Endocrinol. Metab.* **55**, 219.
Stewart, A. F., Insogna, K. L., Goltzman, D., and Broadus, A. E. (1983). *Proc. Natl. Acad. Sci. U.S.A.* **80**, 1454.
Strewler, G. J., Williams, R. D., and Nissenson, R. A. (1983). *J. Clin. Invest.* **71**, 769.
Tashjian, A. H., Jr., Ontjes, D. A., and Munson, P. L. (1964). *Biochemistry* **3**, 1175.
Teitelbaum, A. P., Nissenson, R. A., and Arnaud, C. D. (1982). *Endocrinology* **111**, 1524.
Tregear, G. W., van Rietschoten, J., Green, E., Keutmann, H. T., Niall, H. D., Reit, B., Parsons, J. A., and Potts, J. T., Jr. (1973). *Endocrinology* **93**, 1349.
Warshawsky, H., Goltzman, D., Rouleau, M. F., and Bergeron, J. J. M. (1980). *J. Cell Biol.* **85**, 682.

DISCUSSION

G. D. Aurbach. As you pointed out, in pseudohypoparathyroidism, there is an apparent genetic defect in production of a guanine nucleotide regulatory protein that is required for normal hormonal action. At the same time, you as well as Chayen's group in England have found some abnormalities in biological activity of circulating hormone. In view of the fact that you have found phosphorylation of parathyroid hormone and in view of Dr. Kronenberg's findings concerning the significance of the amino-terminal portion of the molecule for processing, is it possible to reconcile these two apparently divergent sets of observations? Namely, could there be an abnormality in parathyroid hormone secretion attributable also to abnormal cyclic AMP generation in the parathyroid cell?

D. Goltzman. I think that's a very cogent comment. We obviously have to reconcile two apparently disparate sets of observations: that there is an abnormality in the guanine-nucleotide regulatory protein of the adenylate cyclase system, and that there is abnormality of parathyroid gland function and abnormal forms of parathyroid hormone in the circulation. Your group has shown that the parathyroid gland adenylate cyclase system may be involved in secretion, and certainly there is the possibility that it may play a role in hormonal processing as well, whether at the level of phosphorylation, or elsewhere. Consequently the same primary genetic defect involving the adenylate cyclase system could be used to explain both the target tissue resistance to parathyroid hormone and the abnormality in the parathyroid gland and in parathyroid hormone which has been reported both by Loveridge, Fischer, and their co-workers and by our group. I believe we have to do more work in that direction, however, to see if this hypothesis will be sustained.

W. D. Odell. I have two questions. Since you have the tools, it's hard to believe that you haven't assessed the potency of phosphorylated parathyroid hormone by bioassay and immunoassay. Could you tell us about biological potency of the phosphorylated compounds? There are a very small number of patients listed in the world literature that have what might be called a hormonopathy of parathormone, where the serum contains high immunoactivity and very little bioactivity. They don't have pseudohypoparathyroidism. Have you studied any of these patients?

D. Goltzman. In answer to your first question, yes, we have examined the biological activity of the phosphorylated form, but the studies are at the moment too preliminary. It appears as though the phosphorylated hormone has reduced biological activity but we have not yet assigned a relative potency to it. I would like to bioassay the material in several systems first, and not just in the cytochemical bioassay.

The second question dealt with the issue of hypoparathyroidism in which these are elevated immunoreactive PTH levels of presumed low biological activity. We examined plasma from a father and son, both members of a family presumed to have this variant of hypoparathyroidism, and were disappointed to find that in our hands immunoreactive PTH levels were not elevated, so that these patients appeared to have hypoparathyroidism due to a decrease in circulating hormone. I am unaware of any other families or individuals presumed to have this syndrome of dyshormonogenesis. We would of course be interested in studying such patients.

W. D. Odell. Another family from Australia has been described in Sydney.

D. Goltzman. I would be very happy to study these patients if we could gain access to them.

G. T. Campbell. One question that is invariably asked by students concerns the pathogenesis of postmenopausal osteoporosis. Is there any evidence that would indicate that after menopause there's a change in the type of hormone manufactured or released or a change in the mode of degradation of the hormone, so that you would have a circulating biologically active fragment which so far has escaped detection by radioimmunoassays. Second, are there any fragments which have different degrees of biological activity at bone and at the kidney? Third, do you think the binding of the iodinated parathyroid hormone at the glomerulus represents a true site of action of the hormone, or does it reflect the possibility that for some reason these fragments are just being freely filtered?

D. Goltzman. With respect to the first question regarding what the forms of bioactive parathyroid hormone are in osteoporosis, we have not examined that issue as yet. It has been suggested that after menopause, with reduction in estrogen levels, excessive bone loss occurs which could increase extracellular calcium levels and thereby reduce PTH release. It would therefore be of interest to measure bioactive PTH in this state. It has also been reported that with aging, as renal function decreases, immunoreactive parathyroid hormone levels increase. Nevertheless, the production of 1,25-dihydroxyvitamin D in response to increased parathyroid hormone levels has been said to be subnormal in aging patients. Consequently, secondary hyperparathyroidism may develop in association with a decrease in renal function and a relative deficiency of the active form of vitamin D, and may play a pathogenetic role in the development of senile osteoporosis. It would therefore be equally important to examine bioactive PTH forms in senile as well as in postmenopausal osteoporosis.

In response to the second question, it is clear that the information for interaction with bone and kidney is contained within the same parathyroid hormone molecule, i.e., within the major glandular and secreted 84 amino acid form. One possibility for segregating this information within this molecule was suggested by Martin, Slatopolsky, and their co-workers, They reported that the amino-terminal fragment of the intact hormone might be preferentially active in bone. The implication then was that the intact hormone or both fragment and intact hormone would be the active forms in kidney. We found that both in bone and in kidney, at least with respect to the ability to stimulate adenylate cyclase, cleavage of the intact hormone to an amino-terminal fragment was not necessary, i.e., both forms are active in bone and in kidney. Are there other ways that the capacity of the amino-terminal end of the molecule to interact with bone and kidney can be modulated? We demonstrated that mid-region and carboxy-terminal determinants on the intact hormone can bind to specific and saturable sites in kidney and bone tissue. The specificity, saturability, and other studies suggest that these sites are probably not involved in degradation. Furthermore, mid-region and carboxy fragments do not stimulate the adenylate cyclase system and therefore these regions of the molecule are presumably inert or at least do not function via the adenylate cyclase system. It is therefore possible that the function of these regions of the molecule is to modulate the capacity of the amino-terminal region to interact with its receptor, which is cyclase linked. If there are differences in the affinity and/or capacity of the binding sites for the mid-region and the carboxyl sites in kidney versus bone, then one could foresee how this could influence the ability of the amino-terminal and of the molecule to interact with receptors in bone and kidney and therefore influence the potency of this hormone in the two different target tissues.

With respect to the third question, we believe, for the following reasons, that the glomerular binding of iodinated parathyroid hormone does represent a true site of action, at least in the rat. First, the binding *in vivo* is specific and saturable. Second, others have demonstrated PTH-stimulated adenylate cyclase activity in isolated glomeruli. Third, Brenner and his colleagues have reported effects of PTH on glomerular function. Since many chemical and physical factors may presumably influence glomerular function, this role of PTH may not be as obvious as its role in aspects of tubular function.

D. R. Stewart. Going back to pseudohypoparathyroidism, there are three potential explanations. One is a regulatory subunit type of problem, one is an abnormal PTH type of problem, and the third is the circulating inhibitor of PTH in action. Is there any way to tie those three together, particularly the inhibitor? No one seems to want to talk about the inhibitor. Is it clear that these really are all the same patients? Have you looked, for example, at G unit activity in the patients you studied?

D. Goltzman. One of the three patients with pseudohypoparathyroidism we studied was found by Dr. Levine in Dr. Aurbach's laboratory to have a low level of guanine nucleotide regulatory protein activity. Therefore, I believe we are studying the same group of patients. An abnormality in parathryoid gland function with abnormal circulating PTH consequently does appear to coexist with reduced G unit activity, and, as I speculated, may be a result of this reduction in activity. The parathyroid abnormality could then account for the predominance of hypoparathyroidism in this disorder of multihormonal resistance. Parathyroid hormone dysgenesis with the production of a circulating form acting as an antagonist (such forms have been synthesized *in vitro*) could then contribute to the PTH resistance. Nevertheless, I am not certain that the available data, as yet, permit us to conclude that the inhibitor identified *in vitro* is indeed related to PTH. Thus Loveridge and his co-workers have recently reported that after gel filtration of plasma from patients with pseudohypoparathyroidism, the inhibitory activity did not coelute with immunoreactive PTH. We have found that inhibitory activity persists in plasma, unchanged, after producing, with hypercalcemia, a substantial reduction in immunoreactive PTH. Thus the relationship of the *in vitro* inhibitory activity to PTH forms and the *in vivo* significance of the inhibition are at present uncertain.

L. A. Fitzpatrick. I'd like to ask about another clinical syndrome, i.e., familial hypocalciuric and hypercalcemia. It's often difficult by radioimmunoassay alone to distinguish these patients from patients with primary hyperparathyroidism. Have you looked at PTH bioactivity in FHH patients?

D. Goltzman. Not as yet, but we're hoping to.

G. Segre. We have data from a different model which suggest that peripheral metabolism of PTH results in the appearance in the circulation of amino-terminal fragments. As you may remember, several years ago we showed that when intact PTH was incubated with Kupffer cells *in vitro* amino-terminal fragments were generated and released into the medium of those cells and that they were rapidly catabolized. More recently, using internally labeled intact PTH probes injected into normal rats, we found evidence for amino-terminal fragments in the kidney when the blood flow to the kidneys was ligated. Thus evidence for amino-terminal fragments in circulation does exist. Chemically these fragments extend to include position 1 of the hormone and thus are potentially biologically active. To this extent our experiments confirm your results. My first question concerns studies of the phosphorylation of PTH. As you know, in these glands slices, synthesis of parathyroid hormone is essentially calcium independent within the time frame of the experiments. Do you have any evidence that by changing ambient calcium concentration in the medium you can alter the phosphorylation of newly synthesized parathyroid hormone in the gland?

D. Goltzman. Not as yet. This is, however, an important point since a calcium-stimulated, protein kinase C-mediated phosphorylation could be a mechanism for posttranslational modification of PTH.

G. Segre. My second question has to do with your tumor hypercalcemia factor. The gel filtration chromatographs of patient's plasma show the biologic activity eluting in the void volume. I think that you were using BioGel P-100 which has an exclusion limit of about 100,000 Da. Subsequently you demonstrated that the factor stimulating adenylate cyclase from tumor extracts has molecular weight of about 9000–10,000. Do you have a way of reconciling these two observations.

D. Goltzman. The major bioactive forms we found in the rat tumor were approximately 9000 and 28,000 Da. We only found one form in the human tumor which was about 9000 Da. My suspicion is that in fact there is a larger bioactive molecule synthesized in both the human and the rat. We may have only found the smaller moiety because the larger material has been degraded by the time we received the tumor. Therefore I think the profile we see in the plasma, which shows predominantly large-molecular-weight material, is a heterogeneous profile of the material biosynthesized and released by both the human and rat tumors.

G. Segre. Does oxidized intact hormone have any intrinsic biologic activity of its own? If it doesn't is it an effective inhibitor of PTH-stimulated adenylate cyclase?

D. Goltzman. These are interesting questions. We have not tested oxidized intact PTH as either an enhancer or an inhibitor of PTH-stimulated adenylate cyclase. There is a marked reduction in the capacity of oxidized PTH to stimulate adenylate cyclase activity and concomitant with this there is a reduction in the capacity of the oxidized hormone to bind to the amino-terminal site. Therefore, there is a good correlation between binding to the amino-terminal site and reduction of adenylate cyclase activity. We do not know, however, if the mid-region or carboxyl regions of the oxidized hormone can influence the capacity of unoxidized hormone to stimulate adenylate cyclase activity.

G. D. Aurbach. I should point out that Dr. Rice is the one who first described the Leydig cell tumor that produces hypercalcemia in rats, and it's most appropriate that he may want to make a comment at this time.

B. F. Rice. I would like to congratulate the author on having the courage to attempt bioassays on these materials. I would like to confirm that I was also unable to demonstrate any bioassayable activity in terms of raising the serum calcium in thyroparathyroidectomized rats using the classical Munson technique and Munson extraction. Concerning the data you presented in preventing the fall in serum calcium in acutely thyroparathyroidectomized rats, we were never able to be very certain that the material we were injecting was not just very toxic or that we were carrying over a lot of calcium from the serum of the animals. I noticed that the only other bioassay evidence that you have is skeletal cyclase, is that correct?

D. Goltzman. We have shown bioactivity of tumor extracts in a renal cytochemical bioassay, a renal adenylate cyclase-stimulating assay, and skeletal adenylate cyclase-stimulating assay, as well as *in vivo*. We did not use a Munson extraction but a substantial modification of the Munson assay for our *in vivo* work, which involved intravenous infusion of test substances rather than subcutaneous injection. Therefore, our experience and yours are unlikely to be easily comparable. Our purified extract has been purified through cartridges of octadecylsilyl-silica and one stage of reverse-phase HPLC, so I think it's highly unlikely that it was contaminated with calcium. Furthermore our control experiments failed to reveal any "toxic" effects. I think the fact that we observed, after infusion of the extracts, a maintenance of the serum calcium of these parathyroidectomized animals, rather than a rise in serum calcium, indicates it is probably a dose–response phenomenon.

B. F. Rice. There are major differences in the hypercalcemias of malignancies. One model you used was the squamous cell cancer in human, the other, a Leydig cell cancer or Leydig cell localized tumor in the rat. These play very different roles biologically. One responds to corticoids and estrogens, the other does not. One behaves very similarly to primary hyperparathyroidism, the other really does not in biological systems. I wonder if you have any comment on that.

D. Goltzman. Needless to say, tumors produce many factors which can modulate their behavior, but we are examining a subset of features which a variety of different types of tumors have in common. Dr. Stewart will, I believe, go into that in a bit more detail. Certainly with respect to human tumors, for example, renal cell carcinomas behave differently from skin squamous cell carcinomas, yet there are certain biochemical features which they have in common related to the development of hypercalcemia. This is why we have studied them as a group. The rat Leydig cell tumor associated with hypercalcemia does appear to be a reasonable model of the human syndrome in this respect. In other respects, of course, this rat tumor behaves differently from the human tumors, just as human lung squamous cell carcinomas and skin squamous cell carcinomas and renal adenocarcinomas behave differently from one another.

N. Samaan. In patients with pseudohyperparathyroidism or ectopic PTH, did you have a chance to examine some of these tumors by immunostaining, because in some of these tumors the immunoreactive material in the circulation may be normal, but if you immunostain the tumor, using PTH antibody to the amino terminus, you might find it positive. This is not confined to this particular tumor-producing PTH-like substance, but can occur in other nonendocrine hormone-producing tumors.

D. Goltzman. We have not personally performed any immunohistochemical studies. There are, however, a number of reports in the literature of finding positive immunohistochemical staining for PTH in these tumors. The problem is that when one tries to obtain more rigorous evidence that indeed the tumor is producing parathyroid hormone or something immunochemically related to it, it is extremely difficult. Powell and his co-workers showed, using radioimmunoassays, that it was generally difficult or impossible to extract immunologically active parathyroid hormone from the vast majority of tumors associated with hypercalcemia. Attempts to incubate tumors *in vitro* with radiolabeled amino acids and then immunoprecipitate extracted, newly synthesized peptides with a parathyroid hormone antibody, have also generally been unsuccessful. Simpson and her colleagues demonstrated that attempts to extract messenger RNA from these tumors and hybridize it with a cDNA for PTH were difficult if not impossible in the majority of cases. Our own data show that peptides extracted from these tumors with PTH-like bioactivity do not have the amino acid composition of PTH. Therefore, although a variety of attempts have been made to show that in fact these tumors do produce parathyroid hormone ectopically, with only a few exceptions they have failed. I think the immunocytochemical evidence which has been reported is probably disproportionate to the other evidence in that respect.

H. Kulin. I may not have heard correctly, but I believe one of your patients with hyperparathyroidism had X-linked hypophosphotemic rickets. Is that correct? Was that individual studied before or after treatment?

D. Goltzman. That is correct. This was after treatment with phosphates. We have had access to several of the patients whom Dr. Charles Schriver at the Montreal Childrens Hospital has followed for a number of years. They were initially treated with phosphate, but have been recently switched to treatment with 1,25-dihydroxyvitamin D, alone or with phosphate. The patients we have seen, who were initially treated with phosphate only, generally developed secondary hyperthyroidism.

H. Kulin. My understanding is that this group of individuals is heterogeneous in that the majority do not have elevated PTH prior to treatment, but a small select group do. Of the latter type we have treated one such family. I wonder if there isn't a good deal of PTH heterogeneity within this particular disease entity itself?

D. Goltzman. That's an interesting point. We unfortunately have not had the opportunity to study patients prior to treatment to examine this heterogeneity with our methods.

S. Cohen. I remember hearing a talk by Hector Deluca in which he described the action of vitamin D, and in that talk he said that he didn't think that the parathyroid hormone was really a hormone but acted through its vitamin D complex. I have never heard a parathyroid man rebut that. Can you enlighten me on the parathyroid hormone point of view?

D. Goltzman. Parathyroid hormone does appear capable of exerting at least some of its effects, notably those influencing renal function, in the presence of vitamin D deficiency. Nevertheless, it appears less effective in this state, and its mechanism of action does appear to be tightly linked to vitamin D.

T. M. Kelly. I want to go back to the disparity you found between levels of immunoreactive and bioactive PTH in patients with pseudohypothyroidism. One would expect that if normal feedback mechanisms were working, the level of immunoreactive PTH would be within the normal range. Can you speculate on why the levels of bioactive PTH do not reach the normal range? Do parathyroid cells give out before achieving the necessary level of hormone production? Also, does your bioassay involve adenylate cyclase activity or depend on the presence of the enzyme's guanine nucleotide-regulating protein?

D. Goltzman. If, in pseudohypoparathyroidism, immunoreactive PTH is being released which is biologically inert as a result of abnormal synthesis or metabolism of the hormone within the parathyroid gland, then one could conceive of a situation in which no degree of hypocalcemia would correct this error in hormonogenesis and result in the release of bioactive PTH. Structure–function studies *in vitro* have shown that only minor modification of the

ANALYSIS OF BIOACTIVE PARATHYROID HORMONE   703

amino-terminus of the active hormone, i.e., deletion or extension, will result in a peptide which retains immunoreactivity but is biologically inactive. Although there is currently no evidence that the altered PTH we see in the circulation is an amino-terminal-deleted form or an extended precursor form, these are possibilities we must consider. As far as the assay goes, we were using a renal cytochemical bioassay. We did show in previous studies that cyclic AMP and cyclic AMP analogs will mimic the effects of parathyroid hormone in this system; presumably therefore, this is a cyclase-dependent bioassay response.

T. M. Kelly.   So the assay could be affected by the absence of G protein in pseudohyperparathyroid patients?

D. Goltzman.   The assay is an *in vitro* assay performed in guinea pig kidneys which have no evidence of a G unit abnormality.

M. W. Draper.   Have your more recent studies using labeled material, and with your superbly sensitive bioassays, allowed you to more directly address the question of the plasma half-life of bioactive, presumably N-terminal, PTH in chronic renal failure versus other states?

D. Goltzman.   Yes, we found no substantial increase in the plasma half-life of bioactive PTH in patients who have end stage renal failure as opposed to patients who have normal renal function. What appears to occur is that as failure progresses tissues other than the kidney assume a proportionately greater role in the metabolism of the intact hormone. These tissues, such as liver, may then produce greater quantities of circulating bioactive amino-terminal fragments. The overall clearance of biologically active hormone does not decrease substantially, however, because of extrarenal metabolism. However, the clearance of the inert mid-region and carboxy-immunoreactive fragments, which are almost exclusively dependent on renal metabolism, does decrease, resulting in increased plasma concentrations of these moieties.

Nephrogenous Cyclic AMP, Adenylate Cyclase-Stimulating Activity, and the Humoral Hypercalcemia of Malignancy

J. W. GODSALL, W. J. BURTIS, K. L. INSOGNA, A. E. BROADUS, AND A. F. STEWART

Departments of Internal Medicine, Division of Endocrinology and Metabolism, The West Haven VA Medical Center, West Haven, Connecticut, and Yale University School of Medicine, New Haven, Connecticut

I. Introduction

At the time of this writing, there are two predominant postulates regarding the nature of the factor or factors responsible for the humoral hypercalcemia of malignancy (HHM). One group has postulated that the HHM factor is a tumor-derived growth factor (TGF) which is similar in nature if not identical to tumor growth factor α (1,2). Several other groups, including our own, believe that the HHM factor is an adenylate cyclase-stimulating protein which acts through parathyroid hormone receptors but which is unrelated genomically to parathyroid hormone (PTH). The evidence supporting the PTH-like factor postulate, we believe, is compelling. This article summarizes that evidence.

For the purpose of organization, we have come to divide our work into five areas. These include (1) clinical studies on human HHM, (2) the characterization of animal models of the human syndrome, (3) attempts to purify and characterize the putative HHM factor, (4) molecular biological studies, and (5) elucidation of the normal role of the HHM factor. In this article particular emphasis will be given to the first of these areas, which includes our previously unpublished 5-year experience with 133 patients with malignancy-associated hypercalcemia.

II. Clinical Features of HHM

Our initial report on nephrogenous cyclic AMP (NcAMP) excretion in patients with malignancy-associated hypercalcemia (MAHC) contained six major observations which relate to HHM (3). (1) Patients with MAHC can be subdivided into two groups depending upon whether their NcAMP values are elevated or suppressed. (2) Those with elevated NcAMP have predominantly squamous and renal carcinomas, tumors historically linked to the syndrome of HHM. These patients as a group have few or no bone metastases. (3) The groups of patients with suppressed NcAMP excretion have predominantly breast carcinoma or hematologic malignancies, and display evidence of widespread skeletal invasion by tumor. (4) Eighty-two percent ($n=41$) of these 50 unselected patients fell into the "elevated NcAMP" group, suggesting that HHM may be far more common than previously appreciated. (5) Normocalcemic patients with cancer ("cancer controls") have normal NcAMP excretion. (6) As compared to a group of patients with primary hyperparathyroidism (1°HPT) both groups of patients with MAHC have markedly elevated values for fasting or fractional calcium excretion, reduced values for circulating 1,25-dihydroxyvitamin D [1,25(OH)$_2$D], and reduced or undectable values for circulating immunoreactive parathyroid hormone (iPTH).

We interpreted these observations to mean that among patients with MAHC (1) NcAMP measurements can be used as a specific marker for the HHM syndrome in general, and for the presence in the circulation of the putative HHM factor (HHM-F) in particular; (2) the HHM-F interacts with the same proximal tubular PTH receptor/adenylate cyclase complex as does PTH; (3) the development of sensitive *in vitro* PTH–adenylate cyclase assays could serve as detection systems for use in the measurement and purification of the HHM-F; and (4) while the HHM-F is in some respects PTH-like, it is in other respects unlike PTH: it fails to stimulate the proximal tubular 1α-hydroxylase responsible for 1,25(OH)$_2$D synthesis, it fails to stimulate distal tubular calcium reabsorption, and it fails to interact, unlike PTH, with a series of region-specific PTH antisera. Collectively, these findings seem to imply that the proximal tubular PTH receptor which is coupled to adenylate cyclase may recognize a region of the PTH molecule which is different than that recognized by other PTH receptors. Further, they suggest that while the HHM-F has some structural or steric similarity to PTH, it is not identical to PTH.

From August 1, 1978, to June 30, 1984, we studied a total of 133 patients with cancer and hypercalcemia, including the 50 in our initial report. Eight of these patients were discovered to have a combination of metastatic cancer and 1°HPT (1°HPT + cancer group), with the parathyroid disease being responsible for their hypercalcemia. The diagnosis of 1°HPT in these patients was documented by either (1) the reversal of hypercalcemia following parathyroidectomy (3 patients), or (2) a history of hypercalcemia of several months to years duration antedating the discovery of their cancer (5 patients).

The control group of patients with metastatic cancer and normal serum calcium values (cancer controls) was enlarged from the 15 patients in our initial series to 40 patients in the current series. Fifty-two patients with documented 1°HPT were also studied.

Samples for iPTH, NcAMP, plasma vitamin D metabolites, fasting calcium excretion (FCaE), the renal tubular threshold for phosphate (TmP/GFR), alkaline phosphatase, albumin, 5'-nucleotidase, electrolytes, and urinary hydroxyproline were collected and analyzed as previously described (3,4), or using standard clinical chemistry laboratory methods. The only substantial analytical differences as compared to our earlier report are the use of a mid-region PTH radioimmunoassay (5), and the adoption of a simpler plasma $1,25(OH)_2D$ microassay (6). Bone radionuclide scans were performed using ^{99m}Tc-labeled methanediphosphonate.

As previously reported (3), the initial 50 samples were collected under rigorous conditions with respect to fasting, hydration therapy, medication, and sample handling. In contrast to these initial 50 patients, samples on the subsequent 83 were collected by more than 10 different physicians and nurses, and were assayed for NcAMP by two different technicians. While attempts continued to be aimed at studing patients in the fasting state and prior to therapy, and while this is believed to have been achieved in most patients, it has not always been the case. Some of the second 83 patients are known to have eaten meals or to have received furosemide, mithramycin, and/or intravenous saline prior to sample collection.

A. RESULTS

1. NcAMP Excretion

Figure 1 shows the results of NcAMP excretion in the six groups of patients, including 40 cancer controls, 52 patients with 1°HPT, the 8 patients with cancer whose hypercalcemia was due to coexisting 1°HPT, and the remaining 125 patients with MAHC. These latter patients were divided into groups based upon whether these NcAMP excretion was high-normal (2.0-2.5 nmol/dl GF) or frankly elevated (>2.5 nmol/dl GF) (94 patients or 75%), or low-normal (0.5-1.0 nmol/dl GF) or frankly low (0.5 nmol/dl GF) (23 patients or 18%). As in our initial studies, patients with MAHC and elevated NcAMP excretion are considered to have HHM, while those with reduced NcAMP excretion were labeled as having local osteolytic hypercalcemia or "LOH." Eight patients (6%) in the MAHC group had NcAMP values which were squarely in the normal range. These patients were included in a separate "unclassified" group. The cancer control group had a mean NcAMP value indistinguishable from that of healthy individuals, and only six of these 40 were outside the

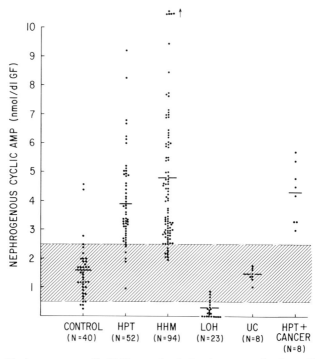

FIG. 1. Nephrogenous cyclic AMP excretion in the six groups of patients. Control refers to patients with cancer and normal serum calcium values; HPT to patients with primary hyperparathyroidism; HHM to patients with humoral hypercalcemia of malignancy; LOH to patients with local osteolytic hypercalcemia; UC to unclassified patients; and HPT + Cancer to patients with concomitant primary hyperparathyroidism and cancer. The numbers in parentheses in this and subsequent figures indicate the numbers on whom this measurement is available.

normal range. Two of the three with elevated NcAMP values subsequently became hypercalcemic. Forty-seven of 52 patients with 1°HPT had elevated NcAMP values, with five patients (10%) displaying values in the normal range. All eight patients with coexisting cancer and 1°HPT had elevated NcAMP values, with a mean value indistinguishable from that in patients with 1°HPT alone.

Urinary cyclic AMP excretion (Fig. 2) mirrored the findings for nephrogenous cyclic AMP excretion in the six groups. Ninety of the 94 patients in the HHM group and 47 of 52 patients in the 1°HPT group had UcAMP values exceeding 4.0 nmol/dl GF, where as 21 of 23 patients in the LOH group and 7 or 8 in the unclassified group had values below 4.0 nmol/dl GF. Mean plasma cyclic AMP (± SD) in the six groups, in nanomoles per liter, was as follows: cancer controls, 18.0 ± 4.3; 1°HPT, 18.5

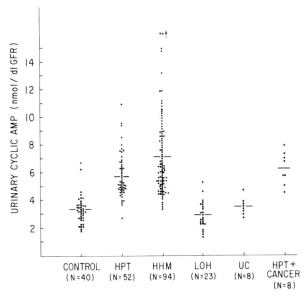

FIG. 2. Urinary cyclic AMP in the six groups described in the legend to Fig. 1.

± 3.5; HHM, 23.6 ± 6.9; LOH, 27.7 ± 8.4; unclassified, 22.6 ± 7.3; and 1°HPT + cancer, 19.4 ± 7.6. The elevations in plasma cAMP presumably reflect the stress of acute illness as well as reductions in the glomerular filtration rate (7).

2. Serum Calcium

Figure 3 shows serum calcium concentrations in the six groups. Mean values for serum calcium in the three groups of patients with cancer-related hypercalcemia (mean ± SD) were 13.2 ± 1.7, 13.0 ± 2.4, and 12.3 ± 1.0. These values were significantly higher than the two groups of patients with 1°HPT, whose corresponding values were 11.4 ± 0.7 and 11.6 ± 0.8.

3. Tumor Histology

Table I details the histologic breakdown of patients in the five groups of patients with cancer. Table II shows the same information given as a percentage of the total in the group. While the contents of the tables are self-explanatory, two important features are worthy of note. First, when patients are classified into groups solely based upon their values for NcAMP excretion, those with elevated NcAMP excretion contain, as their largest single group, patients with squamous carcinomas (46 of 94 patients or 49%); patients with renal carcinomas and transitional cell car-

TABLE I

Tumor Histology in the Five Cancer Groups[a]

Tumor histology	Controls	Elevated NcAMP	Low NcAMP	UC[b]	H + C[c]
Squamous					
Endometrium		2			
Esophagus	2	2			
Head and neck	19	14			
Lung		15	1	2	1
Skin		1			
Cervix		1			
Vulva		4			
Ovary		1			
Bladder		1			
Unknown primary		5			
Total squamous	21	46	1	2	1
Renal	1	11		1	
Bladder, transitional cell		5	1	1	
Miscellaneous					
Prostate					1
Adenocarcinoma					
Lung	2	4	2		
Stomach		1			
Parotid	1	1	1		
Unknown primary		3	1		
Ovary	3	1			2
Endometrium	1	3			
Colon		1			1
Mesothelioma		1			
Apudoma	1	1			
Melanoma	4		2	1	
Small cell, lung		4			
Hepatoma	1	1			
Dysgerminoma, ovary			1		
Breast	2	4	5	2	2
Hematologic					
Hodgkin's lymphoma			1		
Non-Hodgkin's lymphoma	1	1	3		
Acute myelogenous leukemia		1	1		
Chronic lymphocytic leukemia		1			
Multiple myeloma	2	4	4	1	1
Total hematologic	3	7	9	1	1
Totals	40	94	23	8	8

[a] Given in absolute numbers of patients.
[b] Unclassified patients. See text and Fig. 1.
[c] Primary hyperparathyroidism with coexisting cancer. See Fig. 1 and text.

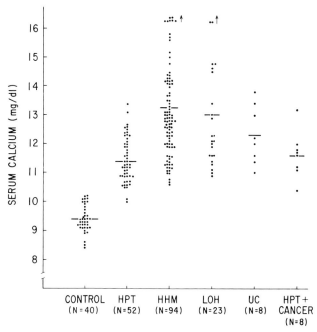

FIG. 3. Serum calcium values in the six groups described in the legend to Fig. 1.

cinomas of the bladder comprise the distant second and third largest groups. Breast and hematologic malignancies account for only 4 and 7 patients, respectively (or 4 and 7%, respectively) of patients in the elevated NcAMP group. In contrast, patients in the low NcAMP group had predominantly breast (5 of 23 patients or 22%) or hematologic (9 of 23 patients or 43%) malignancies. Thus, among patients in the low NcAMP

TABLE II

Tumor Histology in the Two Major MAHC Groups[a]

	Elevated NcAMP (%)	Low NcAMP (%)
Squamous	49	4
Renal	12	0
Bladder	5	5
Miscellaneous	8	14
Adenocarcinomas	15	19
Breast	4	24
Hematologic	7	43

[a] Shown as a percentage of the total number of patients in each group.

group, 14 of 23 or 67% had either breast cancer or a hematologic malignancy.

The second point to note is that, by design, the patients in the cancer control group have tumors with histologies and a distribution which resemble those in the MAHC groups: 21 of the 40 (53%) had squamous carcinomas and 8 (13%) had breast or hematologic tumors. As noted above, these normocalcemic patients as a group have normal values for NcAMP excretion (Fig. 1).

4. Bone Metastases

Fifty-nine of 94 patients in the HHM group had bone scans (63%), as did 19 of 23 patients with low NcAMP (83%). The results of these bone scans are shown in Fig. 4. Twenty-six of 59 (52%) patients in the HHM group had no bone metastases, and an additional 10 had between one and three metastases. Thus, 36 of 59 patients (61%) in this group had three or fewer bone metastases. In contrast, among the patients in the LOH group, 12 of 19 (63%) had seven or more areas of uptake on their bone

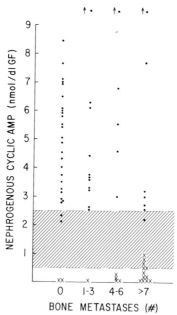

FIG. 4. Nephrogenous cyclic AMP excretion plotted as a function of the number of bone metastases in those patients in the HHM (●) and LOH (×) groups. Note that there is an inverse relation between NcAMP values and the number of bone metastases in patients with MAHC. Also, note that if patients were classified based on the presence or absence of bone metastases, the two groups would overlap with regard to NcAMP values.

scans, and an additional 4 patients had between four and six areas of uptake, so that 16 of 19 (84%) patients in the LOH group had four or more bone metastases. Two patients in the LOH group had no evidence of skeletal involvement, and are of particular interest. One was a 17-year-old female with an ovarian dysgerminoma whose case has previously been published (8). Her hypercalcemia reversed following resection of her tumor, indicating the humoral nature of her hypercalcemia. The other patient in this group is a 57-year-old man with a histocytic lymphoma and dramatic elevation in circulating $1,25(OH)_2D$ levels (see Fig. 8). His case also has been discribed in detail elsewhere (9). One patient in the LOH group had between one and three bone metastases as assessed by bone scan. This patient had multiple myeloma and almost certainly had widespread bone marrow involvement not apparent on radionuclide scanning (10). Thus, with the exception of the patients with the ovarian dysgerminoma, and the $1,25(OH)_2D$-associated lymphoma, all of the patients in the LOH group had widespread skeletal tumor involvement. As can be seen from the figure, there is an inverse relationship between the numbers of bone metastases and values for NcAMP excretion.

5. Serum Phosphorus and TmP/GFR

Values for serum phosphorus are shown in Fig. 5. The major points to emphasize from this figure are that (1) patients in the 1°HPT + cancer and the 1°HPT groups are both hypophosphatemic with respect to the cancer controls ($p < 0.05$). Patients in the HHM group appear hypophosphatemic as compared to controls but this does not reach statistical significance. Those with LOH are similar to the cancer controls.

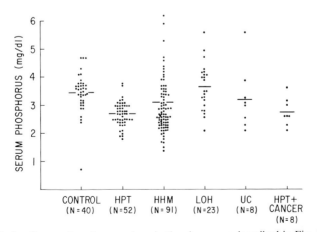

FIG. 5. Serum phosphorus values in the six groups described in Fig. 1.

Values for the renal phosphorus threshold (TmP/GFR) are shown in Fig. 6. Patients in the 1°HPT, HHM, and 1°HPT + cancer groups are all significantly different from controls ($p < 0.05$), and from those with LOH ($p < 0.05$). Patients in the LOH group appear to have mean values somewhat lower than the cancer controls, but this did not attain statistical significance.

6. Fasting Calcium Excretion

Values for fasting calcium excretion are shown in Fig. 7. Patients with 1°HPT alone or 1°HPT + cancer had mean values which were similar and which were mildly but not significantly elevated as compared to the control group. In contrast, patients in both the LOH group and the HHM groups had dramatic increases in fasting calcium excretion. These were significantly different ($p < 0.05$) from both the cancer control group and the 1°HPT groups, with mean values of 0.75 and 1.11 mg/dl GF, respectively. The difference between patients with 1°HPT and HHM remained even when patients with 1°HPT and HHM were matched for both serum total calcium measurements and ionized calcium measurements: when the patients with 1°HPT and HHM with total serum calcium values between 11.0 and 11.9 mg/dl (mean values = 11.4 ± 0.3 and 11.4 ± 0.3, respectively) were compared, fasting calcium excretion values were 0.19 ± 0.08 mg/dl GF in the former and 0.51 ± 0.29 in the latter. Similarly, when patients from our earlier report were matched for ionized serum calcium values between 5.0 and 6.0 mg/dl (mean values = 5.62 ± 0.3 and 5.50 ±

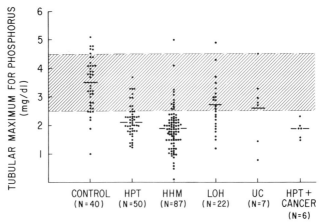

FIG. 6. The renal phosphate threshold (TmP/GFR) values in the six groups described in Fig. 1.

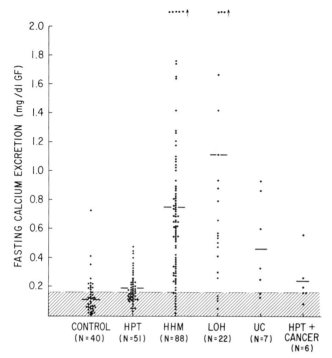

FIG. 7. Fasting calcium excretion in the six groups described in Fig. 1.

0.3, respectively), fasting calcium excretion was 0.30 ± 0.14 mg/dl GF in the former and 0.71 ± 0.61 in the latter.

7. Vitamin D Metabolites

Plasma 1,25(OH)$_2$D concentrations in the six groups are shown in Fig. 8. Mean values were normal in the cancer controls (33 pg/ml) and elevated in the 1°HPT group (72 pg/ml). In the HHM and LOH groups, mean plasma 1,25(OH)$_2$D values were low-normal (22.5 pg/ml in both groups), with 30 and 60% of patients in these two groups having undetectable values. As with the fasting calcium excretion measurements described above, the differences in plasma 1,25(OH)$_2$D values between patients with 1°HPT and HMM remained even when patients were matched for serum calcium concentration. For example, in those patients with 1°HPT and HHM whose serum ionized calcium was between 5.0 and 6.0 mg/dl (mean shown above), the plasma 1,25(OH)$_2$D values were 92 ± 30 and 23 ± 16 pg/ml, respectively. One patient in the LOH group had a dramatically elevated 1,25(OH)$_2$D value (116 pg/ml), a value exceeded by only one

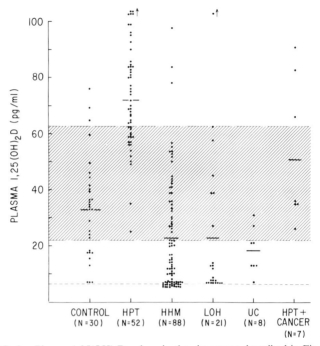

FIG. 8. Plasma 1,25(OH)$_2$D values in the six groups described in Fig. 1.

patient in the 1°HPT group. This was the patient with histocytic lymphoma and presumed 1,25(OH)$_2$D-mediated hypercalcemia. Interestingly, of the seven patients with 1°HPT and coexisting cancer in whom 1,25(OH)$_2$D was measured, only three had elevated values.

Plasma 25-hydroxyvitamin D values were normal in all six groups, with mean (±SD) values as follows: cancer controls, 22.6 ± 16; 1°HPT, 22.2 ± 10.6; HHM, 16.6 ± 9.4; LOH, 15.4 ± 6.1; unclassified, 20.0 ± 18.9; and 1°HPT + cancer, 19.3 ± 5.1.

8. Immunoreactive Parathyroid Hormone

Values for iPTH in the six groups are shown in Fig. 9. In the assay employed, values for iPTH were elevated in 49 of 52 patients with 1°HPT and were normal in 58 of 71 patients with malignancy-associated hypercalcemia. Twelve of 51 (23.5%) patients in the HHM group had elevated values for iPTH, and 1 of 13 (7.6%) in the LOH group had an elevated iPTH value. Sera from all six of the patients with 1°HPT + cancer assayed in this assay had elevated iPTH values, and the other two patients had elevated iPTH values in a commercial PTH assay.

Figure 10 shows serum iPTH values as a function of serum calcium

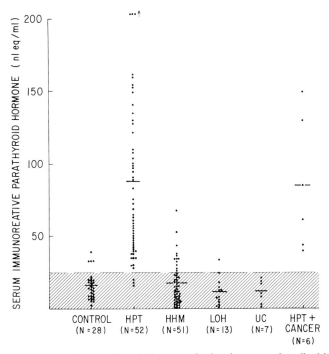

FIG. 9. Immunoreactive parathyroid hormone in the six groups described in Fig. 1.

concentration in the six groups. In contrast to the substantial overlap observed between the 1°HPT and MAHC groups shown in Fig. 9, when iPTH is examined as a function of serum calcium, the two groups are clearly separated.

9. *Serum Chloride:Phosphate Ratio*

Figure 13 displays the chloride:phosphate ratio in the six groups. It can be seen that patients with 1°HPT and 1°HPT + cancer, with only three exceptions, have ratios exceeding 32. Patients with MAHC have widely varying values, with no obvious pattern. Serum chloride concentration in the six groups were as follows (mean ± SD): cancer controls, 100.1 ± 3.8; 1°HPT, 106.1 ± 3.2; HHM, 99.2 ± 6.8; LOH, 95.3 ± 6.1; unclassified, 99.0 ± 6.8; 1°HPT + cancer, 106.0 ± 2.0.

10. *Serum Bicarbonate*

Figure 14 shows the serum bicarbonate values in the six groups. While none of the groups differs in a statistically significant way, serum bicarbonate tended to be lowest in the patients with 1°HPT.

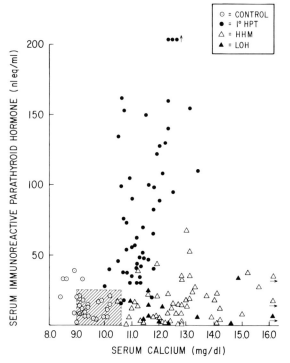

FIG. 10. Immunoreactive parathyroid hormone in the cancer controls, patients with 1°HPT, HHM, and LOH plotted as a function of the serum calcium value. For the 1°HPT group, $r = +0.52$, $p < 0.001$. For the HHM group, $r = +0.13$, $p = NS$.

11. Serum Alkaline Phosphatase

Figure 11 shows the alkaline phosphatase values in the six groups. Alkaline phosphatase was elevated in 10 of 46 patients with 1°HPT, 49 of 86 with HHM, and 8 of 16 in the LOH group. Serum 5'-nucleotidase values were available in 15 patients in the HHM group, and this measure of biliary obstruction was elevated in all 15. Similarly, in the one patient with an elevated alkaline phosphatase in the HHM group for whom a 5'-nucleotidase value was available, this value was elevated. These findings strongly suggest that liver, not the skeleton, is the source of circulating alkaline phosphatase elevations in the minority of patients with HHM in whom this abnormality is found.

12. Urinary Hydroxyproline

Figure 12 shows the results of urinary hydroxyproline excretion in the patients on whom this measurement was performed. In both groups of

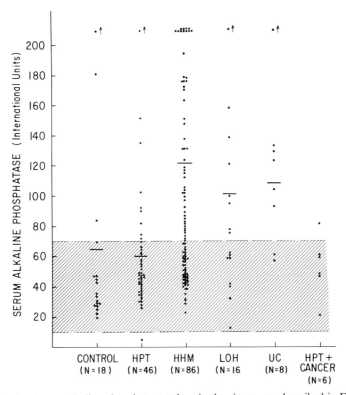

FIG. 11. Serum alkaline phosphatase values in the six groups described in Fig. 1.

patients with malignancy-associated hypercalcemia, urinary hydroxyproline excretion was dramatically elevated. In contrast, mean values in the cancer controls and in the 1°HPT group were at the upper limits of normal.

B. DISCUSSION

We have studied 133 patients with malignancy-associated hypercalcemia. In 8 of these patients (6%), hypercalcemia appeared to result from coexisting 1°HPT. Biochemically, these patients closely resembled the 52 healthy patients with 1°HPT with the single exception being that in 4 of 7 in whom plasma 1,25(OH)$_2$D values were measured, these values were in the normal range instead of being elevated.

The remaining 125 patients had hypercalcemia which resulted from their tumors. These patients were classified into groups depending upon whether their NcAMP excretion was high or high-normal (94 patients or

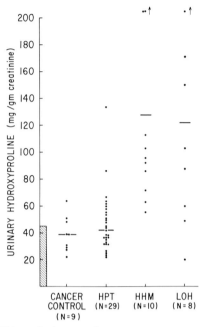

FIG. 12. Urinary hydroxyproline excretion in selected patients.

75%) or low or low-normal (23 patients or 18%). In this series of patients in which samples were collected less carefully than in our initial series, and when samples were analyzed for NcAMP excretion by multiple technicians, 117 of 125 patients (95%) could be reliably categorized. Only 8 patients (6%) were identified who had normal NcAMP excretion. The explanation for these normal values remains uncertain, but may be related to improper sample processing or handling. These patients displayed no distinguishing features when compared to the remaining patients with MAHC. If anything, they most closely resemble, from a histologic standpoint, the patients in the HHM group (Table I).

Despite these allowances in sample collection and processing, the findings in the current series closely mimic those reported earlier (3). The most striking observation, to our mind, continues to be that when patients are grouped based upon NcAMP values, strikingly different profiles emerge with respect to both tumor histology and the number of bone metasases. The group of patients with elevated NcAMP excretion has a striking preponderance of squamous carcinomas as well as a substantial number of urothelial (renal and bladder) carcinomas. As a group, these patients have few or no skeletal metastases. We continue to interpret these findings as indicating that this group of patients with elevated

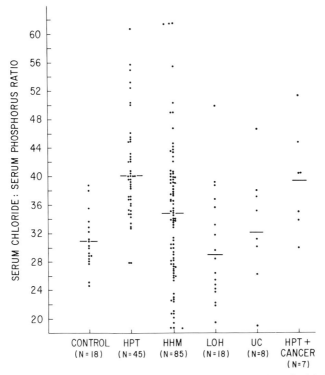

FIG. 13. Serum chloride:phosphorus ratios in the six groups described in Fig. 1.

NcAMP excretion, predominantly squamous and urothelial neoplasmas, and few or no bone metastases, has "prototypical" humoral hypercalcemia of malignancy. The second group of patients with reduced NcAMP excretion also has a preponderance of tumor types, but in this case they are breast carcinoma and hematologic tumors. Also differentiating them from the elevated NcAMP group is the extensive degree of skeletal tumor involvement. In contrast to the patients with elevated NcAMP excretion who appear to have prototypical HHM, these patients, with two exceptions, appear to be hypercalcemic as a result of their extensive skeletal involvement by tumor. Selection of an appropriate designation for this group is problematic, as indicated by the inclusion in this group of the two patients without bone metastases—the woman with the ovarian dysgerminoma whose hypercalcemia reversed with tumor eradication, and the man with the dramatic elevation in plasma $1,25(OH)_2D$ resulting from a lymphoma—who clearly had a form of humoral hypercalcemia of malignancy. What this group fundamentally has in common is a negative biochemical feature—suppression of NcAMP excretion—which relates

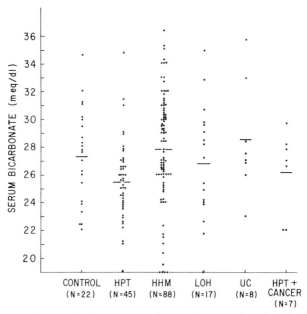

FIG. 14. Serum bicarbonate values in the six groups described in Fig. 1.

to a negative physiologic feature—their hypercalcemia is not due to circulating parathyroid hormone or PTH-like factors. This group illustrates the point that a variety of forms of MAHC exist, most being poorly characterized pathophysiologically. We have designated this group as having local osteolytic hypercalcemia (LOH) to indicate that in the vast majority, hypercalcemia results from local osteolytic events within the skeleton.

In the current series, as in our earlier series, normocalcemic patients with cancer generally displayed normal values for NcAMP excretion. The two cancer controls with the highest NcAMP values both went on to become hypercalcemic within 1 month. These findings are in direct conflict with those of Kukreja et al. (11), who reported that NcAMP excretion in their control group was elevated as compared to their normal population. Part of this "elevation" may be more apparent than real in that their normal population consisted of only 8 patients, yielding a normal range of 0.7–1.9 nmol/dl GF, a range somewhat narrower than ours (0.5–2.5 nmol/dl GF). It may be that as the number of normal individuals studied by Kukreja et al. increases their normal range will expand. Other differences must be present in this study, however, as compared to our own, to explain the facts that most of their controls were in the same range as that reported for hypercalcemic patients, and that their cancer controls have far higher values than ours. We can only guess that differences in sample collection or analysis must exist in the Kukreja study as compared to our own.

Two major differences exist between our study and that of Rude et al. (12): (1) some of their normocalcemic cancer controls had NcAMP values which were elevated; and (2) no definite differences in NcAMP excretion between patients with bone metastases (who were presumed to have local skeletal destruction as the cause of their hypercalcemia) and those without bone metastases (who were presumed to have "pseudohyperparathyroidism" or HHM) were identified. While some methodologic differences between the Rude study and our own must certainly exist (as suggested by their substantially higher NcAMP values in all of their patient groups as compared to our own), the differences between this study and our own are more apparent than real. The apparent elevation in NcAMP excretion in their controls diminishes when interpreted using the normal NcAMP range defined by their 18 normals (mean ± 2SD = 2.2 ± 2.4, or 0.0–4.6). When examined in this light, most of their cancer controls are within or near the normal range. Thus, we believe, allowing for methodologic differences and their relatively wide normal range, that their cancer control group, like ours, has normal or near-normal NcAMP excretion.

The second and most important difference between the Rude study and our own is the apparent failure to identify a distinct pattern of NcAMP excretion in their two MAHC patient groups. The word "apparent" is important here, for our own data, when presented using the criteria of Rude et al., look strikingly similar to their data (vide infra). The major difference between our study and their study emerges as one of patient classification: we used NcAMP to define the patient groups and in so doing achieved clear-cut separation into two pathophysiologic groups. Rude et al. used the presence or absence of bone metastases, as proposed by Lafferty (13), to define these groups, and in so doing achieved no apparent separation with respect to NcAMP excretion. We believe that this is the result of a flaw in their criterion for selecting patients with pseudohyperparathyroidism (or HHM) from the larger group with MAHC. In order for a patient to be included in their HHM group, "no evidence for bone metastases could be found by bone scan and/or autopsy." Using this criterion, only 17 of 91 (18.7%) patients with MAHC were believed to have HHM (in contrast to the 75% in our study), and the remaining 74 patients were presumed to be hypercalcemic as the result of their bone metastases. This latter assumption is not likely to be valid for two reasons. First, many patients in their group with putative local osteolytic hypercalcemia can be assumed to have had only a small number of bone metastases which would be inadequate to cause hypercalcemia.

Unfortunately, no information on the number or extent of skeletal involvement in their "Bone Metastases" group is provided in the paper. Second, even when patients do have bone metastases, humoral mechanisms may be operative, as evidenced by the presence of humorally mediated osteoclastic bone resorption which occurs in portions of the skeleton devoid of bone metastases (14). Thus, we believe that the majority of the patients in the Rude study were incorrectly classified. This belief is strengthened by examination of their Fig. 4 (Fig. 15 in the present article). One can see that the majority of patients in the "GI," "ENT," "Lung," and "Renal" groups have high-normal or elevated values for NcAMP excretion, exactly as occurred in our study, while the majority of patients in the "Breast" and "Myeloma" groups have reduced NcAMP excretion, again mirroring our findings. When our data (Fig. 16) are plotted in a fashion similar to that of Rude *et al.* (Fig. 4 and 15), the two groups of data become almost indistinguishable.

In summary, we feel that methodologic differences and the small number of patients in the Kukreja study explain the differences found in this study as compared to our own. The Rude study, while different in some

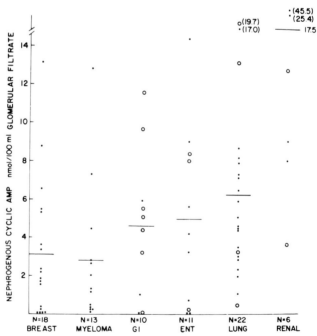

FIG. 15. Nephrogenous cyclic AMP excretion in selected patients with MAHC as reported by Rude *et al*. Reprinted with permission from Rude *et al*. (12).

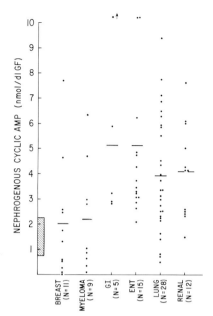

FIG. 16. Nephrogenous cyclic AMP excretion in the 125 patients with MAHC in the current study, displayed using the format of Rude et al. (12). This is the same data shown in Fig. 1. Compare to Fig. 1 and 15. Note that when the NcAMP data are evaluated in this manner, no definite patterns of NcAMP excretion appear.

methodologic respects, and while different (and incorrect, we believe) in their classification of patients, is very similar to our own. Had their patient groups been selected based upon NcAMP excretion rather than bone metastases, the authors would have arrived at conclusions similar to our own.

The other biochemical findings in the current study expand the observations made earlier by many investigators. TmP/GFR values are reduced in HHM and in 1°HPT, reflecting the role of the proximal tubular PTH receptor/adenylate cyclase complex in the inhibition of phosphorus reabsorption. TmP/GFR values are mildly reduced as compared to controls in the LOH group, perhaps reflecting the phosphaturic effects of hypercalcemia per se (15).

As reported previously, fasting calcium excretion is dramatically elevated in both the LOH and HHM groups when compared with the 1°HPT groups (3,16). This remains true even when patients with HHM and 1°HPT who have identical serum calcium values are compared, indicating that the marked hypercalciuria which occurs in patients in the HHM group is not related to an increase in the filtered load of calcium, but to a relative reduction in renal tubular calcium reabsorption in patients with HHM.

Plasma 1,25(OH)$_2$D values are, on the mean, reduced in patients with HHM and LOH as compared to those with 1°HPT, a finding confirmed by Breslau et al. (17), and explaining the observed reduction in intestinal ^{45}Ca absorption in patients with MAHC (18). While it is conceivable that the reduction in 1,25(OH)$_2$D could be due to an increase in the metabolic clearance of 1,25(OH)$_2$D, this seems unlikely. Similarly, differences in levels of a circulating vitamin D binding protein are an unlikely explanation, since (1) levels of plasma 25(OH)D are normal, and (2) these patients display the physiologic correlate of reduced plasma 1,25(OH)$_2$D levels, namely reduced intestinal calcium absorption. If the cause for the reduction in plasma 1,25(OH)$_2$D levels is diminished production, one must ask why the activity of renal 1α-hydroxylase is reduced. Inhibition by elevated serum calcium values seems unlikely in view of the fact that many of the patients with MAHC who have reduced values of 1,25(OH)$_2$D have serum calcium concentrations equivalent to the patients with 1°HPT who have elevated plasma 1,25(OH)$_2$D levels. We have suggested that the PTH receptors involved with 1α-hydroxylase may be different from those associated with adenylate cyclase so that the latter PTH receptors recognize the HHM factor but the former do not. Interesting in this regard is the observation that when tumors from humans with HHM are transplanted into immunodeficient mice, plasma 1,25(OH)$_2$D levels rise in these animals, a finding which suggests that murine and human 1α-hydroxylase/PTH receptors may be different (19). Note should be taken of the extraordinary patient who has MAHC in the setting of a lymphoma, with dramatic elevations in plasma 1,25(OH)$_2$D. To date a total of 7 such patients have been described. (9,17).

Immunoreactive PTH levels in patients with MAHC, depending on the assay employed, are undetectable or inappropriately reduced as compared to patients with 1°HPT (3,16,20–22). These observations are reproduced clearly in Fig. 9 and 10. While some patients may have iPTH values which are above the "normal" range, when iPTH is examined as a function of serum calcium, patients with MAHC and with 1°HPT can be clearly separated. That the apparent "PTH immunoreactivity" in certain HHM sera is not due to interaction of antisera with PTH-like immunoreactive determinants on the HHM factor is apparent from the observation that mean iPTH values in this and other PTH immunoassays are no different in the HHM as compared to LOH groups. The reasons for the presence of circulating immunoreactive "PTH" in the plasma of some patients with MAHC is unclear but probably relates to the combined effects of (1) continued basal PTH secretion by incompletely suppressible normal parathyroid tissue (23), (2) reduced renal clearance of PTH C-terminals in patients with compromised renal function, and (3) nonspecific binding in a given assay.

The observations that patients with HHM display elevated NcAMP excretion and enhanced phosphate excretion suggest that the HHM factor is in some sense "PTH-like." The observations that in HHM as compared to 1°HPT, renal tubular calcium reabsorption is reduced, that renal 1α-hydroxylase activity is reduced, and that PTH immunoreactivity is reduced or absent, suggest that the HHM factor is in other respects "PTH-unlike."

Examination of bone histology in patients with HHM provides a potent extention to the "PTH-like" and "PTH-unlike" nature of the HHM factor (14). Like patients with 1°PTH, osteoclastic bone resorption is accelerated in HHM. In fact, it is strikingly increased beyond that encountered in 1°HPT. Unlike patients with 1°HPT, however, osteoblastic activity is markedly reduced in patients with HHM, indicating that bone cell activity is uncoupled in HHM, and suggesting, again, that the mechanism of action of the HHM factor on a given target organ may be similar to but different from the effects of PTH on that same target organ.

Serum alkaline phosphatase values were elevated in 49 of 86 patients with HHM, a finding which might suggest the presence of enhanced osteoblastic activity in these patients. Yet, in all 15 of the patients in whom serum 5'-nucleotidese, an indicator of biliary obstruction, was measured, it was elevated. These findings suggest that when serum alkaline phosphatase is elevated in patients with HHM, the biliary tree, not the skeleton, is the source of the alkaline phosphatase. These findings correlate with the reduction in osteoblastic activity in HHM mentioned above (14), and agree with the findings of Lafferty (13) who found that while the serum alkaline phosphatase was elevated in 20 of 42 patients with pseudohyperparathyroidism (or HHM), 15 of these 20 patients whose liver was examined at autopsy had hepatic metastases.

Urinary hydroxyproline measurements were available in a minority of patients. Again, as would be predicted from the results of bone biopsy, urinary hydroxyproline values were dramatically elevated in both groups of patients with MAHC, with mean values being some 3-fold higher than those encountered in the 1°HPT group. This 3-fold increment over that seen in 1°HPT is quantitatively identical to the mean 3-fold increment in osteoclast number reported in patients with HHM as compared to those with 1°HPT (14). Increases in urinary hydroxyproline in patients with MAHC have also been reported by Siris *et al.* (24).

In contrast to a prior report by Palmer *et al.* (25), the chloride:phosphate ratio was not clearly discriminant between patients with 1°HPT and patients with MAHC. Like Palmer's patients with 1°HPT, the vast majority of patients with 1°HPT in the current series displayed chloride:phosphate ratios above 33. Unlike the Palmer study, however, patients with MAHC had chloride:phosphate ratios which ran the gamut from high to low values. The prime reason for the difference between the studies would seem to be that the phosphorus values in the "non-1°HPT" groups in the two studies were quite different: in Palmer's study the non-1°HPT patients' mean serum phosphorus value was 4.5 mg/dl, whereas in the current study, the mean serum phosphorus values in the HHM and LOH groups were 3.1 and 3.7, respectively. The higher phosphorus values in Palmer's non-1°HPT patients result from the fact that this group also contained substantial numbers of patients with the milk-alkalai syndrome, vitamin D intoxication, immobilization, and multiple myeloma, all conditions in which the serum phosphorus value would be expected to be high-normal or elevated. In contrast, the serum phosphorus values in the current HHM and LOH groups were similar to or below the normal range, a situation which would tend to raise the chloride:phosphate ratio. Interestingly, in the Palmer study, the one patient with "ectopic hyperparathyroidism" described had a chloride:phosphate ratio in the 1°HPT range. Our data would support the concept that when the chloride:phosphate ratio is below 33, a cause of hypercalcemia other than 1°HPT is likely. A ratio of greater than 33 supports no particular diagnosis.

No significant differences in serum bicarbonate were encountered among any of the groups, although the patients with 1°HPT tended to have lower serum bicarbonate values than those of controls or those with hypercalcemia from other causes. Collection and analysis of serum bicarbonate concentration were not rigorous, however, and it may be that had samples been collected on ice, under oil, and assayed immediately, significant differences would have emerged. These findings support the long-standing observation that PTH is bicarbonatruric but shed no light on the possibility raised by Heinemann (26) that rapid bone resorption in patients with MAHC exposes skeletal proton buffering sites and predisposes to a metabolic alkalosis.

A final clinical or *in vivo* observation regarding human HHM involves the cytochemical bioassay (CBA) for PTH. Using this highly sensitive and specific PTH bioassay, Goltzman, in collaboration with our group, has shown that plasma levels of PTH-like cytochemical bioactivity in 16 patients with HHM (defined on the basis of elevated NcAMP excretion) were dramatically elevated as compared to levels in 10 patients with LOH (defined on the basis of reduced NcAMP excretion (Fig. 17) (27). This PTH-like circulating cytochemical bioactivity was shown to be poorly inhibitable with PTH antisera and to elute in gel filtration studies with a

NcAMP, ACSA, AND MAHC 729

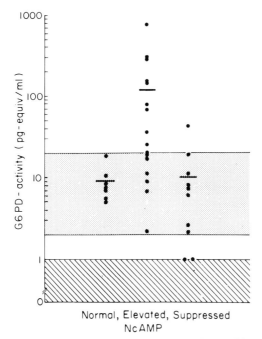

FIG. 17. Plasma cytochemical bioactivity (CBA) in patients with cancer and normal serum calcium values, and from patients with HHM and LOH. Reproduced with permission from Goltzman et al. (27).

molecular weight in excess of that assigned to native PTH (Fig. 18). These findings indicate that (1) patients with HHM display two distinct markers for circulating biologically active PTH: elevated NcAMP excretion and elevated circulating PTH-like cytochemical bioactivity, and (2) this HHM-associated circulating cytochemical bioactivity is unlikely to be related to native PTH.

C. SUMMARY

The information in this portion of the article can be briefly summarized as follows. First, patients with MAHC can be divided into two general categories based upon their values for NcAMP excretion. The discrepancies between our study and those of Kukreja et al. and Rude et al. can be best explained by the limited number of patients examined and methodologic differences in the former study, and by minor methodologic and major organizational differences in the latter. As indicated, our differences with the Rude study are largely interpretive. Second, most normo-

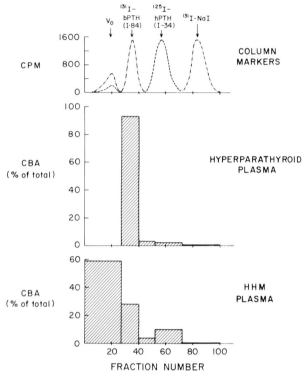

FIG. 18. Elution profiles of plasma CBA in a patient with primary hyperparathyroidism, and a patient with HHM. Reproduced with permission from Goltzman et al. (27).

calcemic patients with cancer, in our hands and in those of Rude et al. (and perhaps in those of Kukreja et al. had they examined larger numbers of normal individuals), have normal NcAMP excretion. Third, given the above, elevated NcAMP excretion serves as a marker for the syndrome of HHM, and presumably for the presence in the circulation of the HHM factor. Fourth, HHM resembles 1°HPT with respect to the presence of accelerated bone resorption, elevated NcAMP excretion, reduced TmP/GFR values, and elevated plasma cytochemical bioactivity, but differs from 1°HPT in that in patients with HHM, renal tubular calcium reabsorption is reduced, renal 1α-hydroxylase activity is reduced, iPTH values are reduced, and osteoblastic bone formation is reduced. Collectively these findings suggest that the HHM factor is structurally or sterically related to PTH, and is capable of binding to and activating certain but not all PTH receptor systems. Fifth, the homogeneity among patients with HHM from a clinical, histological, and biochemical standpoint strongly implies that the mediator of hypercalcemia in this group of patients is likely to be a single substance or closely related family of substances. Sixth, and finally, the findings that patients with HHM are characterized by three pathophysiologic hallmarks—accelerated osteoclastic bone resorption, elevated NcAMP excretion, and elevated plasma PTH-like cytochemical bioactivity—suggest that *in vitro* assays of bone resorption, PTH-sensitive adenylate cyclase assays, and the PTH-sensitive cytochemical bioassay may be useful as detection systems for the presumed HHM factor. The third section will focus on the attempts in our laboratory and others to identify and characterize PTH-like adenylate cyclase-stimulating activity in tumor extracts or tumor-conditioned tissue culture medium derived from patients with HHM.

III. Animal Models of HHM

As will be discussed below, a major difficulty in any attempt at purification of the HHM factor is the limited availability of human HHM-associated tumor which can be used for extraction and purification work. An early and important goal of our group, therefore, was the development and characterization of animal models of human HHM. While a number of potential animal models exist, we have chosen to work most closely with three: (1) a dimethylbenzanthracene (DMBA)-induced murine cutaneous squamous carcinoma (28), (2) the Rice 500 strain of the Fisher rat Leydig cell tumor (29,30), and (3) human HHM-associated tumors transplanted into immunodeficient ("nude") mice. The development and characterization of these models have provided animal and human material for HHM factor protein purification and molecular biological studies, and will permit the physiologic studies on the animal HHM syndrome.

Both the murine and rat HHM syndromes closely resemble human HHM. (1) Both models can be demonstrated to be "humoral" in the sense that hypercalcemia reverses with tumor resection, and histologic evidence of bone resorption is present in the absence of skeletal tumor involvement. (2) Hypercalcemia is associated with hypophosphatemia and evidence of renal phosphorus wasting. (3) Urinary calcium excretion is dramatically increased, even under conditions of dietary calcium restriction. (4) Urinary cAMP or NcAMP excretion is elevated. (5) Immunoreactive PTH levels fall as hypercalcemia develops. The sole discrepancy between the human and rodent HHM syndromes is the circulating $1,25(OH)_2D$ values, which are elevated in rat and murine HHM, but reduced in human HHM. As noted above, when human HHM-associated tumors are transplanted into nude mice, the animals can be shown to demonstrate an increase in plasma $1,25(OH)_2D$ values (19).

We have not examined the nude mouse model in the detail described for the DMBA-induced murine and the Leydig cell rat models. Recently, however, Kukreja and collaborators have demonstrated that a nude mouse model bearing a transplantable human squamous carcinoma shows the biochemical features described above (19).

The striking concordance between NcAMP or UcAMP excretion in the animal and human HHM syndromes, and the findings of normal UcAMP or NcAMP excretion in normocalcemic human controls as well as in animal controls, provides strong support for the thesis that a PTH-like adenylate cyclase-stimulating factor may be responsible for the hypercalcemia associated with HHM.

IV. Adenylate Cyclase-Stimulating Activity (ACSA)

The purpose of this section is to briefly review our attempts at characterizing and purifying the ACSA identified in HHM-derived tumor extracts and tumor-conditioned culture medium. Attention will be given to the choice of detection systems for the HHM factor, sources of HHM factor for purification, and attempts at the characterization and the purification of the HHM factor.

A. DETECTION SYSTEMS

As indicated in Sections II and III, the three dominant characteristics or hallmarks which characterize the HHM syndrome are osteoclastic bone resorption, increases in circulating PTH-like cytochemical bioactivity, and increases in NcAMP excretion. These three characteristics of the *in vivo* syndrome suggest that *in vitro* assays which measure osteoclastic bone resorption, PTH-like cytochemical bioactivity, or PTH-like adenylate cyclase-stimulating activity (ACSA) should be useful in detecting the presence of the presumed HHM factor. While we have employed all three types of assays (in collaboration with Dr. Goltzman in Montreal and Dr. Agnes Vignery at Yale), we have focused on the adenylate cyclase (AC) assays). The reasons for selecting AC assays over *in vitro* bone resorption systems are that the bone resorption assays are (1) more expensive, (2) more labor intensive, (3) less sensitive to PTH, (4) less reproducible, and (5) most importantly in our view, far less specific, in that a long list of specific and nonspecific agonists of *in vitro* bone resorption exists. In contrast, the AC assays we have employed are highly specific in terms of the agonists to which they respond. Nevertheless, the *in vitro* bone resorption systems remain the only means of documenting bone resorption and must therefore be included at some stage in any purported HHM factor purification scheme.

The cytochemical bioassay is recommended by its exquisite sensitivity and specitivity, but given its labor-intensive nature, is not useful for rapidly screening large numbers of column fractions or tumor extracts. We have reserved this assay for situations where sensitivity is the major requirement (see Section V).

We have employed two AC assays. One is a canine renal cortical guanylnucleotide-amplified AC assay (31), similar to that employed by Nissenson *et al.* (32). More recently, through the generosity of Drs. Gideon and Sevgi Rodan, we have established the 17/2.8 rat osteosarcoma (ROS) PTH-sensitive intact cell AC assay (33). These assays are routinely used to assay 200 to 300 samples per week. As indicated above, these assays are highly specific, being stimulated only by PTH, the putative HHM factor, and, in the case of the ROS assay, isoproterenol. The renal AC assay is capable of measuring PTH in concentrations of 10^{-10} M, and the ROS assay 10^{-11} M.

B. SOURCES OF HHM FACTOR

We have used four major types of tumors as source material for the HHM factor. These include (1) human tumors from patients with HHM, (2) murine tumors from the DMBA-induced cutaneous squamous carcinoma, (3) tumors from the Rice 500 Leydig cell tumor line passaged in Fisher 344 rats, and (4) human tumors passaged in "nude" mice. Tumors are either grown in tissue culture and the tumor-conditioned medium examined for bioactivity in the assays described, or are extracted in acid-urea as previously described (31).

Each of these sources has potential advantages and disadvantages. The human tumors we have studied have been primarily squamous carcinomas, which grow poorly in culture. Thus, most of the work we have done with these tumors has been on tumor extracts. The same is true for the DMBA-induced murine squamous carcinomas. In contrast, the Leydig cell tumor grows rapidly and easily in tissue culture. Thus, with the Leydig cell tumor, much of our work has been on tumor-conditioned medium.

C. CHARACTERIZATION AND PURIFICATION

Our initial attempts of ACSA measurement involved acid-urea extracts of human tumors. As indicated in Fig. 19, four of the first five HHM-derived tumors we extracted and assayed contained ACSA in the renal

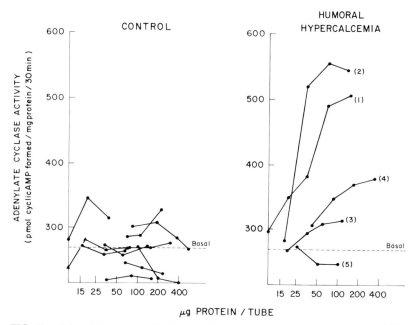

FIG. 19. Adenylate cyclase-stimulating activity in extracts of human control and HHM-associated tumors, using the renal cortical adenylate cyclase assay. Reproduced with permission from Abrahmson et al. (19).

AC assay, while control extracts from tumors derived from normocalcemic patients contained no significant activity (31). Using a similar assay, Strewler et al. demonstrated the presence of ACSA in tissue culture medium conditioned by an HHM-associated renal carcinoma (34). Control tumor-conditioned medium contained no activity. Similarly, Rodan et al. using the ROS assay as a detection system found abundant ACSA in HHM-derived tumor extracts and conditioned medium but not in controls (33). At the time of this writing, we have identified ACSA, using the renal AC, in extracts of 18 of 20 human HHM-associated tumors. This activity is not detectable in 10 samples of nonmalignant human normal tissues, nor is it measurable in seven extracts of tumors derived from patients with hypercalcemia due to mechanisms other than HHM. Twenty extracts of control tumors from patients with normal serum calcium values were also examined. Only four of these contained measurable ACSA.

Acid-urea extracts from the murine DMBA-induced squamous carcinomas also contain potent ACSA in the renal AC and ROS systems (Fig. 20) (28,41). Finally, the Leydig cell tumor extracts and conditioned medium

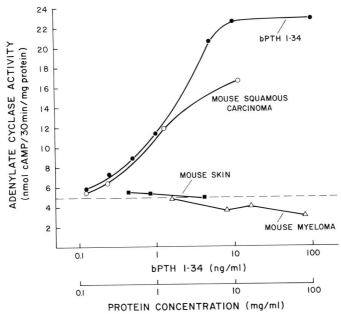

FIG. 20. Dose–response curves of 1–34 bPTH and extracts of the DMBA-induced murine squamous carcinoma, normal murine skin, and a hypercalcemia-associated murine lymphoma, as measured in the renal cortical adenylate cyclase assay. Reproduced with permission from Gkonos et al. (28).

can be shown to contain potent activity in the ROS assay (33,35). When this activity is concentrated and partially purified it can be shown to contain potent activity in the renal AC assay as well.

All three species of ACSA (human, murine, rat) are heat (100°C) stable, basic proteins which are insensitive to reducing agents (28,31,35,36). The ACSA from all three tumor types can be inhibited with the PTH analog and inhibitor $Nle^{8,18}Tyr^{34}$(3–34)bPTH amide (Figs. 21 and 22), but cannot be inhibited by PTH antisera with varying specificities (Fig. 23) (23,28,31,33). This is true as well for the ACSA described by Strewler et al. and Rodan et al.

Efforts in our laboratory are currently focused on complete purification of the various ACSA described above. Material from two human HHM-associated tumors has been purified 20,000- and 5000-fold using a sequence of acid–urea extraction followed by ethanol–sodium chloride precipitation, followed by high-performance liquid chromatography (HPLC). Both of these highly purified extracts dramatically stimulate the fetal bone resorption system. Using the techniques described above, material has

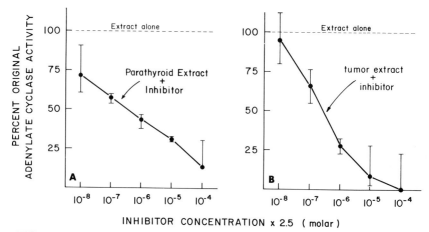

FIG. 21. Inhibition by the PTH analog and inhibitor, Nle^{8,18},Tyr^{34}(3–34)bPTH, of renal cortical adenylate cyclase activity induced by parathyroid extract (A) and HHM tumor extract (B). Reproduced with permission from Abrahmson et al. (19).

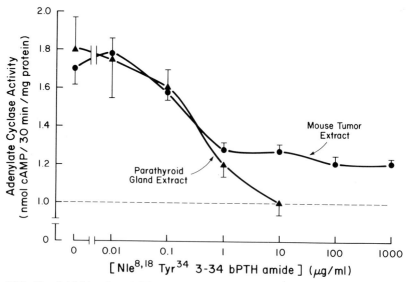

FIG. 22. Inhibition by Nle^{8,18},Tyr^{34}(3–34)bPTH amide of the renal adenylate cyclase activity induced by 1–34 bPTH and by murine squamous carcinoma extract. Note that, unlike Fig. 21, the murine ACSA is incompletely inhibitable, suggesting that more than one adenylate cyclase-stimulating factor is present in this extract. This has been borne out by more recent work (41). Reproduced with permission from Gkonos et al. (28).

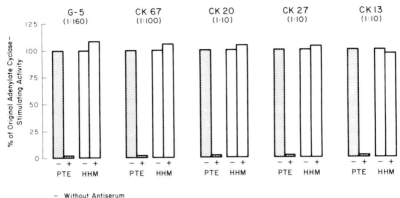

FIG. 23. Effect of preincubating parathyroid extract and human HHM tumor extract with PTH antisera. Note that each antiserum abolishes activity in parathyroid extract, but that identical titers of the same antisera have no effect on the activity produced by HHM-associated tumor extracts. Antisera were generously produced by Drs. G. V. Segre and L. E. Mallette. G5 has mid-region PTH specificity and the CK antisera, amino-terminal specificity.

been purified to apparent homogeneity in microgram quantities from one of the human tumors and can be shown to have a molecular weight of 29,000 as assessed by polyacrylamide gel electrophoresis (36). This apparently pure adenylate cyclase-stimulating protein also displays potent and maximal stimulation in the fetal long bone resorption assay.

Qualitatively similar results have been obtained using the murine and rat tumors. ACSA from both tumors can be prepared in highly purified form and can be shown to have gel filtration and reverse-phase HPLC elution characteristics indistinguishable from the human ACSA.

V. Molecular Biological Approaches

While related to parathyroid hormone in a functional sense, it is uncertain from the above observations whether the HHM-associated adenylate cyclase stimulating protein is related genomically to native PTH. In this era of the recognition of posttranslational processing of parent or precursor proteins into families of functionally distinct peptides, it is reasonable to conder that the PTH-like ACSA observed in HHM-associated tumors might be a product of the PTH gene. With this question in mind, Simpson et al. (37) as well as our own group (38) have examined mRNA extracts from human and animal HHM-associated tumors for the presence of an mRNA species which might hybridize with a bovine PTH cDNA probe.

No detectable mRNA encoding for PTH has been identified by these hybridization studies, indicating that HHM-associated ACSA is not a translation product of PTH mRNA but that the adenylate cyclase-stimulating protein is derived from its own mRNA and DNA.

With this in mind, we have attempted to identify such an mRNA in human and animal HHM-derived tumor extracts (39). Tumors examined were the rat Leydig cell tumor, the DMBA murine squamous carcinoma, and a human HHM-associated squamous carcinoma. Controls included two lymphomas (one human, one murine) associated with hypercalcemia of a mechanism distinct from HHM. Polyadenylated mRNA was extracted and prepared from each tumor, and was microinjected into *Xenopus* oocytes. Oocyte-conditioned media were then examined for the presence of bioactivity as detected using the PTH-sensitive cytochemical bioassay (CBA). Clear-cut bioactivity was present in each of the three HHM-associated oocyte media, but in neither of the control media (Table III, Fig. 24). This cytochemical bioactivity was competitively inhibited by the PTH analog described earlier (Nle^{8,18},Tyr$^{3-34}$bPTH amide). These findings underscore the correlation between the presence of the HHM syndrome as defined using NcAMP measurements, and the presence of PTH-like bioactivity in HHM (but not control) tumors. In addition, they indicate a correlation between ACSA content in HHM-derived tumor

TABLE III

Cytochemical Bioactivity (CBA) in Xenopus Oocyte-Conditioned Medium[a]

mRNA	CBA (pg eq 1–84 bPTH/ml)
Experimental	
Mouse squamous cell tumor	8.0 ± 1.6
Rat Leydig cell tumor	3.5 ± 0.8
Human squamous cell tumor	30.0 ± 1.0
Control	
Mouse myeloma	<1.7 ± 0.3
Human lymphoma	<1.7 ± 0.3

[a] Reproduced from Ref. 39 with permission. Experimental mRNA refers to polyadenylated messenger RNA prepared from three HHM-associated tumors. Control mRNA refers to identically prepared mRNA from two tumors associated with hypercalcemia of a different mechanism. See text and Ref. 19.

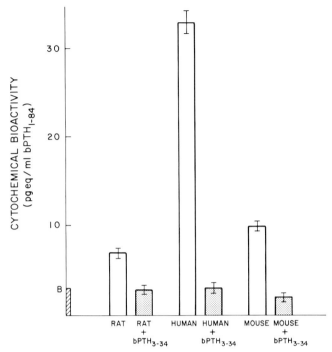

FIG. 24. Cytochemical bioactivity in oocyte-conditioned medium. *Xenopus* oocytes were microinjected with polyadenylated mRNA prepared from the Rice 500 rat Leydig cell tumor (Rat), a human HHM-associated tumor (Human), and the murine DMBA-induced squamous carcinoma (Mouse). Medium conditioned by oocytes injected with control mRNA preparations produced no bioactivity. The stippled bars indicate the effect of Nle^{8,18},Tyr^{34}(3–34)bPTH amide on the CBA produced by each tumor. In each case, the PTH inhibitor blocks the bioactivity. Reproduced with permission from Broadus *et al.* (39).

extracts and the ability of mRNA from these tumors to induce the translation of a closely related or identical protein. Finally, and most importantly from our own perspective, they indicate the feasibility of strategies aimed at cloning the cDNA for what we believe to be the HHM-F.

VI. Normal Role of the HHM Factor

As indicated, the observations that HHM-derived tumors contain a PTH-like AC-stimulating protein and that these tumors contain no measurable PTH mRNA indicate that these HHM-derived tumors must possess a separate species of mRNA encoding for the HHM-associated ACSA. An extension of these observations might suggest that normal tissues should contain PTH-like ACSA (or HHM factor), as well as the corresponding mRNA. With this thought in mind, we have examined cultured normal human cells for their ability to produce ACSA (40). We selected human squamous epithelial (keratinocyte) cultures to examine, since, as noted in Section II, the largest subgroup of patients with HHM is patients with squamous carcinomas. We selected the ROS AC assay as a detection system, given its greater sensitivity as compared to the renal AC system.

All of the keratinocyte cultures we have examined have contained easily detectable ACSA (40). In contrast, none of the control cultures examined contained measurable ACSA. The keratinocyte-derived ACSA, like the rat, murine, and human ACSAs, is protease sensitive, heat stable and has an apparent MW in the 30,000 range. The activity is completely inhibitable with $Nle^{8,18}, Tyr^{34}(3-34)bPTH$ amide. These findings are precisely those associated with the ACSA in animal and human HHM-derived tumor extracts, and provides strong evidence that nonmalignant human keratinocytes secrete an ACSA which is similar or identical to that produced by their malignant counterparts. Further characterization of this material is in progress.

The role of this PTH-like ACSA in nonmalignant human keratinocytes remains uncertain. It is tantalizing to view the epidermis as a separate calcium–PTH–vitamin D endocrine system, complete with vitamin D synthetic capacity and receptors, PTH (HHM factor?) receptors, HHM factor synthetic capacity, and calcium-sensitive mechanisms of differentiation. This now primitive area represents fertile ground for future investigation.

VII. Summary

The observation presented in this article can be briefly summarized as follows.

1. Clinical HHM is a pathophysiologic syndrome which can be readily distinguished from 1°HPT and LOH.

2. Human HHM and animal models of HHM are associated with increases in NcAMP or UcAMP excretion. In contrast, NcAMP or UcAMP excretion is normal in humans and animals with HHM. These observations, and the findings that humans with MAHC can be divided into clinically and pathophysiologically distinct groups based upon NcAMP excretion, lead inescapably to the conclusion that the observed elevations in NcAMP are related mechanistically to hypercalcemia in patients with HHM.

3. The ACSA which is present in human and animal HHM-associated tumors is not typically present in tumors derived from normocalcemic

patients or animals. This ACSA can be prepared in highly purified form from human and animal tumors and can be shown to result from a basic protein with a molecular weight of approximately 30,000. This highly purified material contains potent *in vitro* bone-resorbing activity. These observations provide strong evidence that this PTH-like ACSA is related mechanistically to HHM.

4. Human and animal HHM-associated tumors contain mRNA which encodes for a secretary protein which is closely related or identical to the ACSA found in tumor extracts.

5. Normal, nonmalignant human keratinocytes can be shown to contain measurable PTH-like ACSA using sensitive assays. These findings imply that the PTH-like ACSA which is present in malignant epithelia may play a normal role in the physiology of nonmalignant squamous epithelia.

6. Efforts are currently underway which are aimed at producing this HHM-associated ACSA in amounts sufficient for protein sequencing and antibody production.

ACKNOWLEDGMENTS

The authors wish to thank the following individuals for their help in collecting the data described herein. They include N. D. Ravin, P. J. Gkonos, L. M. Milstone, J. Gertner, J. J. Merendino, E. C. Weir, A. Webb, H. Kronenberg, R. Baron, A. Vignery, A. Aminiafshar, T. Wu, R. Jones, J. Berman, M. Mitnick, B. Dreyer, and J. Wysolmerski. We thank Ms. Theresa Gould and Helen Losnes for expert manuscript preparation. Supported by the Veterans Administration, West Haven, Connecticut, NIH Grants AM 30102 and RR 125, and the General Clinical Research Center of the Yale-New Haven Hospital. Dr. Insogna is the Recipient of a Clinical Associate Physician Award from the NIH.

REFERENCES

1. Mundy, G. R., Ibbotson, K. J., D'Souza, S. M., Simpson, E. L., Jacobs, J. W., and Martin, T. J. (1984). *N. Engl. J. Med.* **310**, 1718–1727.
2. Mundy, G. R., Ibbotson, K. J., and D'Souza, S. M. (1985). *J. Clin. Invest.* **76**, 391–394.
3. Stewart, A. F., Horst, R., Deftos, L. J., Cadman, E. C., Lang, R., and Broadus, A. E. (1980). *N. Engl. J. Med.* **303**, 1377-1383.
4. Kivirikko, K. I., Laitinen, O., and Prockop, D. J. (1967). *Anal. Biochem.* **19**, 249–255.
5. Mallette, L. E., Tuma, S. N., Berger, R. E., and Kirkland, J. L. (1982). *J. Clin. Endocrinol. Metab.* **54**, 1017–24.
6. Reinhardt, T. A., Horst, R. L., Orf, J. W., and Hollis, B. W. (1984). *J. Clin. Endocrinol. Metab.* **58**, 91–98.
7. Broadus, A. E. (1981). *Recent Prog. Horm. Res.* **37**, 667–701.
8. Stewart, A. F., Broadus, A. E., Schwartz, P. E., Kohorn, E. I., and Romero, R. (1982). *Cancer* **49**, 2389–2394.
9. Rosenthal, N. D., Insogna, K. L., Godsall, J. W., Smaldone, L., Waldron, J. W., and Stewart, A. F. (1985). *J. Clin. Endocrinol. Metab.* **60**, 29–33.

10. Wahner, H. W., Kyle, R. A., and Beaubout, J. W. (1980). *Mayo Clin. Proc.* **55,** 739–746.
11. Kukreja, S. C., Shermerdiak, W. P., Lad, T. E., and Johnson, P. A. (1980). *J. Clin. Endocrinol. Metab.* **51,** 167–169.
12. Rude, R. K., Sharp, C. F., Fredericks, R.S., Oldham, S. B., Elbaum, N., Link, J., Irwin, L., and Singer, F. R. (1981). *J. Clin. Endocrinol. Metab.* **52,** 765–771.
13. Lafferty, F. W. (1966). *Medicine (Baltimore)* **45,** 247–260.
14. Stewart, A. F., Vignery, A., Silvergate, A., Ravin, N. D., LiVolsi, V., Broadus, A. E., and Baron, R. (1982). *J. Clin. Endocrinol. Metab.* **55,** 219–227.
15. Schussler, G. C., Verso, M. A., and Nemoto, T. (1972). *J. Clin. Endocrinol. Metab.* **35,** 497–504.
16. Powell, D., Singer, F. R., Murray, T. M., Minkin, C., and Potts, J. T., Jr. (1973). *N. Engl. J. Med.* **289,** 176–81.
17. Breslau, N. A., McGuire, J. L., Zerwekh, J. E., Frenkel, E. P., and Pak, C. Y. C. (1984). *Ann. Intern. Med.* **100,** 1–7.
18. Coombes, R. C., Ward, M. K., Greenberg, P. B. *et al.* (1976). *Cancer* **38,** 2111–2120.
19. Abrahmson, E., Kukla, L. J., Shevrin, D. H., Lad, T. E., McGuire, W. P., and Kukrejas, S. (1984). *Calcif. Tissue Int.* **36,** 563.
20. Riggs, B. L., Arnaud, C. D., Reynolds, J. C., and Smith, L. H. (1971). *J. Clin. Invest.* **50,** 2079–2083.
21. Benson, R. C., Riggs, B. L., Pickard, B. M., and Arnaud, C. D. (1974). *Am. J. Med.* **56,** 821–826.
22. Raisz, L. G., Yajnik, C. H., Bockman, R. S., and Bower, B. F. (1979). *Ann. Intern. Med.* **91,** 739–740.
23. Mayer, G. P., Habener, J. F., and Potts, J. T., Jr. (1976). *J. Clin. Invest.* **57,** 678–683.
24. Siris, E. S., Sherman, W. H., Baquiran, D. C., Schlatterer, J. D., Osserman, E. F., and Canfield, R. E. (1980). *N. Engl. J. Med.* **302,** 310–315.
25. Palmer, F. J., Nelson, J. C., and Bacchus, H. (1974). *Ann. Intern. Med.* **80,** 200–204.
26. Heinemann, H. O. (1965). *Metabolism* **14,** 1137–1152.
27. Goltzman, D., Stewart, A. F., and Broadus, A. E. (1981). *J. Clin. Endocrinol. Metab.* **53,** 899–904.
28. Gkonos, P. J., Hayes, T., Burtis, W. J., McGuire, J., Jacoby, R., and Stewart, A. F. (1984). *Endocrinology* **115,** 2384–2390.
29. Insogna, K. L., Stewart, A. F., Namnum, P. A., Weir, E. C., Vignery, A. M.-C., Baron, R., Kirkwood, J. M., Deftos, L. J., and Broadus, A. E. (1984). *Endocrinology* **114,** 888–896.
30. Martodam, R. R., Thornton, K. S., Sica, D. A., D'Souza, S. M., Flora, L., and Mundy, G. R. (1983). *Calcif. Tissue Int.* **35,** 512.
31. Stewart, A. F., Insogna, K. L., Goltzman, D., and Broadus, A. E. (1983). *Proc. Natl. Acad. Sci. U.S.A.* **80,** 1454–58.
32. Nissenson, R. A., Abott, S. R., Teitelbaum, A. P., Clark, O. H., and Arnaud, C. D. (1981). *J. Clin. Endocrinol. Metab.* **52,** 840–846.
33. Rodan, S. B., Insogna, K. L., Vignery, A. M., Stewart, A. F., Broadus, A. E., D'Souza, S. M., Bertolini, D. R., Mundy, G. R., and Rodan, G. A. (1983). *J. Clin. Invest.* **72,** 1511–1515.
34. Strewler, G. J., Williams, R. D., and Nissenson, R. A. (1983). *J. Clin. Invest.* **71,** 769–774.
35. Insogna, K., Weir, E., Stewart, A., Godsall, W., and Broadus, A. E. (1984). *Proc. Am. Soc. Bone Min. Res.* Abstr. A54.
36. Stewart, A. F., Vignery, A., and Burtis, W. J. (1984). *Clin. Res.* **32,** 410A.

37. Simpson, E. L., Mundy, G. R., D'Souza, S. M., Ibbotson, K. J., Bockman, R., and Jacobs, J. W. (1983). *N. Engl. J. Med.* **309**, 325-330.
38. Insogna, K. L., Stewart, A. F., Vignery, A., Baron, R., Kirkwood, J., Weir, E., Deftos, L., Kronenberg, H., and Broadus, A. E. (1982). *Proc. Am. Soc. Bone Min. Res.* P. S40.
39. Broadus, A. E., Goltzman, A. E., Webb, A. C., and Kronenberg, H. M. (1985). *Endocrinology* **117**, 1661-1666.
40. Merendino, J. J., Insogna, K. L., Milstone, L. M., and Stewart, A. F. (1985). *Proc. Am. Soc. Bone Min. Res.* Abstr. No. 146.
41. Burtis, W. J., Insogna, K. L., Broadus, A. E., and Stewart, A. F. (1985). *Proc. Am. Soc. Bone Min. Res.* Abstr. No. 117.

DISCUSSION

G. D. Aurbach. You now have three different systems producing tumor hypercalcemia factors, the human tumor, the rat Leydig cell tumor, and the skin squamous cell cancer. Do I understand correctly that in each of these systems the adenylate cyclase-stimulating activity coincides during purification procedures with bone resorbing activities?

A. Stewart. It's clearly true in the human. In the rat, such studies are underway, but it is clear that the rat material also comigrates with bone resorption. In the mouse the situation is more complicated. Work with mouse tumor has shown that there are two different kinds of adenylate cyclase-stimulating activity in the tumor that can be separated by reverse-phase HPLC and on gel filtration. There isn't very much activity in these fractions, however, and we've not been able to show convincing bone resorption, but we have been using much less material than we used with the human and the rat.

B. F. Rice. I'd like to congratulate Dr. Stewart for having the courage to propose the existence of a third calcium-regulating hormone. I can remember not too long ago when a very eminent group of endocrinologists told me that we now know all that was needed to be known about calcium regulation because we had the two important hormones, vitamin D and its metabolites, and parathyroid hormone. Because these Proceedings are so widely read, I think it's important that someone speak as a clinician as well as a basic scientist on this subject. You have implied that the breast cancer hypercalcemias by and large are all associated with bone metastasis and therefore are probably due to some osteolytic local type phenomena and not due to a humoral phenomenon. You used as some evidence for that normal tubular resorption of phosphorous. As a clinician who has taken care of many patients with breast cancer hypercalcemia, I can assure you, and I suspect you may have similar data, that when these patients are adequately hydrated, they will have very low tubular resorption levels of phosphorous similar to patients in your other category. Would you care to comment on that?

A. Stewart. Actually, it's a pet interest of mine. It's clear that the majority of patients with breast cancer have become hypercalcemic in the presence of widespread skeletal metastases, but it's also very clear from the experience you're describing, from our own experience, and from published reports from other groups that some patients with breast cancer become hypercalcemic in the absence of bone metastases. Sherwood has described a case in the *New England Journal* (**285**, 154-156, 1971), and two other cases have been reported, one in *Cancer* (**48**, 2383-2385, 1981) and one in the *British Medical Journal* (**2**, 204-205, 1972), of patients who had breast cancer, hypercalcemia, and evidence of humorally mediated bone resorption. In our series of 11 patients with breast cancer hypercalcemia (for some reason there isn't a lot of breast cancer hypercalcemia in New Haven) the breakdown is as follows: about half had elevated NcAMP and about half had suppressed NcAMP. What's fascinating to me about this is that two of the patients who had elevated NcAMP also had no bone metastases. From one of these we were able to get material postmortem; it contained enormous quantities of this cyclase-stimulating activity. That patient had normal parathyroids at autopsy and no bone metastases. We did a bone biopsy which was identical to the one discussed here. So in summary, my view on breast cancer is that the majority of these patients probably are hypercalcemic as the result of bone metastases, but it's very clear to me that there is a minority, maybe even a large minority of patients with breast cancer, who are hypercalcemic predominantly through humoral mechanisms.

B. F. Rice. Now I would like to ask a question as a basic scientist. It appeared to me for a long time that until one is able to clearly understand the mechanism of action of hypercalcemia in some of the tumor models, it's very unlikely that we're going to be able to understand the mechanism of action in the human model. There are some biologic data available in tumor models which I would like to ask you to try to explain. If the tumors, for example, are transplanted into the spleens of thyroparathyroidectomized animals or normal animals, one never sees hypercalcemia unless they've had lesions from the spleen, implying very rapid metabolism of the hypercalcemic factor, since it all goes to the liver.

A. Stewart. Does the tumor grow under those circumstances?

B. F. Rice. It grows very well and these are in castrated animals, so you have high levels of gonadotropins and very large tumors in the spleen (this information has all been published). There's a second set of data that indicate that when the tumor is transplanted into parabiotic animals that are thyroparathyroidectomized, the hypercalcemia does not occur on the other side. This is exactly contrary to what one sees when one injects parathyroid hormone into a parabiotic pair. One sees simultaneous hypercalcemia in both animals.

A. Stewart. With respect to the parabiotic model, I have not worked with such a model, but I don't think there can be any question that hypercalcemia occurs through a humoral mechanism in that model because simple excision of the tumor makes the condition disappear. It's got to be humorally mediated in that model. Also there isn't any bone metastatic disease. I'm simply not familiar enough with the parabiotic model to comment on it.

With regard to your first question about the spleen transplant model, again I was unaware of those data. I suppose what's different about it is that in that model, obviously the spleen drains into the portal system and into the liver, and, as you suggest, maybe something happens in the liver.

R. M. Carey. One of the chief biochemical differences between primary hyperparathyroidism and tumor hypercalcemia is the difference in circulating 1,25-dihydroxyvitamin D. Have you looked at the tumor extract or purified material from human tumor to see whether it stimulates 1-hydroxylation?

A. Stewart. Let me just give you some more information before I refuse to answer that question. The other piece of information that's really interesting in regard to the vitamin D story is that if you take a tumor from a human with HHM who has suppressed $1,25(OH)_2D$ and transplant it into a nude mouse, the mouse then develops the full syndrome; the mouse also has elevated $1,25(OH)_2D$. Thus, the same tumor, making presumably the same factor in a mouse's body that it does in a human, causes elevation of $1,25(OH)_2D$. What that means to me is that rodent kidneys and human kidneys may be different with respect to how they perceive this factor.

The question is whether there is some system in which you could examine 1α-hydroxylase directly from, say, rodent or human tubules? There are preparations; Kiyoshi Kurokawa has meticulously dissected out nephron segments from a variety of different species and has been able to show 1α-hydroxylase activity induced by PTH. It's a lot of work. It's something we talked about a good deal, but have not done because of the work involved.

Finally, those systems are not particularly sensitive to PTH. You'd have to have a lot of this factor to look at. To me, it's one of the most fascinating areas in this field.

R. M. Carey. How does the molecular weight of the purified material differ from that of osteoclast activating factor?

A. Stewart. The material we're talking about, and that David Goltzman also finds, is around 28,000. David also finds a species that's smaller, around 9000. Osteoclast-activating factor is still a mystery. It's probable that there are lots of osteoclast-activating factors. One of them has been recently shown to be identical to interleukin 1, and I don't believe that it's really been clearly shown what the other osteoclast-activating factors are. They have been reported to have molecular weights in the 10,000 range and the 13,000 range, and as low as 140 was reported in one paper. I think that the adenylate cyclase-stimulating activity is different. It's hard to compare it directly to osteoclast-activating factor because neither has clearly been purified and identified. But I think it is different from OAF because patients with myeloma have low cyclics as a rule, and when we've looked at conditioned medium from activated spleen cells that Drs. Baron, Horowitz, and Vignery at Yale have given us; we don't see any cyclase stimulation.

N. Samaan. I have two short comments. One of the causes of hypercalcemia associated with malignancy which should not be ignored is the concomitant presence of primary hyperparathyroidism. In our experience over the last 7 years, we operated, in a Cancer Hospital (I want to specify that), on 272 patients with primary hyperparathyroidism; 26% of these patients had a history of cancer in other parts of the body in the past. Of them 17 had breast cancer, so we have to be aware of this phenomenon. The second cause of hypercalcemia we find in a Cancer Hospital is associated with chemotherapy and infection. We find hypercalcemia associated with fungus infection due to an increase in 1α-25-hydroxyvitamin D. We should be aware of these two causes of hypercalcemia, reports of which have been published (N. A. Samaan, *et al.*, *Surgery* **80**, 232–389, 1976; N. Tannir *et al.*, *Cancer* **55**, 615–617, 1985).

A. Stewart. I'm familiar with your studies and I agree completely with your findings. Our experience has been the same. In a group of 133 patients with cancer and hypercalcemia that we've looked at over the last 5 years, 8 have had coexisting primary hyperparathyroidism. It clearly happens and you're absolutely right in saying that it should be considered.

D. Goltzman. I think the San Francisco group has shown that in dogs, calcium may be an important modulator of 1-hydroxylase, and certainly the 1-hydroxylase is not modulated only by parathyroid hormone, but also phosphate levels and perhaps calcium levels; perhaps the interplay of these factors differs in different species, so that in the human, for example, one would assume that calcium rises to very high levels, but you actually get an inhibition of the 1-hydroxylase, whereas in other species, such as rodents, the calcium level may not be as important.

The other comment relates to the apparent discordance between osteoclastic and osteoblastic activity that you see in the bones of patients with tumor hypercalcemia. Most patients that we see with primary hyperparathyroidism have relatively modest hypercalcemia and probably have relatively modest bone resorption. Patients with tumor hypercalcemia usually have very high levels of serum calcium. Do you think it's reasonable to compare these two groups of patients, or should we be trying in fact to see what bone histology looks like in those few patients with primary hyperparathyroidism, who have serum calciums of 15 or 16 mg%?

A. Stewart. With regard to the question of whether serum calcium or the degree of hypercalcemia and hypophosphatemia could be influencing the 1α-hydroxylase, when we have compared serum calcium and phosphorous values from patients with HHM with patients with hyperparathyroidism, we found that the hyperparathyroid patients had elevated

1,25(OH)$_2$D values. If we just examine the group of HHMs with calcium values in the 11–11.5 range and phosphates in the 3 range, they still have low 1,25(OH)$_2$Ds. Thus, I don't like that as an explanation. It's even true if you match them for ionized serum calciums.

The bone story is more difficult to sort out. The idea that severe hypercalcemia could be inhibiting bone formation is, I suppose, possible, and all I really can say is that it's difficult to biopsy large numbers of these patients, and it's difficult to find patients with hypercalcemia with primary hyperparathyroidism in the range of 14, 15, and 16. But in the small group of patients on whom we reported, in 7 patients, if you include patients who had even mild hypercalcemia, the findings are the same, so I don't think it is hypercalcemia, but clearly that's an incompletely answered question.

G. Segre. I would hate to give the impression that the answers are all in, and that all agree that this syndrome is due to a factor stimulating adenylate cyclase activity. I think it's important to note that Greg Mundy and his group have identified factors in the transforming growth factor class that presumably cause this syndrome. In our laboratory we've been working with a transitional cell bladder cancer which causes marked bone resorption *in vitro*. To add a bit more confusion, we can neither find transforming growth factor activity nor adenylate cyclase-stimulating activity in conditioned media or in urine from this patient. The molecular weight of this factor in urine or medium conditioned by tumor cells *in vitro* seems to be approximately 15,000. Thus, this is probably another hypercalcemia factor that differs from others that have been identified. I'd like you to comment particularly on the issue of the relationship between transforming growth factors and hypercalcemia associated with malignancy.

A. Stewart. Let me make it very clear that I don't believe that this is the only factor. I'm sure that there are many others. I do think that this cyclase-stimulating factor is the most common as evidenced by the fact that the vast majority of these patients have elevated nephrogenous cyclic AMP values *in vivo*, and the majority of these patients are clinically similar and most of them have squamous carcinomas or renal carcinoma. It is difficult for me to believe that such a homogeneous clinical group of patients could have a lot of different mechanisms, but clearly other mechanisms sometimes exist. We've described a patient with an ovarian disgerminoma who is hypercalcemic. It clearly was humorally mediated because when the tumor was resected, the hypercalcemia disappeared. She had none of these HHM-associated biochemical indices. Thus, there's one example, and there are other examples in the literature. Your group has described a patient with melanoma with hypercalcemia. When her metastatic inguinal nodes were removed, her hypercalcemia disappeared, and yet there was none of this adenylate cyclase-stimulating activity or cytochemical bioactivity. I should make it very clear that I am sure that many other things can cause hypercalcemia in patients with cancer, but I strongly believe that this is the most common.

Let me say something about the TGF theory so that you will be aware that there is another major contender, namely α-TGF or TGF-related peptides. This notion arises from the observation that epidermal growth factor is a stimulator of bone resorption in some bone resorption assays. It's clear that the Leydig cell tumor and the Walker 256 carcinoma in rats both produce tumor-derived growth factors. It's been reported by Mundy *et al.* that one of the human tumors that they've examined produces tumor-derived growth factor. It's been reported that messenger RNA for α-TGF can be extracted from a human tumor from a patient with HHM. My problems with these observations are that I think Todaro and his co-workers have shown that TGFs are present in many tumors, but I've yet to see any data surveying (similar to the prevalence figure that I showed) a large group of human tumors from patients who did have hypercalcemia and those who didn't for the presence of α-TGF. My strong suspicion is that the α-TGF would be present in many tumors, not just the patients with hypercalcemia.

My second point of disagreement with the TGF theory is that many things turn on bone resorption *in vitro,* and the finding that something does turn on bone resorption *in vitro* doesn't necessarily make it the hypercalcemic factor. Having said that, it's clear that this is an unresolved issue, and I don't want to come down too hard on the other side. Another possibility is that maybe our material is α-TGF: healthy keratinocytes make it, and maybe it is some sort of growth factor. In that sense it may well be a growth factor, but I don't think it's the same as α-TGF because Mundy and Todaro and co-workers have shown that the growth factors are very sensitive to reducing agents, and in our hands none of these materials is. In addition, α-TGF does not appear to have PTH-like bioactivity in adenylate cyclase assays, and our material, in preliminary experiments, does not appear to have TGF activity in TGF assays. So I think they're different.

G. Segre. You emphasize suppression of bone formation in the biopsies of patients with hypercalcemia in malignancy. I assume that these people were ill enough to be in the hospital and at bed rest for a period of time. Thus, it would seem to me that the proper control population would be other patients, perhaps those with malignancy without the hypercalcemic syndrome who are biopsied after a week or so of bed rest. The sample would be subjected to histomorphometric analysis to quantitate bone resorption, but more specifically bone formation. Bed rest alone might markedly reduce bone formation rates.

A. Stewart. I agree that larger numbers of samples and more care in patient selection would be of interest, but obtaining these biopsies is difficult. These patients with advanced cancer have generally undergone unsuccessful surgery, radiation, and chemotherapy. They are reluctant to undergo bone biopsies so it's very hard to get biopsies in large numbers of people. The same issue is pertinent to Dr. Goltzman's question. However, we did biopsies on 7 patients with the syndrome, 2 of whom were fully ambulatory, had not had chemotherapy, and were clinically well. The findings were the same as for the other five. So to the extent that 2 out of 7 means anything, I believe these findings. Also the mouse and the rat model show very similar findings. I don't believe that there are extraneous factors, such as hypercalcemia, immobilization, or chronic illness.

G. Segre. Have you had any experience using the PTH antagonist, (Nle^8, Nle^{18},Tyr^{34})bPTH(3-34) amide, in your own bone resorption assay. As you know Greg Mundy and his group have reported that this PTH analog inhibits PTH-stimulated bone resorption, but not "factor"-stimulated bone resorption.

A. Stewart. Mundy has shown that in his bone resorption system you can inhibit, using the same analogs that we've talked about, bone resorption induced by PTH. We've not been able to show that in our assay. Have you had any experience with this?

G. Segre. We cannot show inhibition of bone resorption in the mouse calvarial assay. We have not used the long-bone assay.

A. Stewart. I understand that Nissenson, Strewler, and Klein have not been able to show PTH analog inhibition.

T. M. Kelley. You stated that low levels of 1,25-dihydroxyvitamin D and the increased urinary fractional excretion of calcium in patients with humoral hypercalcemia of malignancy were evidence that neither the proximal nor distal renal tubules were seeing PTH-like activity. Yet your PTH-like molecule stimulates renal adenylate cyclase activity. What is the location of this renal adenylate cyclase? Is it linked to any biologic effect since it's not altering vitamin D levels of calcium excretion?

A. Stewart. That's really a good question. What this all means to me is that there are different types of PTH receptors, and that this material is capable of interacting with certain PTH receptors and not others. It's capable of turning on proximal tubular adenylate cyclase and its manifestation. The biological correlate of this is inhibition of phosphate reabsorption. However, 1α-hydroxylase is also proximal tubular and yet it doesn't seem to be activated, so to my mind that means that there are probably two different classes of receptors. The distal tubular calcium reabsorption system has not been worked out completely, but my reading of this would be the same. In fact, the same can be said of bone. Clearly this material is capable of turning on bone resorption and, with the reservations that people have had about formation, it is reduced. Thus, here is another system in which some PTH-like effects are present and others aren't, findings which again suggest to me that there are different PTH receptor classes in bone.

H. Kronenberg. My comment is a follow-up of the discussion on phosphaturia in breast cancer patients. I think it was your *New England Journal* article that showed convincingly that phosphate reabsorption was an absolutely useless index for distinguishing the various types of hypercalcemic patients. Phosphate reabsorption is affected by a hundred different things, not the least of which is blood calcium level, so that it's not at all surprising that phosphaturia might be found in patients with hypercalcemia caused by a variety of different mechanisms. My first question has to do with the fascinating observation that the same tumor that results in low levels of $1,25(OH)_2D$ in patients causes an elevation of the $1,25(OH)_2D$ levels in nude mice. One possibility, of course, is that the differing results reflect a species difference. Another possibility comes from the observation that Bob Neer and others have made that in older people it is difficult to elevate $1,25(OH)_2D$ levels in response to infusions of parathyroid hormone. This observation raises the possibility that in older people or sicker people it might be difficult to elevate $1,25(OH)_2D$ levels in response to the hypercalcemic factor because of their kidney function or their ages, rather than because they are human. Do you have any young people with this cyclase syndrome? What are their $1,25(OH)_2D$ levels? Have you studied old people with hyperparathyroidism? Are their $1,25(OH)_2D$ levels elevated?

A. Stewart. We certainly have seen people in their 60s, 70s, and 80s with hyperparathyroidism and, as a rule, they have elevated $1,25(OH)_2Ds$. I don't think that we have any children in our study but there are some young adults; for example, there's a 17-year-old with a squamous carcinoma of the jaw, and she had reduced plasma $1,25(OH)_2D$. I don't think it's age. Another notion, just to make all the speculation complete, is that people have suggested that PTH secretion is pulsatile. Maybe that's important in the generation or the stimulation of 1α-hydroxylase. In contrast, maybe these tumors release their material tonically, and somehow that may not be sufficient to turn on 1α-hydroxylase. The problem with this theory is that if you take these same tumors and put them in nude mice, the nude mice get elevated $1,25(OH)_2D$ levels. The published record of $1,25(OH)_2D$ in nude mice can be found in an article by Kukreja and collaborators in *Calcified Tissue International* (**36**, 563, 1984), but we've seen the same thing. I know that Buck Strewler has seen the same thing as well.

H. Kronenberg. My last question is a follow-up of this fascinating observation on keratinocytes. The skin is the biggest organ in the body, so that if skin produces the hypercalcemic factor normally, even in small amounts, obviously this would have profound physiological implications. I wondered whether either in normal skin or in abraded skin or in UV-irradiated skin or in any kind of skin in which you might have stimulated the kerotinocytes one way or another you've been able to find this material, that is, in skin taken directly from animals?

A. Stewart. When these observations first appeared in our lab studies with Len Milstone and Jack Merendino, my strong bias was that this was all irrelevant to systemic calcium metabolism but Len disagreed. I thought given the specific activity of this material, that the amount of this material in skin is orders of magnitude less than what's present as PTH in parathyroid glands. His response was "Yes, but there are orders of magnitude more kerotinocytes than parathyroid cells." Thus whether this material could have some systemic importance, I simply have no idea. There clearly are some diseases of skin that are calcium related, but in terms of what the factor's normal role might be, and in terms of experiments designed to look at that, I can only say that we don't have data, but that we're very actively looking for data right now.

H. Kronenberg. Have you looked for it in skin itself?

A. Stewart. We've not yet looked at animal or human skin extracts, as opposed to keratinocyte cultures, in sensitive bioassays.

M. S. Katz. With respect to your earlier discussion with Dr. Segre, I'd like to reiterate the confusing nature of some of the disparate data on the cyclase-stimulating activity from these tumors. In collaboration with Dr. Gregory Mundy, Dr. Gloria Gutierre and I have looked at the cyclase-stimulating activity of the conditioned media from cells in culture from 7 animal and human tumors, using the adenine labeling assay for adenylate cyclase in intact UMR 106 osteosarcoma cells.

A. Stewart. Using adenine labeling?

M. S. Katz. We've been unable to find any correlation between bone resorbing activity of these conditioned media with cyclase-stimulating activity. In fact, one tumor has no cyclase-stimulating activity and substantial bone-resorbing activity, while another tumor shows cyclase-stimulating activity but no bone-resorbing activity. With one tumor, the hypercalcemic Walker rat carcinosarcoma of the breast, in preliminary gel filtration studies we have not found the coelution that you have illustrated for the human tumor. Another point that I would like to comment on is that unlike the results you showed we have been unable to find complete inhibition by the 3–34 PTH analog of the cyclase-stimulating activity from at least two of our tumors. Finally, we have tried purified α-TFG in our cyclase assay and have not found stimulation by the growth factor.

A. Stewart. As far as the UMR line goes, that's the line I believe that Buck Strewler is working with, and the data that he's generated and data that we've generated are basically identical. I don't know why there's a difference in terms of cyclase production in some of the tumors that you've looked at, but I can only say that Strewler is finding exactly the same thing that we are, and so is Gideon Rodan from where the line came. As far as the concordance between bone-resorbing activity *in vitro* and the cyclase stimulation *in vitro*, until these things are purified, it's not going to be clear which is the real thing. If you had to pick two assays and judge them in terms of specificity, there's no contest with respect to the bone resorption in the renal or ROS assay. We simply get no stimulation in the ROS or renal cyclase assay from things other than PTH, this adenylate cyclase-stimulating activity that I've talked about, and in the case of the ROS assay, isoproterenol. Again, that's the experience of Nissenson and collaborators using similar assays. I believe that it is the experience of David Goltzman as well. On the other hand, a myriad of things stimulate bone resorption *in vitro*, and the finding that something stimulates bone resorption *in vitro* needs to be interpreted, I think, extremely cautiously. Again I don't know why your group is finding adenylate cyclase-stimulating activity that's of a smaller molecular weight than that reported from other laboratories.

M. S. Katz. I have no specific comment other than to say that I'm commenting only with respect to our inability to find a correlation between the two responses and not the specificity of what is stimulating the bone resorption.

D. Goltzman. There's one point I would like to make with respect to *in vitro* bone resorption assays. I believe that it is easy to be led astray if one attempts to extrapolate to the *in vivo* situation in *in vitro* bone resorption assays. A classical example might be that of prostaglandin E. Prostaglandins of the E series are known to be very potent stimulators of bone resorption *in vitro*. Unfortunately, when they are administered *in vivo* they are rapidly metabolized, and probably never reach bone in order to produce bone resorption. Consequently, I think one has to be cautious about using only *in vitro* bone resorption assays to assess potential *in vivo* mediators of bone resorption. Since α-TGF is believed to act via the EGF receptor, and since EGF and α-TGF are now freely available, I believe it would be important to perform *in vivo* perfusions to see if in fact these agents, after administration *in vivo*, do produce bone resorption with or without hypercalcemia. Additionally, there is available an antiserum to the EGF receptor which has been reported by Dr. Mundy's group to inhibit the stimulation of bone resorption *in vitro* by extracts of tumors associated with hypercalcemia. If there is sufficient antiserum, it would be very interesting to perform passive immunization experiments to determine if the antiserum could inhibit hypercalcemia in the animal models of the hypercalcemia of malignancy. I think that there are therefore important experiments to be done to verify the relevance of results obtained with *in vitro* bone resorption assays to the *in vivo* situation.

INDEX

A

Acetylcholine
  in arcuate nucleus, 33
  in salivary gland parasympathetic nerves, 15
  VIP coexistence with, 15
Achlya (water mold)
  antheridiol-binding protein, male, 576–577
  comparison with vertebrate steroid receptors, 576–577, 579
  steroid pheromones, male and female, 575, 577
Acromegaly, human
  combined with ectopic GRF secretion, 590–593, 596–599, 611–613
  GH and GRF pulsatile secretion and, 612–613
  GH response to GRF injection, 617–618
  hypothalamic gangliocytomas and, 596, 598–599
ACTH, *see* Adrenocorticotropic hormone
ACTH-like material, *Tetrahymena pyriformis*
  purification and activity, 559, 566–568
Adenylate cyclase
  defects in type IA pseudohypoparathyroidism, 677, 679
  inhibition by ANF, 227–228
Adenylate cyclase, renal cortical, stimulation by
  HHM factor, 733–736
  dose-response curves, 735
  PTH analog and inhibitor effects, 735–736
  keratinocyte cultures, human, 740–741
  parathyroid extract, 736
  PTH analog and inhibitor effects, 735–736
Adrenal cortex
  aldosterone secretion *in vivo*
    ANF inhibitory effect, 223–225
    angiotensin II antagonism with, 223–224
    dopamine inhibition site, 286

aldosterone secretion *in vitro*
  ACTH and, 264–269
  angiotensin II and, 264–268
  ASF and, 263–269
  ANF receptors, 226
  FGF
    physiological role unknown, 188–189
    properties, 170
Adrenaline, *see* Epinephrine
Adrenal medulla
  dopamine β-hydroxylase, 4
  enkephalin-immunoreactive structures, 4–5, 16–17
  multiple messengers, release mechanism, 16–17
β-Adrenergic receptors, ovarian, rat
  juvenile, 413–414
  superior ovarian nerve and, 414–415
  during pubertal phases, 425
Adrenocorticotropic hormone (ACTH)
  aldosterone secretion regulation
    in vitro by adrenocortical cells, 264–269
    K^+, extracellular, and, 265–267
    ouabain-blocked, 266
    receptor sites, 266, 268
  in vivo, 251–252
  release, effects of
    CRF, 44, 49
    dopamine, 49
    enkephalin, 49
    PHI, 49
Aldosterone
  in vitro secretion by adrenocortical cells
    ACTH and, 264–269
    angiotensin II and, 264–268
    ASF and, 263–269
  in vivo secretion, regulation by
    ACTH, 251–252
    angiotensin II, 251, 253–254
    dopamine antagonism, 284–285
    metoclopramide, 255–256
    dopaminergic mechanism, 282–284
    potassium ion, 251–253
    renin, 251, 253–254
    sodium depletion, 253–255

plasma level
  ANF-induced decrease, 223–224
  ASF-induced increase, 270–271
  head-out water immersion and, human, 230–231
  in idiopathic hyperaldosteronism, human, 275–278
    angiotensin III effect, 275–278
  in patients with aldosterone-producing adenoma, 275–278
    unaffected by angiotensin III, 275–278
Aldosterone-producing adenoma, human
  aldosterone, plasma
    angiotensin III infusion and, 275–277
    upright posture and, 275–276
  ASF, plasma, 277–278
    upright posture and, 277–278
  symptoms, 275
Aldosterone-stimulating factor (ASF)
  discovery, 254–256, 258
  glycoprotein nature
    binding to concanavalin A-Sepharose, 262
    neuraminidase treatment and, 260–262
    sugar content and, 262
  in vitro effects on adrenocortical cells, 263–269
    cAMP-independent, 264–265, 267
    dose-dependent, 264
    K^+, extracellular, and, 265–267
    ouabain-blocked, 265–266
    specific receptor sites and, 266, 268
  in vivo effects, chronical, rat
    aldosterone increase in plasma, 270–271
    blood pressure increase, 269–272
    failure after adrenalectomy, 269–270, 272
  isolation and purification
    DEAE-cellulose chromatography, 258–259
    gel filtration on Sephadex G-100, 258–259
    HPLC, 263, 273–274
    paper electrophoresis, 260
  measurement by affinity chromatography/HPLC
    pituitary gland, 272

  plasma, 273–274
  urine, 273–274, 279–280
  pituitary gland origin, 272–274
  plasma level, human
    in idiopathic hyperaldosteronism
      angiotensin III and, 277–279
      dexamethasone and, 279
      dietary sodium restriction and, 281–282
  small fragments
    activities, 263–264
    HPLC chromatogram, 263
  urinary, human
    in idiopathic hyperaldosteronism
      dexamethasone and, 280–281
      excretion increase, 280–281
Alkaline phosphatase, serum, in HHM, 718, 721, 727
γ-Aminobutyric acid (GABA)
  in arcuate nucleus, 31, 33
  as neurotransmitter in CNS, history, 2
Amoeba proteus, opiate receptors, 574, 577
Androgen receptors
  deficiency in testicular feminization, animal, human, 72–73
  KAP high sensitivity to, 94–99
  testosterone effects, murine kidney, 94–99
    kinetics, 94–95
3 α-Androstanediol, ovarian secretion during puberty, rat, 247
ANF, *see* Atrial natriuretic factor
ANF precursor
  amino acid sequence, 210, 211–213
  as ANF storage form, 232
  coding by 3 exons, 213, 214
  processing to ANF, 232–233
  structure, 213–215
ANF receptors, tissue distribution, 226–227
Angiotensin II
  aldosterone secretion regulation
    in vitro by adrenocortical cells, 264–268
      K^+, extracellular, and, 265–267
      ouabain-blocked, 266
      receptor sites, 266, 268
    in vivo, 251, 253–254

… INDEX

Angiotensin III, aldosterone response
 in idiopathic hyperaldosteronism, 275–278
 in patients with aldosterone-producing adenoma, 275–278
Angiotensin converting enzyme
 ANF metabolism and, 233–234
Antheridiol receptor, *Achlya*, male
 comparison with vertebrate steroid receptors, 576–577, 579
Antigens, multiple in single neuron immunohistochemistry
 combined with retrograde tracing, 9–11
 methods, 6–8
 specificity and sensitivity, 8–9
Aqueous two-phase partitioning
 estrogen receptor changes due to ligand binding, 301–305
Arachidonic acid, leukotriene production
 in brain, 50
 in leukocytes, 49
Arcuate nucleus
 acetylcholine neurons, 33
 GABA neurons, 31, 33
 galanin, 34, 35
 GRF, 31, 34, 35
 LHRH, 30
 metorphamide-like peptide, 38–39
 multiple messengers, significance, 42
 neurotensin, 32–34
 NPY, 37–39
 opioid peptides, 36–37
 somatostatin, 30, 39
 TRH, 30
 tuberoinfundibular dopamine neurons, 28–29, 31
Aromatase, rat
 fetal
  hypothalamic, 388–389
 ovarian
  induction by cAMP, forskolin, and VIP, 390
  unresponsive to gonadotropins, 389–390
 in ovarian granulosa cells, FGF and, 188–189
ASF, *see* Aldosterone-stimulating factor
Aspergillus fumigatus, insulin-like material, 559, 561

Atrium natriuretic factor (ANF)
 biological activities
  adenylate cyclase inhibition, 227–228
  antihypertensive, 218–221, 224
  cardiac output and, 220–221
  dose-dependent, 218–219
  mechanism of, 219–221
  Ca^{2+} concentration and, 227–228
  cGMP increase, plasma and urine, 227
  guanylate cyclase activation, 216, 227
  natriuresis and, 217–218
  renal hemodynamic effects, 216–217
  smooth muscle vasorelaxation *in vitro*, 216
 in blood plasma
  experimental increase, 230–232
  head-out water immersion, human, 230–231
  immunoreactivity, 228–230
  low-molecular-weight-peptides, 229–230
 discovery, 207–208
 future research, 234–236
 interactions with
  neurohumoral cardiovascular systems, 235
  renin–angiotensin–aldosterone system, 235
 metabolism
  angiotensin converting enzyme and, 233–234
  processing from precursor, 232–233
 multiple peptides
  amino acid sequence, 208–211
  isolation and purification, 208–209
 secretion inhibition of
  aldosterone, 222, 223–225
  renin, 220, 222–223, 224
  vasopressin, 225
 structure–activity relationships, 215
 tissue distribution
  immunoreactivity, 225–226
  mRNA-coded synthesis, rat, 225

B

Bacillus brevis, gramicidin, spore production and, 571

B. subtilis, somatostatin-like material, 559, 562, 564–566
Bacteria, *see also specific bacteria*
hCG-like material, 559–560
Bicarbonate, serum, in HHM, 717, 720, 728
Biting/scratching, rat
responses to CGRP-tachykinin interactions, 26–28
Blood pressure
ANF effects, 218–220
in hypertensive animals, 218–220, 221, 224
in normotensive animals, 218
ASF-induced rise, 269–272
failure after adrenalectomy, 269–270, 272
Bone
in HHM, metastases, human, 712–713
PTH binding, rat
in vitro to osteoblast-derived osteosarcoma cells, 690–692
in vivo to osteoblasts, 686–688
resorption, in malignancy-associated hypercalcemia, human, 680
Brain, *see also specific regions*
ANF and ANF mRNA detection, 225–226
ANF receptors, distribution, 227
FGF
bovine, 160–165
acidic, 161, 164–165
hippocampal neuron survival *in vitro* and, 191
human, 163–165
physiological role unknown, 191
Bromocriptine, injection into female tammar
corpus luteum reactivation, 488–490
seasonal changes, 489–490
prolactin release inhibition, 489

C

Calcitonin gene-related peptide (CGRP)
injected into spinal cord, interaction with
substance P, 26–28
tachykinins, 26–27

Calcium ion
ANF effects and, 227–228
in HHM, human
fasting excretion, 714–715, 725–726
serum, 709, 711
PTH mRNA synthesis and, 647–648
Candida albicans, corticosterone binding, 577
Cardiovascular system, rat
responses to CGRP–tachykinin interactions, 26–27
Caudate nucleus, cat
dopamine and CCK, presynaptic interaction, 23, 24
CCK, *see* Cholecystokinin
Cell culture
adrenocortical cells, aldosterone secretion, 263–269
GH_3 cells, pituitary tumor-derived, *see* GH_3 cells
H4IIE (rat hepatoma), insulin as PEPCK regulator, 114–132
hippocampal neuron survival, FGF and, 191
pituitary
FGF release, KCl-induced, 180–181
estradiol and, 180–181
prolactin release, FGF effect, 182–185
estradiol synergism, 183–185
vasopressin release, ANF and, 225
Central nervous system (CNS)
neurotransmitters
coexistence with peptides, 11–15
history, 2
Chloride/phosphate ratio, serum, in HHM, 717, 719, 727–728
CGRP, *see* Calcitonin gene-related peptide
Cholecystokinin (CCK), in caudate nucleus
presynaptic interaction with dopamine, 23, 24
Choline acetyltransferase
acetylcholine detection in arcuate nucleus, 33
Chromatin, estrogen receptor binding
transformed, 319–320
unoccupied, 317
Chromatography
affinity/HPCL
ASF assay, 272–274, 279–280

DEAE-cellulose, for ASF, 258–259
heparin-Sepharose affinity, for FGF,
  150–153, 161, 164, 167, 192
high-pressure liquid (HPLC)
  ASF assay
    in plasma and urine, 273–274
    small fragments, 263
  reverse-phase, for FGF, 146–147, 162, 164
immunoaffinity column, for FGF, 149–150
Chromosome 5, mouse
  Gus complex for β-glucuronidase, 87–89
Chromosome X, mouse
  Tfm locus, for β-glucuronidase, 88
Clondine, gonadotropin secretion and, goldfish, 533
Clostridium perfringens, TSH-like material, 559
CNS, see Central nervous system
Corpus luteum, FGF-like angiogenic factor, 165–167
Corpus luteum, tammar
  inhibition
    photoperiod-controlled, 493
    by prolactin, 488–490
  ovulation control by, 490
  pouch young removal effects, 479, 481–482
  during pregnancy and postpartum estrus
    estradiol, content and secretion, 478–480
    progesterone
      content and secretion, 478–480
      postovulatory pulses, frequencies, 480–482
  reactivation by bromocryptine, 488–490
  seasonal changes, 489–490
  role in seasonal breeding, 482–483
Corticosterone receptors, Candida albicans, 577
Corticotropin-releasing factor (CRF)
  ACTH release and, 44, 49
  in paraventricular nucleus, 42, 44–47
  prolactin release and, 44, 48–49
Cortisol, human plasma
  in idiopathic hyperaldosteronism, dexamethasone and, 279
CRF, see Corticotropin-releasing factor

Cyclic AMP (cAMP)
  aromatase induction in fetal ovary, rat, 390
  ASF-unaffected in vitro, 264–265, 267
  in H4IIE cells, effect on PEPCK mRNA
    activity stimulation, 114–117, 119–122
    inhibition by insulin, 114–117, 119–122
    transcription stimulation, 123–125, 128–132
    inhibition by insulin, 123–125, 128–132
    model of, 134–135
  in HHM, human
    nephrogenous excretion, 705–708, 719–725, 729–731
    urinary excretion, 708–709
  LHRH release in pubertal rat and, 430
Cyclic GMP (cGMP), in plasma and urine, increase by ANF, 227
Cycloheximide, PEPCK mRNA synthesis and, 127

D

DARPP-32 phosphoprotein, median eminence
  dopamine neurons and, 40–41
  tanycytes and, 40–41
Deoxycorticosterone
  ANF mRNA in atrial muscle and tissues and, 233
Dexamethasone
  in idiopathic hyperaldosteronism, cortisol increase, 279
  PEPCK mRNA synthesis stimulation, 128
  inhibition by insulin, 128
Diabetes mellitus, FGF increase in vitreous, human, 190
Diacylglycerol, PEPCK gene transcription and, 134–135
5α-Dihydrotestosterone, ornithine decarboxylase mRNA induction, 79–80
DNA
  cytosolic estrogen receptor binding
    transformed, 318
    unoccupied, 315

β-glucuronidase-coding, physical map, mouse
  from genomic DNA clone, 89–90
  from mRNA sequence, 89
  Southern blot analysis, 90–91
ornithine decarboxylase-coding, nucleotide sequences, mouse, 83–84
prepropTH-encoding, transcription and translation
  in cell-free system, from plasmid, 653–654
  in cultured cells, from retroviral vec- 654–655
PTH-coding, isolation and cloning, 643–645
Dopamine
  ACTH release and, 48–49
  aldosterone secretion inhibition
    in adrenal cortex, 286
    interaction with
      angiotensin II, 284–285
      metoclopramide, 282–284
      upright posture-induced angiotensin III, 285
    Na^+ intake and, 284–285
  in arcuate nucleus, 31
  in caudate nucleus, presynaptic interaction with CCK, 23, 24
  gonadotropin release inhibition, goldfish
    in vitro, 521–523
      after LHRH analog-induced increase, 523–524
    in vivo, 521, 523
      after GnRH-induced increase, 523
  GRIF activity, goldfish, 513, 521–532
    dopaminergic receptors in pituitary and, 529–532
    in vitro assay, 532
    mechanism of, scheme, 540–541
    ovarian maturation and, 528–529
    ovulation and, 528
  growth hormone secretion and, 34, 36
  LH release direct inhibition
    in normal and pathological conditions, human, 537–539
    rabbit, 538, 540
    rat, 536–537
  as neurotransmitter in CNS, history, 2
  prolactin release and, 48–49, 283
  renin–angiotensin–aldosterone system and, 284–285
  secretion, somatostatin and, 36
Dopamine β-hydroxylase, in adrenal medulla, 4
Dopaminergic receptors, pituitary
  GRIF activity and, goldfish, 529–532
  in vitro assay, 532
Drosophila melanogaster, insulin receptors, 553–555

E

β-Endorphin-like material, Tetrahymena pyriformis
  purification and activity, 559, 566–568
Enkephalin
  ACTH release and, 48–49
  -immunoreactive structures in adrenal medulla, 4–5, 16–17
  in paraventricular nucleus, 45–47
  prolactin release and, 48–49
Epinephrine
  gonadotropin secretion increase, goldfish, 533, 535
  as neurotransmitter in PNS, history, 1
Escherichia coli
  insulin-like material, 559, 566–568
  plasmids, see Plasmids, E. coli
  somatostatin-like material, 559, 562
  TSH receptors, 574
17β-Estradiol
  binding to
    pituitary receptor in vivo, 299–300
    soluble receptor, 312–316
  GnRH pulse generator inhibition, lamb, 363
    growth and nutrition effects, 368–373
    ovariectomy and, 368–370
    photoperiod effects, 364–366, 370–372
    ovariectomy and, 366
  gonadotropin secretion and, infantile rat
    high dose effects, 401–402
    negative feedback failure, 398–400
    α-fetoprotein and, 398
  gonadotropins in ovariectomized tammar and, 484–485
  LH-induced production, lamb, 354–355

in LH surge, mechanism of action, lamb, 361
in McCune-Albright syndrome, increase in plasma, 459
testolactone treatment, girls, 459
ovarian secretion *in vitro*, rat
during juvenile period
LH pulses and, 408-409
VIP and, 416
during pubertal phases, 425-427
pituitary cell stimulation in release of
FGF, bovine, 180-181
prolactin, FGF synergism, 183-185
thyrotropin, FGF synergism, 182, 184, 185
pituitary FGF content in Fischer rats and, 181-182
in precocious puberty, LHRH analog and, girls, 457
during pregnancy and postpartum estrus, tammar
in Graafian follicles, 478
secretion from, 478-480
plasma, 476-478
pubertal growth in normal boys and, 448-449
Silastic implant in ovariectomized lamb, LH secretion and, 358
uptake by GH_3 cells, 312
Estrogen receptors
cytoplasmic, old model, 289-299
evidences for, 299-230
in GH_3 cells, estradiol binding, noncooperative, 312-313
multiple forms, model, 303, 306
nuclear, new model, 297-298
evidences for, 300-312
in *Paracoccidioides brasiliensis*, 577
physicochemical exclusion from cytoplasm, 317-318
in *Saccharomyces cerevisiae*, 577
soluble, estrogen binding
complex binding equilibria and, 313
cooperative, 312-314
noncooperative in monomeric form bound to hydroxylapatite, 313-316
transformed
cytosolic, binding to DNA, specificity, 318

nuclear, binding to chromatin, 319
estrogen-regulated genes and, 319
nuclear matrix and, 319-320
unoccupied
cytosolic, binding to DNA, 315
immunocytochemical studies, 310-311
in membrane, controversy, 311
nuclear, binding to chromatin, low-affinity, 317
nuclear components, 314-315
nuclear, in GH_3 cells, 306-309
unoccupied and nontransformed, cytosolic, partitioning
coefficients, 301-302
equilibrium stability, time course, 303, 305
pH and salt effects, 301-304
thermodynamic equation, 301, 303
Estrogens
gonadotropin release by ovariectomized infantile rat and, 398, 400
Ethinyl estradiol, pubertal growth and
in normal girls, 445-446
in Turner syndrome, girls
biphasic effect on growth, 445-447
bone age and breast budding unchanged by, 448

F

Fast Blue
in dopamine neurons
arcuate nucleus, 43
mediane eminence, 10
retrograde tracing of neuron pathways, 9-10, 43
Fasting, human
GH pulse frequency and, 605-608
in HHM, Ca^{2+} excretion, 714-715, 725-726
α-Fetoprotein, rat serum and tissues
infantile period, 398
juvenile period, 404
neonatal period, 391
FGF, *see* Fibroblast growth factor
Fibroblast growth factor (FGF)
adrenal gland
biological activity, 170
physiological role unknown, 188-189

purification and identification, 170
brain, physiological role unknown, 191
brain, bovine
  acidic
    amino acid composition, 165
    purification and activity, 161, 164
  amino acid composition, 162–163
  purification and activity, 160–162
brain, human, 163–165
  amino acid sequence, 163
  purification and activity, 163, 164
cellular localization, 177
corpus luteum, 165–167
  amino acid sequence, 166
  angiogenic activity, 166–167
  production by granulosa and luteal cells, 166
  purification, 166–167
future research, 196
history, 143–145
identity from various sources, 195
kidney, 167–170
  biological activity, 168–169
  isolation and identification, 167–169
  microheterogeneity, 168
  physiological role unknown, 189
macrophage
  angiogenic factor and, 190
  biological activity, 172–173
  macrophage-derived growth factor and, 174
  purification and identification, 173
macrophage-derived cell lines, 173–174
molecular forms
  high-molecular-weight precursor, 178–179
  low-molecular-weight product, 178–179
ovary, rat
  angiogenic activity, 186, 188
  aromatase inhibition, 188–189
  in immature females, 186–188
retina, bovine
  basic and acidic forms, 172
  biological activity, 171–172
  purification and identification, 171–172
retina, human
  detection in ocular fluid, 190–191
  increase in diabetes, 190
  isolation from vitreous, 192

serum, bovine, human
  molecular weight determination, 175–177
  radioimmunoassay, 175
serum, diurnal variations, human, rat, 193–195
tumor angiogenic factor and, 190
Fibroblast growth factor, pituitary
  amino acids
    composition, 163
    sequences, 153–156
    hydrophylicity, 156–157
  bovine, KCl-induced release in cell culture, 180–181
    estradiol and, 180–181
  in Fischer 344 rats, content, 181–182
    estradiol and, 181–182
  incubation with rat pituitary cells
    acute, hormone release unchanged by, 182–183
    prolonged, prolactin release stimulation, 182–186
      estradiol synergism, 183–185
    prolonged, thyrotropin release stimulation, 182, 184, 185
  isolation
    under denaturating and nondenaturating conditions, 146–148
    heparin-Sepharose affinity chromatography, 150–153
    immunoaffinity column chromatography, 149–150
    radioimmunoassay, 148–149
  mitogenic activity
    bioassay, 145–146
    with distinct mesodermal cells, 152–153
  immunoneutralization, 147, 149
  synthetic fragments of, 157–160
    antagonistic activity, 158–160
Flutamide, effects on mRNA responses to testosterone, 98–99
Follicle-stimulating hormone (FSH)
  infantile rat, secretion
    estrogen effects after ovariectomy, 398, 400
    LHRH injection and, 398–399
    reaching peak and decline, 395–397
    testosterone effect after ovariectomy, 401

juvenile rat, secretion
  decline, 404
  estrogen inhibitory control, 405–406
  testosterone effect, 405–406
in lactational quiescence period, tammar
  ovariectomy and, 483
  estradiol effect, 484–485
  seasonal changes absence, 483–485
in precocious puberty, LHRH analog
  and, girls, 451–453
Forskolin, aromatase induction in fetal
  ovary, rat, 390

G

GABA, see γ-Aminobutyric acid
GAD, see Glutamate decarboxylase
Galanin
  in arcuate nucleus, 32, 34, 35
  in median eminence, 35
Genes
  androgen-responsive, murine kidney
    β-glucuronidase, 85, 87–93
      Gus complex on chromosome 5, 87–89
      Tfm locus on X chromosome, 88
    KAP, 76–79
    molecular mechanism of action, 100
    ornithine decarboxylase, 79–87
  ANF precursor
    exon actions, 213, 214 (scheme)
    nucleotide sequence, 213
  GRF, structure, human, 593
  PEPCK, see also RNA, messenger,
    PEPCK
    structure, 130
    transcription in H4IIE cells
      cAMC and, 128–132
      insulin and, 128–132
      multihormonal regulation, 128–129, 134–135
  prepropPTH, mutants, structure, 658–659
  PTH
    human, insetred in Escherichia coli
      plasmid, 649
    transcription in rat pituitary GH4
      cells, 650
    location on chromosome 11, human,
      646–647
    genetic map, 647

one transcription initiation site, rat,
  648
structure, bovine, human, rat, 644
  nucleotide sequence homology,
    645–646, 648
  two intron insertion, 644, 645
  two transcription initiation sites,
    bovine, human, 648–649
GH_3 cells
  estradiol uptake, saturation analysis,
    312
  prolactin release
    FGF and estradiol synergism, 184–186
    by intact cells and cytoplasts, 307, 309
  unoccupied nuclear receptors for
    estrogen, 306–309
    glucocorticoid, 309–310
Glucocorticoid receptor, unoccupied,
  nuclear, GH_3 cells, 309–310
Glucocorticoids, see also Dexamethasone
  PEPCK gene transcription and, 128,
    134–135
β-Glucuronidase
  deficiency in mucopolysaccharidosis
    type VII, human, 86–87
  lysosomal form L, 87
  microsomal form X, 87
    egasyn binding, 87
  testosterone effects, murine kidney, 74–75
  gene regulation, 85, 87–93
Glutamate, as neurotransmitter in CNS,
  history, 2
Glutamate decarboxylase (GAD)
  GABA detection in arcuate nucleus, 31, 33
Glycine, as neurotransmitter in CNS,
  history, 2
GnRH, see Gonadotropin-releasing hormone
GnRH pulse generator, female sheep
  in hypothalamic portal blood, 362
  pubertal, frequency increase, 362–363
  LH pulse frequency and, 362–363
  puberty timing and, model, 373–375
  regulation in lamb
    by growth and nutrition, 368–373
    mechanism of, 372–373
    ovariectomy and, 368–370

photoperiod and, 370–372
by photoperiod, 363–367
estradiol inhibition and, 364–366
failure after ovariectomy, 366–367
GnRH receptors, rat
yeast α mating factor binding to, 573
Gonadotropin, goldfish
LHRH analog-induced release inhibition by dopamine, 523
LHRH analog-induced release, stimulation by
haloperidol, 530–531
metoclopramide, 523, 526
norepinephrine
in vitro, 532–533
pimozide, 523, 525–530
spiperone, 530–531
release by pituitary
in GRIF deficiency, temperature effects, 520
spontaneous, 519
inhibition by
dopamine, 521–524
GRIF, 517, 519
regulation by neurotransmitters, scheme, 540–541
stimulation by
clondine, 533
epinephrine, 533, 535
GnRH, 519
6-hydroxydopamine, 521, 526–527
norepinephrine
in vivo, 532–533
reserpine, 521, 526–527
Gonadotropin release inhibitory factor (GRIF) goldfish
dopamine, activity of, 521–532; *see also* Dopamine, GRIF activity, goldfish
epinephrine and, 533, 535
evidence for, 517–520
norepinephrine and, 532–534
ovarian maturation and, 528–529
ovulation and, 528
pathway to pituitary, 526–527
secretion by ventrobasal hypothalamus, 517–520
Gonadotropin-releasing hormone (GnRH), teleost
comparison with LHRH, 515–517
properties, 515

Gonadotropins
secretion ontogeny, rat
fetal, 380
infantile, 395–404
neonatal, 390–391
in teleosts
LHRH-induced secretion, 515–517
two forms, 514
Graafian follicles, tammar
estradiol during pregnancy and postpartum estrus
content, 478
secretion, 478–480
Gramicidin, *Bacillus brevis*
spore production stimulation, 571
GRF, see Growth hormone-releasing factor
GRIF, see Gonadotropin release inhibitory factor
Growth hormone (GH)
dopamine-regulated release, 34–36
GRF-induced release
control *in vivo,* scheme, 605–606
inhibition by somatostatin, 605–606
somatomedin C induction, 605–606
Growth hormone, human
in acromegalic patients
with ectopic GRF secretion, pulses, 611–613
GRF effect, 617–618
during continuous GRF infusion, 609–611
insulin administration and, 610–612
partial desensitization to highest doses, 609–611
pulses, 609–611, 615–616
deficiency, GH treatment, boys, 449–450
in precocious puberty, girls, 456
response to GRF, *see also* Growth hormone-releasing factor
in adults, 599–601, 620–625
in children with short stature, 619–621
secretion pulses, fasting and, 605–608
Growth hormone, rat
GRF-induced release
in pituitary cells *in vitro,* 601–605
during continuous GRF administration, 613–615
juvenile, secretion
diurnal rhythm, 409
ovarian functions and, 412–413
puberty timing and, 410, 412–413

Growth hormone-releasing factor (GRF)
  in arcuate nucleus, 31, 34, 35
  hypothalamic, rat, 589–590
    amino acid sequence, 593–594
    isolation, somatostatin and, 590
  in mediane eminence, 35
Growth hormone-releasing factor, human
  combined with hypothalamic-releasing hormones, 615, 617
    circulating hormone levels in men and women and, 617
    independent actions, 617
  continuous infusion, GH release and, 608–611, 615–616
    insulin administration and, 610–612
    partial desensitization at highest dose, 609–611
  GH response induction
    in acromegalic patients, 617–618
    in adults
      comparative intravenous, subcutaneous, and intranasal administration, 621–625
      with GH deficiency in childhood, 620–623
      men and women, 599–601
    in children with short stature
      with GH deficiency, 619, 621
      without GH deficiency, 619–620
    in rat pituitary cells *in vitro*, 601–605
      during continuous GRF infusion, 613–615
    growth rate response in children with GH deficiency, 623–627
  localization
    in brain, 594–595
    in gut and other tissues, 595
    in neuroendocrine tumors, patients with acromegaly, 590–593, 596–599
  from pancreatic tumor
    amino acid sequence, 593
    isolation, 590–583
    multiple forms, 593
  in patients with acromegaly and ectopic GRF secretion, pulses, 612–613
  pharmacokinetics after injection, 608–609
  somatomedin C increase in adults with GH deficiency, 621, 623
Guanylate cyclase, activation by ANF, 216, 227

H

Haloperidol, gonadotropin release and, goldfish, 530–531
hCG, *see* Human chorionic gonadotropin
hCG receptor, *Pseudomonas maltophilia*
Heart
  atrium
    ANF abundance, 225
    ANF mRNA abundance, 225, 233
    cardiac output, ANF effect, 221
  ventricle
    ANF and ANF mRNA detection, 225
HHM, *see* Humoral hypercalcemia of malignancy
HHM factor
  adenylate cyclase-stimulating activity, 733–735
  distinction from PTH mRNA-coded product, 737–739
  inhibition by PTH analog and inhibitor, 735–726
  PTH-like cytochemical bioactivity, 738–739
  coded by mRNA from HHM-associated tumors, 738–739
  purification, 735, 737
  extraction from HHM-associated tumors, 733
  in nonmalignant keratinocytes, human, 740–741
Human chorionic gonadotropin (hCG)
  in vitro estradiol release induction in ovary, rat, 426
  -like material, in bacteria, 559–560
Humoral hypercalcemia of malignancy (HHM), animal
  cutaneous squamous carcinoma, mouse, 731–732
  human HHM-associated tumor transplanted into nude mouse, 731–732
  Leydig cell tumor, Fischer rat, 731–732
Humoral hypercalcemia of malignancy, human *see also* Hypercalcemia, malignancy-induced
  alkaline phosphatase, serum, 718, 721, 727
  bicarbonate, serum, 717, 720, 728
  bone metastases, 712–713
  calcium
    fasting excretion, 714–715, 725–726

serum, 709, 711
cAMP excretion
  nephrogenous, 705–708, 719–725, 729–731
  urinary, 708–709
chloride/phosphate ratio, serum, 717, 719, 727–728
hydroxyproline, urinary, 718–719, 722, 727
phosphorus
  renal threshold, 714, 725
  serum, 713
PTH, circulating
  cytochemical bioactivity, 728–729
  immunoreactivity, 716–718, 726–727
tumors, histology, 709–712
vitamin D metabolites, 715–716, 726
Hydra
head activator
  amino acid sequence, 557
  detection in mammals, 558
  vertebrate-type neuropeptides, 557–558
6-Hydroxydopamine
  gonadotropin release increase, goldfish, 521
  after LHRH analog-induced stimulation, 526–527
Hydroxylapatite, estrogen receptor immobilization, 313–314
Hydroxyproline, urinary, in HHM, 718–719, 722, 727
5-Hydroxytryptamine, in spinal cord interaction with
  substance P, presynaptically, 24–25
  TRH, postsynaptically, 23, 24
  subcellular distribution in ventral horn, 24
Hypercalcemia, malignancy-associated, human, *see also* Humoral hypercalcemia of malignancy
  biochemical features, 680
  bone resorption, 680
  circulating PTH
    heterogenous PTH-like activity, 681
    low immunoreactivity, 680
  tumor-extracted PTH-like activity
    in vitro, heterogeneity, 682–684
    in vivo effects, 682
    mRNA translation in *Xenopus* oocytes and, 682

Hyperparathyroidism, human
  with diminished renal function
    PTH and its fragment secretion, 672–676
    distinction from HHM, 706, 713–719, 725–728, 730, 740
  with normal renal function
    bioactivity of major glandular PTH form, 671
    PTH secretion increase, 670–671
Hypogonadotropic hypogonadism, boys
  adult height enhancement, 461–462
  puberty delay, 461
Hypopituitarism, children
  GH deficiency, GRF effect, 619, 621
Hypothalamus
  ANF immunoreactivity, 225–226
  ANF mRNA expression, 225
  basal, DARPP-32-containing tanycytes
    dopamine neurons and, 40–41
    LHRH release and, 40–41
  median eminence, *see* Median eminence
  paraventricular nucleus, *see* Paraventricular nucleus
  peptides, *see* Peptides, hypothalamic
  pubertal, rat
    LHRH release, estradiol-induced, 429–432
    norepinephrine release, 427–428
    prostaglandin E_2 release, 428–429
  ventrobasal, goldfish
    GnRH secretion, 518–519
    GRIF secretion, 517–520
      lesions and, 517–519
      pituitary pars distalis transplantation and, 519

I

Idiopathic hyperaldosteronism, human
  aldosterone, plasma
    angiotensin III infusion and, 275–277
    upright posture and, 275–279
  ASF, plasma
    content increase, 277
    upright posture and, 277–279
  ASF, urinary, excretion increase, 280–281
  cortisol, plasma, dexamethasone effect, 279
  symptoms, 275

proopiomelanocortin derivatives and, 279–280
Insulin
during continuous GRF infusion, GH release and, 610–612
function in early chicken embryo, 573 (table)
PEPCK inhibition
in liver and adipose tissues, 112
model of, 134–135
in rat H4IIE cells, 114–115
PEPCK mRNA and, see RNA, messenger, PEPCK
protein synthesis and, 111–113
regulation of mRNAs for enzymes and proteins, 132–133
Insulin receptor
Drosophila melanogaster, 553, 555
H4IIE cells
affinity to porcine and guinea pig insulins, 115–116
PEPCK mRNA regulation and, 115–117
Insulin-related peptides
in insects, immunoreactivity, 552–554
in molluscs, localization, 555–557
prothoracicotropic hormones, silkworm amino acid sequences, 555–556
homology with vertebrate insulins, 555–556
purification and activities
Aspergillus fumigatus, 559, 561
Escherichia coli, 559, 561, 563
Neurospora crassa, 559, 561
Tetrahymena pyriformis, 559–562
Intercellular communication system
biochemical elements, evolution, 549–551
classic concept, 549–550
endocrine and nervous systems overlaps, 550–551

K

KAP, see Kidney androgen-regulated protein
Keratinocytes, nonmalignant, human PTH-like HHM factor activity, 740–741
Kidney
ANF effects
glomerular filtration rate and, 217, 218

hemodynamic, 216–217
natriuretic, 217–218
renin secretion decrease, 220, 222–223
cortex, ANF receptors in glomeruli, 226
FGF
physiological role unknown, 189
properties, 167–170
PTH binding
in vitro, to membranes, canine, 690–692
in vivo, localization, rat, 688–689
testosterone effects, mouse
androgen receptors and, 94–99
β-glucuronidase activation, 74–75
gene regulation and, 85, 87–93
mRNA species induction, 75, 82–100
ornithine decarboxylase activation, 74
gene regulation and, 79–87
protein induction, 74
gene regulation and, 76–79
Kidney androgen-regulated protein (KAP)
amino acid sequence, 76–77
testosterone effects
in renal cytosol, 76–78
in renal mRNA cell-free translation, 76

L

Lemna, somatostatin-like material, 580
Leukotrienes
arachidonic acid derivatives, 49
formation in brain, 50
LTC$_4$
luteinizing hormone release and, 50
Purkinje cell activation, 50
metabolism, 49–50
LH, see Luteinizing hormone
LH receptors, ovarian, rat during pubertal phases, 425
LHRH, see Luteinizing hormone-releasing hormone
LHRH analog
precocious puberty treatment, human, 451–457
LHRH receptors, ovarian, rat during pubertal phases, 425
Liver, PTH binding *in vivo*, rat
by hepatocytes, 686
by Kupffer cells, 686
Local osteolytic hypercalcemia (LOH), human

distinction from HHM, 707–709, 711–722, 725–726, 728–729, 740
LOH, see Local osteolytic hypercalcemia
Luteinizing hormone, human
  in precocious puberty, LHRH analog and, girls, 451–453
Luteinizing hormone, lamb
  pulse frequency, increase before surge, 354–357
    gonadostat hypothesis and, 357–358
    after ovariectomy, 355–357
    sensitivity to estradiol, 359–360
  secretion regulation
    by estradiol during puberty, 358
    by growth and nutrition, 368–373
      mechanism of, 372–373
    ovariectomy and, 368–370
    photoperiod and, 370–372
    by photoperiod, 364–367
      estradiol inhibition and, 364–366
      failure after ovariectomy, 366–367
  surge induction by
    estradiol, 352–353
    LH, 354–355
Luteinizing hormone, mammals
  release inhibition by dopamine, 536–540
  release stimulation by LTC_4, 50
Luteinizing hormone, rat
  infantile, secretion
    estrogen effects after ovariectomy, 398, 400
    LHRH injection and, 398–399
    sporadic increases, 396–398
  juvenile, secretion
    estradiol effect, day 20–34, 404–405
    pulses, morning and afternoon, 406–407
    ovarian hormone release in vitro and, 408–409
    prolactin effect, 411–412
    response to LHRH decline, 418–419
  peripubertal, secretion
    afternoon pulses, 420
    gonadal-independent, 420–421
    minisurges, ovariectomy effect, 422–423
  release during puberty onset, 432–434
Luteinizing hormone, tammar, female
  lactational quiescence period
    lutectomy and, 484, 486

ovariectomy and, 483–484
estradiol effect, 484–485
pouch young removal and, 484
seasonal changes absence, 483–486
preovulatory pulses, 477–478, 480
Luteinizing hormone, tammar, male
  photoperiod-unresponsive, 503
  response to female reactivation, 503
Luteinizing hormone-releasing hormone (LHRH)
  amino acid sequence, 572
  homology with yeast α mating factor, 572
  in arcuate nucleus, 30
  gonadotropin release induction, teleosts, 515–517
  origin of, 30
  in precocious puberty, human, 451–452
  suppression by LHRH analog, 451–453
  release in median eminence
    DARPP-32-containing tanycytes and, 40–41
    dopamine neurons and, 40–41
Luteinizing hormone-releasing hormone, rat
  fetal, 388
  infantile gonadotropin secretion induction, 398–399
  milk, effect on neonatal ovarian receptors, 392–394
  neonatal pulsatile release, 394–395
  pubertal secretion, estrogen-induced, 429–432
  prostaglandin E_2–cAMP pathway, 430
  protein kinase C-dependent pathway, 430–431
  secretion by juvenile hypothalamus in vitro,
  prostaglandin E_2 and, 417–419

M

Macrophage-derived growth factor
  FGF and, comparison, 174
Macrophages
  -derived cell lines, FGF activities, 173–174

FGF
 angiogenic factor and, 190
 properties, 172–174
Marsupials
 reproduction pattern, 472–474
 in tammar wallaby (*Macropus euge-nii*), see Tammar, female
Median eminence
 dopamine neuron projection to, 10–11, 35
 DARPP-32-containing tanycytes and, 40–41
 galanin, 35
 GRF, 35
 LHRH release, 40–41
 tyrosine hydroxylase, 10, 40–41
Melatonin, female lamb
 circulating, diurnal rhythm
  under natural photoperiod, 344–345
  pineal gland denervation and, 344–346
  ontogeny, photoperiod effects, 349–350
  puberty timing and, 349–351
  short and lond day effects, 344
 nightly infusion after pineal gland denervation, 346–347
Melatonin, female tammar
 birth after summer solstice and, 501
 pinealectomy effect, 501–503
 injection before dark
  melatonin secretion and, 498–499
  progesterone pulses and, 497–498
  prolactin pulses and, 498
  reactivation induction, 495–497
 plasma, diurnal rhythm
  photoperiod and, 494–495
  pinealectomy and, 494
Metoclopramide, stimulation of
 aldosterone secretion, human, 255–256
 dopaminergic mechanism, 282–284
 LH release, direct action, human, 537–539
 LHRH analog-induced gonadotropin release, goldfish, 523, 526
Metorphamide-like peptide, in arcuate nucleus, 38–39
Mucopolysaccharidosis type VII, human β-glucuronidase deficiency, 86–87

Muscles, smooth, isolated, ANF effect, 216

N

Natriuresis, ANF action, mechanism of, 217–218
Neovascularization, induction by FGF
 from corpus luteum, 166–167
 from macrophages, 173
 from retina, 171–172
Neuraminidase, ASF treatment by, 260–262
Neurons
 acetylcholine
  in arcuate nucleus, 33
  sympathectomy and, 18–19
 cholinergic, see acetylcholine
 dopamine, tyrosine hydroxylase-positive
  in arcuate nucleus, 28–29, 31
  containing another messengers, 39
  Fast Blue retrograde transportation, 43
  in mediane eminence, 10–11, 35
  DARPP-32-containing tanycytes and, 40–41
  LHRH release and, 40–41
 GABA, in arcuate nucleus, 31, 33
 hippocampal, survival *in vitro*, FGF and, 191
 messenger storage, phylogeny, 51–52
 multimessenger transmission, 52–53
 multiple antigens, immunohistochemistry, 6–9
 neurotransmitters
  coexisting with peptides, 15–18
  –receptor relation, 50–51
  retrograde tracing, 9–11
 parasympathetic in salivary gland, multiple messengers, 15, 20–22
 phenotype plasticity in adult animals, 18–19
 sympathetic, multiple messengers
  in salivary gland, 20–22
  in vas deference, 15, 17–18
Neuropeptides, vertebrate-type, *Hydra*, 557–558
Neuropeptide Y (NPY)
 in arcuate nucleus, 37–39
 in juvenile ovary, rat, 415–416

in sympathetic neurons of
  salivary gland
    blood flow and, 21–22
    noradrenaline secretion and, 20–22
    release by electrical stimulation, 22
  vas deference
    noradrenaline coexistence with, 15, 17–18
    subcellular distribution, 15, 17–18
Neurospora crassa, insulin-like material, 559, 561
Neurotensin, in arcuate nucleus, 32–34
  prolactin secretion and, 33–34
Neurotransmitters
  in CNS
    coexistence with peptides
      examples, 11–14
      species variability, mammalian, 14–15
      history, 2
    frequency-coded release, 15, 18, 19
    interactions with peptides
      postsynaptic
        in salivary gland, 20–23
        in spinal cord ventral horn, 23, 24
      presynaptic
        in caudate nucleus, 24
        in spinal cord, 24–25
        in vas deferens, 23
    in synaptic vesicles, 15–18
Noradrenaline, *see* Norepinephrine
Norepinephrine
  gonadotropin secretion by pituitary and, goldfish
    in vitro increase, 533–534
    in vivo increase, 532–533
    mechanism of, scheme, 540–541
  hypothalamic production during puberty, rat, 427–428
  as neurotransmitter in CNS, history, 2
  subcellular distribution
    in peripheral and central neurons, 15–17
    in vas deferens sympathetic nerves, rat, 15, 17–18
    NPY coexistence with, 15, 17–18
NPY, *see* Neuropeptide Y

O

Opiate receptors, μ-type, *Amoeba proteus*, 574, 577
Ornithine decarboxylase, murine kidney
  induction by
    5 α-dihydrotestosterone, 79–80
    testosterone, 74, 79–80
  gene regulation, 79–87
  purification and heterogeneity, 81–82
  rapid turnover, androgen effects, 81
Ovariectomy
  gonadotropin release and, tammar, 483–484
  LH secretion sensitivity to photoperiod, lamb
    absence at normal diet, 366–367, 369
    maintenance at malnutrition, 369
Ovary
  corpus luteum, *see* Corpus luteum
  Graafian follicles, *see* Graafian follicles
Ovary, rat
  fetal, aromatase, 389–390
  granulosa cells, aromatase inhibition by FGF, 188–189
  immature, FGF detection, 186–188
  juvenile
    β-adrenergic receptors, 413–414
    superior ovarian nerve and, 414–415
    norepinephrine, 413
    superior ovarian nerve and, 414–415
    regulatory mechanisms, scheme, 417–418
    steroid release *in vitro*
      growth hormone and, 412
      hyperprolactinemia and, 410–411
      LH and, 408–409
      VIP and, 416
  neonatal, milk LHRH effects on
    LHRH receptors, 392–393
    steroid induction *in vitro*, 392–393
  during pubertal phases
    3α-androstanediol secretion, 427
    estradiol secretion, 425–427
    LH receptors, 425
    LHRH receptors, 425
    substance P, 426–427
    testosterone secretion, 426–427
    VIP, 425–427

P

Paracoccidioides brasiliensis, estradiol
  binding, 577
Parathyroid hormone (PTH)
  amino acid sequence, 643
  in disease states, heterogeneity, 693
    (table)
  hypercalcemia in malignancies, 680–
    684
  hyperparathyroidism
    with diminished renal function, 672–
      676
    with normal renal function, 670–671
  pseudohypoparathyroidism, 677–679
  extract, adenylate cyclase stimulation,
    736
  inhibition by PTH analog and inhibi-
    tor, 735–736
  in HHM, circulating
    cytochemical bioactivity, 728–729
    immunoreactivity, 716–718, 726–727
  HHM factor distinction from, 737–739
  NH_2-terminal fragments
    binding to receptors in vitro, 690–692
    bioactivity and, 665–666
    serine residue phosphorylation, 667,
      669–670
    amino acid sequence and, 669
  phosphorylation, intraglandular, 667–668
  radiolabeled, binding studies
    in vitro evaluation
      osteosarcoma cells, 690–692
      renal membranes, 690–692
    in vivo detection, rat
      bone, 686–688
      kidney, 688–689
      liver, 686
      methods, 684–686
  synthesis and secretion, scheme, 641–642
Paraventricular nucleus, parvocellular part
  CRF, 42, 44–47
  enkephalin, 45, 47
  PHI, 44–47
PEPCK, see Phosphoenolpyruvate carboxykinase
Peptide hormones, vertebrate-type
  in multicellular invertebrates, 552 (ta-
    ble)–558

  in plants, 578–580
  in unicellular organisms, 559 (table)–570
    evidence against contamination, 568–
      570
  in vertebrates, multiple sites, 549, 551
    (table)
Peptides, hypothalamic, see also specific
  peptides
  coexistence with neurotransmitters in
    CNS, 3–6
    examples, 11–14
    species variability, mammalian, 14–15
  coexistence with neurotransmitters in
    PNS, 3–6
  frequency-coded release, 15, 18, 19
  as messenger molecules, history, 2
  PHI
    ACTH release and, 49
    in paraventricular nucleus, 44–47
    prolactin release and, 44, 48–49
    in synaptic vesicles, 15–18
Peptides, opioid, in arcuate nucleus, 36–37
Peripheral nervous system (PNS)
  adrenaline as neurotransmitter, history,
    1
  neurotransmitters, coexistence with
    peptides, 3–6
PGFM, see Prostaglandin metabolite
PHI, see Peptides, hypothalamic, PHI
Phosphoenolpyruvate carboxykinase
  (PEPCK)
  reactions catalyzed by, 112
  regulation by insulin, 112–114
  model of, 134–135
  synthesis in H4IIE cells
    inhibition by insulin, 114–115
    stimulation by cAMP analog, 114–115
Phosphorus, in HHM, human
  renal threshold, 714, 725
  serum, 713
Photoperiod
  female reactivation and, tammar, 491–
    499
  melatonin diurnal rhythm and, 494–
    495
  melatonin injections and, 495–497
  pinealectomy and, 494
  progesterone secretion and, 492
  sequence of events, 499

information transmission, sheep, 341, 343
melatonin as photic clue, 343–351; see also Melatonin
LH secretion in lamb and, 364–367
malnutrition effect, 370–372
after ovariectomy, 366–367
puberty timing, sheep
artificial long days, 333–341
early exposure to, 347–348
artificial short days, 339–342
seasonal changes, 337–339
seasonal breeding and sheep, 334
tammar, 474–476
Pimozide, gonadotropin release and, goldfish, 523, 525–530
Pinealectomy, female tammar
birth timing and, 501–503
photoperiod-sensitivity and, 494
progesterone release and, 502
Pineal gland, female sheep
denervation
melatonin diurnal rhythm absence, 344–346
puberty induction delay, 344–346
reversal by melatonin nightly infusion, 346–347
in photic information transmission, 343
Pituitary cells, rat
GH4 line
human PTH gene promoter recognition, 650
PTH retrovirus-infected, PTH production, 658
TRH-stimulated PTH secretion, 658
GRF-stimulated GH release
hemolytic plaque assay, 603–605
in perfused dispersed cells
during continuous GRF administration, 613–615
sex differences, 601–603
somatotroph cell content and, 603–605
Pituitary gland
ANF and ANF mRNA detection, 225
ASF localization, 272–274
estrogen receptor localization, 299–300
FGF, see Fibroblast growth factor, pituitary

prolactin release *in vitro*
FGF prolonged stimulation, 182–186
estradiol synergism, 183–185
somatotroph hyperplasia, human, 591
thyrotropin release *in vitro*, FGF and, 182, 184, 185
vasopressin release, ANF and, 225
Plasmids, *Escherichia coli*
with human PTH gene portion, 649
introduction into rat GH4 cells, 650
PTH gene transcription, 650
with prepropPTH-encoding cDNA and *lac* promoter
coding for pre-β-lactamase, 653
PTH and β-lactamase synthesis and transport, 653
Plasmids, *Streptococcus faecalis*
transfer by conjugation, 571
sex pheromones and, 571
PNS, see Peripheral nervous system
Potassium ion, aldosterone secretion and, 251–253
Precocious puberty, human
central, gonadotropin-dependent
early LHRH secretion onset, 451–452
growth hormone increase, girls, 456
LHRH analog treatment
bone maturation normalization, 455
estradiol decrease, girls, 457
growth rate decrease, 454–455
LH and FSH suppression, girls, 451–453
testosterone suppression, boys, 453–454
somatomedin C increase, 456–458
LHRH analog effect, 456
familial male
combined spironolactone and testolactone treatment, 460–461
resistance to LHRH analog, 460
McCune–Albright syndrome
resistance to LHRH analog, 459
testolactone treatment, girls
estradiol and ovarian volume decrease, 459
growth rate and bone maturation decrease, 459
Pre-βlactamase, *Escherichia coli*
processing and transport across membrane *in vitro*, 653–654

Preproparathyroid hormone (prepropTH)
  processing to PTH, 650
  as PTH mRNA primary translation
    product, 650
  signal sequence, 643, 650–651
  binding to signal recognition particle,
    651
  in PTH transport across endoplasmic
    reticulum, 651–654
Progesterone
  juvenile ovarian, secretion *in vitro*, rat
    growth hormone and, 412
  hyperprolactinemia and, 410–411
  LH pulses and, 408–409
  VIP and, 416
  in pubertal transition, female sheep,
    359–361
Progesterone, tammar
  in corpus luteum, 478
  secretion from, 479–480
  during lactation, hypophysectomy and,
    486–488
  photoperiod and, 492
  pinealectomy and, 502
  postovulatory changes, 480–482
  pouch young removal and, 479, 481–482
  during pregnancy and postpartum estrus,
    476–478
  pulses, induction by melatonin injection,
    497–498
Progesterone receptor, unoccupied, nuclear, 309
Prolactin
  corpus luteum inhibition, tammar, 488–490
  suckling by pouch young and, 488
  juvenile rat, secretion
    diurnal rhythm, 409
    puberty timing and, 410–411
  peripubertal secretion, rat
    afternoon surge, 421–422
    gonadal-independent, 422
  plasma, metoclopramide effect, human,
    283
  inhibition by dopamine, 283
  during pregnancy and postpartum estrus,
    tammar, 477
  pulses, female tammar
    melatonin injections and, 498
    photoperiod and, 498–499

release by rat pituitary cells
  increase by
    CRF, 44, 48–49
    enkephalin, 48–49
    FGF, 182–186
    PHI, 44, 48–49
  inhibition by dopamine, 48–49
  release *in vivo*
    neurotensin-inhibited, 33–34
    by tumors in Fischer 344 rats, 181–182
  synthesis in GH_3 cells and cytoplasts,
    307, 309
Proopiomelanocortin, derivatives
  in idiopathic hyperaldosteronism, human, 279–280
Proparathyroid hormone (proPTH)
  processing to PTH, 650–653
  "pro" sequence in PTH secretion, 652–653
Prostaglandin E_2
  hypothalamic production, rat
    during pubertal phases, 428–429
    LHRH release and, 429–432
  LHRH secretion by rat juvenile hypothalamus *in vitro* and, 417–419
Prostaglandin metabolite (PGFM), tammar
  during pregnancy and postpartum estrus,
    477
Proteins
  egasyn, binding to β-glucuronidase X, 87
  KAP, *see* Kidney androgen-regulated protein
  testosterone effects, murine kidney, 74, 76
Prothoracicotropic hormones, silkworm
  insulin-like amino acid sequence, 555–556
  metamorphosis control by, 555–556
  multiple forms, 555
Pseudohypoparathyroidism, human
  PTH secretion
    acceleration, 677–678
    low bioactivity, 677–679
    type IA, symptoms, 677–679
    adenylate cyclase subunit defect, 677, 679
    type II, symptoms, 677
Pseudomonas maltophilia, hCG binding, 574
PTH, *see* Parathyroid hormone

Pubertal growth, human
  central precocious puberty, 451–458; see also Precocious puberty, central
  familial male precocious puberty, 460–461; see also Precocious puberty, familial male
  growth hormone deficiency and treatment, boys, 449–450
  hypogonadotropic hypogonadism, boys, 461
  McCune–Albreight syndrome, 459
  normal boys
    estradiol low doses and, 448–449
    velocity standards, 443–444
  regulation by estrogen doses, girls, 445–446
  Turner syndrome, girls, 443, 445–448
Puberty
  determinants, female sheep, see Sheep, female
  genesis, female rat, see Rat, female
  growth during, human, see Pubertal growth, human

R

Rat, female
  fetal period
    aromatase, hypothalamic and ovarian, 388–390
    LHRH production on day 12, 388
  feto-neonatal period, neuroendocrine events, 396
  growth during sexual development, 385–386
  infantile period
    estradiol negative feedback failure, 398–400
    α-fetoprotein and, 398
    FSH secretion during, 395–401
    gonadotropin secretions, 395–404
    estrogen postovariectomy effects, 400–401
    LHRH injection and, 398–399
    ovarian steriod increase and, 396–397
    testosterone postovariectomy effects, 399–401
    LH sporadic increases, 396–398
    neuroendocrine events, 403–404
  juvenile period
    α-fetoprotein decrease, 404
    FSH secretion decline, 404–406
    growth hormone secretion, 409–410, 412–413
    LH secretion
      estradiol effect, 404–405
      prolactin effect, 411–412
      pulses and surges, 406–408
      ovarian functions, 408–418; see also Ovary, rat
    prolactin secretion, 409–411
    LH release during development, 432–435
  neonatal period
    α-fetoprotein high level, 391
    gonadotropin secretion increase, 390–391
    LHRH pulsatile release, 394–395
    ovarian functions, milk LHRH effects, 392–394
  peripubertal period
    initial events, 420–423
      LH afternoon pulses, regulation, 420–421
      LH minisurges, regulation, 422–423
      prolactin afternoon surges, regulation, 421–422
    precipitation events, 423–432
      estrogen-induced LHRH release, 429–432
      hypothalamic functions, 427–429
      ovarian functions, 425–427
    puberty phases, 424–425
    scheme of, 424
Renin
  aldosterone secretion regulation, 251, 253–254
  plasma activity
    ANF-induced decrease, 220, 222, 224
    head-out water immersion, human, and, 230–231
  secretion rate, ANF effect, 220, 222–223
Renin–angiotensin–aldosterone system, 251
  ANF interactions with, 235
  dopamine interaction with, 284–285
Reserpine, gonadotropin release and, goldfish, 521, 526–527

Retina
  FGF, bovine
    biological activity, 171–172
    purification and identification, 171–172
  FGF, human
    in ocular fluid
      detection, 190–191
      in diabetes, content increase, 190
      in vitreous, isolation from, 192
Retrovirus, carrying PTH cDNA
  infectious retrovirus production, 656–658
  PTH induction in GH4 cells, rat, 658
  PTH synthesis in cultured cells, 654–655
RNA, messenger (mRNA)
  ANF
    in atrial muscle, abundance, 225, 233
    dehydration and, 233
    deoxycorticosterone and, 233
  for enzymes and proteins, insulin-regulated, 132–133
  from hypercalcemia-associated tumors, 682, 737–738
  PTH-like activity translation in *Xenopus* oocytes, 682, 737–738
  PEPCK, H4IIE cells
    activity inhibition by insulin, 114–115
    insulin receptor and, 115–117
    cAMP-induced transcription, insulin effect
      dose-dependent, 123, 125
      time course of, 123–125
    cytoplasmic, insulin effects
      content decrease, 118–119, 122
      degradation unchanged, 120
      reversibility, 119–120
      synthesis and activity inhibition, correlation, 118–119
    dexamethasone-induced synthesis, 128
    inhibition by insulin, 129
    nuclear, 120–123
      egress to cytoplasm, insulin and, 121, 123
      11 species, content of, insulin effect, 121–122
      turnover, insulin and, 121
    transcription without cAMP, insulin effect
      direct action, 126–127
      specificity, 126
      unchanged by cycloheximide, 127

PTH, bovine
  extracellular Ca^{2+} and, 647–648
  in PTH gene assay, 643–644
  testosterone-induced, murine kidney
    β-glucuronidase, 75, 92–99
      Gus complex and, 92–93
      kinetics, 94–95
      nuclear androgen receptors and, 94–99
    KAP, 75–76, 94–99
      cell-free translation, 76–79
      flutamide effects, 98–99
      kinetics, 94–95
      nuclear androgen receptors and, 94–99
      nucleotide sequence, 77
    ornithine decarboxylase, 75, 82–83, 85
      flutamine effects, 98–99
      kinetics, 94–95
      multiple genes for, 83–85
      nuclear androgen receptors and, 94–99
      nucleotide sequences, 82–83
      strain differences, 85–87

S

Saccharomyces cerevisiae
  estradiol receptors, 577
  sex pheromones
    life cycle and, 571–572
    α mating factor, comparison with GnRH
      amino acid sequence homology, 572
      binding to rat GnRH receptors, 573
Salivary gland, cat
  parasympathetic neurons
    multiple messengers, release mechanism, 15
    VIP postsynaptic interaction with acetylcholine, 20–21
  sympathetic neurons
    NPY postsynaptic interaction with noradrenaline, 20–22
Sex pheromones, unicellular organisms
  intercellular coomunications and, 571–573
  Saccharomyces cerevisiae, 571–573
  Streptococcus faecalis, 571

Sheep, female
  growth curve till maturation, 332
  comparison with human female, 332
  LH secretion in lamb, *see* Luteinizing hormone
  pubertal transition
    estradiol positive feedback, 361
    LH surges, 360–361
    progesterone and, 360–361
    short luteal phases, 360–361
  puberty regulation
    GnRH pulse generator and, 362–375
    gonadostat hypothesis, 357–358
    hypothalamic mechanism, models, 351–352, 359–360, 373–375
    LH pulse frequency and, 351–352, 355–357
  puberty timing
    delay by pineal gland denervation, 344–346
    nightly melatonin infusion and, 346–347
    GnRH pulse generator role, 362–375; *see also* GnRH pulse generator
    growth rate and, 335–336
    melatonin as photic clue, 343–357; *see also* Melatonin
    photoperiod effects
      long days, 339–341, 347–348
      seasonal changes, 337–339
      short days, 339–342
      season of birth and, 334–337
    reproductive cycle initiation, hypothesis
      LH pulse sensitivity to estradiol and, 359–360
      progesterone role, 359–360
    seasonal breeding, 333–334
    photoperiod and, 334
Silkworm, prothoracicotropic hormones, 555–556
Sodium ion
  depletion, aldosterone secretion and, 253–255
  dopamine–angiotensin II antagonism and, 284–285
  dopaminergic activity and, 285
  dietary restriction, plasma ASF and, 281–282
Somatomedin C
  GRF-induced in adults with GH deficiency, 621, 623

  induction by GH in peripheral tissues, 605–606
  in precocious puberty, human, 456–458
  LHRH analog effect, 456
Somatostatin
  in arcuate nucleus, 30, 39
  dopamine release and, 36
  GRF isolation from hypothalamus and, rat, 590
  -like material, purification and activity
    Bacillus subtilis, 559, 562, 564–566
    Escherichia coli, 559, 562
    Lemna, 580
    spinach, 578–580
    Tetrahymena pyriformis, 559, 562, 564
  origin of, 30
Spinach, somatostatin-like material, 578–580
Spinal cord, rat
  CGRP, interaction with
    substance P, 26–28
    tachykinins, 26–27
  5-hydroxytryptamine, interaction with
    substance P, presynaptically, 24–25
    TRH, postsynaptically, 23, 24
  ventral horn, multiple messengers, 24
Spiperone
  binding to pituitary homogenate, goldfish, 532
  gonadotropin release and, goldfish, 530–531
Spironolactone, combined with testolactone
  familial male precocious puberty treatment, 460–461
Streptococcus faecalis, sex pheromones
  amino acid sequence, 571
  plasmid transfer during conjugation and, 571
Substance P
  ovarian, rat
    juvenile, detection, 415–417
    secretion *in vitro*, pubertal phases, 426–427
  in spinal cord
    5-hydroxytryptamine and, 24–25
    interaction with CGRP, 26–28
Superior ovarian nerve, rat
  juvenile ovarian function and, 414–415, 417

INDEX

Synaptic vesicles, neurotransmitters and peptides
  in adrenal medulla, 16–17
  in salivary gland parasympathetic nerves, cat, 15
  in spinal cord ventral horn, rat, 24
  in vas deferens sympathetic nerves, rat, 15, 17–18

T

Tachykinins, in spinal cord interaction with CGRP, 26–27
Tammar, female
  corpus luteum
    postovulatory changes, 480–481
    pouch young removal and, 481–482
    role in seasonal breeding, 482–483
  lactational quiescence
    bromocryptine-induced reactivation, 488–490
    corpus luteum
      effect on follicle growth, 486
      hypophysectomy and, 486–488
      inhibition by prolactin, 488–490
    gonadotropins in plasma
      estradiol effect after ovariectomy, 484–485
      lutectomy and, 484, 486
      pouch young removal and, 484
      ovariectomy and, 484
      seasonal changes absence, 483–486
    prolactin release and, 488–490
    suckling by pouch young and, 487–488
  pregnancy and postpartum estrus
    estradiol in Graafian follicles, 478
    secretion from, 479–480
    hormonal secretion, 476–478
    LH preovulatory pulses, 477–478, 480
    progesterone in corpus luteum, 478
    secretion from, 479–480
  reproduction annual cycle, 474–476
    birth after summer solstice, 475, 500–501
    melatonin role, 501
    pinealectomy and, 501–503
    hormonal control, scheme, 500
  seasonal quiescence
    photoperiod-induced reactivation, 491–494
    melatonin injections and, 495–499
    melatonin release and, 494–495
    pinealectomy and, 494
    prolactin release and, 498–499
    progesterone release and, 492
Tammar, male, seasonal hormonal changes
  in response to female presence, 503–504
  unaffected by photoperiod, 503
Tanycytes
  DARPP-32-containing in basal hypothalamus, 40–41
  dopamine neurons in median eminence and, 40
Teleosts
  GnRH, 515–517
  gonadotropins
    goldfish, see Gonadotropin, goldfish
    two forms, 514
  GRIF activity, 513, 535–536
    in goldfish, see Gonadotropin release inhibitory factor, goldfish
    mechanism of, 540–541
Testicular feminization
  androgen receptor deficiency, anumal, human, 72–73
  mRNA species, murine kidney, 97–98
  testosterone effects, 97–98
Testolactone
  combined with spironolactone, familial male precocious puberty treatment, 460–461
  McCune–Albright syndrome treatment, girls, 459
Testosterone, boys
  in precocious puberty, LHRH analog and, 453–454, 457
Testosterone, male tammar
  photoperiod-unresponsive, 503
  response to female reactivation, 503–504
Testosterone, murine kidney
  β-glucuronidase activation, 74–75
  β-glucuronidase mRNA stimulation, 92–99
  KAP induction, 76
  KAP mRNA stimulation, 76–79, 94–99
  mRNA species induction, 75, 82–100
  ornithine decarboxylase activation, 74
  ornithine decarboxylase mRNA induction, 79–80, 94–98
  protein induction, 74

Testosterone, rat
 infantile
  gonadotropin secretion and, 401
  production by ovary, 399
 juvenile
  gonadotropin secretion and, 405–406
  ovarian progesterone release *in vitro* and, 412
 ovarian secretion *in vitro*
  during juvenile period, VIP effect, 416
  during pubertal phases, 426–427
Tetrahymena pyriformis
 ACTH-like material, 559, 566–568
 β-endorphin-like material, 559, 566–568
 insulin-like material, 559–562
 somatostatin-like material, 559, 562, 564
Thyrotropin-releasing hormone (TRH)
 in arcuate nucleus, 30
 origin of, 30
 in spinal cord ventral horn
  cooperation with 5-hydroxytryptamine, 23, 24
  subcellular distribution, 24
Thyroid-stimulating hormone (TSH)
 -like material in *Clostridium perfringens*, 559
TRH, *see* Thyrotropin-releasing hormone
TSH, *see* Thyroid-stimulating hormone
TSH receptors
 Escherichia coli, 574
 Yersinia enterocolitica, 574
Tumors
 angiogenic factor, FGF and, 190
 in HHM, histology, human, 709–712
 neuroendocrine, GRF-secreting, acromegaly and, human, 590–593, 596–599
 pancreatic, GRF-secreting, human
  acromegaly and, 590, 592, 597
  GRF isolation and structure, 592–593
  morphology, 596
  removal causing GH fall, 591
  pituitary somatotroph hyperplasia and, 591
  ultrastructure, 596–597
 prolactin-secreting in Fischer rats
  pituitary FGF increase by estradiol and, 181–182

Turner syndrome, girls, 443
 ethinyl estradiol treatment
  biphasic effect on growth, 445–447
  lack of effect on bone age and breast budding, 448
Tyrosine hydroxylase, *see also* Neurons, dopamine, tyrosine hydroxylase-positive
 coexistence with Fast Blue tracer
  in arcuate nucleus, 43
  in mediane eminence, 10

U

Unicellular organisms
 hormone-binding proteins, 574 (table)–577, 579
 intercellular communication, 570–573
 vertebrate-type peptide hormones, 559–570

V

Vas deference, sympathetic neurons, rat
 noradrenaline and NPY
  coexistence and release, 15, 17–18
  presynaptic interactions, 23
Vasoactive intestinal polypeptide (VIP)
 aromatase induction in fetal ovary, rat, 390
 ovarian, rat
  juvenile, 415–417
  steroid secretion and, 416
  during pubertal phases, 425
  estrogen response to, 425–427
  progesterone response to, 426
 in parasympathetic salivary gland neurons
  acetylcholine secretion and, 20–21
  blood flow and, 20
  release by electrical stimulation, 21
  prolactin-releasing activity, 44, 48
Vasopressin, ANF effects on release, 225
VIP, *see* Vasoactive intestinal polypeptide
Vitamin D, metabolites in HHM, human, 715–716, 726

Y

Yersinia enterocolitica, TSH binding